AF616089

University
of Southampto

SELECTED PAPERS OF

F.W.J. Olver

Part I

World Scientific Series in 20th Century Mathematics

Published

Vol. 1 The Neumann Compendium
edited by F. Bródy and T. Vámos

Vol. 2 40 Years in Mathematical Physics
by L. D. Faddeev

Vol. 3 Selected Papers of Yu I Manin
by Y. Manin

Vol. 4 A Mathematician and His Mathematical Work — Selected Papers of S. S. Chern
edited by S. Y. Cheng, P. Li and G. Tian

Vol. 5 Fields Medallists' Lectures
edited by Sir Michael Atiyah and Daniel Iagolnitzer

Vol. 6 Selected Logic Papers
by G. E. Sacks

Vol. 7 Selected Papers of F. W. J. Olver (in 2 Volumes)
edited by Roderick Wong

World Scientific Series in 20th Century Mathematics Vol. 7

SELECTED PAPERS OF F.W.J. Olver

Part I

Edited by

Roderick Wong

Department of Mathematics
City University of Hong Kong

Published by

World Scientific Publishing Co. Pte. Ltd.

P O Box 128, Farrer Road, Singapore 912805

USA office: Suite 1B, 1060 Main Street, River Edge, NJ 07661

UK office: 57 Shelton Street, Covent Garden, London WC2H 9HE

British Library Cataloguing-in-Publication Data
A catalogue record for this book is available from the British Library.

The author and publisher would like to thank the following publishers of the various journals and books for their assistance and permission to include the selected reprints found in this volume:
Academic Press
American Mathematical Society
Association for Computing Machinery
Birkhäuser
Cambridge University Press
Elsevier Science Ltd
International Press
London Mathematical Society
Marcel Dekker, Inc.
National Bureau of Standards (now National Institute of Standards and Technology)
Oxford University Press (2 letters)
Royal Irish Academy
Royal Society of Edinburgh
Royal Society of London
SIAM Publications
Utilitas Mathematica

ISBN 981-02-4106-2 (Set)
ISBN 981-02-4994-2 (Vol. 1)

Printed in Singapore by Uto-Print

PREFACE

These two volumes contain a selection of my research papers published in the period 1949–1999. The topics are Asymptotic Analysis, Special Functions (also known as Higher Transcendental Functions), and Numerical Analysis. Each paper in the collection falls into one or more of these categories, but because of heavy interactions it would be more harmful than helpful to attempt to classify the papers separately in these categories.

The directions of my research were shaped early in my career. In 1944, at age nineteen and after completing my B.Sc. degree at London University, I was assigned to the British Admiralty Computing Service. I joined a pool of other mathematicians to carry out day-to-day numerical calculations to assist in the successful prosecution of World War II. This was my introduction to numerical analysis.

Soon after the war I accepted a position as Scientific Officer in the newly-formed Mathematics Division of the National Physical Laboratory, reasoning, quite correctly at that time, that this would lead me to the frontiers of research in numerical analysis more effectively than by returning to a university. One of the projects at N.P.L. in which I became very interested was the compilation of numerical tables of zeros of Bessel functions. These tables were published eventually for the Royal Society of London; see [B2] on page 1067. The problems were very challenging mathematically in the desk-machine era of computation, and they ushered in my life-long interest in research in asymptotics and special functions.

I. Asymptotic Analysis

(i) *Asymptotic expansions for second-order linear differential equations containing a large parameter.* The object is to construct expansions of solutions that are uniform with respect to the independent variable. Many different cases of practical importance arise, depending on the nature of the turning points and singularities of the differential equation in the region under consideration. Expansions in terms of elementary functions are treated in [6], [9], [11];* expansions involving Airy functions (simple turning points) are treated in [6], [9], [12], [13]; expansions involving Bessel functions of fixed order (multiple and fractional turning points and simple poles) are treated in [8], [9], [29], [30], [34]; expansions involving parabolic cylinder functions (coalescing simple turning points) are treated in [26], [28], [33]. Connection-formula problems for the asymptotic solutions, that is, the evaluation of Stokes constants, are treated in [16], [17], [30], [31], [34], [53].

*Numbered references without letters refer to the papers included in these volumes. See pages 1061–1068 for complete listings.

(ii) *Solutions of linear differential equations in the neighborhood of an irregular singularity.* Second-order equations are considered in [14], [15], and equations of arbitrary order in [54], [55], [56].

(iii) *Error bounds.* The original objective was to control remainder terms rigorously in numerical applications. Later, I realized that well-constructed bounds lead to important insights in the asymptotic theory. In fact, error bounds tend to reverse the usual relation between a problem and its solution, enabling us to generalize very easily the kind of problem for which a given asymptotic expansion furnishes the solution. Results for second-order differential equations are given in [11], [12], [13], [14], [15], [16], [17], [26], [28], [29], [30], [34], [35]; for differential equations of arbitrary order in [54]; for definite integrals in [20], [21], [24].

(iv) *Re-expansions of remainder terms.* The main object is to increase the accuracy attainable from expansions of standard Poincaré type by identifying contributions that are relatively exponentially small. Other important features include increased regions of validity in the complex plane, greater insight into the Stokes phenomenon, and the phenomenon of resurgence, that is, the re-emergence of the original Poincaré coefficients in the higher re-expansions. See [43], [44], [45], [46], [48], [49], [50], [52], [53].

(v) *Generalized asymptotic expansions.* The sometimes strange nature of generalized asymptotic expansions is illustrated in [22] by an example in which the sum of a convergent series of inverse factorials involving a large parameter is represented asymptotically by twice the partial sums of the same series.

II. Special Functions

(i) *Asymptotic expansions.* A general survey (for 1974) was given in [25].

Some of the general theories constructed in the papers listed in Topic I evolved from particular problems pertaining to special functions. At other times, special functions were chosen to provide illustrations of a new general theory. Either way, my philosophy has been that however comprehensive a newly-constructed general theory may appear to be, hidden shortcomings may emerge on making applications to significant examples. especially ones derived from other branches of mathematics or the natural sciences.

New expansions for Bessel functions, or their zeros or integrals, were supplied in [3], [4], [7], [32], [35]; for confluent hypergeometric functions in [14], [36], [43], [45], [46]; for Gamma and incomplete Gamma functions in [43], [44], [49], [51]; for generalized hypergeometric functions in [5];

for Legendre functions in [22], [27]; and for parabolic cylinder functions and their zeros in [10].

(ii) *Computational algorithms.* Computational algorithms for zeros of Bessel functions are constructed in [2], [3], and for Bessel and related functions in [18], [19], [23], [47]. See also [1] for an application of Bessel functions to the summation of a slowly convergent double series.

III. Numerical Analysis

My career has spanned an exciting half century in which computing aids have undergone radical changes. This is reflected in my publications. Algorithms in the early papers were constructed with desk machines and mathematical tables in mind. More recent papers rely on today's sophisticated automatic computers and associated software. Others address mathematical problems pertaining to computers themselves.

(i) *Boundary-value methods for difference and differential equations.* Boundary-value methods are needed when rogue solutions grow more rapidly than the wanted solution in direct step-by-step methods, leading to disastrous instability. These problems are treated in [18], [19], [23], [47] for second-order equations, and in [55] for equations of arbitrary order. The object is to devise stable configurations of boundary values, supported by error analysis.

(ii) *Error analysis.* In [19], [23], [37], [41] rigorous *a priori* or *a posteriori* bounds are developed for some standard processes in numerical analysis.

(iii) *Computer arithmetic.* The fixed-point and floating-point number systems are only two of many possible ways to represent numbers in a digital computer. The object of the research in [38], [39], [40], [42] is to imbed these systems into a more general system (or systems). Advantages of the new system(s) include: (a) a uniform metric error measure that effectively reduces to absolute precision and relative precision in the right circumstances; (b) closure with respect to arithmetic carried out in finite precision, thereby eliminating the awkward problems of overflow and underflow. (The last-named problems arise frequently in the construction of algorithms for generating the special functions.)

IV. Other Publications

The papers [A1]–[A41], and books [B1]–[B8] listed on pages 1064–1067, contain results of further research work, again on Topics I, II, and III above. Particular mention needs to be made of the frequently-referenced text [B6] that was first published in 1974. This book contains much of my research on Topics I and II prior to this date, often presented in a more comprehensive, or more accessible, form. It also contains a considerable number of results not published previously.

Acknowledgments

I am deeply indebted to my friend Roderick Wong for suggesting that my papers be published in book form, and for making all the necessary arrangements. I also express my thanks to the World Scientific Publishing Company for agreeing to undertake this task. Lastly, I thank all my co-authors, or their executors, for granting permission for their papers to be included in this collection, as well as the publishers of the journals or books in which all the papers appeared originally.

Frank W. J. Olver
University of Maryland
College Park, MD 20742-2431
April 1999

A PERSONAL TRIBUTE

D. S. Jones

When I was asked to write something about Frank Olver, I was reminded of an occasion when I was asked to advise about finance for Frank's research. I wrote "There is no better man in the world for this kind of research. Give him the money." and sent it off. Almost immediately came the reply "Not enough. We need 500 words." Not enough! Did he have to be a member of the Royal Family or a television star as well as being top man in his field? I could try to write the sentence in ten different ways but my computer would quickly inform me that I would be short of target still. I could describe his congenial companionship. I could describe his dancing to strict tempo on a boat in Canada; unusually for him he did not report that an error analysis was necessary. I could mention the happy evening spent with him and his charming wife in a bar in Hong Kong.

But was that enough for the man who revolutionised the evaluation of Bessel functions in the days when calculations were carried out on mechanical or, occasionally, electrical machines? Just because to-day your PC gives the value of a Bessel function in the flick of an eyelid and does not specify Olver's name in the process does not mean that his innovations are not hidden away in the works.

Much of the research on Bessel functions is summarised in the chapter he contributed to the *Handbook of Mathematical Functions* edited by Abramowitz and Stegun. A better idea of the enormous breadth and skill he has displayed is obtained from his classic text *Asymptotics and Special Functions*. While this book was written before his important work on hyperasymptotics it contains a vast amount of information not only on special functions but also on the solutions to general differential equations and approximations to them. Here you will discover how to make reliable approximations, derive error estimates for them, their behaviour near Stokes' lines and connection formulae. In spite of the generality of treatment the needs of practical calculations are always borne in mind.

I have alluded to his expertise at dancing in the ballroom but if you want to see dancing par excellence watch how he manoeuvres on the complex plane to derive his results. Lesser beings would not attempt such intricate steps. They can only stare in amazement and incredulity. The depth of analysis is profound and the details extremely tricky. Yet it all ends up in nice pictures which you can understand easily even if you cannot comprehend how they were obtained.

The evidence is in front of you. Read some of the pages of this book and savour the beautiful clarity of Olver's writing, the tremendous precision of his mathematical reasoning and the brilliance of his technical ability. You will soon be full of admiration for his achievements.

A wise man once said that to achieve something "you have to get your boots on before breakfast" but you have to be up damned early to get them on before Frank.

FRANK OLVER AND ASYMPTOTICS

R. Wong

Asymptotic analysis can mainly be divided into two major areas. In one area, we study the behavior of integrals which depend on a large parameter. In the other area, we investigate the behavior of solutions to some differential equations as the independent variable or an auxiliary parameter tends to infinity. In the first case, there are well-known results such as Laplace's approximation, Kelvin's principle of stationary phase and Debye's method of steepest descent. In the second area, we also have powerful tools such as the JWKB approximation, Turning-point theory and connection formulas. It is in the second area where Frank Olver has made tremendous contributions. In fact, now it is nearly impossible for someone going through the literature in this part of asymptotics not to notice the important results that Frank Olver has given.

First, he established the global existence of uniform asymptotic expansions for solutions to ordinary differential equations with turning points; see [6].* Second, he derived uniform asymptotic expansions for solutions to differential equations with simple or double poles; see [8] and [9]. Third, he constructed numerical bounds for the error terms associated with these expansions; see [12] and [13]. Fourth, he gave uniform asymptotic approximations for solutions to differential equations with two coalescing turning points; see [26]. There are many other important results that are also attributed to him. Here I can mention only a few of them, and will describe briefly how these results have helped me to solve some problems that have arisen in my own work.

As applications of the general theory developed in [6, 8, 9], Frank has given probably the best asymptotic approximations for Bessel functions [7] and parabolic cylinder functions [10]. The delicate estimates in [13] have enabled me to solve two problems proposed by Lee Lorch. One concerns the monotonicity of the inflection points of Bessel function $J_\nu(x)$, and the other is to show that the first two and three terms in the asymptotic expansion of the zeros $j_{\nu,k}$ of the function $J_\nu(x)$, as $\nu \to \infty$, form respectively a lower and upper bound for $j_{\nu,k}$. It is also the concept of computable error bounds, advocated by Frank Olver, that had given me ideas of how to settle a 60-year old conjecture of G. Szegö concerning the monotonicity of Lebesgue constants for the Legendre series, and how to prove a more than 20-year old conjecture of R. A. Askey on the monotonicity of the relative extrema of the Jacobi polynomials $P_0^{(0,-1)}(x)$.

Recently I and my co-workers have been investigating the uniform asymptotics of Meixner, Meixner-Pollaczek and Krawtchouk polynomials. It turns

*The numbers here refer to the articles listed under "Papers Included in this Collection".

out that their asymptotic expansions involve the parabolic cylinder function and its derivative. To show that these expansions are uniform with respect to the argument, we have made tremendous use of Frank's results in [10].

In reading Frank's papers in detail and having to make use of his results, I came to realize the amount of work involved in carrying out all the detailed and rigorous analysis, and how deep and powerful these results are. Many of Frank's papers are very substantial. In addition to those mentioned above, see, for instance, [34] and [52]. Frank's work is always very thorough and complete. He has set such a high standard in asymptotics that we younger generations will find it difficult to follow.

CONTENTS

TRANSFORMATION OF CERTAIN SERIES OCCURRING IN AERODYNAMIC INTERFERENCE CALCULATIONS

By F. W. J. OLVER

(*The National Physical Laboratory, Teddington, Middlesex*)

[Received 18 January 1949]

SUMMARY

A transformation is established which converts a certain slowly convergent series into a rapidly convergent and easily computable form. Numerical values of the coefficients of the new series are given for some cases of practical importance. The method used may be extended in its application; general forms of slowly convergent series which can be similarly transformed are stated.

1. Introduction†

In aerodynamic calculations of interference on lifting surfaces in rectangular wind-tunnels (**1**, **2**), the chordwise variation of induced upwash involves the computation of the following double series:

$$F(\alpha,\beta) = \sum_{n=1}^{\infty}\sum_{m=-\infty}^{\infty}(-)^n[f_n(\alpha-m)-f_n(\beta-m)], \tag{1}$$

where
$$f_n(y) = \frac{y(y^2+2\mu^2n^2)}{n^2(y^2+\mu^2n^2)^{\frac{3}{2}}},$$

α, β are real parameters and μ is a positive parameter. In aerodynamic work μ is the ratio of tunnel height to tunnel breadth.

The series (1) is very slowly convergent and laborious to compute by direct summation. Brown (**1**) gives an estimate for the remainder after a finite number of terms, and though the use of this estimate considerably shortens the work, it remains laborious.

In this paper the following transformation is established,

$$F(\alpha,\beta) = \chi(\beta)-\chi(\alpha), \tag{2}$$

where

$$\chi(x) = \tfrac{1}{6}\pi^2x+\sum_{s=1}^{\infty}k_s\sin 2\pi sx, \tag{3}$$

$$k_s = 8\pi\mu^2s\sum_{n=1}^{\infty}(-)^{n-1}K_0(2\pi\mu sn)+4\mu\sum_{n=1}^{\infty}(-)^{n-1}\frac{K_1(2\pi\mu sn)}{n}, \tag{4}$$

and $K_\nu(z)$ denotes the modified Bessel function (Watson, **3**).

The series (3), (4) are each rapidly convergent unless μ is very small.

† The author desires to acknowledge the collaboration of Mr. H. C. Garner who supplied the aerodynamic information in this section.

It will be seen that a transformation of this type can be established for any convergent series (1) with $f_n(y)$ of the more general form

$$\psi_n(y,\mu)(y^2+\mu^2n^2)^{-(r+\frac{1}{2})},$$

where $\psi_n(y,\mu)$ is a polynomial in y of fixed degree, and r is a fixed positive integer. The transformed series will always converge more rapidly than the original. Convergent series of the form

$$\sum_{m=-\infty}^{\infty}\left[\frac{\psi(\alpha-m,\mu)}{\{(\alpha-m)^2+\mu^2\}^{r+\frac{1}{2}}}-\frac{\psi(\beta-m,\mu)}{\{(\beta-m)^2+\mu^2\}^{r+\frac{1}{2}}}\right],$$

where $\psi(y,\mu)$ is a polynomial in y, may also be transformed into more rapidly convergent series by the same method.

2. Transformation of the series

Where fractional powers of complex quantities occur in the following analysis their principal values are to be understood. Write

$$S_n(\alpha,\beta)=\sum_{m=-\infty}^{\infty}[f_n(\alpha-m)-f_n(\beta-m)]\quad(n\geqslant 1),$$

so that
$$F(\alpha,\beta)=\sum_{n=1}^{\infty}(-)^nS_n(\alpha,\beta). \tag{5}$$

Then by Cauchy's residue theorem,

$$S_n(\alpha,\beta)=\int_C\phi(z)\,dz, \tag{6}$$

where

$$\phi(z)=\left[\frac{(\alpha-z)\{(\alpha-z)^2+2\mu^2n^2\}}{\{(\alpha-z)^2+\mu^2n^2\}^{\frac{3}{2}}}-\frac{(\beta-z)\{(\beta-z)^2+2\mu^2n^2\}}{\{(\beta-z)^2+\mu^2n^2\}^{\frac{3}{2}}}\right]\frac{\cot\pi z}{2in^2},$$

and C is the contour comprising the two straight lines $|\mathrm{Im}\,z|=c$, $0<c<\mu n$ (Fig. 1). The singularities of the function $\phi(z)$ consist of simple poles at the points $z=0,\ \pm1,\ \pm2,\ldots$, and branch points† at $z=\alpha\pm i\mu n,\ \beta\pm i\mu n$. The cuts corresponding to these branch points are parallel to the imaginary axis (Fig. 1).

In the domain $|\mathrm{Im}\,z|\geqslant c$, $\phi(z)=O(|z|^{-2})$ as $z\to\infty$, except between the pair of cuts in the upper half-plane, and also between the pair of cuts in the lower half-plane. In the former of these excepted regions $\phi(z)\to -n^{-2}$, as $z\to\infty$; and in the latter $\phi(z)\to n^{-2}$, as $z\to\infty$. It follows that in (6) the contour C may be deformed, and

$$S_n(\alpha,\beta)=\frac{2}{n^2}(\alpha-\beta)+\lim_{p\to\infty}\int_\Gamma\phi(z)\,dz. \tag{7}$$

† It is assumed in Fig. 1 and in the analysis that $\alpha>\beta$. That the final results remain true if $\alpha\leqslant\beta$ follows by interchange of α, β.

Γ comprises the aggregate of the four closed loop contours Γ_1, Γ_2, Γ_3, Γ_4 (Fig. 1).

Consider the integrals round Γ_1, Γ_2. Make the transformation

$$t = i(z-\alpha)/\mu n,$$

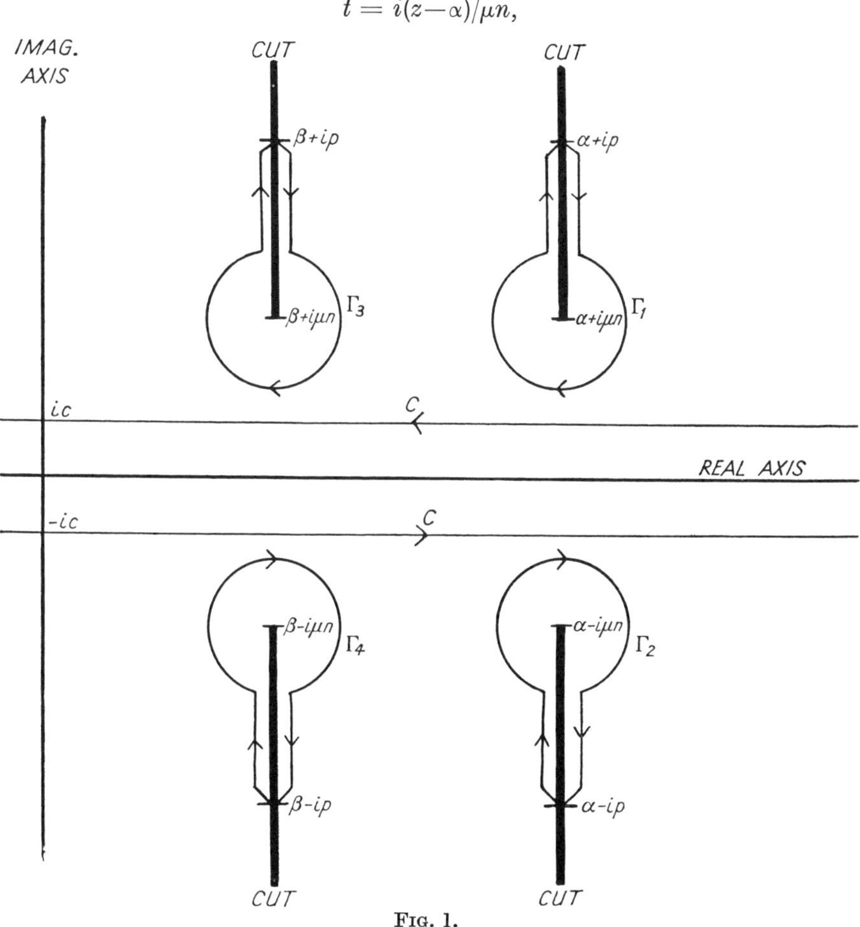

Fig. 1.

so that $\alpha - z = i\mu n t$. Then by use of Cauchy's theorem

$$\int_{\Gamma_1+\Gamma_2} \phi(z)\,dz = -\frac{\mu}{2n}\left[\int_{-q}^{(-1+)} + \int_{q}^{(1+)}\right]\frac{t(2-t^2)}{(1-t^2)^{\frac{3}{2}}}\coth \pi(\mu n t + i\alpha)\,dt, \tag{8}$$

where $q = p/\mu n$.

The paths of integration on the right-hand side of (8) are illustrated in Fig. 2. Similarly

$$\int_{\Gamma_3+\Gamma_4} \phi(z)\,dz = \frac{\mu}{2n}\left[\int_{-q}^{(-1+)} + \int_{q}^{(1+)}\right]\frac{t(2-t^2)}{(1-t^2)^{\frac{3}{2}}}\coth \pi(\mu n t + i\beta)\,dt. \tag{9}$$

Adding (8) and (9), and letting $p \to \infty$, equation (7) can be expressed in the form

$$S_n(\alpha,\beta) = \frac{2}{n^2}(\alpha-\beta)+\frac{\mu i}{2n}\sin\pi(\alpha-\beta)\times$$
$$\times\left[\int\limits_{-\infty}^{(-1+)}+\int\limits_{\infty}^{(1+)}\right]\frac{t(2-t^2)}{(1-t^2)^{\frac{3}{2}}}\frac{dt}{\sinh\pi(\mu nt+i\alpha)\sinh\pi(\mu nt+i\beta)}. \tag{10}$$

It is easily shown that

$$\operatorname{cosech}\pi(\mu nt+i\alpha)\operatorname{cosech}\pi(\mu nt+i\beta)+\operatorname{cosech}\pi(\mu nt-i\alpha)\operatorname{cosech}\pi(\mu nt-i\beta)$$

can be expanded in the form

$$4\operatorname{cosec}\pi(\alpha-\beta)\sum_{s=1}^{\infty}A_s e^{-2\pi\mu snt},\quad \text{where}\quad A_s \equiv \sin 2\pi s\alpha-\sin 2\pi s\beta.$$

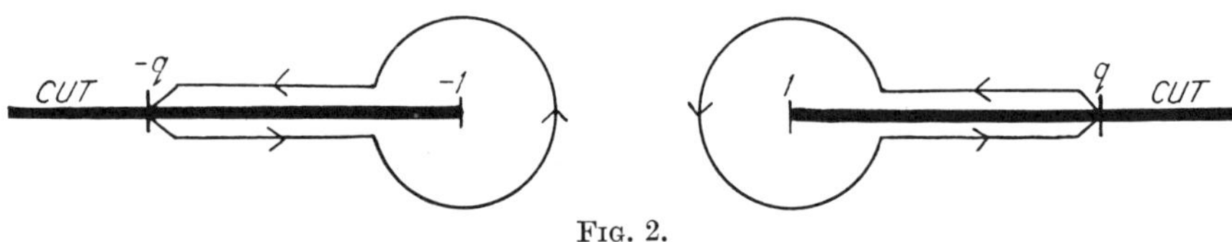

FIG. 2.

Hence on combining the two integrals in (10), and inverting the order of summation and integration, the following expansion is obtained.

$$S_n(\alpha,\beta) = \frac{2}{n^2}(\alpha-\beta)+\frac{2\mu i}{n}\sum_{s=1}^{\infty}A_s\int\limits_{\infty}^{(1+)}\frac{t(2-t^2)}{(1-t^2)^{\frac{3}{2}}}e^{-2\pi\mu snt}\,dt.$$

Integrating by parts,

$$S_n(\alpha,\beta) = \frac{2}{n^2}(\alpha-\beta)+\frac{i}{\pi n^2}\sum_{s=1}^{\infty}\frac{A_s}{s}\int\limits_{\infty}^{(1+)}\left[\frac{3}{(1-t^2)^{\frac{5}{2}}}-\frac{1}{(1-t^2)^{\frac{3}{2}}}\right]e^{-2\pi\mu snt}\,dt. \tag{11}$$

Now (Watson, **3**, 172) if $\mathrm{Rl}\,\nu > -\frac{1}{2}$ and $|\arg z| < \frac{1}{2}\pi$,

$$K_\nu(z) = \frac{\pi^{\frac{1}{2}}(\frac{1}{2}z)^\nu}{\Gamma(\nu+\frac{1}{2})}\int\limits_1^\infty e^{-zt}(t^2-1)^{\nu-\frac{1}{2}}\,dt.$$

Hence
$$\int\limits_{\infty}^{(1+)}e^{-zt}(1-t^2)^{\nu-\frac{1}{2}}\,dt = -\frac{2i\Gamma(\nu+\frac{1}{2})}{\pi^{\frac{1}{2}}(\frac{1}{2}z)^\nu}\cos\nu\pi K_\nu(z).$$

Analytic continuation extends this formula to all values of ν. Putting $\nu = -1, -2$, in turn,

$$\int\limits_{\infty}^{(1+)}e^{-zt}(1-t^2)^{-\frac{3}{2}}\,dt = -2izK_1(z),$$

$$\int\limits_{\infty}^{(1+)}e^{-zt}(1-t^2)^{-\frac{5}{2}}\,dt = -\tfrac{2}{3}iz^2K_2(z) = -\tfrac{2}{3}iz[zK_0(z)+2K_1(z)].$$

Substituting these results in (11), and substituting the above values of A_s,

$$S_n(\alpha,\beta) = \frac{2}{n^2}(\alpha-\beta)+\sum_{s=1}^{\infty}(\sin 2\pi s\alpha-\sin 2\pi s\beta)\times$$
$$\times\left[8\pi\mu^2 sK_0(2\pi\mu sn)+\frac{4\mu}{n}K_1(2\pi\mu sn)\right]. \quad (12)$$

Substitution of this value for $S_n(\alpha,\beta)$ in (5) establishes the final transformation given by equations (2), (3), (4) of the introduction. Since

$$K_\nu(z) \sim (\pi/2z)^{\frac{1}{2}}e^{-z}, \quad \text{as} \quad z\to+\infty,$$

the series (3), (4), (12) each converge very rapidly provided μ is not small. In particular

$$k_s \sim 4\pi\mu^{\frac{3}{2}}s^{\frac{1}{2}}e^{-2\pi\mu s}, \quad \text{as} \quad s\to\infty.$$

These series can be readily computed provided tables of the Bessel functions are available.

3. Numerical tables

Cases of practical importance include the square tunnel ($\mu = 1$) and the duplex tunnel ($\mu = \frac{1}{2}$ or 2). For these cases numerical results have been obtained; the relevant values of k_s are given in Table 1. The computation entailed the subsidiary tabulation of the functions $K_0(n\pi)$ and $K_1(n\pi)$. These are given in Table 2, together with the associated functions $I_0(n\pi)$ and $I_1(n\pi)$, computed incidentally for checking purposes and included for the sake of completeness. With the aid of Table 2, values of k_s can be computed for all rectangular tunnels for which μ is an integer or half an odd integer. All numbers in Tables 1 and 2 are positive. End figures have been correctly rounded off.

The work described here has been carried out as part of the research programme of the National Physical Laboratory, and this paper is published by permission of the Director of the Laboratory.

TABLE 1

s	k_s		
	$\mu = \frac{1}{2}$	$\mu = 1$	$\mu = 2$
1	0·246709	0·026951	0·000133
2	0·013476	0·000066	0·000000
3	0·000682	0·000000	
4	0·000033		
5	0·000002		
6	0·000000		

$\pi^2/6 = 1·644934$

TABLE 2

n	$K_0(n\pi)$	$K_1(n\pi)$	$I_0(n\pi)$	$I_1(n\pi)$
1	0·02950 8684	0·03391 3503	5·47784 50	4·49145 67
2	0·00091 6584	0·00098 6996	87·1085	79·8390
3	0·00003 2532	0·00003 4216	$1·6331 \times 10^3$	$1·5439 \times 10^3$
4	0·00000 1221	0·00000 1269	$3·261 \times 10^4$	$3·128 \times 10^4$
5	0·00000 0047	0·00000 0049	$6·7 \times 10^5$	$6·5 \times 10^5$
6	0·00000 0002	0·00000 0002	$1·4 \times 10^7$	$1·4 \times 10^7$
7	0·00000 0000	0·00000 0000	3×10^8	3×10^8

REFERENCES

1. W. S. BROWN, 'Wind tunnel corrections on ground effect', *Aeronautical Research Committee, Reports and Memoranda*, No. 1865 (1938).

2. H. C. GARNER, 'Wind tunnel interference on two-dimensional and finite wings fitted with controls', *Aeronautical Research Committee Report* No. 10936 (1947).

3. G. N. WATSON, *Theory of Bessel Functions* (Cambridge, 1922).

[Extracted from the *Proceedings of the Cambridge Philosophical Society,*
Vol. 46. Pt. 4.]

A NEW METHOD FOR THE EVALUATION OF ZEROS OF BESSEL FUNCTIONS AND OF OTHER SOLUTIONS OF SECOND-ORDER DIFFERENTIAL EQUATIONS

BY F. W. J. OLVER

Communicated by E. T. GOODWIN

Received 10 March 1950

1. *Introduction.* Methods for the evaluation of zeros of the Bessel functions J, Y have been described in detail in a series of Notes by Bickley, Miller and Jones (1). In the course of work for the British Association Mathematical Tables Committee, they encounted difficulties in computing these zeros for functions of orders other than zero and unity. The two simplest methods of computation are inverse interpolation where the functions are tabulated at a small interval, and the McMahon expansion for the larger zeros. To the accuracy required, the ranges of applicability of these methods overlap for the functions of zero and unit order, but after these a gap appears and widens steadily with increasing order.

Attempts to close this gap were made in the series of Notes referred to above, in particular by a method of interpolation using Taylor's series (Note III) and by use of the 'Miller-Jones expansion' (Note IV). Neither of these methods however, was completely successful, and the gap remained.

The present paper describes a method of obtaining the zeros as a solution of a differential equation. It shares with the McMahon expansion the advantage of being entirely independent of computed values of the functions, and with its aid the gap can be bridged without a prohibitive amount of labour. It is at present being successfully employed to complete the programme of the Mathematical Tables Committee*.

The method is not restricted to Bessel functions, and could be applied to the solution of any linear second-order differential equation provided that it has a sequence of fairly regularly spaced zeros. For this reason the description that follows is in general terms; the particular applications to the zeros and stationary values of Bessel functions are obtained as special cases.

2. *The differential equation for the zeros.* Consider the equation

$$\frac{d^2w}{dz^2}+p(z)\frac{dw}{dz}+q(z)\,w = 0, \tag{2·1}$$

where $p(z)$, $q(z)$ are analytic functions of z. Let $F(z)$, $G(z)$ be linearly independent solutions of (2·1). The function

$$C(z,\alpha) \equiv F(z)\cos\alpha - G(z)\sin\alpha, \tag{2·2}$$

where α is a parameter, also satisfies (2·1). The equation

$$C(z,\alpha) = 0 \tag{2·3}$$

* Now the Royal Society Mathematical Tables Committee.

defines z, α as functions of each other. If $z = z_0$ is an ordinary point of the differential equation (2·1), then the corresponding α_0 is determined, save for an arbitrary additive multiple of π, by the relation
$$\tan\alpha_0 = F(z_0)/G(z_0). \tag{2·4}$$
Now
$$d\alpha/dz = (GF' - FG')/(F^2 + G^2), \tag{2·5}$$
and cannot vanish at $z = z_0$ in virtue of the usual Wronskian properties. Hence a neighbourhood N of α_0 can be found in which there exists a unique analytic solution, say $z = \rho(\alpha)$, of the equation (2·3), with the property $\rho(\alpha_0) = z_0$. It is proposed to find the differential equation satisfied by $\rho(\alpha)$.

Write
$$\left.\begin{aligned} A(z,\alpha) &= F(z)\sin\alpha + G(z)\cos\alpha, \\ B(z,\alpha) &= F'(z)\sin\alpha + G'(z)\cos\alpha, \\ D(z,\alpha) &= F'(z)\cos\alpha - G'(z)\sin\alpha. \end{aligned}\right\} \tag{2·6}$$
Then
$$\frac{\partial A}{\partial\alpha} = C, \quad \frac{\partial B}{\partial\alpha} = D, \quad \frac{\partial C}{\partial\alpha} = -A, \quad \frac{\partial D}{\partial\alpha} = -B, \quad \frac{\partial A}{\partial z} = B, \quad \frac{\partial C}{\partial z} = D. \tag{2·7}$$
Also, by (2·1),
$$\frac{\partial B}{\partial z} = -pB - qA, \quad \frac{\partial D}{\partial z} = -pD - qC. \tag{2·8}$$
Differentiate (2·3) with respect to α, using (2·7). Then
$$\rho' D - A = 0, \quad \text{where} \quad \rho' \equiv d\rho/d\alpha, \quad \text{etc.} \tag{2·9}$$
Differentiating twice again successively using (2·3), (2·7), (2·8), we obtain
$$(\rho'' - p\rho'^2)D - 2\rho' B = 0, \tag{2·10}$$
$$\left[\rho''' - 3p\rho'\rho'' + \left(p^2 - \frac{dp}{d\rho}\right)\rho'^3 - 2\rho'\right]D + 3(p\rho'^2 - \rho'')B + 2q\rho'^2 A = 0. \tag{2·11}$$
Eliminate A, B, D, from the last three equations. Then
$$2\rho'\rho''' - 3\rho''^2 + Q(\rho)\rho'^4 - 4\rho'^2 = 0, \tag{2·12}$$
where
$$Q(\rho) = 4q(\rho) - [p(\rho)]^2 - 2p'(\rho). \tag{2·13}$$
Equation (2·12) is the required differential equation for $\rho(\alpha)$. Starting in the neighbourhood N of α_0, we can integrate it numerically for values of $\rho(\alpha)$ beyond N. Each ρ is a zero $z = \rho$, of the function $C(z,\alpha)$ with the corresponding value of α. In particular, if α is an odd multiple of $\frac{1}{2}\pi$, the corresponding ρ is a zero of $G(z)$. Even multiples of $\frac{1}{2}\pi$ correspond to zeros of $F(z)$.

The values of the derivative of a function at its own zeros are often of interest. These can be obtained incidentally from the numerical solution of (2·12) as follows. From (2·2), (2·6),
$$BC - AD = F(z)G'(z) - F'(z)G(z) = W(z), \tag{2·14}$$
$W(z)$ denoting the Wronskian of $F(z)$, $G(z)$. Putting $z = \rho(\alpha)$ and using (2·3), (2·9), we obtain
$$\rho' D^2 = -W(\rho).$$
Thus
$$\left[\frac{\partial}{\partial z}C(z,\alpha)\right]_{z=\rho} = \pm\left[-\frac{W(\rho)}{\rho'}\right]^{\frac{1}{2}}. \tag{2·15}$$
It may be remarked that the formula (2·15) can be used, quite independently of (2·12), to compute the values of the derivative at the zeros of the function whenever a sequence of zeros are known, ρ' being obtained by numerical differentiation.

3. *Zeros of Bessel functions.* Take $p(\rho) = \rho^{-1}$, $q(\rho) = \rho^{-2}(\rho^2 - n^2)$, $\alpha = \pi t$. Then the results of the previous section show that the zeros (ρ) of the cylinder function

$$\mathscr{C}_n(z) \equiv J_n(z) \cos \pi t - Y_n(z) \sin \pi t, \tag{3·1}$$

satisfy the differential equation

$$2\rho^2\rho'\rho''' - 3\rho^2\rho''^2 + (4\rho^2 + 1 - 4n^2)\rho'^4 - 4\pi^2\rho^2\rho'^2 = 0, \tag{3·2}$$

where here, and henceforth, the derivatives are taken with respect to t and not α.

In order to remove the arbitrary constant in the relation

$$\tan \pi t = J_n(\rho)/Y_n(\rho), \tag{3·3}$$

the values of ρ, t are now supposed to be such that $t = 0$ when $\rho = 0$. Consideration of the graph of $J_n(\rho)/Y_n(\rho)$ shows that $j_{n,s}$, the sth positive zero of $J_n(z)$, is the value of ρ for which $-t = s$. Similarly, $y_{n,s}$ is the value of ρ for which $-t = s - \frac{1}{2}$.

From (2·15)

$$\mathscr{C}'_n(\rho) = \pm[-2/\rho\rho']^{\frac{1}{2}}. \tag{3·4}$$

Another example is furnished by the zeros of $J'_n(z)$, $Y'_n(z)$. These functions satisfy the differential equation

$$z^2(z^2 - n^2)w'' + z(z^2 - 3n^2)w' + [z^4 - (2n^2 + 1)z^2 + n^2(n^2 - 1)]w = 0. \tag{3·5}$$

Hence by (2·12) the zeros of $\mathscr{C}'_n(z)$ satisfy

$$2\rho'\rho''' - 3\rho''^2 + Q(\rho)\rho'^4 - 4\pi^2\rho'^2 = 0, \tag{3·6}$$

where
$$Q(\rho) = [4\rho^6 - 3(4n^2+1)\rho^4 + 2n^2(6n^2 - 5)\rho^2 + n^4(1 - 4n^2)]/\rho^2(\rho^2 - n^2)^2. \tag{3·7}$$

At a zero of $\mathscr{C}'_n(z)$,

$$\mathscr{C}''_n(\rho) = \pm[-2(\rho^2 - n^2)/\rho^3\rho']^{\frac{1}{2}}, \tag{3·8}$$

so that, using Bessel's equation, we find

$$\mathscr{C}_n(\rho) = \pm[-2\rho/\rho'(\rho^2 - n^2)]^{\frac{1}{2}}. \tag{3·9}$$

4. *Numerical integration for Bessel function zeros.* (*a*) *Mathematical details.* The equation (2·12), and the particular case (3·2), is third order and non-linear; its solution by direct methods would be a formidable and laborious task. However, whenever the zeros are fairly regularly spaced, as is the case for all but the smallest zeros of J and Y, ρ'' and ρ''' are small compared with ρ'. In these circumstances these equations can be treated as first-order equations and integrated numerically by a method of successive approximation. In general terms the procedure is as follows:

(i) Neglect ρ'', ρ''' and integrate the resulting first-order equation to find a first approximation to the required solution.

(ii) Obtain from this approximation an estimate of the 'correction terms' involving ρ'' and ρ'''.

(iii) Solve again the first-order equation, in which the estimated correction terms are now included, to obtain a better approximation to the required solution ρ.

(iv) Repeat this iterative cycle until there is no further change in ρ.

In practice this programme is modified in several ways, various artifices being introduced which reduce the labour involved. The most important of these is a procedure whereby the approximation to the correction terms obtained at stage (ii) is materially improved before the next integration at stage (iii) takes place. This reduces the number of cycles of the iterative process.

In the case of the Bessel functions the mathematical details are as follows. Equation (3·2) can be written in the form

$$\rho' = -\frac{2\pi\rho}{(4\rho^2+1-4n^2)^{\frac{1}{2}}}\left[1-\frac{1}{2\pi^2}\frac{\rho'''}{\rho'}+\frac{3}{4\pi^2}\frac{\rho''^2}{\rho'^2}\right]^{\frac{1}{2}}. \qquad (4\cdot1)$$

This equation is to be integrated numerically at an interval h in t, and for this purpose it is conveniently rewritten as

$$\rho_1 = -kR(\rho)\,(1+\Delta), \qquad (4\cdot2)$$

where

$$\rho_1 = 2h\rho', \quad \rho_2 = 2h^2\rho'', \quad \rho_3 = 2h^3\rho''',$$

$$k = 2\pi h, \quad R(\rho) = \rho(\rho^2-a^2)^{-\frac{1}{2}}, \quad a^2 = n^2-\tfrac{1}{4},$$

and

$$\left.\begin{aligned} 1+\Delta &= (1-2\lambda)^{\frac{1}{2}}, \\ \lambda &= \frac{1}{k^2}\left(\frac{\rho_3}{\rho_1}-\frac{3}{2}\frac{\rho_2^2}{\rho_1^2}\right). \end{aligned}\right\} \qquad (4\cdot3)$$

where

The first step is the integration of (4·2), with Δ neglected, to give a first approximation $\rho^{(1)}$ to the required ρ. This equation has the analytical solution

$$t-t_0 = \frac{1}{2\pi}[F_n(\rho_0^{(1)})-F_n(\rho^{(1)})], \qquad (4\cdot4)$$

where $\rho^{(1)} = \rho_0^{(1)}$ at $t = t_0$, and where

$$F_n(\rho^{(1)}) = [4(\rho^{(1)})^2+1-4n^2]^{\frac{1}{2}}-(4n^2-1)^{\frac{1}{2}}\tan^{-1}\left[\frac{4(\rho^{(1)})^2+1-4n^2}{4n^2-1}\right]^{\frac{1}{2}}(4n^2>1). \qquad (4\cdot5)$$

It is easier, however, to obtain $\rho^{(1)}$ by numerical integration, from the central-difference extrapolation formula

$$\rho^{(1)}(t+h) = \rho^{(1)}(t-h)+(1+\tfrac{1}{6}\delta^2-\tfrac{1}{180}\delta^4+\ldots)\,\rho_1^{(1)}(t). \qquad (4\cdot6)$$

When $\rho^{(1)}$ has been obtained, an estimate $\lambda^{(1)}$ for λ, given by the second of (4·3), can be computed from the equation

$$\lambda^{(1)} = \frac{1}{k^2}\left[\frac{\rho_3^{(1)}}{\rho_1^{(1)}}-\frac{3}{2}\left(\frac{\rho_2^{(1)}}{\rho_1^{(1)}}\right)^2\right]. \qquad (4\cdot7)$$

An estimate $\Delta^{(1)}$ for Δ is therefore available, and this could be substituted in (4·2), and a further integration performed.

Now it is clear that the number of cycles of this iterative process necessary for convergence depends on the difference between the estimated Δ and the true Δ; the better the estimate the fewer the cycles. The question therefore arises of the possibility of improving the estimate $\Delta^{(1)}$ without, at this stage, performing another integration.

An answer can be given by following *analytically* the consequences of a second integration, by which a second approximation $\rho^{(2)}$ would be computed from the equation

$$\rho_1^{(2)} = -kR(\rho^{(2)})\,(1+\Delta^{(1)}), \qquad (4\cdot8)$$

and a second estimate $\lambda^{(2)}$ from (4·7), with $\rho^{(2)}$ replacing $\rho^{(1)}$.

The dominant term in (4·7) is that involving ρ_3, and to a close approximation it follows that

$$\lambda^{(2)}-\lambda^{(1)} = \frac{\rho_3^{(2)}}{k^2\rho_1^{(2)}}-\frac{\rho_3^{(1)}}{k^2\rho_1^{(1)}}. \qquad (4\cdot9)$$

Further, since $\rho^{(1)}$ was obtained from an equation like (4·8), but with Δ equal to zero, it follows that, to the first order,

$$\rho_1^{(2)} = (1+\Delta^{(1)})\rho_1^{(1)}, \qquad (4\cdot10)$$

and, on using approximate central-difference formulae for numerical differentiation,

$$\left.\begin{aligned} \rho_3^{(2)} &= (\delta^2 - \tfrac{1}{12}\delta^4)(\rho_1^{(1)} + \rho_1^{(1)}\Delta^{(1)}), \\ \rho_3^{(1)} &= (\delta^2 - \tfrac{1}{12}\delta^4)\rho_1^{(1)}. \end{aligned}\right\} \tag{4·11}$$

Substitution in (4·3), (4·9), and subsequent expansion then yields the approximation

$$\Delta^{(2)} = \Delta^{(1)} - \frac{1}{k^2\rho_1^{(1)}}[(\delta^2 - \tfrac{1}{12}\delta^4)(\rho_1^{(1)}\Delta^{(1)}) - \Delta^{(1)}(\delta^2 - \tfrac{1}{12}\delta^4)\rho_1^{(1)}], \tag{4·12}$$

which can be calculated without difficulty when $\rho^{(1)}$ and $\rho_1^{(1)}$ and hence $\Delta^{(1)}$ have been obtained.

This estimate for $\Delta^{(2)}$ is now used in equation (4·2), and a second integration of this equation is performed to obtain a better approximation $\rho^{(3)}$ to ρ. The $\Delta^{(3)}$ corresponding to this $\rho^{(3)}$ can be calculated and compared with the $\Delta^{(2)}$ actually used in the integration. If the difference

$$d = \Delta^{(3)} - \Delta^{(2)} \tag{4·13}$$

is negligible, the process is ended and the problem solved. Otherwise the method described for deriving a better estimate for Δ is used again, obtaining $\Delta^{(4)}$ from a formula analogous to (4·12), given by

$$\Delta^{(4)} = \Delta^{(3)} - \frac{1}{k^2\rho_1^{(3)}}[(\delta^2 - \tfrac{1}{12}\delta^4)(\rho_1^{(3)}d) - d(\delta^2 - \tfrac{1}{12}\delta^4)\rho_1^{(3)}]. \tag{4·14}$$

Proceeding in this way, we obtain the required solution in fewer cycles than would be necessary without the refinements. Indeed, the functions obtained by numerical integration are $\rho^{(1)}, \rho^{(3)}, \rho^{(5)}, \ldots$, and by other numerical processes $\Delta^{(1)}, \Delta^{(3)}, \ldots$, and good approximations to $\Delta^{(2)}, \Delta^{(4)}, \ldots$; the intermediate quantities $\rho^{(2)}, \rho^{(4)}, \ldots$, are never obtained.

(*b*) *Numerical details.* In the practical application of the method of the preceding subsection, the integration is started with one or more known zeros of fairly high magnitude, corresponding to values of $-t$ of twenty or thirty, and then proceeds with steadily decreasing values of $-t$. The larger zeros of J and Y are very regularly spaced; it may therefore be expected that the method will be very satisfactory for the larger values of $-t$ but become less so as $-t$ decreases and the differences of ρ become greater. Ultimately, the method is no longer practicable, and for the smallest zeros an alternative method, discussed in § 6, is used. In the example quoted later the alternative method was introduced at $t = -4·0$ and used for the computation of the first three zeros of $J_{20}(x)$ and the first four zeros of $Y_{20}(x)$.

As with all methods of successive approximation, it is possible to start with a small number of figures, more being added as the work proceeds. Six or seven decimals might be retained, for example, in the first integration which produces $\rho^{(1)}$. The use of equation (4·6) in this connexion is straightforward, the chosen interval being such that $\delta^2\rho_1^{(1)}$ can be extrapolated without difficulty, and $\delta^4\rho_1^{(1)}$ ignored.

It is then required to obtain $\Delta^{(1)}$ to more decimal places for use in future work. Significant figures are inevitably lost by numerical differentiation, and it is preferable at this stage to obtain $\rho_2^{(1)}$ and $\rho_3^{(1)}$ by differentiating the differential equation. It is found that

$$\rho_2^{(1)} = -\tfrac{1}{2}k^2a^2\rho^{(1)}[(\rho^{(1)})^2 - a^2]^{-2}, \quad \rho_3^{(1)} = \tfrac{1}{4}k^2a^2\rho_1^{(1)}[3(\rho^{(1)})^2 + a^2][(\rho^{(1)})^2 - a^2]^{-3},$$

whence

$$\lambda^{(1)} = \tfrac{1}{8}a^2[6(\rho^{(1)})^2 - a^2][(\rho^{(1)})^2 - a^2]^{-3}. \tag{4·15}$$

When this formula is used, $\lambda^{(1)}$ is usually obtainable to eleven decimals from the six or seven decimal values of $\rho^{(1)}$.

The corresponding eleven decimal values of $\Delta^{(1)}$ may of course agree only in the first few figures with the true Δ, but it is found that the improved estimate $\Delta^{(2)}$, obtained from equation (4·12), is often powerful enough to give four or five more correct figures. It may be remarked here that the term $\frac{1}{12}\Delta^{(1)}\delta^4\rho_1^{(1)}$ is usually negligible, and that if $\delta\rho_1^{(1)}$ and $\Delta^{(1)}$ are very small, a good enough approximation is given by

$$\Delta^{(2)} = \Delta^{(1)} - k^{-2}\,\delta^2\Delta^{(1)}. \qquad (4\cdot16)$$

A similar simplification can be used in the case of equation (4·14).

When subsequent integrations are performed for $\rho^{(3)}$, $\rho^{(5)}$, ..., it is no longer practicable to use an equation like (4·15) for λ, and formulae for numerical differentiation must be used, given by

$$\left.\begin{aligned}\rho_2 &= (\mu\delta - \tfrac{1}{6}\mu\delta^3 + \tfrac{1}{30}\mu\delta^5 - \tfrac{1}{140}\mu\delta^7 + \ldots)\,\rho_1,\\ \rho_3 &= (\delta^2 - \tfrac{1}{12}\delta^4 + \tfrac{1}{90}\delta^6 - \tfrac{1}{560}\delta^8 + \tfrac{1}{3150}\delta^{10} - \ldots)\,\rho_1.\end{aligned}\right\} \qquad (4\cdot17)$$

If ten decimals are retained in $\rho_1^{(3)}$ the loss of significant figures at intervals $h \geqslant \frac{1}{4}$ is not serious enough to prevent accurate computation of $\lambda^{(3)}$ to eleven decimals.

In performing the integrations for $\rho^{(3)}$ it is found that not more than eight or nine decimals need be retained in $\rho^{(3)}$ in order to obtain $\rho_1^{(3)}$ to ten decimals. Ten decimals can subsequently be regained in ρ by use of the central-difference quadrature formula

$$\rho(t+h) = \rho(t) + \tfrac{1}{2}[\mu - \tfrac{1}{12}\mu\delta^2 + \tfrac{11}{720}\mu\delta^4 - \tfrac{191}{60480}\mu\delta^6 + \tfrac{2497}{3628800}\mu\delta^8 - \ldots]\,\rho_1(t+\tfrac{1}{2}h). \quad (4\cdot18)$$

When more than seven decimals in ρ have to be retained in the integration, the differences become so large that the extrapolation required in equation (4·6) is so difficult as to make this method inconvenient. A backward-difference formula of integration, involving no estimation, may then be used. The particular form found most convenient here is given by

$$\rho(t+h) = \rho(t-h) + (1 + a_2\nabla^2 + a_3\nabla^3 + a_4\nabla^4 + \ldots)\,\rho_1(t), \qquad (4\cdot19)$$

where
$$a_s = \frac{(-)^s}{s!}\,[B_s^{(s)} + \tfrac{1}{2}sB_{s-1}^{(s-1)}].$$

Exact values of the generalized Bernoulli numbers $B_s^{(s)}$ for $s = 1\,(1)\,10$ are given by Milne-Thomson ((2), p. 136), and from these the following table has been derived:

s	a_s
2	$0\cdot1\dot{6}$
3	$0\cdot1\dot{6}$
4	$0\cdot16\dot{1}$
5	$0\cdot1\dot{5}$

s	a_s
6	0·15066 13757
7	0·14642 85714
8	0·14275 57319
9	0·13954 14462
10	0·13670 18732

It has already been remarked that a given number of decimals in ρ yields a greater accuracy in the derived value of ρ_1. This comparative insensitivity of ρ_1 to the value of ρ (which is consequent upon the smallness of ρ'' compared with ρ') has the further advantage of ensuring that the usual difficulties associated with integration using a backward-difference formula do not arise in this case.

One further point is of computational interest. It is rarely necessary to go beyond $\Delta^{(4)}$, and the final approximation to ρ could be obtained from the equation

$$\rho_1^{(5)} = -kR(\rho^{(5)})\,(1+\Delta^{(4)}). \tag{4·20}$$

Labour can be saved if the solution is written in the form

$$\rho^{(5)} = \rho^{(3)} + \epsilon.$$

Then, on expansion,

$$\rho_1^{(3)} + \epsilon_1 = -k[R(\rho^{(3)}) + \epsilon R'(\rho^{(3)}) + \ldots]\,(1+\Delta^{(4)}). \tag{4·21}$$

Since
$$\rho_1^{(3)} = -kR(\rho^{(3)})\,(1+\Delta^{(2)}),$$

it follows that
$$\epsilon_1 = kR(\rho^{(3)})\,(\Delta^{(2)}-\Delta^{(4)}) - k(1+\Delta^{(4)})\,[\epsilon R'(\rho^{(3)}) + \ldots]. \tag{4·22}$$

Usually ϵ^2 is negligible, and equation (4·22) can be integrated for ϵ and ϵ_1 by the use of equation (4·6).

5. *Numerical example.* As an illustration of the above method, some details are now given of the computation of the first thirty zeros of $J_{20}(x)$, $Y_{20}(x)$ to ten decimal places. The starting value assumed was the thirty-first zero of $Y_{20}(x)$,

$$y_{20,31} = \rho(-30·5) = 124·84472\;03114,$$

which was obtained from the McMahon asymptotic expansion.

Down to the eighth zero of $Y_{20}(x)$ it was possible to integrate at an interval $h = \frac{1}{2}$. The values of ρ produced at each step are then alternately zeros of $J_{20}(x)$ and $Y_{20}(x)$. From the eighth zero to the fourth the interval had to be halved. Below the fourth zero the differences increase so rapidly that the alternative method, described in § 6, was preferable to further reduction of the interval.

Tables are given of extracts from the computations involved in this example. The following explanations of the tables summarize the various stages described in § 4.

Table 1. Solution of the equation

$$\rho_1^{(1)} = -kR(\rho^{(1)}). \tag{5·1}$$

Table 1

$-t$	$\rho^{(1)}$	$\rho_1^{(1)}$	δ	δ^2	δ^3	δ^4	$\dfrac{10000}{(\rho^{(1)})^2-a^2}$	$R(\rho^{(1)})$
11·0	61·931811	−3·319325		−963		−21	2·9105309	1·0565739
			−10852		−137			
10·5	60·269478	3·330177		1100		29	3·0934277	1·0600282
			11952		166			
10·0	58·601450	3·342129		1266		33	3·2955662	1·0638328
			13218		199			
9·5	56·927138	3·355347		1465		43	3·5199505	1·0680403
			14683		242			
9·0	55·245859	3·370030		1707		55	3·7702345	1·0727139
			16390		297			
8·5	53·556823	3·386420		2004		78	4·0509064	1·0779309
			−18394		−375			
8·0	51·859105	−3·404814		−2379		−92	4·3675431	1·0837862

Table 2. Computation of $\Delta^{(1)}$ by (4·15) and the first of (4·3). Computation of $\Delta^{(2)}$ by (4·12). When $-t > 22·0$, (4·12) can be replaced by (4·16). Before use $\Delta^{(2)}$ is checked by differencing on a National Accounting Machine.

Table 2

$-t$	$\lambda^{(1)}$	$\frac{1}{2}(\lambda^{(1)})^2$	$\Delta^{(1)}$	$\rho_1^{(1)}\Delta^{(1}$	δ	δ^2	δ^3	δ^4	δ^5	δ^6	δ^7	δ^8	$\Delta^{(2)}$
	10^{-11}	10^{-11}	10^{-11}	10^{-10}									10^{-11}
11·0	+2786017	+ 39	−2786056	+ 924783		+19959		+1076		+96		+78	−2780073
					+129116		+4782		+377		+80		
10·5	3164644	50	3164694	1053899		24741		1453		176		−2	3157309
					153857		6235		553		78		
10·0	3613667	65	3613732	1207756		30976		2006		254		+57	3604531
					184833		8241		807		135		
9·5	4150271	86	4150357	1392589		39217		2813		389		127	4138769
					224050		11054		1196		262		
9·0	4796992	115	4797107	1616639		50271		4009		651		116	4782340
					274321		15063		1847		378		
8·5	5583793	156	5583949	1890960		65334		5856		1029		372	5564882
					+339655		+20919		+2876		+750		
8·0	+6551140	+215	−6551355	+2230615		+86253		+8732		+1779		+597	−6526368

Table 3

$-t$	$\rho^{(3)}$	$\dfrac{(\rho^{(3)})^2}{(\rho^{(3)})^2-a^2}$	$\rho_1^{(3)}$	δ	δ^2	δ^3	δ^4	δ^5	δ^6	δ^7	δ^8	δ^9	δ^{10}
12·5	66·88982 5254	1·09811 022833	−3·29203 44474		−66 33777		−1 18268		−5117		−515		+257
				−828 52996		−8 37296		−25196		−1661		−13	
12·0	65·24176 5998	1·10364 976746	3·30031 97470		74 71073		1 43464		6778		528		−389
				903 24069		9 80760		31974		2189		−402	
11·5	63·58938 1069	1·10970 511148	3·30935 21539		84 51833		1 75438		8967		930		+20
				987 75902		11 56198		40941		3119		−382	
11·0	61·93227 3077	1·11634 653718	3·31922 97441		96 08031		2 16379		12086		1312		−223
				1083 83933		13 72577		53027		4431		−605	
10·5	60·26999 1316	1·12365 740640	3·33006 81374		109 80608		2 69406		16517		1917		−540
				1193 64541		16 41983		69544		6348		−1145	
10·0	58·60202 2077	1·13173 734823	3·34200 45915		126 22591		3 38950		22865		3062		−467
				1319 87132		19 80933		92409		9410		−1612	
9·5	56·92777 6540	1·14070 642055	3·35520 33047		146 03524		4 31359		32275		4674		−1312
				1465 90656		24 12292		1 24684		14084		−2924	
9·0	55·24657 5623	1·15071 062360	3·36986 23703		170 15816		5 56043		46359		7598		−2124
				1636 06472		29 68335		1 71043		21682		−5048	
8·5	53·55763 0882	1·16192 930940	3·38622 30175		199 84151		7 27086		68041		12646		−3620
				−1835 90623		−36 95421		−2 39084		−34328		−8668	
8·0	51·86001 9940	1·17458 530063	−3·40458 20798		−236 79572		−9 66170		−102369		−21314		−8517

Table 3. Solution of the equation

$$\rho_1^{(3)} = -kR(\rho^{(3)})\,(1+\Delta^{(2)}), \tag{5·2}$$

using (4·19).

Table 4. Computation of $\lambda^{(3)}$ and d from the solution given in Table 3, by (4·3), (4·13) and (4·17). It is found that d is negligible when $-t \geqslant 10·0$, so that above this point the solution of (4·2), as far as ρ_1 is concerned, has been achieved in Table 3.

Table 5. Computation of $\Delta^{(2)} - \Delta^{(4)}$ by

$$\Delta^{(2)} - \Delta^{(4)} = -(1-k^{-2}\delta^2)\,d, \tag{5·3}$$

which is an adequate approximation to (4·14) when $-t \geqslant 4·0$. Solution of (4·22) for ϵ_1, ϵ. Final values of ρ obtained by applying (4·18) to $\rho_1^{(3)}$ and adding the correction ϵ. $\mathscr{C}_n'(\rho)$ computed by (3·4).

Table 4

$-t$	$\rho_3^{(3)}$	$\rho_2^{(3)}$	$\lambda^{(3)}$	$\frac{1}{2}(\lambda^{(3)})^2$	d	δ	δ^2
	10^{-10}	10^{-8}	10^{-11}	10^{-11}	10^{-11}		
11·0	—	—	—	—	—		—
						—	
10·5	—	—	—	—	—		—
						—	
10·0	−12594595	−1253766	+3604465	+ 65	+ 1		− 6
						− 5	
9·5	14567928	1389263	4138687	86	− 4		+ 2
						− 3	
9·0	16969981	1546550	4782233	114	7		−12
						−15	
8·5	19924305	1730499	5564749	155	22		− 6
						−21	
8·0	−23600160	−1947434	+6526198	+213	−43		−18

Table 5

$-t$	$\Delta^{(2)}-\Delta^{(4)}$	$\dfrac{a^2(1+\Delta^{(4)})}{\rho(\rho^2-a^2)}$	ϵ	ϵ_1	ρ final	$-\dfrac{4h}{\rho\rho_1}$	$\mathscr{C}_n'(\rho)$
	10^{-11}		10^{-10}	10^{-10}		10^{-13}	
11·0	—	—	—	—	61·93227 30728˙	97291 670887	−0·09863 654033
10·5	—	—	—	—	60·26999 13120˙	99649 642962	+0·09982 466777
10·0	0	—	—	—	58·60202 20738	102119 893157	+0·10105 438791
9·5	+ 4	+0·003	+0˙	+ 1	56·92777 65309.	104709 704315	−0·10232 775983
9·0	6	0·003	1˙	2	55·24657 56147	107426 757533	−0·10364 688009
8·5	21	0·003	3˙	7	53·55763 08699.	110279 066352	+0·10501 384021
8·0	+41	+0·003	+9.	+14	51·86001 99283˙	113274 853101	+0·10643 065963

6. *Computation of the early zeros of the Bessel functions.* If the dependent and independent variables in (3·2) are interchanged, the following equation results:

$$2\frac{dt}{d\rho}\frac{d^3t}{d\rho^3} - 3\left(\frac{d^2t}{d\rho^2}\right)^2 - 4\frac{\rho^2-a^2}{\rho^2}\left(\frac{dt}{d\rho}\right)^2 + 4\pi^2\left(\frac{dt}{d\rho}\right)^4 = 0. \tag{6·1}$$

Write $u = (-d\rho/dt)^{\frac{1}{2}}$, so that $dt/d\rho = -1/u^2$. Then (6·1) becomes

$$\frac{d^2u}{d\rho^2} = \frac{\pi^2}{u^3} - \frac{\rho^2-a^2}{\rho^2}u. \tag{6·2}$$

In outline the method of this section is as follows. Equation (6·2) is integrated to give u as a function of the argument ρ, the two initial conditions being obtained by interpolating the results of the previous computations for higher zeros. Quadrature of u^{-2} then gives t, and the zeros are obtained by inverse interpolation as the values of ρ corresponding to $t = -0{\cdot}5, \ -1{\cdot}0, \ldots$. Corresponding values of the derivative are obtainable from the equation

$$\mathscr{C}'_n(\rho) = \pm (2/\rho u^2)^{\frac{1}{2}}. \tag{6·3}$$

The numerical integration of (6·2), notwithstanding the non-linearity of this equation, can be carried out by 'difference-correction' methods of integration very similar to those described by Fox and Goodwin (3). The most convenient method here is the analogue of the powerful Method VII ((3), p. 381). The recurrence relation is

$$u_{-1} + \phi(u_{-1}) = 2u_0 - 10\phi(u_0) - u_1 - \phi(u_1) + \Delta, \tag{6·4}$$

where u_{-1}, u_0 and u_1 are the values of u at the tabular points $\rho - h, \rho, \rho + h$ respectively,

$$\phi(u) = h^2[f_n(\rho)\, u - \tfrac{1}{12}\pi^2 u^{-3}], \quad f_n(\rho) = (\rho^2 - a^2)/12\rho^2,$$

and Δ, the difference correction (not to be confused with the Δ of § 4), is given by

$$\Delta = (-\tfrac{1}{240}\delta^6 + \tfrac{13}{15120}\delta^8 - \ldots)\, u_0. \tag{6·5}$$

Equation (6·4), a quartic for u_{-1}, is most conveniently solved by a method of successive approximation such as Newton's rule. Write

$$K \equiv 2u_0 - 10\phi(u_0) - u_1 - \phi(u_1) + \Delta; \tag{6·6}$$

then if an approximation to u_{-1} is x, a better approximation is

$$x + [K - x - \phi(x)]/[1 + \phi'(x)]. \tag{6·7}$$

Each application of this process approximately doubles the number of correct decimals in u_{-1}. In practice one application generally suffices.

Having solved the recurrence relation (6·4) with $\Delta = 0$, we use the solution obtained to compute an estimated value of Δ by (6·5). The next step is the solution of (6·4) using the new Δ. It is not necessary to use the full number of figures in carrying out the second solution. The following device* saves much labour. Let the solution of (6·4) for $\Delta = 0$ be denoted by $u^{(1)}$ and the difference correction derived from it by $\Delta^{(1)}$. Suppose that the solution of (6·4) for $\Delta = \Delta^{(1)}$ is $u^{(1)} + \eta$. Then, if η^2 is neglected, it is found that η satisfies the *linear* recurrence relation

$$[1 + \phi'(u_{-1})]\,\eta_{-1} = [2 - 10\phi'(u_0)]\,\eta_0 - [1 + \phi'(u_1)]\,\eta_1 + \Delta^{(1)}, \tag{6·8}$$

which can be used to compute η, its two initial values being zero. This process can be repeated as often as necessary, new corrections to Δ (if any) being obtained by (6·5), operating on η instead of u.

By this method the table of Bessel function zeros computed from the thirty-first to the fourth by the method of §§ 4 and 5, was extended down to the first, ten decimals being obtained throughout.

A complete check on the whole of the numerical work is provided by continuing the integration of (6·2) as far as the point $\rho = n$, and by comparing the values obtained at this

* This was suggested by Mr C. W. Clenshaw. It is of general application to the numerical solution of differential equations by difference-correction methods.

point with those given by independent computation. For, by differentiating (3·3), it is seen that

$$u = 2^{-\frac{1}{2}}\pi\rho^{\frac{1}{2}}[J_n^2(\rho)+Y_n^2(\rho)]^{\frac{1}{2}}, \tag{6·9}$$

and its values at $\rho = n$ can therefore be computed from Meissel's asymptotic expansions ((4), p. 247) for $J_n(n)$, $Y_n(n)$. Similarly, t can be computed from these quantities by (3·3). In the above example the discrepancy in t at $\rho = 20$ was less than 4 units in the eleventh decimal place.

It will be noted that, in virtue of (6·9), equation (6·2) can be derived directly from the well-known differential equation ((5), p. 42) for the modulus $M \equiv (J_n^2 + Y_n^2)^{\frac{1}{2}}$. In effect, the method of this section amounts to the computation and inverse interpolation of the phase of the Hankel function $H_n^{(1)}(\rho)$. The advantage of the previous method in obtaining zeros independently of the values of the function has now been lost. However, though the present method would not be considered for use over an extended range, it does provide a practicable way of determining the zeros in the short range where the method of § 4 becomes unwieldy.

7. *Summary.* The zeros of solutions of the general second-order homogeneous linear differential equation are shown to satisfy a certain non-linear differential equation. The method here proposed for their determination is the numerical integration of this differential equation. It has the advantage of being independent of tabulated values of the actual functions whose zeros are being sought. As an example of the application of the method the Bessel functions $J_n(x)$, $Y_n(x)$ are considered. Numerical techniques for integrating the differential equation for the zeros of these Bessel functions are described in detail.

The author wishes to acknowledge the collaboration of Mrs G. E. Olver and Mr C. W. Clenshaw, of the Mathematics Division of the National Physical Laboratory, who carried out most of the computations described in the examples and made practical suggestions for developing the methods.

The work described above has been carried out as part of the research programme of the National Physical Laboratory, and this paper is published by permission of the Director of the Laboratory.

REFERENCES

(1) Bickley, W. G., Miller, J. C. P. and Jones, C. W. Notes on the evaluation of zeros and turning values of Bessel functions. *Phil. Mag.* 36 (1945), 121–33 and 200–10.
(2) Milne-Thomson, L. M. *Calculus of finite differences* (Macmillan, 1933).
(3) Fox, L. and Goodwin, E. T. Some new methods for the numerical integration of ordinary differential equations. *Proc. Cambridge Phil. Soc.* 45 (1949), 373–88.
(4) Watson, G. N. *Theory of Bessel functions* (Cambridge, 1944).
(5) Bickley, W. G. Formulae relating to Bessel functions of moderate or large argument and order. *Phil. Mag.* 34 (1943), 37–49.

National Physical Laboratory
Teddington, Middlesex

Reprinted from the *Proceedings of the Cambridge Philosophical Society*, Volume 47, Part 4, pp. 699–712, 1951.

A FURTHER METHOD FOR THE EVALUATION OF ZEROS OF BESSEL FUNCTIONS AND SOME NEW ASYMPTOTIC EXPANSIONS FOR ZEROS OF FUNCTIONS OF LARGE ORDER

By F. W. J. OLVER

Communicated by E. T. Goodwin

Received 4 December 1950

1. *Introduction.* In a recent paper (1) I described a method for the numerical evaluation of zeros of the Bessel functions $J_n(z)$ and $Y_n(z)$, which was independent of computed values of these functions. The essence of the method was to regard the zeros ρ of the cylinder function

$$\mathscr{C}_n(z) \equiv J_n(z)\cos \pi t - Y_n(z)\sin \pi t, \tag{1·1}$$

as a function of t and to solve numerically the third-order non-linear differential equation satisfied by $\rho(t)$. It has since been successfully used to compute ten-decimal values of $j_{n,s}, y_{n,s}$, the sth positive zeros* of $J_n(z)$, $Y_n(z)$ respectively, in the ranges $n = 10\,(1)\,20$, $s = 1\,(1)\,20$. During the course of this work it was realized that the least satisfactory feature of the new method was the time taken for the evaluation of the first three or four zeros in comparison with that required for the higher zeros; the direct numerical technique for integrating the differential equation satisfied by $\rho(t)$ becomes unwieldy for the small zeros and a different technique (described in the same paper) must be employed. It was also apparent that no mere refinement of the existing methods would remove this defect and that a new approach was required if it was to be eliminated. The outcome has been the development of the method to which the first part (§§ 2–6) of this paper is devoted.

The basis of the new method is the integration of an integro-differential equation. Previously ρ was regarded as a function of t for fixed n. This situation is now reversed; keeping t fixed, we consider the integration of the equation satisfied by ρ as a function of n, taken as a continuous positive variable. This equation is given by Watson (2) in another context but not in a form suitable for immediate numerical application. In order to use his result it is necessary to derive an asymptotic expansion of the equation.

The method has been extended to include the determination of the values of the derivative $\mathscr{C}'_n(z)$ at the zeros of $\mathscr{C}_n(z)$, and also to the determination of the zeros of $\mathscr{C}'_n(z)$ and the corresponding stationary values of $\mathscr{C}_n(z)$. One advantage of these methods is their production of zeros for half-odd-integer orders at the same time as for integer orders. They are speedy and are at present being used extensively in the Royal Society Mathematical Tables Committee's programme for the evaluation of the zeros of J, Y, J' and Y' for integer and half-odd-integer orders.

The second part (§§ 7–10) of this paper is devoted to the development of asymptotic expansions of the zeros and turning values for large values of the order n. These

* In order to fix the relation between ρ and t it is supposed, as in (1), that $t = 0$ when $\rho = 0$. Then $j_{n,s} = \rho(n, -s)$, $y_{n,s} = \rho(n, \frac{1}{2} - s)$.

expansions are obtained from the asymptotic expansion of the integro-differential equation referred to above and form an extension of the work of Watson ((2), Chapter 15) on this subject.

2. *The asymptotic expansion of the integro-differential equation for the zeros.* In the course of the following analysis it is frequently necessary to integrate or differentiate certain asymptotic series. The justification of these processes is usually straightforward; suitable regions can generally be found in which the asymptotic property is uniform with respect to the parameter involved. It will therefore be unnecessary to mention this fact each time such a process occurs.

If $\rho(n,t)$ denotes a zero of $\mathscr{C}_n(z)$ (as given by (1·1)), then from Watson ((2), p. 508)

$$\frac{\partial \rho}{\partial n} = F(\rho, n), \tag{2·1}$$

where
$$F(x,n) = 2x\int_0^\infty K_0(2x\sinh t)\,e^{-2nt}dt. \tag{2·2}$$

It is proposed to determine the asymptotic expansion of $F(x,n)$ for large x and n.

Using an integral representation of the Bessel function K_0 ((2), p. 181), we have

$$\begin{aligned} F(x,n) &= 2x\int_0^\infty e^{-2nt}\,dt\int_0^\infty e^{-2x\cosh u\sinh t}\,du \\ &= 2x\int_0^\infty du\int_0^\infty e^{-2nt-2x\cosh u\sinh t}\,dt. \end{aligned} \tag{2·3}$$

Now ((2), p. 316), if n and x are large

$$\int_0^\infty e^{-2nt-2x\cosh u\sinh t}\,dt \sim \sum_{m=0}^\infty \frac{(2m)!\,a_m(\lambda\,\mathrm{sech}\,u)}{(2x\cosh u)^{2m+1}}, \tag{2·4}$$

where $\lambda = n/x$, and
$$a_m(\cosh\gamma) = \frac{1}{2\pi i}\int^{(0+)}(t\cosh\gamma+\sinh t)^{-2m-1}\,dt. \tag{2·5}$$

Hence the required expansion is

$$F(x,n) \sim \sum_{m=0}^\infty \frac{(-)^m}{n^{2m}} f_m(\lambda), \tag{2·6}$$

where
$$f_m(\lambda) = (-)^m\,(2m)!\,(\tfrac{1}{2}\lambda)^{2m}\int_0^\infty \mathrm{sech}^{2m+1}u\,a_m(\lambda\,\mathrm{sech}\,u)\,du. \tag{2·7}$$

Values of $a_m(\cosh\gamma)$ for $m = 0, 1, 2, 3$ are given by Watson and from these we obtain*

$$\left.\begin{aligned} f_0(\lambda) &= \int_0^\infty \frac{du}{\cosh u+\lambda}, \quad f_1(\lambda) = \frac{\lambda^2}{4}\int_0^\infty \frac{\cosh u\,du}{(\cosh u+\lambda)^4}, \\ f_2(\lambda) &= \frac{\lambda^4}{16}\int_0^\infty \frac{(9\cosh u-\lambda)}{(\cosh u+\lambda)^7}\cosh u\,du, \\ f_3(\lambda) &= \frac{\lambda^6}{64}\int_0^\infty \frac{(225\cosh^2 u-54\,\lambda\cosh u+\lambda^2)}{(\cosh u+\lambda)^{10}}\cosh u\,du, \\ f_4(\lambda) &= \frac{\lambda^8}{256}\int_0^\infty \frac{(11025\cosh^3 u-4131\,\lambda\cosh^2 u+243\lambda^2\cosh u-\lambda^3)}{(\cosh u+\lambda)^{13}}\cosh u\,du. \end{aligned}\right\} \tag{2·8}$$

* The expression for $f_4(\lambda)$ has been obtained by using (2·5).

In the application of (2·6) to (2·1), λ always lies in the range $0<\lambda<1$; consequently, it is advantageous to pretabulate the coefficients $f_m(\lambda)$ as interpolable functions of λ over this range. This is most conveniently effected by expressing them in terms of the functions

$$I_m(\lambda) = \lambda^{m-1}\int_0^\infty (\cosh u+\lambda)^{-m}\,du. \tag{2·9}$$

We obtain the relations

$$\left.\begin{aligned} f_0 &= I_1, \qquad f_1 = \tfrac{1}{4}(I_3-I_4),\\ f_2 &= \tfrac{1}{16}(9I_5-19I_6+10I_7), \quad f_3 = \tfrac{1}{64}(225I_7-729I_8+784I_9-280I_{10}),\\ f_4 &= \tfrac{1}{256}(11025I_9-48231I_{10}+78786I_{11}-56980I_{12}+15400I_{13}).\end{aligned}\right\} \tag{2·10}$$

The functions $I_m(\lambda)$ can be computed from the relations

$$I_1 = (1-\lambda^2)^{-\frac{1}{2}}\cos^{-1}\lambda, \quad I_2 = \lambda(1-\lambda^2)^{-1}(1-\lambda I_1), \tag{2·11}$$

and

$$mI_{m+1} = \lambda^2(1-\lambda^2)^{-1}[(m-1)\,I_{m-1}-(2m-1)\,I_m] \quad (m \geqslant 2). \tag{2·12}$$

3. *Numerical integration for the zeros.* The direct numerical integration of equation (2·1) is only practicable if $\rho(n,t)$ has differences which converge at the selected interval of integration. When $\rho(n,t)$ is regarded as a function of n its singularity occurs at $n=t$, where the derivative becomes infinite, so that we cannot employ direct numerical methods for very low values of n without an excessive reduction of the interval of integration. Again, equation (2·6) can only be used to compute $\partial\rho/\partial n$ from ρ and n if n is large enough to ensure the rapid decrease of the early terms of the asymptotic series. Fortunately, so far as the first few zeros are concerned these restrictions largely coincide; direct numerical integration is generally worth while only in the region where the asymptotic expansion can be used. Thus it is not proposed that the new method be used for low values of n; rather, this region is to be covered beforehand by other methods (such as those described in (1)), and the results so obtained used as a sequence of starting values for the integrations in the n-direction.

The actual integrations are most conveniently performed in step-by-step manner, using Laplace's backward-difference formula:

$$f(n+h)-f(n) = (\tfrac{1}{2}+\tfrac{1}{4}\nabla+\tfrac{5}{24}\nabla^2+\tfrac{3}{16}\nabla^3+\tfrac{251}{1440}\nabla^4+\tfrac{95}{576}\nabla^5+\tfrac{19087}{120960}\nabla^6+\tfrac{5257}{34560}\nabla^7 + \tfrac{1070017}{7257600}\nabla^8+\ldots)\,2hf'(n), \tag{3·1}$$

h being the chosen interval of integration. The usual difficulties associated with the use of a backward-difference formula do not arise here owing to the comparative insensitivity of $F(x,n)$ to small changes in x. From (2·6) it is seen that

$$\left|\frac{\partial F}{\partial x}\right| \doteqdot \frac{\lambda}{x}\,|f_0'(\lambda)| < \frac{1}{3x}, \quad \text{when} \quad 0<\lambda<1.$$

Thus provided $\rho>4$, it is possible to work with at least one decimal less in ρ than in $\partial\rho/\partial n$. The full accuracy in ρ can subsequently be achieved by use of the central-difference quadrature formula

$$f(n+h)-f(n) = \tfrac{1}{2}[\mu-\tfrac{1}{12}\mu\delta^2+\tfrac{11}{720}\mu\delta^4-\tfrac{191}{60480}\mu\delta^6+\tfrac{2497}{3628800}\mu\delta^8-\ldots]\,2hf'(n+\tfrac{1}{2}h). \tag{3·2}$$

4. *Value of the derivative at the zeros.* The values of $\mathscr{C}_n'(z)$ at the zeros $z=\rho$ of $\mathscr{C}_n(z)$ can be evaluated from the solution $\rho(n,t)$ of equation (2·1) as follows.

Regarding ρ as a function of the independent variables n and t we obtain, by differentiating equation (2·1),

$$\frac{\partial^2 \rho}{\partial n \, \partial t} = \frac{1}{\rho}\frac{\partial \rho}{\partial t} G(\rho, n), \tag{4·1}$$

where

$$G(x, t) = x \partial F / \partial x. \tag{4·2}$$

Now ((1), equation (3·4))

$$\mathscr{C}_n'(\rho) = \pm \left(-\frac{\rho}{2}\frac{\partial \rho}{\partial t}\right)^{-\frac{1}{2}}. \tag{4·3}$$

If the relation between ρ and t is defined as in § 1, then the ambiguous sign is positive, and writing $c(n, t) \equiv \mathscr{C}_n'(\rho)$, we have

$$\frac{\partial \rho}{\partial t} = -\frac{2}{\rho c^2}.$$

Differentiating, we obtain

$$\frac{\partial^2 \rho}{\partial n \, \partial t} = \frac{2}{\rho^2 c^2}\frac{\partial \rho}{\partial n} + \frac{4}{\rho c^3}\frac{\partial c}{\partial n}.$$

Substituting these expressions in (4·1), we find that

$$\frac{1}{c}\frac{\partial c}{\partial n} = -\frac{1}{2\rho}\left[\frac{\partial \rho}{\partial n} + G(\rho, n)\right]. \tag{4·4}$$

The asymptotic expansion of $G(x, n)$ may be deduced from (2·6), (4·2) to be

$$G(x, n) \sim \sum_{m=0}^{\infty} \frac{g_m(\lambda)}{n^{2m}}, \tag{4·5}$$

where

$$g_m(\lambda) = (-)^{m-1} \lambda f_m'(\lambda). \tag{4·6}$$

The coefficients $g_m(\lambda)$ can be pretabulated. In terms of the integrals (2·9) they have the form

$$\left.\begin{aligned} &g_0 = I_2, \quad g_1 = \tfrac{1}{2}(I_3 - 3I_4 + 2I_5), \quad g_2 = -\tfrac{1}{8}(18I_5 - 70I_6 + 87I_7 - 35I_8),\\ &g_3 = \tfrac{1}{32}(675I_7 - 3339I_8 + 6052I_9 - 4788I_{10} + 1400I_{11}),\\ &g_4 = -\tfrac{1}{128}(44100I_9 - 266652I_{10} + 635085I_{11} - 746713I_{12} + 434280I_{13} - 100100I_{14}).\end{aligned}\right\} \tag{4·7}$$

After the equation (2·1) has been integrated numerically for ρ, the corresponding values of $\log c$ (and thence c) can be produced by simple quadrature of the quantity on the right-hand side of equation (4·4), using the asymptotic expansion (4·5) to compute the values of $G(\rho, n)$.

5. *Zeros of* $\mathscr{C}_n'(z)$ *and stationary values of* $\mathscr{C}_n(z)$. If the zeros of

$$\mathscr{C}_n'(z) \equiv J_n'(z) \cos \pi t - Y_n'(z) \sin \pi t \tag{5·1}$$

are denoted by $\sigma(n, t)$, then from Watson ((2), p. 510)

$$\frac{\partial \sigma}{\partial n} = H(\sigma, n), \tag{5·2}$$

where

$$H(x, n) = \frac{2x}{x^2 - n^2}\int_0^\infty (x^2 \cosh 2t - n^2) K_0(2x \sinh t) \, e^{-2nt} \, dt. \tag{5·3}$$

The asymptotic expansion of $H(x, n)$ may be deduced from that of $F(x, n)$ as follows.

Using the differential equation satisfied by K_0, we have

$$4x^2\int_0^\infty K_0(2x\sinh t)\sinh^2 t\, e^{-2nt}\,dt$$

$$= 2x\int_0^\infty K_0'(2x\sinh t)\sinh t\, e^{-2nt}\,dt + 4x^2\int_0^\infty K_0''(2x\sinh t)\sinh^2 t\, e^{-2nt}\,dt$$

$$= \left(\frac{x}{2}\frac{\partial}{\partial x}+\frac{x^2}{2}\frac{\partial^2}{\partial x^2}\right)\left(\frac{F}{x}\right),$$

using (2·2). Substituting this result in (5·3) we obtain

$$H = F+\frac{x^2}{2(x^2-n^2)}\left(x\frac{\partial^2}{\partial x^2}+\frac{\partial}{\partial x}\right)\left(\frac{F}{x}\right). \tag{5·4}$$

Hence, from (2·6),

$$H(x,n)\sim\sum_{m=0}^{\infty}\frac{h_m(\lambda)}{n^{2m}}, \tag{5·5}$$

where $\lambda = n/x$ as before, and

$$\left.\begin{aligned}(-)^m h_m(\lambda) &= f_m(\lambda)-\lambda^2(1-\lambda^2)^{-1}h_m^{(1)}(\lambda),\\ h_m^{(1)}(\lambda) &= \tfrac{1}{2}[\lambda^2 f''_{m-1}(\lambda)+3\lambda f'_{m-1}(\lambda)+f_{m-1}(\lambda)].\end{aligned}\right\} \tag{5·6}$$

This is the required expansion.

In terms of the integrals (2·9),

$$\left.\begin{aligned}&h_0^{(1)} = 0,\quad h_1^{(1)} = \tfrac{1}{2}(I_1-3I_2+2I_3),\quad h_2^{(1)} = \tfrac{1}{8}(9I_3-37I_4+48I_5-20I_6),\\ &h_3^{(1)} = \tfrac{1}{32}(225I_5-1179I_6+2242I_7-1848I_8+560I_9),\\ &h_4^{(1)} = \tfrac{1}{128}(11025I_7-70281I_8+175248I_9-214552I_{10}+129360I_{11}-30800I_{12}).\end{aligned}\right\} \tag{5·7}$$

The stationary values of $\mathscr{C}_n(z)$ occur at $z = \sigma(n,t)$. Now ((1), eqn (3·9)),

$$\mathscr{C}_n(\sigma) = \pm\left(-\frac{\sigma^2-n^2}{2\sigma}\frac{\partial\sigma}{\partial t}\right)^{-\frac{1}{2}}. \tag{5·8}$$

If the relation between σ and t is defined as in §7 then the ambiguous sign is positive, and writing $b(n,t)\equiv\mathscr{C}_n(\sigma)$, we have

$$\frac{\partial\sigma}{\partial t} = -\frac{2\sigma}{b^2(\sigma^2-n^2)}. \tag{5·9}$$

Differentiating, we obtain

$$\frac{\partial^2\sigma}{\partial n\,\partial t} = \frac{4\sigma}{b^3(\sigma^2-n^2)}\frac{\partial b}{\partial n}+\frac{2(\sigma^2+n^2)}{b^2(\sigma^2-n^2)^2}\frac{\partial\sigma}{\partial n}-\frac{4n\sigma}{b^2(\sigma^2-n^2)^2}. \tag{5·10}$$

Also from (5·2), (5·9),

$$\frac{\partial^2\sigma}{\partial n\,\partial t} = \frac{\partial\sigma}{\partial t}\left[\frac{\partial H}{\partial x}\right]_{x=\sigma} = -\frac{2\sigma}{b^2(\sigma^2-n^2)}\left[\frac{\partial H}{\partial x}\right]_{x=\sigma}. \tag{5·11}$$

Substituting this result in (5·10), we find that

$$\frac{1}{b}\frac{\partial b}{\partial n} = -\frac{1}{2\sigma}\left[\frac{\sigma^2+n^2}{\sigma^2-n^2}\frac{\partial\sigma}{\partial n}-L(\sigma,n)\right], \tag{5·12}$$

where

$$L(x,n) = \frac{2nx}{x^2-n^2}-x\frac{\partial H}{\partial x}. \tag{5·13}$$

From (5·5), (5·13),
$$L(x, n) \sim \sum_{m=0}^{\infty} \frac{l_m(\lambda)}{n^{2m}}, \tag{5·14}$$
where
$$l_0(\lambda) = \lambda h_0'(\lambda) + 2\lambda(1-\lambda^2)^{-1}, \quad l_m(\lambda) = \lambda h_m'(\lambda) \quad (m \geqslant 1). \tag{5·15}$$
So that from (4·6), (5·6), we obtain
$$\left.\begin{aligned} l_0(\lambda) &= -g_0(\lambda) + 2\lambda(1-\lambda^2)^{-1}, \\ \text{and, if } m \geqslant 1, \quad l_m(\lambda) &= -g_m(\lambda) + (-)^{m-1}\lambda^2(1-\lambda^2)^{-1}[2(1-\lambda^2)^{-1} h_m^{(1)}(\lambda) - l_m^{(1)}(\lambda)], \\ \text{where} \quad l_m^{(1)}(\lambda) &= -\lambda dh_m^{(1)}/d\lambda = -\tfrac{1}{2}[\lambda^3 f_{m-1}'''(\lambda) + 5\lambda^2 f_{m-1}''(\lambda) + 4\lambda f_{m-1}'(\lambda)]. \end{aligned}\right\} \tag{5·16}$$

In terms of the integrals (2·9), we have
$$\left.\begin{aligned} &l_1^{(1)} = 2I_2 - 5I_3 + 3I_4, \quad l_2^{(1)} = -\tfrac{1}{4}(9I_3 - 69I_4 + 170I_5 - 170I_6 + 60I_7), \\ &l_3^{(1)} = -\tfrac{1}{16}(450I_5 - 3510I_6 + 10263I_7 - 14315I_8 + 9632I_9 - 2520I_{10}), \\ &l_4^{(1)} = -\tfrac{1}{64}(33075I_7 - 284571I_8 + 982116I_9 - 1754100I_{10} \\ &\qquad\qquad + 1719560I_{11} - 880880I_{12} + 184800I_{13}). \end{aligned}\right\} \tag{5·17}$$

Provided that the coefficients $h_m(\lambda), l_m(\lambda)$ have been pretabulated, the numerical integration of equation (5·2) for $\sigma(n, t)$ follows the method outlined in §3 for the integration of equation (2·1). The corresponding values of $\log b$ (and thence b) can then be obtained by simple quadrature of the quantity of the right-hand side of (5·12).

6. *Numerical example.* As a typical illustration of the above method, the numerical integration of $\rho(n, -1)$, the first zero of $J_n(x)$, has been carried out to ten decimal places over the range $n = 6\,(\tfrac{1}{2})\,20$. Four or five terms of the series (2·6) are sufficient to give $\partial\rho/\partial n$ to this accuracy. Tables 1 and 2 give extracts from the computations.

Ten-decimal values of $I_m(\lambda)$ have been computed for $m = 1\,(1)\,13$, $\lambda = 0{\cdot}00\,(0{\cdot}05)\,1{\cdot}00$ and from these, tables of $f_m(\lambda)$, $g_m(\lambda)$ for $m = 0, 1, 2, 3, 4$ have been prepared to enable integrations to be performed with ten decimals in $\partial\rho/\partial n$ and c when $n > 5$, for any fixed negative value of t. The range of tabulation of $f_m(\lambda)$ and $g_m(\lambda)$ is $\lambda = 0{\cdot}00\,(0{\cdot}01)\,1{\cdot}00$; at this interval these functions are interpolable either linearly or using modified second differences.

The values of ρ can be interpolated in the n-direction so that, from the results of this example, it is possible to obtain quite simply the value of the first zero of $J_n(x)$ for *any* value of n lying in the range $6 \leqslant n \leqslant 20$.

7. *Asymptotic behaviour of the zeros for large orders.* A natural extension of the preceding work is the development of asymptotic expansions of the zeros ρ, σ and the associated functions c, b (as defined above) for large values of n. The earlier work on this subject has been of a restricted nature; numerical values of the coefficients have been obtained in a few special cases, but the methods employed have not always been sound and the numerical results are largely erroneous. A brief account of this work is given in §10.

We shall take as starting-point the work of Watson ((2), pp. 517–21). In this article Watson proves that the *first* zeros of $J_n(x)$, $Y_n(x)$ and $J_n'(x)$ are each represented asymptotically by expressions of the form
$$x = n + (\text{constant})\, n^{\frac{1}{3}} + O(n^{-\frac{1}{3}}), \tag{7·1}$$

Table 1

n	ρ	$\lambda \equiv n/\rho$	$f_0(\lambda)$	$f_1(\lambda)$	$f_2(\lambda)$	$f_3(\lambda)$
12·0	16·69824 9934	0·71863 81835	1·10579 78883	0·01632 462	0·002739	0·0014
12·5	17·25045 4784	0·72461 85771	1·10326 54805	0·01640 554	0·002770	0·0014
13·0	17·80143 5153	0·73027 81988	1·10088 17571	0·01648 104	0·002798	0·0014
13·5	18·35126 1494	0·73564 42501	1·09863 31259	0·01655 168	0·002825	0·0014
14·0	18·89999 7952	0·74074 08210	1·09650 76543	0·01661 792	0·002850	0·0014
14·5	19·44770 3108	0·74558 93336	1·09449 48181	0·01668 017	0·002874	0·0015
15·0	19·99443 0628	0·75020 89096	1·09258 52962	0·01673 879	0·002897	0·0015

Table 2

n	$\partial\rho/\partial n$	δ	δ^2	δ^3	δ^4	δ^5	δ^6	δ^7	δ^8	Final ρ
12·0	1·10568 46545	−252 40564	+16 32006	−1 54085	+23109	−3718	+903	−185	+ 76	16·69824 99339
12·5	1·10316 05981	237 62643	14 77921	1 34694	19391	3000	718	188	− 3	17·25045 47841
13·0	1·10078 43338	224 19416	13 43227	1 18303	16391	2470	530	85	+103	17·80143 51533
13·5	1·09854 23922	211 94492	12 24924	1 04382	13921	2025	445	95	− 10	18·35126 14960
14·0	1·09642 29430	200 73950	11 20542	92486	11896	1675	350	74	+ 21	18·89999 79532
14·5	1·09441 55480	−190 45894	10 28056	− 82265	10221	−1399	276	− 56	+ 18	19·44770 31081
15·0	1·09251 09586		+ 9 45791		+ 8822		+220		+ 38	19·99443 06298

and gives numerical values for the constants. To obtain this result he considers the function*

$$F_n^{(s)}(x) \equiv f(\theta)[\Gamma(\tfrac{2}{3})(\tfrac{1}{6}n)^{\frac{1}{3}}\mathscr{C}_n(n)\,J_{-\frac{1}{3}}(y)+\Gamma(\tfrac{1}{3})(\tfrac{1}{6}n)^{\frac{2}{3}}A_n J_{\frac{1}{3}}(y)], \tag{7·2}$$

in the two cases

$$\left.\begin{array}{llll}\text{(i) } s=1:\ A_n=\mathscr{C}_n'(n), & f(\theta)=(2\theta)^{\frac{1}{2}}, & y=\tfrac{1}{3}n(2\theta)^{\frac{3}{2}}, & x=ne^{\theta};\\ \text{(ii) } s=2:\ A_n=\mathscr{C}_n'(n)+\dfrac{1}{5n}\mathscr{C}_n(n), & f(\theta)=\{3(1-\theta\cot\theta)\}^{\frac{1}{2}}, & y=n(\tan\theta-\theta), & x=n\sec\theta.\end{array}\right\} \tag{7·3}$$

In both cases it is readily verified that at $x = n$,

$$F_n^{(1)}(x) = F_n^{(2)}(x) = \mathscr{C}_n(x) \quad\text{and}\quad \frac{d}{dx}F_n^{(1)}(x) = \frac{d}{dx}F_n^{(2)}(x) = \mathscr{C}_n'(x).$$

Moreover, in the Sturmian sense, $F_n^{(1)}(x)$ is slightly less oscillatory than $\mathscr{C}_n(x)$ if $x \geqslant n$, whereas $F_n^{(2)}(x)$ is slightly more oscillatory than $\mathscr{C}_n(x)$ provided $1{\cdot}979n > x \geqslant n$. Consequently in the region $x \geqslant n$, for sufficiently large values of n the first zero of $\mathscr{C}_n(x)$ lies between the first zero of $F_n^{(1)}(x)$ and the first zero of $F_n^{(2)}(x)$. The two last-named zeros are *both* of the form

$$x = n+\tfrac{1}{2}(3\xi_n)^{\frac{2}{3}}n^{\frac{1}{3}}+O(n^{-\frac{1}{3}}), \tag{7·4}$$

where $y = \xi_n$ is the first zero of

$$S_n(y) \equiv \Gamma(\tfrac{2}{3})(\tfrac{1}{6}n)^{\frac{1}{3}}\mathscr{C}_n(n)\,J_{-\frac{1}{3}}(y)+\Gamma(\tfrac{1}{3})(\tfrac{1}{6}n)^{\frac{2}{3}}\mathscr{C}_n'(n)\,J_{\frac{1}{3}}(y), \tag{7·5}$$

and is of the form

$$\xi_n = \xi+O(n^{-\frac{2}{3}}), \tag{7·6}$$

ξ being independent of n and depending only on the particular zero considered. It follows that the first zero of $\mathscr{C}_n(x)$ which exceeds n is given by

$$x = n+\tfrac{1}{2}(3\xi)^{\frac{2}{3}}n^{\frac{1}{3}}+O(n^{-\frac{1}{3}}). \tag{7·7}$$

* Watson errs by taking $A_n = \mathscr{C}_n'(n)$ in $F_n^{(2)}(x)$, for with this value $dF_n^{(2)}/dx \neq \mathscr{C}_n'(n)$ at $x = n$. However, this error does not seriously affect his analysis.

In order to extend the expression (7·7) to include *all* the zeros of $\mathscr{C}_n(x)$ and $\mathscr{C}'_n(x)$ which exceed n, we require the following lemma:

Let $u_1(x)$ and $u_2(x)$ be solutions in the interval $a \leqslant x \leqslant b$, of the equations

$$\frac{d^2u_1}{dx^2}+I_1u_1 = 0, \quad \frac{d^2u_2}{dx^2}+I_2u_2 = 0,$$

such that when $x = a$

$$u_1(a) = u_2(a) = \alpha \text{ (say)}, \quad u'_1(a) = u'_2(a) = \beta \text{ (say)}, \quad \alpha^2+\beta^2 \neq 0,$$

and let $I'_1(x)$, $I'_2(x)$ be continuous and $I_1(x) \geqslant I_2(x) > 0$ in the interval $a \leqslant x \leqslant b$. Then

(i) *If $u_2(x)$ has m zeros in the interval $a < x \leqslant b$, $u_1(x)$ has at least m zeros in the same interval, and if $r \leqslant m$ the rth zero of $u_1(x)$ is less then the rth zero of $u_2(x)$.*

(ii) *As (i) but with $u_1(x)$, $u_2(x)$ replaced by $u'_1(x)$, $u'_2(x)$ respectively.*

The first part of the lemma is a particular case of a well-known comparison theorem of Sturm ((3), p. 228). The second part may be deduced from the same theorem by considering the functions $U_1(x) \equiv u'_1(x)$, $U_2(x) \equiv u'_2(x)$. These satisfy the equations

$$\frac{d}{dx}\left(\frac{1}{I_1}\frac{dU_1}{dx}\right)+U_1 = 0, \quad \frac{d}{dx}\left(\frac{1}{I_2}\frac{dU_2}{dx}\right)+U_2 = 0,$$

with the initial conditions

$$U_1(a) = U_2(a) = \beta, \quad U'_1(a) = -\alpha I_1(a), \quad U'_2(a) = -\alpha I_2(a).$$

If the lemma be applied in turn to the pairs of functions

$$u_1(x) = \mathscr{C}_n(x), \quad u_2(x) = F^{(1)}_n(x) \qquad \text{and} \qquad u_1(x) = F^{(2)}_n(x), \quad u_2(x) = \mathscr{C}_n(x),$$

it is seen that if $x \geqslant n$, then for sufficiently large values of n, the rth zero of $\mathscr{C}_n(x)$ lies between the rth zero of $F^{(1)}_n(x)$ and the rth zero of $F^{(2)}_n(x)$, and similarly for the turning values. It now follows as in Watson's article that *every* zero of $\mathscr{C}_n(x)$ or $\mathscr{C}'_n(x)$ which exceeds n is of the form (7·7) for large values of n.

In §§8, 9 we develop asymptotic expansions, the first two terms of which are given by the asymptotic expression (7·7). It is therefore necessary to determine the precise values taken by the constant ξ in the various cases.

Consider
$$\mathscr{C}_n(x) \equiv \cos \pi t J_n(x) - \sin \pi t Y_n(x). \tag{7·8}$$

If $\xi^{(1)}_n$, $\xi^{(2)}_n$ denote values of y (as defined by (7·3)) which correspond to zeros of $F^{(1)}_n(x)$ and $F^{(2)}_n(x)$ respectively, then from (7·2) using Meissel's asymptotic expansions ((2), p. 247) for $J_n(n), Y_n(n)$, $J'_n(n)$ and $Y'_n(n)$, we obtain

$$S_n(\xi^{(1)}_n) = 0, \quad S_n(\xi^{(2)}_n) = O(n^{-\frac{2}{3}}). \tag{7·9}$$

Now if ξ_n is *any* continuous positive value of y satisfying the relation

$$S_n(y) = O(n^{-\frac{2}{3}}), \tag{7·10}$$

then from (7·5), (7·8) and again using Meissel's expansions, we find that

$$\cos(\pi t - \tfrac{1}{3}\pi) J_{-\frac{1}{3}}(\xi_n) + \cos(\pi t + \tfrac{1}{3}\pi)\, J_{\frac{1}{3}}(\xi_n) = O(n^{-\frac{2}{3}}). \tag{7·11}$$

If we write
$$H^{(1)}_{\frac{1}{3}}(x) = M(x)\exp\{i\phi(x)\},$$

where $M(x) > 0$, $\phi(0) = -\frac{1}{2}\pi$ and $\phi(x)$ increases continuously with positive x, then

$$J_{\frac{1}{3}}(x) = M(x)\cos\phi(x), \quad J_{-\frac{1}{3}}(x) = M(x)\cos\{\phi(x)+\tfrac{1}{3}\pi\},$$

and from (7·11) we obtain
$$\cos\{\pi t + \tfrac{1}{6}\pi + \phi(\xi_n)\} = O(n^{-\frac{2}{3}}). \tag{7·12}$$

Thus
$$\phi(\xi_n) = (s-t-\tfrac{2}{3})\pi + O(n^{-\frac{2}{3}}), \tag{7·13}$$
where s is an integer, the value of which depends on the enumeration of ξ_n. If we suppose the relation between t and the zeros $\rho(n,t)$ of $\mathscr{C}_n(x)$ is defined as in §1, we obtain the following result:

If $t < -\frac{1}{6}$, then for fixed t and large n*
$$\rho(n,t) = n + \alpha(t)\, n^{\frac{1}{3}} + O(n^{-\frac{1}{3}}), \tag{7·14}$$
where
$$\alpha(t) = \tfrac{1}{2}(3\xi)^{\frac{2}{3}} \quad \textit{and} \quad \arg H^{(1)}_{\frac{1}{3}}(\xi) = -\pi t - \tfrac{2}{3}\pi, \tag{7·15}$$
it being supposed here that $\arg H^{(1)}_{\frac{1}{3}}(x)$ *increases continuously with positive* x *and equals* $-\frac{1}{2}\pi$ *when* $x = 0$.

In order to deal with the turning points σ of $\mathscr{C}_n(x)$, it is first necessary to fix the relation between σ and t. Consideration of the derivative of $J'_n(\sigma)/Y'_n(\sigma)$ shows that if $\sigma > n$, σ increases steadily as t decreases and so if we now fix $\sigma = j'_{n,1}$ when $t = 0$ then
$$j'_{n,s} = \sigma(n, 1-s), \quad y'_{n,s} = \sigma(n, \tfrac{1}{2}-s).$$

When the expression (7·2) is differentiated it is seen that if $\eta^{(1)}_n, \eta^{(2)}_n$ denote values of y corresponding to zeros of $dF^{(1)}_n/dx$ and $dF^{(2)}_n/dx$ respectively, then
$$T_n(\eta^{(1)}_n) = 0, \quad T_n(\eta^{(2)}_n) = O(n^{-\frac{2}{3}}), \tag{7·16}$$
where
$$T_n(y) \equiv \Gamma(\tfrac{2}{3})\,(\tfrac{1}{6}n)^{\frac{1}{3}}\,\mathscr{C}_n(n)\,J_{\frac{1}{3}}(y) - \Gamma(\tfrac{1}{3})\,(\tfrac{1}{6}n)^{\frac{2}{3}}\,\mathscr{C}'_n(n)\,J_{-\frac{2}{3}}(y). \tag{7·17}$$
If η_n is *any* continuous positive value of y satisfying the relation
$$T_n(y) = O(n^{-\frac{2}{3}}), \tag{7·18}$$
then by reasoning similar to that used to establish (7·12), we find that
$$\sin\{\pi t + \tfrac{1}{3}\pi + \omega(\eta_n)\} = O(n^{-\frac{2}{3}}), \tag{7·19}$$
where $\omega(x) = \arg H^{(1)}_{\frac{2}{3}}(x)$. Thus
$$\omega(\eta_n) = \pi(s-t-\tfrac{1}{3}) + O(n^{-\frac{2}{3}}), \tag{7·20}$$
where s is an integer depending on the enumeration of η_n. We hence obtain the second result:

If $t < \frac{1}{6}$, then for fixed t and large n
$$\sigma(n,t) = n + a(t)\, n^{\frac{1}{3}} + O(n^{-\frac{1}{3}}), \tag{7·21}$$
where
$$a(t) = \tfrac{1}{2}(3\eta)^{\frac{2}{3}} \quad \textit{and} \quad \arg H^{(1)}_{\frac{2}{3}}(\eta) = -\pi t - \tfrac{1}{3}\pi, \tag{7·22}$$
it being supposed that $\arg H^{(1)}_{\frac{2}{3}}(x)$ *increases continuously with positive* x *and equals* $-\frac{1}{2}\pi$ *when* $x = 0$.

8. *Asymptotic expansion of* ρ *and* c. If t is fixed and less than $-\frac{1}{6}$ then $\rho > n$, and from† (2·1), (2·6) we have
$$\frac{d\rho}{dn} \sim \sum_{m=0}^{\infty} \frac{(-)^m}{n^{2m}} f_m\left(\frac{n}{\rho}\right), \quad \text{as} \quad n \to \infty. \tag{8·1}$$
Write $p = \rho/n$ and $u = n^{\frac{1}{3}}$. Then from (7·14) we obtain
$$p = 1 + \alpha u^{-2} + O(u^{-4}), \quad \text{as} \quad u \to \infty, \tag{8·2}$$
and from (8·1),
$$\frac{u}{3}\frac{dp}{du} + p \sim \sum_{m=0}^{\infty} \frac{(-)^m}{u^{6m}} \phi_m(p), \tag{8·3}$$

* This restriction ensures that $\rho(n,t) > n$ if $n > 0$.

† In the expression (2·6) x/n is supposed independent of n, but the reasoning may be easily extended to include the present case.

where $\phi_m(\lambda) = f_m(\lambda^{-1})$. If we write $\lambda = 1+\epsilon$, then for sufficiently small values of ϵ, $\phi_m(\lambda)$ can be expanded in the form

$$\phi_m(1+\epsilon) = \sum_{s=0}^{\infty} \phi_s^{(m)} \epsilon^s. \tag{8·4}$$

From (8·2), (8·3) it is now clear that an expansion of p exists in the form

$$p \sim 1 + \sum_{s=1}^{\infty} \alpha_s u^{-2s}, \tag{8·5}$$

and hence

$$\rho \sim n\left[1+\sum_{s=1}^{\infty} \alpha_s n^{-2s/3}\right]. \tag{8·6}$$

The values of the coefficients α_s may be found by comparing coefficients in the relation

$$1-\sum_{s=1}^{\infty} \frac{2s-3}{3}\frac{\alpha_s}{u^{2s}} \sim \sum_{m=0}^{\infty} \frac{(-)^m}{u^{6m}}\left[\sum_{s=0}^{\infty} \phi_s^{(m)}\left(\frac{\alpha_1}{u^2}+\frac{\alpha_2}{u^4}+\frac{\alpha_3}{u^6}+\dots\right)^s\right]. \tag{8·7}$$

The expansions of $\phi_m(1+\epsilon)$ can be obtained from the relations (2·8). In the cases $m = 0$ and 1 they are

$$\left.\begin{aligned} \phi_0(1+\epsilon) &= 1+\tfrac{1}{3}\epsilon-\tfrac{1}{5}\epsilon^2+\tfrac{13}{105}\epsilon^3-\tfrac{5}{63}\epsilon^4+\tfrac{61}{1155}\epsilon^5-\dots, \\ \phi_1(1+\epsilon) &= \tfrac{2}{105}-\tfrac{2}{315}\epsilon-\tfrac{4}{1155}\epsilon^2+\dots, \end{aligned}\right\} \tag{8·8}$$

and from (8·2), (8·7) we obtain

$$\left.\begin{aligned} &\alpha_1 = \alpha, \quad \alpha_2 = \tfrac{3}{10}\alpha^2, \quad \alpha_3 = -\tfrac{1}{350}\alpha^3+\tfrac{1}{70}, \\ &\alpha_4 = -\tfrac{479}{63000}\alpha^4-\tfrac{1}{3150}\alpha, \quad \alpha_5 = \tfrac{20231}{8085000}\alpha^5-\tfrac{551}{161700}\alpha^2. \end{aligned}\right\} \tag{8·9}$$

The asymptotic expansion of c can be deduced from that of ρ by means of equation (4·3), namely,

$$c = \left(-\frac{\rho}{2}\frac{\partial\rho}{\partial t}\right)^{-\frac{1}{2}}. \tag{8·10}$$

From (8·6) we obtain

$$\frac{\partial\rho}{\partial t} \sim n^{\frac{1}{3}}\frac{d\alpha}{dt}\left[1+\sum_{s=1}^{\infty}\left(\frac{d\alpha_{s+1}}{d\alpha}\right)n^{-2s/3}\right]. \tag{8·11}$$

Hence

$$c \sim \kappa(t) n^{-\frac{2}{3}}\left[1+\sum_{s=1}^{\infty} \beta_s n^{-2s/3}\right], \tag{8·12}$$

where

$$\kappa(t) = \left(-\frac{1}{2}\frac{d\alpha}{dt}\right)^{-\frac{1}{2}}, \tag{8·13}$$

$$\left.\begin{aligned} &\beta_1 = -\tfrac{4}{5}\alpha, && \beta_2 = \tfrac{18}{35}\alpha^2, \\ &\beta_3 = -\tfrac{88}{315}\alpha^3-\tfrac{11}{1575}, && \beta_4 = \tfrac{79586}{606375}\alpha^4+\tfrac{9824}{606375}\alpha. \end{aligned}\right\} \tag{8·14}$$

These coefficients are also obtained when the expressions (8·6), (8·12) are substituted in the differential equation (4·4).

Using (7·15), (8·13) and the Wronskian relation for $J_{\frac{1}{3}}$ and $Y_{\frac{1}{3}}$, we obtain

$$\kappa(t) = \frac{1}{\pi|H_{\frac{1}{3}}^{(1)}(\xi)|}\left(\frac{6}{\alpha}\right)^{\frac{1}{2}}, \tag{8·15}$$

ξ being defined as in (7·15).

9. *Asymptotic expansions of σ and b.* If we define $\psi_m(\lambda)$ by the relation

$$\psi_m(\lambda) = (\lambda-1)\,h_m(\lambda^{-1}),$$

then it follows from equations (5·6) that for sufficiently small ϵ, $\psi_m(\lambda)$ can be expanded in the form

$$\psi_m(1+\epsilon) = \sum_{s=0}^{\infty} \psi_s^{(m)} \epsilon^s. \tag{9·1}$$

By reasoning similar to that indicated in §8 we deduce from (5·2), (5·5) that

$$\frac{d\sigma}{dn} \sim \sum_{m=0}^{\infty} \frac{1}{n^{2m}} h_m\left(\frac{n}{\sigma}\right), \quad \text{as} \quad n \to \infty, \tag{9·2}$$

and hence that

$$\sigma \sim n\left[1 + \sum_{s=1}^{\infty} a_s n^{-2s/3}\right], \tag{9·3}$$

where the values of a_s may be obtained by comparing coefficients in the relation

$$\left(1 - \sum_{s=1}^{\infty} \frac{2s-3}{3} \frac{a_s}{u^{2s}}\right)\left(\sum_{s=1}^{\infty} \frac{a_s}{u^{2s}}\right) \sim \sum_{m=0}^{\infty} \frac{1}{u^{6m}}\left[\sum_{s=0}^{\infty} \psi_s^{(m)}\left(\frac{a_1}{u^2} + \frac{a_2}{u^4} + \frac{a_3}{u^6} + \dots\right)^s\right]. \tag{9·4}$$

The expansions of $\psi_m(1+\epsilon)$ can be obtained from the relations (2·8) and (5·6). For $m = 0$ and 1 they are

$$\left.\begin{aligned} \psi_0(1+\epsilon) &= \epsilon\phi_0(1+\epsilon) = \epsilon + \tfrac{1}{3}\epsilon^2 - \tfrac{1}{5}\epsilon^3 + \tfrac{13}{105}\epsilon^4 - \tfrac{5}{63}\epsilon^5 + \tfrac{61}{1155}\epsilon^6 - \dots, \\ \psi_1(1+\epsilon) &= \tfrac{1}{15} + \tfrac{1}{30}\epsilon - \tfrac{37}{1260}\epsilon^2 + \tfrac{49}{3960}\epsilon^3 + \dots, \end{aligned}\right\} \tag{9·5}$$

and the first four coefficients a_s are given by

$$\left.\begin{aligned} a_1 &= a, \quad a_2 = \tfrac{3}{10}a^2 - \tfrac{1}{10}a^{-1}, \\ a_3 &= -\tfrac{1}{350}a^3 - \tfrac{1}{25} - \tfrac{1}{200}a^{-3}, \\ a_4 &= -\tfrac{479}{63000}a^4 + \tfrac{509}{31500}a + \tfrac{1}{1500}a^{-2} - \tfrac{1}{2000}a^{-5}, \end{aligned}\right\} \tag{9·6}$$

a being given by (7·22).

Likewise the asymptotic expansion of the stationary values b can be derived from equation (5·8) by a similar method to that used in §8 for expanding c. The result is

$$b \sim k(t)\, n^{-\frac{1}{3}}\left[1 + \sum_{s=1}^{\infty} b_s n^{-2s/3}\right], \tag{9·7}$$

where

$$\left.\begin{aligned} b_1 &= -\tfrac{1}{5}a, \quad b_2 = \tfrac{9}{350}a^2 + \tfrac{1}{100}a^{-1}, \\ b_3 &= \tfrac{89}{15750}a^3 - \tfrac{47}{4500} + \tfrac{1}{3000}a^{-3}, \end{aligned}\right\} \tag{9·8}$$

and

$$k(t) = \frac{3^{\frac{1}{2}}}{\pi} \frac{1}{a|H_{\frac{2}{3}}^{(1)}(\eta)|}, \tag{9·9}$$

a and η being given by (7·22).

10. *Numerical values of the coefficients.* The most convenient existing tables for the calculation of α, κ, a and k are those of the Airy Integral (**4**). In terms of the functions F, G, χ and ψ defined there, we obtain

$$\left.\begin{aligned} \alpha &= 2^{-\frac{1}{3}}X, \quad \kappa = 2^{\frac{2}{3}}/\{\pi F(-X)\}, \\ a &= 2^{-\frac{1}{3}}Y, \quad k = 2^{\frac{1}{3}}/\{\pi G(-Y)\}, \end{aligned}\right\} \tag{10·1}$$

where X, Y are such that

$$\chi(-X) = -\pi t, \quad \psi(-Y) = -\pi t. \tag{10·2}$$

In this way the functions α, κ have been computed for the first five zeros of J and Y, and the functions a, k for the first five zeros of J' and Y'. The results are listed in Tables 3, 4, 5 and 6, together with the corresponding values of the first few coefficients of the series (8·6), (9·3) and (9·7). In place of the coefficients β_s of (8·12), the coefficients β'_s of the series

$$c^{-1} \sim \frac{n^{\frac{2}{3}}}{\kappa(t)}\left[1 + \sum_{s=1}^{\infty} \beta'_s n^{-2s/3}\right] \tag{10·3}$$

have been evaluated. In terms of α they are given by

$$\left.\begin{aligned} \beta'_1 &= \tfrac{4}{5}\alpha, & \beta'_2 &= \tfrac{22}{175}\alpha^2, \\ \beta'_3 &= -\tfrac{248}{7875}\alpha^3 + \tfrac{11}{1575}, & \beta'_4 &= \tfrac{346}{144375}\alpha^4 - \tfrac{1016}{202125}\alpha. \end{aligned}\right\} \tag{10·4}$$

Table 3

$-t$	Corresponding zero	α_1	α_2	α_3	α_4	α_5
0·5	$y_{n,1}$	+0·9315768	+ 0·2603506	+0·011976	− 0·00602	− 0·0012
1·0	$j_{n,1}$	+1·8557571	+ 1·0331503	−0·003974	− 0·09076	+ 0·0433
1·5	$y_{n,2}$	+2·5962685	+ 2·0221830	−0·035716	− 0·34628	+ 0·2722
2·0	$j_{n,2}$	+3·2446076	+ 3·1582436	−0·083307	− 0·84367	+ 0·8639
2·5	$y_{n,3}$	+3·8341592	+ 4·4102330	−0·146758	− 1·64436	+ 2·0233
3·0	$j_{n,3}$	+4·3816712	+ 5·7597129	−0·226068	− 2·80395	+ 3·9760
3·5	$y_{n,4}$	+4·8970149	+ 7·1942264	−0·321240	− 4·37396	+ 6·9651
4·0	$j_{n,4}$	+5·3866138	+ 8·7046824	−0·432274	− 6·40286	+11·2490
4·5	$y_{n,5}$	+5·8549399	+10·2840965	−0·559169	− 8·93666	+17·0999
5·0	$j_{n,5}$	+6·3052630	+11·9269025	−0·701926	−12·01933	+24·8020

Table 4

$-t$	Corresponding zero	κ	β'_1	β'_2	β'_3	β'_4
0·5	$y_{n,1}$	+0·9555486	+0·7452614	+0·1090993	−0·018476	−0·00288
1·0	$j_{n,1}$	+1·1131028	+1·4846057	+0·4329392	−0·194279	+0·01910
1·5	$y_{n,2}$	+1·2069171	+2·0770148	+0·8473910	−0·544140	+0·09584
2·0	$j_{n,2}$	+1·2748598	+2·5956861	+1·3234545	−1·068707	+0·24929
2·5	$y_{n,3}$	+1·3286404	+3·0673273	+1·8480976	−1·768070	+0·49865
3·0	$j_{n,3}$	+1·3734258	+3·5053370	+2·4135940	−2·642251	+0·86135
3·5	$y_{n,4}$	+1·4119613	+3·9176119	+3·0147235	−3·691258	+1·35358
4·0	$j_{n,4}$	+1·4458854	+4·3092910	+3·6476764	−4·915095	+1·99058
4·5	$y_{n,5}$	+1·4762570	+4·6839520	+4·3095262	−6·313764	+2·78684
5·0	$j_{n,5}$	+1·5038017	+5·0442104	+4·9979401	−7·887264	+3·75619

Table 5

$-t$	Corresponding zero	a_1	a_2	a_3	a_4
0·0	$j'_{n,1}$	+0·8086165	+ 0·0724902	−0·050967	+ 0·00939
0·5	$y'_{n,1}$	+1·8210980	+ 0·9400074	−0·058084	− 0·05402
1·0	$j'_{n,2}$	+2·5780961	+ 1·9551856	−0·089250	− 0·29413
1·5	$y'_{n,2}$	+3·2328653	+ 3·1044932	−0·136685	− 0·77821
2·0	$j'_{n,3}$	+3·8257153	+ 4·3646903	−0·200071	− 1·56685
2·5	$y'_{n,3}$	+4·3751914	+ 5·7198338	−0·279349	− 2·71528
3·0	$j'_{n,4}$	+4·8918203	+ 7·1585294	−0·374502	− 4·27481
3·5	$y'_{n,4}$	+5·3823170	+ 8·6722214	−0·485524	− 6·29376
4·0	$j'_{n,5}$	+5·8513010	+10·2542267	−0·612411	− 8·81804
4·5	$y'_{n,5}$	+6·3021240	+11·8991624	−0·755163	−11·89156

Table 6

$-t$	Corresponding zero	k	b_1	b_2	b_3
0·0	$j'_{n,1}$	+0·6748851	−0·1617233	+0·0291804	−0·006826
0·5	$y'_{n,1}$	+0·5731940	−0·3642196	+0·0907700	+0·023739
1·0	$j'_{n,2}$	+0·5279264	−0·5156192	+0·1747909	+0·086404
1·5	$y'_{n,2}$	+0·4995875	−0·6465731	+0·2718440	+0·180495
2·0	$j'_{n,3}$	+0·4792821	−0·7651431	+0·3789707	+0·305970
2·5	$y'_{n,3}$	+0·4636121	−0·8750383	+0·4945162	+0·462820
3·0	$j'_{n,4}$	+0·4509358	−0·9783641	+0·6173847	+0·651045
3·5	$y'_{n,4}$	+0·4403413	−1·0764634	+0·7467837	+0·870642
4·0	$j'_{n,5}$	+0·4312725	−1·1702602	+0·8821076	+1·121610
4·5	$y'_{n,5}$	+0·4233665	−1·2604248	+1·0228751	+1·403950

The values of β'_s are much smaller than the corresponding values of β_s, and, taking equivalent numbers of terms, greater accuracy is attainable in c from (10·4) than from (8·12). The extension of all these tables to higher zeros offers no difficulty.

The accuracy attainable by use of the tables may be gauged by way of examples. At $s = 1$, $n = 20$, seven-figure accuracy is attained in $j_{n,s}$, $y_{n,s}$, $j'_{n,s}$ and $y'_{n,s}$, and six-figure accuracy in the associated values of c or b; for $s = 5$ and the same value of n, the corresponding results are five and four-figure accuracy. The accuracy lessens as s increases and of course increases with n. Greater accuracy than that indicated here could be achieved by the evaluation of higher terms of the asymptotic series, since in none of these examples did the numerically smallest term appear to be amongst those which were taken into account.

Summarizing the results of §§7–10, we have proved the existence of asymptotic expansions for large positive values of the order n, for

(i) the positive zeros ρ of the real cylinder function $\mathscr{C}_n(x)$ (vide (8·6)),

(ii) the values of $\mathscr{C}'_n(\rho)$ (vide (8·12), (10·3)),

(iii) the positive zeros σ of $\mathscr{C}'_n(x)$ (vide (9·3)),

(iv) the values of $\mathscr{C}_n(\sigma)$ (vide (9·7)),

and we have also derived explicit expressions and, in certain cases numerical values, for the first few coefficients of each of these series.

The earliest attempt to produce such expansions appears to be that of Meissel (5). He gave in effect* twelve-decimal values of the first three irrational coefficients of the series (8·6) for the first zero of $J_n(x)$, but the values given for the second and third coefficients are wrong. Airey (6) used a method of successive approximation to obtain three-decimal values of the first two irrational coefficients of the series (8·6) for the first three zeros of $J_n(x)$; some of these values possess small errors. He also obtained the corresponding coefficients for the series (9·3) for the first three zeros of $J'_n(x)$, but these results are erroneous. The reason for this is mentioned by Watson ((2), p. 516). Some of Airey's results are quoted by Jahnke and Emde (7).

The work described above has been carried out as part of the research programme of the National Physical Laboratory, and this paper is published by permission of the Director of the Laboratory.

* Meissel's expansions actually give n in terms of a series of fractional powers of the zero ρ or σ.

REFERENCES

(1) OLVER, F. W. J. A new method for the evaluation of zeros of Bessel functions and of other solutions of second-order differential equations. *Proc. Cambridge Phil. Soc.* 46 (1950), 570–80.
(2) WATSON, G. N. *Theory of Bessel Functions* (Cambridge, 1944).
(3) INCE, E. L. *Ordinary Differential Equations* (Dover, New York, 1944).
(4) BRITISH ASSOCIATION MATHEMATICAL TABLES, Part-Vol. B, *The Airy Integral* (Cambridge, 1946).
(5) MEISSEL, E. *Astr. Nach.* CXXVIII (1891), cols. 435–8.
(6) AIREY, J. R. The numerical calculation of the roots of the Bessel function $J_n(x)$ and its first derivate $J_n'(x)$. *Phil. Mag.* 34 (1917), 189–95.
(7) JAHNKE, E. and EMDE, F. *Tables of Functions* (Dover, New York, 1945), p. 143.

NATIONAL PHYSICAL LABORATORY
TEDDINGTON, MIDDLESEX

Reprinted from the *Proceedings of the Cambridge Philosophical Society*, Volume 48, Part 3, pp. 414–427, 1952.

SOME NEW ASYMPTOTIC EXPANSIONS FOR BESSEL FUNCTIONS OF LARGE ORDERS

BY F. W. J. OLVER

Communicated by E. T. GOODWIN

Received 13 July 1951

1. *Introduction and summary.* During the course of recent work (6) on the zeros of the Bessel functions $J_n(x)$ and $Y_n(x)$, it became evident that the theory of the asymptotic expansion of Bessel functions whose arguments and orders are of comparable magnitudes was incomplete. The existing expansions for large orders are those of Debye and Meissel, detailed derivations of both of which are given by Watson ((8), pp. 237–48).

If the order n of the Bessel function is large and positive and α, β are fixed positive numbers, then Debye's expansions may be written as

$$\left.\begin{aligned} J_n(n\operatorname{sech}\alpha) &\sim \frac{\exp\{n(\tanh\alpha-\alpha)\}}{(2\pi n\tanh\alpha)^{\frac{1}{2}}}\sum_{s=0}^{\infty}\frac{D_s(-\coth^2\alpha)}{(n\tanh\alpha)^s}, \\ Y_n(n\operatorname{sech}\alpha) &\sim -\frac{\exp\{n(\alpha-\tanh\alpha)\}}{(\frac{1}{2}\pi n\tanh\alpha)^{\frac{1}{2}}}\sum_{s=0}^{\infty}(-)^s\frac{D_s(-\coth^2\alpha)}{(n\tanh\alpha)^s}, \end{aligned}\right\} \tag{1·1}$$

and

$$\left.\begin{aligned} J_n(n\sec\beta) &= \left(\frac{2}{n\pi\tan\beta}\right)^{\frac{1}{2}}[\cos(n\tan\beta-n\beta-\tfrac{1}{4}\pi)E+\sin(n\tan\beta-n\beta-\tfrac{1}{4}\pi)F], \\ Y_n(n\sec\beta) &= \left(\frac{2}{n\pi\tan\beta}\right)^{\frac{1}{2}}[\sin(n\tan\beta-n\beta-\tfrac{1}{4}\pi)E-\cos(n\tan\beta-n\beta-\tfrac{1}{4}\pi)F], \end{aligned}\right\} \tag{1·2}$$

where
$$E\sim\sum_{s=0}^{\infty}(-)^s\frac{D_{2s}(\cot^2\beta)}{(n\tan\beta)^{2s}},\quad F\sim\sum_{s=0}^{\infty}(-)^s\frac{D_{2s+1}(\cot^2\beta)}{(n\tan\beta)^{2s+1}}, \tag{1·3}$$

$D_s(x)$ being a polynomial in x of degree s. For $s=0$, 1 and 2 we have

$$D_0(x)=1,\quad D_1(x)=\tfrac{5}{24}x+\tfrac{1}{8},\quad D_2(x)=\tfrac{385}{1152}x^2+\tfrac{77}{192}x+\tfrac{9}{128}. \tag{1·4}$$

Meissel's expansions can be written in the form

$$J_n(n+\epsilon)=\frac{1}{(48)^{\frac{1}{3}}}P+(\tfrac{3}{4})^{\frac{1}{3}}Q,\quad Y_n(n+\epsilon)=-(\tfrac{3}{4})^{\frac{1}{3}}P+(\tfrac{9}{2})^{\frac{1}{3}}Q, \tag{1·5}$$

where
$$P\sim\frac{\Gamma(\frac{1}{3})}{\pi(n+\epsilon)^{\frac{1}{3}}}\sum_{s=0}^{\infty}P_s(\epsilon)\left(\frac{6}{n+\epsilon}\right)^s,\quad Q\sim\frac{\Gamma(\frac{2}{3})}{\pi(n+\epsilon)^{\frac{2}{3}}}\sum_{s=0}^{\infty}Q_s(\epsilon)\left(\frac{6}{n+\epsilon}\right)^s, \tag{1·6}$$

ϵ being a fixed number; $P_s(\epsilon)$, $Q_s(\epsilon)$ are polynomials in ϵ of degrees $3s$ and $3s+1$ respectively, the first few of which are*

$$\left.\begin{aligned} &\left.\begin{aligned} P_0(\epsilon)&=1, \\ Q_0(\epsilon)&=\epsilon, \end{aligned}\right\}\quad \left.\begin{aligned} P_1(\epsilon)&=-\tfrac{1}{18}\epsilon^3+\tfrac{1}{45}\epsilon, \\ Q_1(\epsilon)&=-\tfrac{1}{36}\epsilon^4+\tfrac{1}{36}\epsilon^2-\tfrac{1}{420}, \end{aligned}\right\} \\ &\left.\begin{aligned} P_2(\epsilon)&=\tfrac{1}{1620}\epsilon^6-\tfrac{7}{3240}\epsilon^4+\tfrac{1}{648}\epsilon^2-\tfrac{1}{8100}, \\ Q_2(\epsilon)&=\tfrac{1}{4536}\epsilon^7-\tfrac{1}{810}\epsilon^5+\tfrac{19}{11340}\epsilon^3-\tfrac{13}{283500}\epsilon. \end{aligned}\right\} \end{aligned}\right\} \tag{1·7}$$

* In terms of Watson's notation,

$$P_s(\epsilon)=(-)^s\,\Gamma(s+\tfrac{1}{3})\,B_{3s}(\epsilon)/\Gamma(\tfrac{1}{3}),\quad Q_s(\epsilon)=(-)^s\,\Gamma(s+\tfrac{2}{3})\,B_{3s+1}(\epsilon)/\Gamma(\tfrac{2}{3}).$$

Examination of Debye's expansions (1·1) and (1·2) reveals that they are suitable for the numerical computation of $J_n(x)$ and $Y_n(x)$ for large values of n, provided that $|x-n|$ is large compared with $n^{\frac{1}{3}}$, for otherwise the early terms of these expansions do not decrease in magnitude. On the other hand it is evident from (1·6) that Meissel's expansions are of practical use only when $x = n+o(n^{\frac{1}{3}})$. Thus, corresponding to values of $|x-n|$ which are comparable with $n^{\frac{1}{3}}$, there are gaps within which neither type of expansion is suitable. These gaps have been called by Watson the *transition regions*.

With a view to remedying this defect Nicholson (5) applied the method of stationary phase to obtain the asymptotic *forms* of the functions in the transition regions, and Watson ((8), pp. 248–52, also (7)), with the aid of some heavy analysis, put Nicholson's results on a sound basis and gave estimates for the remainder terms. These asymptotic forms cannot be expressed in terms of elementary functions and necessarily involve Bessel functions of order one-third or, equivalently, the Airy functions. Watson's version is given by

$$J_n(n\operatorname{sech}\alpha) = \frac{\tanh\alpha}{3^{\frac{1}{2}}\pi}\exp\{n(\tanh\alpha+\tfrac{1}{3}\tanh^3\alpha-\alpha)\}K_{\frac{1}{3}}(\tfrac{1}{3}n\tanh^3\alpha) + \frac{3\theta_1}{n}\exp\{n(\tanh\alpha-\alpha)\}, \quad (1·8)$$

and $$J_n(n\sec\beta) = \tfrac{1}{3}\tan\beta\cos\{n(\tan\beta-\tfrac{1}{3}\tan^3\beta-\beta)\}\{J_{-\frac{1}{3}}(\xi)+J_{\frac{1}{3}}(\xi)\} + \frac{1}{3^{\frac{1}{2}}}\tan\beta\sin\{n(\tan\beta-\tfrac{1}{3}\tan^3\beta-\beta)\}\{J_{-\frac{1}{3}}(\xi)-J_{\frac{1}{3}}(\xi)\}+\frac{24\theta_2}{n}, \quad (1·9)$$

with a similar expression for $Y_n(n\sec\beta)$, where $0<|\theta_1|, |\theta_2|<1$, $\xi\equiv\frac{1}{3}n\tan^3\beta$ and K denotes the modified Bessel function.

More recently, improved versions of the asymptotic forms in the transition regions have been obtained by Langer (4) and Imai (3), using methods based upon the transformation of differential equations. The remainder terms in Langer's formulae are $O(n^{-\frac{4}{3}})$ and in Imai's formulae they are $O(n^{-\frac{5}{3}})$. No attempt to determine the asymptotic *expansions* appears to have been recorded so far.

The main substance of the present paper is the derivation of asymptotic expansions of the Bessel and Hankel functions which are valid in the transition regions. Using a variation of the method of steepest descents, we shall obtain the asymptotic expansions of $H^{(1)}_\nu(\nu+\tau\nu^{\frac{1}{3}})$ and $H^{(2)}_\nu(\nu+\tau\nu^{\frac{1}{3}})$ for large values of ν, where τ is any fixed number. There is no especial difficulty in supposing the orders and arguments of the functions to be complex, and in §§ 2, 3 the expansions are established in the half-plane $|\arg\nu|<\frac{1}{2}\pi$; the extension to other values of arg ν is given in § 6. We shall adopt the convention that whenever the symbols n, x are employed, *real positive* orders and arguments are implied, otherwise these quantities will appear as ν, z respectively.

In § 4 it will be seen that these expansions have adequate overlap with Debye's expansions (1·1) and (1·2), and also that they are valid throughout the region in which Meissel's expansions (1·5) are useful.

The new expansions are of special interest in the theory of zeros of Bessel functions since it is known that $j_{n,s}$ and $y_{n,s}$, the sth positive zeros of $J_n(x)$ and $Y_n(x)$ respectively, have the asymptotic forms

$$j_{n,s} = n+\mu_s n^{\frac{1}{3}}+O(n^{-\frac{1}{3}}), \quad y_{n,s} = n+\mu'_s n^{\frac{1}{3}}+O(n^{-\frac{1}{3}}),$$

where μ_s and μ_s' are constants depending on s. Thus the complete asymptotic expansions of $j_{n,s}$ and $y_{n,s}$ for large values of n and fixed s, which were derived by the author using a different method (6), are obtainable by suitable reversion of the expansions of $J_n(x)$ and $Y_n(x)$ in the transition regions, and this aspect is developed fully in §5.

2. *Asymptotic expansion of the Hankel functions in the transition regions when* $|\arg \nu| < \frac{1}{2}\pi$. In this section we take as our starting point the integral representation of Sommerfeld's type,

$$H_\nu^{(1)}(z) = \frac{1}{\pi i}\int_{-\infty}^{\infty+\pi i} \exp\{z\sinh t - \nu t\}\,dt, \tag{2·1}$$

which is valid under the single restriction $|\arg z| < \frac{1}{2}\pi$. Writing $u \equiv \nu^{\frac{1}{3}}$, the principal value of the cube root being taken, and putting

$$z = \nu + \tau\nu^{\frac{1}{3}} = u^3 + \tau u, \tag{2·2}$$

where τ is any fixed number (real or complex), we shall show that the integral (2·1) may be transformed and then expanded to yield the asymptotic expansion

$$H_\nu^{(1)}(z) \sim \sum_{s=0}^{\infty} \frac{(-)^s}{u^{2s+1}} \chi_s(\tau), \tag{2·3}$$

where

$$\chi_s(\tau) = \sum_{r=0}^{s} (-)^r f_{sr} \tau^r F_{2s+r}(\tau), \tag{2·4}$$

and

$$F_s(\tau) = \frac{1}{\pi i}\int_{-\infty}^{\infty\exp\{\frac{1}{3}\pi i\}} w^s \exp\{\tfrac{1}{6}w^3 + \tau w\}\,dw. \tag{2·5}$$

The coefficients f_{sr} are numbers independent of τ and u, and the functions $F_s(\tau)$ are readily expressible in terms of Airy integrals.

(*a*) *Transformation of Sommerfeld's integral.* The first step in obtaining the asymptotic expansion is to transform the integral (2·1) into a form more resembling the integral (2·5). If we suppose that $\arg \nu$ is restricted to the range

$$|\arg \nu| \leqslant \tfrac{1}{2}\pi - \delta, \tag{2·6}$$

where δ is an arbitrary positive number, then we can assign a number R, depending only on δ and τ, such that $|\arg z| < \frac{1}{2}\pi$ whenever $|\nu| > R$. In these circumstances we obtain from equations (2·1) and (2·2),

$$H_\nu^{(1)}(z) = \frac{1}{\pi i}\int_{-\infty}^{\infty+\pi i} \exp\{u^3(\sinh t - t) + \tau u \sinh t\}\,dt. \tag{2·7}$$

The contours in the integrals (2·1) and (2·7) may be taken to be any curves which are asymptotic to the lines $\arg t = \pi$ and $\arg(t - \pi i) = 0$. Write $\phi(t) \equiv \sinh t - t$ and consider the contours along which $\Im\phi(t) = 0$. If $t \equiv \xi + i\eta$, where ξ and η are real, then

$$\phi(t) = \sinh\xi\cos\eta - \xi + i(\cosh\xi\sin\eta - \eta), \tag{2·8}$$

and so the curves in which we are interested are given by

$$\cosh\xi\sin\eta = \eta. \tag{2·9}$$

In the strip $|\Im t| \leqslant \pi$ there are three such curves, namely, the real axis (COD) and two curves (AOF, EOB) which are asymptotic to the lines $\Im t = \pm\pi$ and which make angles $\pm\frac{1}{3}\pi$ with the real axis at the origin O (see Fig. 1). It is easily verified that $\Re\phi(t)$ is positive on AO, EO, OD and negative on CO, OB, OF and that $\Re\phi(t)$ changes monotonically on each of these curves. The contour COB is the appropriate one for our purposes, since it joins the end points of the integral (2·7).

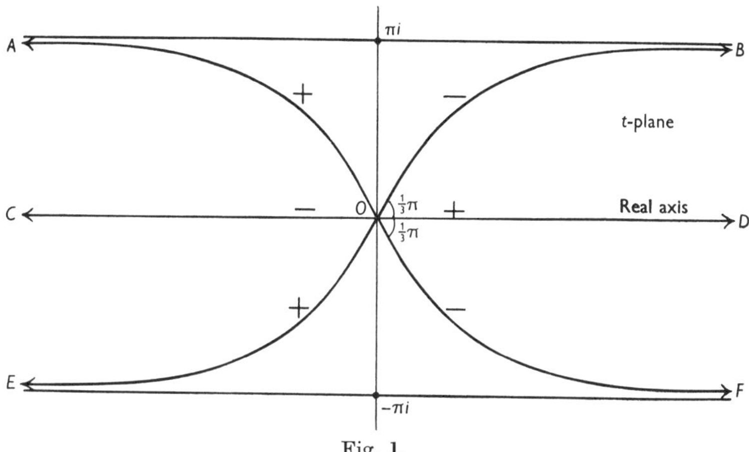

Fig. 1.

Consider a new variable v defined by the equations

$$\tfrac{1}{6}v^3 = \sinh t - t = \phi(t), \quad v = \{6\phi(t)\}^{\frac{1}{3}}, \tag{2·10}$$

where the branch of the cube root is chosen so that $v \sim t$ as $t \to 0$, and is defined by continuity elsewhere on the contour COB. By reversion of (2·10) we obtain the expansion

$$t = \sum_{s=0}^{\infty} (-)^s \lambda_s v^{2s+1}, \tag{2·11}$$

valid within some circle $|v| \leqslant c$, where (cf. (8), p. 232)

$$\lambda_0 = 1, \quad \lambda_1 = 1/60, \quad \lambda_2 = 1/1400, \quad \lambda_3 = 1/25200,$$
$$\lambda_4 = 43/17248000, \quad \lambda_5 = 1213/7207200000, \ldots. \tag{2·12}$$

Making this change of variable in the integral (2·7), we find that

$$H_\nu^{(1)}(z) = \frac{1}{\pi i}\int_{-\infty}^{\infty \exp\{\frac{1}{3}\pi i\}} \exp\{\tfrac{1}{6}u^3v^3 + \tau uv\} f(v, \tau u)\, dv, \tag{2·13}$$

where
$$f(v, \tau u) = e^g \frac{dt}{dv}, \quad g = \tau u(t - v + \tfrac{1}{6}v^3), \tag{2·14}$$

and the contour for the integral (2·13) (which will subsequently be referred to as the contour L) comprises the straight lines $\arg t = \pi$ and $\arg t = \frac{1}{3}\pi$.

(*b*) *Expansion of* $f(v, \tau u)$ *for small values of* $|v|$. From equations (2·11), (2·12) and (2·14) we obtain

$$g = \tau u\left[\tfrac{3}{20}v^3 + \sum_{s=2}^{\infty}(-)^s \lambda_s v^{2s+1}\right] = \tfrac{3}{20}v^2(\tau uv)\left[1 + \sum_{s=1}^{\infty}(-)^{s+1}\tfrac{20}{3}\lambda_{s+1}v^{2s}\right]. \tag{2·15}$$

Hence if $|v|$ is small
$$e^g = 1 + \sum_{s=1}^{\infty} v^{2s} g_s(\tau uv), \tag{2·16}$$

where the coefficients $g_s(x)$ are polynomials in x of degree s, the first four of which are given by

$$\left.\begin{aligned} g_1(x) &= \tfrac{3}{20}x, \quad g_2(x) = \tfrac{9}{800}x^2+\tfrac{1}{1400}x, \quad g_3(x) = \tfrac{9}{16000}x^3+\tfrac{3}{28000}x^2-\tfrac{1}{25200}x,\\ g_4(x) &= \tfrac{27}{1280000}x^4+\tfrac{9}{1120000}x^3-\tfrac{67}{11760000}x^2+\tfrac{43}{17248000}x. \end{aligned}\right\} \tag{2·17}$$

Also, from equation (2·11), we have

$$\frac{dt}{dv} = 1+\sum_{s=1}^{\infty}(-)^s(2s+1)\lambda_s v^{2s} \quad (|v|<c).$$

Using this result and equations (2·14), (2·16), we obtain

$$f(v,\tau u) = \sum_{s=0}^{\infty} v^{2s}\sum_{r=0}^{s}(-)^{s+r}(\tau uv)^r f_{sr}, \tag{2·18}$$

where the coefficients f_{sr} are constants. Using equations (2·12) and (2·17) we readily find that

$$\left.\begin{aligned} &f_{00}=1; \quad f_{10}=\tfrac{1}{20}, \ f_{11}=\tfrac{3}{20}; \quad f_{20}=\tfrac{1}{280}, \ f_{21}=\tfrac{19}{2800}, \ f_{22}=\tfrac{9}{800};\\ &f_{30}=\tfrac{1}{3600}, \ f_{31}=\tfrac{29}{63000}, \ f_{32}=\tfrac{51}{112000}, \ f_{33}=\tfrac{9}{16000};\\ &f_{40}=\tfrac{387}{17248000}, \ f_{41}=\tfrac{5377}{155232000}, \ f_{42}=\tfrac{137}{4704000}, \ f_{43}=\tfrac{9}{448000}, \ f_{44}=\tfrac{27}{1280000}. \end{aligned}\right\} \tag{2·19}$$

If we substitute the expansion (2·18) in the integral (2·13) and invert* the order of integration and summation, then we obtain the expansion given by equations (2·3), (2·4) and (2·5), except that the limits of the integral $F_s(\tau)$ appear as $-\infty\exp(i\arg u)$ and $\infty\exp\{i\arg(u+\frac{1}{3}\pi)\}$. Since $|\arg u|<\frac{1}{6}\pi$, an application of Cauchy's theorem shows that these may be replaced by the limits $-\infty$ and $\infty\exp(\frac{1}{3}\pi i)$ of the integral (2·5).

(c) *Proof of the asymptotic nature of the expansion.* The series (2·18) is a rearrangement of a simple power series in v, and it can be shown without difficulty that both it and the series

$$\sum_{s=0}^{\infty}|v|^{2s}\sum_{r=0}^{s}|\tau uv|^r|f_{sr}| \tag{2·20}$$

certainly converge within some circle $|v|\leqslant V(|\tau u|)$, where $V(|\tau u|)>0$. If we write the series (2·20) in the form

$$\sum_{s=0}^{\infty}|vu^{\frac{1}{3}}|^{2s}\sum_{r=0}^{s}|\tau vu^{\frac{1}{3}}|^r\frac{|f_{sr}|}{|u|^{\frac{1}{3}(2s-2r)}}, \tag{2·21}$$

we see that if $|u|\geqslant 1$ the series (2·20) certainly converges when $|vu^{\frac{1}{3}}|\leqslant V(|\tau|)$, and therefore the expansion (2·18) is also valid under these restrictions.

Hence if $|u|\geqslant 1$, $|v|\leqslant|u|^{-\frac{1}{3}}V(|\tau|)$ and M is an arbitrary positive integer, then

$$\begin{aligned} &\left|f(v,\tau u)-\sum_{s=0}^{M-1}v^{2s}\sum_{r=0}^{s}(-)^{s+r}(\tau uv)^r f_{sr}\right|\\ &\qquad = |v|^{2M}\left|\sum_{s=M}^{\infty}v^{2s-2M}\sum_{r=0}^{s}(-)^{s+r}(\tau uv)^r f_{sr}\right| < \kappa(\tau)|v|^{2M}|u|^{\frac{2}{3}M}, \end{aligned} \tag{2·22}$$

where $\kappa(\tau)$ is independent of both u and v.

Next, consider the behaviour of the quantity on the left-hand side of (2·22) on those parts of L for which

$$|v|>|u|^{-\frac{1}{3}}V(|\tau|). \tag{2·23}$$

* The justification of this operation forms the subject of subsection (c).

L itself may be divided into two parts L_1 and L_2 on which

$$\text{(i) } v = -v_1, \quad t = -t_1, \qquad \text{(ii) } v = e^{\frac{1}{3}\pi i} v_2, \quad t = \pi i + t_2,$$

where t_1, v_1 and v_2 are real numbers varying from 0 to $+\infty$, and where $t_2 \to +\infty$, $\arg t_2 \to 0$ as $v_2 \to +\infty$. Referring to equations (2·10) and (2·14), on L_1 we have $\frac{1}{6}v_1^3 = \sinh t_1 - t_1$, and so as $v_1 \to \infty$,

$$t_1 \sim 3\log v_1, \quad \frac{dt}{dv} = \frac{dt_1}{dv_1} \sim \frac{3}{v_1} \quad \text{and} \quad g \sim \tau u(-3\log v_1 + v_1 - \tfrac{1}{6}v_1^3).$$

Hence on L_1
$$|f(v, \tau u)| < k_1 \exp\{|\tau u|(k_2 v_1 + \tfrac{1}{6}v_1^3)\},$$

where k_1, k_2 are assignable constants. Similarly, on L_2,

$$|f(v, \tau u)| < k_1 \exp\{|\tau u|(k_2 v_2 + \tfrac{1}{6}v_2^3)\}.$$

Hence on those parts of L on which (2·23) holds, we have

$$|f(v, \tau u)| < k_1 \exp\{|\tau u|(k_2|v| + \tfrac{1}{6}|v|^3)\} < k_1 \frac{|v|^{2M}|u|^{\frac{2}{3}M}}{[V(|\tau|)]^{2M}} \exp\{|\tau u|(k_2|v| + \tfrac{1}{6}|v|^3)\}. \tag{2·24}$$

Also, if $|u| \geq 1$ and $\{c_s\}$ denotes a set of suitably assigned constants, then

$$\left|\sum_{s=0}^{M-1} v^{2s} \sum_{r=0}^{s} (-)^{s+r} (\tau u v)^r f_{sr}\right| < \sum_{s=0}^{M-1} |v|^{2s} c_s e^{|\tau u v|} = |v|^{2M} e^{|\tau u v|} \sum_{s=0}^{M-1} \frac{c_s}{|v|^{2M-2s}}$$

$$< |v|^{2M} e^{|\tau u v|} \sum_{s=0}^{M-1} c_s \frac{|u|^{\frac{1}{3}(2M-2s)}}{[V(|\tau|)]^{2M-2s}} < C(\tau)|v|^{2M}|u|^{\frac{2}{3}M} e^{|\tau u v|}, \tag{2·25}$$

in virtue of (2·23), $C(\tau)$ being independent of u and v in the ranges considered.

Combining the results (2·24), (2·25) and using also the result (2·22), we deduce without difficulty that if $|u| \geq 1$ then on the *whole* of L, *no matter whether or not* $|v| \leq |u|^{-\frac{1}{3}} V(|\tau|)$, it is true that

$$\left|f(v, \tau u) - \sum_{s=0}^{M-1} v^{2s} \sum_{r=0}^{s} (-)^{s+r} (\tau u v)^r f_{sr}\right| < K(\tau)|v|^{2M}|u|^{\frac{2}{3}M} \exp\{|\tau u|(k|v| + \tfrac{1}{6}|v|^3)\}, \tag{2·26}$$

where k is an assignable constant and $K(\tau)$ is independent of u and v.

Further, v^3 is negative on L, and hence

$$|\exp\{\tfrac{1}{6}u^3v^3 + \tau u v\}| \leq \exp\{|\tau u v| - \tfrac{1}{6}|v|^3 \Re(u^3)\} \leq \exp\{|\tau u v| - \tfrac{1}{6}|uv|^3 \sin\delta\}, \tag{2·27}$$

since from equation (2·6),

$$\Re u^3 = \Re \nu \geq |\nu| \sin\delta = |u|^3 \sin\delta.$$

Combining the results (2·13), (2·26), (2·27) and using equations (2·4), (2·5), we find without difficulty that

$$H_\nu^{(1)}(z) = \sum_{s=0}^{M-1} \frac{(-)^s}{u^{2s+1}} \chi_s(\tau) + R_M, \tag{2·28}$$

where M is an arbitrary positive integer, and where

$$|R_M| < \frac{2}{\pi} K(\tau)|u|^{\frac{2}{3}M} \int_0^\infty v^{2M} \exp[|\tau u|\{(k+1)v + \tfrac{1}{6}v^3\} - \tfrac{1}{6}|u|^3 v^3 \sin\delta]\, dv, \tag{2·29}$$

provided* $|u| > (|\tau| \operatorname{cosec} \delta + 1)^{\frac{1}{2}}$. Making the change of variable $x = |u| v$ in this integral, we obtain

$$|R_M| < \frac{2}{\pi} K(\tau) |u|^{-\frac{1}{3}(4M+3)} \int_0^\infty x^{2M} \exp[|\tau| \{(k+1)x + \tfrac{1}{6}|u|^{-2} x^3\} - \tfrac{1}{6}x^3 \sin\delta]\, dx$$
$$= O(|u|^{-\frac{1}{3}(4M+3)}). \quad (2{\cdot}30)$$

This estimate for R_M is not quite good enough as it stands to establish the asymptotic property of the series (2·3), but if equation (2·28) is rewritten with M replaced by $M' \equiv$ the integral part of $\frac{1}{2}(3M+1)$, we see that

$$R_M = \sum_{s=M}^{M'-1} \frac{(-)^s}{u^{2s+1}} \chi_s(\tau) + R_{M'} = O(|u|^{-2M-1}), \quad (2{\cdot}31)$$

by use of formula (2·30) with M replaced by M'. This completes the proof.

(*d*) *Evaluation of the coefficients* $\chi_s(\tau)$ *and final form of the asymptotic expansion.* In order to apply the result (2·3) to practical calculations it is necessary to evaluate the

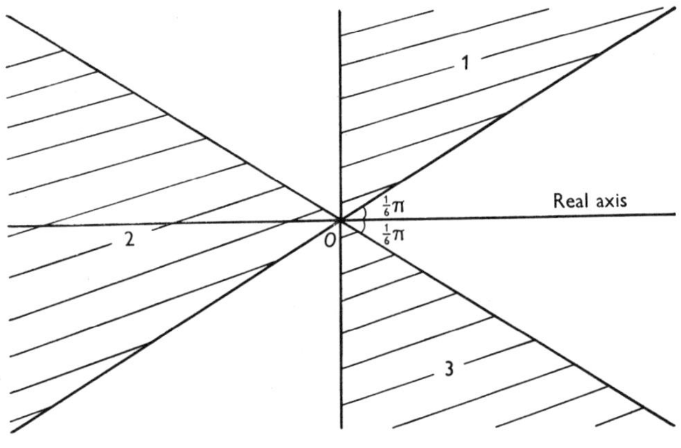

Fig. 2.

coefficients $F_s(\tau)$ and $\chi_s(\tau)$ defined by the expressions (2·4) and (2·5). From (2·5) we obtain

$$F_s(\tau) = \frac{d^s}{d\tau^s} \cdot \frac{1}{\pi i} \int_{-\infty}^{\infty \exp\{\frac{1}{3}\pi i\}} \exp\{\tfrac{1}{6}w^3 + \tau w\}\, dw = F_0^{(s)}(\tau). \quad (2{\cdot}32)$$

To evaluate $F_0(\tau)$, we have ((1), p. B 9) the following integral representations of the Airy functions:

$$\mathrm{Ai}(x) = \tfrac{1}{2} l_{31}, \quad \mathrm{Bi}(x) = \tfrac{1}{2} i (l_{21} + l_{23}), \quad (2{\cdot}33)$$

where

$$l_{rs} = \frac{1}{\pi i} \int_{L_{rs}} \exp\{\tfrac{1}{3}t^3 - xt\}\, dt,$$

and L_{rs} denotes any contour starting at infinity in a shaded sector r of Fig. 2 and finishing at infinity in a shaded sector s.

From Cauchy's theorem $l_{32} + l_{21} + l_{13} = 0$, and so from equations (2·33) we deduce that

$$l_{32} = \mathrm{Ai}(x) + i\,\mathrm{Bi}(x), \quad l_{21} = \mathrm{Ai}(x) - i\,\mathrm{Bi}(x) \quad \text{and} \quad l_{13} = -2\,\mathrm{Ai}(x). \quad (2{\cdot}34)$$

Hence from these equations and (2·32), putting $s = 0$, we obtain

$$F_0(\tau) = 2^{\frac{1}{3}}\{\mathrm{Ai}(-2^{\frac{1}{3}}\tau) - i\,\mathrm{Bi}(-2^{\frac{1}{3}}\tau)\}. \quad (2{\cdot}35)$$

* This restriction ensures that the integral (2·29) converges, and that $|u| > 1$.

The function $F_0(\tau)$ therefore satisfies the differential equation

$$F_0''(\tau) = -2\tau F_0(\tau). \tag{2·36}$$

Consequently

$$F_s(\tau) = \phi_s(\tau) F_0(\tau) - \psi_s(\tau) F_0'(\tau), \tag{2·37}$$

where $\phi_s(\tau)$ and $\psi_s(\tau)$ are polynomials in τ. Differentiating equation (2·36) s times using Leibniz's theorem produces the recurrence relations

$$\left.\begin{aligned} \phi_{s+2} &= -2\tau\phi_s - 2s\phi_{s-1}, \\ \psi_{s+2} &= -2\tau\psi_s - 2s\psi_{s-1}, \end{aligned}\right\} \tag{2·38}$$

which can be used to compute these polynomials, starting from $\phi_0 = 1$, $\psi_0 = 0$; $\phi_1 = 0$, $\psi_1 = -1$. Table 1 gives the values of the first thirteen polynomials of each set.

Table 1

s	$\phi_s(\tau)$	$\psi_s(\tau)$	s	$\phi_s(\tau)$	$\psi_s(\tau)$
0	1	0	7	$-72\tau^2$	$8\tau^3-40$
1	0	-1	8	$16\tau^4-224\tau$	$96\tau^2$
2	-2τ	0	9	$256\tau^3-224$	$-16\tau^4+416\tau$
3	-2	2τ	10	$-32\tau^5+1600\tau^2$	$-320\tau^3+640$
4	$4\tau^2$	4	11	$-800\tau^4+4480\tau$	$32\tau^5-2560\tau^2$
5	16τ	$-4\tau^2$	12	$64\tau^6-8320\tau^3+4480$	$960\tau^4-9600\tau$
6	$-8\tau^3+16$	-24τ			

Substituting the expression (2·37) in equation (2·4) we obtain

$$\chi_s(\tau) = \sum_{r=0}^{s} (-)^r f_{sr} \tau^r \{\phi_{2s+r}(\tau) F_0(\tau) - \psi_{2s+r}(\tau) F_0'(\tau)\},$$

and substitution of this result in the expansion (2·3) and the use of equation (2·35) leads us to the final form of the expansion for the first Hankel function:

$$\begin{aligned} H_\nu^{(1)}(\nu+\tau\nu^{\frac{1}{3}}) \sim \frac{2^{\frac{1}{3}}}{\nu^{\frac{1}{3}}} [\mathrm{Ai}(-2^{\frac{1}{3}}\tau) - i\,\mathrm{Bi}(-2^{\frac{1}{3}}\tau)] \sum_{s=0}^{\infty} \frac{A_s(\tau)}{\nu^{\frac{2}{3}s}} \\ + \frac{2^{\frac{2}{3}}}{\nu^{\frac{1}{3}}} [\mathrm{Ai}'(-2^{\frac{1}{3}}\tau) - i\,\mathrm{Bi}'(-2^{\frac{1}{3}}\tau)] \sum_{s=0}^{\infty} \frac{B_s(\tau)}{\nu^{\frac{2}{3}s}}, \end{aligned} \tag{2·39}$$

where the coefficients $A_s(\tau)$, $B_s(\tau)$ are polynomials in τ given by

$$A_s(\tau) = \sum_{r=0}^{s} (-)^{s+r} f_{sr} \tau^r \phi_{2s+r}(\tau), \quad B_s(\tau) = \sum_{r=0}^{s} (-)^{s+r} f_{sr} \tau^r \psi_{2s+r}(\tau). \tag{2·40}$$

Similarly from Sommerfeld's integral for $H_\nu^{(2)}(z)$ we find that

$$\begin{aligned} H_\nu^{(2)}(\nu+\tau\nu^{\frac{1}{3}}) \sim \frac{2^{\frac{1}{3}}}{\nu^{\frac{1}{3}}} [\mathrm{Ai}(-2^{\frac{1}{3}}\tau) + i\,\mathrm{Bi}(-2^{\frac{1}{3}}\tau)] \sum_{s=0}^{\infty} \frac{A_s(\tau)}{\nu^{\frac{2}{3}s}} \\ + \frac{2^{\frac{2}{3}}}{\nu^{\frac{1}{3}}} [\mathrm{Ai}'(-2^{\frac{1}{3}}\tau) + i\,\mathrm{Bi}'(-2^{\frac{1}{3}}\tau)] \sum_{s=0}^{\infty} \frac{B_s(\tau)}{\nu^{\frac{2}{3}s}}. \end{aligned} \tag{2·41}$$

Using the values of f_{sr} given by equations (2·19) and the values of $\phi_s(\tau)$ and $\psi_s(\tau)$ given in Table 1, we find the first few values of the coefficients to be

$$\left.\begin{aligned} &\left.\begin{aligned} A_0 &= 1, \\ B_0 &= 0, \end{aligned}\right\} \quad \left.\begin{aligned} A_1 &= -\tfrac{1}{5}\tau, \\ B_1 &= \tfrac{3}{10}\tau^2, \end{aligned}\right\} \quad \left.\begin{aligned} A_2 &= -\tfrac{9}{100}\tau^5 + \tfrac{3}{35}\tau^2, \\ B_2 &= -\tfrac{17}{70}\tau^3 + \tfrac{1}{70}, \end{aligned}\right\} \\ &\left.\begin{aligned} A_3 &= \tfrac{957}{7000}\tau^6 - \tfrac{173}{3150}\tau^3 - \tfrac{1}{225}, \\ B_3 &= -\tfrac{9}{1000}\tau^7 + \tfrac{611}{3150}\tau^4 - \tfrac{37}{3150}\tau, \end{aligned}\right\} \\ &\left.\begin{aligned} A_4 &= \tfrac{27}{20000}\tau^{10} - \tfrac{23573}{147000}\tau^7 + \tfrac{5903}{138600}\tau^4 + \tfrac{947}{346500}\tau, \\ B_4 &= \tfrac{549}{28000}\tau^8 - \tfrac{110767}{693000}\tau^5 + \tfrac{79}{12375}\tau^2. \end{aligned}\right\} \end{aligned}\right\} \tag{2·42}$$

3. *Asymptotic expansions of $J_\nu(z)$, $Y_\nu(z)$ and their derivatives in the transition regions.* From the results (2·39) and (2·41) we immediately deduce that

$$J_\nu(\nu+\tau\nu^{\frac{1}{3}}) \sim \frac{2^{\frac{1}{3}}}{\nu^{\frac{1}{3}}} \mathrm{Ai}\,(-2^{\frac{1}{3}}\tau) \sum_{s=0}^{\infty} \frac{A_s(\tau)}{\nu^{\frac{2}{3}s}} + \frac{2^{\frac{2}{3}}}{\nu^{\frac{1}{3}}} \mathrm{Ai}'(-2^{\frac{1}{3}}\tau) \sum_{s=0}^{\infty} \frac{B_s(\tau)}{\nu^{\frac{2}{3}s}}, \tag{3·1}$$

and

$$Y_\nu(\nu+\tau\nu^{\frac{1}{3}}) \sim -\frac{2^{\frac{1}{3}}}{\nu^{\frac{1}{3}}} \mathrm{Bi}\,(-2^{\frac{1}{3}}\tau) \sum_{s=0}^{\infty} \frac{A_s(\tau)}{\nu^{\frac{2}{3}s}} - \frac{2^{\frac{2}{3}}}{\nu^{\frac{1}{3}}} \mathrm{Bi}'(-2^{\frac{1}{3}}\tau) \sum_{s=0}^{\infty} \frac{B_s(\tau)}{\nu^{\frac{2}{3}s}}, \tag{3·2}$$

valid when $|\arg\nu| < \frac{1}{2}\pi$, τ being any fixed number.

By repeated differentiation with respect to τ and use of the differential equation satisfied by $\mathrm{Ai}\,(-2^{\frac{1}{3}}\tau)$ and $\mathrm{Bi}\,(-2^{\frac{1}{3}}\tau)$, we obtain the following expansions for the derivatives, valid under the same conditions:

$$J_\nu^{(m)}(\nu+\tau\nu^{\frac{1}{3}}) \sim \frac{2^{\frac{1}{3}}}{\nu^{\frac{1}{3}(m+1)}} \mathrm{Ai}\,(-2^{\frac{1}{3}}\tau) \sum_{s=0}^{\infty} \frac{A_s^m(\tau)}{\nu^{\frac{2}{3}s}} + \frac{2^{\frac{2}{3}}}{\nu^{\frac{1}{3}(m+1)}} \mathrm{Ai}'(-2^{\frac{1}{3}}\tau) \sum_{s=0}^{\infty} \frac{B_s^m(\tau)}{\nu^{\frac{2}{3}s}}, \tag{3·3}$$

and

$$Y_\nu^{(m)}(\nu+\tau\nu^{\frac{1}{3}}) \sim -\frac{2^{\frac{1}{3}}}{\nu^{\frac{1}{3}(m+1)}} \mathrm{Bi}\,(-2^{\frac{1}{3}}\tau) \sum_{s=0}^{\infty} \frac{A_s^m(\tau)}{\nu^{\frac{2}{3}s}} - \frac{2^{\frac{2}{3}}}{\nu^{\frac{1}{3}(m+1)}} \mathrm{Bi}'(-2^{\frac{1}{3}}\tau) \sum_{s=0}^{\infty} \frac{B_s^m(\tau)}{\nu^{\frac{2}{3}s}}. \tag{3·4}$$

The coefficients $A_s^m(\tau)$, $B_s^m(\tau)$ satisfy the recurrence relations

$$A_s^{m+1}(\tau) = \frac{d}{d\tau} A_s^m(\tau) + 2\tau B_s^m(\tau), \quad B_s^{m+1}(\tau) = -A_s^m(\tau) + \frac{d}{d\tau} B_s^m(\tau). \tag{3·5}$$

For $m = 1, 2$ and $s = 0(1)4$, these coefficients are given by

$$\left.\begin{aligned}
&\left.\begin{aligned} A_0^1 &= 0, \\ B_0^1 &= -1, \end{aligned}\right\} \quad \left.\begin{aligned} A_1^1 &= \tfrac{3}{5}\tau^3 - \tfrac{1}{5}, \\ B_1^1 &= \tfrac{4}{5}\tau, \end{aligned}\right\} \quad \left.\begin{aligned} A_2^1 &= -\tfrac{131}{140}\tau^4 + \tfrac{1}{5}\tau, \\ B_2^1 &= \tfrac{9}{100}\tau^5 - \tfrac{57}{70}\tau^2, \end{aligned}\right\} \\
&\left.\begin{aligned} A_3^1 &= -\tfrac{9}{500}\tau^8 + \tfrac{5437}{4500}\tau^5 - \tfrac{593}{3150}\tau^2, \\ B_3^1 &= -\tfrac{699}{3500}\tau^6 + \tfrac{2617}{3150}\tau^3 - \tfrac{23}{3150}, \end{aligned}\right\} \\
&\left.\begin{aligned} A_4^1 &= \tfrac{369}{7000}\tau^9 - \tfrac{999443}{693000}\tau^6 + \tfrac{31727}{173250}\tau^3 + \tfrac{947}{346500}, \\ B_4^1 &= -\tfrac{27}{20000}\tau^{10} + \tfrac{46631}{147000}\tau^7 - \tfrac{3889}{4620}\tau^4 + \tfrac{1159}{115500}\tau, \end{aligned}\right\}
\end{aligned}\right\} \tag{3·6}$$

and

$$\left.\begin{aligned}
&\left.\begin{aligned} A_0^2 &= -2\tau, \\ B_0^2 &= 0, \end{aligned}\right\} \quad \left.\begin{aligned} A_1^2 &= \tfrac{17}{5}\tau^2, \\ B_1^2 &= -\tfrac{3}{5}\tau^3 + 1, \end{aligned}\right\} \quad \left.\begin{aligned} A_2^2 &= \tfrac{9}{50}\tau^6 - \tfrac{188}{35}\tau^3 + \tfrac{1}{5}, \\ B_2^2 &= \tfrac{97}{70}\tau^4 - \tfrac{64}{35}\tau, \end{aligned}\right\} \\
&\left.\begin{aligned} A_3^2 &= -\tfrac{951}{1750}\tau^7 + \tfrac{48527}{6300}\tau^4 - \tfrac{616}{1575}\tau, \\ B_3^2 &= \tfrac{9}{500}\tau^8 - \tfrac{15161}{6300}\tau^5 + \tfrac{4222}{1575}\tau^2, \end{aligned}\right\} \\
&\left.\begin{aligned} A_4^2 &= -\tfrac{27}{10000}\tau^{11} + \tfrac{163003}{147000}\tau^8 - \tfrac{1193893}{115500}\tau^5 + \tfrac{2349}{4125}\tau^2, \\ B_4^2 &= -\tfrac{927}{14000}\tau^9 + \tfrac{1269133}{346500}\tau^6 - \tfrac{615077}{173250}\tau^3 + \tfrac{23}{3150}. \end{aligned}\right\}
\end{aligned}\right\} \tag{3·7}$$

Certain values of the coefficients for higher values of m appear in §5.

From Bessel's differential equation ((8), p. 38), we find that

$$(u^5 + 2\tau u^3 + \tau^2 u)\, J_\nu''(\nu+\tau\nu^{\frac{1}{3}}) + (u^2+\tau)\, J_\nu'(\nu+\tau\nu^{\frac{1}{3}}) + (2\tau u^3 + \tau^2 u)\, J_\nu(\nu+\tau\nu^{\frac{1}{3}}) = 0, \tag{3·8}$$

where $u \equiv \nu^{\frac{1}{3}}$, as in §2. Substituting the expansions given above for $J_\nu(\nu+\tau\nu^{\frac{1}{3}})$ and its first two derivatives, and equating to zero the coefficient of u^{3-2s}, we deduce that

$$\left.\begin{aligned} A_s^2 + 2\tau A_{s-1}^2 + \tau^2 A_{s-2}^2 + A_{s-1}^1 + \tau A_{s-2}^1 + 2\tau A_s + \tau^2 A_{s-1} &= 0, \\ \text{and} \quad B_s^2 + 2\tau B_{s-1}^2 + \tau^2 B_{s-2}^2 + B_{s-1}^1 + \tau B_{s-2}^1 + 2\tau B_s + \tau^2 B_{s-1} &= 0. \end{aligned}\right\} \tag{3·9}$$

These equations afford checks on the values of the coefficients given by equations (2·42), (3·6) and (3·7).

Another relation useful as a check is obtained from the Wronskian

$$J_\nu(\nu+\tau\nu^{\frac{1}{3}})\,Y'_\nu(\nu+\tau\nu^{\frac{1}{3}})-J'_\nu(\nu+\tau\nu^{\frac{1}{3}})\,Y_\nu(\nu+\tau\nu^{\frac{1}{3}}) = \frac{2}{\pi}(\nu+\tau\nu^{\frac{1}{3}})^{-1}.$$

Substitution of the expansions above leads to the relation

$$\sum_{r=0}^{s}(A^1_r B_{s-r}-A_r B^1_{s-r}) = (-)^s\tau^s \quad (s\geqslant 0). \tag{3·10}$$

4. *Overlap of the new expansions with those of Debye and Meissel.* The highest power of τ occurring in the coefficient $A_s(\tau)$ occurs in the term $f_{ss}\tau^s\phi_{3s}(\tau)$ of the expression (2·40) for $A_s(\tau)$. It is evident from equations (2·38) and Table 1 that $\phi_s(\tau)$ is of degree not exceeding $\frac{1}{2}s$, and hence $A_s(\tau)$ is of degree $\leqslant \frac{5}{2}s$, this bound being attained for even values of s. Similarly $B_s(\tau)$ is of degree $\leqslant \frac{1}{2}(5s-1)$, the bound being attained for odd values of s. The early terms of the series (3·1) and (3·2) are therefore small provided $|\tau^{\frac{5}{2}}\nu^{-\frac{2}{3}}|$ is small, and the new expansions are suitable for the numerical computation of $J_\nu(z)$ and $Y_\nu(z)$ provided

$$z = \nu+o(|\nu|^{\frac{3}{5}}).$$

As we have already seen in §1, Debye's expansions (1·1) and (1·2) are suitable for numerical computation when $|x-n|$ is large compared with $n^{\frac{1}{3}}$. There is thus adequate overlap between the two sets of expansions for large values of n.

It is of interest for a computer to know the stage at which it would be advisable to change from one form of expansion to the other. Roughly speaking, the new expansions behave like power series in $\tau^{\frac{5}{2}}n^{-\frac{2}{3}}$, whereas those of Debye behave like comparable power series in $2n^{-1}|(x/n)^2-1|^{-\frac{3}{2}}$, so that equivalent numbers of terms of the series produce the same accuracy when

$$2n^{-1}|(x/n)^2-1|^{-\frac{3}{2}} = n|(x/n)-1|^{\frac{5}{2}},$$

i.e. when

$$|(x/n)-1| = 2^{-\frac{1}{8}}n^{-\frac{1}{2}}, \text{ approximately.}$$

Thus for practical purposes the change-over should occur roughly when $x = n\pm n^{\frac{1}{2}}$.

Meissel's expansions (1·5) are suitable for numerical application when $x = n+o(n^{\frac{1}{3}})$ and this lies within the region of validity of the new expansions. Indeed, the expansions (1·5) may be obtained from (3·1) and (3·2) by assuming τ to be small and expanding the Airy functions in ascending powers of τ (see (1), p. B 17). The details of this transformation are of no special interest and are not reproduced here.

The minimum accuracy attainable by use of the expansions (3·1) and (3·2) and the expressions (2·39) may be gauged from the fact that four or five accurate significant figures are obtained in the values of $J_n(n\pm n^{\frac{1}{2}})$ when $20\leqslant n\leqslant 100$. For values of the argument x which are such that $|x-n|$ is much less than $n^{\frac{1}{2}}$, the accuracy obtainable is increased considerably, while for values of $|x-n|$ which are greater than $n^{\frac{1}{2}}$ the appropriate expansions to employ are those of Debye, as we have seen above. Greater accuracy than that indicated here can be obtained by the calculation of higher terms of the expansions, since it did not appear that the smallest term was included amongst those which were taken into account.

5. *Asymptotic expansions for the zeros and turning values of cylinder functions of large positive orders.* Consider the real cylinder function

$$\mathscr{C}_n(x)\equiv \cos\pi t J_n(x)-\sin\pi t Y_n(x). \tag{5·1}$$

Using the expansions (3·1) and (3·2), we obtain

$$\mathscr{C}_n(n+\tau n^{\frac{1}{3}}) \sim \frac{2^{\frac{1}{3}}}{u} F \sin(\chi+\pi t)\left[1+\sum_{s=1}^{\infty}\frac{A_s(\tau)}{u^{2s}}\right]+\frac{2^{\frac{2}{3}}}{u} G \sin(\psi+\pi t)\sum_{s=1}^{\infty}\frac{B_s(\tau)}{u^{2s}}, \tag{5·2}$$

where $u \equiv n^{\frac{1}{3}}$ and F, χ, G and ψ are the functions defined in (1), with argument $x = -2^{\frac{1}{3}}\tau$. Thus we have

$$F(x)\, e^{i\chi(x)} = \mathrm{Bi}\,(x)+i\,\mathrm{Ai}\,(x), \quad G(x)\, e^{i\psi(x)} = \mathrm{Bi}'\,(x)+i\,\mathrm{Ai}'\,(x),$$

the phases $\chi(x)$, $\psi(x)$ being continuous functions of x, fixed by $\chi(0) = -\psi(0) = \frac{1}{6}\pi$.

The zeros of the right-hand side of the relation (5·2) must be such that the following equation is satisfied asymptotically:

$$\sin\{\chi(-2^{\frac{1}{3}}\tau)+\pi t\} = 0. \tag{5·3}$$

It follows that the positive zeros of $\mathscr{C}_n(x)$ of fixed enumeration are each of the form

$$x = n+(\text{const.})\, n^{\frac{1}{3}}+O(n^{-\frac{1}{3}}). \tag{5·4}$$

Let us denote the zeros of $\mathscr{C}_n(x)$ by the function $\rho(n,t)$. Then in order to ascribe the value of the constant in the expression (5·4), we shall suppose as in (6) that $\rho(n,t)$ is a continuous function of t, fixed by $\rho(n,0) = 0$. Then it is easily seen that

$$j_{n,s} = \rho(n,-s), \quad y_{n,s} = \rho(n,\tfrac{1}{2}-s). \tag{5·5}$$

With this convention we deduce from equation (5·3) the result that for large n and fixed negative t,

$$\rho(n,t) = n+\alpha n^{\frac{1}{3}}+\epsilon, \tag{5·6}$$

where $\epsilon = O(n^{-\frac{1}{3}})$ and α is such that*

$$\chi(-2^{\frac{1}{3}}\alpha) = -\pi t. \tag{5·7}$$

The higher terms of the expansion of $\rho(n,t)$ can be obtained in the following manner. From Taylor's theorem we have

$$\sum_{m=0}^{\infty}\frac{\epsilon^m}{m!}\mathscr{C}_n^{(m)}(n+\alpha n^{\frac{1}{3}}) = 0, \tag{5·8}$$

and using the expansions (3·3) and (3·4) with $\tau = \alpha$, we find in virtue of equations (5·1) and (5·7) that

$$\mathscr{C}_n^{(m)}(n+\alpha n^{\frac{1}{3}}) \sim \frac{2^{\frac{2}{3}}}{u^{m+1}} G(-2^{\frac{1}{3}}\alpha)\sin\{\psi(-2^{\frac{1}{3}}\alpha)-\chi(-2^{\frac{1}{3}}\alpha)\}\sum_{s=0}^{\infty}\frac{B_s^m(\alpha)}{u^{2s}}. \tag{5·9}$$

It is now clear from the relations (5·8) and (5·9) that an expansion of ϵ exists of the form

$$\epsilon \sim \alpha_2 u^{-1}+\alpha_3 u^{-3}+\alpha_4 u^{-5}+\ldots, \tag{5·10}$$

and hence we obtain the important result that for large n and fixed t ($t<0$)

$$\rho(n,t) \sim n\left[1+\sum_{s=1}^{\infty}\alpha_s n^{-\frac{2}{3}s}\right], \tag{5·11}$$

where $\alpha_1 = \alpha$ and the higher coefficients α_s are obtainable by equating to zero the coefficients of powers of u^{-2} in the relation

$$\sum_{m=0}^{\infty}\frac{1}{m!}\frac{1}{u^{2m}}\left(\alpha_2+\frac{\alpha_3}{u^2}+\frac{\alpha_4}{u^4}+\ldots\right)^m\left[\sum_{s=0}^{\infty}\frac{B_s^m(\alpha)}{u^{2s}}\right] \sim 0. \tag{5·12}$$

* These quantities will not be confused with the α, ϵ appearing in §1.

Values of $A_s^m(\tau)$ and $B_s^m(\tau)$ for $m = 0, 1, 2$ and $s = 0(1)4$, have been given in §§2 and 3. Using the recurrence relations (3·5), we obtain

$$\left.\begin{aligned} &\left.\begin{aligned} A_0^3 &= -2,\\ B_0^3 &= 2\tau, \end{aligned}\right\} \quad \left.\begin{aligned} A_1^3 &= -\tfrac{6}{5}\tau^4+\tfrac{44}{5}\tau,\\ B_1^3 &= -\tfrac{26}{5}\tau^2, \end{aligned}\right\} \quad B_2^3 = -\tfrac{9}{50}\tau^6+\tfrac{382}{35}\tau^3-\tfrac{71}{35};\\ &A_0^4 = 4\tau^2, \quad B_0^4 = 4, \quad B_1^4 = \tfrac{6}{5}\tau^4-\tfrac{96}{5}\tau; \quad B_0^5 = -4\tau^2. \end{aligned}\right\} \tag{5·13}$$

Replacing τ by α and substituting these expressions in the relation (5·12) we find that

$$\alpha_2 = \tfrac{3}{10}\alpha^2, \quad \alpha_3 = -\tfrac{1}{350}\alpha^3+\tfrac{1}{70}, \quad \alpha_4 = -\tfrac{479}{63000}\alpha^4-\tfrac{1}{3150}\alpha, \quad \alpha_5 = \tfrac{20231}{8085000}\alpha^5-\tfrac{551}{161700}\alpha^2. \tag{5·14}$$

The results (5·7), (5·11) and (5·14) give the required expansion for the zeros.

The asymptotic expansion of the value of the derivative $\mathscr{C}_n'(x)$ at the zeros $x = \rho(n, t)$ may be obtained in the following way. From Taylor's theorem we have

$$\mathscr{C}_n'(\rho) = \sum_{m=0}^{\infty} \frac{\epsilon^m}{m!} \mathscr{C}_n^{(m+1)}(n+\alpha n^{\frac{1}{3}}).$$

Substitution of the expansions (5·9) and (5·10) in this equation leads to the result

$$\mathscr{C}_n'(\rho) \sim \frac{\kappa(t)}{n^{\frac{2}{3}}}\left[1+\sum_{s=1}^{\infty}\beta_s n^{-\frac{2}{3}s}\right], \tag{5·15}$$

where
$$\kappa(t) = -2^{\frac{2}{3}} G(-2^{\frac{1}{3}}\alpha)\sin\{\psi(-2^{\frac{1}{3}}\alpha)-\chi(-2^{\frac{1}{3}}\alpha)\},$$

and the coefficients β_s are obtainable by comparing coefficients in the relation

$$1+\sum_{s=1}^{\infty}\frac{\beta_s}{u^{2s}} \sim -\sum_{m=0}^{\infty}\frac{1}{m!}\frac{1}{u^{2m}}\left(\alpha_2+\frac{\alpha_3}{u^2}+\frac{\alpha_4}{u^4}+\ldots\right)^m\left[\sum_{s=0}^{\infty}\frac{B_s^{m+1}(\alpha)}{u^{2s}}\right]. \tag{5·16}$$

Using the Wronskian property of the Airy functions we find that

$$\kappa(t) = 2^{\frac{2}{3}}/\{\pi F(-2^{\frac{1}{3}}\alpha)\}. \tag{5·17}$$

The values of the first four coefficients β_s are readily found to be

$$\beta_1 = -\tfrac{4}{5}\alpha, \quad \beta_2 = \tfrac{18}{35}\alpha^2, \quad \beta_3 = -\tfrac{88}{315}\alpha^3-\tfrac{11}{1575}, \quad \beta_4 = \tfrac{79586}{606375}\alpha^4+\tfrac{9824}{606375}\alpha. \tag{5·18}$$

The expansions (5·11) and (5·15), together with the expressions (5·7), (5·14), (5·17) and (5·18) for the coefficients, have already been given by the present writer in a previous paper (6), having been derived by a method based upon the application of Sturmian techniques to Bessel's differential equation. Previously, however, the results were only established for $t < -\frac{1}{6}$. We now see that they are valid for $t < 0$; in other words, they apply equally to positive zeros of $\mathscr{C}_n(x)$ which are *less* than the order n.

In a similar manner we can derive the asymptotic expansions of the zeros of the derivative $\mathscr{C}_n'(x)$ and of the associated stationary values of $\mathscr{C}_n(x)$. The results so obtained confirm the expansions for these quantities which have already been given in (6).

Numerical values of the early coefficients of the expansions for the first five zeros and turning values of $J_n(x)$ and $Y_n(x)$ are tabulated in (6).

6. *Extension of the region of validity of the asymptotic expansions for the Hankel functions.* The expansions (2·39) and (2·41) have been established in §2 under the restriction $|\arg \nu| < \frac{1}{2}\pi$. We shall now consider the validity of these expansions for other ranges of $\arg \nu$ and also their form for other values of the cube root $\nu^{\frac{1}{3}}$.

Expansion of $H_\nu^{(1)}(\nu+\tau\nu^{\frac{1}{3}})$ when $0<\arg\nu<\pi$. The analytic continuation of $H_\nu^{(1)}(z)$ in the half-plane $0<\arg z<\pi$ is given (see (2)) by

$$H_\nu^{(1)}(z) = \frac{1}{\pi i}\int_{-\infty+\frac{1}{2}\pi i}^{\infty+\frac{1}{2}\pi i} \exp\{z\sinh t-\nu t\}\,dt. \tag{6·1}$$

We take the contour in this integral to be a curve along which $\phi(t)\equiv\sinh t-t$ is purely imaginary. Such curves are given (cf. (2·8)) by

$$\sinh\xi\cos\eta = \xi. \tag{6·2}$$

In the strip $|\Im t|<\frac{1}{2}\pi$ there are three of these curves, namely, part of the imaginary axis, and two curves (POS, ROQ) which are asymptotic to the lines $\Im t=\pm\frac{1}{2}\pi$ and which make angles $\pm\frac{1}{6}\pi$ with the real axis at the origin O (see Fig. 3). It is easily verified that $\Im\phi(t)$ is positive on PO, OQ and negative on RO, OS and that it changes monotonically on each of these curves.

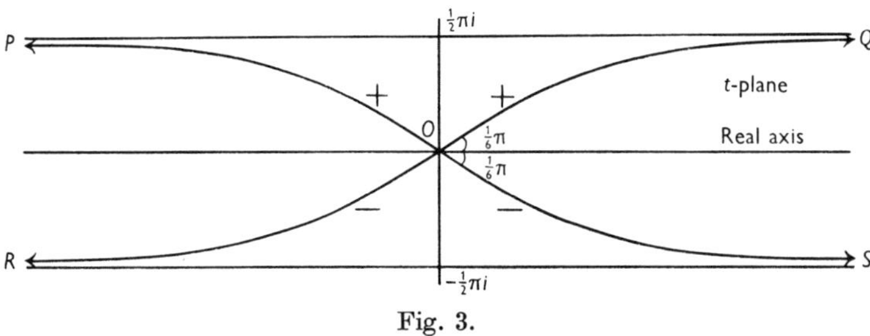

Fig. 3.

Taking the contour in the integral (6·1) to be POQ and using reasoning similar to that of § 2 (a), we find that on transforming to the v-plane (cf. (2·13)),

$$H_\nu^{(1)}(\nu+\tau\nu^{\frac{1}{3}}) = \frac{1}{\pi i}\int_{\infty\exp\{\frac{5}{6}\pi i\}}^{\infty\exp\{\frac{1}{6}\pi i\}} \exp\{\tfrac{1}{6}u^3v^3+\tau uv\}f(v,\tau u)\,dv, \tag{6·3}$$

where the contour of this integral comprises two straight lines through the origin.

The remainder of the derivation of the asymptotic expansion is exactly similar to § 2 (b)–§ 2 (d), and the final form of the expansion in this range of $\arg\nu$ is again given by (2·39).

Expansion of $H_\nu^{(1)}(\nu+\tau\nu^{\frac{1}{3}})$ when $\frac{1}{2}\pi<\arg\nu<\frac{3}{2}\pi$. In this case we use the integral representation

$$H_\nu^{(1)}(z) = \frac{1}{\pi i}\int_{-\infty+\pi i}^{\infty} \exp\{z\sinh t-\nu t\}\,dt, \tag{6·4}$$

taking the contour to be AOD (see Fig. 1). The modification of the analysis is very similar to that indicated above and the expansion (2·39) is also valid in this region.

Summarizing, the expansion (2·39) for $H_\nu^{(1)}(\nu+\tau\nu^{\frac{1}{3}})$ is valid throughout the range $-\frac{1}{2}\pi<\arg\nu<\frac{3}{2}\pi$. Similarly, we can show that the expansion (2·41) for $H_\nu^{(2)}(\nu+\tau\nu^{\frac{1}{3}})$ is valid when $-\frac{3}{2}\pi<\arg\nu<\frac{1}{2}\pi$.

Alternative forms of expansion. Let us write

$u_1\equiv u=\nu^{\frac{1}{3}}$, the principal cube root; $\quad u_2\equiv u\exp\{\frac{2}{3}\pi i\}$, $\quad u_3\equiv u\exp\{-\frac{2}{3}\pi i\}$.

Then if the analysis of §2 is repeated taking $z = \nu + \tau u_2$ and $z = \nu + \tau u_3$ in turn, the only modifications required are to change the contour in the integral (2·5) to

$$\int_{\infty \exp\{-\frac{1}{3}\pi i\}}^{-\infty} \quad \text{and} \quad \int_{\infty \exp\{\frac{1}{3}\pi i\}}^{\infty \exp\{-\frac{1}{3}\pi i\}} \quad \text{respectively,}$$

and to carry through the effect of this change in §2 (d). Using equations (2·34) we obtain the following results:

$$H_\nu^{(1)}(\nu + \tau u_r) \sim \frac{2^{\frac{1}{3}}}{u_r} C_r(-2^{\frac{1}{3}}\tau) \sum_{s=0}^{\infty} \frac{A_s(\tau)}{u_r^{2s}} + \frac{2^{\frac{2}{3}}}{u_r} C_r'(-2^{\frac{1}{3}}\tau) \sum_{s=0}^{\infty} \frac{B_s(\tau)}{u_r^{2s}} \quad (-\tfrac{1}{2}\pi < \arg \nu < \tfrac{3}{2}\pi), \tag{6·5}$$

and

$$H_\nu^{(2)}(\nu + \tau u_r) \sim \frac{2^{\frac{1}{3}}}{u_r} C_{r+1}(-2^{\frac{1}{3}}\tau) \sum_{s=0}^{\infty} \frac{A_s(\tau)}{u_r^{2s}} + \frac{2^{\frac{2}{3}}}{u_r} C_{r+1}'(-2^{\frac{1}{3}}\tau) \sum_{s=0}^{\infty} \frac{B_s(\tau)}{u_r^{2s}} \quad (-\tfrac{3}{2}\pi < \arg \nu < \tfrac{1}{2}\pi), \tag{6·6}$$

where

$$C_1(x) = C_4(x) = \text{Ai}(x) - i\,\text{Bi}(x), \quad C_2(x) = \text{Ai}(x) + i\,\text{Bi}(x) \quad \text{and} \quad C_3(x) = -2\,\text{Ai}(x).$$

The work described above has been carried out as part of the research programme of the National Physical Laboratory, and this paper is published by permission of the Director of the Laboratory.

REFERENCES

(1) *British Association Mathematical Tables*, part-vol. B, *The Airy integral* (Cambridge, 1946).
(2) Copson, E. T. *Theory of functions* (Oxford, 1935), pp. 323–4.
(3) Imai, I. Asymptotic solutions of ordinary differential equations of the second order. *Phys. Rev.* (2), 80 (1950), 1112.
(4) Langer, R. E. On the asymptotic solutions of ordinary differential equations with application to Bessel functions of large order. *Trans. Amer. math. Soc.* 33 (1931), 23–64.
(5) Nicholson, J. W. The asymptotic expansion of Bessel functions. *Phil. Mag.* (6), 19 (1910), 228–49.
(6) Olver, F. W. J. A further method for the evaluation of zeros of Bessel functions and some new asymptotic expansions for zeros of functions of large order. *Proc. Camb. phil. Soc.* 47 (1951), 699–712.
(7) Watson, G. N. Bessel functions of large order. *Proc. Camb. phil. Soc.* 19 (1918), 96–110.
(8) Watson, G. N. *Theory of Bessel functions* (Cambridge, 1944).

National Physical Laboratory
Teddington, Middlesex

NOTE ON THE ASYMPTOTIC EXPANSION OF GENERALIZED HYPERGEOMETRIC FUNCTIONS

F. W. J. OLVER*.

[*Extracted from the Journal of the London Mathematical Society, Vol.* 28, 1953.]

I have recently made a minor extension of a result of Barnes concerning the exponentially infinite asymptotic expansions of linear combinations of generalized hypergeometric functions. In the course of this I found an error in his work which does not appear to have been recorded.

The result is stated by Barnes ([**1**], §48) in the form

$$\frac{\prod_{r=1}^{q}\Gamma(1-\rho_r)}{\prod_{r=1}^{p}\Gamma(1-\alpha_r)}\,{}_pF_q\{\alpha;\ \rho;\ (-)^\mu x\}$$

$$+\sum_{r=1}^{q}\exp\{(\mu-2m)(1-\rho_r)\pi i\}\,x^{1-\rho_r}\,\Gamma(\rho_r-1)$$

$$\times\frac{\prod_{t=1}^{q}{}'\,\Gamma(\rho_r-\rho_t)}{\prod_{t=1}^{p}\Gamma(\rho_r-\alpha_t)}\,{}_pF_q\{1+\alpha-\rho_r;\ 2-\rho_r,\ \ldots,\ \rho_q-\rho_r+1;\ (-)^\mu x\}$$

$$\sim\exp\{-\mu\, e^{(\mu-2m)\pi i/\mu}\,x^{1/\mu}\}(2\pi)^{\frac{1}{2}(\mu-1)}\,\mu^{-\frac{1}{2}}\,x^{\theta/\mu}\,e^{(\mu-2m)\theta\pi i/\mu}$$

$$\times\sum_{n=0}^{\infty}\frac{\lambda_n\exp\{(2m-\mu)n\pi i/\mu\}}{x^{n/\mu}},\tag{1}$$

where $\theta\equiv\Sigma\alpha-\Sigma\rho+\frac{1}{2}(\mu-1)$, $\mu\equiv q+1-p$ is an integer $(\mu>1)$, $|\arg x|<\pi$ and $m=0,\ 1,\ \ldots,\ \mu$.

Barnes derives this by selecting particular phase ranges of x in the following theorem ([**1**], §47)

$$\frac{\prod_{r=1}^{q}\Gamma(1-\rho_r)}{\prod_{r=1}^{p}\Gamma(1-\alpha_r)}\,{}_pF_q\{\alpha\,;\rho\,;(-)^\mu x\}$$

$$+\sum_{r=1}^{q}x^{1-\rho_r}\Gamma(\rho_r-1)\frac{\prod_{t=1}^{q}{}'\,\Gamma(\rho_r-\rho_t)}{\prod_{t=1}^{p}\Gamma(\rho_r-\alpha_t)}\,{}_pF_q\{1+\alpha-\rho_r;\ 2-\rho_r,\ \ldots,\ \rho_q-\rho_r+1\,;(-)^\mu x\}$$

$$\sim\exp\{-\mu x^{1/\mu}\}(2\pi)^{\frac{1}{2}(\mu-1)}\,\mu^{-\frac{1}{2}}\,x^{\theta/\mu}\sum_{n=0}^{\infty}\frac{\lambda_n}{x^{n/\mu}},\tag{2}$$

which is valid when $|\arg x|<(\mu+1)\pi$.

* Received 28 May, 1952; read 12 June, 1952.

In order to extend the result (1) to a wider range of $\arg x$, put $x = \xi e^{2s\pi i}$ in (2), where s is an integer and where

$$(k-1)\pi < \arg \xi < (k+1)\pi,$$

k being an arbitrary real number. The condition $|\arg x| < (\mu+1)\pi$ is satisfied provided that s is restricted to the range

$$-\tfrac{1}{2}\mu - \tfrac{1}{2}k \leqslant s \leqslant \tfrac{1}{2}\mu - \tfrac{1}{2}k.$$

Thus if $\frac{1}{2}\mu - \frac{1}{2}k$ is an integer S' (say), then

$$s = S'-\mu,\ S'-\mu+1,\ \ldots,\ S',$$

but if $\frac{1}{2}\mu - \frac{1}{2}k$ is non-integral, then

$$s = S-\mu+1,\ S-\mu+2,\ \ldots,\ S,$$

where S is the integer such that

$$S < \tfrac{1}{2}\mu - \tfrac{1}{2}k < S+1. \tag{3}$$

On substituting $x = \xi e^{2s\pi i}$ in (2) and replacing ξ, s by x, $\frac{1}{2}\mu - m$ respectively, we obtain the result (1). This is evidently valid when

$$(k-1)\pi < \arg x < (k+1)\pi \tag{4}$$

for the values

$$m = \tfrac{1}{2}\mu - S',\ \tfrac{1}{2}\mu - S'+1,\ \ldots,\ \tfrac{3}{2}\mu - S' \quad (\tfrac{1}{2}\mu - \tfrac{1}{2}k \text{ integral}), \tag{5}$$

or

$$m = \tfrac{1}{2}\mu - S,\ \tfrac{1}{2}\mu - S+1,\ \ldots,\ \tfrac{3}{2}\mu - S-1 \quad (\tfrac{1}{2}\mu - \tfrac{1}{2}k \text{ non-integral}). \tag{6}$$

If we now suppose S to be an *arbitrary* integer, then clearly (1) is true under the conditions (4) and (6) provided that k is restricted by (3), *i.e.*

$$\mu - 2S - 2 < k < \mu - 2S.$$

Hence we deduce that (4) may be replaced by

$$(\mu - 2S - 3)\pi < \arg x < (\mu - 2S + 1)\pi.$$

Summarizing and replacing S by $-N$, we have proved that: (1) *is valid in the range*

$$(\mu + 2N - 3)\pi < \arg x < (\mu + 2N + 1)\pi, \tag{7}$$

where N is an arbitrary integer or zero, and m takes the values

$$m = \tfrac{1}{2}\mu + N,\ \tfrac{1}{2}\mu + N + 1,\ \ldots,\ \tfrac{3}{2}\mu + N - 1. \tag{8}$$

It is now seen that for $|\arg x| < \pi$ (the range considered by Barnes) the values $m = 0, 1, \ldots, \mu$ are admissible only if μ is an *even* integer; they are obtained by putting $N = -\frac{1}{2}\mu$ and $1 - \frac{1}{2}\mu$ in (7) and (8). If μ is an odd integer then (1) is valid when $|\arg x| < \pi$ only for the values $m = \frac{1}{2}, \frac{3}{2}, \ldots, \mu - \frac{1}{2}$ (obtained by putting $N = -\frac{1}{2}\mu + \frac{1}{2}$).

This note is published with the permission of the Director of the National Physical Laboratory.

Reference.

1. Barnes, E. W., "The asymptotic expansion of integral functions defined by generalized hypergeometric series", *Proc. London Math. Soc.* (2), 5 (1906), 59–116.

National Physical Laboratory,
Teddington, Middlesex.

[307]

THE ASYMPTOTIC SOLUTION OF LINEAR DIFFERENTIAL EQUATIONS OF THE SECOND ORDER FOR LARGE VALUES OF A PARAMETER

By F. W. J. OLVER, *The National Physical Laboratory*

(*Communicated by Sir Edward Bullard, F.R.S.—Received* 22 *February* 1954—
Revised 26 *March* 1954)

CONTENTS

Asymptotic solutions of the differential equations

$$\mathrm{d}^2w/\mathrm{d}z^2 = \{uz^n + f(z)\}\, w \quad (n = 0, 1)$$

for large positive values of u, have the formal expansions

$$w = P(z)\left\{1 + \sum_{s=1}^{\infty} \frac{A_s(z)}{u^s}\right\} + \frac{P'(z)}{u} \sum_{s=0}^{\infty} \frac{B_s(z)}{u^s},$$

where P is an exponential or Airy function for $n = 0$ or 1 respectively. The coefficients $A_s(z)$ and $B_s(z)$ are given by recurrence relations. This paper proves that solutions of the differential equations exist whose asymptotic expansions in Poincaré's sense are given by these series, and that the expansions are uniformly valid with respect to the complex variable z. The method of proof differs from those of earlier writers and fewer restrictions are made.

1. Introduction and summary

The purpose of this investigation is to determine asymptotic expansions of solutions of the differential equation

$$\frac{\mathrm{d}^2w}{\mathrm{d}z^2} = \{up(z) + q(z)\}\, w \tag{1·1}$$

for large values of the parameter u. It is supposed that z lies in a simply-connected domain **D** and $p(z)$, $q(z)$ are analytic functions of the complex variable z independent of u. Attention will be confined to positive values of u; if u is a large complex parameter of the form $u = |u|\,\mathrm{e}^{\mathrm{i}\vartheta}$, where ϑ is a fixed real number, then the transformation $z' = \mathrm{e}^{\frac{1}{2}\mathrm{i}\vartheta} z$ converts (1·1) into the same form of equation but with u replaced by $|u|$.

The asymptotic character of the solutions of (1·1) as $u \to \infty$ can take many different forms depending on the number and nature of the *transition points* of (1·1) in **D**, such points being defined here† as ones at which $p(z)$ vanishes, or $p(z)$ or $q(z)$ has a singularity.

Many papers have been published on this subject. Nevertheless, the existing theory does not yet appear to be complete even in the cases when (1·1) has no transition points or only one transition point, a simple zero of $p(z)$. In the more recondite cases previous work has generally been concerned only with the determination of the asymptotic *form* of the solutions in the immediate neighbourhood of various kinds of transition points, and asymptotic *expansions* have not been investigated.

It may be thought that the form of equation (1·1) is unduly restrictive, particularly as other writers have considered more general forms, for example

$$\frac{d^2w}{dz^2} = \left\{up(z) + \sum_{s=0}^{\infty} \frac{q_s(z)}{u^s}\right\} w. \tag{1·2}$$

The principal reason for adopting the present form of equation is that many important second-order differential equations are already in the form (1·1), or may be readily transformed into it. Such equations include those of Bessel, Weber and Legendre, and new asymptotic expansions of standard solutions of these equations for large orders have been obtained, and it is hoped to give these in subsequent papers.

Three cases will be considered here, and they will be referred to as cases A, B and C. They occur respectively when equation (1·1) has, in **D**, (i) no transition points, (ii) one transition point, a simple zero of $p(z)$, (iii) one transition point, a double pole of $p(z)$. The expansions in the three cases are similar and it is convenient to treat them together. Similar expansions also exist when the only transition point is simultaneously a simple pole of $p(z)$ and a double pole of $q(z)$, but consideration of this important case is deferred.

The paper is arranged as follows. In § 2 it is shown that (1·1) may be transformed into a similar equation with $p(z)$ replaced by unity in cases A and C, and by z in case B. In § 3 series are constructed which formally satisfy these standard forms of equation. In § 4 some relevant properties of the exponential and Airy functions are recorded. In § 5 two existence theorems are stated which reveal the asymptotic nature of the series constructed in § 3. These theorems are the main results of the paper; their application is briefly considered in § 6. In § 7 the work of other writers is outlined. The remaining sections, §§ 8 to 11, are devoted to the proofs of the theorems stated in § 5.

2. Transformation to standard form

In (1·1) let us take a new independent variable ζ and a new dependent variable W, related by

$$W = \dot{z}^{-\frac{1}{2}} w, \tag{2·1}$$

dots denoting differentiations with respect to ζ. Then it is readily verified that W satisfies the equation

$$\frac{d^2W}{d\zeta^2} = \{u\dot{z}^2 p(z) + f(\zeta)\} W, \tag{2·2}$$

† Points at which $p(z)$ vanishes have been called 'turning points' by some writers (for example, Langer 1949) and transition points by others (Watson 1944, p. 248; and Cherry 1950). The wider definition of transition point used here is a useful generalization.

where
$$f(\zeta) \equiv \dot{z}^2 q(z) + \dot{z}^{\frac{1}{2}} \frac{d^2}{d\zeta^2}(\dot{z}^{-\frac{1}{2}}) = \dot{z}^2 q(z) + \frac{3\ddot{z}^2 - 2\dot{z}\dddot{z}}{4\dot{z}^2}. \tag{2·3}$$

The relation between z and ζ is now prescribed in the following ways.

Case A. Here $p(z)$ has no zeros in **D**. We take $\dot{z}^2 p(z) = 1$, so that

$$\zeta = \int \{p(z)\}^{\frac{1}{2}} dz. \tag{2·4}$$

Equation (2·2) accordingly becomes

$$\frac{d^2 W}{d\zeta^2} = \{u + f(\zeta)\} W. \tag{2·5}$$

The relation (2·4) maps the domain **D** on a certain domain **G** in the ζ-plane, and it is clear from (2·3) and (2·4) that $f(\zeta)$ is a regular (holomorphic) function of ζ in **G**.

Case B. Here $p(z)$ has a simple zero in **D** at $z = z_0$, say, and the transformation (2·4) is unsuitable because $dz/d\zeta$ becomes infinite there. Instead, we take $\dot{z}^2 p(z) = \zeta$ and

$$\zeta = \left[\frac{3}{2}\int_{z_0}^{z} \{p(z)\}^{\frac{1}{2}} dz\right]^{\frac{2}{3}}. \tag{2·6}$$

Near $z = z_0$ we have

$$\zeta = \text{constant} \times (z - z_0) + (z - z_0)^2 O(1). \tag{2·7}$$

Equation (2·2) becomes
$$\frac{d^2 W}{d\zeta^2} = \{u\zeta + f(\zeta)\} W, \tag{2·8}$$

where $f(\zeta)$ is given by (2·3) and is a regular function of ζ in the transformed domain.

Case C. Here $p(z)$ has a double pole in **D** at $z = z_0$, and we may allow $q(z)$ to have a single or double pole at the same point. The same transformation (2·4) is made as in case A above, so that the transformed equation is again given by (2·5). Near $z = z_0$ we have

$$p(z) = (z - z_0)^{-2}\{k^2 + (z - z_0)\, O(1)\}, \tag{2·9}$$

where k is a non-zero constant, and so

$$\zeta = k \ln (z - z_0) + (z - z_0)\, O(1). \tag{2·10}$$

Thus $z = z_0$ corresponds to $\zeta = -\infty \exp(\mathrm{i} \arg k)$. It is seen from (2·3) that $f(\zeta)$ is again a regular function of ζ in the corresponding ζ-domain. The transition point is now at infinity, but a sufficient condition that the asymptotic expansions of solutions of (2·5) remain uniformly valid at infinity is (see § 5)

$$f(\zeta) = O(|\zeta|^{-2}), \quad \text{as} \quad |\zeta| \to \infty. \tag{2·11}$$

From (2·10) and (2·3) we obtain

$$z - z_0 \sim e^{\zeta/k}, \quad f(\zeta) = \frac{e^{2\zeta/k}}{k^2} q(e^{\zeta/k} + z_0) + \frac{1}{4k^2} + O(e^{\zeta/k}) = c + O(e^{\zeta/k}),$$

as $\zeta \to -\infty \exp(\mathrm{i} \arg k)$, where

$$c = \frac{1}{k^2}[(z - z_0)^2 q(z)]_{z=z_0} + \frac{1}{4k^2}.$$

Thus equation (2·5) itself may violate the condition (2·11), but by making the linear change of variables

$$u_1 = u + c, \quad f_1(\zeta) = f(\zeta) - c,$$

we arrive at the equation
$$\frac{d^2W}{d\zeta^2} = \{u_1 + f_1(\zeta)\} W,$$
which does satisfy the condition.

Thus in all three cases the original differential equation may be transformed into one of the *standard forms* (2·5) or (2·8) in which $f(\zeta)$ is a regular function. Accordingly, we shall confine our attention to these standard forms. It will be convenient to revert to the symbols w, z and **D** in place of W, ζ and **G** respectively.

The foregoing transformation for case B was first given by Langer (1931). That for case A is much older and was used by Liouville (1837).

3. Formal construction of series solutions

We consider now the standard form
$$\frac{d^2w}{dz^2} = \{ug(z) + f(z)\} w, \tag{3·1}$$
where $g(z) = 1$ for cases A and C, and z for case B.

Let $w = P(z)$ be any solution of the equation
$$\frac{d^2w}{dz^2} = ug(z)\, w. \tag{3·2}$$
(When $g(z) = 1$, P is an exponential function, and when $g(z) = z$ it is an Airy function.) Then as a possible solution of (3·1) we consider the series
$$w(z) = P(z)\left\{1 + \sum_{s=1}^{\infty} \frac{A_s(z)}{u^s}\right\} + \frac{P'(z)}{u} \sum_{s=0}^{\infty} \frac{B_s(z)}{u^s}, \tag{3·3}$$
where the coefficients $A_s(z)$ and $B_s(z)$ are independent of u. Term-by-term differentiation, using (3·2), yields
$$w'(z) = P(z) \sum_{s=0}^{\infty} \frac{C_s(z)}{u^s} + P'(z)\left\{1 + \sum_{s=1}^{\infty} \frac{D_s(z)}{u^s}\right\}, \tag{3·4}$$
and
$$w''(z) = uP(z) \sum_{s=0}^{\infty} \frac{E_s(z)}{u^s} + P'(z) \sum_{s=0}^{\infty} \frac{F_s(z)}{u^s}, \tag{3·5}$$
where
$$C_s = A'_s + gB_s, \quad D_s = A_s + B'_{s-1}, \tag{3·6}$$
and
$$\left.\begin{aligned} E_s &= C'_{s-1} + gD_s = A''_{s-1} + gA_s + 2gB'_{s-1} + g'B_{s-1}, \\ F_s &= C_s + D'_s = 2A'_s + B''_{s-1} + gB_s, \end{aligned}\right\} \tag{3·7}$$
primes denoting differentiations with respect to z.

Substituting (3·3) and (3·5) in (3·1), we see that the differential equation is formally satisfied if
$$E_{s+1} = gA_{s+1} + fA_s, \quad F_{s+1} = gB_{s+1} + fB_s. \tag{3·8}$$
Substituting E_{s+1} and F_{s+1} from (3·7), we obtain
$$A''_s - fA_s + 2gB'_s + g'B_s = 0, \quad 2A'_{s+1} + B''_s - fB_s = 0. \tag{3·9}$$
These equations may be integrated to give
$$B_s = \tfrac{1}{2} g^{-\frac{1}{2}} \int g^{-\frac{1}{2}} (fA_s - A''_s)\, dz, \quad A_{s+1} = -\tfrac{1}{2} B'_s + \tfrac{1}{2} \int fB_s\, dz. \tag{3·10}$$

Equations (3·10) may be regarded as recurrence relations for the coefficients $A_s(z)$ and $B_s(z)$. Taking $A_0 = 1$ we can determine all the higher coefficients, apart from the presence of certain arbitrary integration constants.

If the series (3·3) converged uniformly with respect to z then the process of term-by-term differentiation would be valid, and (3·3) would define a solution of (3·1). In general, however, the series diverges and it is natural to examine the manner and circumstances in which (3·3) represents a solution of (3·1). As a preliminary to this purpose some relevant facts about the Airy and exponential functions will now be stated.

4. Some properties of the Airy† and exponential functions

The Airy functions are the solutions of the differential equation

$$\frac{d^2w}{dz^2} = zw. \tag{4·1}$$

One standard solution is the well-known function

$$\mathrm{Ai}\,(z) = \frac{z^{\frac{1}{2}}}{\pi\sqrt{3}} K_{\frac{1}{3}}(\zeta), \tag{4·2}$$

where $\zeta = \frac{2}{3}z^{\frac{3}{2}}$. It is an integral function of z, and for large $|z|$ it takes the asymptotic values (Watson 1944, p. 202)

$$\mathrm{Ai}\,(z) = \frac{z^{-\frac{1}{4}}}{2\sqrt{\pi}}\,e^{-\zeta}\{1+O(|\zeta|^{-1})\}, \quad \mathrm{Ai}'\,(z) = -\frac{z^{\frac{1}{4}}}{2\sqrt{\pi}}\,e^{-\zeta}\{1+O(|\zeta|^{-1})\}, \tag{4·3}$$

when $|\arg z| < \pi$, and

$$\left.\begin{aligned} \mathrm{Ai}\,(-z) &= \frac{z^{-\frac{1}{4}}}{\sqrt{\pi}}\{\cos(\zeta - \tfrac{1}{4}\pi) + e^{|\mathscr{I}\zeta|}\,O(|\zeta|^{-1})\}, \\ \mathrm{Ai}'\,(-z) &= \frac{z^{\frac{1}{4}}}{\sqrt{\pi}}\{\sin(\zeta - \tfrac{1}{4}\pi) + e^{|\mathscr{I}\zeta|}\,O(|\zeta|^{-1})\}, \end{aligned}\right\} \tag{4·4}$$

when $|\arg z| < \frac{2}{3}\pi$. The expressions (4·3) and (4·4) are equivalent in their common regions of validity. From them we may deduce the useful inequalities

$$\left.\begin{aligned} |\mathrm{Ai}\,(z)| &< A(1+|z|^{\frac{1}{4}})^{-1}\,|\exp(-\tfrac{2}{3}z^{\frac{3}{2}})|, \\ |\mathrm{Ai}'\,(z)| &< A(1+|z|^{\frac{1}{4}})\,|\exp(-\tfrac{2}{3}z^{\frac{3}{2}})|, \end{aligned}\right\} \tag{4·5}$$

valid when $|\arg z| \leq \pi$, where A is some constant.

The functions $\mathrm{Ai}\,(z\,e^{\pm\frac{2}{3}\pi i})$ also satisfy (4·1), and we shall find it convenient to use the notation

$$P_1(z) = \mathrm{Ai}\,(z), \quad P_2(z) = \mathrm{Ai}\,(z\,e^{\frac{2}{3}\pi i}), \quad P_3(z) = \mathrm{Ai}\,(z\,e^{-\frac{2}{3}\pi i}). \tag{4·6}$$

Any two of these functions comprise a linearly independent pair of solutions of (4·1), the Wronskians being

$$W(P_1, P_2) = \frac{e^{-\frac{1}{6}\pi i}}{2\pi}, \quad W(P_1, P_3) = \frac{e^{\frac{1}{6}\pi i}}{2\pi}, \quad W(P_2, P_3) = \frac{i}{2\pi}. \tag{4·7}$$

If the sectors $\quad |\arg z| \leq \frac{1}{3}\pi, \quad -\pi \leq \arg z \leq -\frac{1}{3}\pi, \quad \frac{1}{3}\pi \leq \arg z \leq \pi,$

are denoted by $\mathbf{S}_1$, $\mathbf{S}_2$, $\mathbf{S}_3$ respectively (figure 1), then it is seen from (4·3) and (4·6) that $P_j(z)$ is exponentially small in $\mathbf{S}_j$ $(j = 1, 2, 3)$, when $|z|$ is large. Consequently in $\mathbf{S}_j$ any

† A fuller account of the properties of the Airy functions in the complex plane is given in the Appendix to the following paper.

independent pair of solutions of (4·1) must include $P_j(z)$ as a member if it is to be satisfactory from the numerical standpoint.†

Another standard solution of (4·1) is the function

$$\mathrm{Bi}\,(z) = \mathrm{e}^{\frac{1}{6}\pi\mathrm{i}}\,P_2(z) + \mathrm{e}^{-\frac{1}{6}\pi\mathrm{i}}\,P_3(z), \tag{4·8}$$

introduced in the *British Association Mathematical Tables* (1946). Although there is no region in which Bi (z) is exponentially small, it is important because it is real when z is real and is a numerically satisfactory† second solution to Ai (z) for both positive and negative z; its large negative zeros lie asymptotically half-way between those of Ai (z).

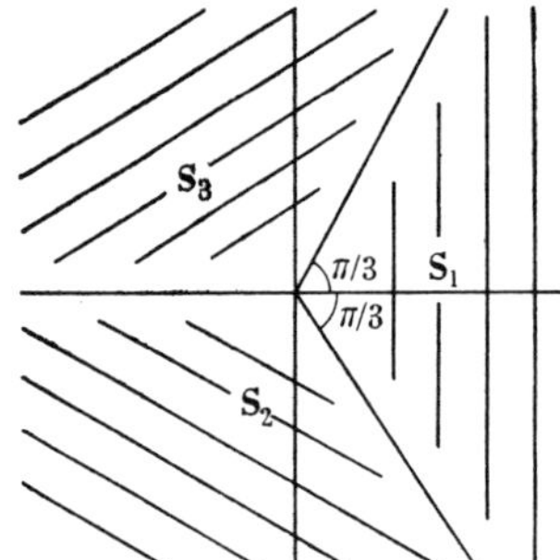

FIGURE 1. Regions in which $P_1(z)$, $P_2(z)$ and $P_3(z)$ are exponentially small.

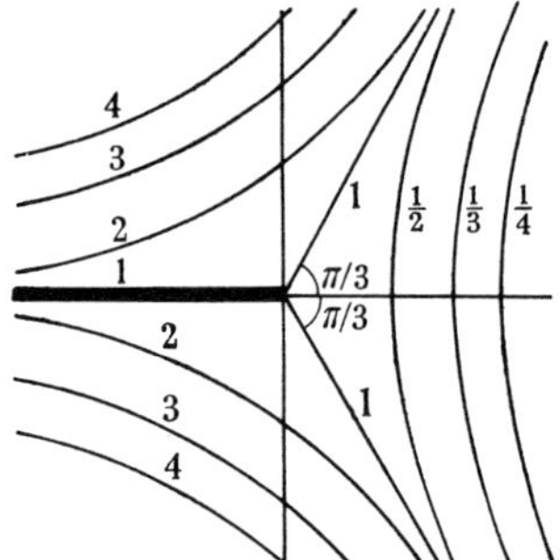

FIGURE 2. Level curves of $\exp(-\frac{2}{3}z^{\frac{3}{2}})$. Values of $|\exp(-\frac{2}{3}z^{\frac{3}{2}})|$ are indicated on the curves.

Level curves. The level curves of a function $F(z)$ are the contours along which

$$|F(z)| = \text{constant}.$$

In later sections we are concerned with the level curves of the functions exp z and $\exp(-\frac{2}{3}z^{\frac{3}{2}})$. Those of the former are straight lines parallel to the imaginary axis; those of $\exp(-\frac{2}{3}z^{\frac{3}{2}})$ have the polar equation

$$r^{\frac{3}{2}}\cos\tfrac{3}{2}\theta = \text{constant}, \tag{4·9}$$

where $z \equiv r\,\mathrm{e}^{\mathrm{i}\theta}$. Typical curves are illustrated in figure 2.

5. Asymptotic nature of the series; theorems A and B

The asymptotic nature of the series given in § 3 may be described by means of two existence theorems. In this and the following two sections the theorems are stated and their application described, the proofs being deferred until §§ 8 to 11.

In the following it is supposed that $f(z)$ is regular in a simply-connected open domain **D**, the boundaries of which consist of a finite number of straight lines; if **D** is unbounded it is supposed that‡

$$f(z) = O(|z|^{-2}), \tag{5·1}$$

as $|z| \to \infty$ in **D**, uniformly with respect to arg z. We denote by **D**$'$ any simply-connected domain lying wholly in **D**, the boundaries of which consist of a finite number of straight lines not intersecting the boundaries of **D**.

† For a discussion of criteria for numerically satisfactory solutions of second-order differential equations see Miller (1950).

‡ There may of course be domains extending arbitrarily near infinity in which (5·1) is not satisfied.

In the first theorem we suppose a_j ($j = 1, 2$) to be an arbitrary fixed point of $\mathbf{D}'$; if $\mathbf{D}'$ is unbounded a_j can be the point at infinity on a straight line lying in $\mathbf{D}'$. In addition, we denote by $\mathbf{D}_j$ the domain comprising those points z of $\mathbf{D}'$ which can be joined to a_j by a contour which lies in $\mathbf{D}'$ and is *wholly to the left* if $j = 1$, or *wholly to the right* if $j = 2$, of the line through z parallel to the imaginary axis.

THEOREM A. *Let sequences of functions $A_s(z)$ and $B_s(z)$ be defined*† *by the relations* $A_0(z) = 1$,

$$A_{s+1}(z) = -\tfrac{1}{2}A_s'(z) + \tfrac{1}{2}\int f(z)\, A_s(z)\, \mathrm{d}z + \text{constant}, \tag{5·2}$$

$$B_s(z) = A_s(z) + A_{s-1}'(z), \tag{5·3}$$

the arbitrary constants in (5·2) *being subject to the condition that there exists a function $\phi(u)$ of u alone, with the asymptotic expansion*

$$\phi(u) \sim 1 + \sum_{s=1}^{\infty} \frac{A_s(c)}{u^s}, \tag{5·4}$$

as $u \to \infty$, for a fixed point c in $\mathbf{D}'$.

Then for large u the equation

$$\frac{\mathrm{d}^2 w}{\mathrm{d}z^2} = \{u^2 + f(z)\}\, w \tag{5·5}$$

has solutions $W_1(z)$ and $W_2(z)$ such that if z lies in $\mathbf{D}_1$

$$W_1(z) = \mathrm{e}^{uz}\left\{\sum_{s=0}^{m-1} \frac{A_s(z)}{u^s} + O\left(\frac{1}{u^m}\right)\right\}, \quad W_1'(z) = u\,\mathrm{e}^{uz}\left\{\sum_{s=0}^{m-1} \frac{B_s(z)}{u^s} + O\left(\frac{1}{u^m}\right)\right\}, \tag{5·6}$$

and if z lies in $\mathbf{D}_2$

$$W_2(z) = \mathrm{e}^{-uz}\left\{\sum_{s=0}^{m-1} (-)^s \frac{A_s(z)}{u^s} + O\left(\frac{1}{u^m}\right)\right\}, \quad W_2'(z) = -u\,\mathrm{e}^{-uz}\left\{\sum_{s=0}^{m-1} (-)^s \frac{B_s(z)}{u^s} + O\left(\frac{1}{u^m}\right)\right\}, \tag{5·7}$$

where each of the O's is uniform with respect to z. Here m is an arbitrary unbounded integer, but the solutions $W_1(z)$ and $W_2(z)$ themselves are independent of m.

Typical domains are illustrated in figures 3 and 4. The thickened lines in figure 3 denote cuts and are the boundaries of $\mathbf{D}$; the broken lines are the boundaries of $\mathbf{D}'$. The domain $\mathbf{D}_1$ is the unshaded region in figure 4.

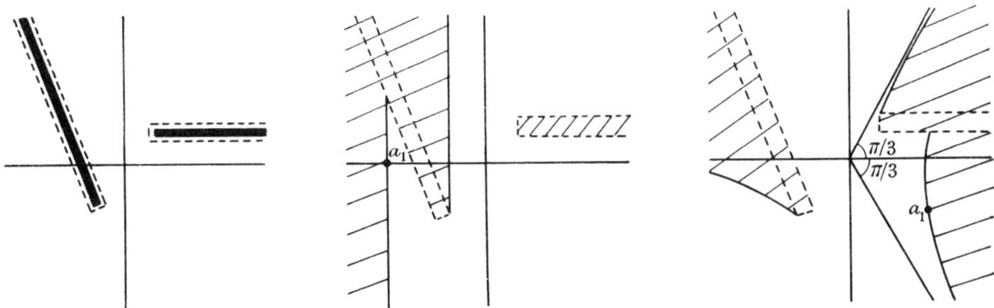

FIGURE 3. Domains $\mathbf{D}$, $\mathbf{D}'$. FIGURE 4. Case A: $\mathbf{D}_1$. FIGURE 5. Case B: $\mathbf{D}_1$.

Before stating the second theorem we introduce the following notation and definitions. It is supposed now that $z = 0$ is an interior point of the domain $\mathbf{D}'$:

(i) $v \equiv u^{\frac{1}{3}}$; $\rho_1 \equiv 1$, $\rho_2 \equiv \mathrm{e}^{\frac{2}{3}\pi\mathrm{i}}$, $\rho_3 \equiv \mathrm{e}^{-\frac{2}{3}\pi\mathrm{i}}$.

† Here A_s corresponds to A_s and B_s of § 3, B_s to C_s and D_s of § 3, and u has been replaced by u^2.

(ii) a_j $(j=1,2,3)$ is an arbitrary fixed point of the region common to the domain $\mathbf{D}'$ and the sector $\mathbf{S}_j$ (figure 1). If this common region is unbounded then a_j can be the point at infinity on a line lying in it.

(iii) $\mathbf{D}_j$ is the domain comprising those points z of $\mathbf{D}'$ for which

$$|\exp\{-\tfrac{2}{3}(\rho_j z)^{\frac{3}{2}}\}| \geqslant |\exp\{-\tfrac{2}{3}(\rho_j a_j)^{\frac{3}{2}}\}|,$$

and a contour can be found joining z and a_j which lies in $\mathbf{D}'$ and at most two of the sectors $\mathbf{S}_1$, $\mathbf{S}_2$, $\mathbf{S}_3$, and does not cross the level curve of $\exp\{-\tfrac{2}{3}(\rho_j z)^{\frac{3}{2}}\}$ through z.

A typical domain $\mathbf{D}_1$ corresponding to domains $\mathbf{D}$ and $\mathbf{D}'$ of figure 3 is the unshaded region in figure 5. The curved boundaries are level curves of $\exp(-\tfrac{2}{3}z^{\frac{3}{2}})$ (cf. figure 2).

Theorem B. *Let sequences of functions $A_s(z)$, $B_s(z)$, $C_s(z)$ and $D_s(z)$ be defined by the relations* $A_0(z)=1$,

$$\left.\begin{aligned} B_s(z) &= \tfrac{1}{2}z^{-\frac{1}{2}}\int_0^z t^{-\frac{1}{2}}\{f(t)\,A_s(t)-A_s''(t)\}\,\mathrm{d}t, \\ A_{s+1}(z) &= -\tfrac{1}{2}B_s'(z)+\tfrac{1}{2}\int f(z)\,B_s(z)\,\mathrm{d}z+\text{constant}, \end{aligned}\right\} \tag{5·8}$$

$$C_s(z) = A_s'(z)+zB_s(z), \quad D_s(z) = A_s(z)+B_{s-1}'(z), \tag{5·9}$$

the arbitrary constants in the second of (5·8) *being subject to the condition that there exists a function $\phi(u)$ having the asymptotic expansion*

$$\phi(u) \sim 1+\sum_{s=1}^{\infty}\frac{A_s(c)}{u^s}, \tag{5·10}$$

as $u\to\infty$, for fixed a point c in $\mathbf{D}'$.

Then the equation

$$\frac{\mathrm{d}^2w}{\mathrm{d}z^2} = \{uz+f(z)\}\,w \tag{5·11}$$

has solutions $W_j(z)$ $(j=1,2,3)$, such that if z lies in $\mathbf{D}_j$

$$W_j(z) = P_j(vz)\left\{1+\sum_{s=1}^{m}\frac{A_s(z)}{u^s}\right\}+\frac{P_j'(vz)}{v^2}\sum_{s=0}^{m-1}\frac{B_s(z)}{u^s}+\frac{\exp\{-\frac{2}{3}(\rho_j vz)^{\frac{3}{2}}\}}{1+|vz|^{\frac{1}{4}}}\,O\left(\frac{1}{u^{m+\frac{1}{2}}}\right), \tag{5·12}$$

$$W_j'(z) = P_j(vz)\sum_{s=0}^{m-1}\frac{C_s(z)}{u^s}+vP_j'(vz)\left\{1+\sum_{s=1}^{m}\frac{D_s(z)}{u^s}\right\}+(1+|vz|^{\frac{1}{4}})\exp\{-\tfrac{2}{3}(\rho_j vz)^{\frac{3}{2}}\}\,O\left(\frac{1}{u^m}\right), \tag{5·13}$$

as $u\to\infty$, where the functions $P_j(z)$ $(j=1,2,3)$ are defined by (4·6) *and the O's are uniform with respect to z. Here m is an arbitrary unbounded integer, and the solutions $W_j(z)$ are independent of m.*

Remarks on the theorems. (i) It is evident that (5·6) and (5·7) are asymptotic expansions in the classical sense of Poincaré. Similarly, it may be verified that (5·12) and (5·13) have the equivalent forms

$$W_j(z) \sim P_j(vz)\sum_{s=0}^{\infty}\frac{A_s(z)}{u^s}+\frac{P_j'(vz)}{v^2}\sum_{s=0}^{\infty}\frac{B_s(z)}{u^s}, \tag{5·14}$$

$$W_j'(z) \sim P_j(vz)\sum_{s=0}^{\infty}\frac{C_s(z)}{u^s}+vP_j'(vz)\sum_{s=0}^{\infty}\frac{D_s(z)}{u^s}. \tag{5·15}$$

(ii) The condition that a function $\phi(u)$ exists having the properties (5·4) or (5·10) is not very restrictive. We could, for example, take any set of values for $A_s(c)$ such that $\Sigma u^{-s}A_s(c)$ converges for large u, or even take $A_{s+1}(c)=0$ $(s\geqslant 0)$.

6. Application of the theorems

The coefficients $A_s(z)$, $B_s(z)$, $C_s(z)$ and $D_s(z)$ defined by equations (5·2), (5·3), (5·8) and (5·9) are regular functions of z in **D**. In these relations they are expressed in the form of integrals, and it may not always be convenient or even possible to perform the integrations analytically and obtain explicit expressions for the coefficients. The integrals may usually be evaluated numerically, however, and tables of coefficients can be prepared for the standard solutions of a given differential equation. An important practical point is that each quantity occurring in the series (5·6), (5·7), (5·14) and (5·15) is a function of a *single* variable, and the general tabulation of the solutions $W_j(z)$, which themselves are functions of two variables z and u, is mainly reduced to the preparation of a few single-entry tables. Such tables have, in fact, already been compiled by the writer for Bessel's and Legendre's equations.

These points are illustrated in the paper immediately following this, in which theorems A and B are used to obtain asymptotic expansions of Bessel functions of large order. In addition, it is shown there how the asymptotic series may be reverted to obtain powerful asymptotic expansions for the zeros and stationary values of the functions.

7. Previous results

The earliest writings on the asymptotic solution of differential equations for large values of a parameter appear to be those of Carlini,† Green (1837) and Liouville (1837). These authors gave, without rigorous investigation, formal asymptotic solutions in intervals free from transition points (defined in § 1). Carlini considered a form of Bessel's equation for large orders, but the equations of Green and Liouville were more general.

Extensive researches have been published on the existence theory of asymptotic solutions in intervals free from transition points. The principal contributors are Horn (1899), Schlesinger (1907), Birkhoff (1908), Tamarkin (1928) and Turrittin (1936). These writers investigated the asymptotic solution of an arbitrary number of simultaneous first-order differential equations for real values of the independent variable z and large complex values of the parameter u. Theorem A, on the other hand (in the case of second-order equations), provides a satisfactory account of the behaviour of solutions in the complex z-plane. There is a further marked difference. The relevant final result of the earlier writers applied to equation (5·5) shows that for each value of m, pairs of solutions exist having the asymptotic forms (5·6) and (5·7). Theorem A gives the more powerful result that there exists a pair of solutions having the properties (5·6) and (5·7) for *all* values of m; in a sense this means we can take $m = \infty$ in the earlier result.

If the range of integration extends from‡ $z = -\infty$ to $z = +\infty$ it can be deduced quite simply from the earlier result that solutions $W_1^*(z)$ and $W_2^*(z)$ exist, independent of m, having the asymptotic forms (5·6) and (5·7) with the arbitrary constants in (5·2) fixed by

$$A_{s+1}(-\infty) = 0 \text{ for } W_1^*(z), \quad A_{s+1}(+\infty) = 0 \text{ for } W_2^*(z) \quad (s \geqslant 0). \tag{7·1}$$

For let $W_{m,1}(z)$ and $W_{m,2}(z)$ be the solutions having the properties

$$W_{m,1}(z) = e^{uz}\left\{1 + \sum_{s=1}^{m-1} \frac{A_s(z)}{u^s} + O\left(\frac{1}{u^m}\right)\right\}, \quad W_{m,2}(z) = e^{-uz}\left\{1 + \sum_{s=1}^{m-1} (-)^s \frac{A_s(z)}{u^s} + O\left(\frac{1}{u^m}\right)\right\}, \tag{7·2}$$

† An account of Carlini's paper is given by Watson (1944, p. 6).

‡ There are complications if the range of integration is finite.

with the values of the coefficients $A_s(z)$ fixed by the first of (7·1). We define $W_1^*(z)$ to be the solution of (5·5) with the condition†

$$W_1^*(z) \sim e^{uz} \quad \text{as} \quad z \to -\infty. \tag{7·3}$$

It can be expressed in the form

$$W_1^*(z) = k_{m,1}\, W_{m,1}(z) + k_{m,2}\, W_{m,2}(z), \tag{7·4}$$

where $k_{m,1}$ and $k_{m,2}$ are independent of z. Letting $z \to -\infty$, we see from (7·2) that for all sufficiently large u

$$k_{m,2} = 0, \quad k_{m,1} = 1 + O(u^{-m}).$$

Substituting this result and (7·2) in (7·4), we obtain

$$W_1^*(z) = e^{uz}\left\{1 + \sum_{s=1}^{m-1} \frac{A_s(z)}{u^s} + O\left(\frac{1}{u^m}\right)\right\}, \tag{7·5}$$

and, since $W_1^*(z)$ is obviously independent of m, this is the result stated. Similarly for $W_2^*(z)$.

If c is any point of the real axis, we may multiply the solution $W_1^*(z)$ by $1/\omega(u)$, where $\omega(u)$ is a function of u alone having the Poincaré asymptotic expansion

$$\omega(u) \sim 1 + \frac{A_1(c)}{u} + \frac{A_2(c)}{u^2} + \ldots, \tag{7·6}$$

and the coefficients A_s in this relation are fixed by the first of (7·1). It would then follow that a solution $W_1(z)$, independent of m, exists having the property (5·6) with a new set of coefficients A_s; the arbitrary constants in (5·2) are now determined by

$$A_{s+1}(c) = 0 \quad (s \geqslant 0).$$

This is essentially the result given by theorem A in the case of real variables. The proof is incomplete, however, until it is established that $\omega(u)$ *does* exist having the property (7·6). This particular result is in fact obtained later in case B (equation (10·3) with $z = c$), and a similar proof of it holds in case A. The proofs of theorems A and B given in this paper do not follow these lines however, and are quite independent of earlier results.

One of the first successful attempts to deal with the case when there is a transition point in the interval of integration‡ is that of Jeffreys§ (1924; or Jeffreys & Jeffreys 1946, pp. 492–4). He reasoned that if the function $p(z)$ in (1·1) has a simple zero at $z = z_0$, then near this point the differential equation is approximately represented by

$$\frac{d^2 w}{dz^2} = u p'(z_0)\,(z - z_0)\, w,$$

solutions of which are Ai (Z) and Bi (Z), where

$$Z \equiv \{u p'(z_0)\}^{\frac{1}{3}}\,(z - z_0).$$

The drawback to this procedure is that the functions Ai (Z) and Bi (Z) are reasonable approximations to solutions of (1·1) only near $z = z_0$, and in any case there is no indication of the error of the approximation.

† That such a solution exists follows from (5·1) and the ordinary asymptotic theory of linear differential equations (Ince 1944, §7·31).

‡ A review of the literature concerned with this case has been published recently by the California Institute of Technology (1953).

§ In a recent paper (1953) Jeffreys points out that this approach was partially anticipated by Gans and Rayleigh.

These disadvantages are largely removed in the method originally given by Langer (1931). He showed how an equation of the form (1·1) may be transformed, as shown in § 2, into (2·8), and he proved that under certain conditions equation (2·8) possesses solutions which for large u are asymptotic to Bessel functions of order one-third. The interval in which these approximations are valid is actually a function of u which shrinks to zero as $u \to \infty$. However, Langer also gave approximations to the same solutions in terms of exponential functions which are valid outside this interval. He later extended these results to the case of complex variables (1932) and to different kinds of transition points (1934, 1935).

The asymptotic form of solution (5·12) was also given, in effect, by Langer (1937, 1949), who proved a result concerning its validity closely analogous to that of Birkhoff and others quoted above for case A; it concerns only real values of the independent variable, and the asymptotic solutions obtained depend on the integer m. An investigation of the solutions when the independent variable is complex has been made by Cherry (1950), who has proved a result resembling theorem B of § 5.

Using the present notation, Cherry shows that solutions $W_{m,j}(z)$ $(j = 1, 2, 3)$ of (5·11) exist for large positive u such that

$$W_{m,j}(z) = P_j(vz)\left\{\sum_{s=0}^{m-1} \frac{A_s(z)}{u^s} + \frac{a_m(z,u)}{u^m}\right\} + \frac{P_j'(vz)}{v^2}\left\{\sum_{s=0}^{m-1} \frac{B_s(z)}{u^s} + \frac{b_m(z,u)}{u^m}\right\},$$

where $|a_m(z,u)|$ and $|b_m(z,u)|$ are bounded; this result is similar to that of Birkhoff for real variables quoted above for case A (see equations (7·2)). By making an additional assumption Cherry then deduces that in certain bounded z-domains solutions $W_j^*(z)$ exist, independent of m, having the asymptotic expansions

$$W_j^*(z) \sim P_j(vz) \sum_{s=0}^{\infty} \frac{A_s^j(z)}{u^s} + \frac{P_j'(vz)}{v^2} \sum_{s=0}^{\infty} \frac{B_s^j(z)}{u^s}, \tag{7·7}$$

as $u \to \infty$, uniformly with respect to z, where the coefficients $A_s^j(z)$, $B_s^j(z)$ are the values of $A_s(z)$, $B_s(z)$ given by (5·8) with the arbitrary constants fixed by

$$A_{s+1}^j(\infty\, \rho_j^{-1}) = 0 \quad (s \geqslant 0),$$

where ρ_j is defined in § 5. The assumption is that $f(z)$ is regular in sectors of non-zero angles extending to infinity centred on the rays $\arg z = 0, \pm\frac{2}{3}\pi$, and in these sectors $f(z) = O(|z|)$ as $|z| \to \infty$ (Cherry 1950, p. 232). The result (7·7) is similar to (7·5) above for case A.

Unlike theorem B Cherry's result does not show that solutions exist when the values of $A_1(z), A_2(z), \ldots$, in (5·8) are prescribed at arbitrary points. Moreover, in this theorem it is not stipulated that the rays $\arg z = 0, \pm\frac{2}{3}\pi$, are included in the domain in which $f(z)$ is regular, or that the domain of uniform validity of the expansion has to be bounded.

The rest of this paper is concerned with the proof of the theorems of § 5. The methods used differ considerably from those of Langer and Cherry.

8. Preliminaries in the proof of the existence theorems

Only a proof for theorem B will be given, since a similar but simpler proof holds for theorem A.

In the next section we consider properties of the coefficients $A_s(z)$ and $B_s(z)$ defined by (5·8), and it is convenient to introduce here some definitions concerning the function $f(z)$.

It was postulated in §5 that $f(z)$ is regular in a simply-connected open domain $\mathbf{D}$, the boundaries of which consist of a finite number of straight lines, and that if $\mathbf{D}$ is unbounded, $f(z) = O(|z|^{-2})$ as $|z|\to\infty$, uniformly with respect to $\arg z$. Consequently in $\mathbf{D}$,

$$|f(z)|\leqslant F_1, \quad |z^2 f(z)|\leqslant F_2, \tag{8·1}$$

where F_1 and F_2 are assignable constants. We shall sometimes use the symbol F generically to denote a combination of these constants.

From Liouville's theorem it follows that $\mathbf{D}$ cannot be the whole z-plane, unless $f(z)\equiv 0$. In other words, a definite boundary always exists.

Auxiliary domains. Let us suppose that straight lines are drawn in $\mathbf{D}$ parallel to the boundaries and at a distance δ from them; if the boundaries include cuts then semicircles of radius δ are described centred on the end-points. Clearly a constant d can be assigned such that for every positive $\delta\leqslant d$, the aggregate of lines and semicircles defines the boundary of another simply-connected domain which also contains $z = 0$ as an interior point. This domain will be denoted by $\mathbf{D}_\delta$, and when $\delta = d$ by $\mathbf{D}_d$.

Neighbourhoods of infinity. There is only a finite number of points at which the boundaries of the domains $\mathbf{D}$ and $\mathbf{D}_d$ change direction abruptly, including the end-points of cuts. Let the one most remote from the origin be of affix R', and let $R\equiv\max(R'+d, 1)$. The simply-connected subdomains of $\mathbf{D}$ lying outside the circle $|z| = R$ will be denoted by $\mathbf{E}^1, \mathbf{E}^2, \ldots$, or, typically, by $\mathbf{E}$. Similarly, the simply-connected subdomains of $\mathbf{D}_\delta$ lying outside the circle $|z| = R+\delta$ will be denoted by $\mathbf{E}^1_\delta, \mathbf{E}^2_\delta, \ldots$, or typically by $\mathbf{E}_\delta$.

Where the symbol δ is used in the following sections it is assumed that

$$0<\delta\leqslant\min(d, \tfrac{1}{2}R, 1). \tag{8·2}$$

9. Properties of the coefficients $A_s(z)$ and $B_s(z)$

These coefficients are defined by $A_0(z) = 1$ and (see (5·8))

$$B_s(z) = \tfrac{1}{2}z^{-\frac{1}{2}}\int_0^z t^{-\frac{1}{2}}\{f(t)\,A_s(t) - A_s''(t)\}\,dt \quad (s\geqslant 0), \tag{9·1}$$

$$A_{s+1}(z) = -\tfrac{1}{2}B_s'(z) + \tfrac{1}{2}\int_0^z f(t)\,B_s(t)\,dt + \text{constant} \quad (s\geqslant 0). \tag{9·2}$$

The integration paths in (9·1) and (9·2) both lie in $\mathbf{D}$, and the branches of the square roots occurring in (9·1) are taken to be the principal ones near the origin and defined by continuity elsewhere. Clearly $A_s(z)$ and $B_s(z)$ are regular functions of z in $\mathbf{D}$.

The arbitrary constant in (9·2) is fixed temporarily by the condition

$$A_{s+1}(c) = 0 \quad (s\geqslant 0), \tag{9·3}$$

for some point c, which may be at infinity, in $\mathbf{D}_d$. Then

$$A_{s+1}(z) = -\tfrac{1}{2}B_s'(z) + \tfrac{1}{2}\int_c^z f(t)\,B_s(t)\,dt + \tfrac{1}{2}B_s'(c). \tag{9·4}$$

This restriction is relaxed in §11, and in any case it is not necessary for the truth of lemma 1 which follows.

LEMMA 1. *If $|z| \to \infty$ in $\mathbf{E}_\delta$, then*

$$A_s(z) = \alpha_s + O(|z|^{-\frac{3}{2}}), \tag{9·5}$$

$$B_s(z) = \beta_s z^{-\frac{1}{2}} + O(|z|^{-\frac{3}{2}}), \tag{9·6}$$

uniformly with respect to arg z. *The quantities α_s and β_s are constants, and the branch of $z^{-\frac{1}{2}}$ in* (9·6) *is continuous in $\mathbf{E}_\delta$.*

This may be proved by induction. Suppose that (9·5) is true for a particular s as $|z| \to \infty$ in $\mathbf{E}_\eta$, where η is an arbitrary number lying in the range $0 < \eta < \delta$. Then in $\mathbf{E}_\eta$

$$A_s(z) = \alpha_s + a_s(z), \quad \text{where} \quad |a_s(z)| < \lambda_s |z|^{-\frac{3}{2}}, \tag{9·7}$$

λ_s being a constant.

Now if z lies in $\mathbf{E}_\epsilon$, where $\eta < \epsilon \leqslant \delta$, we have from Cauchy's integral formula

$$A_s''(z) = \frac{1}{\pi \mathrm{i}} \int_{|t-z|=\epsilon-\eta} \frac{A_s(t)}{(t-z)^3}\, \mathrm{d}t = \frac{1}{\pi \mathrm{i}} \int_{|t-z|=\epsilon-\eta} \frac{a_s(t)}{(t-z)^3}\, \mathrm{d}t,$$

and so, using (9·7),

$$|A_s''(z)| < \frac{2}{(\epsilon-\eta)^2} \frac{\lambda_s}{\{|z|-(\epsilon-\eta)\}^{\frac{3}{2}}} = O(|z|^{-\frac{3}{2}}), \quad \text{as} \quad |z| \to \infty. \tag{9·8}$$

From this result and (5·1), (9·7) it is seen that the integral

$$\int_0^z t^{-\frac{1}{2}} \{f(t)\, A_s(t) - A_s''(t)\}\, \mathrm{d}t$$

converges as $|z| \to \infty$ in $\mathbf{E}_\epsilon$, and hence (9·1) may be written in the form

$$B_s(z) = \beta_s z^{-\frac{1}{2}} + b_s(z), \quad b_s(z) = -\tfrac{1}{2} z^{-\frac{1}{2}} \int_z^\infty t^{-\frac{1}{2}} \{f(t)\, A_s(t) - A_s''(t)\}\, \mathrm{d}t, \tag{9·9}$$

where β_s is a constant, and the upper limit ∞ denotes a point at infinity in $\mathbf{E}_\epsilon$. From (5·1), (9·7) and (9·8) we see that

$$b_s(z) = O(|z|^{-\frac{3}{2}}), \tag{9·10}$$

as $|z| \to \infty$ in $\mathbf{E}_\epsilon$. Thus if (9·5) is true in $\mathbf{E}_\eta$, then (9·6) holds in $\mathbf{E}_\epsilon$.

Next, it follows from (5·1), (9·10) and the first of (9·9) that $\int_0^z f(t)\, B_s(t)\, \mathrm{d}t$ converges as $|z| \to \infty$ in $\mathbf{E}_\epsilon$. We may therefore write (9·2) in the form

$$A_{s+1}(z) = \alpha_{s+1} + a_{s+1}(z),$$

where α_{s+1} is a constant and

$$a_{s+1}(z) = -\tfrac{1}{2} B_s'(z) - \tfrac{1}{2} \int_z^\infty f(t)\, B_s(t)\, \mathrm{d}t. \tag{9·11}$$

From (9·1) we obtain

$$B_s'(z) = \{f(z)\, A_s(z) - A_s''(z) - B_s(z)\}/(2z), \tag{9·12}$$

(cf. (3·9)), and referring to (9·5), (9·8), (9·9) and (9·10) it is seen that $B_s'(z) = O(|z|^{-\frac{3}{2}})$. Hence from (9·11) we deduce that $a_{s+1}(z) = O(|z|^{-\frac{3}{2}})$, as $|z| \to \infty$ in $\mathbf{E}_\epsilon$. This completes the proof of lemma 1.

LEMMA 2. *When z lies in $\mathbf{D}_\delta$ and $|t| \leqslant T$, the series*

$$A(z,t) = \sum_{s=0}^\infty A_s(z) \frac{t^{2s}}{(2s)!}, \quad B(z,t) = \sum_{s=0}^\infty B_s(z) \frac{t^{2s}}{(2s)!}, \tag{9·13}$$

converge uniformly with respect to z and t. Here $T > 0$ is an assignable constant which may depend on δ, but is independent of z.

Let the maximum† moduli in $\mathbf{D}_\delta$ of $A_s(z)$, $A_s'(z)$, $A_s''(z)$, $B_s(z)$ and $B_s'(z)$ be denoted by $M_s(\delta)$, $M_s^{(1)}(\delta)$, $M_s^{(2)}(\delta)$, $P_s(\delta)$ and $P_s^{(1)}(\delta)$ respectively, and let η, ϵ be arbitrary numbers such that $0<\eta<\epsilon\leqslant\delta$.

From (9·1) we obtain

$$|B_s(z)|\leqslant\tfrac{1}{2}|z|^{-\frac{1}{2}}\left|\int_0^z t^{-\frac{1}{2}}\{f(t)\,A_s(t)-A_s''(t)\}\,\mathrm{d}t\right|.$$

If the domain $\mathbf{D}_\epsilon$ were a star-domain the path of integration could be taken to be a straight line, and using (8·1) we would immediately deduce that

$$P_s(\epsilon)\leqslant F_1M_s(\epsilon)+M_s^{(2)}(\epsilon).$$

In general, however, the path cannot be taken to be a single straight line but by considering separately the three parts into which it is divided by the circles $|z|=R$ and $|z|=\rho$, where ρ is the shortest distance between $z=0$ and the boundary of $\mathbf{D}_d$, it may be seen that

$$P_s(\epsilon)\leqslant\kappa\{F_1M_s(\epsilon)+M_s^{(2)}(\epsilon)\}, \tag{9·14}$$

where $\kappa\geqslant1$ is a constant whose value depends on the configuration of the boundaries of the domains $\mathbf{D}$ and $\mathbf{D}_d$.

Similarly, from (9·4) we obtain, using now the symbol κ generically,

$$M_{s+1}(\epsilon)\leqslant P_s^{(1)}(\epsilon)+\kappa FP_s(\epsilon). \tag{9·15}$$

From the maximum-modulus theorem and lemma 1 with δ replaced by ϵ, it follows that $|B_s(z)|$ attains its maximum value in $\mathbf{D}_\epsilon$ on the boundary. Hence from (8·1) and (9·12) we find that

$$P_s^{(1)}(\epsilon)<\{F_1M_s(\epsilon)+M_s^{(2)}(\epsilon)+P_s(\epsilon)\}/(2\rho). \tag{9·16}$$

Combining (9·14), (9·15) and (9·16), we obtain

$$M_{s+1}(\epsilon)<\{\kappa^2F+(1+\kappa)\,(2\rho)^{-1}\}\{F_1M_s(\epsilon)+M_s^{(2)}(\epsilon)\}. \tag{9·17}$$

Now if z lies in $\mathbf{D}_\epsilon$, we have from Cauchy's integral formula

$$A_s''(z)=\frac{1}{\pi\mathrm{i}}\int_{|t-z|=\epsilon-\eta}\frac{A_s(t)}{(t-z)^3}\,\mathrm{d}t,$$

and hence

$$M_s^{(2)}(\epsilon)\leqslant2(\epsilon-\eta)^{-2}M_s(\eta). \tag{9·18}$$

Substituting this result in (9·17) and using the fact that $M_s(\epsilon)\leqslant M_s(\eta)$, we obtain

$$\left.\begin{aligned}&M_{s+1}(\epsilon)<GM_s(\eta),\\ \text{where}\quad &G=\{\kappa^2F+(1+\kappa)\,(2\rho)^{-1}\}\{F_1+2(\epsilon-\eta)^{-2}\}.\end{aligned}\right\} \tag{9·19}$$

This result is the key to the proof, and it may now be completed as follows. Let us write

$$\delta_{n,s}\equiv\frac{s-n}{s}\delta\quad(n=0,1,2,\ldots,s). \tag{9·20}$$

Then from (9·19) we obtain

$$M_s(\delta)\equiv M_s(\delta_{0,s})<G_sM_{s-1}(\delta_{1,s})<G_s^2M_{s-2}(\delta_{2,s})<\ldots<G_s^s, \tag{9·21}$$

$$\left.\begin{aligned}\text{where}\quad &G_s\equiv\{\kappa^2F+(1+\kappa)\,(2\rho)^{-1}\}\,(F_1+2s^2\delta^{-2})\leqslant k_\delta s^2\quad(s\geqslant1),\\ &k_\delta\equiv\{\kappa^2F+(1+\kappa)\,(2\rho)^{-1}\}\,(F_1+2\delta^{-2}).\end{aligned}\right\} \tag{9·22}$$

† The existence of these maxima follows from lemma 1.

Thus if z lies in $\mathbf{D}_\delta$, then

$$|A_s(z)| \leqslant M_s(\delta) < k_\delta^s s^{2s} \quad (s \geqslant 1). \tag{9·23}$$

Further, from (9·18) to (9·22) we deduce that

$$M_{s-1}^{(2)}(\delta) \leqslant 2\delta^{-2} s^2 M_{s-1}(\delta_{1,s}) < 2\delta^{-2} s^2 G_s^{s-1} \leqslant 2\delta^{-2} k_\delta^{s-1} s^{2s}. \tag{9·24}$$

Substituting (9·23) and (9·24) in (9·14), we see that if z lies in $\mathbf{D}_\delta$, then

$$|B_s(z)| \leqslant P_s(\delta) < \kappa k_\delta^s \{F_1 s^{2s} + 2\delta^{-2}(s+1)^{2s+2}\}. \tag{9·25}$$

The truth of lemma 2 is now obvious from the results (9·23), (9·25) and Stirling's formula

$$(2s)! \sim \surd(2\pi)\, s^{2s+\frac{1}{2}}\, 2^{2s+\frac{1}{2}}\, \mathrm{e}^{-2s}. \tag{9·26}$$

The value of T can be taken to be any positive number less than $2\,\mathrm{e}^{-1} k_\delta^{-\frac{1}{2}}$.

Lemma 3. *If z lies in $\mathbf{E}_\delta$ and $|t| \leqslant \tau$, then*

$$\frac{\partial}{\partial z} A(z,t) \equiv \sum_{s=1}^{\infty} A_s'(z) \frac{t^{2s}}{(2s)!} = O(|z|^{-\frac{3}{2}}), \tag{9·27}$$

$$\frac{\partial^2}{\partial z\,\partial t} A(z,t) \equiv \sum_{s=1}^{\infty} A_s'(z) \frac{t^{2s-1}}{(2s-1)!} = O(|z|^{-\frac{3}{2}}), \tag{9·28}$$

as $|z| \to \infty$, uniformly with respect to t and $\arg z$. *Here τ is an assignable number independent of z, and $0 < \tau \leqslant T$.*

To prove this result it is necessary to examine how the constants implied in the O's occurring in the proof of lemma 1 depend on s.

As in the proof of lemmas 1 and 2 it is supposed that η and ϵ are arbitrary numbers such that $0 < \eta < \epsilon \leqslant \delta$. We first investigate the connexion between λ_s and λ_{s+1}, where λ_s is the constant introduced in (9·7).

If z lies in $\mathbf{E}_\epsilon$, then from (9·8)

$$|A_s''(z)| < \frac{2}{(\epsilon-\eta)^2} \frac{\lambda_s}{\{|z| - (\epsilon-\eta)\}^{\frac{3}{2}}} < \frac{2^{\frac{5}{2}} \lambda_s}{(\epsilon-\eta)^2 |z|^{\frac{3}{2}}}, \tag{9·29}$$

since from (8·2), $\qquad |z| - (\epsilon-\eta) > |z| - \delta > |z|(1 - \delta R^{-1}) > \frac{1}{2}|z|.$

Using this result and (8·1), (9·7), we find that

$$|z^{-\frac{1}{2}}\{f(z)\,A_s(z) - A_s''(z)\}| < \frac{F_2}{|z|^{\frac{5}{2}}}\left(|\alpha_s| + \frac{\lambda_s}{|z|^{\frac{3}{2}}}\right) + \frac{2^{\frac{5}{2}}\lambda_s}{(\epsilon-\eta)^2 |z|^2} < \frac{\mu_s}{|z|^2}, \tag{9·30}$$

where

$$\mu_s = F_2(|\alpha_s| + \lambda_s) + 2^{\frac{5}{2}}(\epsilon-\eta)^{-2}\lambda_s, \tag{9·31}$$

since $|z| > 1$ in $\mathbf{E}_\epsilon$ (see § 8).

In the integral (9·9) the path of integration may be taken to be a straight line whose equation is of the form $t = z + p\,\mathrm{e}^{\mathrm{i}\varpi}$, where $0 \leqslant p \leqslant \infty$, ϖ is a constant, and on the path $|t|^2 \geqslant |z|^2 + p^2$. Accordingly, we obtain from (9·9) and (9·30)

$$|b_s(z)| < \frac{1}{2|z|^{\frac{1}{2}}} \int_0^\infty \frac{\mu_s\,\mathrm{d}p}{|z|^2 + p^2} = \frac{\pi\mu_s}{4|z|^{\frac{3}{2}}}. \tag{9·32}$$

Substituting this result and (9·7), (9·9) and (9·29) in (9·12), we obtain

$$|B_s'(z)| < \frac{1}{2|z|}\left\{\frac{F_2}{|z|^2}\left(|\alpha_s| + \frac{\lambda_s}{|z|^{\frac{3}{2}}}\right) + \frac{2^{\frac{5}{2}}\lambda_s}{(\epsilon-\eta)^2|z|^{\frac{3}{2}}} + \frac{|\beta_s|}{|z|^{\frac{1}{2}}} + \frac{\pi\mu_s}{4|z|^{\frac{3}{2}}}\right\} < \frac{|\beta_s| + 2\mu_s}{2|z|^{\frac{3}{2}}}. \tag{9·33}$$

Similarly, from (9·11) using (9·32) and (9·33), we find that

$$|a_{s+1}(z)| < \frac{|\beta_s|+2\mu_s}{4|z|^{\frac{3}{2}}} + \frac{1}{2}\int_0^\infty \frac{F_2(|\beta_s|+\mu_s)}{(|z|^2+p^2)^{\frac{5}{4}}}\,\mathrm{d}p < (F_2+\tfrac{1}{2})(|\beta_s|+\mu_s)|z|^{-\frac{3}{2}}.$$

Substituting for μ_s from (9·31) and using (8·2), we prove that:

$$\left.\begin{array}{ll} \textit{If} & |a_s(z)| < \lambda_s |z|^{-\frac{3}{2}} \quad \textit{in} \quad \mathbf{E}_\eta, \\ \textit{then} & |a_{s+1}(z)| < F\{|\alpha_s|+|\beta_s|+\lambda_s(\epsilon-\eta)^{-2}\}|z|^{-\frac{3}{2}} \quad \textit{in} \quad \mathbf{E}_\epsilon. \end{array}\right\} \tag{9·34}$$

We now consider the magnitudes of $|\alpha_s|$ and $|\beta_s|$. Letting $|z|\to\infty$ in $\mathbf{E}_\delta$, we see from (9·7) and (9·23) that

$$|\alpha_s| < k_\delta^s s^{2s}, \tag{9·35}$$

where k_δ is defined by (9·22). Next we have by definition (see (9·1) and (9·6))

$$\beta_s = \frac{1}{2}\int_0^\infty t^{-\frac{1}{2}}\{f(t)\,A_s(t) - A_s''(t)\}\,\mathrm{d}t. \tag{9·36}$$

If t lies in $\mathbf{E}_\epsilon$, then from (9·29)

$$|A_s''(t)| < 2^{\frac{3}{2}}\lambda_s(\epsilon-\eta)^{-2}|t|^{-\frac{3}{2}},$$

and if t lies anywhere in $\mathbf{D}_\epsilon$, then clearly

$$|A_s(t)| \leqslant M_s(\epsilon), \quad |A_s''(t)| \leqslant M_s^{(2)}(\epsilon).$$

Substituting these estimates in (9·36), we see that

$$|\beta_s| < \kappa F\{M_s(\epsilon)+M_s^{(2)}(\epsilon)+\lambda_s(\epsilon-\eta)^{-2}\},$$

where κ is the generic constant introduced in (9·14) and (9·15). Hence, with the aid of (9·23) and (9·24), we obtain

$$|\beta_s| < \kappa F\epsilon^{-2}\{k_\epsilon^s(s+1)^{2s+2}+\lambda_s(\epsilon-\eta)^{-2}\}, \tag{9·37}$$

where k_ϵ is defined by (9·22) with δ replaced by ϵ. Substituting (9·35) and (9·37) in (9·34) and changing s into $n-1$, we see that:

$$\left.\begin{array}{ll} \textit{If} & |z^{\frac{3}{2}}a_{n-1}(z)| < \lambda_{n-1} \quad \textit{in} \quad \mathbf{E}_\eta, \\ \textit{then} & |z^{\frac{3}{2}}a_n(z)| < \kappa F\epsilon^{-2}\{k_\epsilon^{n-1}n^{2n}+\lambda_{n-1}(\epsilon-\eta)^{-2}\} \quad \textit{in} \quad \mathbf{E}_\epsilon. \end{array}\right\} \tag{9·38}$$

Let ζ be any fixed number in the range $0<\zeta<\delta$, and let us write

$$\zeta_{n,s} \equiv \zeta + \frac{n}{s}(\delta-\zeta) \quad (n = 0, 1, 2, \ldots, s;\ s\geqslant 1).$$

Then taking $\eta = \zeta_{0,s} = \zeta$, $\epsilon = \zeta_{1,s}$ and $n = 1$ in the result (9·38), we see that

$$|z^{\frac{3}{2}}a_1(z)| < \kappa F\zeta_{1,s}^{-2} < \kappa F\zeta^{-2} = \lambda_{1,s},$$

say. Similarly,

$$|z^{\frac{3}{2}}a_2(z)| < \kappa F\zeta_{2,s}^{-2}\{k_\epsilon 2^4+\lambda_{1,s}s^2(\delta-\zeta)^{-2}\} < \kappa F\zeta^{-2}\{k_\zeta 2^4+\lambda_{1,s}s^2(\delta-\zeta)^{-2}\} = \lambda_{2,s},$$

say, and generally,

$$|z^{\frac{3}{2}}a_n(z)| < \lambda_{n,s} \quad (n = 1, 2, \ldots, s), \tag{9·39}$$

where

$$\lambda_{n,s} = \kappa F\zeta^{-2}\{k_\zeta^{n-1}n^{2n}+\lambda_{n-1,s}s^2(\delta-\zeta)^{-2}\}. \tag{9·40}$$

Let us write

$$\lambda_{n,s}^* \equiv K^n s^{2n+2},$$

where

$$K = \max\,[k_\zeta, \kappa F\zeta^{-2}\{1+(\delta-\zeta)^{-2}\}]. \tag{9·41}$$

Then it is evident that

$$\kappa F\zeta^{-2}\{k_\zeta^{n-1}n^{2n}+\lambda^*_{n-1,s}s^2(\delta-\zeta)^{-2}\}<\kappa F\zeta^{-2}\lambda^*_{n-1,s}\{1+s^2(\delta-\zeta)^{-2}\}<Ks^2\lambda^*_{n-1,s}=\lambda^*_{n,s}.$$

From this inequality and (9·40) it follows by induction that

$$\lambda_{n,s}<\lambda^*_{n,s}\quad (n=1,2,\ldots,s),$$

and, in particular,

$$\lambda_{s,s}<\lambda^*_{s,s}=K^s s^{2s+2}.$$

Collecting these results, we see that if z lies in $\mathbf{E}_\delta$ then

$$|A_s(z)-\alpha_s|<K^s s^{2s+2}|z|^{-\frac{3}{2}},$$

where K is independent of z *and* s. Using Cauchy's integral formula we easily deduce from this result that

$$|A'_s(z)|<\text{constant}\times K^s s^{2s+2}|z|^{-\frac{3}{2}},$$

and from this inequality and (9·26) the truth of lemma 3 is established.

10. Proof of theorem B

Let us suppose that the domain $\mathbf{D}_\delta$ (defined in § 8) contains the whole of $\mathbf{D}'$ (defined in § 5); this can always be arranged by taking δ to be sufficiently small. Consider the functions $L_j(z)$ $(j=1,2,3)$ defined by

$$L_j(z)=u^{\frac{1}{2}}P_j(vz)\int_0^\tau \mathrm{e}^{-t\sqrt{u}}A(z,t)\,\mathrm{d}t+u^{-\frac{1}{6}}P'_j(vz)\int_0^\tau \mathrm{e}^{-t\sqrt{u}}B(z,t)\,\mathrm{d}t, \tag{10·1}$$

where $v\equiv u^{\frac{1}{3}}$, $A(z,t)$ and $B(z,t)$ are given by (9·13), and τ is the quantity introduced in lemma 3. More accurately, τ is the least of all such quantities when all the subdomains $\mathbf{E}^1_\delta, \mathbf{E}^2_\delta, \ldots$, are taken into account. The derivative $L'_j(z)$ is given by

$$L'_j(z)=u^{\frac{1}{2}}P_j(vz)\int_0^\tau \mathrm{e}^{-t\sqrt{u}}\left(\frac{\partial A}{\partial z}+zB\right)\mathrm{d}t+u^{-\frac{1}{6}}P'_j(vz)\int_0^\tau \mathrm{e}^{-t\sqrt{u}}\left(uA+\frac{\partial B}{\partial z}\right)\mathrm{d}t. \tag{10·2}$$

Substituting the expansions (9·13) for $A(z,t)$ and $B(z,t)$ in (10·1) and integrating term by term, we may verify without difficulty that

$$L_j(z)\sim P_j(vz)\sum_{s=0}^{\infty}\frac{A_s(z)}{u^s}+\frac{P'_j(vz)}{v^2}\sum_{s=0}^{\infty}\frac{B_s(z)}{u^s}, \tag{10·3}$$

as $u\to\infty$, uniformly with respect to z in $\mathbf{D}'$. Similarly, using (5·9), we obtain from (10·2)

$$L'_j(z)\sim P_j(vz)\sum_{s=0}^{\infty}\frac{C_s(z)}{u^s}+vP'_j(vz)\sum_{s=0}^{\infty}\frac{D_s(z)}{u^s}. \tag{10·4}$$

Comparing (10·3) and (10·4) with (5·14) and (5·15), we see that $L_j(z)$ has the desired asymptotic form. It does not, however, satisfy the differential equation (5·11). For, from (10·1) and (10·2), we derive

$$\frac{\mathrm{d}^2L_j}{\mathrm{d}z^2}-(uz+f)L_j=u^{\frac{1}{2}}P_j(vz)\int_0^\tau \mathrm{e}^{-t\sqrt{u}}\left(\frac{\partial^2 A}{\partial z^2}-fA+2z\frac{\partial B}{\partial z}+B\right)\mathrm{d}t$$

$$+u^{-\frac{1}{6}}P'_j(vz)\int_0^\tau \mathrm{e}^{-t\sqrt{u}}\left(2u\frac{\partial A}{\partial z}+\frac{\partial^2 B}{\partial z^2}-fB\right)\mathrm{d}t.$$

Now, using (9·13) and (3·9) with $g(z) = z$, we find that

$$\frac{\partial^2 A}{\partial z^2} - fA + 2z\frac{\partial B}{\partial z} + B = 0, \quad 2u\frac{\partial A}{\partial z} + \frac{\partial^2 B}{\partial z^2} - fB = 2u\frac{\partial A}{\partial z} - 2\frac{\partial^3 A}{\partial z\,\partial t^2},$$

and so

$$\frac{\mathrm{d}^2 L_j}{\mathrm{d}z^2} - (uz+f)\,L_j = -P_j'(vz)\,R(z), \tag{10·5}$$

where

$$R(z) = 2u^{-\frac{1}{2}}\int_0^\tau \mathrm{e}^{-t\sqrt{u}}\frac{\partial^3 A}{\partial z\,\partial t^2}\,\mathrm{d}t - 2u^{\frac{1}{2}}\int_0^\tau \mathrm{e}^{-t\sqrt{u}}\frac{\partial A}{\partial z}\,\mathrm{d}t.$$

Integrating the last integral twice by parts, we find

$$R(z) = 2v\,\mathrm{e}^{-\tau\sqrt{u}}\{R_1(z) + u^{-\frac{1}{2}}R_2(z)\}, \tag{10·6}$$

where

$$R_1(z) = \left(\frac{\partial A}{\partial z}\right)_{t=\tau} = \sum_{s=1}^{\infty} A_s'(z)\,\frac{\tau^{2s}}{(2s)!}, \quad R_2(z) = \left(\frac{\partial^2 A}{\partial z\,\partial t}\right)_{t=\tau} = \sum_{s=1}^{\infty} A_s'(z)\,\frac{\tau^{2s-1}}{(2s-1)!}. \tag{10·7}$$

Only if $R(z)$ vanished identically would $L_j(z)$ be a solution of (5·11), and this is not the case.

Let us denote by $W_j(z)$ the solution of (5·11) with the conditions

$$\left.\begin{aligned} &W_j(a_j) = L_j(a_j), \quad W_j'(a_j) = L_j'(a_j), \quad \text{if } a_j \text{ is finite,}\\ &W_j(z) \dot{\sim} L_j(z) \quad \text{as} \quad z \to a_j, \quad \text{if } a_j \text{ is finite.}\end{aligned}\right\} \tag{10·8}$$

We shall prove theorem B by showing that for large u the difference $W_j(z) - L_j(z)$ is asymptotically negligible compared with the error term in (5·12). Only a proof for $j = 1$ need be recorded; the other two cases are exactly similar.

From (5·11) and (10·5) we derive

$$\frac{\mathrm{d}^2}{\mathrm{d}z^2}\{W_1(z) - L_1(z)\} - \{uz + f(z)\}\{W_1(z) - L_1(z)\} = P_1'(vz)\,R(z). \tag{10·9}$$

Let us define a sequence of functions $\{h_n(z)\}$ by the relations $h_0(z) = 0$ and

$$\frac{\mathrm{d}^2}{\mathrm{d}z^2}h_n(z) - uzh_n(z) = f(z)\,h_{n-1}(z) + P_1'(vz)\,R(z) \quad (n \geqslant 1), \tag{10·10}$$

with the boundary values

$$h_n(a_1) = h_n'(a_1) = 0. \tag{10·11}$$

Suppose first that z lies in the part of $\mathbf{D}_1$ common to the sector $\mathbf{S}_1 + \mathbf{S}_2$ (figure 1). Applying the principle of variation of parameters to (10·10) and using (4·7), we find that

$$h_n(z) = \frac{2\pi\,\mathrm{e}^{\frac{1}{6}\pi\mathrm{i}}}{v}\int_z^{a_1}\{P_1(vz)\,P_2(vt) - P_2(vz)\,P_1(vt)\}\{f(t)\,h_{n-1}(t) + P_1'(vt)\,R(t)\}\,\mathrm{d}t, \tag{10·12}$$

if $n \geqslant 1$. Putting $n = 1$ and substituting (10·6), we obtain

$$h_1(z) = \mathrm{e}^{-\tau\sqrt{u}}\exp\{-\tfrac{2}{3}(vz)^{\frac{3}{2}}\}\int_z^{a_1}\psi(z,v,t)\,\{R_1(t) + u^{-\frac{1}{2}}R_2(t)\}\,\mathrm{d}t, \tag{10·13}$$

where

$$\psi(z,v,t) = 4\pi\,\mathrm{e}^{\frac{1}{6}\pi\mathrm{i}}\{P_1(vz)\,P_2(vt) - P_2(vz)\,P_1(vt)\}\,P_1'(vt)\exp\{\tfrac{2}{3}(vz)^{\frac{3}{2}}\}. \tag{10·14}$$

In consequence of the definition of the domain $\mathbf{D}_1$ given in § 5 we may select a path of integration in (10·12) and (10·13) lying wholly in $\mathbf{D}'$ and sector $\mathbf{S}_1 + \mathbf{S}_2$, such that if t is any point on the path

$$|\exp(-\tfrac{2}{3}t^{\frac{3}{2}})| \leqslant |\exp(-\tfrac{2}{3}z^{\frac{3}{2}})|. \tag{10·15}$$

Using (4·5), (4·6) and (10·14), we deduce that

$$|\psi(z,v,t)| < \frac{A}{1+|vz|^{\frac{1}{4}}} + \frac{A\,|\exp\{\frac{4}{3}(vz)^{\frac{3}{2}} - \frac{4}{3}(vt)^{\frac{3}{2}}\}|}{1+|vz|^{\frac{1}{4}}},$$

where the symbol A is used generically to denote a constant independent of z, u and t. If (10·15) is substituted in the last result, it is seen that

$$|\psi(z,v,t)| < A(1+|vz|^{\frac{1}{4}})^{-1}, \tag{10·16}$$

and substituting this result in (10·13), we obtain

$$|h_1(z)| < A\,\mathrm{e}^{-\tau\sqrt{u}}\,\frac{|\exp\{-\frac{2}{3}(vz)\}^{\frac{3}{2}}|}{1+|vz|^{\frac{1}{4}}}\int_z^{a_1}|R_1(t)+u^{-\frac{1}{2}}R_2(t)|\,|\mathrm{d}t|. \tag{10·17}$$

From (10·7) and lemma 3 it follows that

$$R_1(z) = O(|z|^{-\frac{3}{2}}), \quad R_2(z) = O(|z|^{-\frac{3}{2}}), \tag{10·18}$$

as $|z| \to \infty$ in $\mathbf{D}'$, uniformly with respect to $\arg z$. The path of integration in (10·12), (10·13) and (10·17) may be confined to the following contours and still satisfy the conditions above:

(i) the level curves of $\exp(-\frac{2}{3}z^{\frac{3}{2}})$ through z and a_1,
(ii) the rays $\arg z = 0, -\frac{2}{3}\pi$,
(iii) the boundary of $\mathbf{D}'$.

With the aid of (4·9) and (10·18), it may be shown that on such paths

$$\int_z^{a_1}|R_1(t)\,\mathrm{d}t| < r_1, \quad \int_z^{a_1}|R_2(t)\,\mathrm{d}t| < r_2, \tag{10·19}$$

where r_1, r_2 are assignable constants independent of z and, of course, u. Substituting these inequalities in (10·17) and absorbing the factor $r_1+u^{-\frac{1}{2}}r_2$ in the generic constant A, we obtain

$$|h_1(z)| < A\,\mathrm{e}^{-\tau\sqrt{u}}(1+|vz|^{\frac{1}{4}})^{-1}|\exp\{-\frac{2}{3}(vz)^{\frac{3}{2}}\}|. \tag{10·20}$$

Next, from (10·12), we obtain

$$h_{n+1}(z)-h_n(z) = \frac{1}{v}\exp\{-\frac{2}{3}(vz)^{\frac{3}{2}}\}\int_z^{a_1}\chi(z,v,t)f(t)\frac{\{h_n(t)-h_{n-1}(t)\}}{\exp\{-\frac{2}{3}(vt)^{\frac{3}{2}}\}}\,\mathrm{d}t, \tag{10·21}$$

provided $n \geqslant 1$, where

$$\chi(z,v,t) = 2\pi\,\mathrm{e}^{\frac{1}{6}\pi\mathrm{i}}\{P_1(vz)\,P_2(vt)-P_2(vz)\,P_1(vt)\}\exp\{\frac{2}{3}(vz)^{\frac{3}{2}}-\frac{2}{3}(vt)^{\frac{3}{2}}\}.$$

Using (4·5) and (10·15) we readily show (cf. (10·16)) that

$$|\chi(z,v,t)| < A(1+|vz|^{\frac{1}{4}})^{-1}. \tag{10·22}$$

Let us assume temporarily that

$$|h_n(z)-h_{n-1}(z)| < k_n(v)\,\mathrm{e}^{-\tau\sqrt{u}}(1+|vz|^{\frac{1}{4}})^{-1}|\exp\{-\frac{2}{3}(vz)^{\frac{3}{2}}\}|,$$

where $k_n(v)$ is independent of z. Then from (10·21) and (10·22) we obtain

$$|h_{n+1}(z)-h_n(z)| < \frac{A}{v}k_n(v)\,\mathrm{e}^{-\tau\sqrt{u}}\frac{|\exp\{-\frac{2}{3}(vz)^{\frac{3}{2}}\}|}{1+|vz|^{\frac{1}{4}}}\int_z^{a_1}|f(t)\,\mathrm{d}t|.$$

Now from (5·1) we have $f(z) = O(|z|^{-2})$ as $|z| \to \infty$, and so $\int_z^{a_1} |f(t)\,dt|$ taken along the admissible paths of integration is a bounded function of z (cf. (10·19)), and may be absorbed in the constant A. Hence, by induction using (10·20), we deduce that

$$|h_{n+1}(z) - h_n(z)| < A\left(\frac{A}{v}\right)^n e^{-\tau\sqrt{u}} \frac{|\exp\{-\frac{2}{3}(vz)^{\frac{3}{2}}\}|}{1+|vz|^{\frac{1}{4}}}. \tag{10·23}$$

Thus when $v > A$, the series

$$\sum_{n=0}^{\infty} \{h_{n+1}(z) - h_n(z)\}$$

converges uniformly with respect to z. From (10·8), (10·9), (10·10) and (10·11) it follows that its sum is $W_1(z) - L_1(z)$. Moreover, if $v \geq 2A$, we have

$$|W_1(z) - L_1(z)| = \left|\sum_{n=0}^{\infty} \{h_{n+1}(z) - h_n(z)\}\right| < A\,e^{-\tau\sqrt{u}} \frac{|\exp\{-\frac{2}{3}(vz)^{\frac{3}{2}}\}|}{1+|vz|^{\frac{1}{4}}} \sum_{n=0}^{\infty} \frac{1}{2^n}$$

$$= 2A\,e^{-\tau\sqrt{u}} \frac{|\exp\{-\frac{2}{3}(vz)^{\frac{3}{2}}\}|}{1+|vz|^{\frac{1}{4}}}, \tag{10·24}$$

and so for large u, $W_1(z) - L_1(z)$ is asymptotically negligible compared with the error term in (5·12). This completes the proof of (5·12) for $j = 1$ when z lies in the part of $\mathbf{D}_1$ common to the sector $\mathbf{S}_1 + \mathbf{S}_2$. A similar proof holds when z lies in the part of $\mathbf{D}_1$ common to $\mathbf{S}_1 + \mathbf{S}_3$, the complementary solutions of (10·10) being taken to be $P_1(vz)$ and $P_3(vz)$, in place of $P_1(vz)$ and $P_2(vz)$.

For the derivative we derive, from (10·12) and (10·6),

$$h_1'(z) = v\,e^{-\tau\sqrt{u}} \exp\{-\tfrac{2}{3}(vz)^{\frac{3}{2}}\} \int_z^{a_1} \psi_1(z, v, t)\,\{R_1(t) + u^{-\frac{1}{2}} R_2(t)\}\,dt, \tag{10·25}$$

and

$$h_{n+1}'(z) - h_n'(z) = \exp\{-\tfrac{2}{3}(vz)^{\frac{3}{2}}\} \int_z^{a_1} \chi_1(z, v, t)\,f(t)\,\frac{\{h_n(t) - h_{n-1}(t)\}}{\exp\{-\frac{2}{3}(vt)^{\frac{3}{2}}\}}\,dt, \tag{10·26}$$

if $n \geq 1$, where

$$\psi_1(z, v, t) = 4\pi\,e^{\frac{1}{6}\pi i} \{P_1'(vz)\,P_2(vt) - P_2'(vz)\,P_1(vt)\}\,P_1'(vt) \exp\{\tfrac{2}{3}(vz)^{\frac{3}{2}}\},$$

$$\chi_1(z, v, t) = 2\pi\,e^{\frac{1}{6}\pi i} \{P_1'(vz)\,P_2(vt) - P_2'(vz)\,P_1(vt)\} \exp\{\tfrac{2}{3}(vz)^{\frac{3}{2}} - \tfrac{2}{3}(vt)^{\frac{3}{2}}\}.$$

We readily show (cf. (10·16) and (10·22)) that

$$|\psi_1(z, v, t)|,\ |\chi_1(z, v, t)| < A(1 + |vz|^{\frac{1}{2}}), \tag{10·27}$$

and thence, using also (10·19) and (10·23), that

$$|h_1'(z)| < Av\,e^{-\tau\sqrt{u}}\,(1 + |vz|^{\frac{1}{4}})\,|\exp\{-\tfrac{2}{3}(vz)^{\frac{3}{2}}\}|,$$

$$|h_{n+1}'(z) - h_n'(z)| < A(A/v)^n\,v\,e^{-\tau\sqrt{u}}\,(1 + |vz|^{\frac{1}{4}})\,|\exp\{-\tfrac{2}{3}(vz)^{\frac{3}{2}}\}|.$$

Thus when $v \geq 2A$

$$|W_1'(z) - L_1'(z)| = \left|\sum_{n=0}^{\infty} \{h_{n+1}'(z) - h_n'(z)\}\right| < 2Av\,e^{-\tau\sqrt{u}}\,(1 + |vz|^{\frac{1}{4}})\,|\exp\{-\tfrac{2}{3}(vz)^{\frac{3}{2}}\}|. \tag{10·28}$$

This result in conjunction with (10·4) establishes (5·13) when z lies in $\mathbf{D}_1$ and $\mathbf{S}_1 + \mathbf{S}_2$. A similar proof holds when z lies in $\mathbf{D}_1$ and $\mathbf{S}_1 + \mathbf{S}_3$.

This completes the proof of theorem B when the arbitrary constants in the second of (5·8) are prescribed by (9·3).

11. Existence of solutions when $A_1(c)$, $A_2(c)$, ..., do not vanish

Let $\phi(u)$ be an arbitrary function of u with the property

$$\phi(u) \sim 1 + \sum_{s=1}^{\infty} \frac{\phi_s}{u^s},$$

as $u \to \infty$, where $\phi_1, \phi_2, \ldots$, are constants. Then $\phi(u)\, W_j(z)$ is a solution of (5·11), and from (5·14) and (5·15) we see that the asymptotic expansions of this function and its derivative for large u are given by

$$\phi(u)\, W_j(z) \sim P_j(vz) \sum_{s=0}^{\infty} \frac{A_s^*(z)}{u^s} + \frac{P_j'(vz)}{v^2} \sum_{s=0}^{\infty} \frac{B_s^*(z)}{u^s},$$

$$\phi(u)\, W_j'(z) \sim P_j(vz) \sum_{s=0}^{\infty} \frac{C_s^*(z)}{u^s} + vP_j'(vz) \sum_{s=0}^{\infty} \frac{D_s^*(z)}{u^s},$$

where

$$A_s^* = A_s + \phi_1 A_{s-1} + \ldots + \phi_{s-1} A_1 + \phi_s, \qquad B_s^* = B_s + \phi_1 B_{s-1} + \ldots + \phi_{s-1} B_1 + \phi_s B_0,$$

$$C_s^* = C_s + \phi_1 C_{s-1} + \ldots + \phi_{s-1} C_1 + \phi_s C_0, \quad D_s^* = D_s + \phi_1 D_{s-1} + \ldots + \phi_{s-1} D_1 + \phi_s.$$

With the aid of (3·6) and (3·9) with $g(z) = z$, we may immediately verify that A_s^*, B_s^*, C_s^* and D_s^* satisfy the same recurrence relations as A_s, B_s, C_s and D_s. Moreover, using (9·3), we obtain

$$A_s^*(c) = \phi_s.$$

In other words, dropping the stars, theorem B remains true if the arbitrary constants in the second of (5·8) have any prescribed values, provided there exists a function $\phi(u)$ with the property

$$\phi(u) \sim 1 + \sum_{s=1}^{\infty} \frac{A_s(c)}{u^s}$$

for some point c of $\mathbf{D}'$. This is the result stated in § 5.

The work described above has been carried out as part of the research programme of the National Physical Laboratory, and this paper is published by permission of the Director of the Laboratory.

References

Birkhoff, G. D. 1908 *Trans. Amer. Math. Soc.* **9**, 219–231.
British Association Mathematical Tables 1946 *The Airy integral.* Cambridge University Press.
California Institute of Technology 1953 *Tech. rep. No.* 1. *Asymptotic solutions of differential equations with turning points.*
Cherry, T. M. 1950 *Trans. Amer. Math. Soc.* **68**, 224–257.
Green, G. 1837 *Trans. Camb. Phil. Soc.* **6**, 457–462.
Horn, J. 1899 *Math. Ann.* **52**, 271–292.
Ince, E. L. 1944 *Ordinary differential equations.* New York: Dover Publications.
Jeffreys, H. 1924 *Proc. Lond. Math. Soc.* **23**, 428–436.
Jeffreys, H. 1953 *Proc. Camb. Phil. Soc.* **49**, 601–611.
Jeffreys, H. & Jeffreys, B. S. 1946 *Methods of mathematical physics.* Cambridge University Press.
Langer, R. E. 1931 *Trans. Amer. Math. Soc.* **33**, 23–64.
Langer, R. E. 1932 *Trans. Amer. Math. Soc.* **34**, 447–480.
Langer, R. E. 1934 *Trans. Amer. Math. Soc.* **36**, 90–106.
Langer, R. E. 1935 *Trans. Amer. Math. Soc.* **37**, 397–416.
Langer, R. E. 1937 *Phys. Rev.* **51**, 669–676.
Langer, R. E. 1949 *Trans. Amer. Math. Soc.* **67**, 461–490.
Liouville, J. 1837 *J. Math. pures appl.* **2**, 16–35.
Miller, J. C. P. 1950 *Quart. J. Mech.* **3**, 225–235.
Schlesinger, L. 1907 *Math. Ann.* **63**, 277–300.
Tamarkin, J. 1928 *Math. Z.* **27**, 1–54.
Turrittin, H. L. 1936 *Amer. J. Math.* **58**, 364–376.
Watson, G. N. 1944 *Theory of Bessel functions.* Cambridge University Press.

[328]

THE ASYMPTOTIC EXPANSION OF BESSEL FUNCTIONS OF LARGE ORDER

By F. W. J. OLVER, *The National Physical Laboratory*

(*Communicated by Sir Edward Bullard, F.R.S.—Received* 22 *February* 1954)

CONTENTS

New expansions are obtained for the functions $I_\nu(\nu z)$, $K_\nu(\nu z)$ and their derivatives in terms of elementary functions, and for the functions $J_\nu(\nu z)$, $Y_\nu(\nu z)$, $H^{(1)}_\nu(\nu z)$, $H^{(2)}_\nu(\nu z)$ and their derivatives in terms of Airy functions, which are uniformly valid with respect to z when $|\nu|$ is large. New series for the zeros and associated values are derived by reversion and used to determine the distribution of the zeros of functions of large order in the z-plane. Particular attention is paid to the complex zeros of $Y_n(z)$ and the Hankel functions when the order n is an integer or half an odd integer, and for this purpose some new asymptotic expansions of the Airy functions are derived. Tables are given of complex zeros of Airy functions and other quantities which facilitate the rapid calculation of the smaller complex zeros of $Y_n(z)$, $Y'_n(z)$, and the Hankel functions and their derivatives, when $2n$ is an integer, to an accuracy of three or four significant figures.

1. Introduction and summary

The theory of the preceding paper (Olver 1954) may be applied to Bessel's differential equation to produce asymptotic expansions of the functions $J_\nu(\nu z)$, $Y_\nu(\nu z)$, $H^{(1)}_\nu(\nu z)$, $H^{(2)}_\nu(\nu z)$ and their derivatives in terms of Airy functions, which are uniformly valid with respect to z when $|\nu|$ is large. In this paper these expansions are considered in detail, together with the expansions of the zeros and associated values which are obtained from them by reversion. Similar expansions for the functions (but not for the derivatives or zeros) have also been given in a recent paper by Cherry (1950).

Other forms of expansion of Bessel functions of large order are those of Debye and Meissel (see, for example, Watson 1944, pp. 235–248) and the present writer (1952), which are valid when $|z-1|\gg|\nu|^{-\frac{2}{3}}$, $|z-1|\ll|\nu|^{-\frac{2}{3}}$ and $|z-1|\ll|\nu|^{-\frac{1}{3}}$, respectively. Because of the property of uniformity there are no such restrictions on the expansions of this paper, which are, moreover, more powerful than the earlier forms, particularly near the transition points $z = \pm 1$. In consequence, the new expansions are of considerable computational importance, and to facilitate their application comprehensive tables of the coefficients have been prepared for real arguments and orders and will be published elsewhere.

The investigation of the zeros of functions of large order has been rather difficult with the existing expansions for the functions. The only explicit series which have been given are those of McMahon (Watson 1944, pp. 505–506; also Bickley & Miller 1945) for the very large zeros, and those of the present writer (1951, 1952) which give the expansion of real zeros of fixed enumeration in the form

$$n + \alpha_1 n^{\frac{1}{3}} + \alpha_2 n^{-\frac{1}{3}} + \dots, \tag{1·1}$$

where† $n \equiv \nu$ and $\alpha_1, \alpha_2, \dots$ are numerical constants. The values of these constants increase for the larger zeros, and expansions of this form are useful only for the smaller zeros. Other series could be obtained by reversion of Debye's expansions, but they would not be valid for indefinitely large orders if the enumeration is kept fixed.

The series given below do not have these disadvantages. The asymptotic expansion of a zero of fixed enumeration is evolved in the form

$$\nu p_0(\nu^{-\frac{2}{3}}a) + \nu^{-1} p_1(\nu^{-\frac{2}{3}}a) + \nu^{-3} p_2(\nu^{-\frac{2}{3}}a) + \dots, \tag{1·2}$$

where a is the corresponding Airy function zero, and the coefficients $p_0(\zeta), p_1(\zeta), \dots$, are transcendental functions of the argument $\zeta \equiv \nu^{-\frac{2}{3}}a$ which may be pretabulated. The expansion (1·2) is uniformly valid with respect to all the zeros and numerically it is very powerful; it is effectively a power series in ν^{-2} in which the early coefficients actually decrease in magnitude. In addition to numerical applications the new series are of theoretical value in determining the asymptotic distribution of complex zeros of Bessel functions of large orders, both real and complex, and lead to some new and interesting results. It is found, for example, that if n is a positive integer, then in the domain $|\arg z| \leqslant \pi$ the function $Y_n(nz)$ has, in addition to its infinite set of real positive zeros, two infinite strings of zeros asymptotically near to the negative real axis, together with $2n$ zeros asymptotically near to the boundary of an eye-shaped domain in the z-plane whose extreme points are at $z = \pm 1$.

For completeness an account is also given of the analogous asymptotic expansions of the modified Bessel functions $I_\nu(\nu z)$ and $K_\nu(\nu z)$. These expansions, which are in terms of exponential and not Airy functions, may be obtained from the Debye expansions for $J_\nu(\nu z)$ and $H_\nu^{(1)}(\nu z)$ for complex arguments and orders (Watson 1944, pp. 262–268), but in this paper they will be derived from the defining differential equation. The only previous account of this approach appears to be that of Lehmer (1945) with corrections by Miller (1945), which deals briefly with $I_n(z)$ for real arguments and orders. The modified Bessel functions are considered first because their asymptotic theory is simpler than that of the Bessel functions themselves.

The arrangement of the paper is as follows. The expansions of $I_n(nz)$, $K_n(nz)$ and their derivatives for large positive orders are given in § 2. The extension of these results to complex orders is considered in § 3.

† Here and elsewhere in this paper we use the convention that when the symbol n is used to denote the order of the Bessel functions it is real and positive.

In § 4 expansions are developed for $J_n(nz)$, $Y_n(nz)$, $H_n^{(1)}(nz)$ and $H_n^{(2)}(nz)$ when the order is large and positive, and in § 5 these results are extended to include complex orders. The coefficients in these expansions are defined by recurrence formulae, and explicit expressions for them are obtained in § 6. This section also contains expansions for the derivatives of the Bessel functions.

Zeros are investigated in §§ 7, 8, 9 and 10. The asymptotic expansions of §§ 4, 5 and 6 are reverted in § 7 to give series for the zeros of $J_\nu(z)$ and $J'_\nu(z)$ and the respective associated values of $J'_\nu(z)$ and $J_\nu(z)$. The corresponding expansions for $Y_n(z)$ and $Y'_n(z)$ are considered in § 8, with particular reference to the complex zeros of these functions when $2n$ is an integer, and a similar investigation of the zeros of the Hankel functions and their derivatives is made in § 9. In § 10 tables are given which facilitate rapid calculation of approximate values of the complex zeros of $Y_n(z)$, $Y'_n(z)$, the Hankel functions and their derivatives, and examples are given of their application.

Certain properties of the Airy functions Ai and Bi are frequently used in the paper. Since they do not all appear to be well known, particularly in the case of Bi, they have been collected together and given in the Appendix.

2. The expansions of $I_n(nz)$ and $K_n(nz)$ for large positive orders

The functions $z^{\frac{1}{2}}I_n(nz)$ and $z^{\frac{1}{2}}K_n(nz)$ satisfy the differential equation

$$\frac{d^2w}{dz^2} = \left\{\left(\frac{1+z^2}{z^2}\right)n^2 - \frac{1}{4z^2}\right\}w. \tag{2·1}$$

This equation has transition points at $z = 0, \pm i$. We shall suppose throughout this section that z is confined to the half-plane $|\arg z| < \frac{1}{2}\pi$; the results obtained may always be extended subsequently to other phase ranges not including the imaginary axis† with the aid of the continuation formulae

$$\left.\begin{aligned} I_\nu(z\,e^{m\pi i}) &= e^{m\nu\pi i}\,I_\nu(z), \\ K_\nu(z\,e^{m\pi i}) &= e^{-m\nu\pi i}\,K_\nu(z) - \pi i \sin m\nu\pi \operatorname{cosec} \nu\pi\, I_\nu(z), \end{aligned}\right\} \tag{2·2}$$

m being an arbitrary integer.

Following the procedure described in the preceding paper (§ 2, cases A and C), we make simultaneous changes of variables in (2·1) from w, z to W, ζ respectively, given by

$$\left(\frac{d\zeta}{dz}\right)^2 = \frac{1+z^2}{z^2}, \qquad W = \left(\frac{dz}{d\zeta}\right)^{-\frac{1}{2}} w = \left(\frac{1+z^2}{z^2}\right)^{\frac{1}{4}} w. \tag{2·3}$$

Then W satisfies the equation

$$\frac{d^2W}{d\zeta^2} = \{n^2 + f(\zeta)\}\,W, \tag{2·4}$$

where

$$f(\zeta) = -\frac{\dot{z}^2}{4z^2} + \dot{z}^{\frac{1}{2}}\frac{d^2}{d\zeta^2}(\dot{z}^{-\frac{1}{2}}), \qquad \dot{z} \equiv \frac{dz}{d\zeta}. \tag{2·5}$$

† Expansions which include the imaginary axis in their region of validity are of a different kind; they may be obtained from the results given later in §§ 4 and 5 by means of the relations

$$I_\nu(z) = e^{-\frac{1}{2}\nu\pi i}\,J_\nu(ze^{\frac{1}{2}\pi i}), \quad K_\nu(z) = \tfrac{1}{2}\pi i\, e^{\frac{1}{2}\nu\pi i}\,H_\nu^{(1)}(ze^{\frac{1}{2}\pi i}).$$

The first of equations (2·3) may be integrated to give

$$\zeta = \surd(1+z^2)+\ln\frac{z}{1+\surd(1+z^2)}, \tag{2·6}$$

it being convenient to take the arbitrary constant of integration equal to zero. This relation maps the half-plane $|\arg z|<\frac{1}{2}\pi$ conformally on the domain **D** comprising the half-plane $\mathscr{R}\zeta>0$ and the half-strip $|\mathscr{I}\zeta|<\frac{1}{2}\pi$, $\mathscr{R}\zeta\leqslant 0$. Corresponding points of the transformation are indicated in figures 1 and 2.

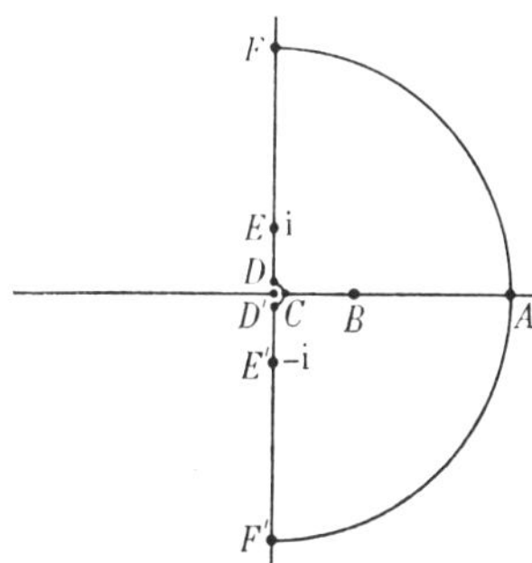

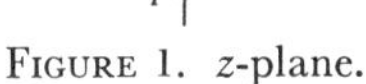

FIGURE 1. z-plane.

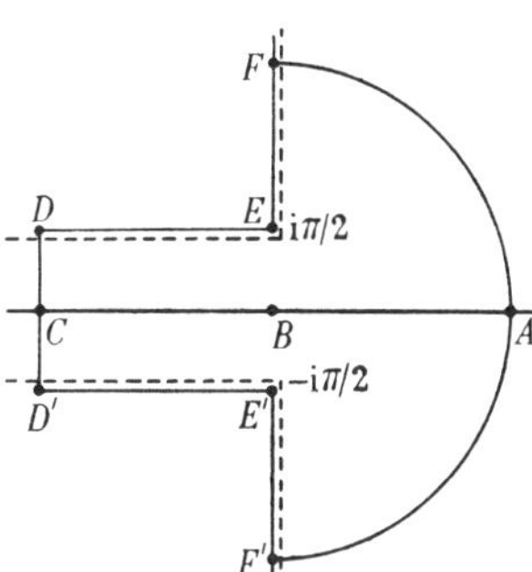

FIGURE 2. ζ-plane.

Substituting the first of (2·3) in (2·5) we find, after a little reduction, that

$$f(\zeta) = \frac{z^2(4-z^2)}{4(1+z^2)^3}. \tag{2·7}$$

Thus $f(\zeta)$ is a regular (holomorphic) function of ζ in **D**. Moreover, from (2·6) it is seen that $z\sim\zeta$ as $|\zeta|\to\infty$ in the right-hand half-plane, and if ζ lies in the strip $|\mathscr{I}\zeta|<\frac{1}{2}\pi$, $\mathscr{R}\zeta\leqslant 0$, then $z\sim 2\,\mathrm{e}^{\zeta-1}$ as $|\zeta|\to\infty$. Hence

$$f(\zeta) = O(|\zeta|^{-2}) \tag{2·8}$$

as $|\zeta|\to\infty$ in **D**, uniformly with respect to $\arg\zeta$.

Thus the preliminary conditions are satisfied for the application of theorem A† to equation (2·4). Let **D**′ denote the part of **D** bounded by the lines

$$\mathscr{I}\zeta = \pm(\tfrac{1}{2}\pi-\delta),\quad \mathscr{R}\zeta\leqslant\delta; \qquad \mathscr{R}\zeta=\delta,\quad |\mathscr{I}\zeta|\geqslant\tfrac{1}{2}\pi-\delta,$$

where $0<\delta\leqslant\frac{1}{2}\pi$; these boundaries are indicated by the broken lines in figure 2. Then by taking a_1, a_2 to be the points at infinity on the negative and positive real axes respectively, the domains $\mathbf{D}_1$ and $\mathbf{D}_2$ of theorem A coincide with **D**′ itself. Accordingly, if a sequence of coefficients‡ $\{U_s(\zeta)\}$ is defined by the equations $U_0(\zeta)=1$ and

$$U_{s+1}(\zeta) = -\tfrac{1}{2}U_s'(\zeta)+\tfrac{1}{2}\int f(\zeta)\,U_s(\zeta)\,\mathrm{d}\zeta \quad (s\geqslant 0), \tag{2·9}$$

then solutions $W_1(\zeta)$ and $W_2(\zeta)$ of (2·4) exist such that, if ζ lies in **D**′,

$$W_1(\zeta)\sim \mathrm{e}^{n\zeta}\sum_{s=0}^{\infty}\frac{U_s(\zeta)}{n^s}, \qquad W_2(\zeta)\sim \mathrm{e}^{-n\zeta}\sum_{s=0}^{\infty}(-)^s\frac{U_s(\zeta)}{n^s}, \tag{2·10}$$

as $n\to\infty$, uniformly with respect to ζ.

† Here and elsewhere the terms 'theorem A' and 'theorem B' refer to the theorems of the preceding paper, §5.

‡ The notation A_s of theorem A has been changed here to U_s for later convenience.

We shall now express $I_n(nz)$ in terms of $W_1(\zeta)$ and $W_2(\zeta)$. From the second of (2·3) it follows that

$$I_n(nz) = \lambda_n(1+z^2)^{-\frac{1}{4}} W_1(\zeta) + \mu_n(1+z^2)^{-\frac{1}{4}} W_2(\zeta), \tag{2·11}$$

where λ_n, μ_n are independent of z. Because of the uniform property of the expansions (2·10), we may keep n fixed and let $z \to 0$ through positive values, or, equivalently, $\zeta \to -\infty$. It is then immediately obvious that $\mu_n = 0$ for all sufficiently large n. Next, substituting (2·10) in (2·11), we see that, for large n,

$$\lambda_n \sim e^{-n\zeta}(1+z^2)^{\frac{1}{4}} I_n(nz) \div \left\{\sum_{s=0}^{\infty} \frac{U_s(\zeta)}{n^s}\right\}.$$

Now as $z \to +\infty$, n being fixed, we have from (2·6) and the well-known asymptotic expansion of $I_n(z)$ for large z,

$$\zeta = z + O(z^{-1}), \qquad I_n(nz) \sim (2\pi nz)^{-\frac{1}{2}} e^{nz},$$

and so

$$e^{-n\zeta}(1+z^2)^{\frac{1}{4}} I_n(nz) \to (2\pi n)^{-\frac{1}{2}}$$

as $z \to +\infty$. Hence if we now fix the arbitrary constants in (2·9) by making $U_{s+1}(+\infty) = 0$ $(s \geq 0)$, so that this relation becomes

$$U_{s+1}(\zeta) = -\tfrac{1}{2}U_s'(\zeta) - \tfrac{1}{2}\int_{\zeta}^{\infty} f(t)\, U_s(t)\, dt, \tag{2·12}$$

then λ_n can be replaced by $(2\pi n)^{-\frac{1}{2}}$ in equation (2·11) provided that the sign of equality is replaced by asymptotic equality. Thus we derive the desired result

$$I_n(nz) \sim \frac{e^{n\zeta}}{\sqrt{(2\pi n)}\,(1+z^2)^{\frac{1}{4}}} \sum_{s=0}^{\infty} \frac{U_s(\zeta)}{n^s} \tag{2·13}$$

as $n \to \infty$, uniformly with respect to z in the half-plane $|\arg z| \leq \frac{1}{2}\pi - \epsilon$, where ϵ is an arbitrary positive number in the range $0 < \epsilon < \frac{1}{2}\pi$.

The corresponding expansion for $K_n(nz)$ may be derived in a similar manner, using the relation

$$K_n(nz) \sim \left(\frac{\pi}{2nz}\right)^{\frac{1}{2}} e^{-nz}, \quad \text{as} \quad z \to +\infty.$$

The result is given by

$$K_n(nz) \sim \sqrt{\left(\frac{\pi}{2n}\right)} \frac{e^{-n\zeta}}{(1+z^2)^{\frac{1}{4}}} \sum_{s=0}^{\infty} (-)^s \frac{U_s(\zeta)}{n^s}, \tag{2·14}$$

as $n \to \infty$, uniformly with respect to z in $|\arg z| \leq \frac{1}{2}\pi - \epsilon$.

The linking of $I_n(nz)$ and $K_n(nz)$ with $W_1(\zeta)$ and $W_2(\zeta)$ may alternatively be effected by using the known behaviour of the functions at $z = 0$ in place of $z = +\infty$. This procedure leads to the result

$$U_s(-\infty) = \gamma_s, \tag{2·15}$$

where $\gamma_0, \gamma_1, \gamma_2, \ldots$ are the coefficients in the asymptotic expansion

$$\frac{\sqrt{(2\pi)}\, n^{n-\frac{1}{2}}}{e^n\, \Gamma(n)} \sim \gamma_0 + \frac{\gamma_1}{n} + \frac{\gamma_2}{n^2} + \ldots. \tag{2·16}$$

The first seven are given by

$$\left.\begin{aligned} &\gamma_0 = 1, \quad \gamma_1 = -\frac{1}{12}, \quad \gamma_2 = \frac{1}{288}, \quad \gamma_3 = \frac{139}{51840}, \\ &\gamma_4 = -\frac{571}{24\,88320}, \quad \gamma_5 = -\frac{1\,63879}{2090\,18880}, \quad \gamma_6 = \frac{52\,46819}{7\,52467\,96800}. \end{aligned}\right\} \tag{2·17}$$

The coefficients $U_s(\zeta)$ can be evaluated explicitly in terms of z. In order to arrive at a convenient practical form we introduce the auxiliary variable

$$u \equiv (1+z^2)^{-\frac{1}{2}}.$$

Then, using (2·3) and (2·7), we obtain

$$\frac{du}{d\zeta} = -u^2(1-u^2), \qquad f(\zeta) = \tfrac{1}{4}u^2(1-u^2)(5u^2-1), \tag{2·18}$$

and the recurrence relation (2·12) becomes

$$U_{s+1} = \tfrac{1}{2}u^2(1-u^2)\frac{dU_s}{du} + \tfrac{1}{8}\int_0^u (1-5u^2)\, U_s\, du. \tag{2·19}$$

From this result it is seen that U_s is a polynomial in u. For $s = 0, 1, 2, 3$, we have

$$\left.\begin{aligned} &U_0 = 1, \quad U_1 = \tfrac{1}{8}u - \tfrac{5}{24}u^3, \quad U_2 = \tfrac{9}{128}u^2 - \tfrac{77}{192}u^4 + \tfrac{385}{1152}u^6, \\ &U_3 = \tfrac{75}{1024}u^3 - \tfrac{4563}{5120}u^5 + \tfrac{17017}{9216}u^7 - \tfrac{85085}{82944}u^9. \end{aligned}\right\} \tag{2·20}$$

When $u = 1$ we see from (2·15) that $U_s = \gamma_s$, and this affords a useful check on the coefficients in (2·20).

The asymptotic expansions of the derivatives of the modified Bessel functions may be obtained by differentiating the expansions (2·13) and (2·14) with respect to z, the legitimacy of this process being a consequence of theorem A. Using (2·18) the expansions are found to be

$$I_n'(nz) \sim \frac{(1+z^2)^{\frac{1}{4}}}{z} \frac{e^{n\zeta}}{\sqrt{(2\pi n)}} \sum_{s=0}^{\infty} \frac{V_s}{n^s}, \qquad K_n'(nz) \sim -\sqrt{\left(\frac{\pi}{2n}\right)} \frac{(1+z^2)^{\frac{1}{4}}}{z} e^{-n\zeta} \sum_{s=0}^{\infty} (-)^s \frac{V_s}{n^s}, \tag{2·21}$$

where the coefficients V_s are polynomials in u, given by

$$V_s = U_s - u(1-u^2)\left(\tfrac{1}{2}U_{s-1} + u\frac{dU_{s-1}}{du}\right). \tag{2·22}$$

The first four are

$$\left.\begin{aligned} &V_0 = 1, \quad V_1 = -\tfrac{3}{8}u + \tfrac{7}{24}u^3, \quad V_2 = -\tfrac{15}{128}u^2 + \tfrac{99}{192}u^4 - \tfrac{455}{1152}u^6, \\ &V_3 = -\tfrac{105}{1024}u^3 + \tfrac{5577}{5120}u^5 - \tfrac{6545}{3072}u^7 + \tfrac{95095}{82944}u^9. \end{aligned}\right\} \tag{2·23}$$

They take the same values as the corresponding U_s at $u = 1$. Another check is given by

$$U_{2s}V_0 - U_{2s-1}V_1 + U_{2s-2}V_2 - \ldots + U_0 V_{2s} = 0 \quad (s \geqslant 1), \tag{2·24}$$

which is derived from the Wronskian relation for $I_n(nz)$ and $K_n(nz)$.

3. The expansion of $I_\nu(\nu z)$ and $K_\nu(\nu z)$ for large complex orders

Let $\nu = n\,e^{i\vartheta}$, where ϑ is real and fixed and n is large and positive. The functions

$$(1+z^2)^{\frac{1}{4}} I_\nu(\nu z) \quad \text{and} \quad (1+z^2)^{\frac{1}{4}} K_\nu(\nu z)$$

satisfy the differential equation

$$\frac{d^2W}{d\zeta^2} = \{n^2 e^{2i\vartheta} + f(\zeta)\}\, W \tag{3·1}$$

(cf. (2·4)), where ζ is given by (2·6). With the substitution $\xi = \zeta e^{i\vartheta}$, this becomes

$$\frac{d^2W}{d\xi^2} = \{n^2 + e^{-2i\vartheta} f(\xi\, e^{-i\vartheta})\}\, W. \tag{3·2}$$

If z lies in the half-plane $|\arg z|<\frac{1}{2}\pi$, then ξ lies in a domain **E** which is the domain **D** of § 2 rotated through an angle ϑ. Let **E**′ be the domain in the ξ-plane corresponding to **D**′ in the ζ-plane. By applying theorem A to (3·2), we find that solutions W_1, W_2 exist such that

$$W_1 \sim e^{n\xi} \sum_{s=0}^{\infty} \frac{u_s(\xi)}{n^s}, \qquad W_2 \sim e^{-n\xi} \sum_{s=0}^{\infty} (-)^s \frac{u_s(\xi)}{n^s}, \tag{3·3}$$

as $n\to\infty$, uniformly with respect to ξ in **E**′. The coefficients $u_s(\xi)$ are given by $u_0(\xi)=1$ and

$$u_{s+1}(\xi) = -\tfrac{1}{2}u_s'(\xi) - \tfrac{1}{2}\int_{\xi}^{\infty e^{i\theta}} e^{-2i\vartheta} f(t\,e^{-i\vartheta})\, u_s(t)\, dt.$$

The function $(1+z^2)^{\frac{1}{4}} I_\nu(\nu z)$ may be expressed in terms of these solutions in a manner similar to that of § 2, using the relations

$$I_\nu(\nu z) \sim \frac{(\frac{1}{2}\nu z)^\nu}{\Gamma(\nu+1)} \quad \text{as} \quad z\to 0, \qquad I_\nu(\nu z) \sim \frac{e^{\nu z}}{(2\pi\nu z)^{\frac{1}{2}}} \quad \text{as} \quad z\to+\infty. \tag{3·4}$$

The first of these is valid for all ν other than a negative integer, and the second holds when $|\arg \nu|<\frac{1}{2}\pi$. We find

$$I_\nu(\nu z) \sim \frac{e^{n\xi}}{\surd(2\pi\nu)\,(1+z^2)^{\frac{1}{4}}} \sum_{s=0}^{\infty} \frac{u_s(\xi)}{n^s}, \tag{3·5}$$

as $n\to\infty$, provided $|\vartheta|<\frac{1}{2}\pi$.

If the original variables ζ and ν are restored in place of n and ξ, it is seen that

$$u_s(\xi) = e^{-si\vartheta}\, U_s(\zeta),$$

where $U_s(\zeta)$ is defined by (2·12), and (3·5) becomes

$$I_\nu(\nu z) \sim \frac{e^{\nu\zeta}}{\surd(2\pi\nu)\,(1+z^2)^{\frac{1}{4}}} \sum_{s=0}^{\infty} \frac{U_s(\zeta)}{\nu^s}. \tag{3·6}$$

In other words, (2·13) *remains valid if n is replaced by the complex variable ν, provided that* $|\arg\nu|<\frac{1}{2}\pi$. *We may, in a similar manner, prove that the same is true of* (2·14).

When $\arg\nu$ lies outside the range $|\arg\nu|<\frac{1}{2}\pi$, the corresponding expansions may be deduced from (3·6) and the analogous expansion for $K_\nu(\nu z)$, by means of (2·2) and the formulae

$$I_{-\nu}(z) = I_\nu(z) + 2\pi^{-1}\sin\nu\pi K_\nu(z), \qquad K_{-\nu}(z) = K_\nu(z). \tag{3·7}$$

In general they are mixtures of both positive and negative exponential-type solutions.

4. The expansion of $J_n(nz)$, $Y_n(nz)$ and the Hankel functions for large positive orders

The functions $z^{\frac{1}{2}}J_n(nz)$ and $z^{\frac{1}{2}}Y_n(nz)$ satisfy

$$\frac{d^2w}{dz^2} = \left\{\left(\frac{1-z^2}{z^2}\right)n^2 - \frac{1}{4z^2}\right\} w. \tag{4·1}$$

This equation has transition points at $z=0, \pm 1$. Taking $z_0=1$ and following the procedure described in the preceding paper (§ 2, case B), we introduce new variables W, ζ given by

$$\zeta\left(\frac{d\zeta}{dz}\right)^2 = \frac{1-z^2}{z^2}, \qquad W = \left(\frac{dz}{d\zeta}\right)^{-\frac{1}{2}} w. \tag{4·2}$$

Then W satisfies the equation

$$\frac{d^2W}{d\zeta^2} = \{n^2\zeta + f(\zeta)\}\,W, \tag{4·3}$$

where

$$f(\zeta) = -\frac{\dot{z}^2}{4z^2} + \dot{z}^{\frac{1}{2}}\frac{d^2}{d\zeta^2}(\dot{z}^{-\frac{1}{2}}), \qquad \dot{z} \equiv \frac{dz}{d\zeta}. \tag{4·4}$$

From the first of (4·2) we obtain

$$\tfrac{2}{3}\zeta^{\frac{3}{2}} = \pm\int_1^z \frac{\sqrt{(1-z^2)}}{z}\,dz. \tag{4·5}$$

We adopt the lower sign here to make ζ real when z is real, and carrying out the integration we find that

$$\tfrac{2}{3}\zeta^{\frac{3}{2}} = \ln\frac{1+\sqrt{(1-z^2)}}{z} - \sqrt{(1-z^2)}. \tag{4·6}$$

In order to see how the z-plane is mapped on the ζ-plane it is helpful to introduce auxiliary variables σ and ρ, related to z and ζ by the equations

$$z = \operatorname{sech}\sigma, \qquad \tfrac{2}{3}\zeta^{\frac{3}{2}} = \rho = \sigma - \tanh\sigma. \tag{4·7}$$

The sector $0 < \arg z < \pi$ is mapped conformally on the half-strip $-\pi < \mathscr{I}\sigma < 0$, $\mathscr{R}\sigma > 0$, and on a domain in the ρ-plane comprising the half-plane $\mathscr{R}\rho < 0$ and the half-strip $-\pi < \mathscr{I}\rho < 0$, $\mathscr{R}\rho \geqslant 0$. This in turn is mapped conformally in the ζ-plane on a domain bounded by the real axis, the ray $\arg\{\zeta e^{\frac{1}{3}\pi i} - (\frac{3}{2}\pi)^{\frac{2}{3}}\} = 0$ and the curve whose parametric equation is given by

$$\zeta = (\tfrac{3}{2})^{\frac{2}{3}}(t - i\pi)^{\frac{2}{3}} \quad (0 \leqslant t < \infty). \tag{4·8}$$

Corresponding points of the transformation of the sector $|\arg z| < \pi$ are indicated in figures 3, 4, 5 and 6. For clarity only the parts of the σ and ρ domains corresponding to $0 < \arg z < \pi$ have been shown in figures 4 and 5. In figure 6 the points E, E' have affixes $(\frac{3}{2}\pi)^{\frac{2}{3}} e^{\mp\frac{1}{3}\pi i}$ and the curve ED has the equation (4·8), $E'D'$ being its conjugate.

For the purpose of applying theorem B† to (4·3) it is not sufficient to confine attention to the interior of the ζ-domain which corresponds to $|\arg z| < \pi$, because the distance between the boundary curves ED and $E'D'$ shrinks to zero as the curves approach infinity. It is necessary to investigate how the remainder of the ζ-plane is mapped on the z-plane. The first point to be noticed in this connexion is that the line BE of the ζ-plane is mapped on curves in the σ- and z-planes whose parametric equations are readily verified to be

$$\sigma = t - \tfrac{1}{2}\pi i \pm i\cos^{-1}\{(1 - t^{-1}\tanh t)^{\frac{1}{2}}\cosh t\}, \tag{4·9}$$

$$z = \pm(t\coth t - t^2)^{\frac{1}{2}} + i(t^2 - t\tanh t)^{\frac{1}{2}}, \tag{4·10}$$

where $t \equiv \mathscr{R}\sigma$ and describes the range $0 \leqslant t \leqslant t_0$, where $t_0 = 1{\cdot}19968\ldots$ is the positive root of $t = \coth t$. These curves are indicated by the broken lines in figures 3 and 4; the points P, P' where they cut the imaginary z-axis (figure 3) have affixes

$$\pm i(t_0^2 - 1)^{\frac{1}{2}} = \pm i\,0{\cdot}66274\ldots$$

and correspond to the points $(\frac{3}{4}\pi)^{\frac{2}{3}} e^{\mp\frac{1}{3}\pi i}$ in the ζ-plane.

Next, we observe from (4·7) that the effect of increasing σ by $-i\pi$ is to increase ρ by $-i\pi$ and to change the sign of z.

† See first footnote on p. 331.

From these results we deduce that the sector $|\arg\zeta|<\frac{1}{3}\pi$ is mapped on the eye-shaped domain **K** in the z-plane bounded by the curve BPE, whose equation is (4·10), and its conjugate $BP'E'$. The transformation is not of course one-to-one and is more properly represented by

$$\tfrac{2}{3}\zeta^{\frac{3}{2}} = \ln\{1+\surd(1-z^2)\} - \operatorname{Ln} z - \surd(1-z^2) \qquad (4\cdot11)$$

(cf. (4·6)), where $\operatorname{Ln} z$ denotes the many-valued function whose imaginary part equals $\mathrm{i}\arg z$. If z lies in **K** and $m\pi<\arg z<(m+1)\pi$, where m is any integer, then ζ lies in the curved half-strip bounded by the curves

$$\zeta = (\tfrac{3}{2})^{\frac{2}{3}}(t-m\pi\mathrm{i})^{\frac{2}{3}}, \qquad \zeta = (\tfrac{3}{2})^{\frac{2}{3}}(t-\overline{m+1}\,\pi\mathrm{i})^{\frac{2}{3}} \quad (t\geqslant 0),$$

and the ray $\arg\zeta=-\frac{1}{3}\pi$ if $m\geqslant 0$, or $\arg\zeta=\frac{1}{3}\pi$ if $m<0$.

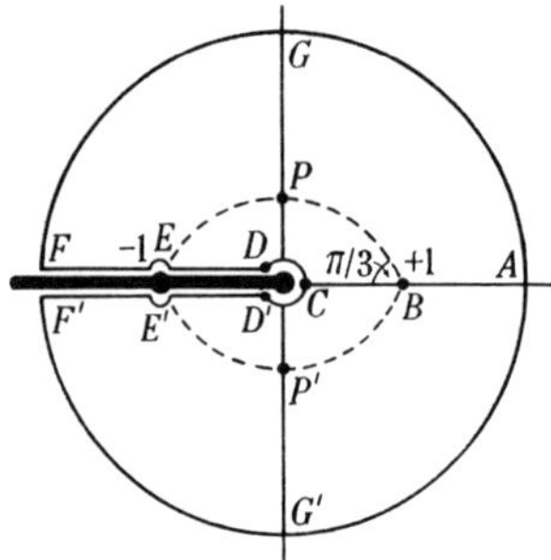

FIGURE 3. z-plane.

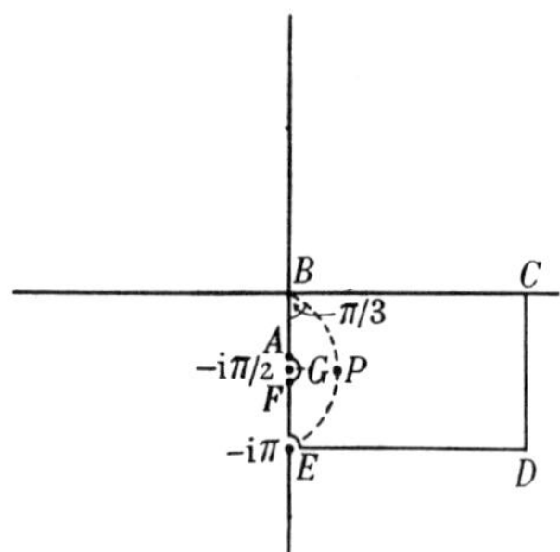

FIGURE 4. σ-plane.

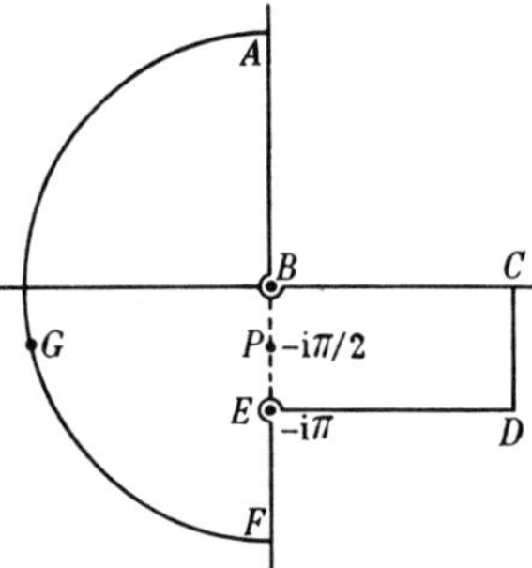

FIGURE 5. ρ-plane.

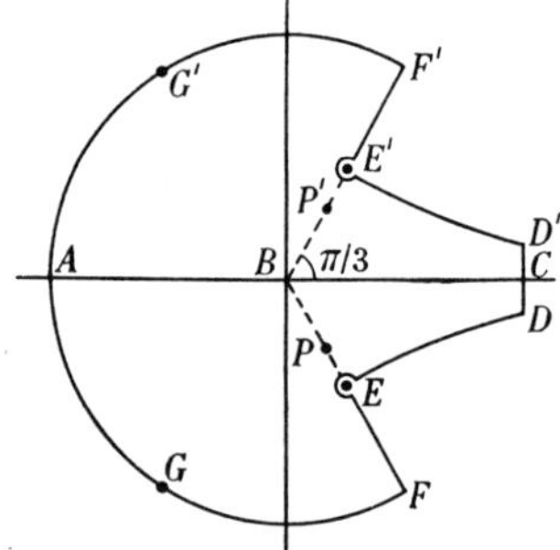

FIGURE 6. ζ-plane.

If z is regarded as a function of ζ it is now clear that $z(\zeta)$ is regular over the whole ζ-plane provided cuts are made along the rays $\arg\zeta=\pm\frac{1}{3}\pi$ from $\zeta=(\frac{3}{2}\pi)^{\frac{2}{3}}\mathrm{e}^{\pm\frac{1}{3}\pi\mathrm{i}}$ to infinity. We shall denote the open domain formed by the ζ-plane cut in this way by the symbol† **D**.

We return to the function $f(\zeta)$. Substituting the first of (4·2) in (4·4) and carrying out some reduction, we find that

$$f(\zeta) = \frac{5}{16\zeta^2} + \frac{\zeta}{4}\frac{z^2(z^2+4)}{(z^2-1)^3}. \qquad (4\cdot12)$$

The only point of **D** at which $z^2=1$ is $\zeta=0$. Near this point, however, we find from (4·4) and the first of (4·2)

$$z(\zeta) = 1-2^{-\frac{1}{3}}\zeta+\tfrac{3}{10}2^{-\frac{2}{3}}\zeta^2+\tfrac{1}{700}\zeta^3+\zeta^4 O(1), \qquad f(\zeta)=\frac{2^{\frac{1}{3}}}{70}+\zeta O(1), \qquad (4\cdot13)$$

and so $f(\zeta)$ is a regular function of ζ throughout **D**.

† No confusion with the **D** of §§ 2 and 3 will arise.

If $|\zeta|\to\infty$ in the sector $|\arg(-\zeta)|<\frac{2}{3}\pi$, then $|z|\to\infty$ and from (4·6) and (4·12) we obtain

$$z=\tfrac{2}{3}(-\zeta)^{\frac{3}{2}}+\tfrac{1}{2}\pi+O(|\zeta|^{-\frac{3}{2}}),\qquad f(\zeta)\sim-\frac{1}{4\zeta^2}. \tag{4·14}$$

On the other hand, if $|\zeta|\to\infty$ in $|\arg\zeta|<\frac{1}{3}\pi$ then $|z|\to0$, and from (4·11) and (4·12) we obtain

$$z\sim2\exp(-\tfrac{2}{3}\zeta^{\frac{3}{2}}-1),\qquad f(\zeta)\sim\frac{5}{16\zeta^2}. \tag{4·15}$$

Thus the preliminary conditions for the applicability of theorem B to equation (4·3) are satisfied. Following the notation of that theorem we define $\mathbf{D}'$ to be the remainder of the ζ-plane after the removal of the half-strips

$$|\mathscr{I}(\zeta e^{\pm\frac{1}{3}\pi i})|<\delta,\qquad \mathscr{R}(\zeta e^{\pm\frac{1}{3}\pi i})>(\tfrac{3}{2}\pi)^{\frac{2}{3}}-\delta$$

enclosing the cuts of $\mathbf{D}$, δ being a small arbitrary positive number. Taking a_1, a_2, a_3 to be the points at infinity on the rays $\arg\zeta=0,\ -\frac{2}{3}\pi,\ \frac{2}{3}\pi$ respectively, we find that the domains $\mathbf{D}_1$, $\mathbf{D}_2$, $\mathbf{D}_3$ all coincide with $\mathbf{D}'$.

Applying theorem B we see that solutions $W_j(\zeta)$ $(j=1,2,3)$ of (4·3) exist such that as $n\to\infty$

$$W_j(\zeta)\sim P_j(n^{\frac{2}{3}}\zeta)\sum_{s=0}^{\infty}\frac{A_s(\zeta)}{n^{2s}}+\frac{P_j'(n^{\frac{2}{3}}\zeta)}{n^{\frac{4}{3}}}\sum_{s=0}^{\infty}\frac{B_s(\zeta)}{n^{2s}}, \tag{4·16}$$

uniformly with respect to ζ in $\mathbf{D}'$. The coefficients are defined by the equations $A_0(\zeta)=1$, and

$$\left.\begin{aligned}B_s(\zeta)&=\tfrac{1}{2}\zeta^{-\frac{1}{2}}\int_0^\zeta t^{-\frac{1}{2}}\{f(t)\,A_s(t)-A_s''(t)\}\,dt,\\ A_{s+1}(\zeta)&=-\tfrac{1}{2}B_s'(\zeta)+\tfrac{1}{2}\int f(\zeta)\,B_s(\zeta)\,d\zeta.\end{aligned}\right\} \tag{4·17}$$

We now express $J_n(nz)$ in terms of W_1, W_2 and W_3. From (4·2) it follows that

$$J_n(nz)=\left(\frac{\zeta}{1-z^2}\right)^{\frac{1}{4}}W(\zeta), \tag{4·18}$$

where W denotes a linear combination of W_1, W_2 and W_3. Letting $\zeta\to+\infty$, $z\to0$, we see immediately that if n is sufficiently large, W is a multiple of W_1.

Next, if we take $j=1$ and $\zeta=n^{-\frac{2}{3}}a_m'$ in (4·16), where a_m' is the mth negative real zero of $\mathrm{Ai}'(z)$, we obtain

$$W_1(\zeta)\sim\mathrm{Ai}(a_m')\left\{1+\sum_{s=1}^{\infty}\frac{A_s(n^{-\frac{2}{3}}a_m')}{n^{2s}}\right\}, \tag{4·19}$$

when n is large, uniformly with respect to m. For large m we have (see Appendix, equations (A 19) and (A 20))

$$a_m'=-\mu^{\frac{2}{3}}\{1+O(m^{-2})\},\qquad \mathrm{Ai}(a_m')=(-)^{m-1}\pi^{-\frac{1}{2}}\mu^{-\frac{1}{6}}\{1+O(m^{-2})\}, \tag{4·20}$$

where $\mu\equiv\frac{3}{8}\pi(4m-3)$. Substituting in the first of (4·14) we obtain

$$z(n^{-\frac{2}{3}}a_m')=\tfrac{2}{3}\mu n^{-1}+\tfrac{1}{2}\pi+O(m^{-1}), \tag{4·21}$$

and substituting this value for z in the well-known asymptotic formula (Watson 1944, p. 199)

$$J_n(nz)=(2/n\pi z)^{\frac{1}{2}}\{\cos(nz-\tfrac{1}{2}n\pi-\tfrac{1}{4}\pi)+O(z^{-1})\},$$

we have

$$J_n(nz)=(3/\mu\pi)^{\frac{1}{2}}\{\cos(\tfrac{2}{3}\mu-\tfrac{1}{4}\pi)+O(m^{-1})\}=(-)^{m-1}(3/\mu\pi)^{\frac{1}{2}}\{1+O(m^{-1})\}. \tag{4·22}$$

Combining the relations (4·18) to (4·22) and letting $m\to\infty$, we see that

$$J_n(nz)\div\left\{\left(\frac{\zeta}{1-z^2}\right)^{\frac{1}{4}}W_1(\zeta)\right\}\sim\frac{\sqrt{2}}{n^{\frac{1}{3}}}\div\left\{1+\sum_{s=1}^{\infty}\frac{A_s(-\infty)}{n^{2s}}\right\}$$

for large n. Hence if we now prescribe

$$A_{s+1}(-\infty) = 0 \quad (s \geqslant 0), \tag{4·23}$$

we derive
$$J_n(nz) \sim \left(\frac{4\zeta}{1-z^2}\right)^{\frac{1}{4}} \left\{ \frac{\mathrm{Ai}\,(n^{\frac{2}{3}}\zeta)}{n^{\frac{1}{3}}} \sum_{s=0}^{\infty} \frac{A_s(\zeta)}{n^{2s}} + \frac{\mathrm{Ai}'\,(n^{\frac{2}{3}}\zeta)}{n^{\frac{5}{3}}} \sum_{s=0}^{\infty} \frac{B_s(\zeta)}{n^{2s}} \right\}, \tag{4·24}$$

as $n \to \infty$.

The result (4·24) is the desired expansion. The coefficients occurring in it are given by (4·17) with the arbitrary constants of integration fixed by (4·23). The expansion is uniform with respect to ζ in $\mathbf{D}'$, and hence it is uniform with respect to z in the sector $|\arg z| \leqslant \pi - \delta$, and if z lies in $\mathbf{K}$ it holds for *any* value of $\arg z$. When z lies outside these regions corresponding expansions may be deduced with the aid of the continuation formula

$$J_n(nz\,\mathrm{e}^{m\pi\mathrm{i}}) = \mathrm{e}^{mn\pi\mathrm{i}}\, J_n(nz),$$

m being any integer.

The significance of the eye-shaped domain $\mathbf{K}$ in relation to $J_n(nz)$ is now apparent. As $n \to \infty$, $|J_n(nz)|$ becomes exponentially small or exponentially large according as z lies inside† or outside $\mathbf{K}$, unless z happens to lie on the real axis outside $\mathbf{K}$ in which event $J_n(nz)$ oscillates boundedly.

The corresponding results for the Hankel functions $H_n^{(1)}(nz)$ and $H_n^{(2)}(nz)$ may be derived in a similar way. If $\zeta \to \infty\,\mathrm{e}^{-\frac{2}{3}\pi\mathrm{i}}$ then $z \to \infty\,\mathrm{e}^{\frac{1}{2}\pi\mathrm{i}}$, and $H_n^{(1)}(nz)$ becomes exponentially small. Accordingly, $H_n^{(1)}(nz)$ must be a multiple of $W_2(\zeta)$ for all sufficiently large n. Similarly, $H_n^{(2)}(nz)$ is a multiple of $W_3(\zeta)$. The precise results may be verified to be

$$H_n^{(1)}(nz) \sim \frac{2\,\mathrm{e}^{-\frac{1}{3}\pi\mathrm{i}}}{n^{\frac{1}{3}}} \left(\frac{4\zeta}{1-z^2}\right)^{\frac{1}{4}} W_2(\zeta), \qquad H_n^{(2)}(nz) \sim \frac{2\,\mathrm{e}^{\frac{1}{3}\pi\mathrm{i}}}{n^{\frac{1}{3}}} \left(\frac{4\zeta}{1-z^2}\right)^{\frac{1}{4}} W_3(\zeta), \tag{4·25}$$

where the coefficients $A_s(\zeta)$ and $B_s(\zeta)$ implied in the expansions for $W_2(\zeta)$ and $W_3(\zeta)$ are the same as in (4·24). In deriving these results use may be made of the fact that $\zeta^{\frac{1}{4}}B_s(\zeta) \to 0$ as $|\zeta| \to \infty$ in the sector $|\arg(-\zeta)| < \frac{2}{3}\pi$, which may itself be proved from (4·24) by setting $\zeta = n^{-\frac{2}{3}}a_m$, where a_m is the mth negative zero of $\mathrm{Ai}\,(z)$, and letting $m \to \infty$.

Combining the two results (4·25), we obtain

$$Y_n(nz) \sim -\left(\frac{4\zeta}{1-z^2}\right)^{\frac{1}{4}} \left\{ \frac{\mathrm{Bi}\,(n^{\frac{2}{3}}\zeta)}{n^{\frac{1}{3}}} \sum_{s=0}^{\infty} \frac{A_s(\zeta)}{n^{2s}} + \frac{\mathrm{Bi}'\,(n^{\frac{2}{3}}\zeta)}{n^{\frac{5}{3}}} \sum_{s=0}^{\infty} \frac{B_s(\zeta)}{n^{2s}} \right\}, \tag{4·26}$$

this expansion having the same region of validity as (4·24) and (4·25).

5. The expansion of $J_\nu(\nu z)$, $Y_\nu(\nu z)$ and the Hankel functions for large complex orders

As in § 3 let us write $\nu = n\,\mathrm{e}^{\mathrm{i}\vartheta}$. Then $(1-z^2)^{\frac{1}{4}}\,\zeta^{-\frac{1}{4}}\,J_\nu(\nu z)$ satisfies the equation

$$\frac{\mathrm{d}^2 W}{\mathrm{d}\xi^2} = \{n^2\xi + \mathrm{e}^{-\frac{4}{3}\mathrm{i}\vartheta} f(\xi\,\mathrm{e}^{-\frac{2}{3}\mathrm{i}\vartheta})\}\, W \tag{5·1}$$

(cf. (4·3)), where $\xi \equiv \zeta\,\mathrm{e}^{\frac{2}{3}\mathrm{i}\vartheta}$ and $\zeta, f(\zeta)$ are defined by (4·11) and (4·12). The z-region comprising the sector $|\arg z| < \pi$ and the domain $\mathbf{K}$ of § 4 with any value of $\arg z$, is mapped on the ξ-plane cut along the rays

$$\arg\xi = \pm\tfrac{1}{3}\pi + \tfrac{2}{3}\vartheta, \qquad |\xi| \geqslant (\tfrac{3}{2}\pi)^{\frac{2}{3}}.$$

† It may be shown (Watson 1944, pp. 268–270) that if z lies inside $\mathbf{K}$ and n is *any* positive integer, then $|J_n(nz)| \leqslant 1$.

These cuts are indicated by the heavy lines in figure 7, the points E, E' having affixes $(\frac{3}{2}\pi)^{\frac{2}{3}}\exp(\mp\frac{1}{3}i\pi+\frac{2}{3}i\vartheta)$.

From theorem B it follows that solutions W_j $(j=1,2,3)$ of (5·1) exist such that

$$W_j \sim P_j(n^{\frac{2}{3}}\xi)\sum_{s=0}^{\infty}\frac{a_s(\xi)}{n^{2s}}+\frac{P_j'(n^{\frac{2}{3}}\xi)}{n^{\frac{4}{3}}}\sum_{s=0}^{\infty}\frac{b_s(\xi)}{n^{2s}}, \tag{5·2}$$

as $n\to\infty$. The regions of validity of these expansions do not, however, extend over the whole of the ξ-plane cut in the manner described. If, for example, $0<\vartheta<\frac{1}{2}\pi$ and the point a_1 of theorem B is taken to be at infinity on the positive real axis, then the definition of $\mathbf{D}_1$ shows that the expansion (5·2) with $j=1$ does not necessarily hold in the domain bounded by the branch of the level curve of $\exp(-\frac{2}{3}\xi^{\frac{3}{2}})$ through E and the cut through E. This excluded domain has been shaded in figure 7.

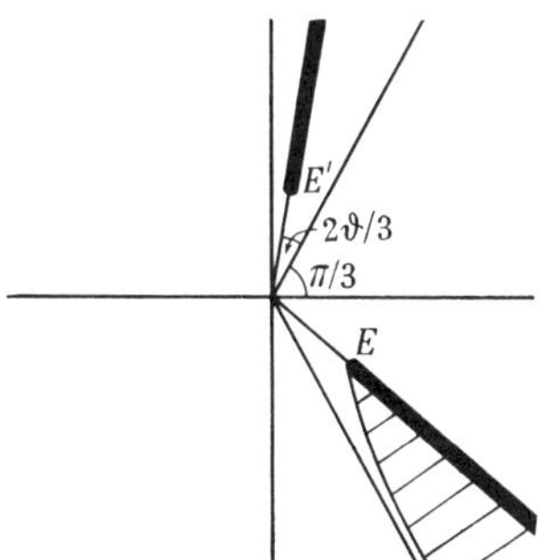

FIGURE 7. ξ-plane.

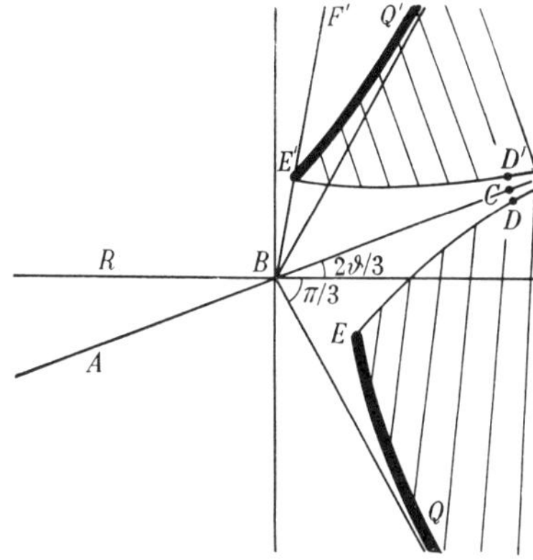

FIGURE 8. ξ-plane.

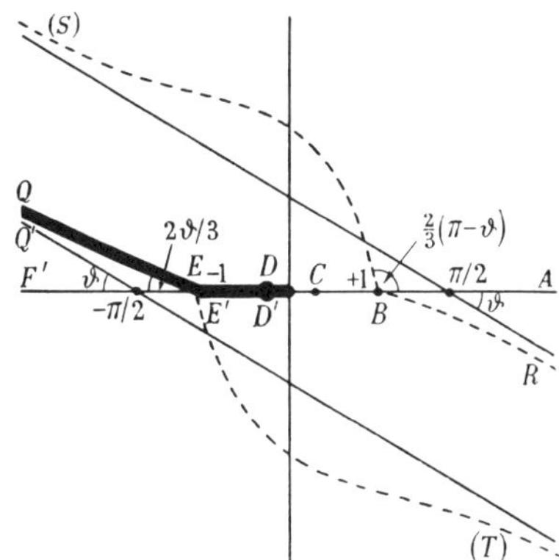

FIGURE 9. z-plane.

We are not, however, bound to make the cuts from E, E' along the rays indicated, and larger regions of validity result if we make them instead along the outward branches of the level curves of $\exp(-\frac{2}{3}\xi^{\frac{3}{2}})$ which pass through E, E'. From (4·7) it is seen that these curves are mapped in the ρ-plane as straight lines parallel to $\arg\rho=\pm\frac{1}{2}\pi-\vartheta$, and using this fact it is fairly easy to determine from (4·7) how they are mapped on the σ- and z-planes.

Figures 8 and 9 show corresponding points of the ξ, z transformation with a value of ϑ between 0 and $\frac{1}{2}\pi$. They should be compared with figures 3 and 6. In figure 8 the curves

EQ, $E'Q'$ are parts of level curves of $\exp(-\frac{2}{3}\xi^{\frac{3}{2}})$, and ED, $E'D'$ are the curves of figure 6 rotated round the origin through an angle $\frac{2}{3}\vartheta$. The unshaded part of the ξ-plane is mapped on the z-plane cut along the join of $z = 0$ and $z = -1$ and the curve EQ whose parametric equation is defined by

$$z = \operatorname{sech}\sigma, \qquad \sigma - \tanh\sigma = -\mathrm{i}\pi + t\,\mathrm{e}^{\mathrm{i}(\frac{3}{2}\pi-\vartheta)} \quad (0 \leqslant t < \infty),$$

and

$$\sigma + \mathrm{i}\pi \sim (3t)^{\frac{1}{3}}\,\mathrm{e}^{\mathrm{i}(\frac{1}{2}\pi-\frac{1}{3}\vartheta)} \quad \text{as} \quad t \to 0.$$

This curve leaves $z = -1$ at an angle $\pi - \frac{2}{3}\vartheta$ with the real axis and has the line

$$\arg(z + \tfrac{1}{2}\pi) = \pi - \vartheta$$

as asymptote. In the cut z-plane, $\arg z$ takes its principal value everywhere except between $E'Q'$ and the negative real axis to the left of $z = -1$. In the theory of zeros (see § 7) the negative real ξ-axis is of interest; it is mapped on the z-plane as the image (BR) in $z = 0$ of the curve EQ.

As in § 4 it is necessary to consider the mapping of the shaded zones of figure 7 on the z-plane. Without going into detail it may be stated that the upper zone is mapped on the interior of the continuous curve given by

$$z = \operatorname{sech}\sigma, \qquad \sigma - \tanh\sigma = \mathrm{i}\pi + t\,\mathrm{e}^{\mathrm{i}(\frac{1}{2}\pi-\vartheta)} \quad (0 \leqslant t < \infty),$$

and

$$\sigma \sim \mathrm{i}\pi + (3t)^{\frac{1}{3}}\,\mathrm{e}^{\mathrm{i}(\frac{1}{6}\pi-\frac{1}{3}\vartheta)} \quad \text{as} \quad t \to 0.$$

This curve starts from $z = -1$ at an angle $\frac{1}{3}\pi - \frac{2}{3}\vartheta$ with the real axis and winds itself round $z = 0$ in the clockwise sense, tending to the origin as $t \to \infty$. The transformation is not one-to-one and $\arg z$ takes all possible values less than $-\pi$.

In the lower shaded zone the points

$$\xi = (\tfrac{3}{2}m\pi)^{\frac{2}{3}}\,\mathrm{e}^{\mathrm{i}(-\frac{1}{3}\pi+\frac{2}{3}\vartheta)} \quad (m = 1, 2, 3, \ldots),$$

are branch points of the function $z(\xi)$. If cuts are made along the outward branches of the level curves of $\exp(-\frac{2}{3}\xi^{\frac{3}{2}})$ through each of these points, then the whole zone is mapped on the interior z-region bounded by the curves EQ, BR, BS and $E'T$ of figure 9, $\arg z$ taking all possible values greater than π. The equation of BS is given by

$$z = \operatorname{sech}\sigma, \qquad \sigma - \tanh\sigma = t\,\mathrm{e}^{-\mathrm{i}(\frac{1}{2}\pi+\vartheta)} \quad (0 \leqslant t < \infty),$$

and

$$\sigma \sim (3t)^{\frac{1}{3}}\,\mathrm{e}^{-\mathrm{i}(\frac{1}{6}\pi+\frac{1}{3}\vartheta)} \quad \text{as} \quad t \to 0,$$

and $E'T$ is its image in $z = 0$. BS leaves $z = 1$ at an angle $\frac{2}{3}\pi - \frac{2}{3}\vartheta$ with the real axis and has the line $\arg(z - \frac{1}{2}\pi) = \pi - \vartheta$ as asymptote.

It has been assumed above that $0 < \vartheta < \frac{1}{2}\pi$. If $-\frac{1}{2}\pi < \vartheta < 0$, the roles of the upper and lower zones are interchanged.

Now that the correspondence between the z- and ξ-planes has been established we are in a position to express the Bessel functions in terms of the solutions W_j of (5·1). This is done in a manner similar to that of § 4 for positive orders, and restoring afterwards the variables ν, ζ in place of n, ξ we find that *if the z-plane is cut in the manner indicated in figure* 9, $\arg z$ *having its principal value on the positive real axis, then the expansions* (4·24), (4·25) *and* (4·26) *hold with n replaced by ν, provided* $|\arg\nu| < \frac{1}{2}\pi$.

The region of validity of the expansions also extends to the z-region corresponding to the shaded ξ-zones of figure 8 which we have just examined. Asymptotic expansions valid for other ranges of $\arg z$ and $\arg \nu$ may be deduced with the aid of standard continuation formulae, given, for example, by Watson (1944, pp. 74–75).

6. Formulae for the coefficients and expansions for the derivatives

The coefficients $A_s(\zeta)$ and $B_s(\zeta)$ in the expansions for $J_\nu(\nu z)$, $Y_\nu(\nu z)$ and the Hankel functions are defined by (4·17) and (4·23), $f(\zeta)$ being given by (4·12). They are regular functions in the cut ζ-plane whether the cuts are made as in § 4 or as in § 5.

The first of (4·17) with $s = 0$ may be integrated directly, using (4·12) and the first of (4·2), to give

$$B_0(\zeta) = -\frac{5}{48\zeta^2} - \frac{1}{\zeta^{\frac{1}{2}}}\left(\frac{v}{8} - \frac{5v^3}{24}\right) \tag{6·1}$$

where

$$v \equiv (1-z^2)^{-\frac{1}{2}}. \tag{6·2}$$

It is not practicable to continue this process, but explicit expressions for the higher coefficients may be obtained by the following procedure which is suggested in a remark of Cherry (1950, p. 250, footnote).

If ζ is taken temporarily to be positive and fixed then $n^{\frac{2}{3}}\zeta \to +\infty$ as $n \to \infty$, and $\mathrm{Ai}\,(n^{\frac{2}{3}}\zeta)$, $\mathrm{Ai}'\,(n^{\frac{2}{3}}\zeta)$ may be replaced in (4·24) by their asymptotic expansions (see Appendix, equations (A 1), (A2), (A3) and (A 6))

$$\mathrm{Ai}\,(n^{\frac{2}{3}}\zeta) \sim \frac{\exp\left(-\frac{2}{3}n\zeta^{\frac{3}{2}}\right)}{2n^{\frac{1}{6}}\zeta^{\frac{1}{4}}\sqrt{\pi}} \sum_{s=0}^{\infty} \frac{(-)^s a_s}{n^s \zeta^{\frac{3}{2}s}}, \qquad \mathrm{Ai}'\,(n^{\frac{2}{3}}\zeta) \sim -\frac{n^{\frac{1}{6}}\zeta^{\frac{1}{4}}}{2\sqrt{\pi}} \exp\left(-\frac{2}{3}n\zeta^{\frac{3}{2}}\right) \sum_{s=0}^{\infty} \frac{(-)^s b_s}{n^s \zeta^{\frac{3}{2}s}}, \tag{6·3}$$

in which $a_0 = 1$, $b_0 = 1$, and

$$a_s = \frac{(2s+1)\,(2s+3)\ldots(6s-1)}{s!\,(144)^s}, \qquad b_s = -\frac{6s+1}{6s-1}a_s. \tag{6·4}$$

The result so obtained must be equivalent to Debye's expansion (Watson 1944, p. 243), which can be written in the form

$$J_n(nz) \sim \frac{\exp\left(-\frac{2}{3}n\zeta^{\frac{3}{2}}\right)}{\sqrt{(2\pi n)}\,(1-z^2)^{\frac{1}{4}}} \sum_{s=0}^{\infty} \frac{\overline{U}_s}{n^s} \tag{6·5}$$

(cf. (2·13)), where $\overline{U}_s$ is given by the same formulae (2·19) and (2·20) as U_s but with u replaced by v, defined in (6·2). In consequence we derive the asymptotic equality

$$\sum_{s=0}^{\infty} \frac{\overline{U}_s}{n^s} \sim \sum_{s=0}^{\infty} \frac{(-)^s a_s}{n^s\zeta^{\frac{3}{2}s}} \sum_{s=0}^{\infty} \frac{A_s(\zeta)}{n^{2s}} - \sum_{s=0}^{\infty} \frac{(-)^s b_s}{n^s\zeta^{\frac{3}{2}s}} \sum_{s=0}^{\infty} \frac{\zeta^{\frac{1}{2}}B_s(\zeta)}{n^{2s+1}}.$$

Equating coefficients of n^{-1}, we confirm the result (6·1), and with the aid of the relation

$$a_{2s}b_0 - a_{2s-1}b_1 + a_{2s-2}b_2 - \ldots + a_0 b_{2s} = 0 \quad (s>0),$$

which may be derived from the Wronskian relation for Ai and Bi, we may show by induction that

$$A_s(\zeta) = \sum_{m=0}^{2s} b_m \zeta^{-\frac{3}{2}m}\,\overline{U}_{2s-m}, \qquad \zeta^{\frac{1}{2}}B_s(\zeta) = -\sum_{m=0}^{2s+1} a_m \zeta^{-\frac{3}{2}m}\,\overline{U}_{2s-m+1}. \tag{6·6}$$

Thus $A_s(\zeta)$ and $\zeta^{\frac{1}{2}}B_s(\zeta)$ are polynomials in $\zeta^{-\frac{3}{2}}$ and v. The validity of these results for all values of ζ in the cut plane follows by the principle of analytic continuation.

Debye's formula may be regarded as a particular case of the present expansions. The same is true also of Meissel's expansions for Bessel functions of nearly equal argument and order (Watson 1944, pp. 245–248) and of the expansions given by the present writer (1952) for $J_\nu(z)$, $Y_\nu(z)$ and the Hankel functions when $z = \nu + \tau\nu^{\frac{1}{3}}$, τ being fixed and $|\nu|$ large. They may all be obtained from the expansions of §§ 4 and 5 by substituting appropriate expansions for the Airy functions.

The systematic tabulation of the coefficients $A_s(\zeta)$ and $B_s(\zeta)$ is most easily performed directly from the defining formulae (4·17), using processes of numerical differentiation and integration. For this purpose it is convenient to have the values of $A_1, A_2, \ldots$ at $\zeta = 0$, and they may be obtained by substituting $z = 1$ in (4·24) and comparing with Meissel's formula for $J_n(n)$ (Watson 1944, pp. 232–233). The first three values are found to be

$$A_1(0) = -\tfrac{1}{225} = -0{\cdot}00444\,44444\ldots, \qquad A_2(0) = +0{\cdot}00069\,3735\ldots,$$
$$A_3(0) = -0{\cdot}00035\,38\ldots.$$

Tables of $A_s(\zeta)$ and $B_s(\zeta)$ have been prepared for $s = 0, 1, 2, 3$ for the case of real positive arguments and orders. It transpires that these early coefficients decrease in magnitude with increasing s; in the range $-\infty < \zeta < \infty$ the maximum values of $|A_3|$ and $|B_3|$ are 0·00040 and 0·0010 approximately. Thus the series (4·24), (4·25) and (4·26) are particularly well suited to numerical applications.

Expansions for the derivatives

The expansion (4·24) may be differentiated term by term; if we write

$$\phi(\zeta) \equiv \left(\frac{4\zeta}{1-z^2}\right)^{\frac{1}{4}} = \left(-\frac{2}{z}\frac{dz}{d\zeta}\right)^{\frac{1}{2}} \tag{6·7}$$

(cf. (4·2)), and replace n by ν, we obtain

$$\nu\frac{dz}{d\zeta}J'_\nu(\nu z) \sim \frac{\phi'(\zeta)}{\phi(\zeta)}J_\nu(\nu z) + \phi(\zeta)\left\{\frac{\mathrm{Ai}\,(\nu^{\frac{2}{3}}\zeta)}{\nu^{\frac{1}{3}}}\sum_{s=0}^{\infty}\frac{A'_s(\zeta)+\zeta B_s(\zeta)}{\nu^{2s}} + \frac{\mathrm{Ai}'\,(\nu^{\frac{2}{3}}\zeta)}{\nu^{\frac{5}{3}}}\sum_{s=-1}^{\infty}\frac{A_{s+1}(\zeta)+B'_s(\zeta)}{\nu^{2s}}\right\}.$$

Hence we derive the desired result

$$J'_\nu(\nu z) \sim -\psi(\zeta)\left\{\frac{\mathrm{Ai}\,(\nu^{\frac{2}{3}}\zeta)}{\nu^{\frac{4}{3}}}\sum_{s=0}^{\infty}\frac{C_s(\zeta)}{\nu^{2s}} + \frac{\mathrm{Ai}'\,(\nu^{\frac{2}{3}}\zeta)}{\nu^{\frac{2}{3}}}\sum_{s=0}^{\infty}\frac{D_s(\zeta)}{\nu^{2s}}\right\}, \tag{6·8}$$

where

$$\psi(\zeta) = 2/\{z\phi(\zeta)\}, \tag{6·9}$$

$$C_s(\zeta) = \chi(\zeta)\,A_s(\zeta) + A'_s(\zeta) + \zeta B_s(\zeta), \qquad D_s(\zeta) = A_s(\zeta) + \chi(\zeta)\,B_{s-1}(\zeta) + B'_{s-1}(\zeta), \tag{6·10}$$

and

$$\chi(\zeta) \equiv \frac{\phi'(\zeta)}{\phi(\zeta)} = \frac{1}{4\zeta} - \frac{z^2}{2}\left\{\frac{\zeta}{(1-z^2)^3}\right\}^{\frac{1}{2}} = \frac{4 - z^2\{\phi(\zeta)\}^6}{16\zeta}. \tag{6·11}$$

The functions $\phi(\zeta)$, $\psi(\zeta)$, $\chi(\zeta)$, $C_s(\zeta)$ and $D_s(\zeta)$ are regular in the same regions as $z(\zeta)$, and (6·8) is valid under the same conditions as the expansion for $J_\nu(\nu z)$. The corresponding expansions for $Y'_\nu(\nu z)$ and the Hankel function derivatives are obtained by replacing Ai by the appropriate Airy functions.

Explicit formulae for $C_s(\zeta)$ and $D_s(\zeta)$ may be derived in a manner similar to that outlined above for $A_s(\zeta)$ and $B_s(\zeta)$; (6·8) is compared with Debye's expansion for $J'_n(nz)$. We find that

$$\zeta^{-\frac{1}{2}}C_s(\zeta) = -\sum_{m=0}^{2s+1} b_m\zeta^{-\frac{3}{2}m}\bar{V}_{2s-m+1}, \qquad D_s(\zeta) = \sum_{m=0}^{2s} a_m\zeta^{-\frac{3}{2}m}\bar{V}_{2s-m}, \tag{6·12}$$

where the functions $\bar{V}_s$ are given by (2·22) and (2·23) with u replaced by v, as defined by (6·2).

7. Zeros of $J_\nu(z)$ and $J'_\nu(z)$

Zeros of $J_n(z)$

We consider first the case when the order is real and positive. From (4·24) it follows that for large n the values of ζ corresponding to zeros of $J_n(nz)$ satisfy the equation (see also Olver 1954, equation (5·12))

$$\mathrm{Ai}\,(n^{\frac{2}{3}}\zeta)+(1+|\,n^{\frac{2}{3}}\zeta\,|^{\frac{1}{4}})^{-1}\exp\left(-\tfrac{2}{3}n\zeta^{\frac{3}{2}}\right)O(n^{-1})=0. \tag{7·1}$$

It is shown in the Appendix that the zeros of Ai (z) are all real and negative. If, in the usual notation, the sth zero is denoted by a_s, we deduce that the corresponding value of ζ is given by

$$\zeta=\alpha+\epsilon, \tag{7·2}$$

where

$$\alpha=n^{-\frac{2}{3}}a_s, \qquad \epsilon=O(n^{-\frac{4}{3}}), \tag{7·3}$$

and so $j_{n,s}$, the sth positive zero of $J_n(z)$, is given by

$$j_{n,s}=nz(\alpha+\epsilon)=nz(\alpha)+O(n^{-\frac{1}{3}}). \tag{7·4}$$

An asymptotic expansion for $j_{n,s}$ may be found as follows. From (4·24) and (7·2) we derive by expansion

$$W_1(\alpha)+\frac{\epsilon}{1!}W'_1(\alpha)+\frac{\epsilon^2}{2!}W''_1(\alpha)+\ldots=0, \tag{7·5}$$

where $W_1(\zeta)$ is defined in § 4. Now from (4·16) with $j=1$ we obtain,† if m is any integer,

$$\left.\begin{aligned}W_1^{(2m)}(\zeta)&\sim n^{2m}\,\mathrm{Ai}\,(n^{\frac{2}{3}}\zeta)\sum_{r=0}^{\infty}\frac{A_r^{2m}(\zeta)}{n^{2r}}+n^{2m-\frac{4}{3}}\,\mathrm{Ai}'\,(n^{\frac{2}{3}}\zeta)\sum_{r=0}^{\infty}\frac{B_r^{2m}(\zeta)}{n^{2r}},\\ W_1^{(2m+1)}(\zeta)&\sim n^{2m}\,\mathrm{Ai}\,(n^{\frac{2}{3}}\zeta)\sum_{r=0}^{\infty}\frac{A_r^{2m+1}(\zeta)}{n^{2r}}+n^{2m+\frac{2}{3}}\,\mathrm{Ai}'\,(n^{\frac{2}{3}}\zeta)\sum_{r=0}^{\infty}\frac{B_r^{2m+1}(\zeta)}{n^{2r}},\end{aligned}\right\} \tag{7·6}$$

where

$$\left.\begin{aligned}A_r^{2m}&=\frac{\mathrm{d}}{\mathrm{d}\zeta}A_{r-1}^{2m-1}+\zeta B_r^{2m-1}, & A_r^{2m+1}&=\frac{\mathrm{d}}{\mathrm{d}\zeta}A_r^{2m}+\zeta B_r^{2m},\\ B_r^{2m}&=A_r^{2m-1}+\frac{\mathrm{d}}{\mathrm{d}\zeta}B_r^{2m-1}, & B_r^{2m+1}&=A_r^{2m}+\frac{\mathrm{d}}{\mathrm{d}\zeta}B_{r-1}^{2m}.\end{aligned}\right\} \tag{7·7}$$

Hence

$$W_1^{(2m)}(\alpha)\sim n^{2m-\frac{4}{3}}\,\mathrm{Ai}'\,(a_s)\sum_{r=0}^{\infty}\frac{B_r^{2m}(\alpha)}{n^{2r}}, \qquad W_1^{(2m+1)}(\alpha)\sim n^{2m+\frac{2}{3}}\,\mathrm{Ai}'\,(a_s)\sum_{r=0}^{\infty}\frac{B_r^{2m+1}(\alpha)}{n^{2r}},$$

and substituting these values in (7·5), we obtain the asymptotic equality

$$\sum_{r=0}^{\infty}\frac{B_r(\alpha)}{n^{2r}}+\frac{\epsilon n^2}{1!}\sum_{r=0}^{\infty}\frac{B_r^1(\alpha)}{n^{2r}}+\frac{\epsilon^2n^2}{2!}\sum_{r=0}^{\infty}\frac{B_r^2(\alpha)}{n^{2r}}+\frac{\epsilon^3n^4}{3!}\sum_{r=0}^{\infty}\frac{B_r^3(\alpha)}{n^{2r}}+\ldots\sim 0. \tag{7·8}$$

From (7·7) we see that $B_0^1=1$, and so an expansion of ϵ exists of the form

$$\epsilon\sim\frac{\alpha_1}{n^2}+\frac{\alpha_2}{n^4}+\frac{\alpha_3}{n^6}+\ldots, \tag{7·9}$$

and substituting this relation in the first of (7·4) and expanding we obtain the desired result

$$j_{n,s}\sim n\sum_{r=0}^{\infty}\frac{p_r(\alpha)}{n^{2r}}, \tag{7·10}$$

† No question of the legitimacy of the repeated differentiation of the asymptotic series need arise; the results are otherwise obtainable from the differential equation (4·3) satisfied by $W_1(\zeta)$, and the recurrence relations (4·17) satisfied by $A_s(\zeta)$, $B_s(\zeta)$.

where α is given by (7·3) and the coefficients $p_r(\alpha)$ by the relation

$$\sum_{r=0}^{\infty} \frac{p_r(\alpha)}{n^{2r}} \sim z(\alpha) + \left(\frac{\alpha_1}{n^2}+\frac{\alpha_2}{n^4}+\ldots\right) z'(\alpha) + \frac{1}{2!}\left(\frac{\alpha_1}{n^2}+\frac{\alpha_2}{n^4}+\ldots\right)^2 z''(\alpha) + \ldots. \tag{7·11}$$

The first three coefficients are readily found to be

$$p_0 = z, \qquad p_1 = \alpha_1 z', \qquad p_2 = \alpha_2 z' + \tfrac{1}{2}\alpha_1^2 z'', \tag{7·12}$$

where from (7·8) and (7·9),

$$\alpha_1 = -B_0, \qquad \alpha_2 = -(B_1 + \alpha_1 B_1^1 + \tfrac{1}{2}\alpha_1^2 B_0^2 + \tfrac{1}{6}\alpha_1^3 B_0^3). \tag{7·13}$$

If α is replaced by the general variable ζ, it is evident that for all values of r, $p_r(\zeta)$ is a polynomial in the functions $B_r^m(\zeta)$ and the derivatives $z^{(r)}(\zeta)$, and so is regular in the cut ζ-plane **D** of § 4. The expansion (7·10) is uniformly valid with respect to s, and the error on curtailing it at the term $p_r(\alpha)/n^{2r-1}$ is $O(n^{-2r-1})$. Thus the error term in (7·4) is really $O(n^{-1})$.

The value of $J_n'(z)$ at the zeros of $J_n(z)$ is another quantity of interest, and its asymptotic expansion may be conveniently obtained with the aid of the formula (Olver 1950, equation (3·4))

$$\pm J_n'(j_{n,s}) = \left(\tfrac{1}{2} j_{n,s} \frac{dj_{n,s}}{ds}\right)^{-\frac{1}{2}}, \tag{7·14}$$

in which s is regarded as a continuous variable. Using (7·3) and (7·10) we obtain

$$\frac{dj_{n,s}}{ds} \sim n^{\frac{1}{3}} \frac{da_s}{ds} \sum_{r=0}^{\infty} \frac{p_r'(\alpha)}{n^{2r}}.$$

The result analogous to (7·14) for the Airy functions is given by (Olver 1950, equation (2·15))

$$\pm \mathrm{Ai}'(a_s) = \left(-\frac{da_s}{ds}\right)^{-\frac{1}{2}}.$$

Also, remembering that $p_0 \equiv z$, we have from (6·7) and (6·9)

$$p_0 p_0' = -\tfrac{1}{2} z^2 \phi^2 = -2/\psi^2.$$

Combining these three results, and using (7·10) and (7·14), we derive

$$J_n'(j_{n,s}) \sim -\mathrm{Ai}'(a_s) \frac{\psi(\alpha)}{n^{\frac{2}{3}}} \left\{1 + \sum_{r=1}^{\infty} \frac{P_r(\alpha)}{n^{2r}}\right\}, \tag{7·15}$$

where the coefficients $P_r(\alpha)$ may be obtained from the relation

$$\left(1 + \sum_{r=1}^{\infty} \frac{P_r}{n^{2r}}\right)^{-2} \sim -\tfrac{1}{2}\psi^2 \left(\sum_{r=0}^{\infty} \frac{p_r}{n^{2r}}\right)\left(\sum_{r=0}^{\infty} \frac{p_r'}{n^{2r}}\right). \tag{7·16}$$

We find, for example, that

$$P_1 = \tfrac{1}{4}\psi^2(p_0 p_1' + p_1 p_0'), \qquad P_2 = \tfrac{3}{2}P_1^2 + \tfrac{1}{4}\psi^2(p_0 p_2' + p_1 p_1' + p_2 p_0'). \tag{7·17}$$

For numerical purposes the coefficients $p_r(\zeta)$, $\psi(\zeta)$ and $P_r(\zeta)$ of (7·10) and (7·15) may be pretabulated, and this has been done by the writer for the first few values of r. In the relevant range $0 < -\zeta < \infty$ the early coefficients *decrease* in magnitude; the values of $|p_1|$, $|p_2|$, $|p_3|$

do not exceed 0·015, 0·0012, 0·00045, and $|P_1|$, $|P_2|$, $|P_3|$ do not exceed 0·0070, 0·00036, 0·00006 4, respectively. Using the first four terms it is found, for example, that the series will give $j_{n,s}$ and $J'_n(j_{n,s})$ correct to at least ten significant figures when $n \geqslant 4$; the first term alone in each series gives four-figure accuracy at $n = 4$.

Zeros of $J'_n(z)$

When n is large the values of ζ corresponding to zeros of $J'_n(nz)$ are given by (cf. (6·8) and Olver 1954, equation (5·13))

$$\mathrm{Ai}'\,(n^{\frac{2}{3}}\zeta)+(1+|\,n^{\frac{2}{3}}\zeta\,|^{\frac{1}{4}})\exp\left(-\tfrac{2}{3}n\zeta^{\frac{3}{2}}\right)O(n^{-\frac{4}{3}})=0.$$

The zeros a'_s of $\mathrm{Ai}'\,(z)$ are all real and negative (see Appendix), and the corresponding values of ζ are given by

$$\zeta=\beta+\eta,$$

where

$$\beta\equiv n^{-\frac{2}{3}}a'_s, \qquad \eta=O(n^{-\frac{4}{3}}). \tag{7·18}$$

Thus $j'_{n,s}$, the sth zero of $J'_n(z)$, is asymptotically given by

$$j'_{n,s}=nz(\beta+\eta)=nz(\beta)+O(n^{-\frac{1}{3}}).$$

Proceeding in a manner similar to that given above for $j_{n,s}$, we define

$$\left.\begin{aligned} C_r^{2m}&=\frac{\mathrm{d}}{\mathrm{d}\zeta}C_r^{2m-1}+\zeta D_r^{2m-1}, & C_r^{2m+1}&=\frac{\mathrm{d}}{\mathrm{d}\zeta}C_{r-1}^{2m}+\zeta D_r^{2m},\\ D_r^{2m}&=C_r^{2m-1}+\frac{\mathrm{d}}{\mathrm{d}\zeta}D_{r-1}^{2m-1}, & D_r^{2m+1}&=C_r^{2m}+\frac{\mathrm{d}}{\mathrm{d}\zeta}D_r^{2m}. \end{aligned}\right\} \tag{7·19}$$

We find that

$$\eta\sim\frac{\beta_1}{n^2}+\frac{\beta_2}{n^4}+\frac{\beta_3}{n^6}+\ldots, \tag{7·20}$$

and

$$j'_{n,s}\sim n\sum_{r=0}^{\infty}\frac{q_r(\beta)}{n^{2r}}, \tag{7·21}$$

where the coefficients β_r and q_r are related by the formulae (7·8), (7·11) and (7·12) with the symbols B, p, ϵ and α, replaced by C, q, η and β, respectively. There is a difference between the two cases, however, because $C_0^1=\zeta$, and in place of (7·13) we have

$$\beta_1=-\zeta^{-1}C_0, \qquad \beta_2=-\zeta^{-1}(C_1+\beta_1C_1^1+\tfrac{1}{2}\beta_1^2C_0^2+\tfrac{1}{6}\beta_1^3C_0^3). \tag{7·22}$$

In consequence, although $q_0(\zeta)$ $(\equiv z)$ is a regular function of ζ throughout $\mathbf{D}$, the higher coefficients $q_r(\zeta)$ have poles at $\zeta=0$ of order $2r-1$ if $r\geqslant 1$. The error on curtailing the expansion (7·21) at the term $q_r(\beta)/n^{2r-1}$ is not $O(n^{-2r-1})$ but $O(n^{-\frac{2}{3}r-\frac{1}{3}})$, uniformly with respect to s.

The expansion for $J_n(j'_{n,s})$ corresponding to (7·15) may be derived in a similar way from the relations (Olver 1950, equations (3·9) and (2·15))

$$\pm J_n(j'_{n,s})=\left(\frac{j'^2_{n,s}-n^2}{2j'_{n,s}}\frac{\mathrm{d}j'_{n,s}}{\mathrm{d}s}\right)^{-\frac{1}{2}}, \qquad \pm\mathrm{Ai}(a'_s)=\left(a'_s\,\frac{\mathrm{d}a'_s}{\mathrm{d}s}\right)^{-\frac{1}{2}}. \tag{7·23}$$

The result is

$$J_n(j'_{n,s})\sim\mathrm{Ai}\,(a'_s)\,\frac{\phi(\beta)}{n^{\frac{1}{3}}}\left\{1+\sum_{r=1}^{\infty}\frac{Q_r(\beta)}{n^{2r}}\right\}, \tag{7·24}$$

where ϕ is defined by (6·7) and the coefficients Q_r are calculable from the asymptotic equality

$$\left(1+\sum_{r=1}^{\infty}\frac{Q_r}{n^{2r}}\right)^{-2} \sim \frac{\phi^2}{2z\zeta}\left\{\left(\sum_{r=0}^{\infty}\frac{q_r}{n^{2r}}\right)^2-1\right\}\left(\sum_{r=0}^{\infty}\frac{q_r'}{n^{2r}}\right)\left(1+\sum_{r=1}^{\infty}\frac{q_r/q_0}{n^{2r}}\right)^{-1}. \tag{7·25}$$

The first two are given by

$$Q_1 = \varpi_0 q_1' + \varpi_1 q_0', \qquad Q_2 = \tfrac{3}{2}Q_1^2 + \varpi_0 q_2' + \varpi_1 q_1' + \varpi_2 q_0', \tag{7·26}$$

where

$$\varpi_0 = -\frac{\phi^2}{4\zeta}\left(q_0-\frac{1}{q_0}\right) = \frac{1}{z\phi^2}, \qquad \varpi_1 = -\frac{\phi^2}{4\zeta}\left(q_1+\frac{q_1}{z^2}\right), \qquad \varpi_2 = -\frac{\phi^2}{4\zeta}\left(q_2+\frac{q_2}{z^2}-\frac{q_1^2}{z^3}\right). \tag{7·27}$$

The functions $Q_r(\zeta)$ have poles of order not exceeding† $2r$ at $\zeta = 0$, and if the series inside the braces of (7·24) is curtailed at the term $Q_r(\beta)/n^{2r}$, the error in the value for $J_n(j_{n,s}')$ is uniformly $O(n^{-\frac{2}{3}r-1})$.

The fact that the forms of the error terms of (7·21) and (7·24) are weaker than those of (7·10) and (7·15) is somewhat surprising. It arises essentially from the process of reversion used, which is not so effective applied to (6·8) as it is to (4·24). However, in numerical applications the difference is only marked for very low values of s. In these exceptional cases the 'lost' accuracy may always be recovered by solving the equation $J_n'(nz) = 0$ by successive approximation, evaluating the trial values of $J_n'(nz)$ by means of (6·8).

Zeros of functions of complex order

The extension of the results of this section to the half-plane $|\arg\nu| < \frac{1}{2}\pi$ presents no difficulty. The expansions (7·10), (7·15), (7·21) and (7·24) remain valid with n replaced by the complex variable ν and

$$\alpha = \nu^{-\frac{2}{3}}a_s, \qquad \beta = \nu^{-\frac{2}{3}}a_s',$$

these quantities now being complex. From § 5 it is seen that the points $z(\alpha)$ and $z(\beta)$ lie on the curve BR of figure 9, and so when $|\arg\nu| < \frac{1}{2}\pi$ and $|\nu|$ is large the zeros of $J_\nu(\nu z)$ and $J_\nu'(\nu z)$ lie asymptotically close to this curve.

8. Zeros of $Y_n(z)$ and $Y_n'(z)$

The analysis of § 7 may be repeated with equation (4·26) in place of (4·24). If, in the usual notation, $y_{n,s}$ and $y_{n,s}'$ denote the sth positive zeros of $Y_n(z)$ and $Y_n'(z)$ respectively, then the expansions (7·10), (7·15), (7·21) and (7·24) evidently remain valid with the symbols j and J replaced by y and Y respectively, provided that Ai and a are replaced by $-$Bi and b; the quantities α, β are now given by

$$\alpha = n^{-\frac{2}{3}}b_s, \qquad \beta = n^{-\frac{2}{3}}b_s',$$

(cf. (7·3) and (7·18)), where b_s, b_s' are the sth negative zeros of Bi (z), Bi′ (z) respectively.

An important difference emerges, however, between the cases of Y_n and J_n. Whereas the zeros of J_n are all real and positive, Y_n has in addition to its real zeros a number which are complex. In this section we investigate the distribution of the complex zeros.

It is shown in the Appendix that the function Bi (z) has a string of zeros $\beta_1, \beta_2, \ldots$, lying in the sector $\frac{1}{3}\pi < \arg z < \frac{1}{2}\pi$, together with a conjugate set $\bar{\beta}_1, \bar{\beta}_2, \ldots$, in the conjugate sector.

† It has been verified that the order is actually $2r-1$ for $r = 1, 2, 3$, and for these cases the truncation error is $O(n^{-\frac{2}{3}r-\frac{5}{3}})$.

If they are arranged in ascending order of modulus magnitude then the asymptotic expansion of β_s for large s is given by (cf. Appendix equations (A19) and (A22))

$$\beta_s \sim e^{\frac{1}{3}\pi i} \lambda^{\frac{2}{3}} (1 + \tfrac{5}{48}\lambda^{-2} - \tfrac{5}{36}\lambda^{-4} + \tfrac{77125}{82944}\lambda^{-6} - \ldots),$$

where
$$\lambda = \tfrac{3}{8}\{(4s-1)\pi + 2i \ln 2\}.$$

Thus the large zeros in this sector lie asymptotically close to $\arg z = \frac{1}{3}\pi$, but even the small zeros all lie very near to this ray; numerical calculation (see § 10) shows, for example, that $\arg \beta_1 = \frac{1}{3}\pi + 0{\cdot}095 \ldots$.

Corresponding to β_s there is, for large n, a zero $\bar{\eta}_{n,s}$ of $Y_n(z)$ whose asymptotic expansion is given by

$$\bar{\eta}_{n,s} \sim \sum_{r=0}^{\infty} \frac{p_r(n^{-\frac{2}{3}}\beta_s)}{n^{2r-1}} \tag{8·1}$$

(cf. (7·10)), where the functions $p_r(\zeta)$ are those of § 7. This expansion is not, however, uniformly valid with respect to all s; if $s \geqslant n$ the point $\zeta = n^{-\frac{2}{3}}\beta_s$ lies asymptotically close to the cut along the ray $\arg \zeta = \frac{1}{3}\pi$ from $\zeta = (\frac{3}{2}\pi)^{\frac{2}{3}} e^{\frac{1}{3}\pi i}$ to infinity. This cut is a boundary of the region of validity of (4·26) and the expansion (8·1) is uniformly valid only when $s \leqslant \kappa n$, where κ is any fixed number in the range $0 < \kappa < 1$.

The approximate distribution of $\bar{\eta}_{n,s}$ may be found by curtailing (8·1) at its first term, given by

$$\bar{\eta}_{n,s} = nz(n^{-\frac{2}{3}}\beta_s) + O(n^{-1}) \tag{8·2}$$

(see (7·12)), where $z(\zeta)$ is the function defined in § 4. If $s < n$, the point $\zeta = n^{-\frac{2}{3}}\beta_s$ lies close to the segment $BP'E'$ (figure 6) of the ray $\arg \zeta = \frac{1}{3}\pi$; this is mapped in the z-plane on the curve $BP'E'$ (figure 3), the lower boundary of the domain **K**. In consequence, the zeros $\bar{\eta}_{n,s}$ lie close to, and in fact just outside, the lower boundary of the domain† **nK**.

The relations (8·1) and (8·2) naturally break down when the value of s approaches n, that is, for zeros lying in the neighbourhood of the transition point $z = -1$. In order to achieve a more complete account we must consider the expansions of $Y_n(nz)$ corresponding to (4·26) for general phase ranges of z. Before proceeding to this we record some properties of a relevant Airy function.

The Airy function $\mathrm{Di}_m(z)$

This is defined by the equation

$$\mathrm{Di}_m(z) = mi\,\mathrm{Ai}(z) - \mathrm{Bi}(z), \tag{8·3}$$

where m is a constant, which we restrict here to be real. Other Airy functions may be expressed in terms of $\mathrm{Di}_m(z)$. Using the equation (*British Association Mathematical Tables* 1946, p. B 9)

$$\mathrm{Ai}(e^{\frac{2}{3}\pi i} z) = \tfrac{1}{2} e^{\frac{1}{3}\pi i}\{\mathrm{Ai}(z) - i\,\mathrm{Bi}(z)\},$$

we readily verify that

$$\mathrm{Di}_0(z) = -\mathrm{Bi}(z), \qquad \mathrm{Di}_1(z) = 2e^{\frac{5}{6}\pi i}\mathrm{Ai}(e^{-\frac{2}{3}\pi i} z), \qquad \mathrm{Di}_3(z) = 2e^{\frac{2}{3}\pi i}\mathrm{Bi}(e^{\frac{2}{3}\pi i} z). \tag{8·4}$$

Relations connecting functions having different values of m are

$$\mathrm{Di}_{-m}(z) = \overline{\mathrm{Di}_m(\bar{z})}, \qquad \mathrm{Di}_m(e^{\frac{2}{3}\pi i} z) = \tfrac{1}{2}(m-1)\, e^{-\frac{2}{3}\pi i}\, \mathrm{Di}_{(3+m)/(1-m)}(z). \tag{8·5}$$

† The 'domain **nK**' means the domain **K** magnified by the factor n.

The asymptotic expansions of $\mathrm{Di}_m(z)$ and $\mathrm{Di}'_m(z)$ for large $|z|$ may be obtained without difficulty from those of $\mathrm{Ai}(z)$ and $\mathrm{Bi}(z)$ given by equations (A 1) to (A 10) in the Appendix. Using the same notation we find that

$$\left.\begin{aligned}\mathrm{Di}_m(-z) &\sim \sqrt{\left(\frac{m^2-1}{\pi}\right)}\,\mathrm{i}z^{-\frac{1}{4}}\left\{\cos\left(\xi-\tfrac{1}{4}\pi+\frac{\mathrm{i}}{2}\ln\frac{m+1}{m-1}\right)P(\xi)+\sin\left(\xi-\tfrac{1}{4}\pi+\frac{\mathrm{i}}{2}\ln\frac{m+1}{m-1}\right)Q(\xi)\right\} \quad (m^2>1),\\ \mathrm{Di}'_m(-z) &\sim \sqrt{\left(\frac{m^2-1}{\pi}\right)}\,\mathrm{i}z^{\frac{1}{4}}\left\{\cos\left(\xi-\tfrac{3}{4}\pi+\frac{\mathrm{i}}{2}\ln\frac{m+1}{m-1}\right)R(\xi)+\sin\left(\xi-\tfrac{3}{4}\pi+\frac{\mathrm{i}}{2}\ln\frac{m+1}{m-1}\right)S(\xi)\right\} \quad (m^2>1),\end{aligned}\right\} \quad (8{\cdot}6)$$

$$\left.\begin{aligned}\mathrm{Di}_m(z\,\mathrm{e}^{\frac{1}{3}\pi\mathrm{i}}) &\sim \sqrt{\left(\frac{2(m-1)}{\pi}\right)}\,\mathrm{e}^{\frac{1}{6}\pi\mathrm{i}}\,z^{-\frac{1}{4}}\left\{\cos\left(\xi+\tfrac{1}{4}\pi+\frac{\mathrm{i}}{2}\ln\frac{m-1}{2}\right)P(\xi)+\sin\left(\xi+\tfrac{1}{4}\pi+\frac{\mathrm{i}}{2}\ln\frac{m-1}{2}\right)Q(\xi)\right\} \quad (m>1),\\ \mathrm{Di}'_m(z\,\mathrm{e}^{\frac{1}{3}\pi\mathrm{i}}) &\sim \sqrt{\left(\frac{2(m-1)}{\pi}\right)}\,\mathrm{e}^{-\frac{1}{6}\pi\mathrm{i}}\,z^{\frac{1}{4}}\left\{\cos\left(\xi-\tfrac{1}{4}\pi+\frac{\mathrm{i}}{2}\ln\frac{m-1}{2}\right)R(\xi)+\sin\left(\xi-\tfrac{1}{4}\pi+\frac{\mathrm{i}}{2}\ln\frac{m-1}{2}\right)S(\xi)\right\} \quad (m>1),\end{aligned}\right\} \quad (8{\cdot}7)$$

$$\left.\begin{aligned}\mathrm{Di}_m(z\,\mathrm{e}^{\frac{1}{3}\pi\mathrm{i}}) &\sim \sqrt{\left(\frac{2(1-m)}{\pi}\right)}\,\mathrm{e}^{\frac{1}{6}\pi\mathrm{i}}\,z^{-\frac{1}{4}}\left\{\cos\left(\xi+\tfrac{3}{4}\pi+\frac{\mathrm{i}}{2}\ln\frac{1-m}{2}\right)P(\xi)+\sin\left(\xi+\tfrac{3}{4}\pi+\frac{\mathrm{i}}{2}\ln\frac{1-m}{2}\right)Q(\xi)\right\} \quad (m<1),\\ \mathrm{Di}'_m(z\,\mathrm{e}^{\frac{1}{3}\pi\mathrm{i}}) &\sim \sqrt{\left(\frac{2(1-m)}{\pi}\right)}\,\mathrm{e}^{-\frac{1}{6}\pi\mathrm{i}}\,z^{\frac{1}{4}}\left\{\cos\left(\xi-\tfrac{3}{4}\pi+\frac{\mathrm{i}}{2}\ln\frac{1-m}{2}\right)R(\xi)+\sin\left(\xi-\tfrac{3}{4}\pi+\frac{\mathrm{i}}{2}\ln\frac{1-m}{2}\right)S(\xi)\right\} \quad (m<1),\end{aligned}\right\} \quad (8{\cdot}8)$$

each of these expansions being valid in the sector $|\arg z|<\frac{2}{3}\pi$. Expansions for $\mathrm{Di}_m(z\,\mathrm{e}^{-\frac{1}{3}\pi\mathrm{i}})$ and $\mathrm{Di}'_m(z\,\mathrm{e}^{-\frac{1}{3}\pi\mathrm{i}})$ corresponding to (8·7) and (8·8) may be obtained by changing the signs of i and m in the right-hand sides of these relations (see the first of (8·5)).

From these results it is seen that $\mathrm{Di}_m(z)$ and $\mathrm{Di}'_m(z)$ have strings of zeros lying asymptotically close to the rays $\arg z=\pi$, $\frac{1}{3}\pi$ and $-\frac{1}{3}\pi$. We denote those of $\mathrm{Di}_m(z)$ by $d_{m,s}$, $\delta_{m,s}$ and $\epsilon_{m,s}$ $(s=1,2,\ldots)$, respectively, and those of $\mathrm{Di}'_m(z)$ by $d'_{m,s}$, $\delta'_{m,s}$ and $\epsilon'_{m,s}$ $(s=1,2,\ldots)$, respectively, and suppose each string to be arranged in ascending order of modulus.

The asymptotic expansions of these zeros when s is large may be found by reversion of (8·6) to (8·8). Using the notation $T(\lambda)$, $U(\mu)$, $V(\lambda)$ and $W(\mu)$ defined by equation (A 19) of the Appendix, we find for the group of zeros near $\arg z=\pi$

$$\left.\begin{aligned}&d_{m,s}\sim -T(\lambda), && d'_{m,s}\sim -U(\mu),\\ &\mathrm{Di}'_m(d_{m,s})\sim(-)^{s-1}(m^2-1)^{\frac{1}{2}}\,\mathrm{i}V(\lambda), && \mathrm{Di}_m(d'_{m,s})\sim(-)^{s-1}(m^2-1)^{\frac{1}{2}}\,\mathrm{i}W(\mu),\\ &\text{where }\lambda=\frac{3}{8}\left\{(4s-1)\pi-2\mathrm{i}\ln\frac{m+1}{m-1}\right\}, && \mu=\frac{3}{8}\left\{(4s-3)\pi-2\mathrm{i}\ln\frac{m+1}{m-1}\right\}\quad(m^2>1).\end{aligned}\right\} \quad (8{\cdot}9)$$

For the zeros near $\arg z=\frac{1}{3}\pi$, we find

$$\left.\begin{aligned}&\delta_{m,s}\sim\mathrm{e}^{\frac{1}{3}\pi\mathrm{i}}\,T(\lambda), && \delta'_{m,s}\sim\mathrm{e}^{\frac{1}{3}\pi\mathrm{i}}\,U(\mu),\\ &\mathrm{Di}'_m(\delta_{m,s})\sim(-)^s\{2(m-1)\}^{\frac{1}{2}}\,\mathrm{e}^{\frac{1}{6}\pi\mathrm{i}}\,V(\lambda), && \mathrm{Di}_m(\delta'_{m,s})\sim(-)^s\{2(m-1)\}^{\frac{1}{2}}\,\mathrm{e}^{\frac{1}{6}\pi\mathrm{i}}\,W(\mu)\quad(m>1),\\ &\mathrm{Di}'_m(\delta_{m,s})\sim(-)^s\{2(1-m)\}^{\frac{1}{2}}\,\mathrm{e}^{\frac{2}{3}\pi\mathrm{i}}\,V(\lambda), && \mathrm{Di}_m(\delta'_{m,s})\sim(-)^s\{2(1-m)\}^{\frac{1}{2}}\,\mathrm{e}^{\frac{2}{3}\pi\mathrm{i}}\,W(\mu)\quad(m<1),\\ &\text{where }\lambda=\frac{3}{8}\left\{(4s-3)\pi-2\mathrm{i}\ln\frac{m-1}{2}\right\}, && \mu=\frac{3}{8}\left\{(4s-1)\pi-2\mathrm{i}\ln\frac{m-1}{2}\right\}\quad(m>1),\\ &\phantom{\text{where }}\lambda=\frac{3}{8}\left\{(4s-1)\pi-2\mathrm{i}\ln\frac{1-m}{2}\right\}, && \mu=\frac{3}{8}\left\{(4s-3)\pi-2\mathrm{i}\ln\frac{1-m}{2}\right\}\quad(m<1).\end{aligned}\right\} \quad (8{\cdot}10)$$

Finally, for the zeros near $\arg z = -\frac{1}{3}\pi$, we find

$$\left.\begin{array}{ll}
\epsilon_{m,s} \sim \mathrm{e}^{-\frac{1}{3}\pi\mathrm{i}}\, T(\lambda), & \epsilon'_{m,s} \sim \mathrm{e}^{-\frac{1}{3}\pi\mathrm{i}}\, U(\mu), \\
\mathrm{Di}'_m(\epsilon_{m,s}) \sim (-)^s \{2(1+m)\}^{\frac{1}{4}} \mathrm{e}^{-\frac{5}{6}\pi\mathrm{i}}\, V(\lambda), & \mathrm{Di}_m(\epsilon'_{m,s}) \sim (-)^s \{2(1+m)\}^{\frac{1}{4}} \mathrm{e}^{-\frac{1}{6}\pi\mathrm{i}}\, W(\mu) \quad (m>-1), \\
\mathrm{Di}'_m(\epsilon_{m,s}) \sim (-)^s \{2(-1-m)\}^{\frac{1}{4}} \mathrm{e}^{-\frac{1}{6}\pi\mathrm{i}}\, V(\lambda), & \mathrm{Di}_m(\epsilon'_{m,s}) \sim (-)^s \{2(-1-m)\}^{\frac{1}{4}} \mathrm{e}^{-\frac{5}{6}\pi\mathrm{i}}\, W(\mu) \quad (m<-1), \\
\text{where} \quad \lambda = \frac{3}{8}\left\{(4s-1)\,\pi + 2\mathrm{i}\ln\frac{1+m}{2}\right\}, & \mu = \frac{3}{8}\left\{(4s-3)\,\pi + 2\mathrm{i}\ln\frac{1+m}{2}\right\} \quad (m>-1), \\
\lambda = \frac{3}{8}\left\{(4s-3)\,\pi + 2\mathrm{i}\ln\frac{(-1-m)}{2}\right\}, & \mu = \frac{3}{8}\left\{(4s-1)\,\pi + 2\mathrm{i}\ln\frac{(-1-m)}{2}\right\} \quad (m<-1).
\end{array}\right\} \tag{8·11}$$

It can be verified by contour integration that $\mathrm{Di}_m(z)$ and $\mathrm{Di}'_m(z)$ have no zeros other than those enumerated above.

Zeros of $Y_n(z)$ and $Y'_n(z)$ for general phase ranges of z, when n is an integer

If m is any integer we have (Watson 1944, p. 75),

$$Y_n(z\,\mathrm{e}^{m\pi\mathrm{i}}) = \mathrm{e}^{-mn\pi\mathrm{i}}\, Y_n(z) + 2\mathrm{i}\sin mn\pi\, \cot n\pi J_n(z). \tag{8·12}$$

Hence, if n is an integer,

$$Y_n(z\,\mathrm{e}^{m\pi\mathrm{i}}) = (-)^{mn}\{Y_n(z) + 2\mathrm{i}mJ_n(z)\}. \tag{8·13}$$

Combining this result with (4·24) and (4·26), we see that

$$Y_n(nz\,\mathrm{e}^{m\pi\mathrm{i}}) \sim (-)^{mn}\left(\frac{4\zeta}{1-z^2}\right)^{\frac{1}{4}}\left\{\frac{\mathrm{Di}_{2m}(n^{\frac{2}{3}}\zeta)}{n^{\frac{1}{3}}}\sum_{s=0}^{\infty}\frac{A_s(\zeta)}{n^{2s}} + \frac{\mathrm{Di}'_{2m}(n^{\frac{2}{3}}\zeta)}{n^{\frac{5}{3}}}\sum_{s=0}^{\infty}\frac{B_s(\zeta)}{n^{2s}}\right\}, \tag{8·14}$$

for large n, where ζ, $A_s(\zeta)$ and $B_s(\zeta)$ are the same functions as in § 4, and $\mathrm{Di}_m(z)$ is defined by (8·3). The expansion (8·14) is valid when $|\arg z| < \pi$, but we shall use it only in the half-plane $|\arg z| \leqslant \frac{1}{2}\pi$.

Corresponding to a zero δ of $\mathrm{Di}_{2m}(z)$, there is a zero $z = \eta$ of $Y_n(z\,\mathrm{e}^{m\pi\mathrm{i}})$ given by (cf. (8·2))

$$\eta = nz(n^{-\frac{2}{3}}\delta) + O(n^{-1}). \tag{8·15}$$

Consider first the zeros corresponding to $d_{2m,s}$ $(s = 1, 2, \ldots)$. If s is large, and $m \neq 0$, we have from (8·9),

$$n^{-\frac{2}{3}}d_{2m,s} = -\left(\frac{3}{2n}\right)^{\frac{2}{3}}\left\{(s-\tfrac{1}{4})\,\pi - \frac{\mathrm{i}}{2}\ln\frac{2m+1}{2m-1}\right\}^{\frac{2}{3}}\{1 + O(s^{-2})\}.$$

If $s \gg n$, so that $n^{-\frac{2}{3}}d_{2m,s}$ is large, we deduce from this result, (8·15) and the first of (4·14) that

$$\eta = (s-\tfrac{1}{4})\,\pi + \tfrac{1}{2}n\pi - \frac{\mathrm{i}}{2}\ln\frac{2m+1}{2m-1} + O(n^2/s) + O(n^{-1}).$$

Letting $s \to \infty$, we see that the curve of zeros has the asymptote

$$\mathscr{I}z = -\tfrac{1}{2}\ln\frac{2m+1}{2m-1} + O(n^{-1}) \quad (m \neq 0). \tag{8·16}$$

This result may be otherwise obtained from (8·13) and Hankel's expansions (Watson 1944, p. 199) for $J_n(z)$ and $Y_n(z)$. In this way we find that if n is a positive integer or zero then the asymptote is given by

$$\mathscr{I}z = -\tfrac{1}{2}\ln\left|\frac{2m+1}{2m-1}\right|. \tag{8·17}$$

Thus the error term $O(n^{-1})$ in (8·16) is in fact zero.

Consider next the zeros corresponding to $\delta_{2m,s}$ $(s = 1, 2, \ldots)$. The points $\zeta = n^{-\frac{2}{3}}\delta_{2m,s}$, like the points $\zeta = n^{-\frac{2}{3}}\beta_s$ in (8·1), all lie near to the ray $\arg\zeta = \frac{1}{3}\pi$, and hence the corresponding zeros of $Y_n(nz\,\mathrm{e}^{m\pi\mathrm{i}})$ lie near to the lower boundary $BP'E'$ (figure 3) of the domain **K**. These zeros may be regarded as lying on a continuous curve, and it is of interest to know where this curve meets the imaginary z-axis, and how many zeros lie to the right of this axis.

If $m \geqslant 1$, we have from (8·10),

$$n^{-\frac{2}{3}}\delta_{2m,s} = \mathrm{e}^{\frac{1}{3}\pi\mathrm{i}}\left(\frac{3}{2n}\right)^{\frac{2}{3}}\{(s-\tfrac{3}{4})\,\pi - \tfrac{1}{2}\mathrm{i}\ln|m-\tfrac{1}{2}|\}^{\frac{2}{3}}\{1+O(s^{-2})\}. \tag{8·18}$$

The curve in the ζ-plane passing through the zeros is obtained by allowing s to be a continuous real variable. Now from § 4, the points

$$\zeta = \mathrm{e}^{\frac{1}{3}\pi\mathrm{i}}\,(\tfrac{3}{4}\pi)^{\frac{2}{3}} \quad \text{and} \quad z = \bar{z}_0 \equiv -\mathrm{i}(t_0^2-1)^{\frac{1}{2}} = -\mathrm{i}\,0·66274\ldots, \tag{8·19}$$

where t_0 is the positive root of $\coth t = t$, are corresponding points. At $z = \bar{z}_0$ we have, from (4·2),

$$\left(\frac{\mathrm{d}z}{\mathrm{d}\zeta}\right)_{z=\bar{z}_0} = -z\left(\frac{\zeta}{1-z^2}\right)^{\frac{1}{2}} = \mathrm{e}^{\frac{2}{3}\pi\mathrm{i}}\,(\tfrac{3}{4}\pi)^{\frac{1}{3}}\,(1-t_0^{-2})^{\frac{1}{2}}.$$

Setting $s = \frac{1}{2}n + \frac{3}{4}$ in (8·18), we find that the imaginary z-axis and the curve of zeros of $Y_n(nz\,\mathrm{e}^{m\pi\mathrm{i}})$ intersect at

$$z = \bar{z}_0 + \frac{\mathrm{i}}{2n}(1-t_0^{-2})^{\frac{1}{2}}\ln|m-\tfrac{1}{2}| + O(n^{-2}). \tag{8·20}$$

The zeros lying to the right of the imaginary z-axis are seen from (8·18) to be those for which $s - \frac{3}{4} < \frac{1}{2}n$; there are thus $[\frac{1}{2}n + \frac{1}{2}]$ such zeros.

When $m \leqslant 0$ the equation corresponding to (8·18) is obtained by replacing $\frac{3}{4}$ by $\frac{1}{4}$; the equation (8·20) remains valid, but the number of zeros to the right of the imaginary axis is now $[\frac{1}{2}n]$.

The analogous results concerning the zeros corresponding to $\epsilon_{2m,s}$ may be deduced by the use of the conjugate property (see the first of (8·5)),

$$\epsilon_{m,s} = \bar{\delta}_{-m,s}.$$

The curve of these zeros lies near the upper boundary of **K** and intersects the imaginary axis at the point

$$z = z_0 - \frac{\mathrm{i}}{2n}(1-t_0^{-2})^{\frac{1}{2}}\ln|m+\tfrac{1}{2}| + O(n^{-2}). \tag{8·21}$$

The number of zeros to the right of this axis is $[\frac{1}{2}n]$ or $[\frac{1}{2}n + \frac{1}{2}]$ according as $m \geqslant 0$ or $m \leqslant -1$.

We are now in a position to assess the asymptotic distribution of zeros of $Y_n(z\,\mathrm{e}^{m\pi\mathrm{i}})$ in the wider range $|\arg z| \leqslant \pi$. In the quadrant $\frac{1}{2}\pi \leqslant \arg z \leqslant \pi$ the zeros are the same as those of $Y_n(z\,\mathrm{e}^{(m+1)\pi\mathrm{i}})$ rotated through an angle π. There is an infinite string of zeros here if $m \geqslant -1$, having the asymptote

$$\mathscr{I}z = \tfrac{1}{2}\ln\left|\frac{2m+3}{2m+1}\right|$$

(cf. (8·17)), and the curve of zeros near the upper boundary of **nK** continues from the point given by (8·21) towards $z = -n$; the number of zeros on this segment is $[\frac{1}{2}n + \frac{1}{2}]$ or $[\frac{1}{2}n]$ according as $m \geqslant 0$ or $m \leqslant -1$. Combining this with the result for $|\arg z| \leqslant \frac{1}{2}\pi$, we see that for all values of m the total number of zeros on the upper curve is n.

Similarly, there is an infinite string of zeros in the quadrant $-\pi \leqslant \arg z \leqslant -\frac{1}{2}\pi$ if $m \leqslant 1$, and for any m the lower curve from $z = n$ to $z = -n$ also contains n zeros.

These results are illustrated in figures 10, 11 and 12. Figures 10 and 12 are conjugate to each other. The continuous curves are the boundaries of **nK**; they intersect the imaginary axis at the points nz_0 and $n\bar{z}_0$, where $\bar{z}_0$ is given by (8·19). The broken lines indicate the

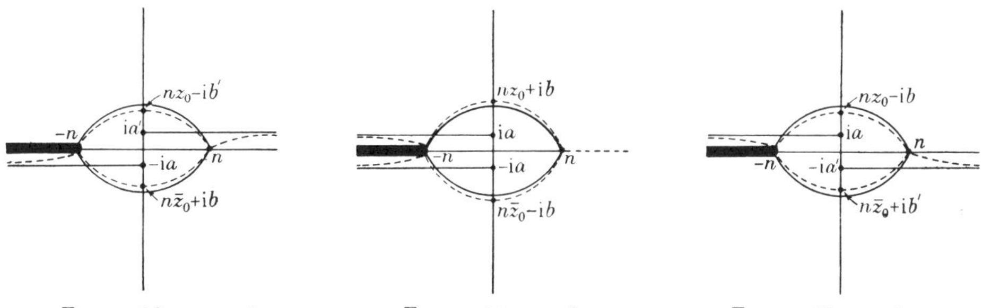

FIGURE 10. $m \leqslant -1$. FIGURE 11. $m = 0$. FIGURE 12. $m \geqslant 1$.

Distribution of the zeros ---- of $Y_n(z)$ in $(2m-1)\pi \leqslant \arg z \leqslant (2m+1)\pi$, when n is an integer.

curves on which the zeros lie. The displacements of the asymptotes of the infinite branches from the real axis are a and a', and the distances between the broken and continuous curves where they cross the imaginary axis are b and b', where, when $m \neq 0$, a, a', b and b' are given by

$$\left.\begin{aligned} a &= \tfrac{1}{2}\ln\frac{4\,|\,m\,|+3}{4\,|\,m\,|+1}, \qquad & b &= \tfrac{1}{2}(1-t_0^{-2})^{\frac{1}{2}}\ln(2\,|\,m\,|+\tfrac{1}{2})+O(n^{-1}), \\ a' &= \tfrac{1}{2}\ln\frac{4\,|\,m\,|+1}{4\,|\,m\,|-1}, \qquad & b' &= \tfrac{1}{2}(1-t_0^{-2})^{\frac{1}{2}}\ln(2\,|\,m\,|-\tfrac{1}{2})+O(n^{-1}). \end{aligned}\right\} \tag{8·22}$$

For $m = 0$ we have

$$a = \tfrac{1}{2}\ln 3 = 0{\cdot}54931\ldots, \quad b = \tfrac{1}{2}(1-t_0^{-2})^{\frac{1}{2}}\ln 2+O(n^{-1}) = 0{\cdot}19146\ldots+O(n^{-1}). \tag{8·23}$$

There are n zeros on each of the finite curves.

The zeros of $Y_n'(z)$ may be investigated in a similar manner, and the results are very similar. Figures 10, 11 and 12 are also applicable to this function, and the number of zeros on each of the finite curves is again n. The significant change is in the form of the error term (cf. (7·22)); in the formulae corresponding to (8·2), (8·15), (8·22) and (8·23) the error term is $O(n^{-\frac{2}{3}})$ and not $O(n^{-1})$.

The results obtained in this section are valid for large values of n. We can, however, determine by contour integration the number of zeros of $Y_n(z\,e^{m\pi i})$ and $Y_n'(z\,e^{m\pi i})$ in the strip $0 \leqslant \mathscr{R}z \leqslant (s+\frac{1}{2}n)\pi$ when n is any positive integer or zero, and s is an integer. We find, without difficulty, that if s is sufficiently large $Y_n(z\,e^{m\pi i})$ and $Y_n'(z\,e^{m\pi i})$ each have $n+s$ zeros in this strip, unless $m = 0$ and n is odd, in which event $Y_n(z)$ has $n+s-1$ zeros and $Y_n'(z)$ has $n+s+1$ zeros. Thus when n is a positive integer $Y_n(z)$ *and* $Y_n'(z)$ *have no zeros other than those indicated by the asymptotic formula* (8·15).

Tables for the numerical evaluation of the first term in the various formulae for the zeros and associated values are given in § 10.

Zeros of $Y_n(z)$ and $Y_n'(z)$ when n is half an odd integer

If we substitute $n = n_0+\epsilon$ in (8·12), where n_0 is an integer and $0\leqslant\epsilon<1$, we obtain

$$Y_n(z\,e^{m\pi i}) = (-)^{mn_0}\{e^{-m\epsilon\pi i}\,Y_n(z)+2i\sin m\epsilon\pi\,\cot\epsilon\pi\,J_n(z)\}. \tag{8·24}$$

The pattern of the zeros of $Y_n(z)$ for general phase ranges of z evidently depends on ϵ, the non-integral part of n. We have considered above the case of integer n; we now deal briefly with the case when n is half an odd integer.

If we put $\epsilon = \frac{1}{2}$ in (8·24), we obtain

$$Y_n(z\,e^{m\pi i}) = (-)^{mn_0}\,e^{-\frac{1}{2}m\pi i}\,Y_n(z).$$

The zeros of $Y_n(z\,e^{m\pi i})$ are thus the same as those of $Y_n(z)$. Because of this and the conjugate property it is sufficient to examine the region $-\frac{1}{2}\pi\leqslant\arg z\leqslant 0$. In this quadrant $Y_n(z)$ has a string of real zeros $y_{n,1}, y_{n,2}, \ldots$, and a number of complex zeros $\bar{\eta}_{n,1}, \bar{\eta}_{n,2}, \ldots$, lying near the lower boundary of **nK** and having the expansion (8·1). The curve on which they lie meets the imaginary z-axis at the point

$$z = n\bar{z}_0-\tfrac{1}{2}i(1-t_0^{-2})^{\frac{1}{2}}\ln 2+O(n^{-1}) \tag{8·25}$$

(cf. (8·20)), where $\bar{z}_0$ is given by (8·19). The number of such zeros lying to the right of this axis is $[\frac{1}{2}n_0]$; if n_0 is odd there is, in addition, a zero on the imaginary axis.

Figure 13 shows the distribution of zeros over the whole plane. The continuous curves are the boundaries of **nK**. The zeros lie on the real axis outside **nK** and on the broken curves. The latter intersect the imaginary axis at the point $n\bar{z}_0-ib$ and its conjugate, where $\bar{z}_0$ and b are given by (8·19) and (8·23). The total number of zeros on each broken curve is $n_0\equiv n-\frac{1}{2}$.

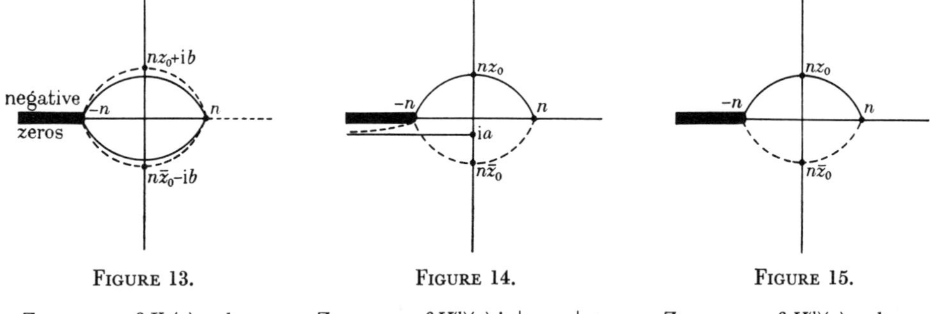

FIGURE 13. Zeros ---- of $Y_n(z)$, when n is half an odd integer.

FIGURE 14. Zeros ---- of $H_n^{(1)}(z)$ in $|\arg z|\leqslant\pi$, when n is an integer.

FIGURE 15. Zeros ---- of $H_n^{(1)}(z)$, when n is half an odd integer.

The same diagram is applicable to the zeros of $Y_n'(z)$, but the number of zeros on each broken curve is now $n_0+1\equiv n+\frac{1}{2}$, and the error term in the asymptotic formulae has to be modified.

As in the case of integer n, it can be shown by contour integration that $Y_n(z)$ and $Y_n(z)$ have no zeros other than those indicated by the asymptotic formulae.

9. Zeros of the Hankel functions and their derivatives

Zeros of $H_n^{(1)}(z)$ and $H_n^{(1)\prime}(z)$ when n is an integer

In this section we extend the analysis of § 8 to the Hankel functions. Only $H_n^{(1)}(z)$ will be considered, the zeros of $H_n^{(2)}(z)$ are the conjugates of those of $H_n^{(1)}(z)$.

If m is any integer we have (Watson 1944, p. 75),

$$H_n^{(1)}(z\,e^{m\pi i}) = e^{-mn\pi i}\,H_n^{(1)}(z) - 2\,e^{-n\pi i}\sin mn\pi \operatorname{cosec} n\pi J_n(z). \tag{9·1}$$

Hence if n is an integer

$$H_n^{(1)}(z\,e^{m\pi i}) = (-)^{mn}\{H_n^{(1)}(z) - 2mJ_n(z)\} = (-)^{mn}\,i\{Y_n(z) + (2m-1)\,iJ_n(z)\}. \tag{9·2}$$

Combining this result with (4·24) and (4·26), we obtain

$$H_n^{(1)}(nz\,e^{m\pi i}) \sim (-)^{mn}\,i\left(\frac{4\zeta}{1-z^2}\right)^{\frac{1}{4}}\left\{\frac{\mathrm{Di}_{2m-1}(n^{\frac{2}{3}}\zeta)}{n^{\frac{1}{3}}}\sum_{s=0}^{\infty}\frac{A_s(\zeta)}{n^{2s}} + \frac{\mathrm{Di}'_{2m-1}(n^{\frac{2}{3}}\zeta)}{n^{\frac{5}{3}}}\sum_{s=0}^{\infty}\frac{B_s(\zeta)}{n^{2s}}\right\}, \tag{9·3}$$

(cf. (8·14)), valid when $|\arg z| < \pi$, where $\mathrm{Di}_m(z)$ is defined by (8·3).

When $m \neq 0$ or 1, $\mathrm{Di}_{2m-1}(z)$ has a string of zeros $d_{2m-1,s}$ near the negative real axis (see § 8). Corresponding to these there is a string of zeros of $H_n^{(1)}(z\,e^{m\pi i})$ lying on a curve whose asymptote is given by

$$\mathscr{I} z = -\tfrac{1}{2}\ln\frac{m}{m-1} + O(n^{-1}) \tag{9·4}$$

(cf. (8·16)). With the aid of Hankel's expansions for $J_n(z)$ and $Y_n(z)$, we can show that the error term $O(n^{-1})$ in this equation is in fact zero. When $m = 0$ or 1, $\mathrm{Di}_{2m-1}(z)$ is a multiple of $\mathrm{Ai}(e^{\pm\frac{2}{3}\pi i}z)$ (see (8·4) and (8·5)), and has no zeros $d_{2m-1,s}$.

Next, if $m \neq 1$, $\mathrm{Di}_{2m-1}(z)$ has a set of zeros $\delta_{2m-1,s}$ near the ray $\arg z = \frac{1}{3}\pi$. Corresponding to these there are zeros of $H_n^{(1)}(z\,e^{m\pi i})$ lying near the lower boundary of **nK**; the curve on which they lie intersects the imaginary z-axis at the point

$$z = n\bar{z}_0 + \tfrac{1}{2}i(1-t_0^{-2})^{\frac{1}{2}}\ln|m-1| + O(n^{-1}) \tag{9·5}$$

(cf. (8·20)). The number of zeros to the right of this axis is $[\frac{1}{2}n+\frac{1}{2}]$ or $[\frac{1}{2}n]$ according as $m \gtrless 1$. If $m = 1$ there are no $\delta_{2m-1,s}$ zeros.

Lastly, if $m \neq 0$, $H_n^{(1)}(z\,e^{m\pi i})$ has a set of zeros lying near the upper boundary of **nK** corresponding to the zeros $\epsilon_{2m-1,s}$ of $\mathrm{Di}_{2m-1}(z)$. The curve of these zeros intersects the imaginary z-axis at the point

$$z = nz_0 - \tfrac{1}{2}i(1-t_0^{-2})^{\frac{1}{2}}\ln|m| + O(n^{-1}), \tag{9·6}$$

and the number of zeros to the right of this axis is $[\frac{1}{2}n]$ or $[\frac{1}{2}n+\frac{1}{2}]$ according as $m \gtrless 0$. If $m = 0$ there are no $\epsilon_{2m-1,s}$ zeros.

The asymptotic distribution of the zeros of $H_n^{(1)}(z)$ in the range

$$(2m-1)\,\pi \leqslant \arg z \leqslant (2m+1)\,\pi$$

may be deduced from these results in a manner similar to that used in § 8 for $Y_n(z)$. For the cases $m \leqslant -1$ and $m \geqslant 1$, the pattern of the zeros is given by figures 10 and 12 respectively, with the quantities a, a', b and b' now given by

$$\left.\begin{aligned} a &= \tfrac{1}{2}\ln\frac{|4m-1|+3}{|4m-1|+1}, & b &= \tfrac{1}{2}(1-t_0^{-2})^{\frac{1}{2}}\ln(|2m-\tfrac{1}{2}|+\tfrac{1}{2}) + O(n^{-1}),\\ a' &= \tfrac{1}{2}\ln\frac{|4m-1|+1}{|4m-1|-1}, & b' &= \tfrac{1}{2}(1-t_0^{-2})^{\frac{1}{2}}\ln(|2m-\tfrac{1}{2}|-\tfrac{1}{2}) + O(n^{-1}),\end{aligned}\right\} \tag{9·7}$$

in place of (8·22). There are n zeros on each of the finite curves.

Figure 14 corresponds to the case $m = 0$, for which there exist only one finite set and one infinite string of zeros. The former are asymptotically situated on the lower boundary of **nK**, and the asymptote of the latter is

$$\mathscr{I} z = -a = -\tfrac{1}{2}\ln 2 = -0{\cdot}34657\ldots.$$

There are n zeros on the finite curve.

Figures 10, 12 and 14 also hold for $H_n^{(1)\prime}(z)$, and on each of the finite curves this function has n zeros also.

It can be verified by contour integration that $H_n^{(1)}(z)$ and $H_n^{(1)\prime}(z)$ have no zeros other than those indicated by the asymptotic formulae.

Zeros of $H_n^{(1)}(z)$ and $H_n^{(1)\prime}(z)$ when n is half an odd integer

If we replace m by $2m$ in (9·1) and substitute $n = n_0+\frac{1}{2}$, where n_0 is an integer, we obtain

$$H_n^{(1)}(z\,\mathrm{e}^{2m\pi\mathrm{i}}) = (-)^m H_n^{(1)}(z).$$

The zeros of $H_n^{(1)}(z\,\mathrm{e}^{2m\pi\mathrm{i}})$ are thus the same as those of $H_n^{(1)}(z)$.

The expansion of $H_n^{(1)}(nz)$ in the range $|\arg z| < \pi$ is given by (9·3) with $m = 0$. The only zeros in the half-plane $|\arg z| \leqslant \frac{1}{2}\pi$ are those corresponding to $\delta_{-1,s} \equiv \mathrm{e}^{-\frac{2}{3}\pi\mathrm{i}}\,a_s$, and they lie asymptotically close to the lower boundary of **K**. The number of these zeros is $[\frac{1}{2}n_0]$, but if n_0 is odd there is also a zero on the imaginary axis.

If we substitute $m = \pm 1$, $n = n_0+\frac{1}{2}$ in (9·1), we obtain

$$H_n^{(1)}(z\,\mathrm{e}^{\pm\pi\mathrm{i}}) = \mp(-)^{n_0}\mathrm{i}\{H_n^{(1)}(z)-2J_n(z)\} = \pm(-)^{n_0}\mathrm{i}H_n^{(2)}(z) = \pm(-)^{n_0}\mathrm{i}\overline{H_n^{(1)}(\bar{z})},$$

and so the zeros of $H_n^{(1)}(z)$ in the quadrants $\frac{1}{2}\pi \leqslant \arg z \leqslant \pi$ and $-\pi \leqslant \arg z \leqslant -\frac{1}{2}\pi$ are the images in the imaginary axis of the zeros in the half-plane $|\arg z| \leqslant \frac{1}{2}\pi$.

Figure 15 illustrates the distribution of zeros over the z-plane. They lie asymptotically close to the lower boundary of **nK** and their total number is $n_0 \equiv n-\frac{1}{2}$. The zeros of $H_n^{(1)\prime}(z)$ are similarly distributed but their total number is $n_0+1 \equiv n+\frac{1}{2}$. It is evident from the form of the explicit formulae for $H_{n_0+\frac{1}{2}}^{(1)}$ and $H_{n_0+\frac{1}{2}}^{(1)\prime}$ (see, for example, Watson 1944, p. 80, equation (12)) that there are no zeros other than those indicated by the asymptotic formulae.

10. Tables for the calculation of zeros and associated quantities

For convenient numerical application of the asymptotic series of §§ 7, 8 and 9 we need tables of the coefficients $p_s(\zeta)$, $q_s(\zeta)$, $P_s(\zeta)$, $Q_s(\zeta)$, $\psi(\zeta)$ and $\phi(\zeta)$, and also of Airy function zeros. For real zeros this initial preparation is not prohibitive. The *British Association Mathematical Tables* (1946) give the first fifty zeros of Ai (z) and Ai′ (z) and the first twenty real zeros of Bi (z) and Bi′ (z) to eight decimal places. The writer has prepared comprehensive tables of coefficients which enable the corresponding Bessel function zeros and associated quantities to be computed to ten significant figures when $n \geqslant 5$, and they will be published elsewhere.

For the purpose of evaluating the complex zeros of $Y_n(z)$, $Y_n'(z)$, the Hankel functions and their derivatives in all phase ranges when n is a positive integer, the coefficients $p_s(\zeta)$, $q_s(\zeta)$, ..., need to be tabulated over a substantial part of the ζ-plane. Tables would also be required of the zeros of the Airy functions $\mathrm{Di}_m(z)$ and $\mathrm{Di}_m'(z)$, defined by (8·3), for all integer values of m. The preparation of comprehensive tables would be a large undertaking. In this section we give a useful set of skeleton tables which can be used, even for moderately large n, to calculate to three- or four-figure accuracy the smaller complex zeros of Bessel functions and associated quantities in the more important phase ranges.

If the asymptotic formulae of § 8 are curtailed at their first term and the substitution $\psi = 2/z\phi$ (cf. (6·9)) is made, we obtain the following equations, n being a positive integer,

$$\eta = nz + O(n^{-1}), \qquad Y_n'(\eta\, e^{m\pi i}) = (-)^{mn+m+1}\, 2n^{-\frac{2}{3}}(z\phi)^{-1}\, \mathrm{Di}_{2m}'(\gamma_{2m})\,\{1+O(n^{-2})\}, \tag{10·1}$$

$$\eta' = nz + O(n^{-\frac{1}{3}}), \qquad Y_n(\eta'\, e^{m\pi i}) = (-)^{mn}\, n^{-\frac{1}{3}}\phi\, \mathrm{Di}_{2m}(\gamma_{2m}')\,\{1+O(n^{-\frac{4}{3}})\}, \tag{10·2}$$

where γ_{2m}, γ_{2m}' are zeros of $\mathrm{Di}_{2m}(z)$, $\mathrm{Di}_{2m}'(z)$ respectively, and η, η' the corresponding roots of

$$Y_n(\eta\, e^{m\pi i}) = 0, \quad Y_n'(\eta'\, e^{m\pi i}) = 0.$$

The argument ζ of the functions z and ϕ is $n^{-\frac{2}{3}}\gamma_{2m}$ in (10·1) and $n^{-\frac{2}{3}}\gamma_{2m}'$ in (10·2). The analogous formulae for the Hankel function zeros are (see § 9)

$$\eta = nz + O(n^{-1}), \quad H_n^{(1)\prime}(\eta\, e^{m\pi i}) = \mathrm{i}(-)^{mn+m+1}\, 2n^{-\frac{2}{3}}(z\phi)^{-1}\, \mathrm{Di}_{2m-1}'(\gamma_{2m-1})\,\{1+O(n^{-2})\}, \tag{10·3}$$

$$\eta' = nz + O(n^{-\frac{1}{3}}), \quad H_n^{(1)}(\eta'\, e^{m\pi i}) = \mathrm{i}(-)^{mn}\, n^{-\frac{1}{3}}\phi\, \mathrm{Di}_{2m-1}(\gamma_{2m-1}')\,\{1+O(n^{-\frac{4}{3}})\}, \tag{10·4}$$

where η, η' are now roots of

$$H_n^{(1)}(\eta\, e^{m\pi i}) = 0, \qquad H_n^{(1)\prime}(\eta'\, e^{m\pi i}) = 0,$$

respectively, and the argument ζ of z and ϕ is $n^{-\frac{2}{3}}\gamma_{2m-1}$ in (10·3) and $n^{-\frac{2}{3}}\gamma_{2m-1}'$ in (10·4). The tables in this section give three-decimal values of γ_m, γ_m', $\mathrm{Di}_m'(\gamma_m)$, $\mathrm{Di}_m(\gamma_m')$, $z(\zeta)$ and $\phi(\zeta)$.

The modulus and phase of the first five complex zeros $\beta_1, \beta_2, \ldots, \beta_5$ of $\mathrm{Bi}(z) = -\mathrm{Di}_0(z)$ and $\beta_1', \beta_2', \ldots, \beta_5'$ of $\mathrm{Bi}'(z)$ are given in table 1, together with the modulus and phase of $\mathrm{Bi}'(\beta_s)$ and $\mathrm{Bi}(\beta_s')$. These zeros lie in the upper half-plane; the corresponding zeros in the lower half-plane are their conjugates.

Tables $2a$, $2b$ and $2c$ give zeros of the function

$$\mathrm{Di}_2(z) \equiv 2\mathrm{i}\,\mathrm{Ai}(z) - \mathrm{Bi}(z).$$

The first five members of each set $\delta_{2,s}$, $\delta_{2,s}'$, $\epsilon_{2,s}$, $\epsilon_{2,s}'$, $d_{2,s}$ and $d_{2,s}'$, defined in § 8, are given, together with the associated values of Di_2' or Di_2.

Larger zeros of the sets given in tables 1 to $2c$ may readily be found by use of the asymptotic series (8·9) to (8·11). To four-figure accuracy the quantities $T(\lambda)$, $U(\mu)$, $V(\lambda)$ and $W(\mu)$ reduce to a single term when $s \geqslant 6$.

With the aid of tables 1 to $2c$ and the *British Association Mathematical Tables* (1946) giving the real zeros of Ai, Ai′, Bi and Bi′, we may trivially obtain the corresponding zeros and associated values of Di_m and Di_m' for any of the values $m = 0, \pm 1, \pm 2, \pm 3, \pm 5$. From (8·4) and (8·5) it is seen that the zeros of $\mathrm{Di}_{\pm 1}$ are $e^{\pm\frac{2}{3}\pi i}$ times those of Ai, the zeros of $\mathrm{Di}_{\pm 3}$ are $e^{\mp\frac{2}{3}\pi i}$ times those of Bi, and the zeros of $\mathrm{Di}_{\pm 5}$ are $e^{\pm\frac{2}{3}\pi i}$ times those of $\mathrm{Di}_{\mp 2}$, the last relation following from (8·5) with $m = 2, 5$. The same relations hold for the zeros of the derivatives.

We now consider tables of $z(\zeta)$ and $\phi(\zeta)$. In the application of (10·1), (10·2), (10·3) and (10·4) it will be noticed that $\arg\zeta = \arg\gamma_m$ or $\arg\gamma_m'$. In consequence, for the values of m contemplated in tables 1 to $2c$, $\arg\zeta$ is approximately $\pm\frac{1}{3}\pi$ or π. Thus $z(\zeta)$ and $\phi(\zeta)$ are needed only near the rays $\arg\zeta = \frac{1}{3}\pi$ and $\arg\zeta = \pi$; the values near $\arg\zeta = -\frac{1}{3}\pi$ are obtainable from the former by means of the conjugate property $z(\bar{\zeta}) = \overline{z(\zeta)}$.

Table 3 gives the modulus and phase of the function $z(\zeta)$ for

$$\arg(\zeta e^{-\frac{1}{3}\pi i}) = -0\cdot15\,(0\cdot05)+0\cdot10, \qquad |\zeta| = 0\,(0\cdot1)\,l,$$

where l is just greater than the value for which $\arg z = -\frac{1}{2}\pi$. It is linearly interpolable in both directions with a possible error of 2 units in the end figure for the higher values of $|\zeta|$. This table covers all the values of ζ near $\arg\zeta = \frac{1}{3}\pi$ which are required in the application of (10·1) to (10·4) using the zeros of the functions $\mathrm{Di}_m(z)$, $\mathrm{Di}'_m(z)$ $(m = 0, \pm1, \pm2, \pm3, \pm5)$, with the exception of β'_1, $\delta_{2,1}$, $\delta'_{-2,1}$, $\delta'_{-3,1}$, $\delta_{5,1}$ and $\delta'_{-5,1}$. These exceptions are covered by table 3*a* which gives $z(\zeta)$ for values of $\arg\zeta$ equal to the phases of these particular zeros. This table is linearly interpolable in the $|\zeta|$ direction; the question of interpolation in the $\arg\zeta$ direction does not, of course, arise.

Tables 4 and 4*a* give the values of $\phi(\zeta)$ corresponding to $z(\zeta)$ in tables 3 and 3*a*. They are linearly interpolable.

Tables 5, 5*a*, 6 and 6*a* give, in a similar manner, values of $z(\zeta)$ and $\phi(\zeta)$ near $\arg\zeta = \pi$. They too are linearly interpolable.

Application of the tables: numerical example

When the order n is large and positive, integer or not, we can, with the aid of the tables, calculate the zeros of $Y_n(z)$, $Y'_n(z)$, $H^{(1)}_n(z)$ and $H^{(1)\prime}_n(z)$ in the sector $|\arg z| \leqslant \frac{1}{2}\pi$. When n is a positive integer we can also calculate the zeros in $|\arg z| \leqslant \frac{1}{2}\pi$ of the functions $Y_n(z\,e^{m\pi i})$, $Y'_n(z\,e^{m\pi i})$ $(m = 0, \pm1)$, and $H^{(1)}_n(z\,e^{m\pi i})$, $H^{(1)\prime}_n(z\,e^{m\pi i})$ $(m = -2\,(1)\,3)$. In terms of general phase ranges this means we can, for integer n, evaluate the zeros of $Y_n(z)$, $Y'_n(z)$ in $|\arg z| \leqslant \frac{3}{2}\pi$, and the zeros of $H^{(1)}_n(z)$, $H^{(1)\prime}_n(z)$ in $-\frac{5}{2}\pi \leqslant \arg z \leqslant \frac{7}{2}\pi$; the corresponding range for $H^{(2)}_n(z)$, $H^{(2)\prime}_n(z)$ is $-\frac{7}{2}\pi \leqslant \arg z \leqslant \frac{5}{2}\pi$.

As an example we evaluate a few complex zeros of $Y_3(z)$. If $n = 3$, then $n^{-\frac{2}{3}} = 0\cdot4807$, and using table 1 we obtain

$$|n^{-\frac{2}{3}}\beta_1| = 1\cdot132, \qquad \arg(n^{-\frac{2}{3}}\beta_1) = \tfrac{1}{3}\pi + 0\cdot095.$$

Entering tables 3 and 4 with $|\zeta| = 1\cdot132$, $\arg\zeta = \frac{1}{3}\pi + 0\cdot095$, we find by interpolation

$$z = 0\cdot783\exp(-0\cdot934i), \qquad \phi = 1\cdot362\exp(0\cdot171i),$$

and so for the corresponding zero $\bar{\eta}_{3,1}$ of $Y_3(z)$ we have approximately (see (10·1))

$$\bar{\eta}_{3,1} = nz = 2\cdot349\exp(-0\cdot934i), \qquad Y'_3(\bar{\eta}_{3,1}) = 2n^{-\frac{2}{3}}(z\phi)^{-1}\,\mathrm{Bi}'(\beta_1) = 0\cdot895\exp(-2\cdot879i),$$

the value of $\mathrm{Bi}'(\beta_1)$ being obtained from table 1.

There are two other zeros $\bar{\eta}_{3,2}$ and $\bar{\eta}_{3,3}$, say, near the lower boundary of **nK** (figure 11), but the same procedure with β_2 or β_3 in place of β_1 fails because $|\zeta|$ lies outside the range of table 3. This means that $\bar{\eta}_{3,2}$ and $\bar{\eta}_{3,3}$ lie in the quadrant $-\pi < \arg z < -\frac{1}{2}\pi$. Accordingly, we enter tables 3*a* and 4*a* with the following values obtained from table 2*a*:

$$|\zeta| = |n^{-\frac{2}{3}}\delta_{2,1}| = 0\cdot590, \qquad \arg\zeta = \arg\delta_{2,1},$$

and we find that

$$z = 0\cdot917\exp(-0\cdot472i), \qquad \phi = 1\cdot291\exp(0\cdot091i).$$

These correspond to a zero of $Y_3(3z\,e^{\pi i})$; the related zero $\eta_{3,3}$ of $Y_3(z)$ is given by

$$\eta_{3,3} = 3\,e^{\pi i}z = 2\cdot751\exp(2\cdot670i),$$

$$Y'_3(\eta_{3,3}) = -2n^{-\frac{2}{3}}(z\phi)^{-1}\,\mathrm{Di}'_2(\delta_{2,1}) = 0\cdot696\exp(1\cdot480i).$$

This actually lies on the upper branch; the corresponding zero on the lower branch is of course its conjugate. Similarly, using $\delta_{2,2}$ in place of $\delta_{2,1}$ we find

$$\eta_{3,2} = 2{\cdot}202 \exp(1{\cdot}779\mathrm{i}), \qquad Y_3'(\eta_{3,2}) = 0{\cdot}986 \exp(-0{\cdot}958\mathrm{i}).$$

The zeros on the infinite dotted curves of figure 11 may be found in a similar manner. For the first zero on the upper branch, $y_{3,1}^{(1)}$ say, we find that

$$|\zeta| = |n^{-\frac{2}{3}} d_{2,1}| = 1{\cdot}143, \qquad \arg\zeta = \arg d_{2,1} = \pi - 0{\cdot}150 \quad \text{(table } 2c\text{)},$$

$$z = 2{\cdot}138 \exp(-0{\cdot}097\mathrm{i}), \qquad \phi = 1{\cdot}062 \exp(0{\cdot}024\mathrm{i}) \quad \text{(tables 5, 6)},$$

$$\left.\begin{aligned} y_{3,1}^{(1)} &= 3\,\mathrm{e}^{\pi\mathrm{i}}\,z = 6{\cdot}414 \exp(3{\cdot}045\mathrm{i}) \\ Y_3'(y_{3,1}^{(1)}) &= -2n^{-\frac{2}{3}}(z\phi)^{-1}\,\mathrm{Di}_2'(d_{2,1}) = 0{\cdot}516 \exp(-1{\cdot}534\mathrm{i}) \end{aligned}\right\} \quad \text{(table } 2c\text{)}.$$

For the purpose of comparison accurate values of these zeros of $Y_3(z)$ and the corresponding $Y_3'(z)$ have been computed by solving the equation $Y_3(z) = 0$ by successive approximation, the values of $Y_3(z)$ and $Y_3'(z)$ required being evaluated from their ascending series (Copson 1944, p. 329). The accurate values for the upper zeros are as follows:

$$\eta_{3,1} = 2{\cdot}352 \exp(0{\cdot}932\mathrm{i}), \qquad Y_3'(\eta_{3,1}) = 0{\cdot}895 \exp(2{\cdot}881\mathrm{i}),$$

$$\eta_{3,2} = 2{\cdot}205 \exp(1{\cdot}781\mathrm{i}), \qquad Y_3'(\eta_{3,2}) = 0{\cdot}986 \exp(-0{\cdot}959\mathrm{i}),$$

$$\eta_{3,3} = 2{\cdot}754 \exp(2{\cdot}671\mathrm{i}), \qquad Y_3'(\eta_{3,3}) = 0{\cdot}696 \exp(1{\cdot}478\mathrm{i}),$$

$$y_{3,1}^{(1)} = 6{\cdot}416 \exp(3{\cdot}046\mathrm{i}), \qquad Y_3'(y_{3,1}^{(1)}) = 0{\cdot}516 \exp(-1{\cdot}534\mathrm{i}).$$

The greatest error in the approximate values derived above is about 0·2 %. The discrepancies arise from neglect of $n^{-1}p_1(\zeta)$ and higher terms in (8·1), and behave relatively as $O(n^{-2})$ for large n. We may infer that if $n \geq 4$ the values obtained by using formulae (10·1) and (10·3) are accurate to the number of figures provided in the tables. The zeros of the derivatives are a little less accurately determined however, because the relative error in their case is only $O(n^{-\frac{4}{3}})$.

Previous numerical information on the complex zeros of Bessel functions is almost non-existent, save for orders 0 and 1. Complex zeros of Y_0 and Y_1 are given by the National Bureau of Standards Computation Laboratory (1950, pp. 405–406), for example,

$$\eta_{1,1} = -0{\cdot}50274\ 3273 + 0{\cdot}78624\ 3714\mathrm{i}, \qquad \eta_{1,1}' = 0{\cdot}57678\ 5129 + 0{\cdot}90398\ 4792\mathrm{i}.$$

The tables given here yield the approximate values

$$\eta_{1,1} = -0{\cdot}486 + 0{\cdot}786\mathrm{i}, \qquad \eta_{1,1}' = 0{\cdot}604 + 0{\cdot}784\mathrm{i}.$$

It is remarkable that formulae (10·1) and (10·2), which were derived on the assumption that n is large, should give reasonable results even for $n = 1$.

Computation of the tables

The method used for computing tables 1 to $2c$ is not without interest. The asymptotic expansions (8·9) to (8·11) were used to obtain all the quantities required as accurately as the series permitted. The values so obtained were then checked by using ascending series to recalculate the values of Di_m and Di'_m at the zeros. When this check failed, particularly for $s=1$ and 2 where the asymptotic series yielded insufficient accuracy, new values were calculated by use of Newton's rule and the ascending series evaluated afresh.

The series used for the zeros near the rays $\arg z = \frac{1}{3}\pi$ are based on the fact that Airy functions with argument $z\,\mathrm{e}^{\pm\frac{1}{3}\pi\mathrm{i}}$ can be expressed in terms of Airy functions with argument $-z$; see (8·5). Thus we have

$$\mathrm{Di}_m(z\,\mathrm{e}^{\frac{1}{3}\pi\mathrm{i}}) = \tfrac{1}{2}\mathrm{e}^{-\frac{1}{3}\pi\mathrm{i}}\{(m-3)\,\mathrm{i}\,\mathrm{Ai}(-z)-(m+1)\,\mathrm{Bi}(-z)\}. \tag{10·5}$$

Putting $z=x+h$, and expanding with the aid of Taylor's theorem and the differential equation satisfied by $\mathrm{Ai}(-x)$ and $\mathrm{Bi}(-x)$, we obtain

$$\mathrm{Di}_m\{(x+h)\,\mathrm{e}^{\frac{1}{3}\pi\mathrm{i}}\} = \tfrac{1}{2}\mathrm{e}^{-\frac{1}{3}\pi\mathrm{i}}\left\{F_m(-x)\sum_{r=0}^{\infty}\frac{h^r}{r!}\phi_r + F'_m(-x)\sum_{r=0}^{\infty}\frac{h^r}{r!}\psi_r\right\}, \tag{10·6}$$

$$\mathrm{Di}'_m\{(x+h)\,\mathrm{e}^{\frac{1}{3}\pi\mathrm{i}}\} = \tfrac{1}{2}\mathrm{e}^{-\frac{2}{3}\pi\mathrm{i}}\left\{F_m(-x)\sum_{r=0}^{\infty}\frac{h^r}{r!}\phi_{r+1} + F'_m(-x)\sum_{r=0}^{\infty}\frac{h^r}{r!}\psi_{r+1}\right\}, \tag{10·7}$$

where $$F_m(-x) \equiv (m-3)\,\mathrm{i}\,\mathrm{Ai}(-x)-(m+1)\,\mathrm{Bi}(-x),$$

and ϕ_r, ψ_r are polynomials in x, given by

$$\phi_0 = 1,\quad \phi_1 = 0,\quad \phi_2 = -x,\quad \phi_3 = -1,\quad \phi_4 = x^2,$$
$$\psi_0 = 0,\quad \psi_1 = -1,\quad \psi_2 = 0,\quad \psi_3 = x,\quad \psi_4 = 2,$$

higher members being readily obtainable from the equations

$$\phi_{r+1} = \phi'_r + x\psi_r, \qquad \psi_{r+1} = \psi'_r - \phi_r.$$

In the numerical application of (10·6) and (10·7), the value of x was taken to be a convenient number approximately equal to $\mathscr{R}(\delta_{m,s}\,\mathrm{e}^{-\frac{1}{3}\pi\mathrm{i}})$ or $\mathscr{R}(\delta'_{m,s}\,\mathrm{e}^{-\frac{1}{3}\pi\mathrm{i}})$, and the values of $\mathrm{Ai}(-x)$, $\mathrm{Bi}(-x)$, $\mathrm{Ai}'(-x)$ and $\mathrm{Bi}'(-x)$ were obtained from the *British Association Mathematical Tables* (1946). Seven terms of each series ensured six-decimal accuracy.

Tables 3 to 6*a* were computed using equations (4·7) and (6·7), the values of σ being found by successive approximation using Newton's rule. All the tables were originally calculated to six decimal places.

The author wishes to acknowledge the assistance of Miss P. M. Sivyer, Miss J. Staton and Mrs O. E. Taylor who computed the tables given in this section.

The work described above has been carried out as part of the research programme of the National Physical Laboratory, and this paper is published by permission of the Director of the Laboratory.

TABLE 1

s	$e^{-\frac{1}{3}\pi i}\,\beta_s$ mod.	phase	$\mathrm{Bi}'(\beta_s)$ mod.	phase	$e^{-\frac{1}{3}\pi i}\,\beta'_s$ mod.	phase	$\mathrm{Bi}(\beta'_s)$ mod.	phase
1	2·354	0·095	0·993	2·641	1·121	0·331	0·750	0·466
2	4·093	0·042	1·136	−0·513	3·257	0·059	0·592	−2·632
3	5·524	0·027	1·224	2·625	4·824	0·033	0·538	0·515
4	6·789	0·020	1·288	−0·519	6·166	0·023	0·506	−2·624
5	7·946	0·015	1·340	2·622	7·374	0·017	0·484	0·519

TABLE 2*a*

s	$e^{-\frac{1}{3}\pi i}\,\delta_{2,s}$ mod.	phase	$\mathrm{Di}'_2(\delta_{2,s})$ mod.	phase	$e^{-\frac{1}{3}\pi i}\,\delta'_{2,s}$ mod.	phase	$\mathrm{Di}_2(\delta'_{2,s})$ mod.	phase
1	1·227	0·249	0·857	−2·043	2·312	0·100	0·642	−1·070
2	3·279	0·058	1·076	1·061	4·079	0·042	0·561	2·084
3	4·835	0·033	1·184	−2·086	5·515	0·027	0·520	−1·054
4	6·172	0·023	1·258	1·053	6·783	0·020	0·494	2·090
5	7·378	0·017	1·315	−2·090	7·942	0·015	0·475	−1·051

TABLE 2*b*

s	$e^{\frac{1}{3}\pi i}\,\epsilon_{2,s}$ mod.	phase	$\mathrm{Di}'_2(\epsilon_{2,s})$ mod.	phase	$e^{\frac{1}{3}\pi i}\,\epsilon'_{2,s}$ mod.	phase	$\mathrm{Di}_2(\epsilon'_{2,s})$ mod.	phase
1	2·344	0·056	1·718	0·537	1·057	0·208	1·307	2·583
2	4·090	0·024	1·967	−2·612	3·251	0·035	1·026	−0·532
3	5·522	0·016	2·119	0·527	4·821	0·019	0·932	2·613
4	6·787	0·011	2·231	−2·615	6·164	0·013	0·877	−0·527
5	7·945	0·009	2·321	0·526	7·373	0·010	0·838	2·615

TABLE 2*c*

s	$-d_{2,s}$ mod.	phase	$\mathrm{Di}'_2(d_{2,s})$ mod.	phase	$-d'_{2,s}$ mod.	phase	$\mathrm{Di}_2(d'_{2,s})$ mod.	phase
1	2·377	−0·150	1·219	1·535	1·241	−0·463	0·907	1·657
2	4·101	−0·066	1·392	−1·587	3·270	−0·094	0·725	−1·548
3	5·528	−0·042	1·499	1·560	4·830	−0·052	0·659	1·584
4	6·792	−0·031	1·578	−1·579	6·169	−0·036	0·620	−1·562
5	7·948	−0·025	1·641	1·565	7·376	−0·027	0·593	1·578

Table 3. $|z|$ and $\arg z$

$[\theta \equiv \arg(\zeta\, e^{-\frac{1}{3}\pi i})]$

θ / $\lvert\zeta\rvert$	−0·15 $\lvert z\rvert$	−0·10 $\lvert z\rvert$	−0·05 $\lvert z\rvert$	0·00 $\lvert z\rvert$	0·05 $\lvert z\rvert$	0·10 $\lvert z\rvert$	−0·15 $\arg z$	−0·10 $\arg z$	−0·05 $\arg z$	0·00 $\arg z$	0·05 $\arg z$	0·10 $\arg z$
0·0	1·000	1·000	1·000	1·000	1·000	1·000	−0·000	−0·000	−0·000	−0·000	−0·000	−0·000
0·1	0·952	0·955	0·958	0·962	0·965	0·969	0·063	0·066	0·068	0·070	0·072	0·073
0·2	0·907	0·913	0·919	0·926	0·933	0·940	0·129	0·134	0·138	0·142	0·145	0·148
0·3	0·864	0·874	0·883	0·893	0·903	0·914	0·197	0·204	0·210	0·216	0·221	0·225
0·4	0·825	0·837	0·850	0·863	0·876	0·890	0·268	0·277	0·285	0·292	0·299	0·304
0·5	0·788	0·803	0·818	0·835	0·852	0·869	−0·342	−0·353	−0·362	−0·371	−0·378	−0·385
0·6	0·753	0·771	0·789	0·809	0·829	0·850	0·418	0·431	0·442	0·452	0·460	0·467
0·7	0·720	0·741	0·763	0·785	0·809	0·833	0·497	0·511	0·524	0·535	0·544	0·551
0·8	0·690	0·713	0·738	0·764	0·791	0·819	0·579	0·594	0·608	0·620	0·630	0·637
0·9	0·662	0·688	0·716	0·745	0·775	0·806	0·663	0·680	0·695	0·707	0·717	0·725
1·0	0·635	0·664	0·695	0·727	0·761	0·796	−0·751	−0·769	−0·785	−0·797	−0·807	−0·814
1·1	0·611	0·643	0·677	0·712	0·750	0·789	0·841	0·860	0·876	0·889	0·898	0·905
1·2	0·588	0·623	0·660	0·699	0·740	0·783	0·935	0·955	0·971	0·984	0·992	0·997
1·3	0·567	0·605	0·645	0·688	0·733	0·780	1·031	1·052	1·068	1·080	1·088	1·091
1·4	0·548	0·589	0·632	0·678	0·728	0·779	1·132	1·153	1·169	1·180	1·186	1·187
1·5	0·530	0·574	0·621	0·671	0·725	0·781	−1·235	−1·257	−1·272	−1·281	−1·286	−1·285
1·6	0·514	0·561	0·612	0·666	0·724	0·786	1·343	1·364	1·378	1·386	1·388	1·384
1·7	0·499	0·549	0·604	0·663	0·727	0·794	1·454	1·475	1·487	1·493	1·492	1·485
1·8	0·485	0·539	0·599	0·663	0·732	0·805	1·570	1·589	1·601	1·603	1·599	1·587
1·9	0·473	—	—	—	—	—	1·689	—	—	—	—	—

Table 3*a*. $|z|$ and $\arg z$

$\arg\zeta$	$\arg\beta'_1$ $(\frac{1}{3}\pi+0·33073)$		$\arg\delta_{2,1}$ $(\frac{1}{3}\pi+0·24909)$		$\arg\delta'_{-2,1}$ $(\frac{1}{3}\pi-0·20756)$		$\arg\delta'_{-3,1}$ $(\frac{1}{3}\pi-0·33073)$		$\arg\delta_{5,1}$ $(\frac{1}{3}\pi-0·24909)$		$\arg\delta'_{-5,1}$ $(\frac{1}{3}\pi-0·46335)$	
$\lvert\zeta\rvert$	$\lvert z\rvert$	$\arg z$	$\lvert z\rvert$	$\arg z$	$\lvert z\rvert$	$\arg z$	$\lvert z\rvert$	$\arg z$	$\lvert z\rvert$	$\arg z$	$\lvert z\rvert$	$\arg z$
0·0	1·000	−0·000	1·000	−0·000	1·000	−0·000	1·000	−0·000	1·000	−0·000	1·000	−0·000
0·1	0·986	0·078	0·980	0·077	0·949	0·060	0·942	0·053	0·946	0·058	0·935	0·045
0·2	0·975	0·158	0·962	0·155	0·900	0·123	0·887	0·109	0·895	0·119	0·874	0·092
0·3	0·966	0·238	0·947	0·235	0·854	0·189	0·835	0·168	0·847	0·182	0·816	0·142
0·4	0·959	0·318	0·934	0·315	0·811	0·257	0·785	0·229	0·802	0·248	0·761	0·195
0·5	0·955	−0·400	0·924	−0·397	0·771	−0·328	0·739	−0·293	0·760	−0·317	0·710	−0·250
0·6	0·954	0·482	0·916	0·480	0·733	0·402	0·695	0·360	0·720	0·389	0·661	0·307
0·7	0·955	0·564	0·910	0·563	0·698	0·478	0·654	0·430	0·682	0·463	0·614	0·368
0·8	0·958	0·647	0·907	0·648	0·664	0·558	0·615	0·503	0·647	0·541	0·571	0·432
0·9	0·965	0·730	0·907	0·733	0·633	0·640	0·579	0·580	0·614	0·621	0·530	0·498
1·0	0·974	−0·813	0·909	−0·819	0·604	−0·726	0·544	−0·659	0·583	−0·705	0·492	−0·568
1·1	0·986	0·897	0·913	0·906	0·577	0·814	0·512	0·742	0·553	0·792	0·456	0·641
1·2	1·001	0·980	0·921	0·993	0·551	0·906	0·482	0·828	0·526	0·883	0·422	0·717
1·3	1·020	1·063	0·931	1·081	0·527	1·002	0·453	0·918	0·500	0·977	0·390	0·797
1·4	1·042	1·146	0·945	1·169	0·505	1·101	0·426	1·011	0·476	1·074	0·360	0·880
1·5	1·068	−1·228	0·963	−1·257	0·484	−1·204	0·401	−1·108	0·454	−1·176	0·332	−0·966
1·6	1·099	1·310	0·984	1·345	0·465	1·310	0·377	1·209	0·433	1·281	0·306	1·056
1·7	1·134	1·391	1·009	1·433	0·447	1·421	0·354	1·315	0·413	1·390	0·282	1·150
1·8	1·175	1·470	1·040	1·521	0·430	1·536	0·333	1·424	0·394	1·504	0·259	1·247
1·9	1·221	1·548	1·076	1·608	0·414	1·655	0·313	1·537	0·376	1·622	0·237	1·347
2·0	1·274	−1·625	—	—	—	—	0·294	−1·655	—	—	0·217	−1·451
2·1	—	—	—	—	—	—	—	—	—	—	0·199	1·558
2·2	—	—	—	—	—	—	—	—	—	—	0·181	1·669

TABLE 4. $|\phi|$ AND $\arg \phi$

$[\theta \equiv \arg (\zeta\, e^{-\frac{1}{3}\pi i})]$

θ / $\lvert\zeta\rvert$	−0·15 $\lvert\phi\rvert$	−0·10 $\lvert\phi\rvert$	−0·05 $\lvert\phi\rvert$	0·00 $\lvert\phi\rvert$	0·05 $\lvert\phi\rvert$	0·10 $\lvert\phi\rvert$	−0·15 arg ϕ	−0·10 arg ϕ	−0·05 arg ϕ	0·00 arg ϕ	0·05 arg ϕ	0·10 arg ϕ
0·0	1·260	1·260	1·260	1·260	1·260	1·260	0·000	0·000	0·000	0·000	0·000	0·000
0·1	1·272	1·272	1·271	1·270	1·269	1·268	0·012	0·013	0·013	0·014	0·014	0·014
0·2	1·285	1·283	1·282	1·280	1·278	1·276	0·025	0·026	0·027	0·028	0·028	0·029
0·3	1·298	1·295	1·293	1·290	1·287	1·285	0·037	0·039	0·040	0·042	0·043	0·044
0·4	1·311	1·308	1·304	1·301	1·297	1·293	0·050	0·052	0·054	0·056	0·057	0·058
0·5	1·324	1·320	1·316	1·311	1·306	1·302	0·063	0·065	0·067	0·070	0·071	0·073
0·6	1·338	1·333	1·327	1·322	1·316	1·310	0·075	0·078	0·081	0·084	0·086	0·088
0·7	1·352	1·346	1·339	1·333	1·326	1·319	0·088	0·092	0·095	0·098	0·101	0·103
0·8	1·366	1·359	1·352	1·344	1·336	1·328	0·101	0·105	0·109	0·112	0·116	0·119
0·9	1·381	1·373	1·365	1·356	1·347	1·338	0·113	0·118	0·123	0·127	0·131	0·134
1·0	1·396	1·387	1·378	1·368	1·358	1·347	0·126	0·132	0·137	0·141	0·146	0·150
1·1	1·412	1·402	1·392	1·381	1·369	1·357	0·139	0·145	0·151	0·156	0·161	0·166
1·2	1·429	1·418	1·406	1·394	1·381	1·368	0·151	0·158	0·165	0·171	0·177	0·182
1·3	1·447	1·435	1·422	1·408	1·394	1·379	0·164	0·172	0·179	0·186	0·193	0·198
1·4	1·466	1·452	1·438	1·423	1·407	1·391	0·176	0·185	0·194	0·202	0·209	0·216
1·5	1·485	1·471	1·456	1·439	1·421	1·403	0·189	0·199	0·209	0·218	0·226	0·233
1·6	1·507	1·491	1·475	1·456	1·437	1·416	0·201	0·212	0·223	0·234	0·243	0·251
1·7	1·529	1·513	1·495	1·475	1·453	1·430	0·213	0·226	0·238	0·250	0·261	0·270
1·8	1·554	1·537	1·517	1·496	1·472	1·446	0·224	0·239	0·253	0·267	0·279	0·290
1·9	1·580	—	—	—	—	—	0·235	—	—	—	—	—

TABLE 4*a*. $|\phi|$ AND $\arg \phi$

arg ζ / $\lvert\zeta\rvert$	arg β_1' ($\frac{1}{3}\pi+0{\cdot}33073$) $\lvert\phi\rvert$	arg ϕ	arg $\delta_{2,1}$ ($\frac{1}{3}\pi+0{\cdot}24909$) $\lvert\phi\rvert$	arg ϕ	arg $\delta'_{-2,1}$ ($\frac{1}{3}\pi-0{\cdot}20756$) $\lvert\phi\rvert$	arg ϕ	arg $\delta'_{-3,1}$ ($\frac{1}{3}\pi-0{\cdot}33073$) $\lvert\phi\rvert$	arg ϕ	arg $\delta_{5,1}$ ($\frac{1}{3}\pi-0{\cdot}24909$) $\lvert\phi\rvert$	arg ϕ	arg $\delta'_{-5,1}$ ($\frac{1}{3}\pi-0{\cdot}46335$) $\lvert\phi\rvert$	arg ϕ
0·0	1·260	0·000	1·260	0·000	1·260	0·000	1·260	0·000	1·260	0·000	1·260	0·000
0·1	1·264	0·016	1·265	0·015	1·273	0·012	1·275	0·010	1·274	0·011	1·277	0·009
0·2	1·267	0·031	1·271	0·031	1·287	0·024	1·291	0·021	1·288	0·023	1·294	0·018
0·3	1·271	0·047	1·276	0·046	1·301	0·036	1·306	0·032	1·303	0·034	1·311	0·026
0·4	1·275	0·063	1·281	0·062	1·315	0·048	1·322	0·042	1·317	0·046	1·329	0·035
0·5	1·278	0·079	1·287	0·077	1·329	0·060	1·338	0·053	1·332	0·057	1·347	0·044
0·6	1·282	0·095	1·292	0·093	1·344	0·072	1·355	0·063	1·348	0·069	1·365	0·053
0·7	1·286	0·111	1·298	0·109	1·359	0·084	1·372	0·073	1·363	0·080	1·384	0·061
0·8	1·289	0·128	1·303	0·125	1·374	0·096	1·389	0·084	1·379	0·092	1·403	0·070
0·9	1·293	0·144	1·309	0·142	1·390	0·108	1·407	0·094	1·396	0·103	1·422	0·078
1·0	1·296	0·161	1·315	0·158	1·406	0·119	1·425	0·104	1·413	0·114	1·442	0·086
1·1	1·300	0·179	1·321	0·176	1·423	0·131	1·444	0·114	1·431	0·126	1·463	0·094
1·2	1·303	0·197	1·327	0·193	1·441	0·143	1·464	0·123	1·449	0·136	1·483	0·101
1·3	1·307	0·215	1·333	0·211	1·460	0·154	1·484	0·133	1·469	0·147	1·504	0·108
1·4	1·310	0·234	1·339	0·230	1·480	0·166	1·505	0·142	1·489	0·158	1·526	0·115
1·5	1·313	0·253	1·345	0·249	1·500	0·177	1·527	0·150	1·510	0·168	1·548	0·121
1·6	1·316	0·273	1·352	0·269	1·522	0·187	1·549	0·158	1·532	0·177	1·571	0·127
1·7	1·318	0·294	1·358	0·289	1·545	0·198	1·573	0·165	1·556	0·187	1·594	0·132
1·8	1·320	0·315	1·364	0·311	1·570	0·207	1·597	0·172	1·581	0·195	1·617	0·136
1·9	1·322	0·338	1·370	0·334	1·597	0·216	1·622	0·178	1·607	0·203	1·640	0·140
2·0	1·322	0·361	—	—	—	—	1·648	0·182	—	—	1·663	0·143
2·1	—	—	—	—	—	—	—	—	—	—	1·686	0·146
2·2	—	—	—	—	—	—	—	—	—	—	1·709	0·148

TABLE 5. $|z|$ AND $\arg z$

$[\tau \equiv \arg(-\zeta)]$

τ / $\lvert\zeta\rvert$	0·00 $\lvert z\rvert$	−0·05 $\lvert z\rvert$	−0·10 $\lvert z\rvert$	−0·15 $\lvert z\rvert$	0·00 $\arg z$	−0·05 $\arg z$	−0·10 $\arg z$	−0·15 $\arg z$
0·0	1·000	1·000	1·000	1·000	0†	−0·000	−0·000	−0·000
0·1	1·081	1·081	1·081	1·080		0·004	0·008	0·011
0·2	1·166	1·166	1·165	1·164		0·007	0·015	0·022
0·3	1·255	1·255	1·254	1·252		0·011	0·022	0·032
0·4	1·348	1·347	1·346	1·344		0·014	0·028	0·042
0·5	1·444	1·443	1·442	1·439		−0·017	−0·034	−0·051
0·6	1·544	1·543	1·541	1·538		0·020	0·039	0·059
0·7	1·647	1·646	1·644	1·641		0·022	0·045	0·067
0·8	1·754	1·753	1·751	1·747		0·025	0·050	0·074
0·9	1·865	1·864	1·861	1·856		0·027	0·054	0·081
1·0	1·979	1·978	1·975	1·970		−0·029	−0·059	−0·088
1·1	2·097	2·095	2·092	2·086		0·031	0·063	0·094
1·2	2·218	2·216	2·213	2·207		0·033	0·066	0·100
1·3	2·342	2·341	2·337	2·330		0·035	0·070	0·105
1·4	2·470	2·468	2·464	2·457		0·037	0·073	0·110
1·5	2·601	2·599	2·595	2·587		−0·038	−0·077	−0·115
1·6	2·735	2·734	2·729	2·721		0·040	0·079	0·119
1·7	2·873	2·871	2·866	2·858		0·041	0·082	0·123
1·8	3·013	3·012	3·006	2·998		0·042	0·085	0·127
1·9	3·157	3·155	3·150	3·141		0·044	0·087	0·131
2·0	3·304	3·302	3·296	3·287		−0·045	−0·090	−0·135
2·1	3·454	3·452	3·446	3·436		0·046	0·092	0·138
2·2	3·607	3·605	3·599	3·588		0·047	0·094	0·141
2·3	3·763	3·760	3·754	3·744		0·048	0·096	0·144
2·4	3·921	3·919	3·913	3·902		0·049	0·098	0·147
2·5	4·083	4·081	4·074	4·063		−0·050	−0·100	−0·150

τ / $\lvert\zeta\rvert$	0·00 $\lvert z\rvert$	−0·05 $\lvert z\rvert$	−0·05 $\arg z$
2·5	4·08	4·08	−0·050
2·6	4·25	4·25	0·051
2·7	4·41	4·41	0·052
2·8	4·58	4·58	0·052
2·9	4·76	4·76	0·053
3·0	4·93	4·93	−0·054
3·1	5·11	5·11	0·054
3·2	5·29	5·29	0·055
3·3	5·48	5·47	0·056
3·4	5·66	5·66	0·056
3·5	5·85	5·85	−0·057
3·6	6·04	6·04	0·057
3·7	6·24	6·23	0·058
3·8	6·43	6·43	0·058
3·9	6·63	6·63	0·059
4·0	6·83	6·83	−0·059
4·1	7·03	7·03	0·060
4·2	7·24	7·24	0·060
4·3	7·45	7·45	0·060
4·4	7·66	7·66	0·061
4·5	7·87	7·87	−0·061
4·6	8·09	8·08	0·061
4·7	8·30	8·30	0·062
4·8	8·52	8·52	0·062
4·9	8·74	8·74	0·062
5·0	8·97	8·97	−0·063

† $\arg z$ is zero in this column.

TABLE 5a. $|z|$ AND $\arg z$

$\arg\zeta$ / $\lvert\zeta\rvert$	$\arg d'_{2,1}$ $(\pi - 0{\cdot}46335)$ $\lvert z\rvert$	$\arg z$	$\arg d'_{3,1}$ $(\pi - 0{\cdot}33073)$ $\lvert z\rvert$	$\arg z$	$\arg d'_{5,1}$ $(\pi - 0{\cdot}20756)$ $\lvert z\rvert$	$\arg z$
0·0	1·000	−0·000	1·000	−0·000	1·000	−0·000
0·1	1·073	0·034	1·077	0·025	1·080	0·016
0·2	1·149	0·067	1·157	0·049	1·163	0·031
0·3	1·229	0·098	1·242	0·071	1·250	0·045
0·4	1·313	0·127	1·330	0·092	1·340	0·058
0·5	1·400	−0·154	1·421	−0·111	1·435	−0·070
0·6	1·491	0·180	1·517	0·130	1·533	0·082
0·7	1·586	0·204	1·615	0·147	1·635	0·093
0·8	1·684	0·227	1·718	0·163	1·740	0·103
0·9	1·786	0·248	1·824	0·178	1·849	0·112
1·0	1·891	−0·269	1·934	−0·193	1·961	−0·121
1·1	2·001	0·288	2·047	0·206	2·077	0·130
1·2	2·114	0·306	2·164	0·219	—	—
1·3	2·230	0·323	—	—	—	—

TABLE 6. $|\phi|$ AND $\arg \phi$

$[\tau \equiv \arg(-\zeta)]$

τ / $\lvert\zeta\rvert$	0·00 $\lvert\phi\rvert$	−0·05 $\lvert\phi\rvert$	−0·10 $\lvert\phi\rvert$	−0·15 $\lvert\phi\rvert$	0·00 $\arg\phi$	−0·05 $\arg\phi$	−0·10 $\arg\phi$	−0·15 $\arg\phi$
0·0	1·260	1·260	1·260	1·260	0†	0·000	0·000	0·000
0·1	1·240	1·240	1·240	1·240		0·001	0·002	0·002
0·2	1·221	1·221	1·221	1·221		0·002	0·003	0·005
0·3	1·202	1·202	1·202	1·202		0·002	0·005	0·007
0·4	1·183	1·183	1·184	1·184		0·003	0·006	0·009
0·5	1·165	1·165	1·166	1·166		0·004	0·008	0·011
0·6	1·148	1·148	1·148	1·149		0·005	0·009	0·014
0·7	1·131	1·131	1·131	1·132		0·005	0·010	0·016
0·8	1·114	1·114	1·115	1·115		0·006	0·012	0·018
0·9	1·098	1·098	1·099	1·099		0·007	0·013	0·020
1·0	1·082	1·082	1·083	1·084		0·007	0·014	0·021
1·1	1·067	1·067	1·068	1·068		0·008	0·016	0·023
1·2	1·052	1·052	1·053	1·054		0·008	0·017	0·025
1·3	1·038	1·038	1·038	1·039		0·009	0·018	0·027
1·4	1·024	1·024	1·024	1·025		0·009	0·019	0·028
1·5	1·010	1·010	1·011	1·012		0·010	0·020	0·030
1·6	0·997	0·997	0·998	0·999		0·010	0·021	0·031
1·7	0·984	0·984	0·985	0·986		0·011	0·022	0·033
1·8	0·972	0·972	0·972	0·973		0·011	0·023	0·034
1·9	0·959	0·960	0·960	0·961		0·012	0·024	0·035
2·0	0·948	0·948	0·949	0·949		0·012	0·024	0·037
2·1	0·936	0·937	0·937	0·938		0·013	0·025	0·038
2·2	0·925	0·925	0·926	0·927		0·013	0·026	0·039
2·3	0·914	0·915	0·915	0·916		0·013	0·027	0·040
2·4	0·904	0·904	0·905	0·906		0·014	0·027	0·041
2·5	0·894	0·894	0·895	0·895		0·014	0·028	0·042

τ / $\lvert\zeta\rvert$	0·00 $\lvert\phi\rvert$	−0·05 $\lvert\phi\rvert$	−0·05 $\arg\phi$
2·5	0·894	0·894	0·014
2·6	0·884	0·884	0·014
2·7	0·874	0·874	0·015
2·8	0·865	0·865	0·015
2·9	0·856	0·856	0·015
3·0	0·847	0·847	0·016
3·1	0·838	0·838	0·016
3·2	0·830	0·830	0·016
3·3	0·821	0·822	0·016
3·4	0·813	0·814	0·017
3·5	0·806	0·806	0·017
3·6	0·798	0·798	0·017
3·7	0·791	0·791	0·017
3·8	0·783	0·784	0·017
3·9	0·776	0·776	0·018
4·0	0·769	0·770	0·018
4·1	0·763	0·763	0·018
4·2	0·756	0·756	0·018
4·3	0·750	0·750	0·018
4·4	0·743	0·743	0·018
4·5	0·737	0·737	0·019
4·6	0·731	0·731	0·019
4·7	0·725	0·725	0·019
4·8	0·720	0·720	0·019
4·9	0·714	0·714	0·019
5·0	0·708	0·708	0·019

† $\arg \phi$ is zero in this column.

TABLE 6a. $|\phi|$ AND $\arg \phi$

$\arg\zeta$ / $\lvert\zeta\rvert$	$\arg d'_{2,1}$ $(\pi - 0·46335)$ $\lvert\phi\rvert$	$\arg\phi$	$\arg d'_{3,1}$ $(\pi - 0·33073)$ $\lvert\phi\rvert$	$\arg\phi$	$\arg d'_{5,1}$ $(\pi - 0·20756)$ $\lvert\phi\rvert$	$\arg\phi$
0·0	1·260	0·000	1·260	0·000	1·260	0·000
0·1	1·242	0·007	1·241	0·005	1·241	0·003
0·2	1·225	0·014	1·223	0·010	1·222	0·006
0·3	1·208	0·021	1·205	0·015	1·203	0·010
0·4	1·191	0·028	1·187	0·020	1·185	0·013
0·5	1·174	0·034	1·170	0·025	1·167	0·016
0·6	1·158	0·041	1·153	0·030	1·150	0·019
0·7	1·142	0·047	1·137	0·034	1·133	0·022
0·8	1·127	0·053	1·121	0·038	1·117	0·024
0·9	1·111	0·059	1·105	0·043	1·101	0·027
1·0	1·097	0·065	1·089	0·047	1·085	0·030
1·1	1·082	0·071	1·075	0·051	1·070	0·032
1·2	1·068	0·076	1·060	0·055	—	—
1·3	1·054	0·081	—	—	—	—

Appendix. Some properties of the Airy functions Ai (z) and Bi (z) in the complex plane

Asymptotic expansions

Let us write
$$\xi \equiv \tfrac{2}{3} z^{\frac{3}{2}}, \tag{A 1}$$

$$u_s \equiv \frac{(2s+1)(2s+3)(2s+5)\dots(6s-1)}{s!\,(216)^s}, \qquad v_s \equiv -\frac{6s+1}{6s-1} u_s, \tag{A 2}$$

and
$$\left.\begin{aligned} L(\xi) &\equiv \sum_{s=0}^{\infty} \frac{u_s}{\xi^s} = 1 + \frac{3.5}{1!\,216}\frac{1}{\xi} + \frac{5.7.9.11}{2!\,(216)^2}\frac{1}{\xi^2} + \frac{7.9.11.13.15.17}{3!\,(216)^3}\frac{1}{\xi^3} + \dots, \\ M(\xi) &\equiv \sum_{s=0}^{\infty} \frac{v_s}{\xi^s} = 1 - \frac{3.7}{1!\,216}\frac{1}{\xi} - \frac{5.7.9.13}{2!\,(216)^2}\frac{1}{\xi^2} - \frac{7.9.11.13.15.19}{3!\,(216)^3}\frac{1}{\xi^3} - \dots, \end{aligned}\right\} \tag{A 3}$$

$$\left.\begin{aligned} P(\xi) &\equiv \sum_{s=0}^{\infty} (-)^s \frac{u_{2s}}{\xi^{2s}} = 1 - \frac{5.7.9.11}{2!\,(216)^2}\frac{1}{\xi^2} + \frac{9.11.13.15.17.19.21.23}{4!\,(216)^4}\frac{1}{\xi^4} - \dots, \\ Q(\xi) &\equiv \sum_{s=0}^{\infty} (-)^s \frac{u_{2s+1}}{\xi^{2s+1}} = \frac{3.5}{1!\,216}\frac{1}{\xi} - \frac{7.9.11.13.15.17}{3!\,(216)^3}\frac{1}{\xi^3} + \dots, \end{aligned}\right\} \tag{A 4}$$

$$\left.\begin{aligned} R(\xi) &\equiv \sum_{s=0}^{\infty} (-)^s \frac{v_{2s}}{\xi^{2s}} = 1 + \frac{5.7.9.13}{2!\,(216)^2}\frac{1}{\xi^2} - \frac{9.11.13.15.17.19.21.25}{4!\,(216)^4}\frac{1}{\xi^4} + \dots, \\ S(\xi) &\equiv \sum_{s=0}^{\infty} (-)^s \frac{v_{2s+1}}{\xi^{2s+1}} = -\frac{3.7}{1!\,216}\frac{1}{\xi} + \frac{7.9.11.13.15.19}{3!\,(216)^3}\frac{1}{\xi^3} - \dots. \end{aligned}\right\} \tag{A 5}$$

Then if $|z|$ is large

$$\mathrm{Ai}\,(z) \sim \tfrac{1}{2}\pi^{-\frac{1}{2}} z^{-\frac{1}{4}}\, \mathrm{e}^{-\xi} L(-\xi), \qquad \mathrm{Ai}'\,(z) \sim -\tfrac{1}{2}\pi^{-\frac{1}{2}} z^{\frac{1}{4}}\, \mathrm{e}^{-\xi} M(-\xi), \quad (|\arg z| < \pi), \tag{A 6}$$

$$\left.\begin{aligned} \mathrm{Ai}\,(-z) &\sim \pi^{-\frac{1}{2}} z^{-\frac{1}{4}} \{\cos(\xi - \tfrac{1}{4}\pi)\, P(\xi) + \sin(\xi - \tfrac{1}{4}\pi)\, Q(\xi)\} \\ \mathrm{Ai}'\,(-z) &\sim \pi^{-\frac{1}{2}} z^{\frac{1}{4}} \{\cos(\xi - \tfrac{3}{4}\pi)\, R(\xi) + \sin(\xi - \tfrac{3}{4}\pi)\, S(\xi)\} \end{aligned}\right\} \quad (|\arg z| < \tfrac{2}{3}\pi), \tag{A 7}$$

$$\mathrm{Bi}\,(z) \sim \pi^{-\frac{1}{2}} z^{-\frac{1}{4}}\, \mathrm{e}^{\xi} L(\xi), \qquad \mathrm{Bi}'\,(z) \sim \pi^{-\frac{1}{2}} z^{\frac{1}{4}}\, \mathrm{e}^{\xi} M(\xi), \quad (|\arg z| < \tfrac{1}{3}\pi), \tag{A 8}$$

$$\left.\begin{aligned} \mathrm{Bi}\,(-z) &\sim \pi^{-\frac{1}{2}} z^{-\frac{1}{4}} \{\cos(\xi + \tfrac{1}{4}\pi)\, P(\xi) + \sin(\xi + \tfrac{1}{4}\pi)\, Q(\xi)\} \\ \mathrm{Bi}'\,(-z) &\sim \pi^{-\frac{1}{2}} z^{\frac{1}{4}} \{\cos(\xi - \tfrac{1}{4}\pi)\, R(\xi) + \sin(\xi - \tfrac{1}{4}\pi)\, S(\xi)\} \end{aligned}\right\} \quad (|\arg z| < \tfrac{2}{3}\pi), \tag{A 9}$$

$$\left.\begin{aligned} \mathrm{Bi}\,(z\,\mathrm{e}^{\pm\frac{1}{3}\pi\mathrm{i}}) &\sim (2/\pi)^{\frac{1}{2}}\, \mathrm{e}^{\pm\frac{1}{6}\pi\mathrm{i}}\, z^{-\frac{1}{4}} \{\cos(\xi - \tfrac{1}{4}\pi \mp \tfrac{1}{2}\mathrm{i}\ln 2)\, P(\xi) + \sin(\xi - \tfrac{1}{4}\pi \mp \tfrac{1}{2}\mathrm{i}\ln 2)\, Q(\xi)\} \\ \mathrm{Bi}'\,(z\,\mathrm{e}^{\pm\frac{1}{3}\pi\mathrm{i}}) &\sim (2/\pi)^{\frac{1}{2}}\, \mathrm{e}^{\mp\frac{1}{6}\pi\mathrm{i}}\, z^{\frac{1}{4}} \{\cos(\xi + \tfrac{1}{4}\pi \mp \tfrac{1}{2}\mathrm{i}\ln 2)\, R(\xi) + \sin(\xi + \tfrac{1}{4}\pi \mp \tfrac{1}{2}\mathrm{i}\ln 2)\, S(\xi)\} \end{aligned}\right\}$$
$$(|\arg z| < \tfrac{2}{3}\pi), \tag{A 10}$$

Expansions (A 6) and (A 7) for Ai and Ai′ may be derived from the well-known asymptotic expansions of Bessel functions of large argument (Watson 1944, chap. 7), using the relations

$$\mathrm{Ai}\,(z) = \frac{z^{\frac{1}{2}}}{\pi\sqrt{3}} K_{\frac{1}{3}}(\xi), \qquad \mathrm{Ai}\,(-z) = \frac{z^{\frac{1}{2}}}{2\sqrt{3}} \{\mathrm{e}^{\frac{1}{6}\pi\mathrm{i}} H^{(1)}_{\frac{1}{3}}(\xi) + \mathrm{e}^{-\frac{1}{6}\pi\mathrm{i}} H^{(2)}_{\frac{1}{3}}(\xi)\},$$

$$\mathrm{Ai}'\,(z) = -\frac{z}{\pi\sqrt{3}} K_{\frac{2}{3}}(\xi), \qquad \mathrm{Ai}'\,(-z) = \frac{z}{2\sqrt{3}} \{\mathrm{e}^{-\frac{1}{6}\pi\mathrm{i}} H^{(1)}_{\frac{2}{3}}(\xi) + \mathrm{e}^{\frac{1}{6}\pi\mathrm{i}} H^{(2)}_{\frac{2}{3}}(\xi)\}.$$

The expansions (A 8), (A 9) and (A 10) for Bi and Bi′ may be derived from (A 6) and (A 7) by means of the relations

$$\mathrm{Bi}\,(z) = \mathrm{e}^{\frac{1}{6}\pi\mathrm{i}}\, \mathrm{Ai}\,(\mathrm{e}^{\frac{2}{3}\pi\mathrm{i}} z) + \mathrm{e}^{-\frac{1}{6}\pi\mathrm{i}}\, \mathrm{Ai}\,(\mathrm{e}^{-\frac{2}{3}\pi\mathrm{i}} z), \quad \mathrm{Bi}'\,(z) = \mathrm{e}^{\frac{5}{6}\pi\mathrm{i}}\, \mathrm{Ai}'\,(\mathrm{e}^{\frac{2}{3}\pi\mathrm{i}} z) + \mathrm{e}^{-\frac{5}{6}\pi\mathrm{i}}\, \mathrm{Ai}'\,(\mathrm{e}^{-\frac{2}{3}\pi\mathrm{i}} z).$$

Distribution of zeros

This may be investigated by Lommel's method (Watson 1944, p. 482). Using the differential equation satisfied by Ai we may verify that

$$\frac{d}{dz}\{b\,\mathrm{Ai}\,(az)\,\mathrm{Ai}'\,(bz)-a\,\mathrm{Ai}\,(bz)\,\mathrm{Ai}'\,(az)\} = (b^3-a^3)\,z\,\mathrm{Ai}\,(az)\,\mathrm{Ai}\,(bz),$$

$$\frac{d}{dz}\{a^2\,\mathrm{Ai}\,(az)\,\mathrm{Ai}'\,(bz)-b^2\,\mathrm{Ai}\,(bz)\,\mathrm{Ai}'\,(az)\} = (a^3-b^3)\,\mathrm{Ai}'\,(az)\,\mathrm{Ai}'\,(bz),$$

where a, b are constants. Hence if $b^3 \neq a^3$

$$\int_0^1 t\,\mathrm{Ai}\,(at)\,\mathrm{Ai}\,(bt)\,dt = \frac{b\,\mathrm{Ai}\,(a)\,\mathrm{Ai}'\,(b)-a\,\mathrm{Ai}\,(b)\,\mathrm{Ai}'\,(a)}{b^3-a^3}-\frac{b-a}{b^3-a^3}\mathrm{Ai}\,(0)\,\mathrm{Ai}'\,(0),$$

$$\int_0^1 \mathrm{Ai}'\,(at)\,\mathrm{Ai}'\,(bt)\,dt = \frac{a^2\,\mathrm{Ai}\,(a)\,\mathrm{Ai}'\,(b)-b^2\,\mathrm{Ai}\,(b)\,\mathrm{Ai}'\,(a)}{a^3-b^3}-\frac{a^2-b^2}{a^3-b^3}\mathrm{Ai}\,(0)\,\mathrm{Ai}'\,(0).$$

Suppose now that a is a non-real zero of $\mathrm{Ai}\,(z)$ or $\mathrm{Ai}'\,(z)$, and b is its conjugate $\bar{a}$. Then

$$\int_0^1 t\,\mathrm{Ai}\,(at)\,\mathrm{Ai}\,(\bar{a}t)\,dt = -\frac{\bar{a}-a}{\bar{a}^3-a^3}\mathrm{Ai}\,(0)\,\mathrm{Ai}'\,(0) = -\frac{1}{r^2}\frac{\sin\theta}{\sin 3\theta}\mathrm{Ai}\,(0)\,\mathrm{Ai}'\,(0), \tag{A 11}$$

$$\int_0^1 \mathrm{Ai}'\,(at)\,\mathrm{Ai}'\,(\bar{a}t)\,dt = -\frac{a^2-\bar{a}^2}{a^3-\bar{a}^3}\mathrm{Ai}\,(0)\,\mathrm{Ai}'\,(0) = -\frac{1}{r}\frac{\sin 2\theta}{\sin 3\theta}\mathrm{Ai}\,(0)\,\mathrm{Ai}'\,(0), \tag{A 12}$$

where $a \equiv r\,e^{i\theta}$. The integrals in (A 11) and (A 12) are necessarily positive and

$$\mathrm{Ai}\,(0)\,\mathrm{Ai}'\,(0) < 0.$$

In order to avoid a contradiction $\sin\theta/\sin 3\theta$ and $\sin 2\theta/\sin 3\theta$ must each be positive and finite. Thus non-real zeros can only occur in the sector $|\arg z| < \frac{1}{3}\pi$.

Next, we may readily show that for all sufficiently large R there are no zeros in the sector $|\arg z| \leqslant \frac{2}{3}\pi$, $|z| \leqslant R$, by examining the changes in $\arg \mathrm{Ai}\,(z)$ and $\arg \mathrm{Ai}'\,(z)$ as z traverses the boundary; on the curved part we use the asymptotic expressions given by the leading terms in (A 6), and on the rays $\arg z = \pm\frac{2}{3}\pi$ we use the relations (*British Association Mathematical Tables* 1946, p. B 9)

$$\mathrm{Ai}\,(x\,e^{\pm\frac{2}{3}\pi i}) = \tfrac{1}{2}e^{\pm\frac{1}{3}\pi i}\{\mathrm{Ai}\,(x) \mp i\,\mathrm{Bi}\,(x)\}, \qquad \mathrm{Ai}'\,(x\,e^{\pm\frac{2}{3}\pi i}) = \tfrac{1}{2}e^{\mp\frac{1}{3}\pi i}\{\mathrm{Ai}'\,(x) \mp i\,\mathrm{Bi}'\,(x)\}.$$

Thus the zeros of $\mathrm{Ai}\,(z)$ *and* $\mathrm{Ai}'\,(z)$ *are all real and negative.*

Equations (A 11) and (A 12) remain valid with Ai replaced by Bi, but in this case $\mathrm{Bi}\,(0)\,\mathrm{Bi}'\,(0) > 0$. Hence $\sin\theta/\sin 3\theta$ and $\sin 2\theta/\sin 3\theta$ must each be negative, and so non-real zeros can only occur in the sectors $\frac{1}{3}\pi < |\arg z| < \frac{1}{2}\pi$. That $\mathrm{Bi}\,(z)$ and $\mathrm{Bi}'\,(z)$ do have zeros in these sectors may be established by examining the changes in $\arg \mathrm{Bi}\,(z)$ and $\arg \mathrm{Bi}'\,(z)$ as z traverses the contour $OABCDEO$ in figure 16. On this diagram AB and DE are arcs of the circle $|z| = R$, where R is arbitrarily large, and B, D are the points $R\,e^{(\frac{1}{3}\pi\mp\delta)i}$, δ being an arbitrary number in the range $0 < \delta < \frac{1}{3}\pi$. The curve BCD has the equation

$$\mathscr{I}\{\tfrac{2}{3}z^{\frac{3}{2}}\} = \text{constant} = \tfrac{2}{3}R^{\frac{3}{2}}\cos\tfrac{3}{2}\delta \equiv p, \tag{A 13}$$

say; C is thus the point $R(\cos\frac{3}{2}\delta)^{\frac{2}{3}}\,e^{\frac{1}{3}\pi i}$.

On OA we have

$$\mathop{\triangle}_{OA} \arg \mathrm{Bi}\,(z) = \mathop{\triangle}_{OA} \arg \mathrm{Bi}'\,(z) = 0. \tag{A 14}$$

On AB and DE we use the asymptotic formulae given by the first terms in (A 8) and (A 9) respectively. We find without difficulty

$$\left.\begin{aligned} \mathop{\triangle}_{AB} \arg \mathrm{Bi}\,(z) &= -\tfrac{1}{4}(\tfrac{1}{3}\pi-\delta)+p+o(1), & \mathop{\triangle}_{AB} \arg \mathrm{Bi}'\,(z) &= \tfrac{1}{4}(\tfrac{1}{3}\pi-\delta)+p+o(1),\\ \mathop{\triangle}_{DE} \arg \mathrm{Bi}\,(z) &= -\tfrac{1}{4}(\tfrac{1}{3}\pi-\delta)+p+o(1), & \mathop{\triangle}_{DE} \arg \mathrm{Bi}'\,(z) &= \tfrac{1}{4}(\tfrac{1}{3}\pi-\delta)+p+o(1),\end{aligned}\right\} \tag{A 15}$$

where p is defined by (A 13).

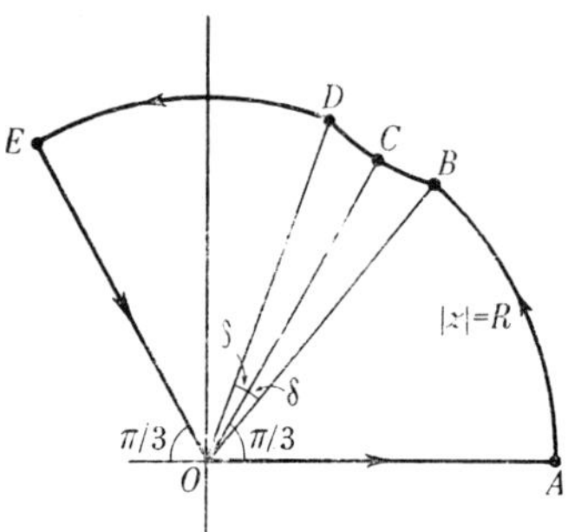

FIGURE 16.

On EO we use the relations

$$\mathrm{Bi}\,(x\,\mathrm{e}^{\frac{2}{3}\pi\mathrm{i}}) = \tfrac{1}{2}\,\mathrm{e}^{-\frac{1}{6}\pi\mathrm{i}}\{3\,\mathrm{Ai}\,(x)+\mathrm{i}\,\mathrm{Bi}\,(x)\}, \qquad \mathrm{Bi}'\,(x\,\mathrm{e}^{\frac{2}{3}\pi\mathrm{i}}) = \tfrac{1}{2}\,\mathrm{e}^{-\frac{5}{6}\pi\mathrm{i}}\{3\,\mathrm{Ai}'\,(x)+\mathrm{i}\,\mathrm{Bi}'\,(x)\}.$$

At $x=0$ we have (*British Association Mathematical Tables* 1946, p. 139)

$$3\,\mathrm{Ai}\,(x)+\mathrm{i}\,\mathrm{Bi}\,(x) = \frac{3^{\frac{1}{3}}}{\Gamma(\frac{2}{3})}\left(1+\frac{\mathrm{i}}{\sqrt{3}}\right), \qquad 3\mathrm{Ai}'\,(x)+\mathrm{i}\,\mathrm{Bi}'\,(x) = \frac{3^{\frac{2}{3}}}{\Gamma(\frac{1}{3})}\left(-1+\frac{\mathrm{i}}{\sqrt{3}}\right),$$

and so

$$\mathop{\triangle}_{EO} \arg \mathrm{Bi}\,(z) = -\tfrac{1}{3}\pi+o(1), \qquad \mathop{\triangle}_{EO} \arg \mathrm{Bi}'\,(z) = \tfrac{1}{3}\pi+o(1). \tag{A 16}$$

On BCD we have from (A 10)

$$\mathrm{Bi}\,(z) \sim (2/\pi)^{\frac{1}{2}}\,\mathrm{e}^{\frac{1}{6}\pi\mathrm{i}}\,z^{-\frac{1}{4}}\cos\phi, \qquad \mathrm{Bi}'\,(z) = (2/\pi)^{\frac{1}{2}}\,\mathrm{e}^{\frac{5}{6}\pi\mathrm{i}}\,z^{\frac{1}{4}}\sin\phi,$$

where

$$\phi \equiv -\tfrac{2}{3}\mathrm{i}z^{\frac{3}{2}}-\tfrac{1}{4}\pi-\tfrac{1}{2}\mathrm{i}\ln 2.$$

From (A 13) we see that

$$\mathscr{R}\phi = p-\tfrac{1}{4}\pi.$$

Hence if R is chosen so that $p-\frac{1}{4}\pi$ is a multiple of π, $n\pi$ say, then $\cos\phi$ is real and one-signed on the contour, and so

$$\mathop{\triangle}_{BCD} \arg \mathrm{Bi}\,(z) = -\tfrac{1}{2}\delta+o(1). \tag{A 17}$$

Similarly, if $p+\frac{1}{4}\pi = n\pi$ then

$$\mathop{\triangle}_{BCD} \arg \mathrm{Bi}'\,(z) = \tfrac{1}{2}\delta+o(1). \tag{A 18}$$

Combining the results (A 14) to (A 18), we find that

$$\mathop{\triangle}_{OABCDEO} \arg \mathrm{Bi}\,(z) = 2p-\tfrac{1}{2}\pi+o(1) = 2n\pi+o(1),$$

$$\mathop{\triangle}_{OABCDEO} \arg \mathrm{Bi}'\,(z) = 2p+\tfrac{1}{2}\pi+o(1) = 2n\pi+o(1).$$

Thus if n is a sufficiently large positive integer, the contour $OABCDEO$ has inside it n zeros of Bi (z) *if* $R = \{\frac{3}{2}(n+\frac{1}{4})\,\pi\sec\frac{3}{2}\delta\}^{\frac{2}{3}}$, *and n zeros of* Bi$'$ (z) *if* $R = \{\frac{3}{2}(n-\frac{1}{4})\,\pi\sec\frac{3}{2}\delta\}^{\frac{2}{3}}$.

Asymptotic expansions for the large zeros

Let us write†

$$\left.\begin{aligned}
T(\lambda) &\equiv \lambda^{\frac{2}{3}}\left(1+\tfrac{5}{48}\lambda^{-2}-\tfrac{5}{36}\lambda^{-4}+\tfrac{77125}{82944}\lambda^{-6}-\ldots\right),\\
U(\mu) &\equiv \mu^{\frac{2}{3}}\left(1-\tfrac{7}{48}\mu^{-2}+\tfrac{35}{288}\mu^{-4}-\tfrac{181223}{207360}\mu^{-6}+\ldots\right),\\
V(\lambda) &\equiv \frac{\lambda^{\frac{1}{6}}}{\sqrt{\pi}}\left(1+\tfrac{5}{48}\lambda^{-2}-\tfrac{1525}{4608}\lambda^{-4}+\tfrac{2397875}{663552}\lambda^{-6}-\ldots\right),\\
W(\mu) &\equiv \frac{\mu^{-\frac{1}{6}}}{\sqrt{\pi}}\left(1-\tfrac{7}{96}\mu^{-2}+\tfrac{1673}{6144}\mu^{-4}-\tfrac{84394709}{26542080}\mu^{-6}+\ldots\right),
\end{aligned}\right\} \tag{A 19}$$

and let the sth negative zeros of Ai (z), Ai′ (z), Bi (z) and Bi′ (z) be denoted by a_s, a'_s, b_s and b'_s respectively. Then if s is large

$$\left.\begin{aligned}
a_s &\sim -T(\lambda), & a'_s &\sim -U(\mu),\\
\mathrm{Ai}'(a_s) &\sim (-)^{s-1}V(\lambda), & \mathrm{Ai}(a'_s) &\sim (-)^{s-1}W(\mu),\\
\text{where}\qquad \lambda &= \tfrac{3}{8}\pi(4s-1), & \mu &= \tfrac{3}{8}\pi(4s-3),
\end{aligned}\right\} \tag{A 20}$$

and

$$\left.\begin{aligned}
b_s &\sim -T(\lambda), & b'_s &\sim -U(\mu),\\
\mathrm{Bi}'(b_s) &\sim (-)^{s-1}V(\lambda), & \mathrm{Bi}(b'_s) &\sim (-)^{s}W(\mu),\\
\text{where}\qquad \lambda &= \tfrac{3}{8}\pi(4s-3), & \mu &= \tfrac{3}{8}\pi(4s-1).
\end{aligned}\right\} \tag{A 21}$$

Similarly, if the complex zeros of Bi (z) and Bi′ (z) in the sector $\frac{1}{3}\pi<\arg z<\frac{1}{2}\pi$ arranged in ascending order of modulus are denoted by β_s and β'_s respectively $(s=1,2,\ldots)$, then for large s

$$\left.\begin{aligned}
\beta_s &\sim \mathrm{e}^{\frac{1}{3}\pi\mathrm{i}}\,T(\lambda), & \beta'_s &\sim \mathrm{e}^{\frac{1}{3}\pi\mathrm{i}}\,U(\mu),\\
\mathrm{Bi}'(\beta_s) &\sim (-)^{s}\,2^{\frac{1}{2}}\,\mathrm{e}^{-\frac{1}{6}\pi\mathrm{i}}\,V(\lambda), & \mathrm{Bi}(\beta'_s) &\sim (-)^{s-1}\,2^{\frac{1}{2}}\,\mathrm{e}^{\frac{1}{6}\pi\mathrm{i}}\,W(\mu),\\
\text{where}\qquad \lambda &= \tfrac{3}{8}\{(4s-1)\,\pi+2\mathrm{i}\ln 2\}, & \mu &= \tfrac{3}{8}\{(4s-3)\,\pi+2\mathrm{i}\ln 2\}.
\end{aligned}\right\} \tag{A 22}$$

The zeros in the sector $-\frac{1}{2}\pi<\arg z<-\frac{1}{3}\pi$ are the conjugates $\bar{\beta}_s$, $\bar{\beta}'_s$.

The expansions (A 20), (A 21) and (A 22) may be derived by reversion of the expansions (A 7), (A 9) and (A 10) respectively. The higher coefficients in formula (A 19) for $V(\lambda)$ and $W(\mu)$ are most conveniently obtained from those of $T(\lambda)$ and $U(\mu)$ by means of the relations (Olver 1950, equation (2·15))

$$\mathrm{Ai}'(a_s)=\pm\left(-\frac{\mathrm{d}a_s}{\mathrm{d}s}\right)^{-\frac{1}{2}}, \qquad \mathrm{Ai}(a'_s)=\pm\left(a'_s\,\frac{\mathrm{d}a'_s}{\mathrm{d}s}\right)^{-\frac{1}{2}}, \tag{A 23}$$

in which s is regarded as a continuous variable.

† Additional terms in the formulae for $T(\lambda)$ and $U(\mu)$ are given in *British Association Mathematical Tables* (1946, p. B48).

References

Bickley, W. G. & Miller, J. C. P. 1945 *Phil. Mag.* **36**, 124–131.

British Association Mathematical Tables 1946 *The Airy integral.* Cambridge University Press.

Cherry, T. M. 1950 *Trans. Amer. Math. Soc.* **68**, 224–257.

Copson, E. T. 1944 *Theory of functions of a complex variable.* Oxford University Press.

Lehmer, D. H. 1945 *Math. Tab., Wash.*, **1**, 133–135.

Miller, J. C. P. 1945 *Math. Tab., Wash.*, **1**, 309–312.

National Bureau of Standards Computation Laboratory 1950 *Table of the Bessel functions* $Y_0(z)$ *and* $Y_1(z)$ *for complex arguments.* New York: Columbia University Press.

Olver, F. W. J. 1950 *Proc. Camb. Phil. Soc.* **46**, 570–580.

Olver, F. W. J. 1951 *Proc. Camb. Phil. Soc.* **47**, 699–712.

Olver, F. W. J. 1952 *Proc. Camb. Phil. Soc.* **48**, 414–427.

Olver, F. W. J. 1954 *Phil. Trans.* A, **247**, 307–327.

Watson, G. N. 1944 *Theory of Bessel functions.* Cambridge University Press.

[65]

THE ASYMPTOTIC SOLUTION OF LINEAR DIFFERENTIAL EQUATIONS OF THE SECOND ORDER IN A DOMAIN CONTAINING ONE TRANSITION POINT

By F. W. J. OLVER, *The National Physical Laboratory*

(*Communicated by Sir Edward Bullard, F.R.S.—Received* 14 *July* 1955)

CONTENTS

The differential equation
$$\frac{d^2w}{dz^2}=\left\{uz^n+\frac{\gamma}{z^2}+g(z)\right\}w,$$
where n is an integer ($\geqslant -1$), u a parameter and γ a constant, has the formal solution
$$w=P(z)\left\{1+\sum_{s=1}^{\infty}\frac{A_s(z)}{u^s}\right\}+\frac{P'(z)}{u}\sum_{s=0}^{\infty}\frac{B_s(z)}{u^s},$$
where $P(z)$ is a solution of the equation
$$\frac{d^2P}{dz^2}=\left(uz^n+\frac{\gamma}{z^2}\right)P.$$
The coefficients $A_s(z)$ and $B_s(z)$ are given by recurrence relations. It is shown that they are analytic at $z=0$ if, and only if, the differential equation for w can be transformed into a similar equation with $n=0$, $\gamma=0$, or $n=1$, $\gamma=0$, or $n=-1$. The first two cases (for which P is an exponential and Airy function respectively) have been treated in a previous paper. The third case, for which P is a Bessel function of order $\pm(1+4\gamma)^{\frac{1}{2}}$, is examined in detail in the present paper. It is proved that for large positive u, solutions exist whose asymptotic expansions in Poincaré's sense are given by the formal series, and that these expansions are uniformly valid with respect to the complex variable z.

1. Introduction and summary

In a recent paper (Olver 1954) I considered asymptotic expansions of solutions of the differential equation
$$\frac{d^2w}{dz^2}=\{up(z)+q(z)\}w, \tag{1·1}$$
for large positive values of the parameter u, which are uniform with respect to z when z lies in a simply-connected domain **D** in which $p(z)$, $q(z)$ are analytic functions of z independent

of u. Three cases A, B and C were considered. In case A the equation had no transition points† in **D**. In case B there was one transition point in **D**, a simple zero of $p(z)$. In case C there was again one transition point, this time a double pole of $p(z)$. It was shown that case C can be transformed into case A.

The principal purpose of the present paper is to consider a fourth case, case D, in which the only transition point in **D** is simultaneously a single pole of $p(z)$ and a double pole of $q(z)$. We shall show that the solutions of the differential equation have asymptotic expansions which are uniform with respect to z and similar in character to those in cases A and B, but whereas the expansions in these cases are in terms of exponential and Airy functions respectively, the present expansions are in terms of Bessel functions whose order depends on the behaviour of $q(z)$ at the transition point. The only previous treatment of this case appears to be that of Langer (1935). This derives the dominant term in the asymptotic expansion, but the result obtained is uniformly valid only in a z-domain which shrinks to zero as $u \to \infty$.

The results given in this paper can be applied to the hypergeometric and confluent hypergeometric equations, and in particular to the determination of new asymptotic expansions of Legendre, Laguerre and Coulomb-wave functions of large orders.

It is natural to inquire whether the solutions of (1·1) have asymptotic expansions, similar in character to those in cases A and B, for any other kind of transition point in **D**. The first part of this paper (§§ 2 to 6) is devoted to the investigation of this question. The basis of the method used for deriving asymptotic solutions of (1·1) is to compare (1·1) with a similar differential equation

$$\frac{d^2w}{dz^2} = \{up(z)+r(z)\}\,w, \tag{1·2}$$

with solutions $w = P(u,z)$, say, and then to satisfy (1·1) (or a transformed form of (1·1)) formally by a series

$$w = P(u,z)\left\{1+\sum_{s=1}^{\infty}\frac{A_s(z)}{u^s}\right\}+\frac{1}{u}\left\{\frac{\partial}{\partial z}P(u,z)\right\}\sum_{s=0}^{\infty}\frac{B_s(z)}{u^s}, \tag{1·3}$$

in which the coefficients $A_s(z)$, $B_s(z)$ are independent of u.

The success of the method depends on two conditions. First, the coefficients $A_s(z)$ and $B_s(z)$ must all be analytic functions of z at the transition point; we cannot otherwise expect the expansion (1·3) to be uniformly valid with respect to z. Secondly, $P(u,z)$ must be a less recondite function of the two variables u, z than the solutions of (1·1), otherwise we have achieved little. It is not easy to ascribe a precise meaning to the second condition. From the computational point of view we mean that $P(u,z)$ can be computed easily from a number of single-entry (or argument) tables. We shall in fact take the exact form of this criterion to be that $P(u,z)$ is of the form $\chi\{z\psi(u)\}$, where χ, ψ are functions of a single variable.

Given these two conditions, we can determine (§§ 4 and 5) all the cases which admit formal series expansions of the desired form. We find, however (§ 6), that apart from certain special or 'non-typical' cases, all the admissible cases can be transformed either into case A or case B of the previous paper, or into case D of the present paper. Thus there can be no

† Here a transition point means one at which $p(z)$ vanishes or $p(z)$ or $q(z)$ has a singularity (see Olver 1954, § 1).

further types of isolated transition point which can be treated by the same method and present new features.

The second part of the paper (§§ 7 to 14) is devoted to the investigation in case D of the asymptotic nature of the formal series (1·3). The main result is stated in § 8 in the form of an existence theorem, theorem D, which is analogous to theorems A and B of the previous paper. The proof of theorem D is given in §§ 9 to 13, and in the final section (§ 14) alternative arrangements of the series (1·3) are briefly considered.

At the end of the paper there is an Appendix in which it is indicated how the proofs of theorems A and B of the previous paper (Olver 1954) may be brought more into line with that given for theorem D in the present paper. In this way it may be shown that a condition which was assumed in order to prove the uniform validity of the expansions at $z = \infty$ may, in fact, be relaxed considerably.

2. Preliminary transformation

We suppose that the transition point of (1·1) is at $z = z_0$ and that in the neighbourhood of $z = z_0$

$$p(z) = \varpi(z)\,(z-z_0)^n, \tag{2·1}$$

where n is an integer, $\varpi(z)$ is a regular (holomorphic) function, $\varpi(z_0) \neq 0$. We may assume that $n \geqslant -1$, for if $n < -2$ the differential equation has an irregular singularity at $z = z_0$, and if $n = -2$ we have a case treated in a previous paper (Olver 1954, case C). We first show that we may transform (1·1) into the same form of equation but with $p(z)$ replaced by z^n.

In (1·1) we take new dependent and independent variables W, ζ related by

$$W = \dot{z}^{-\frac{1}{2}} w, \tag{2·2}$$

dots denoting differentiation with respect to ζ. Then it is readily verified that W satisfies the equation

$$\frac{d^2W}{d\zeta^2} = \{u\dot{z}^2 p(z) + f(\zeta)\}\, W, \tag{2·3}$$

where

$$f(\zeta) = \dot{z}^2 q(z) + \dot{z}^{\frac{1}{2}} \frac{d^2}{d\zeta^2}(\dot{z}^{-\frac{1}{2}}). \tag{2·4}$$

Equation (2·3) has the desired form

$$\frac{d^2W}{d\zeta^2} = \{u\zeta^n + f(\zeta)\}\, W, \tag{2·5}$$

if z and ζ are related by

$$\dot{z}^2 p(z) = \zeta^n. \tag{2·6}$$

On integration the last equation yields

$$\zeta = \left[\frac{n+2}{2}\int_{z_0}^{z} \{p(t)\}^{\frac{1}{2}}\, dt\right]^{2/(n+2)}. \tag{2·7}$$

Substituting (2·1) in (2·7), we find that

$$\zeta = \{\varpi(z_0)\}^{1/(n+2)}\,(z-z_0)\,\{1+(z-z_0)\,O(1)\}, \tag{2·8}$$

as $z \to z_0$, and so $\zeta(z)$ is analytic at $z = z_0$.

Thus the relation (2·7) maps the domain **D** on a certain domain **G** in the ζ-plane, and the point $z = z_0$ corresponds to $\zeta = 0$. Moreover, we observe from (2·4) and (2·8) that if $q(z)$

is analytic at $z = z_0$ then $f(\zeta)$ is analytic at $\zeta = 0$, and if $q(z)$ has a single or double pole at $z = z_0$, then $f(\zeta)$ has the same kind of singularity at $\zeta = 0$.

In the rest of the paper, with the exception of § 4, we shall consider equation (2·5) instead of (1·1), and revert to the symbols z, q, w, $\mathbf{D}$, in place of $\zeta, f, W, \mathbf{G}$, respectively.

3. Formal construction of series solutions

The principle on which we construct asymptotic solutions of the equation

$$\frac{d^2w}{dz^2} = \{uz^n + q(z)\} w \tag{3·1}$$

for large u, is to compare this equation with a similar but less recondite equation

$$\frac{d^2w}{dz^2} = \{uz^n + r(z)\} w, \tag{3·2}$$

which we shall refer to as the *basic equation*, and its solutions as the *basic functions*.

Let $P(u, z)$ be a solution of (3·2). Then as a possible solution of (3·1) we consider the series

$$w(z) = P(u, z) \sum_{s=0}^{\infty} \frac{A_s(z)}{u^s} + \frac{1}{u}\left\{\frac{\partial}{\partial z} P(u, z)\right\} \sum_{s=0}^{\infty} \frac{B_s(z)}{u^s}, \tag{3·3}$$

where $A_0(z) = 1$ and the coefficients $A_s(z)$, $B_s(z)$ are independent of u. Term-by-term differentiation, using (3·2), yields

$$w'(z) = P(u, z) \sum_{s=0}^{\infty} \frac{C_s(z)}{u^s} + \frac{1}{u}\left\{\frac{\partial}{\partial z} P(u, z)\right\} \sum_{s=-1}^{\infty} \frac{D_s(z)}{u^s}, \tag{3·4}$$

$$w''(z) = P(u, z) \sum_{s=-1}^{\infty} \frac{E_s(z)}{u^s} + \frac{1}{u}\left\{\frac{\partial}{\partial z} P(u, z)\right\} \sum_{s=-1}^{\infty} \frac{F_s(z)}{u^s}, \tag{3·5}$$

where
$$C_s = A_s' + z^n B_s + rB_{s-1}, \qquad D_s = A_{s+1} + B_s', \tag{3·6}$$

$$E_s = C_s' + z^n D_s + rD_{s-1} = A_s'' + z^n A_{s+1} + rA_s + 2z^n B_s' + nz^{n-1} B_s + 2rB_{s-1}' + r' B_{s-1}, \tag{3·7}$$

$$F_s = C_{s+1} + D_s' = 2A_{s+1}' + B_s'' + z^n B_{s+1} + rB_s, \tag{3·8}$$

primes denoting differentiations with respect to z.

Substituting (3·3) and (3·5) in (3·1), we see that (3·1) is formally satisfied if

$$E_s = z^n A_{s+1} + qA_s, \qquad F_s = z^n B_{s+1} + qB_s.$$

From the last equations and (3·7), (3·8), we obtain

$$2z^n B_s' + nz^{n-1} B_s = (q-r) A_s - A_s'' - 2rB_{s-1}' - r' B_{s-1}, \tag{3·9}$$

$$2A_{s+1}' = (q-r) B_s - B_s''. \tag{3·10}$$

On integration, these equations yield the recurrence relations

$$B_s = \tfrac{1}{2} z^{-\frac{1}{2}n} \int z^{-\frac{1}{2}n} \left\{(q-r) A_s - A_s'' - 2r^{\frac{1}{2}} \frac{d}{dz}(r^{\frac{1}{2}} B_{s-1})\right\} dz, \tag{3·11}$$

$$A_{s+1} = \tfrac{1}{2} \int (q-r) B_s \, dz - \tfrac{1}{2} B_s'. \tag{3·12}$$

We are interested in the circumstances in which (3·3) represents an asymptotic expansion of a solution of (3·1). In order that this expansion may be uniformly valid with respect to z it is obviously necessary that $A_s(z)$ and $B_s(z)$ are analytic functions of z at the transition point $z = 0$. This condition imposes restrictions on $q(z)$ and $r(z)$. Before proceeding to consider these we shall, in the next section, examine the circumstances in which the solutions of the basic equation (3·2) are in the form of a function of a single variable (cf. § 1).

4. The basic equation

Lemma 1. *Let $z^2p(z)$, $z^2q(z)$ $(p(z) \neq 0)$ be regular (holomorphic) functions of z in a domain which includes $z = 0$ as an interior point. Necessary and sufficient conditions that the equation*

$$\frac{d^2w}{dz^2} = \{up(z) + r(z)\}\, w \qquad (4\cdot1)$$

possesses a non-trivial solution of the form

$$w = \chi\{z\psi(u)\}, \qquad (4\cdot2)$$

are that

$$p(z) = \alpha z^n, \qquad r(z) = \beta z^n + \frac{\gamma}{z^2}, \qquad (4\cdot3)$$

where $n \geqslant -1$ is an integer, and α, β, γ are constants.

To prove this result we write $t = z\psi(u)$ and substitute (4·2) in (4·1). We obtain

$$\{\psi(u)\}^2\chi''(t) = \{up(z) + r(z)\}\chi(t),$$

and hence

$$\frac{up(z)+r(z)}{\{\psi(u)\}^2} = \frac{\chi''(t)}{\chi(t)}, \quad \text{a function of } t. \qquad (4\cdot4)$$

For small $|z|$ we can expand $p(z)$ and $r(z)$ in the form

$$p(z) = z^{-2}\sum_{s=0}^{\infty} p_s z^s, \qquad r(z) = z^{-2}\sum_{s=0}^{\infty} r_s z^s,$$

where p_s, r_s are constants, and so

$$\frac{up(z)+r(z)}{\{\psi(u)\}^2} = \frac{1}{\{\psi(u)\}^2}\left[up\left(\frac{t}{\psi(u)}\right) + r\left(\frac{t}{\psi(u)}\right)\right] = \sum_{s=0}^{\infty}\frac{up_s + r_s}{\{\psi(u)\}^s}\, t^{s-2}. \qquad (4\cdot5)$$

From (4·4) and (4·5) we see that

$$\frac{up_s + r_s}{\{\psi(u)\}^s} = c_s, \quad \text{a constant for all } u. \qquad (4\cdot6)$$

Taking $s = 0$ in this result we see immediately that $p_0 = 0$. Let p_j be the first non-vanishing member of the sequence $p_1, p_2, \ldots$. Then from (4·6) with $s = j$ we have

$$\psi(u) = c_j^{-1/j}\,(up_j + r_j)^{1/j},$$

and substituting this in (4·6) we obtain

$$\frac{up_s + r_s}{(up_j + r_j)^{s/j}} = \frac{c_s}{c_j^{s/j}}.$$

For this equation to hold $(up_j + r_j)^{s/j}$ must be a linear function of u, and since this can only happen if $s = 0$ or j it follows that

$$p_s = r_s = 0 \quad (s \neq 0, j), \qquad p_0 = 0.$$

Thus

$$p(z) = p_j z^{j-2}, \qquad r(z) = r_0 z^{-2} + r_j z^{j-2} \quad (j > 0).$$

This is essentially the result (4·3) of the lemma with $n = j - 2$.

5. Conditions that $A_s(z)$ and $B_s(z)$ are analytic at $z=0$

In accordance with the result of lemma 1 we take

$$r(z) = \beta z^n + \frac{\gamma}{z^2} \quad (n \geqslant -1) \tag{5·1}$$

in equation (3·2) and the results of § 3. We consider now the admissible forms of $q(z)$ for which the recurrence relations (3·11) and (3·12) define coefficients $A_s(z)$ and $B_s(z)$ which are analytic at $z = 0$.

We may note first that the term βz^n in the basic equation given by (3·2) and (5·1) may be absorbed by replacing u by $u+\beta$. There is accordingly no loss of generality in supposing β to be zero. If we write

$$r(z) = \frac{\gamma}{z^2}, \qquad q(z) = \frac{\gamma}{z^2} + g(z), \tag{5·2}$$

the differential equations (3·1) and (3·2) become

$$\frac{d^2w}{dz^2} = \left\{uz^n + \frac{\gamma}{z^2} + g(z)\right\} w, \tag{5·3}$$

$$\frac{d^2w}{dz^2} = \left\{uz^n + \frac{\gamma}{z^2}\right\} w, \tag{5·4}$$

respectively, and the recurrence relations (3·11) and (3·12) can be written in the form

$$A_{s+1} = \tfrac{1}{2}\int gB_s\,dz - \tfrac{1}{2}B_s', \tag{5·5}$$

$$B_s = \tfrac{1}{2}z^{-\frac{1}{2}n}\int z^{-\frac{1}{2}n}G_s\,dz, \tag{5·6}$$

where

$$G_s = gA_s - A_s'' - \frac{2\gamma}{z}\frac{d}{dz}\left(\frac{B_{s-1}}{z}\right). \tag{5·7}$$

We now consider the cases† $n = -1, 0, 1, 2, \ldots$, in turn.

Case $n = -1$. From (5·6) and (5·7) we have

$$B_0 = \tfrac{1}{2}z^{\frac{1}{2}}\int z^{\frac{1}{2}}g\,dz.$$

If $g \sim z^{-2}$ as $z \to 0$, then clearly $B_0 \to -1$. Substituting in (5·5) with $s = 0$, we see that this implies that A_1 has a singularity. Hence it is necessary that $zg(z)$ be analytic at $z = 0$. It may be verified by induction that this condition is also sufficient, provided that the lower limit of integration in (5·6) is taken to be zero; every B_s vanishes at $z = 0$ and every G_s has a simple pole at this point.

Thus in this case the admissible form of (5·3) is given by

$$\frac{d^2w}{dz^2} = \left\{\frac{u}{z} + \frac{\gamma}{z^2} + \frac{h(z)}{z}\right\} w, \tag{5·8}$$

where $h(z)$ denotes an arbitrary function regular in the domain of integration.

† The case $n = -2$ has been dealt with in the previous paper (Olver 1954, case C). It is of interest to note that without transformation this case does not satisfy the conditions of lemma 1.

Equation (5·8) is, in fact, the transformed form of case D (cf. § 1), and the investigation of its solutions is the subject of the second part (§§ 7 to 14) of this paper.

Before passing on to higher values of n we note the following facts common to all such cases. First, from (5·6) and (5·7) we obtain

$$B_0 = \tfrac{1}{2}z^{-\frac{1}{2}n}\int z^{-\frac{1}{2}n} g \, dz. \tag{5·9}$$

Hence if $n \geqslant 0$ it is necessary (though not always sufficient) that $g(z)$ be analytic at $z = 0$, otherwise B_0 will certainly have a singularity at this point. Secondly, if $n \geqslant 1$ it is necessary that the lower limit of integration in (5·6) and (5·9) is zero. Thirdly, we see from (5·5) that A_{s+1} is automatically analytic at $z = 0$ if the same is true of B_s (because g is analytic), and so we need only consider the way in which the behaviour of B_{s-1} is propagated to B_s. For this purpose it is convenient to have the following formulae for G_s which are obtained by substituting (5·5) in (5·7):

$$G_s = \tfrac{1}{2}g\int gB_{s-1}\,dz - gB'_{s-1} - \tfrac{1}{2}g'B_{s-1} + \tfrac{1}{2}B'''_{s-1} - \frac{2\gamma}{z}\frac{d}{dz}\left(\frac{B_{s-1}}{z}\right) \quad (s>0), \qquad G_0 = g. \tag{5·10}$$

Case $n = 0$, $\gamma = 0$. We see at once from (5·6) and (5·10) that the coefficients B_s are all analytic at $z = 0$. Equation (5·3) becomes

$$\frac{d^2w}{dz^2} = \{u + h(z)\}\, w, \tag{5·11}$$

where $h(z)$ has the meaning assigned above (cf. (5·8)). This is case A of the previous paper (Olver 1954).

Case $n = 0$, $\gamma \neq 0$. From (5·6) it is seen that every G_s must be analytic at $z = 0$. Now if B_{s-1} is analytic all the terms in the expression (5·10) for G_s are analytic, except possibly the last. Let the expansions of B_{s-1} and g in the neighbourhood of $z = 0$ be denoted by

$$B_{s-1} = b_0^{(s-1)} + b_1^{(s-1)}z + b_2^{(s-1)}z^2 + \ldots, \tag{5·12}$$

$$g = g_0 + g_1 z + g_2 z^2 + \ldots, \tag{5·13}$$

respectively. Then
$$\frac{2\gamma}{z}\frac{d}{dz}\left(\frac{B_{s-1}}{z}\right) = \sum_{j=0}^{\infty} 2\gamma(j-1)\, b_j^{(s-1)} z^{j-3}.$$

Hence $b_0^{(s-1)} = 0$ and $b_2^{(s-1)} = 0$, for all s. The first of these equations means that the lower integration limit in (5·6) must be zero; the implications of the second are as follows. From (5·6) we see that the expansion of G_s must contain no term in z; in particular, we have $g_1 = 0$. Setting the coefficient of z in (5·10) equal to zero, we obtain

$$12b_4^{(s-1)} - 6\gamma b_4^{(s-1)} = 0,$$

and so either $b_4^{(s-1)} = 0$ or $\gamma = 2$.

Taking the first of these alternatives we see from (5·6) that the expansion of G_s must contain no term in z^3. Hence (cf. (5·10)) $g_3 = 0$ and

$$\tfrac{1}{2}\,.\,6\,.\,5\,.\,4\,.\,b_6^{(s-1)} - 2\gamma\,.\,5b_6^{(s-1)} = 0.$$

Thus $b_6^{(s-1)} = 0$ or $\gamma = 6$. Continuing the argument we see that either $g(z)$ is an even function (in which event the B_s are odd and the G_s are even functions), or $\gamma = j(j+1)$, where $2j+1$ is the first non-vanishing odd power in the expansion of g.

Except when $j = 0$, the latter alternative is a rather special possibility and is unlikely to have much practical significance. We shall call such a case *non-typical* and not regard it as admissible.

Thus the admissible form of (5·3) is given by

$$\frac{d^2w}{dz^2} = \left\{u + \frac{\gamma}{z^2} + h(z^2)\right\} w. \tag{5·14}$$

Case $n = 1, \gamma = 0$. We see from (5·6) and (5·10) that the B_s are analytic at $z = 0$. Equation (5·3) becomes

$$\frac{d^2w}{dz^2} = \{uz + h(z)\} w. \tag{5·15}$$

This is case B of the previous paper.

Cases $n \geqslant 1, \gamma \neq 0$. From (5·6) we see that a term in z^j in the expansion of G_s corresponds to a term in z^{j-n+1}, if $j \neq \frac{1}{2}n - 1$, or $z^{j-n+1} \ln z$, if $j = \frac{1}{2}n - 1$, in the expansion of B_s. Hence G_s must contain no terms in $1, z, z^2, \ldots, z^{n-2}$; in particular, since $G_0 = g$, it follows that

$$g = g_{n-1} z^{n-1} + g_n z^n + g_{n+1} z^{n+1} + \ldots. \tag{5·16}$$

Substituting (5·12) and (5·16) in (5·10), we obtain

$$G_s = 2\gamma\left(\frac{b_0^{(s-1)}}{z^3} - \frac{b_2^{(s-1)}}{z}\right) + \tfrac{1}{4}\{2(1.3 - 4\gamma)\, b_3^{(s-1)} + 3(2.4 - 4\gamma)\, b_4^{(s-1)} z + \ldots\}$$
$$- \tfrac{1}{2}(n-1)\, g_{n-1} b_0^{(s-1)} z^{n-2} + z^{n-1} O(1), \tag{5·17}$$

if $s > 0$. All terms here up to and including z^{n-2} must vanish. Accordingly, we have

$$b_0^{(s-1)} = b_2^{(s-1)} = b_3^{(s-1)} = b_4^{(s-1)} = \ldots = b_{n+1}^{(s-1)} = 0,$$

for all s, the possibilities $4\gamma = 1.3,\ 2.4,\ \ldots$, being rejected as *non-typical* (see above, case $n = 0, \gamma \neq 0$). Thus we have

$$B_{s-1} = b_1^{(s-1)} z + b_{n+2}^{(s-1)} z^{n+2} + b_{n+3}^{(s-1)} z^{n+3} + \ldots \tag{5·18}$$

(cf. (5·12)). Moreover, referring again to (5·6), we now see that G_s must contain no terms in $z^{n-1}, z^{n+1}, z^{n+2}, z^{n+3}, \ldots, z^{2n}$; in particular, it follows that

$$g = g_n z^n + g_{2n+1} z^{2n+1} + g_{2n+2} z^{2n+2} + \ldots. \tag{5·19}$$

Substituting (5·18) and (5·19) in (5·10), and equating to zero the coefficients of $z^{n-1}, z^{n+1}, z^{n+2}, \ldots, z^{2n}$, we find that

$$b_{n+2}^{(s-1)} = b_{n+4}^{(s-1)} = b_{n+5}^{(s-1)} = b_{n+6}^{(s-1)} = \ldots = b_{2n+3}^{(s-1)} = 0,$$

on excluding non-typical possibilities. Repeating the argument, we see that in the expansions of B_{s-1} and g, all coefficients except $b_1^{(s-1)}, b_{n+3}^{(s-1)}, b_{2n+5}^{(s-1)}, \ldots, g_n, g_{2n+2}, g_{3n+4}, \ldots$, must vanish. The admissible form of (5·3) is accordingly

$$\frac{d^2w}{dz^2} = \left\{uz^n + \frac{\gamma}{z^2} + z^n h(z^{n+2})\right\} w \quad (n = 1, 2, 3, \ldots). \tag{5·20}$$

Cases $n \geqslant 2, \gamma = 0$.† Equations (5·16) and (5·17) remain valid. The coefficients of $1, z, \ldots, z^{n-2}$ in the latter equation must vanish and so

$$b_3^{(s-1)} = b_4^{(s-1)} = \ldots = b_n^{(s-1)} = 0, \tag{5·21}$$

$$\tfrac{1}{4}n(n-1)(n+1)\, b_{n+1}^{(s-1)} - \tfrac{1}{2}(n-1)\, g_{n-1} b_0^{(s-1)} = 0. \tag{5·22}$$

† The arguments given also apply to the same values of n when $\gamma \neq 0$.

Now from (5·9) we have

$$b_{n+1}^{(0)} = \frac{g_{2n}}{3n+2}, \qquad b_0^{(0)} = \frac{g_{n-1}}{n}.$$

Substituting these values in (5·22) with $s = 1$, and rejecting the possibility that g_{2n} is a multiple of g_{n-1}^2 as non-typical, we deduce that $g_{2n} = g_{n-1} = 0$ and hence that $b_{n+1}^{(0)} = 0$, $b_0^{(0)} = 0$, and, also from (5·22), that $b_{n+1}^{(s-1)} = 0$.

Consider now (5·6) and (5·10). A term $b_{j+n+2}^{(s-1)} z^{j+n+2}$ (j arbitrary) in the expansion of B_{s-1} becomes $\frac{1}{2}(j+n+2)(j+n+1)(j+n)\, b_{j+n+2}^{(s-1)} z^{j+n-1}$ in $\frac{1}{2}B_{s-1}'''$, and this in turn gives rise to a term $\frac{1}{2}(j+n+2)(j+n+1)(j+n)(2j+n)^{-1}\, b_{j+n+2}^{(s-1)} z^j$ in the expansion of B_s. There are other terms in z^j in B_s arising from the first three terms in the expression (5·10) for G_s (the last term is, of course, zero), but these involve products of the g_j and $b_j^{(s-1)}$. Hence, on rejecting non-typical possibilities, we see that if $b_j^{(s)} = 0$ then $b_{j+n+2}^{(s-1)} = 0$. From this result, (5·21), and the fact that $b_{n+1}^{(s-1)} = 0$, we deduce that

$$b_j^{(s)} = 0 \quad (j = 3, 4, 5, \ldots, n+1 \ (\text{modulo}\ n+2)). \tag{5·23}$$

Hence the expansion of B_0 may be written in the form

$$B_0 = \psi_0 + z\psi_1 + z^2\psi_2, \tag{5·24}$$

where ψ_0, ψ_1, ψ_2 are power series in z^{n+2}, and $\psi_0 = 0$ when $z = 0$. From (5·9) we deduce that

$$g = z^{n-1}\rho_0 + z^n\rho_1 + z^{n+1}\rho_2, \tag{5·25}$$

and differentiating (5·24) and (5·25), we obtain

$$B_0' = z^{-1}\chi_0 + \chi_1 + z\chi_2,$$

$$g' = z^{n-2}\sigma_0 + z^{n-1}\sigma_1 + z^n\sigma_2,$$

where ρ_j, χ_j, σ_j ($j = 0, 1, 2$) are power series in z^{n+2} whose coefficients are positive multiples of the corresponding coefficients of ψ_j. Hence

$$gB_0' + \tfrac{1}{2}g'B_0 = z^{n-2}(\rho_0\chi_0 + \tfrac{1}{2}\sigma_0\psi_0) + \ldots + z^{n+2}(\rho_2\chi_2 + \tfrac{1}{2}\sigma_2\psi_2). \tag{5·26}$$

Now from (5·23) and (5·6) we see that the expansion of the right-hand side of the expression (5·10) for G_s must contain only powers of z equal to $n-1$, n, $n+1$ (modulo $n+2$). In particular the terms $-gB_{s-1}' - \frac{1}{2}g'B_{s-1}$ must be of this form because they are the only terms whose coefficients are quadratic functions of the g_j and $b_j^{(s-1)}$. Therefore the coefficients of z^{n-2} and z^{n+2} in (5·26) must vanish†, and so $\rho_0 = \chi_0 = \sigma_0 = \psi_0$, because in each of these functions the coefficients of the same powers of z^{n+2} are positive multiples of one another. Similarly $\rho_2 = \chi_2 = \sigma_2 = \psi_2 = 0$.

Thus (5·25) reduces to $\qquad g = z^n\rho_1 = z^n h(z^{n+2}),$

where h has the meaning assigned above, and the corresponding form of (5·3) is given by (5·20) with $\gamma = 0$.

† When $n = 2$ the condition is that the *sum* of the coefficients of z^{n-2} and z^{n+2} must vanish. However, rejection of non-typical possibilities leads to the same conclusion.

Summary. In this and the previous section we have shown that in addition to cases A and B of the previous paper (Olver 1954), and case D of the present paper, only the following form of equation (3·1)

$$\frac{d^2w}{dz^2} = \left\{uz^n + \frac{\gamma}{z^2} + z^n h(z^{n+2})\right\} w \quad (n = 0, 1, 2, \ldots), \tag{5·27}$$

where $h(z^{n+2})$ is an arbitrary analytic function of z^{n+2}, satisfies the conditions:

(i) The corresponding basic equation satisfies the conditions of lemma 1.

(ii) The coefficients $A_s(z)$ and $B_s(z)$ in the formal expansion (3·3) are regular functions of z throughout the domain of integration, including the transition point.

(iii) It is not non-typical in the sense that it involves special relations either between γ and the coefficients in the expansion of h in powers of z, or between the coefficients themselves.

6. Transformation of the admissible cases

We now show that equation (5·27) can be transformed into the form (5·8) and so comes under case D. We take new variables W, ζ related by

$$W = \dot{z}^{-\frac{1}{2}} w,$$

dots denoting differentiation with respect to ζ. Then (5·27) becomes

$$\frac{d^2W}{d\zeta^2} = \left[\dot{z}^2\left\{uz^n + \frac{\gamma}{z^2} + z^n h(z^{n+2})\right\} + \dot{z}^{\frac{1}{2}}\frac{d^2}{d\zeta^2}(\dot{z}^{-\frac{1}{2}})\right] W, \tag{6·1}$$

(cf. (2·3) and (2·4)). The relation between z and ζ is now prescribed by

$$\dot{z}^2 z^n = 1/\zeta.$$

On integration this yields

$$\zeta = z^{n+2}/(n+2)^2. \tag{6·2}$$

Substituting these results in (6·1), we find after reduction that

$$\frac{d^2W}{d\zeta^2} = \left[\frac{u}{\zeta} + \left\{\frac{\gamma + \frac{1}{4}}{(n+2)^2} - \frac{1}{4}\right\}\frac{1}{\zeta^2} + \frac{1}{\zeta} h\{(n+2)^2\zeta\}\right] W, \tag{6·3}$$

which is of the form (5·8) with γ replaced by $(\gamma + \frac{1}{4})(n+2)^{-2} - \frac{1}{4}$.

Thus all admissible cases can be transformed either into case A or case B of the previous paper (Olver 1954), or into case D of the present paper.

7. Case D; formal series solutions

We consider now the equation

$$\frac{d^2w}{dz^2} = \left\{\frac{u}{z} + \frac{\gamma}{z^2} + \frac{h(z)}{z}\right\} w, \tag{7·1}$$

(see (5·8)), in which γ is a constant and $h(z)$ a regular function of z in the domain of integration. The corresponding basic equation (cf. (3·2), (5·4)) is given by

$$\frac{d^2w}{dz^2} = \left\{\frac{u}{z} + \frac{\gamma}{z^2}\right\} w, \tag{7·2}$$

solutions of which are $z^{\frac{1}{2}} I_\mu\{2(uz)^{\frac{1}{2}}\}$ and $z^{\frac{1}{2}} K_\mu\{2(uz)^{\frac{1}{2}}\}$, where $\mu^2 = 1 + 4\gamma$ and I_μ, K_μ are the modified Bessel functions of order μ.

Owing to the presence of the square roots in the arguments of the Bessel functions it is convenient to take new variables. We replace z, u by z^2, $\frac{1}{4}u^2$ respectively, and accordingly equation (7·1) becomes

$$\frac{d^2w}{dz^2} = \frac{1}{z}\frac{dw}{dz} + \left\{u^2 + \frac{\mu^2-1}{z^2} + f(z)\right\} w, \tag{7·3}$$

which we shall take as the standard form of differential equation for this case. Here $f(z) = 4h(z^2)$ and so is an even function of z. The solutions of the basic equation are now $zI_\mu(uz)$ and $zK_\mu(uz)$.

A further modification we shall make concerns the form of the expansion (3·3). Instead of using the derivatives of I_μ and K_μ we shall use $I_{\mu+1}$ and $K_{\mu+1}$. The advantages of this change are first that it leads to simpler recurrence relations for the coefficients, and secondly that derivatives of Bessel functions are not normally given in numerical tables and have to be computed from functions of orders μ and $\mu+1$ by means of recurrence relations.

Thus we take the formal series solutions of (7·3) in the form

$$w(z) = z\mathscr{Z}_\mu(uz) \sum_{s=0}^{\infty} \frac{A_s(z)}{u^{2s}} + \frac{z}{u}\mathscr{Z}_{\mu+1}(uz) \sum_{s=0}^{\infty} \frac{B_s(z)}{u^{2s}}, \tag{7·4}$$

where $A_0 = 1$, and $\mathscr{Z}_\mu$ denotes the modified cylinder function

$$\mathscr{Z}_\mu = k_1 I_\mu + k_2 e^{\mu\pi i} K_\mu, \tag{7·5}$$

k_1, k_2 being arbitrary constants. The A_s and B_s occurring in (7·4) are not the same as the A_s and B_s of § 3, and the recurrence relations they satisfy may be found directly as follows.

Differentiating (7·4), and using the relations (Watson 1944, p. 79)

$$\frac{d}{dz}\mathscr{Z}_\mu(uz) = \frac{\mu}{z}\mathscr{Z}_\mu(uz) + u\mathscr{Z}_{\mu+1}(uz), \qquad \frac{d}{dz}\mathscr{Z}_{\mu+1}(uz) = u\mathscr{Z}_\mu(uz) - \frac{\mu+1}{z}\mathscr{Z}_{\mu+1}(uz), \tag{7·6}$$

we obtain

$$w'(z) = \mathscr{Z}_\mu(uz) \sum_{s=0}^{\infty} \frac{C_s(z)}{u^{2s}} + uz\mathscr{Z}_{\mu+1}(uz) \sum_{s=0}^{\infty} \frac{D_s(z)}{u^{2s}}, \tag{7·7}$$

$$w''(z) = \frac{u^2}{z}\mathscr{Z}_\mu(uz) \sum_{s=0}^{\infty} \frac{E_s(z)}{u^{2s}} + u\mathscr{Z}_{\mu+1}(uz) \sum_{s=0}^{\infty} \frac{F_s(z)}{u^{2s}}, \tag{7·8}$$

where

$$C_s = (\mu+1)A_s + zA_s' + zB_s, \qquad D_s = A_s - \mu z^{-1}B_{s-1} + B_{s-1}', \tag{7·9}$$

$$E_s = \mu C_{s-1} + zC_{s-1}' + z^2 D_s, \qquad F_s = C_s - \mu D_s + zD_s'. \tag{7·10}$$

Substituting (7·9) in (7·10), we obtain

$$\left.\begin{aligned} E_s &= z^2A_s + \mu(\mu+1)A_{s-1} + 2(\mu+1)zA_{s-1}' + z^2A_{s-1}'' + zB_{s-1} + 2z^2B_{s-1}', \\ F_s &= A_s + 2zA_s' + zB_s + \mu(\mu+1)z^{-1}B_{s-1} - 2\mu B_{s-1}' + zB_{s-1}''. \end{aligned}\right\} \tag{7·11}$$

From (7·4), (7·7) and (7·8) we see that equation (7·3) is formally satisfied if

$$\left.\begin{aligned} E_{s+1} &= C_s + z^2A_{s+1} + (\mu^2 - 1 + z^2 f)A_s, \\ F_{s+1} &= D_{s+1} + zB_{s+1} + (\mu^2 - 1 + z^2 f)z^{-1}B_s. \end{aligned}\right\} \tag{7·12}$$

Substituting (7·9) and (7·11) in (7·12), we find on reduction that

$$\left.\begin{aligned} \frac{2\mu+1}{z}A_s'+A_s''-fA_s+2B_s' &= 0,\\ 2A_{s+1}'+\frac{2\mu+1}{z^2}B_s-\frac{2\mu+1}{z}B_s'+B_s''-fB_s &= 0.\end{aligned}\right\} \tag{7·13}$$

These equations may be integrated to give

$$\left.\begin{aligned} 2B_s(z) &= -A_s'(z)+\int_0^z\left\{f(t)\,A_s(t)-\frac{2\mu+1}{t}A_s'(t)\right\}\mathrm{d}t,\\ 2A_{s+1}(z) &= \frac{2\mu+1}{z}B_s(z)-B_s'(z)+\int f(z)\,B_s(z)\,\mathrm{d}z,\end{aligned}\right\} \tag{7·14}$$

the integration constant in the second equation being arbitrary. These are the recurrence relations for A_s and B_s. Since $f(z)$ is even, A_s and B_s are regular functions of z in the domain of integration, A_s being an even function and B_s an odd function.

We shall proceed to consider the circumstances in which the formal series (7·4) is an asymptotic expansion (in Poincaré's sense) of a solution of the differential equation (7·3). Throughout the rest of the paper we shall suppose, without loss of generality, that

$$m\equiv\mathscr{R}\mu\geqslant 0. \tag{7·15}$$

We shall, of course, need to establish the desired results for two solutions of (7·3), and the question arises which pair of functions $\mathscr{Z}_\mu$ are we to take? The natural choice is I_μ and K_μ, and these are indeed a numerically satisfactory pair of solutions in the range $|\arg z|\leqslant\frac{1}{2}\pi$. When $|z|$ is small I_μ is small (or bounded) and K_μ is large; when $|z|$ is large I_μ is exponentially large and K_μ exponentially small. Because $f(z)$ is an even function it is adequate to confine our attention to solutions of (7·3) in the range $|\arg z|\leqslant\frac{1}{2}\pi$; solutions in other phase ranges can always be expressed in terms of solutions in this range by connecting formulae. On the other hand, in applications the actual process of connexion may well be tedious. For this reason, and for the sake of completeness, it is desirable to establish the asymptotic property in as wide a phase range as possible.

In the case of the series having the basic function I_μ we shall prove that the asymptotic property holds in all phase ranges; this is a particularly desirable result, since in all phase ranges I_μ is an essential member of any numerically satisfactory pair of solutions of the basic equation.

In the case of the series with the basic function K_μ we establish the asymptotic property in the range $|\arg z|\leqslant\frac{3}{2}\pi$. It is, in fact, unlikely that the valid range exceeds this, at any rate not without restriction on μ. If, for example, μ is half an odd integer, then $K_\mu(z)$ is exponentially small in the ranges $\frac{3}{2}\pi<|\arg z|<\frac{5}{2}\pi$. In the nature of the case we should not expect this property to extend also to the solution of (7·3) which is exponentially small in the sector $|\arg z|<\frac{1}{2}\pi$.

8. Asymptotic nature of the series; theorem D

In this section we reveal the asymptotic nature of the series (7·4) by stating an existence theorem analogous to theorems A and B of the previous paper (Olver 1954). The proof is given in §§ 10 to 13.

In the following it is supposed that $f(z)$ is an even function of z which is regular in a simply-connected open domain **D**, the boundaries of which consist of a finite number of straight lines. If **D** is unbounded it is supposed that

$$f(z) = O(|z|^{-1-\alpha}), \tag{8·1}$$

as $|z| \to \infty$ in **D** uniformly with respect to $\arg z$, where $\alpha > 0$ is a constant. We denote by **D′** any simply-connected domain lying wholly in **D**, the boundaries of which consist of a finite

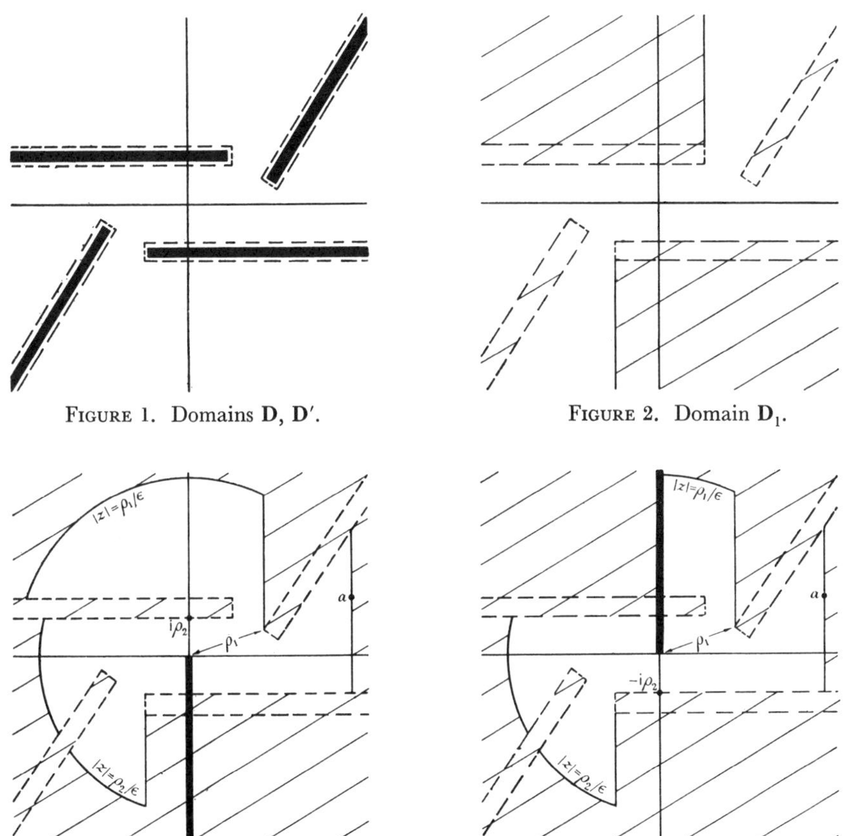

FIGURE 1. Domains **D**, **D′**.

FIGURE 2. Domain $\mathbf{D}_1$.

FIGURE 3. Domain $\mathbf{D}_2$, $-\frac{1}{2}\pi \leqslant \arg z \leqslant \frac{3}{2}\pi$.

FIGURE 4. Domain $\mathbf{D}_2$, $-\frac{3}{2}\pi \leqslant \arg z \leqslant \frac{1}{2}\pi$.

number of straight lines not intersecting the boundaries of **D**. It is supposed that $z = 0$ is an interior point of **D′**.

We define two further domains $\mathbf{D}_1$, $\mathbf{D}_2$ as follows. The domain $\mathbf{D}_1$ comprises those points z of **D′** which can be joined to the origin by a contour which lies in **D′** and does not cross either the imaginary axis, or the line through z parallel to the imaginary axis.

In the second definition we suppose a to be an arbitrary point of the sector $|\arg z| < \frac{1}{2}\pi$ (which may be at infinity), and ϵ to be an arbitrary positive number. Then $\mathbf{D}_2$ comprises those points z of $\mathbf{D}'$ for which $|\arg z| \leqslant \frac{3}{2}\pi$, $\mathscr{R}z \leqslant \mathscr{R}a$, and a contour can be found joining z and a which satisfies the following conditions:

(i) It lies in $\mathbf{D}'$.

(ii) It lies wholly to the right of the line through z parallel to the imaginary axis.

(iii) It does not cross the negative imaginary axis if $\frac{1}{2}\pi \leqslant \arg z \leqslant \frac{3}{2}\pi$, and does not cross the positive imaginary axis if $-\frac{3}{2}\pi \leqslant \arg z \leqslant -\frac{1}{2}\pi$.

(iv) It lies outside the circle $|t| = \epsilon |z|$.

Typical domains are illustrated in figures 1 to 4. The thickened lines in figure 1 denote cuts and are the boundaries of $\mathbf{D}$; they must, of course, be symmetrical about the origin. The broken lines everywhere are boundaries of $\mathbf{D}'$. The domain $\mathbf{D}_1$ is the unshaded region of figure 2. The domain $\mathbf{D}_2$ is represented by the unshaded regions in figures 3 and 4; the former refers to the range $-\frac{1}{2}\pi \leqslant \arg z \leqslant \frac{3}{2}\pi$ and has a cut along the negative imaginary axis, and the latter refers to the range $-\frac{3}{2}\pi \leqslant \arg z \leqslant \frac{1}{2}\pi$ and has a cut along the positive imaginary axis. The curved boundaries are arcs of the circles $|z| = \rho_1/\epsilon$, $|z| = \rho_2/\epsilon$, where ρ_1, ρ_2 are distances from the origin of points on the boundaries of $\mathbf{D}'$, as indicated on the diagrams.

THEOREM D. *Let sequences of functions $A_s(z)$, $B_s(z)$, $C_s(z)$, $D_s(z)$ $(s = 0, 1, 2, \ldots)$ be defined by the relations $A_0(z) = 1$*

$$\left.\begin{aligned} 2B_s(z) &= -A_s'(z) + \int_0^z \left\{f(t)\, A_s(t) - \frac{2\mu+1}{t} A_s'(t)\right\} \mathrm{d}t, \\ 2A_{s+1}(z) &= \frac{2\mu+1}{z} B_s(z) - B_s'(z) + \int f(z)\, B_s(z)\, \mathrm{d}z + \text{constant}, \end{aligned}\right\} \tag{8·2}$$

$$C_s(z) = (\mu+1)\, A_s(z) + zA_s'(z) + zB_s(z), \qquad D_s(z) = A_s(z) - \frac{\mu}{z} B_{s-1}(z) + B_{s-1}'(z). \tag{8·3}$$

Then if $\mathscr{R}\mu \geqslant 0$ the equation

$$\frac{\mathrm{d}^2w}{\mathrm{d}z^2} = \frac{1}{z}\frac{\mathrm{d}w}{\mathrm{d}z} + \left\{u^2 + \frac{\mu^2-1}{z^2} + f(z)\right\} w \tag{8·4}$$

has solutions $W_1(z)$, $W_2(z)$ such that:

(i) *If z lies in $\mathbf{D}_1$*

$$W_1(z) = zI_\mu(uz)\left\{\sum_{s=0}^{N-1} \frac{A_s(z)}{u^{2s}} + O\left(\frac{1}{u^{2N}}\right)\right\} + \frac{z}{u} I_{\mu+1}(uz)\left\{\sum_{s=0}^{N-1} \frac{B_s(z)}{u^{2s}} + \frac{z}{1+|z|} O\left(\frac{1}{u^{2N}}\right)\right\}, \tag{8·5}$$

$$W_1'(z) = I_\mu(uz)\left\{\sum_{s=0}^{N-1} \frac{C_s(z)}{u^{2s}} + (1+|z|)\, O\left(\frac{1}{u^{2N}}\right)\right\} + uzI_{\mu+1}(uz)\left\{\sum_{s=0}^{N-1} \frac{D_s(z)}{u^{2s}} + O\left(\frac{1}{u^{2N}}\right)\right\}, \tag{8·6}$$

as $u \to \infty$, where the O's are uniform with respect to z.

(ii) *If z lies in $\mathbf{D}_2$*

$$W_2(z) = zK_\mu(uz)\left\{\sum_{s=0}^{N-1} \frac{A_s(z)}{u^{2s}} + O\left(\frac{1}{u^{2N}}\right)\right\} - \frac{z}{u} K_{\mu+1}(uz)\left\{\sum_{s=0}^{N-1} \frac{B_s(z)}{u^{2s}} + \frac{z}{1+|z|} O\left(\frac{1}{u^{2N}}\right)\right\}, \tag{8·7}$$

$$W_2'(z) = K_\mu(uz)\left\{\sum_{s=0}^{N-1} \frac{C_s(z)}{u^{2s}} + (1+|z|)\, O\left(\frac{1}{u^{2N}}\right)\right\} - uzK_{\mu+1}(uz)\left\{\sum_{s=0}^{N-1} \frac{D_s(z)}{u^{2s}} + O\left(\frac{1}{u^{2N}}\right)\right\}, \tag{8·8}$$

as $u \to \infty$, where the O's are uniform with respect to z.

Remarks. (i) In (8·5) to (8·8) N is an arbitrary unbounded integer, but the solutions $W_1(z)$ and $W_2(z)$ are themselves independent of N. In the usual notation for asymptotic series we can write (8·5) and (8·7) in the respective forms

$$W_1(z) \sim zI_\mu(uz) \sum_{s=0}^{\infty} \frac{A_s(z)}{u^{2s}} + \frac{z}{u} I_{\mu+1}(uz) \sum_{s=0}^{\infty} \frac{B_s(z)}{u^{2s}}, \tag{8·9}$$

$$W_2(z) \sim zK_\mu(uz) \sum_{s=0}^{\infty} \frac{A_s(z)}{u^{2s}} - \frac{z}{u} K_{\mu+1}(uz) \sum_{s=0}^{\infty} \frac{B_s(z)}{u^{2s}}. \tag{8·10}$$

(ii) It may be noticed that in theorems A and B (Olver 1954) it was stipulated that the arbitrary constants of integration in the recurrence relations for the coefficients $A_s(z)$ and $B_s(z)$ must be such that there exists a function $\phi(u)$ with the property

$$\phi(u) \sim 1 + \sum_{s=1}^{\infty} \frac{A_s(c)}{u^s},$$

as $u \to \infty$. This restriction is in fact unnecessary, because it can be proved that there always exists a $\phi(u)$ with this property whatever the values of $A_1(c)$, $A_2(c)$, ..., (Erdélyi 1955, § 1·7). Accordingly, the corresponding restriction has been omitted from the statement of theorem D.

9. Relevant properties of Bessel functions

In the following we use the symbol A generically to denote a constant which is independent of z but which may depend on μ. It is supposed in this section, as elsewhere, that $\mathscr{R}\mu \geqslant 0$ (cf. (7·15)).

Properties of I_μ. We have (Watson 1944, chapters 3 and 7)

$$I_\mu(z\,e^{p\pi i}) = e^{p\mu\pi i} I_\mu(z), \quad p \text{ an integer}, \tag{9·1}$$

$$I_\mu(z) \sim (\tfrac{1}{2}z)^\mu / \Gamma(\mu+1) \quad \text{as} \quad z \to 0, \tag{9·2}$$

$$I_\mu(z) \sim (2\pi z)^{-\frac{1}{2}} \{e^z + e^{\pm(\mu+\frac{1}{2})\pi i} e^{-z}\} \quad (|\arg z| \leqslant \tfrac{3}{2}\pi - \epsilon) \quad \text{as} \quad |z| \to \infty, \tag{9·3}$$

uniformly with respect to arg z, the upper ambiguous sign applying to the range $-\frac{1}{2}\pi + \epsilon \leqslant \arg z \leqslant \frac{3}{2}\pi - \epsilon$, and the lower to the range $-\frac{3}{2}\pi + \epsilon \leqslant \arg z \leqslant \frac{1}{2}\pi - \epsilon$.

From (9·2) and (9·3) we deduce that

$$|I_\mu(z)| < AV_\mu(z) \quad (|\arg z| \leqslant \tfrac{1}{2}\pi), \tag{9·4}$$

where

$$V_\mu(z) \equiv \frac{|z^m e^z|}{1 + |z|^{m+\frac{1}{2}}} \quad (m \equiv \mathscr{R}\mu). \tag{9·5}$$

Properties of K_μ. Corresponding to (9·1), (9·2) and (9·3), we have

$$K_\mu(z\,e^{-p\pi i}) = e^{p\mu\pi i} K_\mu(z) + \pi i \sin p\mu\pi \operatorname{cosec} \mu\pi \,.\, I_\mu(z), \quad p \text{ an integer}, \tag{9·6}$$

$$\left.\begin{aligned} &K_\mu(z) \sim \tfrac{1}{2}\Gamma(\mu)\,(\tfrac{1}{2}z)^{-\mu} && (\mathscr{R}\mu > 0) \\ &K_\mu(z) \sim \tfrac{1}{2}\Gamma(\mu)\,(\tfrac{1}{2}z)^{-\mu} + \tfrac{1}{2}\Gamma(-\mu)\,(\tfrac{1}{2}z)^{\mu} && (\mathscr{R}\mu = 0,\ \mu \neq 0) \\ &K_\mu(z) \sim -\ln z && (\mu = 0) \end{aligned}\right\} \quad \text{as} \quad z \to 0, \tag{9·7}$$

$$K_\mu(z) \sim \left(\frac{\pi}{2z}\right)^{\frac{1}{2}} e^{-z} \quad (|\arg z| \leqslant \tfrac{3}{2}\pi - \epsilon) \quad \text{as} \quad |z| \to \infty, \tag{9·8}$$

uniformly with respect to arg z.

Let us write
$$l_0(z) \equiv \ln \frac{1+2|z|}{|z|}, \qquad l_\mu(z) \equiv 1 \quad (\mu \neq 0), \tag{9·9}$$

$$X_\mu(z) \equiv l_\mu(z) \frac{1+|z|^m |e^{-z}|}{1+|z|^{\frac{1}{2}} \; |z|^m} \quad (m \equiv \mathscr{R}\mu), \tag{9·10}$$

Then
$$|K_\mu(z)| < A X_\mu(z) \quad (|\arg z| \leqslant \tfrac{3}{2}\pi). \tag{9·11}$$

This result follows immediately from (9·7) and (9·8) when $|\arg z| \leqslant \frac{3}{2}\pi - \epsilon$ $(\epsilon > 0)$; its validity in the wider range of $\arg z$ may be established with the aid of (9·6) and (9·3).

Inequalities for products. From (9·4), (9·5), (9·9), (9·10) and (9·11) we derive

$$\left.\begin{aligned} |I_\mu(z)\, K_\mu(z)| &< \frac{A l_\mu(z)}{1+|z|}, & |I_{\mu+1}(z)\, K_{\mu+1}(z)| &< \frac{A}{1+|z|}, \\ |I_{\mu+1}(z)\, K_\mu(z)| &< \frac{A\,|z|\, l_\mu(z)}{1+|z|^2}, & |I_\mu(z)\, K_{\mu+1}(z)| &< \frac{A}{|z|}, \end{aligned}\right\} \tag{9·12}$$

valid when $|\arg z| \leqslant \frac{1}{2}\pi$. These inequalities remain true if I is replaced by V, or K replaced by X.

Wronskian relations. From the relation
$$\mathscr{W}\{K_\mu(z), I_\mu(z)\} = 1/z \tag{9·13}$$
(Watson 1944, p. 80), we deduce that
$$\mathscr{W}\{zK_\mu(z), zI_\mu(z)\} = z, \tag{9·14}$$
and, with the aid of (9·6),
$$\mathscr{W}\{zK_\mu(z\,e^{-p\pi i}), zI_\mu(z)\} = e^{p\mu\pi i}\, z. \tag{9·15}$$

10. Properties of the coefficients $A_s(z)$ and $B_s(z)$

These coefficients are defined by $A_0 = 1$, and (see (8·2))
$$2B_s(z) = -A_s'(z) + \int_0^z \left\{ f(t)\, A_s(t) - \frac{2\mu+1}{t} A_s'(t) \right\} dt, \tag{10·1}$$

$$2A_{s+1}(z) = \frac{2\mu+1}{z} B_s(z) - B_s'(z) + \int f(z)\, B_s(z)\, dz + \text{constant}. \tag{10·2}$$

The integration paths in (10·1) and (10·2) both lie in $\mathbf{D}$, and $A_s(z)$, $B_s(z)$ are regular functions of z in $\mathbf{D}$, $A_s(z)$ being an even function and $B_s(z)$ being an odd function.

In this section we shall establish properties of $A_s(z)$ and $B_s(z)$ analogous to those given in § 9 of the previous paper (Olver 1954) for the coefficients in case B. For this purpose we define constants d, R, auxiliary domains $\mathbf{D}_\delta$ and neighbourhoods of infinity $\mathbf{E}_\delta^1, \mathbf{E}_\delta^2, \ldots, \mathbf{E}_\delta$, exactly as in § 8 of that paper, and we again suppose (*ibid.*, equation (8·2))
$$0 < \delta \leqslant \min(d, \tfrac{1}{2}R, 1). \tag{10·3}$$

We introduce a further constant σ, defined by
$$\sigma = \min(\alpha, 1),$$
where α is the constant occurring in (8·1). From (8·1) we deduce that
$$|f(z)| \leqslant F_1, \qquad |z^{1+\sigma} f(z)| \leqslant F_2, \tag{10·4}$$
when z lies in $\mathbf{D}$, where F_1, F_2 are assignable constants. We shall use the symbol F generically to denote a combination of these constants, and also in a wider sense to denote any constant

which may depend on σ, μ, the properties of f and the configuration of the boundaries of $\mathbf{D}$ and $\mathbf{D}_d$, but is independent of z, u, s (or n) and δ.

We fix the arbitrary constants in (10·2) by the condition

$$A_{s+1}(c) = 0 \quad (s \geqslant 0), \tag{10·5}$$

for some point c, which may be at infinity, in $\mathbf{D}_d$. This restriction can subsequently be removed in a manner similar to that given in § 11 of the previous paper (Olver 1954) in the proof of theorem B, and we need not dwell further on this point. Accordingly, (10·2) becomes

$$2A_{s+1}(z) = \frac{2\mu+1}{z} B_s(z) - B_s'(z) + \int_c^z f(t)\, B_s(t)\, \mathrm{d}t - \frac{2\mu+1}{c} B_s(c) + B_s'(c) \quad (s \geqslant 0). \tag{10·6}$$

Lemma 2. *If* $|z| \to \infty$ *in* $\mathbf{E}_\delta$, *then*

$$A_s(z) = \alpha_s + O(|z|^{-\sigma}), \quad B_s(z) = \beta_s + O(|z|^{-\sigma}), \quad A_s'(z) = O(|z|^{-1-\sigma}), \tag{10·7}$$

uniformly with respect to $\arg z$, *where* α_s *and* β_s *are constants.*

We prove this by induction. Let η, ϵ be arbitrary numbers such that $0 \leqslant \eta < \epsilon \leqslant \delta$, and suppose that when z lies in $\mathbf{E}_\eta$

$$|A_s'(z)| < \lambda_s |z|^{-1-\sigma}, \tag{10·8}$$

where λ_s is a constant. Then we may write

$$A_s(z) = \alpha_s + a_s(z), \tag{10·9}$$

where α_s is a constant and

$$a_s(z) = -\int_z^\infty A_s'(t)\, \mathrm{d}t. \tag{10·10}$$

The path of integration can be taken to be a straight line having the parametric equation

$$t = z + q|z|\, \mathrm{e}^{\mathrm{i}\varpi}, \tag{10·11}$$

where ϖ is a constant, $0 \leqslant q < \infty$, and on the path

$$|t|^2 \geqslant |z|^2 (1+q^2). \tag{10·12}$$

We deduce that

$$|a_s(z)| < \int_z^\infty \frac{\lambda_s |\mathrm{d}t|}{|t|^{1+\sigma}} \leqslant \frac{\lambda_s}{|z|^\sigma} \int_0^\infty \frac{\mathrm{d}q}{(1+q^2)^{\frac{1}{2}+\frac{1}{2}\sigma}} = \frac{F\lambda_s}{|z|^\sigma}, \tag{10·13}$$

when z lies in $\mathbf{E}_\eta$, where F is the generic constant introduced at the beginning of this section.

It is evident from (8·1), (10·8), (10·9) and (10·13) that $\int\left(fA_s - \frac{2\mu+1}{z} A_s'\right) \mathrm{d}z$ converges at infinity in $\mathbf{E}_\eta$. Accordingly, we may write (10·1) in the form

$$B_s(z) = \beta_s + b_s(z), \tag{10·14}$$

where β_s is a constant, and

$$b_s(z) = -\tfrac{1}{2}A_s'(z) - \tfrac{1}{2}\int_z^\infty \left\{f(t)\, A_s(t) - \frac{2\mu+1}{t} A_s'(t)\right\} \mathrm{d}t. \tag{10·15}$$

The path of integration may be taken to be that defined by (10·11). With the aid of (10·4), (10·8), (10·9) and (10·13) we deduce that

$$\left|f(z)\, A_s(z) - \frac{2\mu+1}{z} A_s'(z)\right| < \frac{F_2}{|z|^{1+\sigma}}\left(|\alpha_s| + F\frac{\lambda_s}{|z|^\sigma}\right) + \frac{|2\mu+1|}{|z|^{2+\sigma}} \lambda_s < \frac{F}{|z|^{1+\sigma}} (|\alpha_s| + \lambda_s), \tag{10·16}$$

since $|z|>R\geqslant 1$. Substituting this result and (10·8) in (10·15) and using (10·12), we deduce that when z lies in $\mathbf{E}_\eta$

$$|b_s(z)|<\frac{\lambda_s}{2|z|^{1+\sigma}}+\frac{1}{2}\int_z^\infty \frac{F}{|t|^{1+\sigma}}(|\alpha_s|+\lambda_s)\,|\mathrm{d}t|\leqslant\frac{\lambda_s}{2|z|^{1+\sigma}}+\frac{F}{|z|^{\sigma}}(|\alpha_s|+\lambda_s),$$

and so (since $|z|>1$) we have

$$|b_s(z)|<F(|\alpha_s|+\lambda_s)\,|z|^{-\sigma}. \tag{10·17}$$

Next, we obtain by differentiation of (10·2) and (10·1),

$$2A'_{s+1}(z)=\frac{2\mu+1}{z}B'_s(z)-\frac{2\mu+1}{z^2}B_s(z)-B''_s(z)+f(z)\,B_s(z), \tag{10·18}$$

$$2B'_s(z)=-A''_s(z)+f(z)\,A_s(z)-\frac{2\mu+1}{z}A'_s(z), \tag{10·19}$$

$$2B''_s(z)=-A'''_s(z)+\frac{\mathrm{d}}{\mathrm{d}z}\left\{f(z)\,A_s(z)-\frac{2\mu+1}{z}A'_s(z)\right\}. \tag{10·20}$$

If, now, z lies in $\mathbf{E}_\epsilon$ we have by Cauchy's integral formula

$$A''_s(z)=\frac{1}{2\pi\mathrm{i}}\int_{|t-z|=\epsilon-\eta}\frac{A'_s(t)}{(t-z)^2}\mathrm{d}t. \tag{10·21}$$

From this result and (10·8), we see that

$$|A''_s(z)|<\frac{1}{\epsilon-\eta}\frac{\lambda_s}{\{|z|-(\epsilon-\eta)\}^{1+\sigma}}=O(|z|^{-1-\sigma}), \tag{10·22}$$

as $z\to\infty$ in $\mathbf{E}_\epsilon$. Substituting this result and (8·1), (10·8), (10·9), (10·13) in (10·19), we obtain

$$B'_s(z)=O(|z|^{-1-\sigma})\quad\text{as}\quad z\to\infty\text{ in }\mathbf{E}_\epsilon. \tag{10·23}$$

Similarly from (10·20) we deduce that

$$B''_s(z)=O(|z|^{-1-\sigma})\quad\text{as}\quad z\to\infty\text{ in }\mathbf{E}_\epsilon. \tag{10·24}$$

Substituting the last two results and (8·1), (10·14), (10·17) in (10·18), we see that

$$A'_{s+1}(z)=O(|z|^{-1-\sigma})\quad\text{as}\quad z\to\infty\text{ in }\mathbf{E}_\epsilon, \tag{10·25}$$

(since $\sigma\leqslant 1$).

The truth of lemma 2 now follows from the results (10·8), (10·13), (10·17) and (10·25) by induction.

LEMMA 3. *When z lies in $\mathbf{D}_\delta$ and $|t|\leqslant T$, the series*

$$A(z,t)=\sum_{s=0}^{\infty}A_s(z)\frac{t^{2s}}{(2s)!},\qquad B(z,t)=\sum_{s=0}^{\infty}B_s(z)\frac{t^{2s}}{(2s)!}, \tag{10·26}$$

converge uniformly with respect to z and t. Here $T>0$ is an assignable constant which may depend on δ, but is independent of z.

Let the maximum† moduli in $\mathbf{D}_\delta$ of $A_s(z)$, $A'_s(z)$, $A''_s(z)$, $B_s(z)$ and $B'_s(z)$ be denoted by $M_s(\delta)$, $M_s^{(1)}(\delta)$, $M_s^{(2)}(\delta)$, $P_s(\delta)$ and $P_s^{(1)}(\delta)$ respectively, and let η, ϵ be arbitrary numbers such that $0\leqslant\eta<\epsilon\leqslant\delta$.

† The existence of these maxima follows from lemma 2.

From (10·1) we obtain

$$|2B_s(z)| \leqslant |A_s'(z)| + \left| \int_0^z \left\{ f(t)\, A_s(t) - \frac{2\mu+1}{t} A_s'(t) \right\} \mathrm{d}t \right|. \tag{10·27}$$

If z lies in $\mathbf{D}_\epsilon$ and inside the circle $|z| = R+\epsilon$ (the boundary of $\mathbf{E}_\epsilon$), the path of integration can be made to consist of parts of the ray $\arg t = \arg z$, the circle $|z| = R+\epsilon$ and the boundary of $\mathbf{D}_\epsilon$. The length of such a path is bounded, $\leqslant l$, where l depends on the configuration of the boundaries of $\mathbf{D}$ and $\mathbf{D}_d$, but is independent of ϵ. Again, in this region $A_s'(z)/z$ is a regular function of z (since $A_s(z)$ is even) and therefore attains its maximum modulus on the boundary, by the maximum modulus theorem. Hence we have

$$\left| \int_0^z \left\{ f(t)\, A_s(t) - \frac{2\mu+1}{t} A_s'(t) \right\} \mathrm{d}t \right| \leqslant l \left\{ F_1 M_s(\epsilon) + \frac{|2\mu+1|}{\rho} M_s^{(1)}(\epsilon) \right\}, \tag{10·28}$$

when z lies in $\mathbf{D}_\epsilon$ and the circle $|z| = R+\epsilon$, where F_1 is defined by (10·4), and ρ is the shortest distance between $z = 0$ and the boundary of $\mathbf{D}_d$.

Next, if z lies in the subdomain $\mathbf{E}_\epsilon$, we may divide the path at the point z_1 where it crosses the circle $|z| = R+\epsilon$. For $\int_0^{z_1}$ the inequality (10·28) holds, and for $\int_{z_1}^z$ the path may be taken to be the straight line with the parametric equation

$$t = z_1 + q\,|z_1| \exp\{\mathrm{i} \arg (z - z_1)\} \quad (0 \leqslant q \leqslant |z-z_1|/|z_1|),$$

and the property

$$|t|^2 \geqslant |z_1|^2 (1+q^2)$$

(cf. (10·12)). Therefore we have

$$\left| \int_{z_1}^z f(t)\, A_s(t)\, \mathrm{d}t \right| \leqslant M_s(\epsilon) \int_{z_1}^z \frac{F_2\, |\mathrm{d}t|}{|t|^{1+\sigma}} < \frac{M_s(\epsilon)\, F_2}{|z_1|^\sigma} \int_0^\infty \frac{\mathrm{d}q}{(1+q^2)^{\frac{1}{2}+\frac{1}{2}\sigma}} < \frac{F}{R^\sigma} M_s(\epsilon), \tag{10·29}$$

since $|z_1| > R$. Also, on integrating by parts we have

$$\left| \int_{z_1}^z \frac{A_s'(t)}{t} \mathrm{d}t \right| = \left| \left[\frac{A_s(t)}{t} \right]_{z_1}^z + \int_{z_1}^z \frac{A_s(t)}{t^2} \mathrm{d}t \right| < \frac{2M_s(\epsilon)}{R} + M_s(\epsilon) \int_{z_1}^z \left| \frac{\mathrm{d}t}{t^2} \right|$$
$$\leqslant M_s(\epsilon) \left\{ \frac{2}{R} + \frac{1}{|z_1|} \int_0^\infty \frac{\mathrm{d}q}{1+q^2} \right\} < (2 + \tfrac{1}{2}\pi) \frac{M_s(\epsilon)}{R}. \tag{10·30}$$

Combining (10·28), (10·29) and (10·30), we see that

$$\left| \int_0^z \left\{ f(t)\, A_s(t) - \frac{2\mu+1}{t} A_s'(t) \right\} \mathrm{d}t \right| < F\{M_s(\epsilon) + M_s^{(1)}(\epsilon)\},$$

when z lies in $\mathbf{D}_\epsilon$, where F is the generic constant defined at the beginning of this section. Substituting this result in (10·27), we obtain

$$P_s(\epsilon) < F\{M_s(\epsilon) + M_s^{(1)}(\epsilon)\}. \tag{10·31}$$

We may apply similar arguments to (10·6); like $A_s'(z)/z$ the function $B_s(z)/z$ is regular in $\mathbf{D}_\epsilon$ and attains its maximum modulus on the boundary. We obtain the result

$$M_{s+1}(\epsilon) < FP_s(\epsilon) + P_s^{(1)}(\epsilon). \tag{10·32}$$

Next, we obtain immediately from (10·19)

$$2P_s^{(1)}(\epsilon) \leqslant M_s^{(2)}(\epsilon) + F_1 M_s(\epsilon) + |2\mu+1|\, \rho^{-1} M_s^{(2)}(\epsilon). \tag{10·33}$$

Combining (10·31), (10·32) and (10·33) to eliminate $P_s(\epsilon)$, $P_s^{(1)}(\epsilon)$, we obtain

$$M_{s+1}(\epsilon) < F\{M_s(\epsilon) + M_s^{(1)}(\epsilon) + M_s^{(2)}(\epsilon)\}. \tag{10·34}$$

From Cauchy's integral formulae

$$A'_s(z) = \frac{1}{2\pi\mathrm{i}}\int_{|t-z|=\epsilon-\eta} \frac{A_s(t)}{(t-z)^2}\,\mathrm{d}t, \qquad A''_s(z) = \frac{1}{\pi\mathrm{i}}\int_{|t-z|=\epsilon-\eta} \frac{A_s(t)}{(t-z)^3}\,\mathrm{d}t,$$

valid when z lies in $\mathbf{D}_\epsilon$, we deduce that

$$M_s^{(1)}(\epsilon) \leqslant (\epsilon-\eta)^{-1} M_s(\eta), \qquad M_s^{(2)}(\epsilon) \leqslant 2(\epsilon-\eta)^{-2} M_s(\eta). \tag{10·35}$$

Substituting these inequalities in (10·34), and using the facts that $M_s(\epsilon) \leqslant M_s(\eta)$ and $\epsilon - \eta < 1$ (cf. (10·3)), we obtain the key result

$$M_{s+1}(\epsilon) < F(\epsilon-\eta)^{-2} M_s(\eta). \tag{10·36}$$

To complete the proof we define

$$\delta_{n,s} = \frac{s-n}{s}\delta \quad (n = 0, 1, 2, \ldots, s). \tag{10·37}$$

Then from (10·36) we deduce that

$$M_s(\delta) \equiv M_s(\delta_{0,s}) < \frac{Fs^2}{\delta^2} M_{s-1}(\delta_{1,s}) < \left(\frac{Fs^2}{\delta^2}\right)^2 M_{s-2}(\delta_{2,s}) < \ldots < \left(\frac{Fs^2}{\delta^2}\right)^s. \tag{10·38}$$

Further, from this result and the first of (10·35), we deduce that

$$M_{s-1}^{(1)}(\delta) \leqslant \frac{s}{\delta} M_{s-1}(\delta_{1,s}) < \frac{s}{\delta}\left(\frac{Fs^2}{\delta^2}\right)^{s-1},$$

and so with the aid of (10·38) and (10·31) with ϵ replaced by δ, we obtain

$$P_s(\delta) < F\{M_s(\delta) + M_s^{(1)}(\delta)\} < F\left(\frac{Fs^2}{\delta^2}\right)^s + F\frac{s+1}{\delta}\left\{\frac{F(s+1)^2}{\delta^2}\right\}^s < F\left(1 + \frac{s+1}{\delta}\right)\left\{\frac{F(s+1)^2}{\delta^2}\right\}^s. \tag{10·39}$$

The truth of lemma 3 follows from the results (10·38), (10·39) and Stirling's formula

$$(2s)! \sim (2\pi)^{\frac{1}{2}}(2s)^{2s+\frac{1}{2}}\mathrm{e}^{-2s}. \tag{10·40}$$

LEMMA 4. *If z lies in $\mathbf{E}_\delta$ and $|t| \leqslant \tau$, then*

$$\frac{\partial}{\partial z}A(z,t) \equiv \sum_{s=1}^{\infty} A'_s(z)\frac{t^{2s}}{(2s)!} = O(|z|^{-1-\sigma}), \tag{10·41}$$

$$\frac{\partial^2}{\partial z\,\partial t}A(z,t) \equiv \sum_{s=1}^{\infty} A'_s(z)\frac{t^{2s-1}}{(2s-1)!} = O(|z|^{-1-\sigma}), \tag{10·42}$$

as $|z| \to \infty$, uniformly with respect to t and $\arg z$. *Here τ is an assignable number independent of z, and $0 < \tau \leqslant T$.*

As in the proof of lemmas 2 and 3 it is supposed that η, ϵ are arbitrary numbers such that $0 \leqslant \eta < \epsilon \leqslant \delta$. We first investigate the connexion between λ_s and λ_{s+1}, where λ_s is the constant introduced in (10·8).

If z lies in $\mathbf{E}_\epsilon$ then from (10·22) it is seen that

$$|A''_s(z)| < \frac{1}{\epsilon-\eta}\frac{\lambda_s}{\{|z|-(\epsilon-\eta)\}^{1+\sigma}} < \frac{2^{1+\sigma}\lambda_s}{(\epsilon-\eta)\,|z|^{1+\sigma}}, \tag{10·43}$$

since $\epsilon-\eta<\frac{1}{2}R<\frac{1}{2}|z|$ (cf. (10·3)). Substituting this result in (10·19) and using also (10·4), (10·8), (10·9) and (10·13), we find that

$$|B_s'(z)|<\frac{2^\sigma\lambda_s}{(\epsilon-\eta)|z|^{1+\sigma}}+\frac{F_2}{2|z|^{1+\sigma}}\left(|\alpha_s|+F\frac{\lambda_s}{|z|^\sigma}\right)+\frac{|2\mu+1|\lambda_s}{2|z|^{2+\sigma}}$$

$$<\left\{\frac{2^\sigma\lambda_s}{\epsilon-\eta}+F(|\alpha_s|+\lambda_s)\right\}\frac{1}{|z|^{1+\sigma}},\tag{10·44}$$

since $|z|>1$. Similarly we have

$$|A_s'''(z)|=\left|\frac{1}{\pi\mathrm{i}}\int_{|t-z|=\epsilon-\eta}\frac{A_s'(t)}{(t-z)^3}\mathrm{d}t\right|<\frac{2}{(\epsilon-\eta)^2}\frac{\lambda_s}{\{|z|-(\epsilon-\eta)\}^{1+\sigma}}<\frac{2^{2+\sigma}\lambda_s}{(\epsilon-\eta)^2|z|^{1+\sigma}},\tag{10·45}$$

and $$\left|\frac{\mathrm{d}}{\mathrm{d}z}\left\{f(z)\,A_s(z)-\frac{2\mu+1}{z}A_s'(z)\right\}\right|=\left|\frac{1}{2\pi\mathrm{i}}\int_{|t-z|=\epsilon-\eta}\frac{f(t)\,A_s(t)-(2\mu+1)\,t^{-1}A_s'(t)}{(t-z)^2}\mathrm{d}t\right|$$

$$<\frac{F(|\alpha_s|+\lambda_s)}{(\epsilon-\eta)|z|^{1+\sigma}}.\tag{10·46}$$

Combining (10·20), (10·45) and (10·46), we find that

$$|B_s''(z)|<\frac{2^{1+\sigma}\lambda_s}{(\epsilon-\eta)^2|z|^{1+\sigma}}+\frac{F(|\alpha_s|+\lambda_s)}{2(\epsilon-\eta)|z|^{1+\sigma}}<\frac{F(|\alpha_s|+\lambda_s)}{(\epsilon-\eta)^2|z|^{1+\sigma}},\tag{10·47}$$

since $\epsilon-\eta<1$.

Substituting the results (10·14), (10·17), (10·44) and (10·47) in (10·18), we obtain on reduction

$$|A_{s+1}'(z)|<\frac{F(|\alpha_s|+|\beta_s|+\lambda_s)}{(\epsilon-\eta)^2|z|^{1+\sigma}}.\tag{10·48}$$

This result holds in $\mathbf{E}_\epsilon$ provided that (10·8) holds in $\mathbf{E}_\eta$.

The magnitudes of $|\alpha_s|$, $|\beta_s|$ may be estimated by letting $|z|\to\infty$ in $\mathbf{E}_\delta$. We see immediately from (10·38) and (10·39) that

$$|\alpha_s|<\left(\frac{Fs^2}{\delta^2}\right)^s,\qquad |\beta_s|<F\left(1+\frac{s+1}{\delta}\right)\left\{\frac{F(s+1)^2}{\delta^2}\right\}^s,$$

and so $$|\alpha_s|,|\beta_s|<\{F(s+1)^2\delta^{-2}\}^{s+1}.\tag{10·49}$$

Substituting this result in (10·48) and changing s into $n-1$, we obtain the result:

$$\left.\begin{aligned}&\textit{If} && |z^{1+\sigma}A_{n-1}'(z)|<\lambda_{n-1}\quad \textit{in}\quad \mathbf{E}_\eta,\\ &\textit{then} && |z^{1+\sigma}A_n'(z)|<F(\epsilon-\eta)^{-2}\{(Fn^2\delta^{-2})^n+\lambda_{n-1}\}\quad \textit{in}\quad \mathbf{E}_\epsilon.\end{aligned}\right\}\tag{10·50}$$

Taking $n=1$, $\eta=0$, $\epsilon=\delta_{s-1,s}$, where $\delta_{n,s}$ is defined by (10·37), we obtain†

$$|z^{1+\sigma}A_1'(z)|\leqslant Fs^2\delta^{-2}F\delta^{-2}=\lambda_{1,s},\tag{10·51}$$

say. Similarly $$|z^{1+\sigma}A_2'(z)|<Fs^2\delta^{-2}\{(F2^2\delta^{-2})^2+\lambda_{1,s}\}=\lambda_{2,s},\tag{10·52}$$

say, and generally $$|z^{1+\sigma}A_n'(z)|<\lambda_{n,s}\quad(n=1,2,\ldots,s),\tag{10·53}$$

where $$\lambda_{n,s}=Fs^2\delta^{-2}\{(Fn^2\delta^{-2})^n+\lambda_{n-1,s}\}.\tag{10·54}$$

Let us write $$\lambda_{n,s}^*\equiv\{2Fs^2\delta^{-2}\}^{n+1}.$$

† Throughout the rest of the proof the symbol F denotes the same constant.

Then from (10·51) we see that $\lambda_{1,s}<\lambda^*_{1,s}$. Moreover, if $1\leqslant n\leqslant s$ we have

$$Fs^2\delta^{-2}\{(Fn^2\delta^{-2})^n+\lambda^*_{n-1,s}\}<Fs^2\delta^{-2}\{(2Fs^2\delta^{-2})^n+\lambda^*_{n-1,s}\}=\lambda^*_{n,s}.$$

Hence by induction $\qquad \lambda_{n,s}<\lambda^*_{n,s} \quad (n=1,2,\ldots,s),$

and in particular $\qquad \lambda_{s,s}<\lambda^*_{s,s}=(2Fs^2\delta^{-2})^{s+1}.$

Collecting these results, we have shown that when z lies in $\mathbf{E}_\delta$

$$|A_s'(z)|<(2F\delta^{-2})^{s+1}s^{2s+2}|z|^{-1-\sigma}.$$

The truth of lemma 4 is now apparent from this inequality and Stirling's formula (10·40).

11. Method of proof of theorem D

Let us suppose that the domain $\mathbf{D}_\delta$ (defined in § 10) contains the whole of the domain $\mathbf{D}'$ (defined in § 8). This can always be arranged by taking δ to be sufficiently small. Consider the function

$$L(z)=uz\mathscr{Z}_\mu(uz)\int_0^\tau e^{-ut}A(z,t)\,dt+z\mathscr{Z}_{\mu+1}(uz)\int_0^\tau e^{-ut}B(z,t)\,dt, \tag{11·1}$$

where $A(z,t)$, $B(z,t)$ are given by (10·26) and τ is the quantity introduced in lemma 4. More accurately, τ is the least of all such quantities when all the subdomains $\mathbf{E}^1_\delta$, $\mathbf{E}^2_\delta$, ... are taken into account.

Differentiating (11·1) and using (7·6), we find that

$$L'(z)=uz\mathscr{Z}_\mu(uz)\int_0^\tau e^{-ut}\left\{\frac{\mu+1}{z}A+\frac{\partial A}{\partial z}+B\right\}dt+z\mathscr{Z}_{\mu+1}(uz)\int_0^\tau e^{-ut}\left\{u^2A-\frac{\mu}{z}B+\frac{\partial B}{\partial z}\right\}dt, \tag{11·2}$$

$$L''(z)=uz\mathscr{Z}_\mu(uz)\int_0^\tau e^{-ut}\left\{\left(\frac{\mu+1}{z}+\frac{\partial}{\partial z}\right)\left(\frac{\mu+1}{z}A+\frac{\partial A}{\partial z}+B\right)+u^2A-\frac{\mu}{z}B+\frac{\partial B}{\partial z}\right\}dt$$
$$+z\mathscr{Z}_{\mu+1}(uz)\int_0^\tau e^{-ut}\left\{\frac{\mu+1}{z}u^2A+u^2\frac{\partial A}{\partial z}+u^2B+\left(\frac{\partial}{\partial z}-\frac{\mu}{z}\right)\left(u^2A-\frac{\mu}{z}B+\frac{\partial B}{\partial z}\right)\right\}dt. \tag{11·3}$$

Substituting the expansions (10·26) for $A(z,t)$ and $B(z,t)$ in (11·1) and integrating term by term, we may verify without difficulty that

$$L(z)=z\mathscr{Z}_\mu(uz)\left\{\sum_{s=0}^{N-1}\frac{A_s(z)}{u^{2s}}+O\left(\frac{1}{u^{2N}}\right)\right\}+\frac{z}{u}\mathscr{Z}_{\mu+1}(uz)\left\{\sum_{s=0}^{N-1}\frac{B_s(z)}{u^{2s}}+\frac{z}{1+|z|}O\left(\frac{1}{u^{2N}}\right)\right\}, \tag{11·4}$$

as $u\to\infty$, uniformly with respect to z in $\mathbf{D}'$, where N is an arbitrary integer. Similarly, using (8·3), we obtain from (11·2)

$$L'(z)=\mathscr{Z}_\mu(uz)\left\{\sum_{s=0}^{N-1}\frac{C_s(z)}{u^{2s}}+(1+|z|)\,O\left(\frac{1}{u^{2N}}\right)\right\}+uz\mathscr{Z}_{\mu+1}(uz)\left\{\sum_{s=0}^{N-1}\frac{D_s(z)}{u^{2s}}+O\left(\frac{1}{u^{2N}}\right)\right\}. \tag{11·5}$$

Comparing (11·4) and (11·5) with (8·5) to (8·8), we see that $L(z)$ has the desired asymptotic form.

Next, on substituting (11·1), (11·2) and (11·3) in the differential equation (8·4), we find on reduction

$$\frac{d^2L}{dz^2}-\frac{1}{z}\frac{dL}{dz}-\left\{u^2+\frac{\mu^2-1}{z^2}+f(z)\right\}L=S(z), \tag{11·6}$$

where
$$S(z) = uz\mathscr{Z}_\mu(uz)\int_0^\tau e^{-ut} S_1\,dt + z\mathscr{Z}_{\mu+1}(uz)\int_0^\tau e^{-ut} S_2\,dt, \tag{11·7}$$

$$S_1 \equiv \frac{2\mu+1}{z}\frac{\partial A}{\partial z} + \frac{\partial^2 A}{\partial z^2} - fA + 2\frac{\partial B}{\partial z},$$

$$S_2 \equiv 2u^2\frac{\partial A}{\partial z} + \left(\frac{2\mu+1}{z^2} - f\right)B - \frac{2\mu+1}{z}\frac{\partial B}{\partial z} + \frac{\partial^2 B}{\partial z^2}.$$

From (7·13) and (10·26) we see that S_1 vanishes identically, and

$$S_2 = 2u^2\frac{\partial A}{\partial z} - \frac{2\partial^3 A}{\partial z\,\partial t^2}.$$

Substituting these results in (11·7) and integrating by parts, we find that

$$S(z) = -z\mathscr{Z}_{\mu+1}(uz)\,R(z), \tag{11·8}$$

where
$$R(z) = 2\int_0^\tau e^{-ut}\left(\frac{\partial^3 A}{\partial z\,\partial t^2} - u^2\frac{\partial A}{\partial z}\right)dt = 2e^{-u\tau}\{uR_1(z) + R_2(z)\}, \tag{11·9}$$

$$R_1(z) = \left(\frac{\partial A}{\partial z}\right)_{t=\tau} = \sum_{s=1}^{\infty} A_s'(z)\frac{\tau^{2s}}{(2s)!}, \qquad R_2(z) = \left(\frac{\partial^2 A}{\partial z\,\partial t}\right)_{t=\tau} = \sum_{s=1}^{\infty} A_s'(z)\frac{\tau^{2s-1}}{(2s-1)!}. \tag{11·10}$$

Thus $L(z)$ does not itself satisfy the differential equation (8·4). We shall prove theorem D by showing that, for large u, the difference between $L(z)$ and an appropriately defined solution of (8·4) is asymptotically negligible compared with the series (11·4). In the next section we consider the solution having the basic function $\mathscr{Z}_\mu = I_\mu$, and in § 13 we consider the solution having the basic function $\mathscr{Z}_\mu = e^{\mu\pi i}K_\mu$.

12. Proof for the solution $W_1(z)$

We define $W_1(z)$ to be the solution of (8·4) with the condition

$$z^{-(\mu+1)}\{W_1(z) - L_1(z)\} \to 0 \quad \text{as} \quad z \to 0, \tag{12·1}$$

where $L_1(z) = L(z)$ is given by (11·1) with $\mathscr{Z} = I$. From (8·4), (11·6) and (11·8) we obtain

$$\frac{d^2}{dz^2}\{W_1(z) - L_1(z)\} - \frac{1}{z}\frac{d}{dz}\{W_1(z) - L_1(z)\} - \left\{u^2 + \frac{\mu^2-1}{z^2} + f(z)\right\}\{W_1(z) - L_1(z)\} = zI_{\mu+1}(uz)\,R(z). \tag{12·2}$$

We define a sequence of functions $\{h_n(z)\}$ by the relation $h_0(z) = 0$ and

$$\frac{d^2}{dz^2}h_n(z) - \frac{1}{z}\frac{d}{dz}h_n(z) - \left(u^2 + \frac{\mu^2-1}{z^2}\right)h_n(z) = f(z)\,h_{n-1}(z) + zI_{\mu+1}(uz)\,R(z) \quad (n \geq 1), \tag{12·3}$$

with the condition
$$z^{-(\mu+1)}h_n(z) \to 0 \quad \text{as} \quad z \to 0. \tag{12·4}$$

Let us suppose that z lies in the sector

$$(p - \tfrac{1}{2})\,\pi \leq \arg z \leq (p + \tfrac{1}{2})\,\pi,$$

where p is an arbitrary integer. Then taking the complementary solutions of (12·3) to be $zI_\mu(uz)$ and $zK_\mu(uz\,e^{-p\pi i})$, and applying the principle of variation of parameters, we find, with the aid of (9·15) and (12·4),

$$h_n(z) = z\,e^{-p\mu\pi i}\int_0^z \{I_\mu(uz)\,K_\mu(ut\,e^{-p\pi i}) - K_\mu(uz\,e^{-p\pi i})\,I_\mu(ut)\}\{f(t)\,h_{n-1}(t) + tI_{\mu+1}(ut)\,R(t)\}\,dt, \tag{12·5}$$

when $n \geqslant 1$. If we write
$$z' = z\,\mathrm{e}^{-p\pi\mathrm{i}}, \qquad t' = t\,\mathrm{e}^{-p\pi\mathrm{i}},$$
so that $|\arg z'| \leqslant \frac{1}{2}\pi$, we obtain on substituting (11·9) in (12·5) with $n = 1$,
$$h_1(z) = \mathrm{e}^{-u\tau}\, zV_\mu(uz') \int_0^z \psi(z,u,t)\,\{R_1(t) + u^{-1}R_2(t)\}\,\mathrm{d}t, \tag{12·6}$$
where V_μ is defined by (9·5), and
$$\psi(z,u,t) = 2u\,\mathrm{e}^{-p\mu\pi\mathrm{i}}\{I_\mu(uz)\,K_\mu(ut') - K_\mu(uz')\,I_\mu(ut)\}\frac{tI_{\mu+1}(ut)}{V_\mu(uz')}. \tag{12·7}$$

In consequence of the definition of the domain $\mathbf{D}_1$ given in § 8, we may choose a path of integration for (12·5) and (12·6) which lies wholly in $\mathbf{D}'$ and has the properties
$$|\arg t'| \leqslant \tfrac{1}{2}\pi, \qquad \mathscr{R}z' \geqslant \mathscr{R}t', \qquad |t| \leqslant \kappa |z|, \tag{12·8}$$
where† κ is independent of z, u, t and p. From (9·1), (9·4) and (9·12) we deduce that
$$\left| u\,\mathrm{e}^{-p\mu\pi\mathrm{i}}\, I_\mu(uz)\,K_\mu(ut')\frac{tI_{\mu+1}(ut)}{V_\mu(uz')}\right| = \left| u\,\mathrm{e}^{p\mu\pi\mathrm{i}}\, I_\mu(uz')\,K_\mu(ut')\frac{tI_{\mu+1}(ut')}{V_\mu(uz')}\right|$$
$$< A\,|\mathrm{e}^{p\mu\pi\mathrm{i}}|\frac{|ut|^2\, l_\mu(ut)}{1+|ut|^2} < A\,|\mathrm{e}^{p\mu\pi\mathrm{i}}|, \tag{12·9}$$
where l_μ is defined by (9·9) and A denotes a generic constant independent of z, u, t and p. Similarly from (9·1), (9·4) and (9·11) we deduce that
$$\left| u\,\mathrm{e}^{-p\mu\pi\mathrm{i}}\, K_\mu(uz')\,I_\mu(ut)\frac{tI_{\mu+1}(ut)}{V_\mu(uz')}\right| < A\,|ut\,\mathrm{e}^{p\mu\pi\mathrm{i}}|\, V_\mu(ut')\,V_{\mu+1}(ut')\frac{X_\mu(uz')}{V_\mu(uz')}$$
$$< A\,|\mathrm{e}^{p\mu\pi\mathrm{i}}|\frac{|ut|^{2m}}{1+|ut|^{2m}}\,\frac{1+|uz|^{2m}}{|uz|^{2m}}\,|\mathrm{e}^{2u(t'-z')}|\, l_\mu(uz). \tag{12·10}$$
From (12·8) we see that
$$|\mathrm{e}^{2u(t'-z')}| \leqslant 1, \tag{12·11}$$
and that
$$\frac{|ut|^{2m}}{1+|ut|^{2m}}\,\frac{1+|uz|^{2m}}{|uz|^{2m}} \leqslant \begin{cases} 2\kappa^{2m} & (|uz| \leqslant 1),\\ 2 & (|uz| \geqslant 1).\end{cases} \tag{12·12}$$
Also, $l_\mu(x)$ is a non-increasing function of the real variable $|x|$ (see (9·9)), and so
$$l_\mu(uz) \leqslant l_\mu(ut/\kappa) \leqslant l_\mu(t), \tag{12·13}$$
provided that $u \geqslant \kappa$. Combining (12·11), (12·12) and (12·13), we see that the right-hand side of (12·10) is less than $A|\mathrm{e}^{p\mu\pi\mathrm{i}}|\, l_\mu(t)$, and substituting this result and (12·9) in (12·7), we obtain
$$|\psi(z,u,t)| < A\,|\mathrm{e}^{p\mu\pi\mathrm{i}}|\, l_\mu(t). \tag{12·14}$$

Applying (12·14) to (12·6), we find that
$$|h_1(z)| < A\,\mathrm{e}^{-u\tau}\,|z\,\mathrm{e}^{p\mu\pi\mathrm{i}}|\, V_\mu(uz') \int_0^z l_\mu(t)\,|R_1(t) + u^{-1}R_2(t)|\,|\mathrm{d}t|. \tag{12·15}$$

† For if the symbols ρ and R have the meanings assigned in §9, then when $|z| \leqslant \rho$ and $|z| \geqslant R$ we can obviously choose a contour such that $|t| \leqslant z$, and when $\rho < |z| < R$ we can choose a contour such that $|t| \leqslant R$. Thus we can take $\kappa = R/\rho$.

From (11·10) and lemma 4 it follows that

$$R_1(z) = O(|z|^{-1-\sigma}), \qquad R_2(z) = O(|z|^{-1-\sigma}), \tag{12·16}$$

as $|z| \to \infty$ in $\mathbf{D}'$, uniformly with respect to $\arg z$. The path of integration in (12·15) may be confined to the ray $\arg t = \arg z$ and the boundary of $\mathbf{D}'$, and still satisfy the conditions above. Hence

$$\int_0^z l_\mu(t)\, |R_1(t)\,\mathrm{d}t| < A, \qquad \int_0^z l_\mu(t)\, |R_2(t)\,\mathrm{d}t| < A, \tag{12·17}$$

and substituting these inequalities in (12·15), we obtain

$$|h_1(z)| < A\,\mathrm{e}^{-u\tau}\, |z\,\mathrm{e}^{p\mu\pi\mathrm{i}}|\, V_\mu(uz'). \tag{12·18}$$

Next, from (12·5) it is seen that

$$h_{n+1}(z) - h_n(z) = \frac{z}{u} V_\mu(uz') \int_0^z \chi(z,u,t)\, f(t) \frac{h_n(t) - h_{n-1}(t)}{t V_\mu(ut')}\,\mathrm{d}t \quad (n \geqslant 1), \tag{12·19}$$

where
$$\chi(z,u,t) = u\,\mathrm{e}^{-p\mu\pi\mathrm{i}} \{I_\mu(uz)\, K_\mu(ut') - K_\mu(uz')\, I_\mu(ut)\} \frac{t V_\mu(ut')}{V_\mu(uz')}. \tag{12·20}$$

Using arguments similar to those above for $\psi(z,u,t)$, we may show that

$$\left| u\,\mathrm{e}^{-p\mu\pi\mathrm{i}}\, I_\mu(uz)\, K_\mu(ut') \frac{t V_\mu(ut')}{V_\mu(uz')} \right| < A \frac{|ut|\, l_\mu(ut)}{1 + |ut|} < A,$$

$$\left| u\,\mathrm{e}^{-p\mu\pi\mathrm{i}}\, K_\mu(uz')\, I_\mu(ut) \frac{t V_\mu(ut')}{V_\mu(uz')} \right| < A \frac{|ut|^{2m+1}}{1 + |ut|^{2m+1}} \frac{1 + |uz|^{2m}}{|uz|^{2m}} |\mathrm{e}^{2u(t'-z')}|\, l_\mu(uz) < A,$$

(since $l_\mu(uz) \leqslant l_\mu(ut/\kappa)$), and thence

$$|\chi(z,u,t)| < A. \tag{12·21}$$

Let us assume temporarily that

$$|h_n(z) - h_{n-1}(z)| < q_n(u)\, |z|\, V_\mu(uz'),$$

where $q_n(u)$ is independent of z. Then, from (12·19) and (12·21), we obtain

$$|h_{n+1}(z) - h_n(z)| < A \frac{|z|}{u} V_\mu(uz') \int_0^z |f(t)|\, q_n(u)\, |\mathrm{d}t| < \frac{A}{u} q_n(u)\, |z|\, V_\mu(uz'),$$

in virtue of (8·1), where A is independent of z, u, p and n (cf. (12·16) and (12·17)). Hence by induction, using (12·18), we have

$$|h_{n+1}(z) - h_n(z)| < A (A/u)^n\, \mathrm{e}^{-u\tau}\, |z\,\mathrm{e}^{p\mu\pi\mathrm{i}}|\, V_\mu(uz') \quad (n \geqslant 0). \tag{12·22}$$

Thus when $u > A$ the series
$$\sum_{n=0}^{\infty} \{h_{n+1}(z) - h_n(z)\}$$

converges uniformly with respect to z. From (12·1), (12·2), (12·3) and (12·4), we see that its sum is $W_1(z) - L_1(z)$. Moreover, if $u > 2A$ we have

$$|W_1(z) - L_1(z)| < \sum_{n=0}^{\infty} A (\tfrac{1}{2})^n\, \mathrm{e}^{-u\tau}\, |z\,\mathrm{e}^{p\mu\pi\mathrm{i}}|\, V_\mu(uz') = 2A\,\mathrm{e}^{-u\tau}\, |z\,\mathrm{e}^{p\mu\pi\mathrm{i}}|\, V_\mu(uz'). \tag{12·23}$$

The proof of (8·5) will be complete if we show this quantity to be asymptotically negligible compared with the series (11·4) with $\mathscr{Z}_\mu = I_\mu$.

Suppose that r is the zero of $I_\mu(uz)$ nearest to $z=0$ (other than $z=0$ itself). Then in consequence of (9·1) and the relation

$$I_\mu(z)\,K_{\mu+1}(z)+I_{\mu+1}(z)\,K_\mu(z) = 1/z \tag{12·24}$$

(Watson 1944, p. 80), we can write

$$z\,\mathrm{e}^{p\mu\pi\mathrm{i}}V_\mu(uz') = V_\mu^{(1)}.zI_\mu(uz)+V_\mu^{(2)}.uz^2(1+|z|)^{-1}I_{\mu+1}(uz), \tag{12·25}$$

where

$$\left.\begin{aligned} V_\mu^{(1)} &= uz'K_{\mu+1}(uz')\,V_\mu(uz') \\ V_\mu^{(2)} &= (1+|z|)\,K_\mu(uz')\,V_\mu(uz') \end{aligned}\right\}(|uz|>\tfrac{1}{2}|r|), \qquad \left.\begin{aligned} V_\mu^{(1)} &= V_\mu(uz')/I_\mu(uz') \\ V_\mu^{(2)} &= 0 \end{aligned}\right\}(|uz|\leqslant\tfrac{1}{2}|r|). \tag{12·26}$$

With the aid of (9·2), (9·5) and (9·12), we may verify that $|V_\mu^{(1)}|$, $|V_\mu^{(2)}|$ are bounded functions of u and z, and the desired result follows immediately.

To establish the expansion (8·6) for the derivative $W_1'(z)$, we differentiate (12·5). Using (7·6), we find

$$h_n'(z) = \frac{\mu+1}{z}h_n(z)+uz\,\mathrm{e}^{-p\mu\pi\mathrm{i}}\int_0^z\{I_{\mu+1}(uz)\,K_\mu(ut\,\mathrm{e}^{-p\pi\mathrm{i}})+\mathrm{e}^{-p\pi\mathrm{i}}K_{\mu+1}(uz\,\mathrm{e}^{-p\pi\mathrm{i}})I_\mu(ut)\}$$
$$\times\{f(t)\,h_{n-1}(t)+tI_{\mu+1}(ut)\,R(t)\}\,\mathrm{d}t. \tag{12·27}$$

Taking $n=1$ and substituting (11·9), we may verify that

$$h_1'(z) = \frac{\mu+1}{z}h_1(z)+uz\,\mathrm{e}^{-u\tau}\,V_\mu(uz')\int_0^z\psi_1(z,u,t)\,\{R_1(t)+u^{-1}R_2(t)\}\,\mathrm{d}t, \tag{12·28}$$

where $$\psi_1(z,u,t) = 2u\,\mathrm{e}^{-p\mu\pi\mathrm{i}}\{I_{\mu+1}(uz)\,K_\mu(ut')+\mathrm{e}^{-p\pi\mathrm{i}}K_{\mu+1}(uz')\,I_\mu(ut)\}\frac{tI_{\mu+1}(ut)}{V_\mu(uz')}. \tag{12·29}$$

Similarly

$$h_{n+1}'(z)-h_n'(z) = \frac{\mu+1}{z}\{h_{n+1}(z)-h_n(z)\}+zV_\mu(uz')\int_0^z\chi_1(z,u,t)\,f(t)\,\frac{h_n(t)-h_{n-1}(t)}{tV_\mu(ut')}\,\mathrm{d}t, \tag{12·30}$$

if $n\geqslant 1$, where

$$\chi_1(z,u,t) = u\,\mathrm{e}^{-p\mu\pi\mathrm{i}}\{I_{\mu+1}(uz)\,K_\mu(ut')+\mathrm{e}^{-p\pi\mathrm{i}}K_{\mu+1}(uz')\,I_\mu(ut)\}\frac{tV_\mu(ut')}{V_\mu(uz')}. \tag{12·31}$$

We readily show that

$$|\psi_1(z,u,t)|<A\,|\mathrm{e}^{p\mu\pi\mathrm{i}}|, \qquad |\chi_1(z,u,t)|<A, \tag{12·32}$$

(cf. (12·14) and (12·21)), and thence using also (12·18) and (12·22), that

$$|h_1'(z)|<A\,\mathrm{e}^{-u\tau}\,|\mathrm{e}^{p\mu\pi\mathrm{i}}|\,(1+|uz|)\,V_\mu(uz')<Au\,\mathrm{e}^{-u\tau}\,|\mathrm{e}^{p\mu\pi\mathrm{i}}|\,(1+|z|)\,V_\mu(uz'), \tag{12·33}$$

$$|h_{n+1}'(z)-h_n'(z)|<A(A/u)^n\,\mathrm{e}^{-u\tau}\,|\mathrm{e}^{p\mu\pi\mathrm{i}}|\,V_\mu(uz')+A(A/u)^{n-1}\,\mathrm{e}^{-u\tau}\,|\mathrm{e}^{p\mu\pi\mathrm{i}}|\,|z|\,V_\mu(uz')$$
$$<A(A/u)^n\,u\,\mathrm{e}^{-u\tau}\,|\mathrm{e}^{p\mu\pi\mathrm{i}}|\,(1+|z|)\,V_\mu(uz'), \tag{12·34}$$

provided that $u>1$. Thus when $u>2A$

$$|W_1'(z)-L_1'(z)| = \left|\sum_{n=0}^{\infty}\{h_{n+1}'(z)-h_n'(z)\}\right|<2Au\,\mathrm{e}^{-u\tau}\,|\mathrm{e}^{p\mu\pi\mathrm{i}}|\,(1+|z|)\,V_\mu(uz'). \tag{12·35}$$

This result corresponds to (12·23). From (12·25) we see that we may write

$$\mathrm{e}^{p\mu\pi\mathrm{i}}\,(1+|z|)\,V_\mu(uz') = V_\mu^{(1)}.(1+|z|)\,I_\mu(uz)+V_\mu^{(2)}.uzI_{\mu+1}(uz), \tag{12·36}$$

where $V_\mu^{(1)}$, $V_\mu^{(2)}$ are defined by (12·26). Hence $|W_1'(z)-L_1'(z)|$ is asymptotically negligible compared with the error terms in (8·6). This completes the proof of (8·6).

13. Proof for the solution $W_2(z)$

We define $W_2(z)$ to be the solution of (8·4) with the conditions

$$\left.\begin{aligned} &W_2(a) = L_2(a), \quad W_2'(a) = L_2'(a), \quad \text{if } a \text{ is finite,} \\ &\lim_{z\to a}[z^{-\frac{1}{2}}\,e^{uz}\{W_2(z)-L_2(z)\}] = 0, \quad \text{if } a \text{ is infinite,}\end{aligned}\right\} \tag{13·1}$$

where $L_2(z) = L(z)$ is given by (11·1) with $\mathscr{L}_\mu = K_\mu$, $\mathscr{L}_{\mu+1} = -K_{\mu+1}$, and a is an arbitrary point of the sector $|\arg z| < \frac{1}{2}\pi$ (cf. § 8). We shall now prove that with this choice of $W_2(z)$ the relations (8·7) and (8·8) hold in $\mathbf{D}_2$. Only a proof for the part of $\mathbf{D}_2$ which lies in the sector $-\frac{1}{2}\pi \leqslant \arg z \leqslant \frac{3}{2}\pi$ will be given; the proof for the part which lies in $-\frac{3}{2}\pi \leqslant \arg z \leqslant \frac{1}{2}\pi$ is exactly similar.

Corresponding to (12·2), we have

$$\frac{d^2}{dz^2}\{W_2(z)-L_2(z)\} - \frac{1}{z}\frac{d}{dz}\{W_2(z)-L_2(z)\} - \left\{u^2 + \frac{\mu^2-1}{z^2} + f(z)\right\}\{W_2(z)-L_2(z)\}$$
$$= -zK_{\mu+1}(uz)\,R(z). \tag{13·2}$$

We define a sequence $\{j_n(z)\}$ by $j_0(z) = 0$,

$$\frac{d^2}{dz^2}j_n(z) - \frac{1}{z}\frac{d}{dz}j_n(z) - \left(u^2 + \frac{\mu^2-1}{z^2}\right)j_n(z) = f(z)\,j_{n-1}(z) - zK_{\mu+1}(uz)\,R(z) \quad (n \geqslant 1), \tag{13·3}$$

with

$$\left.\begin{aligned} &j_n(a) = j_n'(a) = 0 \quad (a \text{ finite}), \\ &\lim_{z\to a}\{z^{-\frac{1}{2}}\,e^{uz} j_n(z)\} = 0 \quad (a \text{ infinite}).\end{aligned}\right\} \tag{13·4}$$

Thus

$$j_n(z) = \int_z^a z\{K_\mu(uz)\,I_\mu(ut) - I_\mu(uz)\,K_\mu(ut)\}\{f(t)\,j_{n-1}(t) - tK_{\mu+1}(ut)\,R(t)\}\,dt \quad (n \geqslant 1) \tag{13·5}$$

(cf. (9·14) and (12·5)), and so

$$j_1(z) = -e^{-u\tau}\,zX_\mu(uz)\int_z^a \omega(z,u,t)\,\{R_1(t) + u^{-1}R_2(t)\}\,dt \tag{13·6}$$

(cf. (12·6) and (11·9)), where

$$\omega(z,u,t) = 2u\{K_\mu(uz)\,I_\mu(ut) - I_\mu(uz)\,K_\mu(ut)\}\,\frac{tK_{\mu+1}(ut)}{X_\mu(uz)}. \tag{13·7}$$

In consequence of the definition of the domain $\mathbf{D}_2$ given in § 8 we may choose a path of integration for (13·5) and (13·6) which lies wholly in $\mathbf{D}'$ and has the properties

$$\mathscr{R}z \leqslant \mathscr{R}t, \qquad |t| \geqslant \epsilon|z|. \tag{13·8}$$

Suppose first that $|\arg z| \leqslant \frac{1}{2}\pi$. Then $|\arg t| \leqslant \frac{1}{2}\pi$, and we deduce from (9·11) and (9·12) that

$$\left|2uK_\mu(uz)\,I_\mu(ut)\,\frac{tK_{\mu+1}(ut)}{X_\mu(uz)}\right| < A, \tag{13·9}$$

and, from (9·4), (9·5), (9·10) and (9·11), that

$$\left|2uI_\mu(uz)\,K_\mu(ut)\,\frac{tK_{\mu+1}(ut)}{X_\mu(uz)}\right| < A\,\frac{1+|ut|^{2m}}{|ut|^{2m}}\,\frac{|uz|^{2m}}{1+|uz|^{2m}}\,|e^{2u(z-t)}|\,\frac{l_\mu(ut)}{l_\mu(uz)}, \tag{13·10}$$

where A denotes a generic constant independent of z, u, t. From (9·9) and (13·8), we see that

$$l_\mu(ut)/l_\mu(uz) \leqslant l_\mu(u\epsilon z)/l_\mu(uz) \leqslant A, \tag{13·11}$$

where A may now depend on ϵ, but is still independent of z, u, t. Using this result and (13·8) we may show that the right-hand side of (13·10) is bounded (cf. (12·12)), and hence we obtain

$$|\omega(z,u,t)| < A \quad (|\arg z| \leqslant \tfrac{1}{2}\pi). \tag{13·12}$$

Alternatively, suppose that $\frac{1}{2}\pi \leqslant \arg z \leqslant \frac{3}{2}\pi$. Then we take $\omega(z,u,t)$ in the form†

$$\omega(z,u,t) = 2u\,\mathrm{e}^{-\mu\pi\mathrm{i}}\{K_\mu(uz\,\mathrm{e}^{-\pi\mathrm{i}})\,I_\mu(ut) - I_\mu(uz)\,K_\mu(ut\,\mathrm{e}^{-\pi\mathrm{i}})\}\frac{tK_{\mu+1}(ut)}{X_\mu(uz)}, \tag{13·13}$$

which may be obtained from (13·7) by use of (9·6) with $p=1$. It is clear, from condition (iii) of the definition of $\mathbf{D}_2$ given in § 8, that t must lie in the sector $-\frac{1}{2}\pi \leqslant \arg t \leqslant \frac{3}{2}\pi$. With the aid of (9·1), (9·4), (9·5), (9·10) and (9·11), we deduce that

$$\left|2u\,\mathrm{e}^{-\mu\pi\mathrm{i}}\,I_\mu(uz)\,K_\mu(ut\,\mathrm{e}^{-\pi\mathrm{i}})\frac{tK_{\mu+1}(ut)}{X_\mu(uz)}\right| = \left|2uI_\mu(uz\,\mathrm{e}^{-\pi\mathrm{i}})\,K_\mu(ut\,\mathrm{e}^{-\pi\mathrm{i}})\frac{tK_{\mu+1}(ut)}{X_\mu(uz)}\right|$$
$$< A\frac{1+|ut|^{2m}}{|ut|^{2m}}\,\frac{|uz|^{2m}}{1+|uz|^{2m}}\,\frac{l_\mu(ut)}{l_\mu(uz)} < A \tag{13·14}$$

(cf. (13·10) and (13·11)). Further, if $|\arg t| \leqslant \frac{1}{2}\pi$ we deduce from (9·10), (9·11) and (9·12) that

$$\left|2u\,\mathrm{e}^{-\mu\pi\mathrm{i}}\,K_\mu(uz\,\mathrm{e}^{-\pi\mathrm{i}})\,I_\mu(ut)\frac{tK_{\mu+1}(ut)}{X_\mu(uz)}\right| < A\,|\mathrm{e}^{2uz}| \leqslant A, \tag{13·15}$$

and if $\frac{1}{2}\pi \leqslant \arg t \leqslant \frac{3}{2}\pi$ we deduce from (9·1), (9·4), (9·5), (9·10), (9·11) and (13·8) that

$$\left|2u\,\mathrm{e}^{-\mu\pi\mathrm{i}}\,K_\mu(uz\,\mathrm{e}^{-\pi\mathrm{i}})\,I_\mu(ut)\frac{tK_{\mu+1}(ut)}{X_\mu(uz)}\right| = \left|2uK_\mu(uz\,\mathrm{e}^{-\pi\mathrm{i}})\,I_\mu(ut\,\mathrm{e}^{-\pi\mathrm{i}})\frac{tK_{\mu+1}(ut)}{X_\mu(uz)}\right| < A\,|\mathrm{e}^{2u(z-t)}| \leqslant A. \tag{13·16}$$

Substitution of (13·14), (13·15), (13·16) in (13·13), and combination of the result with (13·12), shows that

$$|\omega(z,u,t)| < A \tag{13·17}$$

throughout the sector $-\frac{1}{2}\pi \leqslant \arg z \leqslant \frac{3}{2}\pi$. Substituting this inequality in (13·6) and using (12·16), we obtain

$$|j_1(z)| < A\,\mathrm{e}^{-u\tau}\,|z\,X_\mu(uz)\int_z^a|R_1(t)+u^{-1}R_2(t)|\,|\mathrm{d}t| < A\,\mathrm{e}^{-u\tau}\,|z\,X_\mu(uz), \tag{13·18}$$

since the contour can be deformed to consist of parts of the real axis, the boundary of $\mathbf{D}'$, the circle $|t| = \epsilon|z|$ and the lines through z and a parallel to the imaginary axis, and still satisfy the conditions already imposed.

Next, we have from (13·5)

$$j_{n+1}(z) - j_n(z) = \frac{z}{u}X_\mu(uz)\int_z^a \theta(z,u,t)\,f(t)\,\frac{j_n(t)-j_{n-1}(t)}{tX_\mu(ut)}\,\mathrm{d}t \quad (n \geqslant 1), \tag{13·19}$$

where
$$\theta(z,u,t) = u\{K_\mu(uz)\,I_\mu(ut) - I_\mu(uz)\,K_\mu(ut)\}\,tX_\mu(ut)/X_\mu(uz)$$
$$= u\,\mathrm{e}^{-\mu\pi\mathrm{i}}\{K_\mu(uz\,\mathrm{e}^{-\pi\mathrm{i}})\,I_\mu(ut) - I_\mu(uz)\,K_\mu(ut\,\mathrm{e}^{-\pi\mathrm{i}})\}\,tX_\mu(ut)/X_\mu(uz). \tag{13·20}$$

Proceeding in a manner similar to that used for the proof of (13·17), we readily verify that

$$|\theta(z,u,t)| < A, \tag{13·21}$$

† The use of this formula was suggested to the writer by Mr R. C. Thorne.

and thence, from (13·18) and (13·19), by induction that

$$|j_{n+1}(z)-j_n(z)|<A(A/u)^n\,\mathrm{e}^{-u\tau}\,|z|\,X_\mu(uz). \tag{13·22}$$

Thus the series $\sum_{n=0}^{\infty}\{j_{n+1}(z)-j_n(z)\}$ converges uniformly with respect to z to the sum $W_2(z)-L_2(z)$ when $u>A$. When $u>2A$ we deduce that

$$|W_2(z)-L_2(z)|<2A\,\mathrm{e}^{-u\tau}\,|z|\,X_\mu(uz). \tag{13·23}$$

Let us suppose that r' is the zero of $K_0(z)$ in $|\arg z|\leqslant\frac{3}{2}\pi$ of smallest modulus. We set

$$zX_\mu(uz)=X_\mu^{(1)}.zK_\mu(uz)+X_\mu^{(2)}.uz^2(1+|z|)^{-1}K_{\mu+1}(uz), \tag{13·24}$$

where

(i) if $|uz|\leqslant\frac{1}{2}r'$

$$\left.\begin{aligned}X_\mu^{(1)}&=X_\mu(uz)/K_\mu(uz)\\ X_\mu^{(2)}&=0;\end{aligned}\right\}(\mu=0)\qquad \left.\begin{aligned}X_\mu^{(1)}&=uzI_{\mu+1}(uz)\,X_\mu(uz)\\ X_\mu^{(2)}&=(1+|z|)I_\mu(uz)\,X_\mu(uz)\end{aligned}\right\}(\mu\neq 0), \tag{13·25}$$

(cf. (12·24));

(ii) if $|uz|>\frac{1}{2}r'$

$$X_\mu^{(1)}=\frac{uz}{\pi\mathrm{i}}K_{\mu+1}(uz\,\mathrm{e}^{-\pi\mathrm{i}})\,X_\mu(uz),\quad X_\mu^{(2)}=\frac{1+|z|}{\pi\mathrm{i}}K_\mu(uz\,\mathrm{e}^{-\pi\mathrm{i}})\,X_\mu(uz). \tag{13·26}$$

The relations (13·26) may be verified by use of the identity

$$K_\mu(z)\,K_{\mu+1}(z\,\mathrm{e}^{-\pi\mathrm{i}})+K_{\mu+1}(z)\,K_\mu(z\,\mathrm{e}^{-\pi\mathrm{i}})=\pi\mathrm{i}/z, \tag{13·27}$$

which can be obtained from (9·6) with $p=1$, and (12·24).

With the aid of the inequalities of § 9 we may verify that $|X_\mu^{(1)}|$, $|X_\mu^{(2)}|$ are bounded functions for all u,z ($u>1$, $-\frac{1}{2}\pi\leqslant\arg z\leqslant\frac{3}{2}\pi$). Accordingly, $W_2(z)-L_2(z)$ is negligible compared with the error terms in (8·7). This completes the proof of (8·7).

For the derivative, we obtain from (13·5), (7·6) and (11·9)

$$j_1'(z)=\frac{\mu+1}{z}j_1(z)+\mathrm{e}^{-u\tau}uzX_{\mu+1}(uz)\int_z^a\omega_1(z,u,t)\,\{R_1(t)+u^{-1}R_2(t)\}\,\mathrm{d}t, \tag{13·28}$$

$$j_{n+1}'(z)-j_n'(z)=\frac{\mu+1}{z}\{j_{n+1}(z)-j_n(z)\}-zX_{\mu+1}(uz)\int_z^a\theta_1(z,u,t)\,f(t)\,\frac{j_n(t)-j_{n-1}(t)}{tX_\mu(ut)}\,\mathrm{d}t\quad(n\geqslant 1), \tag{13·29}$$

where

$$\begin{aligned}\omega_1(z,u,t)&=2u\{K_{\mu+1}(uz)\,I_\mu(ut)+I_{\mu+1}(uz)\,K_\mu(ut)\}\,tK_{\mu+1}(ut)/X_{\mu+1}(uz)\\&=2u\,\mathrm{e}^{-\mu\pi\mathrm{i}}\{I_{\mu+1}(uz)\,K_\mu(ut\,\mathrm{e}^{-\pi\mathrm{i}})-K_{\mu+1}(uz\,\mathrm{e}^{-\pi\mathrm{i}})\,I_\mu(ut)\}\,tK_{\mu+1}(ut)/X_{\mu+1}(uz),\end{aligned} \tag{13·30}$$

$$\begin{aligned}\theta_1(z,u,t)&=u\{K_{\mu+1}(uz)\,I_\mu(ut)+I_{\mu+1}(uz)\,K_\mu(ut)\}\,tX_\mu(ut)/X_{\mu+1}(uz)\\&=u\,\mathrm{e}^{-\mu\pi\mathrm{i}}\{I_{\mu+1}(uz)\,K_\mu(ut\,\mathrm{e}^{-\pi\mathrm{i}})-K_{\mu+1}(uz\,\mathrm{e}^{-\pi\mathrm{i}})\,I_\mu(ut)\}\,tX_\mu(ut)/X_{\mu+1}(uz).\end{aligned} \tag{13·31}$$

We may verify that $\quad|\omega_1(z,u,t)|<Al_\mu(t),\qquad|\theta_1(z,u,t)|<Al_\mu(t),\quad$ (13·32)

(cf. (13·17) and (13·21)), and thence, with the aid of (13·28), (13·29), (13·18) and (13·22), that

$$|j_1'(z)|<A\,\mathrm{e}^{-u\tau}\{X_\mu(uz)+u\,|z|\,X_{\mu+1}(uz)\}, \tag{13·33}$$

$$|j_{n+1}'(z)-j_n'(z)|<A(A/u)^n\,\mathrm{e}^{-u\tau}\{X_\mu(uz)+u\,|z|\,X_{\mu+1}(uz)\}. \tag{13·34}$$

Thus when $u > 2A$

$$|W_2'(z) - L_2'(z)| = \left|\sum_{n=0}^{\infty}\{j_{n+1}'(z) - j_n'(z)\}\right| < 2A\,\mathrm{e}^{-u\tau}\{X_\mu(uz) + u\,|z|\,X_{\mu+1}(uz)\}. \quad (13\cdot35)$$

From the definition (9·10) of X, we may verify that

$$u|z|\,X_{\mu+1}(uz) < A(1+|uz|)\,X_\mu(uz).$$

Hence we have

$$|W_2'(z) - L_2'(z)| < A\,\mathrm{e}^{-u\tau}(1+|uz|)\,X_\mu(uz) < Au\,\mathrm{e}^{-u\tau}(1+|z|)\,X_\mu(uz), \quad (13\cdot36)$$

provided that $u > 1$. From (13·24), we see that

$$(1+|z|)\,X_\mu(uz) = X_\mu^{(1)}.(1+|z|)\,K_\mu(uz) + X_\mu^{(2)}.uz\,K_{\mu+1}(uz), \quad (13\cdot37)$$

where $X_\mu^{(1)}$, $X_\mu^{(2)}$ are defined by (13·25), (13·26), and are bounded.

Hence $|W_2'(z) - L_2'(z)|$ is negligible compared with the error terms in (8·8). This completes the proof.

14. Other forms of expansion

In applications of theorem D, the form of expansion for $W_1(z)$, $W_1'(z)$, $W_2(z)$ and $W_2'(z)$ given by (8·5) to (8·8) may not always be the most convenient. We mention here two other useful forms.

Instead of using Bessel functions of orders μ and $\mu+1$, we can use orders μ and $\mu-1$. Using the relation

$$\mathscr{Z}_{\mu+1}(uz) = \mathscr{Z}_{\mu-1}(uz) - 2\mu(uz)^{-1}\mathscr{Z}_\mu(uz) \quad (14\cdot1)$$

(Watson 1944, p. 79), we obtain, for example,

$$W_1(z) = zI_\mu(uz)\left\{\sum_{s=0}^{N-1}\frac{A_s^*(z)}{u^{2s}} + O\left(\frac{1}{u^{2N}}\right)\right\} + \frac{z}{u}I_{\mu-1}(uz)\left\{\sum_{s=0}^{N-1}\frac{B_s(z)}{u^{2s}} + \frac{z}{1+|z|}O\left(\frac{1}{u^{2N}}\right)\right\}, \quad (14\cdot2)$$

$$W_1'(z) = I_\mu(uz)\left\{\sum_{s=0}^{N-1}\frac{C_s^*(z)}{u^{2s}} + (1+|z|)\,O\left(\frac{1}{u^{2N}}\right)\right\} + uzI_{\mu-1}(uz)\left\{\sum_{s=0}^{N-1}\frac{D_s(z)}{u^{2s}} + O\left(\frac{1}{u^{2N}}\right)\right\}, \quad (14\cdot3)$$

where $$A_s^*(z) = A_s(z) - 2\mu z^{-1}B_{s-1}(z), \qquad C_s^*(z) = C_s(z) - 2\mu D_s(z). \quad (14\cdot4)$$

If we allow the argument of the Bessel function to be an asymptotic series we need use only one such function. For example, we can write

$$W_1(z) \sim z\left(1 + \sum_{s=1}^{\infty}\frac{a_s(z)}{u^{2s}}\right)I_\mu\left(uz + uz\sum_{s=1}^{\infty}\frac{b_s(z)}{u^{2s}}\right). \quad (14\cdot5)$$

The coefficients $a_s(z)$ and $b_s(z)$ may be related to the $A_s(z)$ and $B_s(z)$ of § 8 by expanding the Bessel function in Taylor's series, and expressing the derivatives $I_\mu^{(r)}(uz)$ in terms of $I_\mu(uz)$ and $I_{\mu+1}(uz)$ by means of recurrence relations. It is found that they are all analytic at $z = 0$, and the first few are given by

$$\left.\begin{aligned} b_1 &= B_0 z^{-1}, \qquad a_1 = A_1 - \mu B_0 z^{-1} - \tfrac{1}{2}(B_0)^2, \\ b_2 &= \{B_1 - A_1 B_0 + \tfrac{1}{3}(B_0)^3\}z^{-1} + (\mu + \tfrac{1}{2})\,(B_0)^2 z^{-2}. \end{aligned}\right\} \quad (14\cdot6)$$

This form of expansion may sometimes be useful in calculating series for the zeros of $W_1(z)$.

In a similar way we have

$$W_1'(z) \sim uz\left(1 + \sum_{s=1}^{\infty}\frac{c_s(z)}{u^{2s}}\right)I_{\mu+1}\left(uz + uz\sum_{s=1}^{\infty}\frac{d_s(z)}{u^{2s}}\right), \quad (14\cdot7)$$

where
$$\left.\begin{aligned} d_1 &= C_0 z^{-2}, \quad c_1 = D_1 + \{(\mu+1)\,C_0 - \tfrac{1}{2}(C_0)^2\}\,z^{-2}, \\ d_2 &= \{C_1 - C_0 D_1\}\,z^{-2} + \{\tfrac{1}{3}(C_0)^3 - (\mu+\tfrac{1}{2})\,(C_0)^2\}\,z^{-4}. \end{aligned}\right\} \tag{14·8}$$

This expansion is not so convenient as (14·5) because the coefficients $c_s(z)$ and $d_s(z)$ have poles at $z = 0$. Nor can this difficulty be overcome by using I'_μ in place of $I_{\mu+1}$.

Appendix. Note on theorems A and B

In a previous paper (Olver 1954) we investigated the behaviour of solutions of the differential equations
$$\frac{d^2w}{dz^2} = \{u^2 + f(z)\}\,w \quad \text{(case A)},$$
$$\frac{d^2w}{dz^2} = \{uz + f(z)\}\,w \quad \text{(case B)},$$
for large positive values of the parameter u. We showed that under certain conditions these equations possess solutions $W_j(z)$ whose asymptotic expansions are given by† (*5·6*), (*5·7*) and (*5·12*), and are uniformly valid with respect to z.

If the domain of integration **D** is taken to be bounded, then no restriction need be made on the behaviour of $f(z)$ at $z = \infty$. However, if **D** is unbounded then in order to prove that the asymptotic expansions are uniformly valid at $z = \infty$, we assumed that
$$f(z) = O(|z|^{-2}),$$
as $|z| \to \infty$ in **D**, uniformly with respect to $\arg z$ (*ibid.*, equation (*5·1*)). The purpose of this Appendix is to point out that this restriction is unnecessarily severe, and that theorems A and B remain valid if equation (*5·1*) is replaced by
$$f(z) = O(|z|^{-1-\alpha}) \quad \text{(for theorem A)}, \tag{A 1}$$
and
$$f(z) = O(|z|^{-\frac{1}{2}-\alpha}) \quad \text{(for theorem B)}, \tag{A 2}$$
where $\alpha > 0$ is a constant.

Theorems A and B may be established under these conditions by bringing their proofs more into line with that of theorem D. We indicate below the principal modifications to the proof of theorem B; the treatment of theorem A is similar and simpler. We introduce a constant σ similar to that defined in § 10, given by
$$\sigma = \min(\alpha, \tfrac{3}{2}).$$

In the statement of *lemma 1*, equations (*9·5*) and (*9·6*) are replaced by
$$A_s(z) = \alpha_s + O(|z|^{-\sigma}), \quad B_s(z) = \beta_s z^{-\frac{1}{2}} + O(|z|^{-\frac{1}{2}-\sigma}), \quad A'_s(z) = O(|z|^{-1-\sigma}). \tag{A 3}$$
To prove the new lemma, we assume that for some s
$$|A'_s(z)| < \lambda_s |z|^{-1-\sigma} \quad \text{in} \quad \mathbf{E}_\eta. \tag{A 4}$$
We may then show that
$$A_s(z) = \alpha_s + a_s(z), \qquad a_s(z) = O(|z|^{-\sigma}) \quad \text{in} \quad \mathbf{E}_\eta \tag{A 5}$$
(cf. (10·9) and (10·13)), and that
$$A''_s(z) = O(|z|^{-1-\sigma}) \quad \text{in} \quad \mathbf{E}_\epsilon \tag{A 6}$$

† Equation numbers, section numbers and lemmas in *italics* refer to the previous paper (Olver 1954).

(cf. (10·22)). Hence

$$z^{-\frac{1}{2}}\{f(z)\,A_s(z)-A_s''(z)\} = O(|z|^{-1-\sigma}) \quad \text{in} \quad \mathbf{E}_\epsilon,$$

and so the integral (*9·1*) converges at $z=\infty$. Accordingly, $B_s(z)$ may be written in the form (*9·9*), where

$$b_s(z) = O(|z|^{-\frac{1}{2}-\sigma}) \quad \text{in} \quad \mathbf{E}_\epsilon. \tag{A 7}$$

Substituting (A 2), (A 5), (A 6), (A 7) in (*9·12*), we may show that $B_s'(z) = O(|z|^{-\frac{3}{2}})$ in $\mathbf{E}_\epsilon$. Similarly, from the equation

$$2zB_s'' = \frac{\mathrm{d}}{\mathrm{d}z}(fA_s)-A_s'''-3B_s',$$

obtained by differentiation of (*9·12*), we deduce that $B_s''(z) = O(|z|^{-\frac{3}{2}-\sigma_1})$ in $\mathbf{E}_\epsilon$ where $\sigma_1 = \min(\sigma, 1)$. Substitution of this result and (A 2), (A 7) in the equation

$$A_{s+1}'(z) = -\tfrac{1}{2}B_s''(z)+\tfrac{1}{2}f(z)\,B_s(z),$$

obtained by differentiation of (*9·2*), shows that in $\mathbf{E}_\epsilon$

$$A_{s+1}'(z) = O(|z|^{-\frac{3}{2}-\sigma_1})+O(|z|^{-1-\sigma}) = O(|z|^{-1-\sigma}),$$

since $\sigma_1 \geqslant \sigma-\frac{1}{2}$. *Lemma 1* is now established by induction.

The statement of *lemma 2* is unaffected, but the proof breaks down at (*9·15*) and needs revision. In order to estimate $|z^{\frac{1}{2}}B_s(z)|$ and $|A_{s+1}(z)|$, we subdivide the paths of integration in (*9·1*) and (*9·4*) at the points where they cross the circles $|z|=\rho$ and $|z|=R+\epsilon$. We may then show, by use of partial integration, that

$$|z^{\frac{1}{2}}B_s(z)|<F\{M_s(\epsilon)+M_s^{(1)}(\epsilon)+M_s^{(2)}(\epsilon)\} \quad \text{in} \quad \mathbf{D}_\epsilon, \tag{A 8}$$

and then, by substitution of this result in (*9·4*), that

$$M_{s+1}(\epsilon)<P_s^{(1)}(\epsilon)+F\{M_s(\epsilon)+M_s^{(1)}(\epsilon)+M_s^{(2)}(\epsilon)\}, \tag{A 9}$$

where F is the generic constant defined in § 10. Combination of (A 9) with (*9·14*) and (*9·16*) (which remain valid), shows that

$$M_{s+1}(\epsilon)<F\{M_s(\epsilon)+M_s^{(1)}(\epsilon)+M_s^{(2)}(\epsilon)\}. \tag{A 10}$$

This result corresponds to (10·34), and the proof may be completed as for lemma 3 of the present paper, except that in place of (10·39) we use

$$|z^{\frac{1}{2}}B_{s-1}(z)|<Fs^2\delta^{-2}M_{s-1}(\delta_{1,s})<(Fs^2\delta^{-2})^s \quad \text{in} \quad \mathbf{D}_\delta, \tag{A 11}$$

obtained from (A 8), (10·35) and (10·38).

Next, in the statement of *lemma 3* we have to replace the $O(|z|^{-\frac{3}{2}})$ by $O(|z|^{-1-\sigma})$. To prove the result, we examine how the constants occurring in the O's in the proof of *lemma 1* depend on s. We may show without difficulty that:

$$\left.\begin{array}{ll} \text{If} & |z^{1+\sigma}A_s'(z)|<\lambda_s \quad \text{in} \quad \mathbf{E}_\eta, \\ \text{then} & |z^{1+\sigma}A_{s+1}'(z)|<F(\epsilon-\eta)^{-2}\{|\alpha_s|+|\beta_s|+\lambda_s\} \quad \text{in} \quad \mathbf{E}_\epsilon \end{array}\right\} \tag{A 12}$$

(cf. (10·48)). The values of $|\alpha_s|$, $|\beta_s|$ are estimated by letting $|z|\to\infty$ in $\mathbf{E}_\delta$. We have at once, from (10·38) and (A 11),

$$|\alpha_s|<(Fs^2\delta^{-2})^s, \qquad |\beta_s|<\{F(s+1)^2\delta^{-2}\}^{s+1}. \tag{A 13}$$

Substituting (A 13) in (A 12) and changing s into $n-1$, we have:

$$\left.\begin{array}{ll}\text{If} & |z^{1+\sigma}A'_{n-1}(z)|<\lambda_{n-1} \quad \text{in} \quad \mathbf{E}_\eta, \\ \text{then} & |z^{1+\sigma}A'_n(z)|<F(\epsilon-\eta)^{-2}\{(Fn^2\delta^{-2})^n+\lambda_{n-1}\} \quad \text{in} \quad \mathbf{E}_\epsilon.\end{array}\right\}$$

This result is exactly the same as (10·50), and the proof may be completed as for lemma 4 of the present paper.

The only other modifications to the proof of theorem B occur in §*10*. In (*10·18*) the $O(|z|^{-\frac{3}{2}})$ are replaced by $O(|z|^{-1-\sigma})$, but (*10·19*) remains valid. The integral occurring in the last inequality on p. *325* no longer converges at infinity. However, we can replace (*10·22*) by the sharper inequality

$$|\chi(z,v,t)|<\frac{A}{(1+|vz|^{\frac{1}{4}})(1+|vt|^{\frac{1}{4}})}\leqslant\frac{A}{(1+|vz|^{\frac{1}{4}})(1+|t|^{\frac{1}{4}})},$$

when $u\geqslant 1$, and substitution of this in (*10·21*) enables us to show that

$$|h_{n+1}(z)-h_n(z)|<\frac{A}{v}k_n(v)\,\mathrm{e}^{-\tau\sqrt{u}}\frac{|\exp\{-\frac{2}{3}(vz)^{\frac{3}{2}}\}|}{1+|vz|^{\frac{1}{4}}}\int_z^a\frac{|f(t)\,\mathrm{d}t|}{(1+|t|^{\frac{1}{4}})^2},$$

which does converge, and we establish (*10·23*) and (*10·24*) as before. In a similar way we may prove that (*10·28*) remains valid.

The work described above has been carried out as part of the research program of the National Physical Laboratory, and this paper is published by permission of the Director of the Laboratory.

References

Erdélyi, A. 1955 *Asymptotic expansions.* Tech. Rep. no. 3, California Institute of Technology.
Langer, R. E. 1935 *Trans. Amer. Math. Soc.* **37**, 397–416.
Olver, F. W. J. 1954 *Phil. Trans.* A, **247**, 307–327.
Watson, G. N. 1944 *Theory of Bessel functions.* Cambridge University Press.

[479]

UNIFORM ASYMPTOTIC EXPANSIONS OF SOLUTIONS OF LINEAR SECOND-ORDER DIFFERENTIAL EQUATIONS FOR LARGE VALUES OF A PARAMETER

By F. W. J. OLVER

The National Physical Laboratory, Teddington, Middx.

(*Communicated by G. B. B. M. Sutherland, F.R.S.—Received* 4 *September* 1957)

CONTENTS

An investigation is made of the differential equations

$$\frac{d^2w}{dz^2} = \{u^2+f(u,z)\}w, \qquad \frac{d^2w}{dz^2} = \{u^2z+f(u,z)\}w,$$

$$\frac{d^2w}{dz^2} = \frac{1}{z}\frac{dw}{dz}+\left\{u^2+\frac{\mu^2-1}{z^2}+f(u,\mu,z)\right\}w,$$

in which u is a large complex parameter, μ is a real or complex parameter independent of u, and z is a complex variable whose domain of variation may depend on arg u and μ, and need not be bounded.

General conditions are obtained under which solutions exist having the formal series

$$w = P(z)\sum_{s=0}^{\infty}\frac{A_s}{u^{2s}}+\frac{P'(z)}{u^2}\sum_{s=0}^{\infty}\frac{B_s}{u^{2s}}$$

as their asymptotic expansions for large $|u|$, uniformly valid with respect to z, arg u and μ. Here $P(z)$ is respectively an exponential function, Airy function or Bessel function of order μ, and the coefficients A_s and B_s are given by recurrence relations.

1. Introduction and summary

In recent papers (Olver 1954*a*, 1956) I investigated the asymptotic expansion of solutions of differential equations of the form

$$\frac{d^2w}{dz^2} = \{u^2p(z)+q(z)\}w \tag{1·1}$$

for large values of the real positive parameter† u, which are uniformly valid with respect to all values of the complex variable z lying in a domain **D**, bounded or otherwise. The asymptotic character of the solutions depends on the number and nature of the *transition points* in **D**, that is, points at which $p(z)$ has a zero, or $p(z)$ or $q(z)$ has a singularity. Zeros of $p(z)$ are also called *turning points*.

Four principal cases were distinguished, denoted by A, B, C and D. In case A, **D** is free from transition points; in case B, **D** contains a simple turning point; in case C, **D** contains a double pole of $p(z)$; in case D, **D** contains a simple pole of $p(z)$. In the last two cases $q(z)$ is permitted to have a single or double pole at the pole of $p(z)$. By simultaneous change of dependent and independent variables equation (1·1) can be transformed into one of the following forms

$$\frac{d^2w}{dz^2} = \{u^2+f(z)\}w \quad \text{(cases A and C)}, \tag{1·2}$$

$$\frac{d^2w}{dz^2} = \{u^2z+f(z)\}w \quad \text{(case B)}, \tag{1·3}$$

$$\frac{d^2w}{dz^2} = -\frac{1}{z}\frac{dw}{dz}+\left\{u^2+\frac{\mu^2-1}{z^2}+f(z)\right\}w \quad \text{(case D)}, \tag{1·4}$$

in which μ is a constant and $f(z)$ is a regular (holomorphic) function in the transformed domain. In (1·4) $f(z)$ is an even function of z.

Each of the equations (1·2), (1·3) and (1·4) can be satisfied formally by series of the form

$$P_j(z)\left\{1+\sum_{s=1}^{\infty}\frac{A_s(z)}{u^{2s}}\right\}+\frac{P_j'(z)}{u^2}\sum_{s=0}^{\infty}\frac{B_s(z)}{u^{2s}}, \tag{1·5}$$

in which $P_j(z)$ is a solution of the corresponding equation with $f(z) = 0$, the suffix j being associated with the particular choice of solution. The $P_j(z)$ are called the *basic functions* and are respectively exponential functions, Airy functions and Bessel functions of order μ for the three equations.

Existence theorems designated A, B and D were proved, which establish that solutions of the equations exist for which the formal series (1·5) is an asymptotic expansion, in the sense of Poincaré, for large positive u, provided that z lies in a certain subdomain $\mathbf{D}_j$ of the transformed domain corresponding to **D**. $\mathbf{D}_j$ is independent of u and can extend to infinity in regions where the condition

$$\left.\begin{aligned} f(z) &= O(|z|^{-1-\sigma}) \quad \text{(equations (1·2) and (1·4))},\\ f(z) &= O(|z|^{-\frac{1}{2}-\sigma}) \quad \text{(equation (1·3))}, \end{aligned}\right\} \tag{1·6}$$

is satisfied; σ being a positive constant.

Any other equation of the form (1·1) which has an isolated transition point and which is amenable to the same general treatment can be transformed into either (1·2), (1·3) or (1·4). In addition, Thorne (1957*a*) has shown that the case in which **D** contains both a simple turning point and a double pole of $p(z)$ can be transformed into (1·3); the pole becomes a point at infinity at which the condition (1·6) is satisfied. He has also indicated that the case in which **D** contains both a simple and a double pole of $p(z)$ can likewise be transformed into the form (1·4).

† For convenience the u of the previous papers is replaced here by u^2.

A further case, requiring a new existence theorem, has been investigated by Thorne (1957*b*), using similar methods. The standard form of equation is

$$\frac{d^2w}{dz^2} = \left\{u^2\left(1+\frac{\alpha^2}{z^2}\right) - \frac{1}{4z^2} + f(z)\right\} w, \tag{1·7}$$

where α is a constant, $f(z)$ is even and $f(z) = O(|z|^{-1-\sigma})$ at infinity. The basic functions here are Bessel functions of order αu. This equation corresponds to (1·1) when $p(z)$ has a double pole and a simple zero in **D**. The branch cut for the solutions which emanates from the pole is made to pass through the zero, and the regions of validity of the corresponding expansions include the pole and the *two* turning points, one on either side of the cut.

Although in the papers cited it is supposed that u is a real positive parameter, it is observed that if u is complex and arg u is fixed then a trivial transformation will convert any of (1·2), (1·3), (1·4) or (1·7) into the same form of equation with u replaced by $|u|$. The existence theorems accordingly apply to the situation when u is large and complex and arg u is fixed.

The principal object of the present paper is to extend theorems A, B and D by establishing conditions under which the asymptotic expansions are uniformly valid over any prescribed range of arg u. The need for this extension in cases A and B has arisen in developing an asymptotic theory for the case in which equation (1·1) contains *two* simple turning points in **D**. The basic functions for this case are the Weber parabolic cylinder functions, being the solutions of the differential equation

$$\frac{d^2w}{dz^2} = (\tfrac{1}{4}z^2 + a)\, w.$$

It is desirable to know asymptotic representations of these functions for large values of the parameter a which are uniformly valid with respect to z and arg a. Such representations can be found by application of the new forms of theorems A and B given in the present paper.

The desired extension of the existence theorems can be achieved by introducing an extra parameter θ in (1·2), (1·3) and (1·4), replacing $f(z)$ by $f(\theta, z)$. The equations become respectively

$$\frac{d^2w}{dz^2} = \{u^2 + f(\theta, z)\}\, w, \tag{1·8}$$

$$\frac{d^2w}{dz^2} = \{u^2 z + f(\theta, z)\}\, w, \tag{1·9}$$

$$\frac{d^2w}{dz^2} = \frac{1}{z}\frac{dw}{dz} + \left\{u^2 + \frac{\mu^2-1}{z^2} + f(\theta, z)\right\} w, \tag{1·10}$$

and the coefficients $A_s(z)$ and $B_s(z)$ in the series (1·5) are replaced by $A_s(\theta, z)$ and $B_s(\theta, z)$, respectively. We seek conditions under which the asymptotic property of this series *for large positive u* is uniform with respect to θ as well as z.

The connexion of this approach with the theory of (1·2), (1·3) and (1·4) for large complex u is as follows. Consider, for example, equation (1·2). If u is complex and $\theta = \arg u$, the substitutions

$$u = |u|\, e^{i\theta}, \qquad z = e^{-i\theta} z_1, \tag{1·11}$$

transform (1·2) into

$$\frac{d^2w}{dz_1^2} = \{|u|^2 + e^{-2i\theta} f(e^{-i\theta} z_1)\}\, w. \tag{1·12}$$

This equation is clearly of the form (1·8) with u positive and $f(\theta, z_1) = e^{-2i\theta}f(e^{-i\theta}z_1)$. In fact, it would be adequate for the immediate purpose if we were to replace $f(\theta, z)$ in (1·8) by $e^{-2i\theta}f(e^{-i\theta}z)$. Greater generality is achieved however, with no extra complication, by taking f to be a general function of the two variables θ and z. And although the parameter θ in (1·8), (1·9) and (1·10) corresponds to arg u in (1·2), (1·3) and (1·4) with complex u, when we consider the asymptotic theory of (1·8), (1·9) and (1·10) we merely regard θ as another parameter, real or complex, either independent of or dependent on the real positive parameter u, or even as a set of such parameters.

In establishing conditions under which the formal series (1·5), with $A_s(z)$ and $B_s(z)$ replaced by $A_s(\theta, z)$ and $B_s(\theta, z)$, represents uniform asymptotic expansions of solutions of equations (1·8), (1·9) and (1·10), it is essential that we permit the boundaries of the z-domain to depend on θ. This is evident from the transformation (1·11) which shows that the z_1-domain is obtained from the z-domain by rotation through an angle θ. We generalize this relationship and allow the z-domain $\mathbf{D}(\theta)$ associated with (1·8), (1·9) and (1·10) to be a general function of θ. This generalization is of great value in applications since it allows the boundary of $\mathbf{D}(\theta)$ to be arranged in the most convenient manner for each θ. In particular, it is usually advantageous to arrange that cuts from branch points of $f(\theta, z)$ are parallel to the imaginary z-axis in the case of (1·8) and (1·10), and lie along level curves of the function $\exp(\frac{2}{3}z^{\frac{3}{2}})$ in the case of (1·9); the regions of validity of the asymptotic expansions are then likely to be maximal.

The proofs given in this paper are simpler in principle than the proofs previously given of the original forms of the existence theorems; two of the three lemmas concerning the magnitude of the coefficients $A_s(\theta, z)$ and $B_s(\theta, z)$ are avoided. The simpler method of proof permits an easy extension to be made to meet the situation in which $f(\theta, z)$ is replaced by a function $f(u, \theta, z)$ having a uniform asymptotic expansion of the form

$$f(u, \theta, z) \sim f_0(\theta, z) + \frac{f_1(\theta, z)}{u} + \frac{f_2(\theta, z)}{u^2} + \dots, \tag{1·13}$$

for large u.

Another extension is the easing of restrictions placed upon the boundaries of the regions of validity. We no longer suppose that these consist of a finite number of straight lines. This is of particular value in case B, for here we often wish to arrange that branch cuts bounding $\mathbf{D}(\theta)$ are level curves of the function $\exp(\frac{2}{3}z^{\frac{3}{2}})$. In this case, moreover, the previous restriction that the distances of all points of the regions $\mathbf{D}_j(\theta)$ from the boundary of $\mathbf{D}(\theta)$ must have a positive lower bound, is replaced by the weaker condition: if z_0 is a boundary point of $\mathbf{D}_j(\theta)$ then $|z_0|^{\frac{1}{2}}$ times the shortest distance between z_0 and the boundary of $\mathbf{D}(\theta)$ must have a positive lower bound. This too simplifies applications as we shall see in § 10.

As our final extension, the existence theorem in case D now establishes uniformity with respect to the variable μ occurring in (1·10), when μ lies in a bounded region in the half-plane $\mathrm{Re}\,\mu \geqslant 0$. The function f and the z-domains are permitted to depend on μ as well as θ.

Uniform asymptotic solutions of second-order differential equations with a large complex parameter have been derived by many authors, though usually only for real values of the independent variable. Cherry (1950), however, develops uniform asymptotic series for case B with u complex and z lying in a given bounded star-domain. The z-regions of validity which result from these conditions are unnecessarily restricted and the greater flexibility in the choice of boundaries permitted in the present paper yields larger regions in applications, even in the case of bounded star-domains. Another condition of Cherry's paper is that the function $f(u, \theta, z)$, defined above, should have a convergent expansion in descending powers of u^2; here we merely suppose that $f(u, \theta, z)$ has an asymptotic expansion in descending powers of u. Other comparisons between the results of Cherry and those of the present writer have been discussed in Olver 1954*a*, § 7.

Kazarinoff & McKelvey (1956) have investigated an extended form of case C with both u and z complex. The z-domain is severely restricted however, being the interior of a fixed simple closed contour.

The present paper is divided into three parts, corresponding to cases A, B and D. The arrangement of each part follows the same pattern. In the opening section (§§ 2, 9 and 14) the problem is stated together with its solution, the corresponding existence theorem. The postulated conditions are of a rather general form, and to facilitate their application the next section (§§ 3, 10 and 15) is mainly devoted to an account of common circumstances in which the general conditions are fulfilled. The proof of each theorem then begins. First (§§ 5, 11 and 17 (i)) bounds are established for the coefficients of the series (1·5) which are uniform with respect to θ and, in case D, μ. Next (§§ 6, 12 and 17 (ii)), a differential equation satisfied by the truncated series is obtained. Combination of this equation with the original equation yields a differential equation for the error term of the asymptotic series, and in the final sections (§§ 7, 8, 13, 18 and 19) an integral equation corresponding to the last-named differential equation is solved by the iterative method.

In part 3 there is an additional section (§ 16) in which bounds are established for the modified Bessel functions $I_\mu(z)$ and $K_\mu(z)$ which are uniform with respect to bounded μ. These may be of use in other contexts.

PART 1. CASE A

2. Statement of conditions and theorem A

We consider the differential equation

$$\frac{d^2w}{dz^2} = \{u^2 + f(u, \theta, z)\} w, \tag{2·1}$$

in which u is a large positive parameter; θ is a set of real or complex parameters ranging over a sct of values $\mathbf{\Theta}$; z is a complex variable ranging over a simply-connected complex domain $\mathbf{D}(\theta)$, bounded or otherwise.

We suppose that for given u and θ, $f(u, \theta, z)$ is a regular (holomorphic) function of z in $\mathbf{D}(\theta)$, and that

$$\left| f(u, \theta, z) - \sum_{s=0}^{m-1} \frac{f_s(\theta, z)}{u^s} \right| < \frac{1}{1+|z|^{1+\sigma}} \frac{k_m}{u^m} \quad \text{(for every } m = 0, 1, 2, \ldots), \tag{2·2}$$

when $z \in \mathbf{D}(\theta)$; $\theta \in \mathbf{\Theta}$; $u \geq u_0 (>0)$, independent of z and θ. Each of the coefficients $f_s(\theta, z)$ is independent of u† and a regular function of z in $\mathbf{D}(\theta)$, and σ is a positive constant independent of u, θ, z and m. The number k_m is independent of u, θ and z, but may depend on m; we shall

† Except in as much as θ may depend on u.

Purchase suggestions to Hartley Library

Produced Monday, 31 January, 2000 at 1:10 AM

9) Request ID: HL-66219 Status: NEWPURCH
Request Type: PURCHASE Operator Access: DLR
Date Placed: 28/1/2000 Date Modified: NEVER
Reply Date: NEVER Date Viewed: NEVER
Origin Library: HARTLEY Service Library: HARTLEY

User ID: 11362704
User Name: Forster, Jonathan, J

REQUEST INFORMATION ---
ISBN/ISSN:9810241062
Author/Editor:R Wong (Ed)
Title:Selected papers of FWJ Olver
Volume/Part:Parts 1 and 2
Publisher:World Scientific
Date:1999
Price:~~$159~~ £83
Number of copies:1
Library:Hartley
Fund code:472211

BBS

BBS ✓

use the symbol k_m generically in this sense, and also the symbol k generically to denote a positive number independent of u, θ and z.

The condition (2·2) implies that

$$f(u,\theta,z) \sim \sum_{s=0}^{\infty} \frac{f_s(\theta,z)}{u^s} \tag{2·3}$$

as $u \to +\infty$, uniformly valid with respect to θ and z. We can construct a formal solution of (2·1) in the form

$$w = \mathrm{e}^{uz} \sum_{s=0}^{\infty} \frac{A_s(\theta,z)}{u^s}. \tag{2·4}$$

Term-by-term differentiation yields

$$\frac{\mathrm{d}w}{\mathrm{d}z} = u\,\mathrm{e}^{uz} \sum_{s=0}^{\infty} \frac{B_s(\theta,z)}{u^s}, \tag{2·5}$$

$$\frac{\mathrm{d}^2w}{\mathrm{d}z^2} = u^2\,\mathrm{e}^{uz} \sum_{s=0}^{\infty} \frac{C_s(\theta,z)}{u^s}, \tag{2·6}$$

where $$B_s = A_s + A'_{s-1}, \qquad C_s = B_s + B'_{s-1} = A_s + 2A'_{s-1} + A''_{s-2}, \tag{2·7}$$

primes denoting differentiations with respect to z. Substituting (2·3), (2·4) and (2·6) in (2·1) and equating coefficients, we see that the last-named equation is formally satisfied if

$$C_{s+2} = A_{s+2} + f_0 A_s + f_1 A_{s-1} + \ldots + f_s A_0, \tag{2·8}$$

that is, if $$2A'_{s+1} = -A''_s + f_0 A_s + f_1 A_{s-1} + \ldots + f_s A_0. \tag{2·9}$$

On integration, this yields A_0 = constant, and

$$A_{s+1} = -\tfrac{1}{2}A'_s + \tfrac{1}{2}\int (f_0 A_s + f_1 A_{s-1} + \ldots + f_s A_0)\,\mathrm{d}z \quad (s \geqslant 0). \tag{2·10}$$

This determines $A_0, A_1, A_2, \ldots$, apart from arbitrary constants of integration, which may depend on θ. Each A_s is a regular function of z in $\mathbf{D}(\theta)$, and single-valued since $\mathbf{D}(\theta)$ is simply-connected.

We can easily verify that a second formal solution of (2·1) is given by

$$w = \mathrm{e}^{-uz} \sum_{s=0}^{\infty} (-)^s \frac{A_s^*(\theta,z)}{u^s}, \tag{2·11}$$

$$\frac{\mathrm{d}w}{\mathrm{d}z} = -u\,\mathrm{e}^{-uz} \sum_{s=0}^{\infty} (-)^s \frac{B_s^*(\theta,z)}{u^s}, \tag{2·12}$$

where the coefficients are determined by A_0^* = constant,

$$A_{s+1}^* = -\tfrac{1}{2}A_s^{*\prime} + \tfrac{1}{2}\int (f_0 A_s^* - f_1 A_{s-1}^* + \ldots + (-)^s f_s A_0^*)\,\mathrm{d}z \quad (s \geqslant 0), \tag{2·13}$$

and $$B_s^* = A_s^* + A_{s-1}^{*\prime}. \tag{2·14}$$

If the functions $f_s(\theta,z)$ of odd suffix s all vanish, then A_s^* and B_s^* are the same as A_s and B_s, respectively; in particular, this happens when $f(u,\theta,z)$ is independent of u.

We define $\mathbf{G}(\theta)$ to be an arbitrary closed subdomain of $\mathbf{D}(\theta)$, subject to the following conditions.

(i) The distance between each point of $\mathbf{G}(\theta)$ and each boundary point of $\mathbf{D}(\theta)$ has a positive lower bound which is independent of θ.

(ii) For each θ, a point $c(\theta)$ and a path joining $c(\theta)$ and z lying wholly in $\mathbf{G}(\theta)$ can be found such that

$$\int_{c(\theta)}^{z} \frac{|\mathrm{d}t|}{1+|t|^{1+\sigma}} < k \quad (z \in \mathbf{G}(\theta)). \tag{2·15}$$

The only restriction we impose on the arbitrary constants associated with the determination of the coefficients $A_s(\theta, z)$ and $A_s^*(\theta, z)$ is that $A_s\{\theta, c(\theta)\}$ and $A_s^*\{\theta, c(\theta)\}$ are bounded functions of θ; thus

$$|A_s\{\theta, c(\theta)\}| < k_s, \qquad |A_s^*\{\theta, c(\theta)\}| < k_s. \tag{2·16}$$

In these circumstances, the asymptotic nature of the formal series may be expressed in the following theorem:

THEOREM A. *The differential equation* (2·1) *possesses solutions* $W_1(u, \theta, z)$ *and* $W_2(u, \theta, z)$ *with the properties*

$$\left.\begin{aligned} W_1(u,\theta,z) &= \mathrm{e}^{uz}\left[\sum_{s=0}^{m} \frac{A_s(\theta,z)}{u^s} + O\left(\frac{1}{u^{m+1}}\right)\right] \\ \frac{\mathrm{d}}{\mathrm{d}z} W_1(u,\theta,z) &= u\,\mathrm{e}^{uz}\left[\sum_{s=0}^{m} \frac{B_s(\theta,z)}{u^s} + O\left(\frac{1}{u^{m+1}}\right)\right] \end{aligned}\right\} \quad (z \in \mathbf{H}_1(\theta), \theta \in \mathbf{\Theta}), \tag{2·17}$$

$$\left.\begin{aligned} W_2(u,\theta,z) &= \mathrm{e}^{-uz}\left[\sum_{s=0}^{m} (-)^s \frac{A_s^*(\theta,z)}{u^s} + O\left(\frac{1}{u^{m+1}}\right)\right] \\ \frac{\mathrm{d}}{\mathrm{d}z} W_2(u,\theta,z) &= -u\,\mathrm{e}^{-uz}\left[\sum_{s=0}^{m} (-)^s \frac{B_s^*(\theta,z)}{u^s} + O\left(\frac{1}{u^{m+1}}\right)\right] \end{aligned}\right\} \quad (z \in \mathbf{H}_2(\theta), \theta \in \mathbf{\Theta}), \tag{2·18}$$

as $u \to +\infty$, *where each of the O's is uniform with respect to z and* θ. *Here m is an arbitrary positive integer or zero, and* $W_1(u, \theta, z)$, $W_2(u, \theta, z)$ *are independent of m.*

The regions of validity $\mathbf{H}_j(\theta)$ $(j = 1, 2)$ are defined as follows. We suppose $a_j(\theta)$ $(j = 1, 2)$ to be any prescribed point of $\mathbf{G}(\theta)$, or the point at infinity on a straight line $\mathscr{L}$ lying in $\mathbf{G}(\theta)$. If $a_j(\theta)$ is at infinity we suppose that $|\arg\{-a_j(\theta)\}| < \frac{1}{2}\pi$ $(j = 1)$ and $|\arg a_j(\theta)| < \frac{1}{2}\pi$ $(j = 2)$. Then $\mathbf{H}_j(\theta)$ comprises those points z of $\mathbf{G}(\theta)$ which can be joined to $a_j(\theta)$ by a path $\mathscr{P}$ having the following properties, t being a typical point of $\mathscr{P}$.

(i) $\mathscr{P}$ lies in $\mathbf{G}(\theta)$.

(ii) $\mathscr{P}$ comprises a finite number of Jordan arcs, each with parametric equations of the form $t = t(\tau)$, where τ is the real parameter of the arc; $t''(\tau)$ is continuous and $t'(\tau)$ does not vanish. If $a_j(\theta)$ is at infinity, $\mathscr{P}$ coincides with $\mathscr{L}$ for all sufficiently large $|t|$.

(iii) $$\int_{\mathscr{P}} \frac{|\mathrm{d}t|}{1+|t|^{1+\sigma}} < k. \tag{2·19}$$

(iv) As t traverses $\mathscr{P}$ from $a_j(\theta)$ to z, $\mathrm{Re}\, t$ is monotonic increasing if $j = 1$, and monotonic decreasing if $j = 2$.

3. REMARKS ON THEOREM A

First, the common circumstances in which some of the conditions given in § 2 are fulfilled are noted.

(i) Condition (2·15) is relatively weak. It is satisfied, for example, if the path joining $c(\theta)$ and z is composed entirely of straight lines, the total number of which is a bounded function of z and θ.

The parametric equation of a typical straight line may be expressed in the form

$$t = (\tau + i\lambda_1)\, e^{i\lambda_2} \quad (\lambda_3 \leqslant \tau \leqslant \lambda_4), \tag{3·1}$$

where λ_1, λ_2, λ_3 and λ_4 are real constants which may depend on θ. Clearly $|t| \geqslant |\tau|$ and

$$\int \frac{|dt|}{1+|t|^{1+\sigma}} \leqslant \int_{\lambda_3}^{\lambda_4} \frac{d\tau}{1+|\tau|^{1+\sigma}} \leqslant 2\int_0^\infty \frac{d\tau}{1+\tau^{1+\sigma}} = k, \tag{3·2}$$

the left-hand integral being taken over the straight line (3·1).

(ii) Similarly, if $\mathscr{P}$ is composed entirely of straight lines, the total number of which is a bounded function of z and θ, then condition (iii) on $\mathscr{P}$ is satisfied. In this event, moreover, condition (ii) on $\mathscr{P}$ is obviously satisfied as well.

We may note in passing that condition (ii) on $\mathscr{P}$ demands rather more than that each arc of $\mathscr{P}$ be a *regular* Jordan arc, for which only $t'(\tau)$ need be continuous.

(iii) It may happen that the inequality (2·2) holds only for $m \leqslant M$, where M is a positive integer or zero and independent of u and θ. In this event it can be deduced from the proof of theorem A given in §§ 5 to 8 that the relations (2·17) and (2·18) hold provided that $m \leqslant M$.

(iv) The method of proof of theorem A, given in §§ 5 to 8, is not immediately applicable to the extended case in which the asymptotic expansion (2·3) contains the additional term $uf_{-1}(\theta, z)$. Langer (1949) has derived formal series solutions of the differential equation in this case and indicated their asymptotic nature when z is real and bounded. An extension of Langer's result to the equation

$$\frac{d^2w}{dz^2} = u^{2p} f w; \qquad f \sim 1 + \frac{f_1}{u} + \frac{f_2}{u^2} + \dots,$$

where p is a positive integer, has been given by Erdélyi (1956), again for real bounded z.

4. Lemma on asymptotic sums

We shall require the following result.

Lemma. *Let $\phi_0(\theta)$, $\phi_1(\theta)$ … be a set of functions satisfying the condition*

$$|\phi_s(\theta)| < c_s \quad (\theta \in \mathbf{\Theta}, s = 0, 1, \dots), \tag{4·1}$$

where $c_0, c_1, \dots$ are independent of u and θ. Then there exists a function $\Phi(u, \theta)$ with the property

$$\Phi(u, \theta) \sim \sum_{s=0}^{\infty} \frac{\phi_s(\theta)}{u^s} \tag{4·2}$$

as $u \to +\infty$, uniformly valid with respect to θ.

This result is known from the general theory of asymptotic sums (van der Corput 1956, p. 391). It suffices here to remark that the function

$$\Phi(u, \theta) = \sum_{s=0}^{\nu(u)} \frac{\phi_s(\theta)}{u^s}, \tag{4·3}$$

where $\nu(u)$ is the largest integer such that

$$c_0 + c_1 + \dots + c_{\nu(u)} + \nu(u) \leqslant u, \tag{4·4}$$

has the required property.

5. Bounds for the coefficients

Inequality (2·2), with $m = 0$, means that

$$|f(u,\theta,z)| < \frac{k}{1+|z|^{1+\sigma}} \quad (z \in \mathbf{D}(\theta), u \geqslant u_0), \tag{5·1}$$

and for $m \geqslant 1$ the same inequality establishes the equation

$$f_{m-1}(\theta,z) = u_0^{m-1}\left\{f(u_0,\theta,z) - \sum_{s=0}^{m-2} \frac{f_s(\theta,z)}{u_0^s}\right\} + \frac{\varpi k_m}{(1+|z|^{1+\sigma})\, u_0}, \tag{5·2}$$

where ϖ is a number such that $|\varpi| < 1$. If $m = 1$ the sum is absent, and with the aid of (5·1) we deduce that

$$|f_0(\theta,z)| < \frac{k}{1+|z|^{1+\sigma}} \quad (z \in \mathbf{D}(\theta)). \tag{5·3}$$

From these three results we readily show by induction that

$$|f_s(\theta,z)| < \frac{k_s}{1+|z|^{1+\sigma}} \quad (z \in \mathbf{D}(\theta), s = 0, 1, \ldots), \tag{5·4}$$

where k_s is the generic symbol denoting a quantity independent of z, θ and u, but which may depend on s.

We use these results to establish the following properties of the set of coefficients $A_s(\theta, z)$ defined by (2·10).

Lemma

$$|A_s(\theta,z)| < k_s \qquad |A_s'(\theta,z)| < \frac{k_s}{1+|z|^{1+\sigma}} \qquad |A_s''(\theta,z)| < \frac{k_s}{1+|z|^{1+\sigma}} \quad (z \in \mathbf{G}(\theta)). \tag{5·5}$$

This, also, may be proved by induction. Let $d(>0)$ be the shortest distance between the boundaries of $\mathbf{D}(\theta)$ and $\mathbf{G}(\theta)$ for all θ (cf. §2, condition (i) on $\mathbf{G}(\theta)$). We define $\mathbf{G}(\theta,\delta)$ $(0<\delta<d)$ to be $\mathbf{G}(\theta)$ plus the aggregate of all points whose distance from a boundary point of $\mathbf{G}(\theta)$ does not exceed δ. Clearly $\mathbf{G}(\theta,\delta)$ is contained in $\mathbf{D}(\theta)$.

We note that if $z \in \mathbf{G}(\theta,\delta)$ there exists a path joining $c(\theta)$ and z, lying in $\mathbf{G}(\theta,\delta)$ and having the property

$$\int_{c(\theta)}^{z} \frac{|\mathrm{d}t|}{1+|t|^{1+\sigma}} < k. \tag{5·6}$$

For if $z \in \mathbf{G}(\theta)$ this is one of the conditions on $\mathbf{G}(\theta)$. If z is any other point of $\mathbf{G}(\theta,\delta)$ then z_0 exists on the boundary of $\mathbf{G}(\theta)$ such that $|z-z_0| \leqslant \delta$; we take the path in (5·6) to be the known path joining $c(\theta)$ and z_0, plus the join of z_0 and z.

Let us suppose, temporarily, that

$$|A_n'(\theta,z)| < \frac{k_n}{1+|z|^{1+\sigma}} \quad (z \in \mathbf{G}(\theta,\delta), n = 0, 1, \ldots, s). \tag{5·7}$$

Clearly

$$A_n(\theta,z) = \int_{c(\theta)}^{z} A_n'(\theta,t)\,\mathrm{d}t + A_n\{\theta, c(\theta)\}. \tag{5·8}$$

If $z \in \mathbf{G}(\theta,\delta)$ we may take the path of integration to be that of the integral on the left of (5·6). Then substituting (5·7) and the first of (2·16) in (5·8) and using (5·6), we find that

$$|A_n(\theta,z)| < k_n \int_{c(\theta)}^{z} \frac{|\mathrm{d}t|}{1+|t|^{1+\sigma}} + k_n = kk_n + k_n = k_n. \tag{5·9}$$

Let η be an arbitrary number, $0<\eta<\delta$. Then $\mathbf{G}(\theta,\eta)$ is contained in $\mathbf{G}(\theta,\delta)$. Cauchy's integral formula shows that

$$A''_s(\theta,z)=\frac{1}{2\pi i}\int_{|t-z|=\delta-\eta}\frac{A'_s(\theta,t)}{(t-z)^2}\,dt. \tag{5·10}$$

If, now, $z\in\mathbf{G}(\theta,\eta)$, then $t\in\mathbf{G}(\theta,\delta)$. [This is obviously true if $z\in\mathbf{G}(\theta)$. If z is any other point of $\mathbf{G}(\theta,\eta)$ then there is a point z_0 on the boundary of $\mathbf{G}(\theta)$ such that $|z-z_0|<\eta$. Hence $|t-z_0|=|(t-z)+(z-z_0)|<\delta-\eta+\eta=\delta$.] We also have

$$\frac{1+|z|^{1+\sigma}}{1+|t|^{1+\sigma}}\leqslant\frac{1+(|t|+\delta-\eta)^{1+\sigma}}{1+|t|^{1+\sigma}}<k. \tag{5·11}$$

Substituting (5·7) with $n=s$, in (5·10) and using (5·11), we deduce that

$$|A''_s(\theta,z)|<\frac{k_s}{2\pi}\int_{|t-z|=\delta-\eta}\frac{|dt|}{|t-z|^2(1+|t|^{1+\sigma})}$$

$$<\frac{kk_s}{2\pi(1+|z|^{1+\sigma})}\int_{|t-z|=\delta-\eta}\frac{|dt|}{|t-z|^2}=\frac{kk_s}{(\delta-\eta)(1+|z|^{1+\sigma})}, \tag{5·12}$$

provided that $z\in\mathbf{G}(\theta,\eta)$. Finally, substituting this result, (5·9) and (5·4) in (2·9), we obtain

$$|2A'_{s+1}(\theta,z)|<\frac{(\delta-\eta)^{-1}kk_s+k_0k_s+k_1k_{s-1}+\ldots+k_sk_0}{1+|z|^{1+\sigma}}=\frac{k_{s+1}}{1+|z|^{1+\sigma}}\quad(z\in\mathbf{G}(\theta,\eta)).$$

The truth of the lemma is now evident.

6. Equation satisfied by the truncated series

We define

$$L_m(u,\theta,z)=e^{uz}\sum_{s=0}^{m}\frac{A_s(\theta,z)}{u^s}. \tag{6·1}$$

Differentiating with respect to z and suppressing arguments on the right-hand side, we obtain

$$\frac{d}{dz}L_m(u,\theta,z)=u\,e^{uz}\left(\sum_{s=0}^{m}\frac{B_s}{u^s}+\frac{A'_m}{u^{m+1}}\right), \tag{6·2}$$

$$\frac{d^2}{dz^2}L_m(u,\theta,z)=u^2\,e^{uz}\left(\sum_{s=0}^{m}\frac{C_s}{u^s}+\frac{2A'_m+A''_{m-1}}{u^{m+1}}+\frac{A''_m}{u^{m+2}}\right). \tag{6·3}$$

Hence we have

$$\frac{d^2}{dz^2}L_m(u,\theta,z)-\{u^2+f(u,\theta,z)\}L_m(u,\theta,z)=e^{uz}R_m(u,\theta,z), \tag{6·4}$$

where

$$R_m(u,\theta,z)=u^2\left\{\sum_{s=0}^{m}\frac{C_s}{u^s}+\frac{2A'_m+A''_{m-1}}{u^{m+1}}+\frac{A''_m}{u^{m+2}}\right\}-\{u^2+f(u,\theta,z)\}\sum_{s=0}^{m}\frac{A_s}{u^s}, \tag{6·5}$$

and is a regular function of z in $\mathbf{D}(\theta)$.

The inequality (2·2) implies that

$$f(u,\theta,z)=\sum_{s=0}^{m-1}\frac{f_s}{u^s}+\frac{\varpi k_m}{(1+|z|^{1+\sigma})\,u^m}\quad(z\in\mathbf{D}(\theta)), \tag{6·6}$$

when $u\geqslant u_0$, where $|\varpi|<1$. Substituting (6·6) in (6·5), we find that the coefficients of $u^2, u, \ldots, u^{-m+1}$, vanish as a consequence of (2·8) and (2·9), and with the aid also of (5·4) and (5·5) we easily show that

$$|R_m(u,\theta,z)|<\frac{1}{1+|z|^{1+\sigma}}\frac{k_m}{u^m}\quad(z\in\mathbf{G}(\theta)), \tag{6·7}$$

provided that $u\geqslant u_0$; henceforth we accept this restriction.

7. Proof of theorem A: $a_j(\theta)$ finite

A proof only for the case $j = 1$ will be recorded. An exactly similar proof holds for $j = 2$; alternatively, the result for $j = 2$ can be deduced from that for $j = 1$ by means of the transformation $z' = -z$.

In this section we suppose that $a_1 \equiv a_1(\theta)$, defined in § 2, is a point of $\mathbf{G}(\theta)$ which is not at infinity. The necessary modifications when a_1 is at infinity are considered in § 8.

From (5·5) and the first of (2·7), we see that

$$|A_s(\theta, a_1)| < k_s, \qquad |B_s(\theta, a_1)| < k_s. \tag{7·1}$$

The lemma of § 4 shows that functions $A(u, \theta)$ and $B(u, \theta)$ exist having the asymptotic expansions

$$A(u,\theta) \sim \sum_{s=0}^{\infty} \frac{A_s(\theta, a_1)}{u^s}, \qquad B(u,\theta) \sim \sum_{s=0}^{\infty} \frac{B_s(\theta, a_1)}{u^s}, \tag{7·2}$$

as $u \to +\infty$, uniformly valid with respect to θ. We define $W_1(u, \theta, z)$ to be the solution of the differential equation (2·1) satisfying the conditions

$$W_1(u, \theta, a_1) = \mathrm{e}^{ua_1} A(u, \theta), \qquad \left[\frac{\mathrm{d}}{\mathrm{d}z} W_1(u, \theta, z)\right]_{z=a_1} = u\,\mathrm{e}^{ua_1} B(u, \theta). \tag{7·3}$$

From (2·1) and (6·4) we obtain

$$\frac{\mathrm{d}^2}{\mathrm{d}z^2}\{W_1(z) - L_m(z)\} - \{u^2 + f(z)\}\{W_1(z) - L_m(z)\} = -\mathrm{e}^{uz} R_m(z), \tag{7·4}$$

where $W_1(z) = W_1(u, \theta, z)$ and the arguments u and θ of the other functions have similarly been suppressed.

Let z_1 be a point of the region $\mathbf{H}_1(\theta)$ defined in § 2, and let ζ be a typical point of the path $\mathscr{P}$ joining z_1 and a_1. In order to establish the asymptotic properties (2·17) we solve, by successive approximation, an integral equation equivalent to the differential equation (7·4). We introduce a sequence of functions $h_{m,n}(u, \theta, z_1, \zeta) \equiv h_{m,n}(\zeta)$ defined by $h_{m,0}(\zeta) = 0$ and

$$h_{m,n}(\zeta) = \frac{1}{2u}\int_{a_1}^{\zeta} \{\mathrm{e}^{u(\zeta - t)} - \mathrm{e}^{u(t-\zeta)}\}\{f(t)\, h_{m,n-1}(t) - \mathrm{e}^{ut} R_m(t)\}\,\mathrm{d}t + \alpha_m \mathrm{e}^{u\zeta} + \beta_m \mathrm{e}^{-u\zeta} \quad (n \geqslant 1), \tag{7·5}$$

where the path of integration is the part of $\mathscr{P}$ between ζ and a_1, and α_m, β_m are determined by the conditions

$$h_{m,n}(a_1) = W_1(a_1) - L_m(a_1), \qquad h'_{m,n}(a_1) = W'_1(a_1) - L'_m(a_1). \tag{7·6}$$

By differentiating (7·5), we readily verify that

$$(\mathrm{d}^2/\mathrm{d}\zeta^2)\, h_{m,n}(\zeta) - u^2 h_{m,n}(\zeta) = f(\zeta)\, h_{m,n-1}(\zeta) - \mathrm{e}^{u\zeta} R_m(\zeta) \quad (n \geqslant 1) \tag{7·7}$$

(cf. (7·4)).

The quantities α_m and β_m are independent of n; in fact from (7·5), (7·6), (7·3), (6·1) and (6·2) we obtain

$$\left.\begin{aligned} \alpha_m \mathrm{e}^{ua_1} + \beta_m \mathrm{e}^{-ua_1} &= \mathrm{e}^{ua_1}\left\{A(u,\theta) - \sum_{s=0}^{m} \frac{A_s(\theta, a_1)}{u^s}\right\} = \mathrm{e}^{ua_1} O(u^{-m-1}), \\ \alpha_m \mathrm{e}^{ua_1} - \beta_m \mathrm{e}^{-ua_1} &= \mathrm{e}^{ua_1}\left\{B(u,\theta) - \sum_{s=0}^{m} \frac{B_s(\theta, a_1)}{u^s} - \frac{A'_m(\theta, a_1)}{u^{m+1}}\right\} = \mathrm{e}^{ua_1} O(u^{-m-1}), \end{aligned}\right\} \tag{7·8}$$

where the O's are uniform with respect to θ. Solving these equations, we see that

$$\alpha_m = O(u^{-m-1}), \qquad \beta_m = e^{2ua_1}O(u^{-m-1}). \tag{7·9}$$

Hence

$$|\alpha_m e^{u\zeta}| < k_m u^{-m-1} |e^{u\zeta}|, \qquad |\beta_m e^{-u\zeta}| = |e^{2ua_1 - u\zeta}| O(u^{-m-1}) < k_m u^{-m-1} |e^{u\zeta}|, \tag{7·10}$$

since

$$\operatorname{Re} z_1 \geqslant \operatorname{Re} \zeta \geqslant \operatorname{Re} t \geqslant \operatorname{Re} a_1, \tag{7·11}$$

in consequence of condition (iv) on $\mathscr{P}$, given in § 2.

Next, we have from (7·5)

$$h_{m,1}(\zeta) = -\frac{1}{2u}\int_{a_1}^{\zeta} \{e^{u\zeta} - e^{u(2t-\zeta)}\} R_m(t)\, dt + \alpha_m e^{u\zeta} + \beta_m e^{-u\zeta}. \tag{7·12}$$

From (7·11) we deduce that

$$|e^{u(2t-\zeta)}| \leqslant |e^{u\zeta}|, \tag{7·13}$$

and using this inequality and (6·7), we see that

$$\left|\int_{a_1}^{\zeta} \{e^{u\zeta} - e^{u(2t-\zeta)}\} R_m(t)\, dt\right| < \frac{2k_m}{u^m} |e^{u\zeta}| \int_{\mathscr{P}} \frac{|dt|}{1+|t|^{1+\sigma}} < \frac{2kk_m}{u^m} |e^{u\zeta}|, \tag{7·14}$$

in virtue of (2·19). Substituting this result and (7·10) in (7·12), we obtain

$$|h_{m,1}(\zeta)| < k_m u^{-m-1} |e^{u\zeta}|. \tag{7·15}$$

From (7·5) we have also

$$h_{m,n+1}(\zeta) - h_{m,n}(\zeta) = \frac{1}{2u}\int_{a_1}^{\zeta} \{e^{u(\zeta-t)} - e^{u(t-\zeta)}\} f(t) \{h_{m,n}(t) - h_{m,n-1}(t)\}\, dt \quad (n \geqslant 1). \tag{7·16}$$

Taking $n = 1$ and substituting (5·1) with z replaced by t, (7·15) with ζ replaced by t and (7·13), we deduce that

$$|h_{m,2}(\zeta) - h_{m,1}(\zeta)| < kk_m u^{-m-2} |e^{u\zeta}|.$$

Continuing the argument by induction, we establish the key result

$$|h_{m,n+1}(\zeta) - h_{m,n}(\zeta)| < k_m k^n u^{-m-n-1} |e^{u\zeta}| \quad (n = 0, 1, \ldots), \tag{7·17}$$

where k_m and k are of course independent of n as well as of u, θ, z and ζ.

Thus the series

$$T = \sum_{n=0}^{\infty} \{h_{m,n+1}(\zeta) - h_{m,n}(\zeta)\}$$

converges if, as we shall now suppose, $u > k$. We shall prove that its sum is $W_1(\zeta) - L_m(\zeta)$. In the method of proof used previously (Olver 1954*a*, § 10; 1956, §§ 12 and 13) this was achieved quite easily by tacitly supposing that the region corresponding to $\mathbf{H}_1(\theta)$ is, in fact, a complex domain, i.e. an open connex set of points. Term-by-term differentiation of the series T is then valid within the domain, and it follows immediately from (7·7) that T satisfies the same differential equation (7·4) as $W_1(z) - L_m(z)$. Therefore, because of the boundary conditions (7·6) imposed at a_1, the two functions are equal in $\mathbf{H}_1(\theta)$. The present proof does not depend on assumptions of this kind concerning the nature of the region $\mathbf{H}_1(\theta)$.

Differentiating (7·12) and (7·16) with respect to ζ, we find that

$$h'_{m,1}(\zeta) = -\frac{1}{2}\int_{a_1}^{\zeta} \{e^{u\zeta} + e^{u(2t-\zeta)}\} R_m(t)\, dt + \alpha_m u\, e^{u\zeta} - \beta_m u\, e^{-u\zeta}, \tag{7·18}$$

and

$$h'_{m,n+1}(\zeta) - h'_{m,n}(\zeta) = \frac{1}{2}\int_{a_1}^{\zeta} \{e^{u(\zeta-t)} + e^{u(t-\zeta)}\} f(t) \{h_{m,n}(t) - h_{m,n-1}(t)\}\, dt \quad (n \geqslant 1). \tag{7·19}$$

Using (2·19), (5·1), (6·7), (7·10), (7·13) and (7·17), we may show that

$$|h'_{m,n+1}(\zeta)-h'_{m,n}(\zeta)|<k_m k^n u^{-m-n}|e^{u\zeta}| \quad (n=0,1,\ldots). \tag{7·20}$$

And from (7·17), (6·7), (5·1) and the relations

$$h''_{m,1}(\zeta)=u^2h_{m,1}(\zeta)-e^{u\zeta}R_m(\zeta), \tag{7·21}$$

$$h''_{m,n+1}(\zeta)-h''_{m,n}(\zeta)=u^2\{h_{m,n+1}(\zeta)-h_{m,n}(\zeta)\}+f(\zeta)\{h_{m,n}(\zeta)-h_{m,n-1}(\zeta)\} \quad (n\geqslant 1), \tag{7·22}$$

obtained from (7·7), we deduce that

$$|h''_{m,n+1}(\zeta)-h''_{m,n}(\zeta)|<k_m k^n u^{-m-n+1}|e^{u\zeta}| \quad (n=0,1,\ldots). \tag{7·23}$$

Thus the series $\sum_{n=0}^{\infty}\{h''_{m,n+1}(\zeta)-h''_{m,n}(\zeta)\}$ converges, and from (7·21), (7·22) it is seen that

$$\sum_{n=0}^{\infty}\{h''_{m,n+1}(\zeta)-h''_{m,n}(\zeta)\}=\{u^2+f(\zeta)\}\sum_{n=0}^{\infty}\{h_{m,n+1}(\zeta)-h_{m,n}(\zeta)\}-e^{u\zeta}R_m(\zeta). \tag{7·24}$$

Condition (ii) on $\mathscr{P}$, given in § 2, affirms that $\mathscr{P}$ has a parametric equation of the form

$$\zeta=\zeta(\tau) \quad (\tau_0\leqslant\tau\leqslant\tau_1), \tag{7·25}$$

where τ_0 and τ_1 correspond to a_1 and z_1, respectively. The function $\zeta(\tau)$ is continuous and $\zeta'(\tau)$, $\zeta''(\tau)$ have only a finite number of discontinuities in the interval $\tau_0\leqslant\tau\leqslant\tau_1$. The series

$$S(\tau)=\sum_{n=0}^{\infty}[h''_{m,n+1}\{\zeta(\tau)\}-h''_{m,n}\{\zeta(\tau)\}] \tag{7·26}$$

is accordingly a uniformly convergent series of continuous functions when $\tau_0\leqslant\tau\leqslant\tau_1$, and we may multiply both sides of this equation by $\zeta'(\tau)$ and integrate term by term to obtain

$$\int_{\tau_0}^{\tau}S(\tau)\,\zeta'(\tau)\,d\tau+h'_{m,1}(a_1)=\sum_{n=0}^{\infty}[h'_{m,n+1}\{\zeta(\tau)\}-h'_{m,n}\{\zeta(\tau)\}]. \tag{7·27}$$

Similarly

$$\int_{\tau_0}^{\tau}\left\{\int_{\tau_0}^{\tau}S(\tau)\,\zeta'(\tau)\,d\tau+h'_{m,1}(a_1)\right\}\zeta'(\tau)\,d\tau+h_{m,1}(a_1)=\sum_{n=0}^{\infty}[h_{m,n+1}\{\zeta(\tau)\}-h_{m,n}\{\zeta(\tau)\}]=T(\tau), \tag{7·28}$$

say. Differentiation of the last equation with respect to τ yields

$$\begin{aligned}T''(\tau)-\{\zeta''(\tau)/\zeta'(\tau)\}\,T'(\tau)&=\{\zeta'(\tau)\}^2S(\tau)\\&=\{\zeta'(\tau)\}^2[u^2+f\{\zeta(\tau)\}]\,T(\tau)-\{\zeta'(\tau)\}^2e^{u\zeta(\tau)}R_m\{\zeta(\tau)\},\end{aligned} \tag{7·29}$$

on substituting for $S(\tau)$ by means of (7·26) and (7·24).

Equation (7·29) is a second-order differential equation for $T(\tau)$. Since, by hypothesis, $\zeta'(\tau)$ does not vanish, all the coefficients are continuous except at the discontinuities of $\zeta'(\tau)$ and $\zeta''(\tau)$, i.e. the junctions of the Jordan arcs. Substituting $z=\zeta(\tau)$ in (7·4), we see that the function $W_1\{\zeta(\tau)\}-L_m\{\zeta(\tau)\}$ satisfies exactly the same differential equation. Moreover, from (7·6) we note that at $\tau=\tau_0$, i.e. at $\zeta=a_1$, $T(\tau)$ and $T'(\tau)$ are respectively equal to the values of $W_1\{\zeta(\tau)\}-L_m\{\zeta(\tau)\}$ and its τ-derivative. Since a linear differential equation with continuous coefficients has a unique solution corresponding to given initial conditions (Ince 1944, § 3·32), it follows that

$$T(\tau)=W_1\{\zeta(\tau)\}-L_m\{\zeta(\tau)\}, \qquad T'(\tau)=(d/d\tau)\,[W_1\{\zeta(\tau)\}-L_m\{\zeta(\tau)\}], \tag{7·30}$$

along the arc with terminal point τ_0. At the junction of this arc with the next, $T(\tau)$ is continuous but $T'(\tau)$ is discontinuous. However, $T'(\tau)/\zeta'(\tau)$ and $W_1'\{\zeta(\tau)\}-L_m'\{\zeta(\tau)\}$ are both continuous. Accordingly, equations (7·30) hold across the junction and therefore also along the second arc. Extending the argument, we see that equations (7·30) hold along the whole of $\mathscr{P}$.

In particular, at $\tau = \tau_1$, i.e. at $\zeta = z_1$, we have

$$\left.\begin{aligned} W_1(z_1)-L_m(z_1) &= T(\tau_1) = \sum_{n=0}^{\infty}\{h_{m,n+1}(z_1)-h_{m,n}(z_1)\},\\ W_1'(z_1)-L_m'(z_1) &= T'(\tau_1)/\zeta'(\tau_1) = \sum_{n=0}^{\infty}\{h'_{m,n+1}(z_1)-h'_{m,n}(z_1)\}.\end{aligned}\right\} \tag{7·31}$$

If we suppose that $u>2k$, we deduce immediately from these results and (7·17), (7·20) that

$$\left.\begin{aligned} |W_1(z_1)-L_m(z_1)| &< k_m u^{-m-1}|e^{uz_1}|\sum_{n=0}^{\infty}2^{-n} = 2k_m u^{-m-1}|e^{uz_1}|,\\ |W_1'(z_1)-L_m'(z_1)| &< k_m u^{-m}|e^{uz_1}|\sum_{n=0}^{\infty}2^{-n} = 2k_m u^{-m}|e^{uz_1}|.\end{aligned}\right\} \tag{7·32}$$

These are the desired results.

8. Proof of theorem A: $a_j(\theta)$ at infinity

We now suppose that $a_1\equiv a_1(\theta)$ is the point at infinity on a straight line $\mathscr{L}$ lying in $\mathbf{G}(\theta)$. Relations (5·8) and (5·5) show that $A_s(\theta,z)$ tends to a limit $A_s(\theta,a_1)$, say, as $z\to a_1$ along $\mathscr{L}$, and that

$$|A_s(\theta,a_1)|\leqslant k_s. \tag{8·1}$$

Therefore, by the lemma of § 4, a function $A(u,\theta)$ exists having the asymptotic expansion

$$A(u,\theta)\sim\sum_{s=0}^{\infty}\frac{A_s(\theta,a_1)}{u^s}, \tag{8·2}$$

as $u\to+\infty$, uniformly valid with respect to θ.

We define $W_1(u,\theta,z)\equiv W_1(z)$ to be the solution of the differential equation (2·1) with the properties

$$\lim\{e^{-uz}W_1(u,\theta,z)\} = A(u,\theta), \qquad \lim\{e^{-uz}(d/dz)\,W_1(u,\theta,z)\} = uA(u,\theta), \tag{8·3}$$

as $z\to a_1$ along $\mathscr{L}$. At this stage we do not know whether such a solution necessarily exists. There can be at most one such solution, however. For suppose $W_1^*(z)$ were a second solution with the same properties. Since $|\arg(-a_1)|<\frac{1}{2}\pi$ (cf. § 2), it follows that W_1, W_1^*, W_1' and $W_1^{*\prime}$ all tend to zero as $z\to a_1$; accordingly the Wronskian of W_1 and W_1^* vanishes.

We define the sequence $h_{m,n}(\zeta)$ by $h_{m,0}(\zeta) = 0$ and

$$h_{m,n}(\zeta) = \frac{1}{2u}\int_{a_1}^{\zeta}\{e^{u(\zeta-t)}-e^{u(t-\zeta)}\}\{f(t)\,h_{m,n-1}(t)-e^{ut}R_m(t)\}\,dt+\alpha_m e^{u\zeta} \quad (n\geqslant 1) \tag{8·4}$$

(cf. (7·5)), where

$$\alpha_m = A(u,\theta)-\sum_{s=0}^{m}\frac{A_s(\theta,a_1)}{u^s}. \tag{8·5}$$

Here ζ is again a typical point of the path $\mathscr{P}$ joining a_1 and z_1, and the path of integration in (8·4) is the part of $\mathscr{P}$ between ζ and a_1; it coincides with $\mathscr{L}$ for all sufficiently large $|\zeta|$ (cf. § 2, condition (ii)).

Equation (8·4) corresponds to (7·5) and the analysis of § 7 between (7·5) and (7·24) can be repeated here with trivial changes; in particular the relations (7·17) and (7·20) to (7·24) are again satisfied.

The parametric equation of $\mathscr{P}$ is of the form (7·25) but with $\tau_0 = -\infty$ of necessity, since $\zeta(\tau)$ is continuous and therefore finite with τ. Term-by-term integration of a uniformly convergent series over an infinite interval is permissible if the same operation on the moduli of the terms yields a convergent series (Titchmarsh 1939, § 1·77). If $u>k$ this condition is satisfied by the series on the right of (7·26) multiplied by $\zeta'(\tau)$, because $\int_{-\infty}^{T} |e^{u\zeta(\tau)}\zeta'(\tau)|\,d\tau$ converges. Therefore we have

$$\int_{-\infty}^{T} S(\tau)\,\zeta'(\tau)\,d\tau = \sum_{n=0}^{\infty} [h'_{m,\,n+1}\{\zeta(\tau)\} - h'_{m,\,n}\{\zeta(\tau)\}], \tag{8·6}$$

corresponding to (7·27), and similarly

$$\int_{-\infty}^{T}\left\{\int_{-\infty}^{T} S(\tau)\,\zeta'(\tau)\,d\tau\right\}\zeta'(\tau)\,d\tau = \sum_{n=0}^{\infty} [h_{m,\,n+1}\{\zeta(\tau)\} - h_{m,\,n}\{\zeta(\tau)\}] = T(\tau), \tag{8·7}$$

corresponding to (7·28). Differentiation of (8·7) again yields (7·29).

We now examine the behaviour of $T(\tau)$ as $\tau \to -\infty$. Since $\mathscr{P}$ coincides with the straight line $\mathscr{L}$ for all sufficiently large $|\zeta|$, it follows that if ϵ is an arbitrary positive number then

$$\int_{a_1}^{\zeta(\tau)} \frac{|dt|}{1+|t|^{1+\sigma}} < \epsilon, \tag{8·8}$$

for all $\tau \leqslant \tau(\epsilon)$ (assignable). With this restriction, we derive from (8·4), with $n = 1$,

$$|h_{m,1}(\zeta) - \alpha_m e^{u\zeta}| = \left|\frac{1}{2u}\int_{a_1}^{\zeta} \{e^{u\zeta} - e^{u(2t-\zeta)}\} R_m(t)\,dt\right| \leqslant \frac{\epsilon k_m}{u^{m+1}} |e^{u\zeta}|, \tag{8·9}$$

in consequence of (6·7), (7·13) and (8·8). Similarly, if $n \geqslant 1$ we have

$$\begin{aligned} |h_{m,\,n+1}(\zeta) - h_{m,\,n}(\zeta)| &= \left|\frac{1}{2u}\int_{a_1}^{\zeta} \{e^{u(\zeta-t)} - e^{u(t-\zeta)}\} f(t)\,\{h_{m,\,n}(t) - h_{m,\,n-1}(t)\}\,dt\right| \\ &< \epsilon k_m k^n u^{-m-n-1} |e^{u\zeta}|, \end{aligned} \tag{8·10}$$

in consequence of (5·1), (7·13), (7·17) and (8·8). Therefore, if $u>2k$, it follows that

$$\left|\sum_{n=0}^{\infty} \{h_{m,\,n+1}(\zeta) - h_{m,\,n}(\zeta)\} - \alpha_m e^{u\zeta}\right| < 2\epsilon k_m u^{-m-1} |e^{u\zeta}|. \tag{8·11}$$

Again, equations (7·18) and (7·19) still hold here provided that we set $\beta_m = 0$ in the former, and from them we deduce that

$$\left|\sum_{n=0}^{\infty} \{h'_{m,\,n+1}(\zeta) - h'_{m,\,n}(\zeta)\} - \alpha_m u\, e^{u\zeta}\right| < 2\epsilon k_m u^{-m} |e^{u\zeta}| \tag{8·12}$$

(cf. (8·11)). The inequalities (8·11) and (8·12) show that

$$e^{-u\zeta(\tau)} T(\tau) \to \alpha_m, \qquad e^{-u\zeta(\tau)}\{T'(\tau)/\zeta'(\tau)\} \to \alpha_m u, \tag{8·13}$$

as $\tau \to -\infty$.

Let $W_1^*(z)$ be the solution of the differential equation (2·1) satisfying the conditions

$$W_1^*(z_1) = T(\tau_1)+L_m(z_1), \qquad W_1^{*\prime}(z_1) = \{T'(\tau_1)/\zeta'(\tau_1)\}+L'_m(z_1), \tag{8·14}$$

Then $W_1^*\{\zeta(\tau)\}-L_m\{\zeta(\tau)\}$ satisfies the differential equation (7·29), and since at $\tau=\tau_1$ this function and its τ-derivative are respectively equal to $T(\tau)$ and $T'(\tau)$ it follows as in § 7 that

$$T(\tau) = W_1^*\{\zeta(\tau)\}-L_m\{\zeta(\tau)\}, \qquad T'(\tau) = (\mathrm{d}/\mathrm{d}\tau)\,[W_1^*\{\zeta(\tau)\}-L_m\{\zeta(\tau)\}], \tag{8·15}$$

for all τ in the range $-\infty<\tau\leqslant\tau_1$ (cf. (7·30)).

Letting $\tau\to-\infty$, we obtain from these equations and (8·13)

$$\left.\begin{aligned}\lim\{\mathrm{e}^{-u\zeta}W_1^*(\zeta)\} &= \alpha_m+\lim\{\mathrm{e}^{-u\zeta}L_m(\zeta)\} = A(u,\theta),\\ \lim\{\mathrm{e}^{-u\zeta}W_1^{*\prime}(\zeta)\} &= \alpha_m u+\lim\{\mathrm{e}^{-u\zeta}L'_m(\zeta)\} = uA(u,\theta),\end{aligned}\right\} \tag{8·16}$$

on substituting (8·5), (6·1), (6·2), and using the equations $A'_s(\theta,a_1)=0$, $B_s(\theta,a_1)=A_s(\theta,a_1)$, obtained from (5·5) and the first of (2·7).

Referring to (8·3), we now see that the solution $W_1(z)$ does exist; in fact, $W_1(z)=W_1^*(z)$. Equations (8·15) show that equations (7·31) again hold, and from them we deduce the inequalities (7·32) as in § 7. This completes the proof.

PART 2. CASE B

9. Statement of conditions and theorem B

Where possible we shall shorten statements and proofs by reference to the corresponding results for case A given in part 1.

The standard form of differential equation for this case is

$$\frac{\mathrm{d}^2w}{\mathrm{d}z^2} = \{u^2z+f(u,\theta,z)\}\,w \tag{9·1}$$

(cf. (1·9)), where (as in part 1) u is a large positive parameter, $\theta\in\mathbf{\Theta}$, $z\in\mathbf{D}(\theta)$, a simply-connected complex domain bounded or otherwise. We suppose in addition that $\mathbf{D}(\theta)$ contains the circle $|z|\leqslant b$, where b is positive and independent of u and θ.

We assume that $f(u,\theta,z)$ is a regular function of z in $\mathbf{D}(\theta)$ and that

$$\left|f(u,\theta,z)-\sum_{s=0}^{m-1}\frac{f_s(\theta,z)}{u^s}\right| < \frac{1}{1+|z|^{\frac{1}{2}+\sigma}}\frac{k_m}{u^m} \quad (m=0,1,2,\ldots) \tag{9·2}$$

(cf. (2·2)), for all $z\in\mathbf{D}(\theta)$, $\theta\in\mathbf{\Theta}$, $u\geqslant u_0$. Each coefficient $f_s(\theta,z)$ is independent of u† and regular in $\mathbf{D}(\theta)$, and the symbols σ and k_m have the meanings assigned in § 2.

Formal solutions of (9·1) in inverse powers of u are given by

$$w = \mathrm{Ci}(u^{\frac{2}{3}}z)\sum_{s=0}^{\infty}\frac{A_s(\theta,z)}{u^s}+\frac{\mathrm{Ci}'(u^{\frac{2}{3}}z)}{u^{\frac{4}{3}}}\sum_{s=0}^{\infty}\frac{B_s(\theta,z)}{u^s} \tag{9·3}$$

(cf. (2·4)), where $\mathrm{Ci}(u^{\frac{2}{3}}z)$ is any solution of the Airy equation

$$\frac{\mathrm{d}^2}{\mathrm{d}z^2}\mathrm{Ci}(u^{\frac{2}{3}}z) = u^2z\,\mathrm{Ci}(u^{\frac{2}{3}}z), \tag{9·4}$$

† Except in as much as θ may depend on u.

and the coefficients $A_s(\theta, z)$ and $B_s(\theta, z)$ are to be determined. Term-by-term differentiation of (9·3) and use of (9·4) yields

$$\frac{dw}{dz} = \mathrm{Ci}(u^{\frac{2}{3}}z)\sum_{s=0}^{\infty}\frac{C_s(\theta, z)}{u^s} + u^{\frac{2}{3}}\mathrm{Ci}'(u^{\frac{2}{3}}z)\sum_{s=0}^{\infty}\frac{D_s(\theta, z)}{u^s}, \tag{9·5}$$

$$\frac{d^2w}{dz^2} = u^2\mathrm{Ci}(u^{\frac{2}{3}}z)\sum_{s=0}^{\infty}\frac{E_s(\theta, z)}{u^s} + u^{\frac{4}{3}}\mathrm{Ci}'(u^{\frac{2}{3}}z)\sum_{s=0}^{\infty}\frac{F_s(\theta, z)}{u^s}, \tag{9·6}$$

where

$$C_s = A_s' + zB_s, \qquad D_s = A_s + B_{s-2}', \tag{9·7}$$

$$\left.\begin{aligned} E_s &= C_{s-2}' + zD_s = A_{s-2}'' + zA_s + 2zB_{s-2}' + B_{s-2}, \\ F_s &= C_s + D_s' = 2A_s' + B_{s-2}'' + zB_s. \end{aligned}\right\} \tag{9·8}$$

Equation (9·1) is formally satisfied if

$$\left.\begin{aligned} E_{s+2} &= zA_{s+2} + f_0A_s + f_1A_{s-1} + \ldots + f_sA_0, \\ F_{s+2} &= zB_{s+2} + f_0B_s + f_1B_{s-1} + \ldots + f_sB_0. \end{aligned}\right\} \tag{9·9}$$

Substituting (9·8) in (9·9), we obtain

$$2zB_s' + B_s = -A_s'' + f_0A_s + f_1A_{s-1} + \ldots + f_sA_0, \tag{9·10}$$

$$2A_{s+2}' = -B_s'' + f_0B_s + f_1B_{s-1} + \ldots + f_sB_0. \tag{9·11}$$

Integrating these equations, we find that $A_0 =$ constant, $A_1 =$ constant, and

$$B_s = \tfrac{1}{2}z^{-\frac{1}{2}}\int_0^z z^{-\frac{1}{2}}(-A_s'' + f_0A_s + f_1A_{s-1} + \ldots + f_sA_0)\,dz, \tag{9·12}$$

$$A_{s+2} = -\tfrac{1}{2}B_s' + \tfrac{1}{2}\int (f_0B_s + f_1B_{s-1} + \ldots + f_sB_0)\,dz. \tag{9·13}$$

Except for the presence of arbitrary constants of integration in (9·13) these two equations determine a set of functions $A_s(\theta, z)$ and $B_s(\theta, z)$ which are regular and single-valued in $\mathbf{D}(\theta)$. We note that if the coefficients $f_s(\theta, z)$ of odd suffix all vanish, then we can arrange that the same is true of $A_s(\theta, z)$ and $B_s(\theta, z)$, i.e. the series occurring in (9·3) contain only even powers of u^{-1}; in particular, this happens when $f(u, \theta, z)$ is independent of u.

We prescribe $\mathbf{G}(\theta)$ here to be a closed subdomain of $\mathbf{D}(\theta)$, having the properties;

(i) $\mathbf{G}(\theta)$ contains the circle $|z| \leqslant b$.

(ii) The distance between each boundary point z_0 of $\mathbf{G}(\theta)$ and each boundary point of $\mathbf{D}(\theta)$ is not less than $d/|z_0|^{\frac{1}{2}}$, where d is a positive constant, assignable independently of θ.

(iii) $$\int_0^z \frac{|dt|}{1+|t|^{1+\sigma_1}} < k \quad (z \in \mathbf{G}(\theta)), \tag{9·14}$$

for some path lying wholly in $\mathbf{G}(\theta)$, where k is the generic symbol denoting a number independent of u, θ and z (cf. § 2), and

$$\sigma_1 = \min(\sigma, \tfrac{3}{2}). \tag{9·15}$$

We suppose that the arbitrary constants associated with the determination of the coefficients are such that

$$|A_s\{\theta, c(\theta)\}| < k_s, \tag{9·16}$$

where $c(\theta)$ is a prescribed point of $\mathbf{G}(\theta)$ (cf. (2·16)).

Using the notation

$$P_1(z) = \mathrm{Ai}\,(z), \qquad P_2(z) = \mathrm{Ai}\,(e^{\frac{2}{3}\pi i}z), \qquad P_3(z) = \mathrm{Ai}\,(e^{-\frac{2}{3}\pi i}z), \tag{9·17}$$

for the Airy functions (Olver 1954*a*, equations (4·6)), we can now state the existence theorem:

Theorem B. *The differential equation* (9·1) *possesses solutions* $W_j(u, \theta, z)$ $(j = 1, 2, 3)$ *with the properties*

$$W_j(u,\theta,z) = P_j(u^{\frac{2}{3}}z)\left[\sum_{s=0}^{m}\frac{A_s(\theta,z)}{u^s} + O\left(\frac{1}{u^{m+1}}\right)\right] + \frac{P_j'(u^{\frac{2}{3}}z)}{u^{\frac{4}{3}}}\left[\sum_{s=0}^{m-1}\frac{B_s(\theta,z)}{u^s} + \frac{1}{1+|z|^{\frac{1}{2}}}O\left(\frac{1}{u^m}\right)\right], \tag{9·18}$$

$$\frac{d}{dz}W_j(u,\theta,z) = P_j(u^{\frac{2}{3}}z)\left[\sum_{s=0}^{m-1}\frac{C_s(\theta,z)}{u^s} + (1+|z|^{\frac{1}{2}})\,O\left(\frac{1}{u^m}\right)\right] + u^{\frac{2}{3}}P_j'(u^{\frac{2}{3}}z)\left[\sum_{s=0}^{m}\frac{D_s(\theta,z)}{u^s} + O\left(\frac{1}{u^{m+1}}\right)\right], \tag{9·19}$$

as $u \to +\infty$, *valid when* $z \in \mathbf{H}_j(\theta)$ *and* $\theta \in \mathbf{\Theta}$, *each of the O's being uniform with respect to z and* θ. *Here m is an arbitrary positive integer or zero and* $W_j(u,\theta,z)$ *is independent of m.*

The regions of validity $\mathbf{H}_j(\theta)$ $(j = 1, 2, 3)$ are defined as follows. We take $a_j(\theta)$ to be any prescribed point of $\mathbf{G}(\theta)$, or the point at infinity on a straight line $\mathscr{L}$ lying in $\mathbf{G}(\theta)$. If $a_j(\theta)$ is at infinity we make the restriction

$$|\arg\{\rho_j a_j(\theta)\}| < \tfrac{1}{3}\pi, \qquad \text{where} \qquad \rho_1 \equiv 1, \qquad \rho_2 \equiv e^{\frac{2}{3}\pi i}, \qquad \rho_3 \equiv e^{-\frac{2}{3}\pi i}. \tag{9·20}$$

Then $\mathbf{H}_j(\theta)$ comprises those points z of $\mathbf{G}(\theta)$ which can be joined to $a_j(\theta)$ by a path $\mathscr{P}$ having the following properties; t being a typical point of $\mathscr{P}$.

(i) $\mathscr{P}$ lies in $\mathbf{G}(\theta)$.

(ii) $\mathscr{P}$ comprises a finite number of Jordan arcs, each with a parametric equation of the form $t = t(\tau)$, where τ is the real parameter of the arc; $t''(\tau)$ is continuous and $t'(\tau)$ does not vanish. If $a_j(\theta)$ is at infinity, $\mathscr{P}$ coincides with $\mathscr{L}$ for all sufficiently large $|t|$.

(iii) $$\int_{\mathscr{P}}\frac{|dt|}{1+|t|^{1+\sigma_1}} < k. \tag{9·21}$$

(iv) As t traverses $\mathscr{P}$ from $a_j(\theta)$ to z, $|\exp\{\frac{2}{3}(\rho_j t)^{\frac{3}{2}}\}|$ is monotonic decreasing, where ρ_j is defined by (9·20).

10. Remarks on theorem B

In applications of the theorem many of the conditions are conveniently interpreted by transformation to the x-plane, where

$$x = \tfrac{2}{3}z^{\frac{3}{2}}, \tag{10·1}$$

the principal value being taken; examples of this interpretation are given below. It is less convenient, however, to develop the *theory* with x as primary variable in place of z, because the transformed differential equation has a singularity at $x = 0$; moreover, it would be necessary to consider the two-sheeted region $|\arg x| \leq \frac{3}{2}\pi$.

(i) If the distance between the boundary points of $\mathbf{G}(\theta)$ and $\mathbf{D}(\theta)$ has a positive lower bound d', condition (ii) on $\mathbf{G}(\theta)$ is fulfilled with $d = b^{\frac{1}{2}}d'$; this is a consequence of condition (i) on $\mathbf{G}(\theta)$. Condition (ii) is less restrictive than the corresponding condition given in § 5 of Olver 1954*a*. Effectively the new condition means that the distance between the boundaries of the x-maps of $\mathbf{G}(\theta)$ and $\mathbf{D}(\theta)$ has a positive lower bound, but because of the two-sheeted nature of the x-maps the precise formulation of the condition in this form is complicated.

A typical example of the greater usefulness of the new form of condition occurs in Olver 1954*b* in an application to Bessel functions of large order. If $\mathbf{D}(\theta)$ is taken to be the ζ-domain corresponding to $|\arg z| < \pi$ (1954*b*, equation (4·6) and figures 3 and 6), then the ζ-domain corresponding to $|\arg z| \leqslant \pi - \epsilon < \pi$ fulfils condition (ii) on $\mathbf{G}(\theta)$ but not the conditions of Olver 1954*a*, § 5.

(ii) Condition (iii) on $\mathbf{G}(\theta)$ is satisfied if the path joining 0 and z is composed of straight lines, the total number of which is a bounded function of z and θ. This is clear from § 3 (i).

The same condition is also satisfied if the x-map of the path is composed of a bounded number of straight lines. For the parametric equation of the curve in the z-plane corresponding to a typical straight line in the x-plane may be expressed in the form

$$z = z(\tau) = \{\tfrac{3}{2}(\tau + i\lambda_1)\, e^{i\lambda_2}\}^{\frac{2}{3}} \quad (\lambda_3 \leqslant \tau \leqslant \lambda_4), \tag{10·2}$$

where $\lambda_1, \lambda_2, \lambda_3$ and λ_4 are real constants. Clearly

$$|z| \geqslant (\tfrac{3}{2}|\tau|)^{\frac{2}{3}} \geqslant |\tau|^{\frac{2}{3}} \quad \text{and} \quad |dz| = |z|^{-\frac{1}{2}}\, d\tau \leqslant |\tau|^{-\frac{1}{3}}\, d\tau.$$

Therefore we have

$$\int \frac{|dz|}{1+|z|^{1+\sigma_1}} \leqslant \int_{\lambda_3}^{\lambda_4} \frac{d\tau}{|\tau|^{\frac{1}{3}}(1+|\tau|^{\frac{2}{3}+\frac{2}{3}\sigma_1})} \leqslant 2\int_0^\infty \frac{d\tau}{\tau^{\frac{1}{3}}(1+\tau^{\frac{2}{3}+\frac{2}{3}\sigma_1})} = k, \tag{10·3}$$

which proves the statement.

(iii) Similarly, condition (iii) on $\mathscr{P}$ is satisfied if $\mathscr{P}$ consists of a bounded number of straight lines, or if the x-map of $\mathscr{P}$ consists of a bounded number of straight lines plus the x-map of $\mathscr{L}$ (should $a_j(\theta)$ be at infinity).

In these circumstances moreover, condition (ii) on $\mathscr{P}$ is satisfied as well. This is immediately clear from (10·2) unless $\lambda_1 = 0$ when $z'(\tau)$ and $z''(\tau)$ each become infinite at $\tau = 0$. In this event, however, the path passes through the origin and the corresponding z-curve comprises two rays emanating from the origin, in directions differing by an angle $\frac{2}{3}\pi$. We have already seen that the condition is satisfied on each of these rays.

(iv) With $j = 1$, condition (iv) on $\mathscr{P}$ implies that in the x-plane $\operatorname{Re} x$ is monotonic decreasing as we proceed along $\mathscr{P}$ away from the map of $a_1(\theta)$. Similar interpretations hold for $j = 2$ and 3.

(v) If the inequality (9·2) holds only for $m \leqslant M$, where M is a positive integer or zero and independent of u and θ, then (9·18) and (9·19) hold unchanged for $m \leqslant M-1$. They also hold for $m = M$, provided that the error terms $O(u^{-m-1})$ and $O(u^{-m})$ in (9·18) are changed to $O(u^{-M-\frac{2}{3}})$ and $O(u^{-M+\frac{2}{3}})$, respectively, and the error terms $O(u^{-m})$ and $O(u^{-m-1})$ in (9·19) are changed to $O(u^{-M+\frac{1}{3}})$ and $O(u^{-M-\frac{2}{3}})$ respectively. This result follows from the inequalities (13·26) given later. Other forms of the error terms are indicated by relations (13·23) and (13·24).

(vi) Langer (1949) has developed formal series solutions of (9·1) in the extended case in which the asymptotic representation (9·2) contains the additional term $uf_{-1}(\theta, z)$, and has established their asymptotic nature for real bounded z.

11. Bounds for the coefficients

From (9·2) we deduce that

$$|f(u,\theta,z)|<\frac{k}{1+|z|^{\frac{1}{2}+\sigma}}\quad (z\in\mathbf{D}(\theta),u\geqslant u_0), \tag{11·1}$$

and

$$|f_s(\theta,z)|<\frac{k_s}{1+|z|^{\frac{1}{2}+\sigma}}\quad (z\in\mathbf{D}(\theta),s=0,1,\ldots), \tag{11·2}$$

(cf. (5·1) and (5·4)).

In order to establish bounds for the coefficients $A_s(\theta,z)$ and $B_s(\theta,z)$ analogous to (5·5), we introduce an auxiliary region $\mathbf{G}(\theta,\delta)$ defined as $\mathbf{G}(\theta)$ plus the aggregate of all points whose distance from a boundary point z_0 of $\mathbf{G}(\theta)$ is less than $\delta|z_0|^{-\frac{1}{2}}$. We restrict

$$0<\delta<\min(d,\tfrac{1}{2}b^{\frac{3}{2}}). \tag{11·3}$$

Clearly $\mathbf{G}(\theta,\delta)$ is contained in $\mathbf{D}(\theta)$, and if η is an arbitrary number such that $0<\eta<\delta$ then $\mathbf{G}(\theta,\delta)$ itself contains $\mathbf{G}(\theta,\eta)$.

As a preliminary we prove that *if* $z\in\mathbf{G}(\theta,\eta)$ *and* $|z|\geqslant b$, *then* $\mathbf{G}(\theta,\delta)$ *contains the circle*

$$|t-z|\leqslant\kappa(\delta-\eta)|z|^{-\frac{1}{2}}, \tag{11·4}$$

where

$$\kappa=(1+db^{-\frac{3}{2}})^{-\frac{1}{2}}. \tag{11·5}$$

Suppose first that z is a point of $\mathbf{G}(\theta,\eta)$ not contained in $\mathbf{G}(\theta)$. There then exists z_0 on the boundary of $\mathbf{G}(\theta)$ such that $|z-z_0|<\eta|z_0|^{-\frac{1}{2}}$, and for points t satisfying (11·4) we have

$$|t-z_0|=|(t-z)+(z-z_0)|<\kappa(\delta-\eta)|z|^{-\frac{1}{2}}+\eta|z_0|^{-\frac{1}{2}}. \tag{11·6}$$

Now

$$|z_0|=|(z_0-z)+z|<\eta|z_0|^{-\frac{1}{2}}+|z|<|z|(db^{-\frac{3}{2}}+1)=|z|\kappa^{-2}.$$

Thus $\kappa|z|^{-\frac{1}{2}}<|z_0|^{-\frac{1}{2}}$, and substituting this inequality in the right of (11·6) we see that $|t-z_0|<\delta|z_0|^{-\frac{1}{2}}$, and hence that $t\in\mathbf{G}(\theta,\delta)$.

Alternatively, suppose that $z\in\mathbf{G}(\theta)$. Then if t is a point of the circle (11·4) which is not itself contained in $\mathbf{G}(\theta)$, there must be at least one boundary point z_0 of $\mathbf{G}(\theta)$ on the join of t and z. In this event

$$|z_0|=|(z_0-z)+z|\leqslant|t-z|+|z|\leqslant\kappa(\delta-\eta)|z|^{-\frac{1}{2}}+|z|<|z|\{1+(\delta-\eta)|z|^{-\frac{3}{2}}\},$$

since $\kappa<1$. Therefore $|z_0|<|z|\kappa^{-2}$. From this result and (11·4) we obtain

$$|t-z_0|\leqslant|t-z|\leqslant\kappa(\delta-\eta)|z|^{-\frac{1}{2}}<(\delta-\eta)|z_0|^{-\frac{1}{2}}<\delta|z_0|^{-\frac{1}{2}}.$$

This completes the proof of the statement that $\mathbf{G}(\theta,\delta)$ contains the circle (11·4).

A further observation concerning points z of the region $\mathbf{G}(\theta,\delta)$ is that a path joining z and $c(\theta)$ exists lying in $\mathbf{G}(\theta,\delta)$ and having the property

$$\int_{c(\theta)}^{z}\frac{|\mathrm{d}t|}{1+|t|^{1+\sigma_1}}<k. \tag{11·7}$$

For if $z\in\mathbf{G}(\theta)$ we set $\int_{c(\theta)}^{z}=\int_{0}^{z}-\int_{0}^{c(\theta)}$, taking the paths of integration in the right-hand integrals to be those postulated by (9·14). If z is any other point of $\mathbf{G}(\theta,\delta)$, then z_0 exists on the boundary of $\mathbf{G}(\theta)$ such that $|z-z_0|<\delta|z_0|^{-\frac{1}{2}}$; we take the path in (11·7) to be the known path joining $c(\theta)$ and z_0 plus the join of z_0 and z; on the latter

$$\int_{z_0}^{z}\frac{|\mathrm{d}t|}{1+|t|^{1+\sigma_1}}<|z-z_0|<db^{-\frac{1}{2}}=k.$$

LEMMA

$$\left.\begin{array}{lll}|A_s(\theta,z)|<k_s, & |A_s'(\theta,z)|<\dfrac{k_s}{1+|z|^{1+\sigma_1}}, & |A_s''(\theta,z)|<\dfrac{k_s}{1+|z|^{\frac{1}{2}+\sigma_1}},\\ |B_s(\theta,z)|<\dfrac{k_s}{1+|z|^{\frac{1}{2}}}, & |B_s'(\theta,z)|<\dfrac{k_s}{1+|z|^{\frac{3}{2}}}, & |B_s''(\theta,z)|<\dfrac{k_s}{1+|z|^{1+\sigma_1}},\end{array}\right\} \tag{11·8}$$

when $z\in\mathbf{G}(\theta)$.

This corresponds to the lemma of § 5 and we prove it in a similar manner. Let us suppose temporarily that

$$|A_n'(\theta,z)|<\frac{k_n}{1+|z|^{1+\sigma_1}}\quad(z\in\mathbf{G}(\theta,\delta),n=0,1,\ldots,s). \tag{11·9}$$

Clearly

$$A_n(\theta,z)=\int_{c(\theta)}^{z}A_n'(\theta,t)\,\mathrm{d}t+A_n\{\theta,c(\theta)\}. \tag{11·10}$$

If $z\in\mathbf{G}(\theta,\delta)$ we may take the path of integration to be the same as that for the integral (11·7), and substituting (9·16) with $s=n$ and (11·9) in (11·10), we deduce that

$$|A_n(\theta,z)|<k_n\int_{c(\theta)}^{z}\frac{|\mathrm{d}t|}{1+|t|^{1+\sigma_1}}+k_n<k_n\quad(z\in\mathbf{G}(\theta,\delta)) \tag{11·11}$$

(cf. the first of (11·8)).

From Cauchy's integral formula we have

$$A_n''(\theta,z)=\frac{1}{2\pi i}\int\frac{A_n'(\theta,t)}{(t-z)^2}\,\mathrm{d}t, \tag{11·12}$$

in which we may take the path of integration to be the circumference of the circle (11·4). If, now, $z\in\mathbf{G}(\theta,\eta)$ and $|z|\geqslant b$, then as we have seen above $t\in\mathbf{G}(\theta,\delta)$. Moreover,

$$|t|=|z+(t-z)|\geqslant|z|-\kappa(\delta-\eta)|z|^{-\frac{1}{2}}>|z|(1-\kappa\delta|z|^{-\frac{3}{2}})\geqslant|z|(1-\kappa\delta b^{-\frac{3}{2}})>\tfrac{1}{2}|z|, \tag{11·13}$$

since $\kappa<1$ and $\delta\leqslant\frac{1}{2}b^{\frac{3}{2}}$ (cf. (11·3) and (11·5)). Substituting (11·9) in (11·12) and using (11·13), we find that

$$|A_n''(\theta,z)|<\frac{1}{2\pi}\int\frac{k_n}{1+|t|^{1+\sigma_1}}\frac{|\mathrm{d}t|}{|t-z|^2}<\frac{k_n}{1+|\frac{1}{2}z|^{1+\sigma_1}}\frac{|z|^{\frac{1}{2}}}{\kappa(\delta-\eta)}<\frac{k_n}{1+|z|^{\frac{1}{2}+\sigma_1}} \tag{11·14}$$

(cf. the third of (11·8)). This result has been proved for $z\in\mathbf{G}(\theta,\eta)$ on the assumption that $|z|\geqslant b$. Application of the maximum-modulus theorem shows that it must also hold for $|z|<b$.

From (11·14), (11·11) and (11·2) we deduce immediately that

$$|-A_n''(\theta,z)+f_0(\theta,z)\,A_n(\theta,z)+\ldots+f_n(\theta,z)\,A_0(\theta,z)|<\frac{k_n}{1+|z|^{\frac{1}{2}+\sigma_1}}. \tag{11·15}$$

Now consider formula (9·12) for $B_s(\theta,z)$, with s replaced by n. If $|z|\leqslant b$ the path of integra- can be taken to be the join of z and the origin. Then clearly

$$|B_n(\theta,z)|<\tfrac{1}{2}|z|^{-\frac{1}{2}}\int_0^z|t|^{-\frac{1}{2}}\frac{k_n|\mathrm{d}t|}{1+|t|^{\frac{1}{2}+\sigma_1}}\leqslant\tfrac{1}{2}k_n|z|^{-\frac{1}{2}}\int_0^z|t^{-\frac{1}{2}}\mathrm{d}t|=k_n. \tag{11·16}$$

Alternatively, suppose that z is a point of $\mathbf{G}(\theta,\eta)$ outside the circle $|z|=b$. Then (11·7), with $c(\theta)$ replaced by 0, shows that a path exists joining 0 and z and lying in $\mathbf{G}(\theta,\eta)$, along which

$$\int_0^z \frac{|dt|}{1+|t|^{1+\sigma_1}}<k.$$

This path, deformed if necessary so that the part inside $|z|=b$ is a straight line, is now taken as the path of integration in (9·12). Let z_1 be its meet with $|z|=b$. Then we have

$$|B_n(\theta,z)|<\tfrac{1}{2}k_n|z|^{-\frac{1}{2}}\left\{\int_0^{z_1}+\int_{z_1}^{z}\right\}\frac{|t^{-\frac{1}{2}}\,dt|}{1+|t|^{\frac{1}{2}+\sigma_1}}<\frac{k_n b^{\frac{1}{2}}}{|z|^{\frac{1}{2}}}+\frac{k_n}{|z|^{\frac{1}{2}}}\int_{z_1}^{z}\frac{|dt|}{1+|t|^{1+\sigma_1}}<\frac{k_n}{|z|^{\frac{1}{2}}}. \tag{11·17}$$

Combining (11·16) and (11·17) and again adjusting k_n, we find

$$|B_n(\theta,z)|<\frac{k_n}{1+|z|^{\frac{1}{2}}}\quad (z\in\mathbf{G}(\theta,\eta)) \tag{11·18}$$

(cf. the fourth of (11·8)).

Next, from (9·10) we have

$$2zB_n'=f_0A_n+f_1A_{n-1}+\dots+f_nA_0-A_n''-B_n. \tag{11·19}$$

Substituting (11·2), (11·11), (11·14) and (11·18), we find that

$$|2zB_n'(\theta,z)|<\frac{k_n}{1+|z|^{\frac{1}{2}}}\quad (z\in\mathbf{G}(\theta,\eta)), \tag{11·20}$$

and with the aid of the maximum-modulus theorem we deduce that

$$|B_n'(\theta,z)|<\frac{k_n}{1+|z|^{\frac{3}{2}}}\quad (z\in\mathbf{G}(\theta,\eta)) \tag{11·21}$$

(cf. the fifth of (11·8)).

Again, differentiation of (11·19) yields

$$2zB_n''=\frac{d}{dz}(f_0A_n+f_1A_{n-1}+\dots+f_nA_0)-A_n'''-3B_n'. \tag{11·22}$$

Since

$$|f_0A_n+f_1A_{n-1}+\dots+f_nA_0|<\frac{k_n}{1+|z|^{\frac{1}{2}+\sigma_1}}\quad (z\in\mathbf{G}(\theta,\delta)), \tag{11·23}$$

we deduce by application of Cauchy's integral formula and the maximum-modulus theorem that

$$\left|\frac{d}{dz}(f_0A_n+f_1A_{n-1}+\dots+f_nA_0)\right|<\frac{k_n}{1+|z|^{\sigma_1}}\quad (z\in\mathbf{G}(\theta,\eta)) \tag{11·24}$$

(cf. (11·9) and the analysis from (11·12) to (11·14)). Also, we have

$$A_n'''(\theta,z)=\frac{1}{\pi i}\int_{|t-z|=\kappa(\delta-\eta)|z|^{-\frac{1}{2}}}\frac{A_n'(\theta,t)}{(t-z)^3}\,dt. \tag{11·25}$$

Hence

$$|A_n'''(\theta,z)|<\frac{k_n}{1+|z|^{\sigma_1}}\quad (z\in\mathbf{G}(\theta,\eta)). \tag{11·26}$$

Substituting the last two inequalities and (11·21) in (11·22), we obtain

$$|2zB_n''(\theta,z)|<\frac{k_n}{1+|z|^{\sigma_1}}+\frac{k_n}{1+|z|^{\frac{3}{2}}}<\frac{k_n}{1+|z|^{\sigma_1}}\quad (z\in\mathbf{G}(\theta,\eta)), \tag{11·27}$$

since $\sigma_1 \leqslant \frac{3}{2}$ (cf. (9·15)). Hence

$$|\, B_n''(\theta, z) \,| < \frac{k_n}{1+|\, z \,|^{1+\sigma_1}} \quad (z \in \mathbf{G}(\theta, \eta)) \tag{11·28}$$

(cf. the sixth of (11·8)).

Finally, we substitute (11·2), (11·18) and (11·28) in (9·11), with s replaced by $s-1$, to obtain

$$|\, A_{s+1}'(\theta, z) \,| < \frac{k_{s+1}}{1+|\, z \,|^{1+\sigma_1}} \quad (z \in \mathbf{G}(\theta, \eta)). \tag{11·29}$$

The lemma is now established from the relations (11·9) and (11·29) by induction.

12. Equation satisfied by the truncated series

For brevity, we shall write $v \equiv u^{\frac{2}{3}}$. We define

$$L_m(u, \theta, z) = P_1(vz) \sum_{s=0}^{m} \frac{A_s(\theta, z)}{u^s} + \frac{P_1'(vz)}{v^2} \sum_{s=0}^{m-1} \frac{B_s(\theta, z)}{u^s} \tag{12·1}$$

(cf. (6·1)). Then we have

$$\frac{\mathrm{d}}{\mathrm{d}z} L_m(u, \theta, z) = P_1(vz) \left\{ \sum_{s=0}^{m-1} \frac{C_s(\theta, z)}{u^s} + \frac{A_m'(\theta, z)}{u^m} \right\} + v P_1'(vz) \left\{ \sum_{s=0}^{m} \frac{D_s(\theta, z)}{u^s} + \frac{B_{m-1}'(\theta, z)}{u^{m+1}} \right\} \tag{12·2}$$

(cf. (6·2)), and

$$\frac{\mathrm{d}^2}{\mathrm{d}z^2} L_m(u, \theta, z) - \{u^2 z + f(u, \theta, z)\} L_m(u, \theta, z) = R_m(u, \theta, z) \tag{12·3}$$

(cf. (6·4)), where $R_m(u, \theta, z)$ is regular in $\mathbf{D}(\theta)$ and is given by

$$R_m(u, \theta, z) = P_1(vz)\, R_m^{(1)}(u, \theta, z) + v^{-2} P_1'(vz)\, R_m^{(2)}(u, \theta, z), \tag{12·4}$$

$$R_m^{(1)}(u, \theta, z) = u^2 \left(\sum_{s=0}^{m} \frac{E_s}{u^s} + \frac{A_{m-1}'' + 2z B_{m-1}' + B_{m-1}}{u^{m+1}} + \frac{A_m''}{u^{m+2}} \right) - (u^2 z + f) \sum_{s=0}^{m} \frac{A_s}{u^s}, \tag{12·5}$$

$$R_m^{(2)}(u, \theta, z) = u^2 \left(\sum_{s=0}^{m-1} \frac{F_s}{u^s} + \frac{2A_m' + B_{m-2}''}{u^m} + \frac{B_{m-1}''}{u^{m+1}} \right) - (u^2 z + f) \sum_{s=0}^{m-1} \frac{B_s}{u^s}. \tag{12·6}$$

The inequality (9·2) implies that

$$f(u, \theta, z) = \sum_{s=0}^{m-1} \frac{f_s}{u^s} + \frac{\varpi k_m}{(1+|\, z \,|^{\frac{1}{2}+\sigma})\, u^m} \quad (z \in \mathbf{D}(\theta), u \geqslant u_0), \tag{12·7}$$

where $|\, \varpi \,| \leqslant 1$. Substituting (12·7) in (12·5) we find that the coefficients of $u^2, u, \ldots, u^{-m+1}$ all vanish as a consequence of (9·10) and the first of (9·9), and with the aid of (11·2) and (11·8) we may verify that

$$|\, R_m^{(1)}(u, \theta, z) \,| < \frac{1}{1+|\, z \,|^{\frac{1}{2}+\sigma_1}} \frac{k_m}{u^m} \quad (z \in \mathbf{G}(\theta), u \geqslant u_0). \tag{12·8}$$

Similarly

$$|\, R_m^{(2)}(u, \theta, z) \,| < \frac{1}{1+|\, z \,|^{1+\sigma_1}} \frac{k_m}{u^{m-1}} \quad (z \in \mathbf{G}(\theta), u \geqslant u_0). \tag{12·9}$$

Substituting (12·8) and (12·9) in (12·4) and using the inequalities

$$|P_1(z)\,| < \frac{k\,|\exp(-\frac{2}{3} z^{\frac{3}{2}})\,|}{1+|\, z \,|^{\frac{1}{4}}} \qquad |\, P_1'(z) \,| < k(1+|\, z \,|^{\frac{1}{4}})\,|\exp(-\frac{2}{3} z^{\frac{3}{2}})\,| \quad (|\arg z\,| \leqslant \pi), \tag{12·10}$$

(Olver 1954*a*, (4·5)), we deduce (cf. (6·7)) that

$$|R_m(u,\theta,z)| < |\exp(-\tfrac{2}{3}uz^{\frac{3}{2}})| \left\{\frac{1}{1+|vz|^{\frac{1}{4}}}\frac{1}{1+|z|^{\frac{1}{2}+\sigma_1}}+\frac{1}{v^{\frac{1}{2}}}\frac{1+|vz|^{\frac{1}{4}}}{1+|z|^{1+\sigma_1}}\right\}\frac{k_m}{u^m}$$

$$< \frac{|\exp(-\tfrac{2}{3}uz^{\frac{3}{2}})|}{1+|vz|^{\frac{1}{4}}}\frac{1}{1+|z|^{\frac{1}{2}+\sigma_1}}\frac{k_m}{u^m} \quad (z\in \mathbf{G}(\theta),\ u\geqslant u_0). \tag{12·11}$$

13. Proof of theorem B

A proof only for $j=1$ will be recorded. Exactly similar proofs can be formulated for the cases $j=2,3$; alternatively, the results for these cases can be deduced from that for $j=1$ by appropriate changes of the variables. We first suppose that $a_1 \equiv a_1(\theta)$ is a point of $\mathbf{G}(\theta)$ which is not at infinity.

Let us write

$$B_s^*(\theta,z) = (1+|z|^{\frac{1}{2}})\,B_s(\theta,z), \qquad C_s^*(\theta,z) = (1+|z|^{\frac{1}{2}})^{-1}C_s(\theta,z). \tag{13·1}$$

Setting $z=a_1$, we see from (11·8) and (9·7) that

$$|A_s(\theta,a_1)|<k_s, \qquad |B_s^*(\theta,a_1)|<k_s, \qquad |C_s^*(\theta,a_1)|<k_s, \qquad |D_s(\theta,a_1)|<k_s, \tag{13·2}$$

Therefore by the lemma of § 4 functions $A(u,\theta)$, $B^*(u,\theta)$, $C^*(u,\theta)$ and $D(u,\theta)$ exist having the asymptotic expansions

$$\left.\begin{aligned} A(u,\theta) &\sim \sum_{s=0}^{\infty}\frac{A_s(\theta,a_1)}{u^s}, & B^*(u,\theta) &\sim \sum_{s=0}^{\infty}\frac{B_s^*(\theta,a_1)}{u^s},\\ C^*(u,\theta) &\sim \sum_{s=0}^{\infty}\frac{C_s^*(\theta,a_1)}{u^s}, & D(u,\theta) &\sim \sum_{s=0}^{\infty}\frac{D_s(\theta,a_1)}{u^s}, \end{aligned}\right\} \tag{13·3}$$

as $u\to+\infty$, uniformly valid with respect to θ. We define $W_1(u,\theta,z)$ to be the solution of the differential equation (9·1) satisfying the conditions

$$\left.\begin{aligned} W_1(u,\theta,a_1) &= P_1(va_1)\,A(u,\theta)+\frac{P_1'(va_1)}{v^2}\frac{B^*(u,\theta)}{1+|a_1|^{\frac{1}{2}}},\\ \left[\frac{\mathrm{d}}{\mathrm{d}z}W_1(u,\theta,z)\right]_{z=a_1} &= P_1(va_1)\,(1+|a_1|^{\frac{1}{2}})\,C^*(u,\theta)+vP_1'(va_1)\,D(u,\theta), \end{aligned}\right\} \tag{13·4}$$

(cf. (7·3)). From (9·1) and (12·3) we have at once

$$\frac{\mathrm{d}^2}{\mathrm{d}z^2}\{W_1(z)-L_m(z)\}-\{u^2z+f(z)\}\{W_1(z)-L_m(z)\} = -R_m(z), \tag{13·5}$$

the arguments u and θ being suppressed.

Let z_1 be a point of the region common to $\mathbf{H}_1(\theta)$, defined in § 9, and the sector $\mathbf{S}_1+\mathbf{S}_2$, defined in Olver 1954*a*, § 4. Then as a consequence of condition (iv) given at the end of § 9, the path $\mathscr{P}$ joining z_1 and a_1 must also lie in $\mathbf{S}_1+\mathbf{S}_2$; in particular, $a_1\in\mathbf{S}_1+\mathbf{S}_2$. This is clear from figures 1 and 2 of Olver 1954*a*. Let ζ be a typical point of the path. By analogy with (7·5) we define the sequence of functions $h_{m,n}(u,\theta,z_1,\zeta)\equiv h_{m,n}(\zeta)$ by $h_{m,0}(\zeta)=0$ and

$$h_{m,n}(\zeta) = \frac{2\pi\,\mathrm{e}^{\frac{1}{6}\pi i}}{v}\int_{a_1}^{\zeta}\{P_2(v\zeta)\,P_1(vt)-P_1(v\zeta)\,P_2(vt)\}\{f(t)\,h_{m,n-1}(t)-R_m(t)\}\,\mathrm{d}t$$
$$+\alpha_m P_1(v\zeta)+\beta_m P_2(v\zeta) \quad (n\geqslant 1), \tag{13·6}$$

where the path of integration is the part of $\mathscr{P}$ between ζ and a_1, and α_m, β_m are determined by the conditions

$$h_{m,n}(a_1) = W_1(a_1) - L_m(a_1), \qquad h'_{m,n}(a_1) = W'_1(a_1) - L'_m(a_1). \tag{13·7}$$

Differentiation of (13·6) and use of the Wronksian relation

$$P_1(v\zeta)\,P'_2(v\zeta) - P'_1(v\zeta)\,P_2(v\zeta) = e^{-\frac{1}{6}\pi i}/(2\pi) \tag{13·8}$$

yield the differential equation

$$\frac{d^2}{d\zeta^2}h_{m,n}(\zeta) - u^2\zeta h_{m,n}(\zeta) = f(\zeta)\,h_{m,n-1}(\zeta) - R_m(\zeta) \quad (n \geqslant 1). \tag{13·9}$$

From (13·6) and (13·7) it is seen that

$$\left.\begin{aligned} \alpha_m P_1(va_1) + \beta_m P_2(va_1) &= W_1(a_1) - L_m(a_1), \\ \alpha_m P'_1(va_1) + \beta_m P'_2(va_1) &= v^{-1}\{W'_1(a_1) - L'_m(a_1)\}. \end{aligned}\right\} \tag{13·10}$$

These equations show that α_m and β_m are independent of n. Solving with the aid of (13·8), we find

$$\left.\begin{aligned} \alpha_m &= 2\pi\, e^{\frac{1}{6}\pi i}[P'_2(va_1)\{W_1(a_1) - L_m(a_1)\} - v^{-1}P_2(va_1)\{W'_1(a_1) - L'_m(a_1)\}], \\ \beta_m &= 2\pi\, e^{\frac{1}{6}\pi i}[v^{-1}P_1(va_1)\{W'_1(a_1) - L'_m(a_1)\} - P'_1(va_1)\{W_1(a_1) - L_m(a_1)\}]. \end{aligned}\right\} \tag{13·11}$$

From (12·1), (12·2) and (13·4) we obtain

$$W_1(a_1) - L_m(a_1) = P_1(va_1)\left\{A(u,\theta) - \sum_{s=0}^{m}\frac{A_s(\theta,a_1)}{u^s}\right\} + \frac{P'_1(va_1)}{v^2}\left\{\frac{B^*(u,\theta)}{1+|a_1|^{\frac{1}{2}}} - \sum_{s=0}^{m-1}\frac{B_s(\theta,a_1)}{u^s}\right\},$$

$$W'_1(a_1) - L'_m(a_1) = P_1(va_1)\left\{(1+|a_1|^{\frac{1}{2}})\,C^*(u,\theta) - \sum_{s=0}^{m-1}\frac{C_s(\theta,a_1)}{u^s} - \frac{A'_m(\theta,a_1)}{u^m}\right\}$$
$$+ vP'_1(va_1)\left\{D(u,\theta) - \sum_{s=0}^{m}\frac{D_s(\theta,a_1)}{u^s} - \frac{B'_{m-1}(\theta,a_1)}{u^{m+1}}\right\}.$$

Using these equations, the inequalities (12·10), the expansions (13·3) and the lemma of § 11, we may verify without difficulty that

$$|W_1(a_1) - L_m(a_1)| < \frac{k_m}{u^{m+1}}\frac{|\exp(-\frac{2}{3}ua_1^{\frac{3}{2}})|}{1+|va_1|^{\frac{1}{4}}} + \frac{k_m}{u^m}\frac{1}{v^2}\frac{1+|va_1|^{\frac{1}{4}}}{1+|a_1|^{\frac{1}{2}}}|\exp(-\frac{2}{3}ua_1^{\frac{3}{2}})| < \frac{k_m}{u^{m+1}}\frac{|\exp(-\frac{2}{3}ua_1^{\frac{3}{2}})|}{1+|va_1|^{\frac{1}{4}}}, \tag{13·12}$$

and

$$|W'_1(a_1) - L'_m(a_1)| < \frac{k_m}{u^m}\frac{1+|a_1|^{\frac{1}{2}}}{1+|va_1|^{\frac{1}{4}}}|\exp(-\frac{2}{3}ua_1^{\frac{3}{2}})| + \frac{k_m}{u^{m+1}}v(1+|va_1|^{\frac{1}{4}})\,|\exp(-\frac{2}{3}ua_1^{\frac{3}{2}})|$$
$$< \frac{k_m}{u^m}(1+|va_1|^{\frac{1}{4}})\,|\exp(-\frac{2}{3}ua_1^{\frac{3}{2}})|. \tag{13·13}$$

Substituting (13·12) and (13·13) in (13·11), and using (12·10) and the corresponding inequalities

$$|P_2(z)| < k\frac{|\exp\frac{2}{3}z^{\frac{3}{2}}|}{1+|z|^{\frac{1}{4}}}, \qquad |P'_2(z)| < k(1+|z|^{\frac{1}{4}})\,|\exp\frac{2}{3}z^{\frac{3}{2}}|, \tag{13·14}$$

the region of validity of which includes $\mathbf{S}_1 + \mathbf{S}_2$, we find that

$$|\alpha_m| < k_m u^{-m-\frac{2}{3}}, \qquad |\beta_m| < k_m u^{-m-\frac{2}{3}}\,|\exp(-\frac{4}{3}ua_1^{\frac{3}{2}})|. \tag{13·15}$$

With the aid of (13·14), (12·10) and the relations

$$|\exp(-\tfrac{2}{3}u\zeta^{\frac{3}{2}})| \geqslant |\exp(-\tfrac{2}{3}ut^{\frac{3}{2}})| \geqslant |\exp(-\tfrac{2}{3}ua_1^{\frac{3}{2}})|, \tag{13·16}$$

obtained from condition (iv) on $\mathscr{P}$ given in § 9, we deduce from (13·15) that

$$|\alpha_m P_1(v\zeta)+\beta_m P_2(v\zeta)| < \frac{k_m}{u^{m+\frac{2}{3}}} \frac{|\exp(-\frac{2}{3}u\zeta^{\frac{3}{2}})|}{1+|v\zeta|^{\frac{1}{4}}} \tag{13·17}$$

(cf. (7·10)).

Next, we have from (13·6)

$$h_{m,1}(\zeta) = -\frac{2\pi\,\mathrm{e}^{\frac{1}{6}\pi\mathrm{i}}}{v}\int_{a_1}^{\zeta}\{P_2(v\zeta)\,P_1(vt)-P_1(v\zeta)\,P_2(vt)\}\,R_m(t)\,\mathrm{d}t+\alpha_m P_1(v\zeta)+\beta_m P_2(v\zeta). \tag{13·18}$$

Using (12·10), (12·11), (13·14) and (13·16), we readily show that

$$|P_2(v\zeta)\,P_1(vt)\,R_m(t)| < \frac{k_m}{u^m}\frac{|\exp(\frac{2}{3}u\zeta^{\frac{3}{2}}-\frac{4}{3}ut^{\frac{3}{2}})|}{(1+|v\zeta|^{\frac{1}{4}})\,(1+|vt|^{\frac{1}{4}})^2}\frac{1}{1+|t|^{\frac{1}{2}+\sigma_1}} < \frac{k_m}{u^m}\frac{|\exp(-\frac{2}{3}u\zeta^{\frac{3}{2}})|}{(1+|v\zeta|^{\frac{1}{4}})\,(1+|t|^{1+\sigma_1})},$$

and

$$|P_1(v\zeta)\,P_2(vt)\,R_m(t)| < \frac{k_m}{u^m}\frac{|\exp(-\frac{2}{3}u\zeta^{\frac{3}{2}})|}{(1+|v\zeta|^{\frac{1}{4}})\,(1+|vt|^{\frac{1}{4}})^2}\frac{1}{1+|t|^{\frac{1}{2}+\sigma_1}} < \frac{k_m}{u^m}\frac{|\exp(-\frac{2}{3}u\zeta^{\frac{3}{2}})|}{(1+|v\zeta|^{\frac{1}{4}})\,(1+|t|^{1+\sigma_1})}.$$

Substitution of the last two results and (13·17) in (13·18) yields

$$|h_{m,1}(\zeta)| < \frac{k_m}{u^{m+\frac{2}{3}}}\frac{|\exp(-\frac{2}{3}u\zeta^{\frac{3}{2}})|}{1+|v\zeta|^{\frac{1}{4}}}\int_{a_1}^{\zeta}\frac{|\mathrm{d}t|}{1+|t|^{1+\sigma_1}}+\frac{k_m}{u^{m+\frac{2}{3}}}\frac{|\exp(-\frac{2}{3}u\zeta^{\frac{3}{2}})|}{1+|v\zeta|^{\frac{1}{4}}} < \frac{k}{v}\frac{k_m}{u^m}\frac{|\exp(-\frac{2}{3}u\zeta^{\frac{3}{2}})|}{1+|v\zeta|^{\frac{1}{4}}}, \tag{13·19}$$

as a consequence of (9·21).

From (13·6) we also obtain

$$h_{m,2}(\zeta)-h_{m,1}(\zeta) = \frac{2\pi\mathrm{e}^{\frac{1}{6}\pi\mathrm{i}}}{v}\int_{a_1}^{\zeta}\{P_2(v\zeta)\,P_1(vt)-P_1(v\zeta)\,P_2(vt)\}f(t)\,h_{m,1}(t)\,\mathrm{d}t. \tag{13·20}$$

From (11·1) and (13·19) we can show that $|vf(t)\,h_{m,1}(t)|$ satisfies the inequality (12·11) which bounds $|R_m(t)|$. It follows immediately that

$$|h_{m,2}(\zeta)-h_{m,1}(\zeta)| < \frac{k^2}{v^2}\frac{k_m}{u^m}\frac{|\exp(-\frac{2}{3}u\zeta^{\frac{3}{2}})|}{1+|v\zeta|^{\frac{1}{4}}},$$

and continuing the argument by induction we obtain

$$|h_{m,n+1}(\zeta)-h_{m,n}(\zeta)| < \left(\frac{k}{v}\right)^{n+1}\frac{k_m}{u^m}\frac{|\exp(-\frac{2}{3}u\zeta^{\frac{3}{2}})|}{1+|v\zeta|^{\frac{1}{4}}} \quad (n=0,1,\ldots), \tag{13·21}$$

where k_m and k are independent of n.

Differentiation of (13·6) yields

$$h'_{m,1}(\zeta) = -2\pi\,\mathrm{e}^{\frac{1}{6}\pi\mathrm{i}}\int_{a_1}^{\zeta}\{P'_2(v\zeta)\,P_1(vt)-P'_1(v\zeta)\,P_2(vt)\}\,R_m(t)\,\mathrm{d}t+\alpha_m vP'_1(v\zeta)+\beta_m vP'_2(v\zeta),$$

and $h'_{m,n+1}(\zeta)-h'_{m,n}(\zeta)$

$$= 2\pi\,\mathrm{e}^{\frac{1}{6}\pi\mathrm{i}}\int_{a_1}^{\zeta}\{P'_2(v\zeta)\,P_1(vt)-P'_1(v\zeta)\,P_2(vt)\}f(t)\,\{h_{m,n}(t)-h_{m,n-1}(t)\}\,\mathrm{d}t \quad (n\geqslant 1).$$

Substituting by means of the inequalities (11·1), (12·10), (12·11), (13·14), (13·15), (13·21) and using (9·21) and (13·16), we readily show that

$$|h'_{m,n+1}(\zeta)-h'_{m,n}(\zeta)|<\left(\frac{k}{v}\right)^n\frac{k_m}{u^m}(1+|v\zeta|^{\frac{1}{4}})\,|\exp(-\tfrac{2}{3}u\zeta^{\frac{3}{2}})|\quad(n=0,1,\ldots).\quad(13\cdot22)$$

The inequalities (13·21) and (13·22) are the key results for this case and correspond to (7·17) and (7·20), respectively, in case A. The series $\Sigma\{h_{m,n+1}(\zeta)-h_{m,n}(\zeta)\}$ and $\Sigma\{h'_{m,n+1}(\zeta)-h'_{m,n}(\zeta)\}$ clearly converge if $v>k$, and carrying out an analysis exactly similar to that between (7·20) and (7·31) we can prove that the sums of these series are $W_1(\zeta)-L_m(\zeta)$ and $W'_1(\zeta)-L'_m(\zeta)$, respectively. Setting $\zeta=z_1$ and taking $v>2k$ in (13·21) and (13·22), we obtain

$$|W_1(z_1)-L_m(z_1)|<X_m,\qquad |W'_1(z_1)-L'_m(z_1)|<Y_m,\quad(13\cdot23)$$

where
$$X_m=\frac{k_m}{u^{m+\frac{2}{3}}}\frac{|\exp(-\frac{2}{3}uz_1^{\frac{3}{2}})|}{1+|vz_1|^{\frac{1}{4}}},\qquad Y_m=\frac{k_m}{u^m}(1+|vz_1|^{\frac{1}{4}})\,|\exp(-\tfrac{2}{3}uz_1^{\frac{3}{2}})|,\quad(13\cdot24)$$

(cf. (7·32)).

In order to express the error terms in the forms implied by the equations (9·18) and (9·19), we set

$$\left.\begin{aligned}X_m&=X_m^{(1)}.P_1(vz_1)+X_m^{(2)}.v^{-2}(1+|z_1|^{\frac{1}{2}})^{-1}P'_1(vz_1),\\ Y_m&=Y_m^{(1)}.P_1(vz_1)(1+|z_1|^{\frac{1}{2}})+Y_m^{(2)}.vP'_1(vz_1),\end{aligned}\right\}\quad(13\cdot25)$$

where (cf. (13·8))

$$X_m^{(1)}=2\pi\,e^{\frac{1}{6}\pi i}P'_2(vz_1)\,X_m,\qquad X_m^{(2)}=-2\pi\,e^{\frac{1}{6}\pi i}v^2(1+|z_1|^{\frac{1}{2}})\,P_2(vz_1)\,X_m,$$
$$Y_m^{(1)}=2\pi\,e^{\frac{1}{6}\pi i}(1+|z_1|^{\frac{1}{2}})^{-1}P'_2(vz_1)\,Y_m,\qquad Y_m^{(2)}=-2\pi\,e^{\frac{1}{6}\pi i}v^{-1}P_2(vz_1)\,Y_m.$$

With the aid of (13·14) and (13·24) we may show that

$$\left.\begin{aligned}|X_m^{(1)}|&<k_mu^{-m-\frac{2}{3}},\qquad |X_m^{(2)}|<k_mu^{-m+\frac{2}{3}},\\ |Y_m^{(1)}|&<k_mu^{-m+\frac{1}{3}},\qquad |Y_m^{(2)}|<k_mu^{-m-\frac{1}{3}}.\end{aligned}\right\}\quad(13\cdot26)$$

The results (9·18) and (9·19) (with $z=z_1$) now follow on changing m into $m+1$ and using the lemma of §11. In an exactly similar manner, or by an appeal to symmetry, we can establish the same results when z_1 is a point of the region common to $\mathbf{H}_1(\theta)$ and $\mathbf{S}_1+\mathbf{S}_3$. This completes the proof of theorem B for finite $a_1(\theta)$.

When $a_1(\theta)$ is at infinity the proof needs modification. The principal change is that $W_1(u,\theta,z)$ is now defined to be the solution of (9·1) with the conditions

$$\lim\{2\pi^{\frac{1}{2}}(vz)^{\frac{1}{4}}\exp(\tfrac{2}{3}uz^{\frac{3}{2}})\,W_1(u,\theta,z)\}=A(u,\theta)-u^{-1}\exp(\tfrac{1}{2}i\arg a_1)\,B^*(u,\theta),$$
$$\lim\{2\pi^{\frac{1}{2}}(vz)^{-\frac{1}{4}}\exp(\tfrac{2}{3}uz^{\frac{3}{2}})\,(d/dz)\,W_1(u,\theta,z)\}=-v\{A(u,\theta)-u^{-1}\exp(\tfrac{1}{2}i\arg a_1)\,B^*(u,\theta)\},$$

as $z\to a_1(\theta)$ along $\mathscr{L}$ (cf. (8·3)). The sequence $h_{m,n}(\zeta)$ is defined by (13·6) with $\beta_m=0$ and

$$\alpha_m=A(u,\theta)-\sum_{s=0}^{m}\frac{A_s(\theta,a_1)}{u^s}-\frac{\exp(\frac{1}{2}i\arg a_1)}{u}\left\{B^*(u,\theta)-\sum_{s=0}^{m-1}\frac{B_s^*(\theta,a_1)}{u^s}\right\}$$

(cf. (8·5)). Other changes are similar to those given in §8 for case A and do not need to be recorded here.

PART 3. CASE D

14. Statement of conditions and theorem D

The standard form of differential equation for this case is taken to be

$$\frac{d^2w}{dz^2} = \frac{1}{z}\frac{dw}{dz} + \left\{u^2 + \frac{\mu^2-1}{z^2} + f(u,\theta,\mu,z)\right\} w \tag{14·1}$$

(cf. (1·10)), in which u is a large positive parameter, θ is a set of real or complex parameters ranging over a set of values $\mathbf{\Theta}$, and μ is a real or complex parameter which ranges over a bounded region $\mathbf{M}$ lying in the half-plane $\operatorname{Re}\mu \geq 0$. We suppose that the variable z ranges over a simply-connected complex domain $\mathbf{D}(\theta,\mu)$, bounded or otherwise. Further, $\mathbf{D}(\theta,\mu)$ must contain the circle $|z| \leq b$, where b is positive and *fixed*, i.e. independent of u, θ and μ.

We assume that $f(u,\theta,\mu,z)$ is a regular *even* function of z in $\mathbf{D}(\theta,\mu)$ and that

$$\left| f(u,\theta,\mu,z) - \sum_{s=0}^{m-1} \frac{f_s(\theta,\mu,z)}{u^s} \right| < \frac{1}{1+|z|^{1+\sigma}} \frac{k_m}{u^m} \quad (m = 0,1,2,\ldots), \tag{14·2}$$

valid when $z \in \mathbf{D}(\theta,\mu)$, $\mu \in \mathbf{M}$, $\theta \in \mathbf{\Theta}$ and $u \geq u_0$, where u_0 is fixed. Here σ is fixed and positive, and each $f_s(\theta,\mu,z)$ is a regular even function of z in $\mathbf{D}(\theta,\mu)$. As in parts 1 and 2 the symbols k and k_m are used generically to denote numbers independent of u, θ, z and also, in this case, μ.

Following the analysis given in Olver 1956, § 7, we readily verify that formal solutions of (14·1) are given by

$$w = z\mathscr{Z}_\mu(uz) \sum_{s=0}^{\infty} \frac{A_s(\theta,\mu,z)}{u^s} + \frac{z}{u}\mathscr{Z}_{\mu+1}(uz) \sum_{s=0}^{\infty} \frac{B_s(\theta,\mu,z)}{u^s}, \tag{14·3}$$

$$\frac{dw}{dz} = \mathscr{Z}_\mu(uz) \sum_{s=0}^{\infty} \frac{C_s(\theta,\mu,z)}{u^s} + uz\mathscr{Z}_{\mu+1}(uz) \sum_{s=0}^{\infty} \frac{D_s(\theta,\mu,z)}{u^s}, \tag{14·4}$$

where $\mathscr{Z}_\mu$ denotes the modified cylinder function of order μ. Here $A_0 =$ constant, $A_1 =$ constant, and the higher coefficients are given by

$$2B_s = -A_s' + \int_0^z \left(f_0 A_s + f_1 A_{s-1} + \ldots + f_s A_0 - \frac{2\mu+1}{z} A_s'\right) dz, \tag{14·5}$$

$$2A_{s+2} = \frac{2\mu+1}{z} B_s - B_s' + \int (f_0 B_s + f_1 B_{s-1} + \ldots + f_s B_0)\, dz + \text{constant}. \tag{14·6}$$

In these equations the arguments θ, μ, z of A_s, B_s and f_s have been suppressed and primes used to denote differentiations with respect to z. These two equations determine a set of functions which are regular functions of z in $\mathbf{D}(\theta,\mu)$, A_s being even and B_s being odd. If the f_s of odd suffix vanish we can arrange that the same is true of both A_s and B_s. The coefficients C_s and D_s appearing in (14·4) are given by

$$C_s = (\mu+1) A_s + zA_s' + zB_s, \qquad D_s = A_s - \mu z^{-1} B_{s-2} + B_{s-2}', \tag{14·7}$$

and are regular even functions of z in $\mathbf{D}(\theta,\mu)$.

We prescribe $\mathbf{G}(\theta,\mu)$ to be a closed subdomain of $\mathbf{D}(\theta,\mu)$ which has the properties:

(i) $\mathbf{G}(\theta,\mu)$ contains the circle $|z| \leq b$.

(ii) The distance between each point of $\mathbf{G}(\theta,\mu)$ and each boundary point of $\mathbf{D}(\theta,\mu)$ has a positive lower bound which is independent of θ and μ.

(iii) $\int_0^z \frac{|\mathrm{d}t|}{1+|t|^{1+\sigma_1}} < k \quad (z \in \mathbf{G}(\theta,\mu)),$ (14·8)

for some path lying wholly in $\mathbf{G}(\theta,\mu)$, where

$$\sigma_1 = \min(\sigma, 1). \tag{14·9}$$

The arbitrary constants occurring in (14·6) are subjected to the condition

$$|A_s\{\theta,\mu,c(\theta,\mu)\}| < k_s, \tag{14·10}$$

where $c(\theta,\mu)$ is a prescribed point of $\mathbf{G}(\theta,\mu)$. The asymptotic nature of the formal series (14·3) and (14·4) can now be expressed in the following theorem:

THEOREM D. *The differential equation* (14·1) *possesses solutions* $W_j(u,\theta,\mu,z)$ $(j = 1,2)$ *with the properties*

$$W_1(u,\theta,\mu,z) = zI_\mu(uz)\left\{\sum_{s=0}^{m}\frac{A_s}{u^s}+O\left(\frac{1}{u^{m+1}}\right)\right\}+\frac{z}{u}I_{\mu+1}(uz)\left\{\sum_{s=0}^{m-1}\frac{B_s}{u^s}+\frac{z}{1+|z|}O\left(\frac{1}{u^m}\right)\right\}, \tag{14·11}$$

$$\frac{\mathrm{d}}{\mathrm{d}z}W_1(u,\theta,\mu,z) = I_\mu(uz)\left\{\sum_{s=0}^{m-1}\frac{C_s}{u^s}+(1+|z|)\,O\left(\frac{1}{u^m}\right)\right\}+uzI_{\mu+1}(uz)\left\{\sum_{s=0}^{m}\frac{D_s}{u^s}+O\left(\frac{1}{u^{m+1}}\right)\right\}, \tag{14·12}$$

$$W_2(u,\theta,\mu,z) = zK_\mu(uz)\left\{\sum_{s=0}^{m}\frac{A_s}{u^s}+O\left(\frac{1}{u^{m+1}}\right)\right\}-\frac{z}{u}K_{\mu+1}(uz)\left\{\sum_{s=0}^{m-1}\frac{B_s}{u^s}+\frac{z}{1+|z|}O\left(\frac{1}{u^m}\right)\right\}, \tag{14·13}$$

$$\frac{\mathrm{d}}{\mathrm{d}z}W_2(u,\theta,\mu,z) = K_\mu(uz)\left\{\sum_{s=0}^{m-1}\frac{C_s}{u^s}+(1+|z|)\,O\left(\frac{1}{u^m}\right)\right\}-uzK_{\mu+1}(uz)\left\{\sum_{s=0}^{m}\frac{D_s}{u^s}+O\left(\frac{1}{u^{m+1}}\right)\right\}, \tag{14·14}$$

as $u\to+\infty$, *valid when* $\theta\in\mathbf{\Theta}$, $\mu\in\mathbf{M}$ *and* $z\in\mathbf{H}_j(\theta,\mu)$. *Here* m *denotes an arbitrary positive integer or zero; the solutions are independent of* m. *Each of the* O*'s is uniform with respect to* θ,μ *and* z, *except that in the case* $j = 2$ *the part of* $\mathbf{M}$ *common to the annulus* $0<|\mu|<\delta$ *is excluded,* δ *being an arbitrary positive number.*

The regions of validity $\mathbf{H}_j(\theta,\mu)$ and the branches of the modified Bessel functions are defined as follows.

$\mathbf{H}_1(\theta,\mu)$ comprises those points z of $\mathbf{G}(\theta,\mu)$ which can be joined to the origin by a path $\mathscr{P}$ having the following properties, t being a typical point of $\mathscr{P}$.

(i) $\mathscr{P}$ lies in $\mathbf{G}(\theta,\mu)$.

(ii) $\mathscr{P}$ comprises a finite number of Jordan arcs, each with parametric equation of the form $t = t(\tau)$, where τ is the real parameter of the arc; $t''(\tau)$ is continuous and $t'(\tau)$ does not vanish.

(iii) $\int_{\mathscr{P}} \frac{|\mathrm{d}t|}{1+|t|^{1+\sigma_1}} < k.$ (14·15)

(iv) $\operatorname{Re} t$ is monotonic as t traverses $\mathscr{P}$ from 0 to z.

(v) If t, ζ are any two points of $\mathscr{P}$ arranged in the order 0, t, ζ, z along $\mathscr{P}$, then

$$|t| < k|\zeta|. \tag{14·16}$$

Equations (14·11) and (14·12) then hold for $z\in\mathbf{H}_1(\theta,\mu)$ uniformly with respect to *all* values of arg z.

In the definition of $\mathbf{H}_2(\theta,\mu)$ we suppose $a(\theta,\mu)$ to be any prescribed point of $\mathbf{G}(\theta,\mu)$ which is outside or on the circle $|z| = b$, or the point at infinity on a straight line $\mathscr{L}$ lying in $\mathbf{G}(\theta,\mu)$. In either event we impose the restriction

$$|\arg a(\theta,\mu)| < \tfrac{1}{2}\pi. \tag{14·17}$$

Then $\mathbf{H}_2(\theta,\mu)$ comprises those points z of $\mathbf{G}(\theta,\mu)$ which can be joined to $a(\theta,\mu)$ by a path $\mathscr{P}$ having the properties (i), (ii), (iii) listed above for the case $j=1$; in addition $\mathscr{P}$ must coincide with $\mathscr{L}$ for all sufficiently large $|t|$, should $a(\theta,\mu)$ be at infinity. In place of (iv) and (v) we have:

(iv) Re t is monotonic decreasing as t traverses $\mathscr{P}$ from $a(\theta,\mu)$ to z.

(v) If t, ζ are any two points of $\mathscr{P}$ arranged in the order a, t, ζ, z along $\mathscr{P}$, then

$$|\zeta|<k|t|. \tag{14·18}$$

If $\operatorname{Re} z\geqslant 0$ the branches of the modified Bessel functions $K_\mu(uz)$ and $K_{\mu+1}(uz)$ in (14·13) and (14·14) are the principal ones. If $\operatorname{Re} z<0$ the branches are determined by $\frac{1}{2}\pi<\arg z<\frac{3}{2}\pi$ or $-\frac{1}{2}\pi>\arg z>-\frac{3}{2}\pi$ according as $\mathscr{P}$ intersects the positive or negative imaginary axis.

15. Remarks on theorem D

Remarks (i) and (ii) made in § 3 in connexion with case A apply equally well here to the conditions (14·8) and (14·15) provided, of course, that we stipulate uniformity with respect to μ. Corresponding to § 3 (iii) we have:

(iii) If the inequality (14·2) holds only for $m\leqslant M$, where M is a fixed positive integer or zero, then equations (14·11) to (14·14) hold for $m\leqslant M-1$. They also hold for $m=M$ provided that the error terms $O(1/u^m)$ appearing in (14·11) and (14·13) are changed to $O(1/u^{M-1})$.

We also make the following observations:

(iv) For $j=2$, condition (v) implies that $\mathscr{P}$ cannot pass through the origin; this is seen by taking $t=0$ in (14·18). From this result and condition (iv) we see that $\mathscr{P}$ cannot intersect both the positive and the negative imaginary axes. Thus there is no ambiguity in the choice of branch of $K_\mu(uz)$ and $K_{\mu+1}(uz)$. If, however, for a given point $z(\operatorname{Re} z<0)$ two alternative paths $\mathscr{P}$ exist which satisfy conditions (i) to (v) and intersect the imaginary axis on opposite sides of the origin, then equations (14·13) and (14·14) hold for both $\frac{1}{2}\pi<\arg z<\frac{3}{2}\pi$ and $-\frac{1}{2}\pi>\arg z>-\frac{3}{2}\pi$.

(v) An investigation has not been made of the necessity of the condition that the annulus $0<|\mu|<\delta$ be excluded from the μ-region of *uniform* validity when $j=2$. This would appear to require the use of a sharper inequality for the Bessel function $K_\mu(z)$ than (16·2) given below. We may note however that since $\mu=0$ is not excluded we can show, by proper choice of δ, that equations (14·13) and (14·14) hold for any *fixed* value of μ. Alternative forms of error terms which hold uniformly throughout $\mathbf{M}$, including the annulus, are given by equations (19·14) and (19·15).

16. Bounds for the Bessel functions

We require bounds for the modified Bessel functions $I_\mu(z)$ and $K_\mu(z)$ which are uniformly valid with respect to bounded μ in the half-plane $\operatorname{Re}\mu\geqslant 0$. These are afforded by the following results which are extensions of the inequalities for fixed μ given in Olver 1956, § 9:

$$|I_\mu(z)|<kV_\mu(z) \quad (\mu\in\mathbf{M}, |\arg z|\leqslant\tfrac{1}{2}\pi), \tag{16·1}$$

$$|K_\mu(z)|<kX_\mu(z) \quad (\mu\in\mathbf{M}, |\arg z|\leqslant\tfrac{3}{2}\pi), \tag{16·2}$$

Here
$$V_\mu(z) = \frac{|z^\alpha e^z|}{1+|z|^{\alpha+\frac{1}{2}}}, \qquad X_\mu(z) = l_\mu(z)\frac{1+|z|^\alpha |e^{-z}|}{1+|z|^{\frac{1}{2}}\,|z|^\alpha}, \tag{16·3}$$

$$l_\mu(z) = \ln\frac{1+2|z|}{|z|} \quad (|\mu|<\delta), \qquad l_\mu(z) = 1 \quad (|\mu|\geqslant\delta), \tag{16·4}$$

where
$$\alpha = \operatorname{Re}\mu\,(\geqslant 0), \tag{16·5}$$

and δ is an arbitrary positive number in the range $0<\delta<\frac{1}{2}$. In (16·1) and (16·2), k denotes a generic number independent of μ and z; the k of (16·2) depends on δ.

These results may be proved as follows.

Large z. An examination of the analysis given in Watson 1944, §7·2 reveals that the asymptotic expansions of Hankel for the Hankel functions $H_\nu^{(1)}(z)$ and $H_\nu^{(2)}(z)$ when $|z|$ is large, are uniformly valid with respect to ν and $\arg z$ when $\nu \in \mathbf{M}$ and

$$-\pi+\epsilon \leqslant \arg z \leqslant 2\pi-\epsilon \quad (\text{in the case of } H_\nu^{(1)}(z)),$$
$$-2\pi+\epsilon \leqslant \arg z \leqslant \pi-\epsilon \quad (\text{in the case of } H_\nu^{(2)}(z)),$$

where ϵ is an arbitrary fixed positive number.

Hence we derive (Watson 1944, §7·23)

$$I_\mu(z) = \frac{e^z}{(2\pi z)^{\frac{1}{2}}}O(1) + \frac{e^{-z\pm(\mu+\frac{1}{2})\pi i}}{(2\pi z)^{\frac{1}{2}}}O(1) \quad \begin{Bmatrix} -\frac{1}{2}\pi+\epsilon \leqslant \arg z \leqslant \frac{3}{2}\pi-\epsilon \text{ (upper sign)} \\ -\frac{3}{2}\pi+\epsilon \leqslant \arg z \leqslant \frac{1}{2}\pi-\epsilon \text{ (lower sign)} \end{Bmatrix}, \tag{16·6}$$

and
$$K_\mu(z) = \left(\frac{\pi}{2z}\right)^{\frac{1}{2}} e^{-z}\,O(1) \quad (|\arg z| \leqslant \tfrac{3}{2}\pi-\epsilon), \tag{16·7}$$

as $|z|\to\infty$, where the O's are uniform with respect to μ and $\arg z$.

Equation (16·6) enables us to prove (16·1) when $|z|>\kappa$, where $\kappa(>0)$ is assignable independently of μ and $\arg z$. Similarly, (16·7) establishes (16·2) when $|z|>\kappa$ and $|\arg z|\leqslant\frac{3}{2}\pi-\epsilon$; the extension to $|\arg z|\leqslant\frac{3}{2}\pi$ is achieved with the aid of the formula

$$K_\mu(z) = e^{\pm\mu\pi i}K_\mu(z\,e^{\pm\pi i}) \pm \pi i I_\mu(z\,e^{\pm\pi i}). \tag{16·8}$$

Bounded z. When $|z|\leqslant\kappa$ we use the series definitions

$$I_\mu(z) = \frac{(\frac{1}{2}z)^\mu}{\Gamma(1+\mu)}\left[1+\frac{(\frac{1}{2}z)^2}{1!(1+\mu)}+\frac{(\frac{1}{2}z)^4}{2!(1+\mu)(2+\mu)}+\dots\right], \tag{16·9}$$

$$I_{-\mu}(z) = \frac{(\frac{1}{2}z)^{-\mu}}{\Gamma(1-\mu)}\left[1+\frac{(\frac{1}{2}z)^2}{1!(1-\mu)}+\frac{(\frac{1}{2}z)^4}{2!(1-\mu)(2-\mu)}+\dots\right], \tag{16·10}$$

and
$$K_\mu(z) = \tfrac{1}{2}\pi \operatorname{cosec}\mu\pi\{I_{-\mu}(z)-I_\mu(z)\}, \tag{16·11}$$

the limit being taken in (16·11) when μ is an integer.

Since $\operatorname{Re}\mu\geqslant 0$ it follows that $|s+\mu|\geqslant s$ $(s=1,2,\dots)$, and from (16·9) we obtain

$$|I_\mu(z)| \leqslant \left|\frac{(\frac{1}{2}z)^\mu}{\Gamma(1+\mu)}\right|\left[1+\frac{(\frac{1}{2}\kappa)^2}{(1!)^2}+\frac{(\frac{1}{2}\kappa)^4}{(2!)^2}+\dots\right] < k|z|^\alpha, \tag{16·12}$$

when $|\arg z|$ is bounded. This enables us to complete the proof of (16·1).

The proof of (16·2) is more complicated and we divide the region **M** into three parts.

(i) Suppose first that μ lies on or outside the circles

$$|\mu - n| = \delta \quad (n = 0, 1, 2, \ldots). \tag{16·13}$$

Only a finite number of these circles lie in **M**, and since $\delta < \frac{1}{2}$ they do not overlap.

Clearly $|s-\mu| \geqslant \delta$ for all positive integer values of s. Hence, from (16·10) we obtain

$$|I_{-\mu}(z)| < \left|\frac{(\frac{1}{2}z)^{-\mu}}{\Gamma(1-\mu)}\right| \left[1 + \frac{(\frac{1}{2}\kappa)^2}{1!\,\delta} + \frac{(\frac{1}{2}\kappa)^4}{2!\,\delta^2} + \ldots\right] < k\,|z|^{-\alpha},$$

since $|\arg z|$ is bounded. Using this result, (16·12) and the fact that $|\operatorname{cosec}\mu\pi|$ is bounded here, we see from (16·11) that

$$|K_\mu(z)| < k\,|z|^{-\alpha}. \tag{16·14}$$

(ii) Next suppose that, for some positive integer n,

$$|\mu - n| \leqslant \delta. \tag{16·15}$$

For fixed z, other than zero, the function $z^\mu K_\mu(z)$ is a regular function of μ inside the circle (16·15), whilst from (16·14) we see that its modulus is bounded (independently of z) on the circumference. Hence by the maximum-modulus theorem

$$|K_\mu(z)| < k\,|z^{-\mu}| < k\,|z|^{-\alpha} \tag{16·16}$$

throughout the circle.

(iii) Finally, suppose that μ lies in the semi-circle

$$|\mu| \leqslant \delta, \quad \operatorname{Re}\mu \geqslant 0. \tag{16·17}$$

We write

$$K_\mu(z) = \tfrac{1}{2}\pi\{z^{-\mu}K_\mu^{(1)}(z) + z^\mu K_\mu^{(2)}(z) + K_\mu^{(3)}(z)\}, \tag{16·18}$$

where

$$K_\mu^{(1)}(z) = \frac{z^\mu I_{-\mu}(z) - I_0(z)}{\sin\mu\pi}, \qquad K_\mu^{(2)}(z) = \frac{I_0(z) - z^{-\mu}I_\mu(z)}{\sin\mu\pi}, \qquad K_\mu^{(3)}(z) = \frac{z^{-\mu} - z^\mu}{\sin\mu\pi} I_0(z).$$

Clearly $K_\mu^{(1)}(z)$ and $K_\mu^{(2)}(z)$ are regular functions of μ inside the circle $|\mu| \leqslant \delta$, whilst their moduli are bounded on the circumference. Applying the maximum-modulus theorem again, we deduce that

$$|K_\mu^{(1)}(z)|,\ |K_\mu^{(2)}(z)| < k \tag{16·19}$$

throughout the circle.

For the remaining function we write

$$K_\mu^{(3)}(z) = \ln z\, I_0(z)\, \frac{\mu}{\sin\mu\pi}\, \frac{e^{-x} - e^x}{x}, \tag{16·20}$$

where $x = \mu \ln z$. Clearly

$$\left|\frac{e^{-x} - e^x}{x}\right| < k\, e^{|\operatorname{Re} x|} < k\,|z|^{-\alpha}, \tag{16·21}$$

when μ lies in the semi-circle (16·17). Hence we have

$$|K_\mu^{(3)}(z)| < k\,|\ln z|\,|z|^{-\alpha} < k\,l(z)\,|z|^{-\alpha}, \tag{16·22}$$

where

$$l(z) = \ln\frac{1 + 2|z|}{|z|} \tag{16·23}$$

(cf. (16·4)), and is a decreasing function of $|z|$. From this result and (16·18), (16·19), we obtain

$$|K_\mu(z)| < k\,|z|^{-\alpha} + k\,|z|^{\alpha} + k\,l(z)\,|z|^{-\alpha} < k\,l(z)\,|z|^{-\alpha}. \tag{16·24}$$

All regions of **M** have now been covered. Combining (16·14), (16·16) and (16·24), we see that

$$|K_\mu(z)| < k\, l_\mu(z)\, |z|^{-\alpha} \quad (\mu \in \mathbf{M}, |z| \leqslant \kappa, |\arg z| \leqslant \tfrac{3}{2}\pi), \tag{16·25}$$

where $l_\mu(z)$ is defined by (16·4). This completes the proof of (16·2).

Inequalities for products. Using (16·1) and (16·2) we can prove that the inequalities (9·12) given in Olver 1956 are uniformly valid with respect to $\mu \in \mathbf{M}$, provided that $l_\mu(z)$ is defined as in the present paper.

17. Preliminaries in the proof of theorem D

The proof of theorem D follows the pattern of the proofs of theorems A and B given in parts 1 and 2, and we shall record only the principal steps.

(i) *Bounds for the coefficients*

Lemma

$$\left.\begin{aligned} &|A_s(\theta,\mu,z)| < k_s, \quad |A_s'(\theta,\mu,z)| < \frac{k_s}{1+|z|^{1+\sigma_1}}, \quad |A_s''(\theta,\mu,z)| < \frac{k_s}{1+|z|^{1+\sigma_1}}, \\ &|B_s(\theta,\mu,z)| < k_s, \quad |B_s'(\theta,\mu,z)| < \frac{k_s}{1+|z|^{1+\sigma_1}}, \quad |B_s''(\theta,\mu,z)| < \frac{k_s}{1+|z|^{1+\sigma_1}}, \end{aligned}\right\} \tag{17·1}$$

when $z \in \mathbf{G}(\theta,\mu)$.

This is the extended form of lemma 2 given in Olver 1956, and may be proved by the same method (cf. also §§ 5 and 11 of the present paper).

(ii) *Equation satisfied by the truncated series*

Let $L_m(u,\theta,\mu,z) \equiv L_m(z)$ be defined by the equation

$$L_m(z) = z\mathscr{Z}_\mu(uz) \sum_{s=0}^{m} \frac{A_s}{u^s} + \frac{z}{u} \mathscr{Z}_{\mu+1}(uz) \sum_{s=0}^{m-1} \frac{B_s}{u^s}, \tag{17·2}$$

in which the arguments θ, μ, z of A_s and B_s have been suppressed. Then

$$L_m'(z) = \mathscr{Z}_\mu(uz)\left\{\sum_{s=0}^{m-1} \frac{C_s}{u^s} + \frac{(\mu+1)\,A_m + zA_m'}{u^m}\right\} + uz\mathscr{Z}_{\mu+1}(uz)\left\{\sum_{s=0}^{m} \frac{D_s}{u^s} + \frac{B_{m-1}' - \mu z^{-1} B_{m-1}}{u^{m+1}}\right\}, \tag{17·3}$$

and by analogy with §§ 6 and 12, $L_m(z)$ satisfies a differential equation of the form

$$L_m''(z) - \frac{1}{z} L_m'(z) - \left\{u^2 + \frac{\mu^2-1}{z^2} + f(z)\right\} L_m(z) = R_m(z), \tag{17·4}$$

where $f(z) \equiv f(u,\theta,\mu,z)$. The function $R_m(z) \equiv R_m(u,\theta,\mu,z)$ can be expressed in the form

$$R_m(z) = z\mathscr{Z}_\mu(uz)\, R_m^{(1)}(z) + z\mathscr{Z}_{\mu+1}(uz)\, R_m^{(2)}(z) \tag{17·5}$$

(cf. (12·4)), where $R_m^{(1)}(z) \equiv R_m^{(1)}(u,\theta,\mu,z)$ and $R_m^{(2)}(z) \equiv R_m^{(2)}(u,\theta,\mu,z)$ are regular functions of z in $\mathbf{D}(\theta,\mu)$, satisfying

$$|R_m^{(1)}(z)|, |R_m^{(2)}(z)| < \frac{1}{1+|z|^{1+\sigma_1}} \frac{k_m}{u^m} \quad (z \in \mathbf{G}(\theta,\mu), u \geqslant u_0). \tag{17·6}$$

18. Proof for the solution W_1

We define $W_1(u,\theta,\mu,z) \equiv W_1(z)$ to be the solution of the differential equation (14·1) satisfying the conditions

$$\lim \{z^{-\mu-1} W_1(z)\} = \frac{(\frac{1}{2}u)^\mu}{\Gamma(1+\mu)} A(u,\theta,\mu), \tag{18·1}$$

$$\lim \{z^{-\mu} W_1'(z)\} = (1+\mu) \frac{(\frac{1}{2}u)^\mu}{\Gamma(1+\mu)} A(u,\theta,\mu), \tag{18·2}$$

as $z \to 0$ along *any* straight line, where $A(u,\theta,\mu)$ is a function having the asymptotic expansion

$$A(u,\theta,\mu) \sim \sum_{s=0}^{\infty} \frac{A_s(\theta,\mu,0)}{u^s} \tag{18·3}$$

as $u \to +\infty$, uniformly valid with respect to θ and μ (cf. § 4).

For a given $A(u,\theta,\mu)$ the solution $W_1(z)$ exists; moreover, no other solution of (14·1) can exist with the properties (18·1) and (18·2) as $z \to 0$ along a *given* straight line. These statements can be proved by considering the forms of a fundamental pair of solutions $w_1(z)$, $w_2(z)$ of (14·1) in the neighbourhood of the origin, given by

$$\left.\begin{aligned} w_1(z) &= z^{1+\mu}\phi_\mu(z), \\ w_2(z) &= z^{1-\mu}\psi_\mu(z) \quad (2\mu \neq 0,1,2,\ldots), \\ w_2(z) &= z^{1-\mu}\psi_\mu(z) + z^{1+\mu}\ln z\,\chi_\mu(z) \quad (2\mu = 0,1,2,\ldots), \end{aligned}\right\}$$

where $\phi_\mu(z)$, $\psi_\mu(z)$ and $\chi_\mu(z)$ are regular functions of z with the properties

$$\phi_\mu(0) \neq 0, \qquad \psi_\mu(0) \neq 0 \quad (\mu \neq 0), \qquad \chi_0(0) \neq 0,$$

(cf. Copson 1944, §§ 10·12 and 10·15). Since $\operatorname{Re}\mu \geqslant 0$ it is clear that $z^{-\mu-1}w_2(z)$ cannot tend to a finite limit unless $\operatorname{Re}\mu = 0$ and $\mu \neq 0$. In the latter event we can easily show that $w_2(z)$ cannot satisfy both (18·1) and (18·2) as $z \to 0$ along a given straight line.

From (14·1) and (17·4) we have

$$\frac{d^2}{dz^2}\{W_1(z) - L_m(z)\} - \frac{1}{z}\frac{d}{dz}\{W_1(z) - L_m(z)\} - \left\{u^2 + \frac{\mu^2-1}{z^2} + f(z)\right\}\{W_1(z) - L_m(z)\} = -R_m(z), \tag{18·4}$$

with $\mathscr{X} = I$ in the definitions of $L_m(z)$ and $R_m(z)$.

Let z_1 be a point of the region $\mathbf{H}_1(\theta,\mu)$, defined in § 14, and suppose that

$$(p-\tfrac{1}{2})\pi \leqslant \arg z_1 \leqslant (p+\tfrac{1}{2})\pi, \tag{18·5}$$

where p is an arbitrary integer. Then condition (iv) of § 14 shows that the postulated path $\mathscr{P}$ joining z_1 and the origin must also lie in the sector (18·5). We define the sequence $h_{m,n}(u,\theta,\mu,z_1,\zeta) \equiv h_{m,n}(\zeta)$ by $h_{m,0}(\zeta) = 0$ and

$$h_{m,n}(\zeta) = \zeta e^{-p\mu\pi i}\int_0^\zeta \{I_\mu(u\zeta)K_\mu(ut e^{-p\pi i}) - K_\mu(u\zeta e^{-p\pi i})I_\mu(ut)\}\{f(t)h_{m,n-1}(t) - R_m(t)\}dt + \alpha_m \zeta I_\mu(u\zeta) \quad (n \geqslant 1), \tag{18·6}$$

where the path of integration is the part of $\mathscr{P}$ between ζ and 0, and

$$\alpha_m = A(u,\theta,\mu) - \sum_{s=0}^{m} \frac{A_s(\theta,\mu,0)}{u^s}. \tag{18·7}$$

The branches of the Bessel functions occurring in (18·6) are determined by continuity and the condition $(p-\frac{1}{2})\pi \leqslant \arg \zeta \leqslant (p+\frac{1}{2})\pi$.

Extending the analysis given in Olver 1956, § 12, we may show that

$$|h_{m,n+1}(\zeta)-h_{m,n}(\zeta)| < \left(\frac{k}{u}\right)^n \frac{k_m}{u^{m+1}} |\zeta e^{p\mu\pi i}| V_\mu(u\zeta') \quad (n=0,1,\ldots), \tag{18·8}$$

and

$$|h'_{m,n+1}(\zeta)-h'_{m,n}(\zeta)| < \left(\frac{k}{u}\right)^n \frac{k_m}{u^{m+1}} |e^{p\mu\pi i}| (1+|u\zeta|) V_\mu(u\zeta') \quad (n=0,1,\ldots), \tag{18·9}$$

where V_μ is defined by (16·3), $\zeta' \equiv \zeta e^{-p\pi i}$, and here and elsewhere in this section the generic numbers k and k_m are independent of p as well as of u, θ, μ and z. From these inequalities and the equation

$$h''_{m,n}(\zeta)-\frac{1}{\zeta}h'_{m,n}(\zeta)-\left(u^2+\frac{\mu^2-1}{\zeta^2}\right)h_{m,n}(\zeta) = f(\zeta)\,h_{m,n-1}(\zeta)-R_m(\zeta) \tag{18·10}$$

obtained by differentiation of (18·6), we deduce that

$$|h''_{m,n+1}(\zeta)-h''_{m,n}(\zeta)| < \left(\frac{k}{u}\right)^n \frac{k_m}{u^{m+1}} |e^{p\mu\pi i}| \frac{1+|u\zeta|^2}{|\zeta|} V_\mu(u\zeta'). \tag{18·11}$$

The parametric equation of $\mathscr{P}$ is of the form

$$\zeta = \zeta(\tau) \quad (\tau_0 \leqslant \tau \leqslant \tau_1), \tag{18·12}$$

where τ_0, τ_1 correspond to 0, z_1, respectively. Term-by-term integration of the series

$$S(\tau) = \sum_{n=0}^{\infty} [h''_{m,n+1}\{\zeta(\tau)\}-h''_{m,n}\{\zeta(\tau)\}] \tag{18·13}$$

multiplied by $\zeta'(\tau)$ is valid over the interval (τ, τ_1) if $\tau > \tau_0$ and $u > k$. Hence we obtain

$$\int_\tau^{\tau_1}\left[\int_\tau^{\tau_1} S(\tau)\,\zeta'(\tau)\,d\tau - \sum_{n=0}^{\infty}\{h'_{m,n+1}(z_1)-h'_{m,n}(z_1)\}\right]\zeta'(\tau)\,d\tau + \sum_{n=0}^{\infty}\{h_{m,n+1}(z_1)-h_{m,n}(z_1)\}$$

$$= \sum_{n=0}^{\infty} [h_{m,n+1}\{\zeta(\tau)\}-h_{m,n}\{\zeta(\tau)\}] = T(\tau),$$

say. Differentiation of this result and use of (18·10) enable us to show that $T(\tau)$ is a solution of the transformed form of the differential equation (18·4) which results from the substitution $z=\zeta(\tau)$. Hence we deduce that

$$T(\tau) = W_1^*\{\zeta(\tau)\}-L_m\{\zeta(\tau)\}, \qquad T'(\tau)/\zeta'(\tau) = W_1^{*\prime}\{\zeta(\tau)\}-L'_m\{\zeta(\tau)\}, \tag{18·14}$$

when $\tau_0 < \tau \leqslant \tau_1$, where $W_1^*(z)$ is the solution of (14·1) satisfying the conditions

$$W_1^*(z_1) = L_m(z_1)+T(\tau_1), \qquad W_1^{*\prime}(z_1) = L'_m(z_1)+\{T'(\tau_1)/\zeta'(\tau_1)\}. \tag{18·15}$$

We now examine the limiting form of $T(\tau)$ as $\tau \to \tau_0$. The part of $\mathscr{P}$ lying inside the circle $|z| \leqslant b$ may be deformed into a straight line without infringing conditions (i) to (v) of § 14. Accordingly, for all $\tau_0 < \tau < \tau(\epsilon)$, assignable, we have

$$\int_0^{\zeta(\tau)} \frac{|dt|}{1+|t|^{1+\sigma_1}} < \epsilon, \tag{18·16}$$

where ϵ is an arbitrary positive number. With this restriction on τ we may repeat the analysis used to establish (18·8) and (18·9) from (18·6), and thence deduce that

$$\left|\sum_{n=0}^{\infty}\{h_{m,n+1}(\zeta)-h_{m,n}(\zeta)\}-\alpha_m\zeta I_\mu(u\zeta)\right|<2\epsilon\frac{k_m}{u^{m+1}}\,|\,\zeta\,e^{p\mu\pi i}\,|\,V_\mu(u\zeta')$$

and

$$\left|\sum_{n=0}^{\infty}\{h'_{m,n+1}(\zeta)-h'_{m,n}(\zeta)\}-\alpha_m\{(\mu+1)\,I_\mu(u\zeta)+u\zeta I_{\mu+1}(u\zeta)\}\right|<2\epsilon\frac{k_m}{u^{m+1}}\,|\,e^{p\mu\pi i}\,|\,(1+|\,u\zeta\,|)\,V_\mu(u\zeta'),$$

when $u>2k$, (cf. (8·11) and (8·12)). Therefore

$$\lim_{\tau\to\tau_0}\{\zeta(\tau)\}^{-\mu-1}T(\tau)=\frac{(\frac{1}{2}u)^\mu}{\Gamma(1+\mu)}\alpha_m,\qquad \lim_{\tau\to\tau_0}\{\zeta(\tau)\}^{-\mu}\frac{T'(\tau)}{\zeta'(\tau)}=(1+\mu)\frac{(\frac{1}{2}u)^\mu}{\Gamma(1+\mu)}\alpha_m. \tag{18·17}$$

Substituting these equations and the limiting forms of $L_m\{\zeta(\tau)\}$ and $L'_m\{\zeta(\tau)\}$, obtained from (17·2) and (17·3), in (18·14), we see that $W_1^*\{\zeta(\tau)\}$ and $W_1^{*\prime}\{\zeta(\tau)\}$ satisfy the conditions (18·1) and (18·2) with z replaced by $\zeta(\tau)$. Therefore $W_1^*(z)=W_1(z)$.

If $u>2k$, we derive from (18·8) and (18·9)

$$|\,W_1(z_1)-L_m(z_1)\,|<2k_m u^{-m-1}\,|\,z_1\,e^{p\mu\pi i}\,|\,V_\mu(uz'_1), \tag{18·18}$$

and

$$|\,W'_1(z_1)-L'_m(z_1)\,|<2k_m u^{-m-1}\,|\,e^{p\mu\pi i}\,|\,(1+|\,uz_1\,|)\,V_\mu(uz'_1), \tag{18·19}$$

where $z'_1\equiv z_1\,e^{-p\pi i}$. The error terms can be expressed in the forms given in (14·11) and (14·12) by setting

$$z_1\,e^{p\mu\pi i}V_\mu(uz'_1)=V_\mu^{(1)}.z_1I_\mu(uz_1)+V_\mu^{(2)}\frac{z_1^2}{u(1+|\,z_1\,|)}I_{\mu+1}(uz_1), \tag{18·20}$$

$$e^{p\mu\pi i}(1+|\,uz_1\,|)\,V_\mu(uz'_1)=V_\mu^{(3)}.(1+|\,z_1\,|)\,I_\mu(uz_1)+V_\mu^{(4)}.uz_1I_{\mu+1}(uz_1), \tag{18·21}$$

where

$$\left.\begin{aligned}&V_\mu^{(1)}=V_\mu(uz'_1)\,/I_\mu(uz'_1) && V_\mu^{(2)}=0\quad(|\,uz_1\,|\leqslant 1),\\ &V_\mu^{(1)}=uz'_1K_{\mu+1}(uz'_1)\,V_\mu(uz'_1), && V_\mu^{(2)}=u^2(1+|\,z_1\,|)\,K_\mu(uz'_1)\,V_\mu(uz'_1)\quad(|uz_1|>1),\end{aligned}\right\} \tag{18·22}$$

and

$$V_\mu^{(3)}=\frac{1+|\,uz_1\,|}{1+|\,z_1\,|}V_\mu^{(1)},\qquad V_\mu^{(4)}=\frac{1+|\,uz_1\,|}{1+|\,z_1\,|}\frac{V_\mu^{(2)}}{u^2}. \tag{18·23}$$

Using (16·1), and (16·2) and the inequality

$$\frac{1}{|\,I_\mu(z)\,|}<k\,|\,z^{-\mu}\,|\quad(\mu\in\mathbf{M},\ |\,z\,|\leqslant 1) \tag{18·24}$$

which can readily be established by use of (16·9), we can prove that

$$|\,V_\mu^{(1)}\,|,\,u^{-2}\,|\,V_\mu^{(2)}\,|,\,u^{-1}|\,V_\mu^{(3)}|\ \text{and}\ |\,V_\mu^{(4)}\,|$$

are bounded. We see immediately that (14·11) and (14·12) hold if the term $O(1/u^m)$ occurring in the former is replaced by $O(1/u^{m-1})$. The final form is obtained on changing m into $m+1$.

19. Proof for the solution W_2

We suppose first that $a\equiv a(\theta,\mu)$ is a point of $\mathbf{G}(\theta,\mu)$ which is not at infinity.

We write

$$C_s^*(\theta,\mu,z)=\frac{C_s(\theta,\mu,z)}{1+|\,z\,|}, \tag{19·1}$$

and define $A(u,\theta,\mu), B(u,\theta,\mu), C^*(u,\theta,\mu)$ and $D(u,\theta,\mu)$ to be functions with uniform expansions for large positive u typified by

$$A(u,\theta,\mu) \sim \sum_{s=0}^{\infty} \frac{A_s(\theta,\mu,a)}{u^s}. \qquad (19\cdot2)$$

Then $W_2(u,\theta,\mu,z) \equiv W_2(z)$ is defined to be the solution of the differential equation (14·1) satisfying the conditions

$$\left.\begin{aligned} W_2(a) &= aK_\mu(ua)\,A(u,\theta,\mu) - (a/u)\,K_{\mu+1}(ua)\,B(u,\theta,\mu),\\ W_2'(a) &= (1+|a|)\,K_\mu(ua)\,C^*(u,\theta,\mu) - uaK_{\mu+1}(ua)\,D(u,\theta,\mu).\end{aligned}\right\} \qquad (19\cdot3)$$

Let z_1 be a point of the region $\mathbf{H}_2(\theta,\mu)$, and let ζ be a typical point of the path $\mathscr{P}$ joining z_1 and a. In the expressions for L_m and R_m given in § 17 we take $\mathscr{L}_\mu = K_\mu$, $\mathscr{L}_{\mu+1} = -K_{\mu+1}$. The sequence $h_{m,n}(\zeta)$ is then defined by $h_{m,0}(\zeta) = 0$ and

$$h_{m,n}(\zeta) = \zeta\int_\zeta^a \{K_\mu(u\zeta)\,I_\mu(ut) - I_\mu(u\zeta)\,K_\mu(ut)\}\{f(t)\,h_{m,n-1}(t) - R_m(t)\}\,\mathrm{d}t$$
$$+\alpha_m\zeta I_\mu(u\zeta) + \beta_m\zeta K_\mu(u\zeta) \quad (n\geqslant 1), \qquad (19\cdot4)$$

where the path of integration is the part of $\mathscr{P}$ between ζ and a, and α_m, β_m are constants. The branches of the modified Bessel functions have their principal values at a, and elsewhere are determined by continuity.

The constants α_m and β_m are determined by the conditions

$$h_{m,n}(a) = W_2(a) - L_m(a), \qquad h'_{m,n}(a) = W_2'(a) - L_m'(a), \qquad (19\cdot5)$$

when $n\geqslant 1$. This implies that

$$\left.\begin{aligned}\alpha_m &= \{uK_{\mu+1}(ua) - (\mu+1)\,a^{-1}K_\mu(ua)\}\{W_2(a) - L_m(a)\} + K_\mu(ua)\,\{W_2'(a) - L_m'(a)\},\\ \beta_m &= \{uI_{\mu+1}(ua) + (\mu+1)\,a^{-1}I_\mu(ua)\}\{W_2(a) - L_m(a)\} - I_\mu(ua)\,\{W_2'(a) - L_m'(a)\}.\end{aligned}\right\} \qquad (19\cdot6)$$

Substituting by means of the inequalities (16·1) and (16·2), and using the fact that $|a|\geqslant b$, we can show that

$$|\alpha_m| < k_m u^{-m-1}\,|ua|\,\{X_\mu(ua)\}^2, \qquad |\beta_m| < k_m u^{-m-1}. \qquad (19\cdot7)$$

Using (19·7) and extending the analysis given in Olver 1956, § 13, we may show that

$$|h_{m,n+1}(\zeta) - h_{m,n}(\zeta)| < \left(\frac{k}{u}\right)^n \frac{k_m}{u^{m+1}}\,|\zeta|\,X_\mu(u\zeta), \qquad (19\cdot8)$$

$$|h'_{m,n+1}(\zeta) - h'_{m,n}(\zeta)| < \left(\frac{k}{u}\right)^n \frac{k_m}{u^{m+1}}\{X_\mu(u\zeta) + |u\zeta|\,X_{\mu+1}(u\zeta)\}, \qquad (19\cdot9)$$

and hence that
$$|h''_{m,n+1}(\zeta) - h''_{m,n}(\zeta)| < \left(\frac{k}{u}\right)^n \frac{k_m}{u^{m+1}}\,\frac{1+|u\zeta|^2}{|\zeta|}\,X_\mu(u\zeta). \qquad (19\cdot10)$$

These are the key results. Using them we deduce that

$$W_2\{\zeta(\tau)\} - L_m\{\zeta(\tau)\} = T(\tau) \equiv \sum_{n=0}^{\infty}[h_{m,n+1}\{\zeta(\tau)\} - h_{m,n}\{\zeta(\tau)\}], \qquad (19\cdot11)$$

$$W_2'\{\zeta(\tau)\} - L_m'\{\zeta(\tau)\} = T'(\tau)/\zeta'(\tau) = \sum_{n=0}^{\infty}[h'_{m,n+1}\zeta(\tau) - h'_{m,n}\{\zeta(\tau)\}], \qquad (19\cdot12)$$

when $u>k$, where
$$\zeta = \zeta(\tau) \quad (\tau_0\leqslant\tau\leqslant\tau_1) \qquad (19\cdot13)$$

is the parametric equation of $\mathscr{P}$.

Thus if $u > 2k$ we have

$$|W_2(z_1) - L_m(z_1)| < 2k_m u^{-m-1} |z_1| X_\mu(uz_1), \tag{19·14}$$

$$|W_2'(z_1) - L_m'(z_1)| < 2k_m u^{-m-1}(1+|uz_1|) X_\mu(uz_1). \tag{19·15}$$

To express the error terms in the form given by equations (14·13) and (14·14), we set

$$z_1 X_\mu(uz_1) = X_\mu^{(1)}.z_1 K_\mu(uz_1) + X_\mu^{(2)} \frac{z_1^2}{u(1+|z_1|)} K_{\mu+1}(uz_1), \tag{19·16}$$

$$(1+|uz_1|) X_\mu(uz_1) = X_\mu^{(3)}.(1+|z_1|) K_\mu(uz_1) + X_\mu^{(4)}.uz_1 K_{\mu+1}(uz_1), \tag{19·17}$$

where $X_\mu^{(1)}$, $X_\mu^{(2)}$, $X_\mu^{(3)}$ and $X_\mu^{(4)}$ are defined as follows. Suppose first $|\mu| \geqslant \delta$, where δ is the arbitrary positive number associated with the definition of $l_\mu(z)$ (cf. (16·4)). Then

$$\left.\begin{aligned} X_\mu^{(1)} &= uz_1 I_{\mu+1}(uz_1) X_\mu(uz_1), \\ X_\mu^{(2)} &= u^2(1+|z_1|) I_\mu(uz_1) X_\mu(uz_1), \end{aligned}\right\} \quad (|uz_1| \leqslant 1), \tag{19·18}$$

$$\left.\begin{aligned} X_\mu^{(1)} &= \pm \frac{uz_1}{\pi i} K_{\mu+1}(uz_1 e^{\mp\pi i}) X_\mu(uz_1), \\ X_\mu^{(2)} &= \pm \frac{u^2(1+|z_1|)}{\pi i} K_\mu(uz_1 e^{\mp\pi i}) X_\mu(uz_1), \end{aligned}\right\} \quad (|uz_1| > 1), \tag{19·19}$$

and

$$X_\mu^{(3)} = \frac{1+|uz_1|}{1+|z_1|} X_\mu^{(1)}, \qquad X_\mu^{(4)} = \frac{1+|uz_1|}{1+|z_1|} \frac{X_\mu^{(2)}}{u^2}. \tag{19·20}$$

The upper signs are taken in (19·19) if $0 \leqslant \arg z_1 < \frac{3}{2}\pi$ and the lower if $-\frac{3}{2}\pi < \arg z_1 < 0$.

Alternatively suppose that $\mu = 0$. Then we define $X_\mu^{(1)}$ and $X_\mu^{(2)}$ by equations (13·25) and (13·26) of Olver 1956, replacing $X_\mu^{(2)}$ by $u^{-2}X_\mu^{(2)}$, and $X_\mu^{(3)}$, $X_\mu^{(4)}$ by equations (19·20) above.

Using (16·1) and (16·2) we can prove that $|X_\mu^{(1)}|$, $u^{-2}|X_\mu^{(2)}|$, $u^{-1}|X_\mu^{(3)}|$ and $|X_\mu^{(4)}|$ are bounded. Changing m into $m+1$ we complete the proof of (14·13) and (14·14) when a is not at infinity.

When a is at infinity modifications have to be made to the proof which are similar to those given in § 8 for case A. The solution $W_2(z)$ is determined by the conditions

$$\left.\begin{aligned} \lim\{(2u/\pi z)^{\frac{1}{2}} e^{uz} W_2(z)\} &= A(u,\theta,\mu) - u^{-1} B(u,\theta,\mu), \\ \lim\{(2u/\pi z)^{\frac{1}{2}} e^{uz} W_2'(z)\} &= -uA(u,\theta,\mu) + B(u,\theta,\mu), \end{aligned}\right\} \tag{19·21}$$

as $z \to a$ along the straight line $\mathscr{L}$. The sequence $h_{m,n}(\zeta)$ is defined by (19·4) with $\alpha_m = 0$ and

$$\beta_m = A(u,\theta,\mu) - \sum_{s=0}^{m} \frac{A_s(\theta,\mu,a)}{u^s} - \frac{1}{u}\left\{B(u,\theta,\mu) - \sum_{s=0}^{m-1} \frac{B_s(\theta,\mu,a)}{u^s}\right\}. \tag{19·22}$$

Other modifications are straightforward. Equations (19·14) and (19·15) again hold and the remainder of the proof is unchanged.

The author wishes to acknowledge the assistance of his colleague Mr G. F. Miller who checked the manuscript of this paper and made many valuable suggestions.

The work described above has been carried out as part of the research programme of the National Physical Laboratory, and this paper is published by permission of the Director of the Laboratory.

References

Cherry, T. M. 1950 *Trans. Amer. Math. Soc.* **68**, 224–257.
Copson, E. T. 1944 *Theory of functions of a complex variable.* Oxford: University Press.
Erdélyi, A. 1956 *Asymptotic expansions.* New York: Stechert-Hafner.
Ince, E. L. 1944 *Ordinary differential equations.* New York: Dover.
Kazarinoff, N. D. & McKelvey, R. 1956 *Canad. J. Math.* **8**, 97–104.
Langer, R. E. 1949 *Trans. Amer. Math. Soc.* **67**, 461–490.
Olver, F. W. J. 1954*a* *Phil. Trans.* A, **247**, 307–327.
Olver, F. W. J. 1954*b* *Phil. Trans.* A, **247**, 328–368.
Olver, F. W. J. 1956 *Phil. Trans.* A, **249**, 65–97.
Thorne, R. C. 1957*a* *Proc. Camb. Phil. Soc.* **53**, 382–398.
Thorne, R. C. 1957*b* *Phil. Trans.* A, **249**, 585–596.
Titchmarsh, E. C. 1939 *Theory of functions.* Oxford: University Press.
van der Corput, J. G. 1956 *J. d'Analyse Math.* **4**, 341–418.
Watson, G. N. 1944 *Theory of Bessel functions.* Cambridge: University Press.

Journal of Research of the National Bureau of Standards—B. Mathematics and Mathematical Physics
Vol. 63B, No. 2, October–December 1959

Uniform Asymptotic Expansions for Weber Parabolic Cylinder Functions of Large Orders

F. W. J. Olver*

(April 6, 1959)

Expansions of solutions of the differential equation

$$\frac{d^2w}{dt^2}=\mu^4(t^2-1)w,$$

are sought for large values of $|\mu|$, which are uniformly valid with respect to arg μ and unrestricted values of the complex variable t. Two types of expansion are found. Those of the first type are in terms of elementary functions and are valid outside the neighborhoods of the points $t=\pm 1$, the turning points of the differential equation. The second are in terms of Airy functions and hold in unbounded regions containing one of the turning points.

The special forms of the expansions when the variables are real are considered in detail, and asymptotic expansions for the zeros of solutions of the differential equation are found by reversion. Numerical examples are included.

Contents

Part 1. Introduction

1. Introduction and General Summary

The purpose of this paper is to determine asymptotic expansions of solutions of the differential equation

$$\frac{d^2w}{dt^2}=\mu^4(t^2-1)w \tag{1.1}$$

for large values of the complex parameter μ which are uniformly valid with respect to arg μ and the complex variable t.

The immediate reason for this investigation is that the solutions of eq (1.1) are basic functions in the theory of the asymptotic solution of the differential equation

$$\frac{d^2w}{dt^2}=\{\mu^4p(t)+q(t)\}w \tag{1.2}$$

for large values of $|\mu|$ in a region in which $p(t)$ has two simple zeros, just as Airy functions are

*Present address: National Physical Laboratory, Teddington, Middlesex, England.

basic functions in the theory of eq (1.2) in a region containing one simple zero of $p(t)$. A knowledge of the uniform asymptotic behavior of the solutions of (1.1) is required in developing the asymptotic theory of (1.2).

Equation (1.1) is of interest, however, in other connections, for example in the study of wave motion inside or outside parabolic cylinders, and wave propagation in an inhomogeneous atmosphere. Hermite polynomials of order n are expressible in terms of the solutions of the equation when $\mu^2=2n+1$. Accordingly, for the sake of completeness many results are derived in this paper which go beyond those needed for the immediate purpose stated in the preceding paragraph.

The desired asymptotic expansions are obtained by application of the general theory of the asymptotic solution of linear differential equations of the second order as developed by the present writer in [9].[1] From the standpoint of the general theory, eq (1.1) is characterized by having turning points at $t=\pm 1$. Asymptotic expansions in terms of elementary functions can be found in any region bounded away from these points and expansions of this kind are obtained in part 2 by application of theorem A of [9].

In order to determine the behavior near the turning points, expansions in terms of Airy functions need to be used. In part 3 these expansions are obtained by application of theorem B of [9]; they are valid in unbounded t-regions which include one (but not both) of the turning points. All combinations of μ and t are covered.

When the variables are real, eq (1.1) has important special solutions. Expansions of these solutions are derived in part 4 from the results of parts 2 and 3. Also included in part 4 are numerical examples illustrating the powerful nature of the expansions.

In part 5 uniform asymptotic expansions for the zeros and associated values are obtained by reversion and illustrated by numerical examples. Much of the analysis of this part and some in earlier parts is similar to that used in [8] in the theory of Bessel functions of large order, and advantage is taken of this analogy whenever possible.

Investigations of the asymptotic solution of eq (1.1) for large $|\mu|$ have been made by several writers. Their work on expansions in terms of elementary functions is described in section 6 of part 2, and on expansions in terms of Airy functions in section 10 of part 3. At this stage it suffices to remark that the expansions obtained in this paper include practically all of the earlier results and extend them in various ways.

2. Relevant Properties of Parabolic Cylinder Functions and the Gamma Function

A comprehensive account of the properties of parabolic cylinder functions is given by Miller in the Introduction to [4]. For the purpose of reference we collect here the properties which will be required in this paper. We use Miller's notation rather than that of Whittaker [13] and adopt Miller's choice of solutions when the variables are real.

The standard form of differential equation for the parabolic cylinder functions is

$$\frac{d^2w}{dz^2}=\left(\frac{1}{4}z^2+a\right)w. \tag{2.1}$$

The principal solution $U(a,z)$ is determined by the condition

$$U(a,z)\sim z^{-a-\frac{1}{2}}e^{-\frac{1}{4}z^2} \qquad \text{as } z\to+\infty. \tag{2.2}$$

It is an integral (entire) function of z and an integral function of a. In terms of Whittaker's notation $D_n(z)$ for parabolic cylinder functions and $W_{k,m}(z)$ for the confluent hypergeometric function, we have

$$U(a,z)=D_{-a-\frac{1}{2}}(z)=2^{-\frac{1}{2}a}z^{-\frac{1}{2}}W_{-\frac{1}{2}a,-\frac{1}{4}}\left(\frac{1}{2}z^2\right). \tag{2.3}$$

[1] Figures in brackets indicate the literature references at the end of this paper.

Other solutions of (2.1) are $U(a,-z)$, $U(-a,iz)$ and $U(-a,-iz)$. The connection formulas are

$$U(-a,\pm iz)=(2\pi)^{-\frac{1}{2}}\Gamma\left(\frac{1}{2}+a\right)\{e^{-i\pi(\frac{1}{2}a-\frac{1}{4})}U(a,\pm z)+e^{i\pi(\frac{1}{2}a-\frac{1}{4})}U(a,\mp z)\}, \tag{2.4}$$

$$U(a,\pm z)=(2\pi)^{-\frac{1}{2}}\Gamma\left(\frac{1}{2}-a\right)\{e^{-i\pi(\frac{1}{2}a+\frac{1}{4})}U(-a,\pm iz)+e^{i\pi(\frac{1}{2}a+\frac{1}{4})}U(-a,\mp iz)\}. \tag{2.5}$$

For fixed a and large $|z|$

$$U(a,z)=z^{-a-\frac{1}{2}}e^{-\frac{1}{4}z^2}\{1+O(|z|^{-2})\}\qquad\left(|\arg z|\leq\frac{3}{4}\pi-\epsilon\right), \tag{2.6}$$

where[2] the O is uniform with respect to arg z. Here and elsewhere ϵ denotes an arbitrary positive number which is independent of all other variables. From (2.4) and (2.6), we obtain

$$U(a,-z)=e^{-i\pi(a+\frac{1}{2})}z^{-a-\frac{1}{2}}e^{-\frac{1}{4}z^2}\{1+O(|z|^{-2})\}+\frac{(2\pi)^{\frac{1}{2}}}{\Gamma(\frac{1}{2}+a)}z^{a-\frac{1}{2}}e^{\frac{1}{4}z^2}\{1+O(|z|^{-2})\}, \tag{2.7}$$

valid when $-\frac{3}{4}\pi+\epsilon\leq\arg z\leq\frac{1}{4}\pi-\epsilon$. The same result with the sign of i changed holds when $-\frac{1}{4}\pi+\epsilon\leq\arg z\leq\frac{3}{4}\pi-\epsilon$; the apparent discrepancy between the formulas in their common region of validity is merely an example of the Stokes phenomenon.

Let the z-plane be divided into four sectors $\mathbf{M}_0$, $\mathbf{M}_1$, $\mathbf{M}_2$, and $\mathbf{M}_3$, defined by $|\arg z|\leq\frac{1}{4}\pi$, $\frac{1}{4}\pi\leq\arg z\leq\frac{3}{4}\pi$, $|\arg(-z)|\leq\frac{1}{4}\pi$ and $-\frac{3}{4}\pi\leq\arg z\leq-\frac{1}{4}\pi$, respectively. Then eq (2.6) shows that for large $|z|$, $U(a,z)$ is exponentially small in $\mathbf{M}_0$ and exponentially large in $\mathbf{M}_1$ and $\mathbf{M}_3$. Equation (2.7) shows that $U(a,z)$ is also exponentially large in $\mathbf{M}_2$, unless a has one of the values $-\frac{1}{2}, -\frac{3}{2}, -\frac{5}{2}, \ldots$, in which case it is exponentially small. In fact in this event

$$U(-n-\tfrac{1}{2},-z)=(-)^nU(-n-\tfrac{1}{2},z)\qquad(n=0,1,2,\ldots). \tag{2.8}$$

The solutions $U(a,\pm z)$, $U(-a,\pm iz)$ clearly form a numerically satisfactory set in the complex plane, because in each sector $\mathbf{M}_j$ one of the solutions is exponentially small and at least two others exponentially large. Accordingly, we may expect that a knowledge of the asymptotic behavior of $U(a,z)$ for large $|a|$ in the sector $|\arg z|\leq\frac{3}{4}\pi$, or even the half-plane $|\arg z|\leq\frac{1}{2}\pi$, will enable us to determine the behavior of any solution of (2.1) over the whole z-plane by use of connection formulas.

Solutions when the variables are real. There are two forms of eq (2.1) of importance when the variables are real, namely

$$\frac{d^2w}{dx^2}=\left(\frac{1}{4}x^2+a\right)w, \tag{2.9}$$

and

$$\frac{d^2w}{dx^2}=\left(a-\frac{1}{4}x^2\right)w, \tag{2.10}$$

in which a and x are real. Equation (2.10) is obtained from (2.1) by setting $z=xe^{-\frac{1}{4}\pi i}$ and replacing a by ia.

Real solutions of (2.9) are $U(a,x)$ and $U(a,-x)$ but they do not form a satisfactory pair for all values of a; in particular they are linearly dependent when $a=-\frac{1}{2}, -\frac{3}{2}, -\frac{5}{2}, \ldots$. The fundamental solutions are taken to be $U(a,x)$ and $V(a,x)$, where

$$V(a,x)=(1/\pi)\Gamma\left(\frac{1}{2}+a\right)\{\sin\pi a\cdot U(a,x)+U(a,-x)\}. \tag{2.11}$$

[2] Miller [4, p. 27], who is more concerned with asymptotic expansions fulfilling the uniqueness conditions of Watson [11] than the Poincaré condition, gives the range $|\arg z|<\frac{1}{2}\pi$. The validity, in Poincaré's sense, for the wider range is established in [14, p. 347].

For all values of a, $U(a,x)$ and $V(a,x)$ are linearly independent. When a is negative it is often convenient to use in place of $V(a,x)$ its multiple $\overline{U}(a,x)$, given by

$$\overline{U}(a,x)=\Gamma\left(\frac{1}{2}-a\right)V(a,x)=\tan \pi a\cdot U(a,x)+\sec \pi a\cdot U(a,-x). \tag{2.12}$$

When $a+\frac{1}{2}$ is a positive integer $\overline{U}(a,x)$ is infinite.

Wronskian relations for these solutions are

$$\mathscr{W}\{U(a,x),V(a,x)\}=(2/\pi)^{\frac{1}{2}}, \qquad \mathscr{W}\{U(a,x),\overline{U}(a,x)\}=(2/\pi)^{\frac{1}{2}}\Gamma\left(\frac{1}{2}-a\right). \tag{2.13}$$

The Hermite polynomials

$$H_n(x)=(-)^n e^{x^2}\frac{d^n}{dx^n}e^{-x^2} \tag{2.14}$$

are related to the function $U(a,x)$ by

$$H_n(x)=2^{\frac{1}{2}n}e^{\frac{1}{2}x^2}U\left(-n-\frac{1}{2},x\sqrt{2}\right). \tag{2.15}$$

For eq (2.10), the fundamental pair of real solutions is taken to be $W(a,x)$ and $W(a,-x)$, defined by

$$W(a,x)=(2k)^{\frac{1}{2}}e^{\frac{1}{4}\pi a}\mathrm{Re}\{e^{i(\frac{1}{2}\phi_2+\frac{1}{8}\pi)}U(ia,xe^{-\frac{1}{4}\pi i})\}, \tag{2.16}$$

$$W(a,-x)=(2/k)^{\frac{1}{2}}e^{\frac{1}{4}\pi a}\,\mathrm{Im}\{e^{i(\frac{1}{2}\phi_2+\frac{1}{8}\pi)}U(ia,xe^{-\frac{1}{4}\pi i})\}. \tag{2.17}$$

where

$$k=(1+e^{2\pi a})^{\frac{1}{2}}-e^{\pi a}, \qquad 1/k=(1+e^{2\pi a})^{\frac{1}{2}}+e^{\pi a} \qquad \phi_2=\arg\Gamma\left(\frac{1}{2}+ia\right). \tag{2.18}$$

The value of $\arg\Gamma(\frac{1}{2}+ia)$ here is not the principal one but is determined by the conditions that it is a continuous function of a and equals $a\ln a-a+O(a^{-1})$ as $a\to+\infty$.

The Wronskian relation is

$$\mathscr{W}\{W(a,x),W(a,-x)\}=1. \tag{2.19}$$

Gamma function expansions. We shall need the asymptotic expansion of the function $\Gamma\left(\frac{1}{2}+z\right)$ and its reciprocal for large $|z|$. From Stirling's series [5, chapter 9], we have

$$\ln\Gamma\left(\frac{1}{2}+z\right)=\frac{1}{2}\ln 2\pi+z\ln z-z+Z(z), \tag{2.20}$$

where

$$Z(z)\sim\sum_{s=1}^{\infty}\frac{B_{2s}(\frac{1}{2})}{2s(2s-1)}\frac{1}{z^{2s-1}}\equiv-\frac{1}{24}\frac{1}{z}+\frac{7}{2880}\frac{1}{z^3}-\frac{31}{40320}\frac{1}{z^5}+\ldots \qquad (|\arg z|\leq\pi-\epsilon) \tag{2.21}$$

uniformly with respect to $\arg z$, B_{2s} denoting the Bernoulli polynomial of degree $2s$. Hence

$$\Gamma\left(\frac{1}{2}+z\right)\sim(2\pi)^{\frac{1}{2}}e^{-z}z^z\sum_{s=0}^{\infty}\frac{\gamma_s}{z^s} \qquad (|\arg z|\leq\pi-\epsilon), \tag{2.22}$$

where the constants are determined by the asymptotic identity

$$\sum_{s=0}^{\infty}\frac{\gamma_s}{z^s}\sim\exp\{Z(z)\}, \tag{2.23}$$

or equivalently,

$$\sum_{s=1}^{\infty}\frac{s\gamma_s}{z^{s+1}}\sim\left\{\sum_{s=1}^{\infty}\frac{B_{2s}(\frac{1}{2})}{2s}\frac{1}{z^{2s}}\right\}\left\{\sum_{s=0}^{\infty}\frac{\gamma_s}{z^s}\right\}. \tag{2.24}$$

Numerical calculation yields

$$\gamma_0=1, \quad \gamma_1=-\frac{1}{24}, \quad \gamma_2=\frac{1}{1152}, \quad \gamma_3=\frac{1003}{414720}, \quad \gamma_4=-\frac{4027}{39813120}. \tag{2.25}$$

Since $Z(z)$ is an odd function of z, it follows from (2.23) that

$$\left\{\sum_{s=0}^{\infty}\frac{\gamma_s}{z^s}\right\}\left\{\sum_{s=0}^{\infty}(-)^s\frac{\gamma_s}{z^s}\right\}\sim 1, \tag{2.26}$$

identically, and hence that

$$\frac{1}{\Gamma(\frac{1}{2}+z)}\sim\frac{e^z z^{-z}}{(2\pi)^{\frac{1}{2}}}\sum_{s=0}^{\infty}(-)^s\frac{\gamma_s}{z^s} \qquad (|\arg z|\leq\pi-\epsilon). \tag{2.27}$$

Part 2. Expansions in Terms of Elementary Functions

3. Asymptotic Solutions of the Differential Equation

In order to bring the differential eq (1.1) into the standard form required for the application of theorem A of [9], we take new variables $\xi=\xi(t)$ and W (not the $W(a,x)$ of sec. 2), defined by

$$\left(\frac{d\xi}{dt}\right)^2=t^2-1, \qquad W=(t^2-1)^{\frac{1}{4}}\,w \tag{3.1}$$

(cf. [7] sec. 2). Equation (1.1) then becomes

$$\frac{d^2W}{d\xi^2}=\{\mu^4+\mathscr{F}(\xi)\}W, \tag{3.2}$$

where

$$\mathscr{F}(\xi)=(t^2-1)^{-\frac{1}{4}}\frac{d^2}{d\xi^2}(t^2-1)^{\frac{1}{4}}=-\frac{3t^2+2}{4(t^2-1)^3}. \tag{3.3}$$

For convenience we prescribe that $\xi=0$ when $t=1$. From the first of eqs (3.1) we then obtain

$$\xi=\int_1^t(t^2-1)^{\frac{1}{2}}\,dt=\tfrac{1}{2}t(t^2-1)^{\frac{1}{2}}-\tfrac{1}{2}\ln\{t+(t^2-1)^{\frac{1}{2}}\}, \tag{3.4}$$

and on expansion

$$\xi=\tfrac{1}{2}t^2-\tfrac{1}{2}\ln 2t-\tfrac{1}{4}+\tfrac{1}{4}\sum_{s=1}^{\infty}\frac{1.3.5\ldots(2s-1)}{s(s+1)!(2t^2)^s} \qquad (|t|>1). \tag{3.5}$$

Temporarily we suppose that the branches of the many-valued functions occurring in (3.4) and (3.5) are positive when $t>1$ and determined by continuity elsewhere. The relation

$$\xi(te^{\pm i\pi})=\xi(t)\mp\tfrac{1}{2}i\pi \tag{3.6}$$

follows immediately from (3.5) when $|t|>1$, and thence by analytic continuation throughout the t-plane cut along the join of $t=-1$ and $t=+1$.

The mapping of the t-plane, cut along the real axis from $-\infty$ to $+1$, is illustrated in figure 1. The center diagram corresponds to the upper half of the t-plane and the right-hand diagram to the lower half. They are of course conjugate to each other. Features of the mapping are that the upper side of the real t-axis is mapped on three straight lines BA (Im $\xi=0$, Re $\xi\geq 0$), AE (Re $\xi=0$, $0\geq$ Im $\xi\geq-\frac{1}{2}\pi$), ED (Im $\xi=-\frac{1}{2}\pi$, Re $\xi\geq 0$), and that the positive imaginary t-axis is mapped on the straight line OC (Im $\xi=-\frac{1}{4}\pi$, Re $\xi\leq 0$).

Next, we make the substitutions (cf. [9], sec. 1)

$$u=|\mu|^2, \qquad \theta=\arg\mu, \qquad \xi=e^{-2i\theta}x \tag{3.7}$$

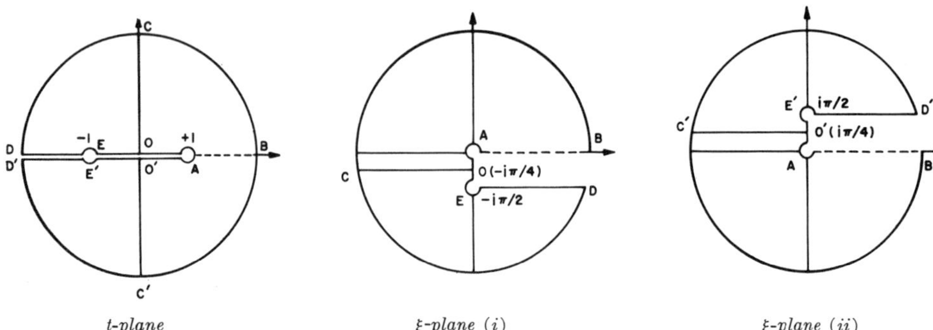

t-plane *ξ-plane (i)* *ξ-plane (ii)*

FIGURE 1. *Mapping of the function ξ(t).*

(the x here is not to be confused with the real variable of sec. 2). Then (3.2) becomes

$$\frac{d^2W}{dx^2}=\{u^2+e^{-4i\theta}\mathscr{F}(e^{-2i\theta}x)\}W, \tag{3.8}$$

which has the form of eq (2.1) of [9]. The relationship between x and t is given by

$$x=x(t)=e^{2i\theta}\xi(t)=e^{2i\theta}\int_1^t (t^2-1)^{\frac{1}{2}}dt; \tag{3.9}$$

the x-map of the t-plane is obtained from the ξ-map by rotation through an angle 2θ. The function $\mathscr{F}$ is expressed in terms of t by (3.3). For large $|x|$ we deduce from (3.3) and (3.5) that

$$e^{-4i\theta}\mathscr{F}(e^{-2i\theta}x)\sim-\frac{3}{4}\,e^{-4i\theta}t^{-4}\sim-\frac{3}{16x^2}, \tag{3.10}$$

uniformly with respect to bounded arg x and bounded θ.

In consequence of the last result we could, if we wished, take the x-domain $\mathbf{D}(\theta)$ of [9], section 2 to be the map of the t-plane cut along the real axis from $-\infty$ to $+1$ with the circles $|t\pm 1|<\delta'$ removed, δ' being a fixed small positive number. The preliminary condition (2.2) of [9] would then be satisfied with $\sigma=1$. Greater regions of validity in the desired expansions will result however, if we arrange that the boundaries of $\mathbf{D}(\theta)$ are parallel to the positive imaginary x-axis where possible. Accordingly, we shall now investigate what we shall call the *principal curves* in the t-plane, namely the t-maps of the straight lines in the ξ-plane passing through A ($\xi=0$) and E ($\xi=\mp\frac{1}{2}i\pi$) and inclined at an angle -2θ to the imaginary ξ-axis. In other words, the principal curves are the level curves of the function $e^{x(t)}$ which pass through $t=\pm 1$.

Near $t=1$ we find from (3.9) that

$$x(t)\sim\frac{1}{3}\,2^{\frac{3}{2}}e^{2i\theta}(t-1)^{\frac{3}{2}}. \tag{3.11}$$

Hence three principal curves emanate from $t=1$, and have as tangents the rays

$$\arg(t-1)=\pm\frac{1}{3}\pi-\frac{4}{3}\theta,\quad \pi-\frac{4}{3}\theta.$$

Equation (3.6) shows that three principal curves also emanate from $t=-1$.

For large $|t|$ we have from (3.5)

$$x(t)\sim\frac{1}{2}\,e^{2i\theta}t^2. \tag{3.12}$$

Hence the six principal curves are asymptotic to the four rays arg $t=\pm\frac{1}{4}\pi-\theta$, $\pm\frac{3}{4}\pi-\theta$.

Figure 2 illustrates the principal curves when $0\leq\theta<\pi$. Those through A are the continuous curves AP, AQ, AS when $0<\theta<\frac{1}{2}\pi$, and AP, AQ, AR when $\frac{1}{2}\pi<\theta<\pi$. When $\theta=0$, AQ degenerates into the curve EQ plus the join AE, and when $\theta=\frac{1}{2}\pi$, AS, AR degenerate into AE+ES, AE+ER, respectively. The principal curves through E are the images in the origin of those through A; this can be verified by using the continuation formula (3.6).

We can now define the branch of the many-valued function $\xi(t)$ which will be the most convenient for our purposes. When $0\leq\theta<\pi$ we cut the t-plane along the principal curves AQ and ER and select the branch of $\xi(t)$ which is continuous in the cut plane and for which

$$\xi(0)=\frac{1}{4}i\pi \quad (0<\theta<\pi), \qquad \xi(0-0i)=\frac{1}{4}i\pi \quad (\theta=0). \tag{3.13}$$

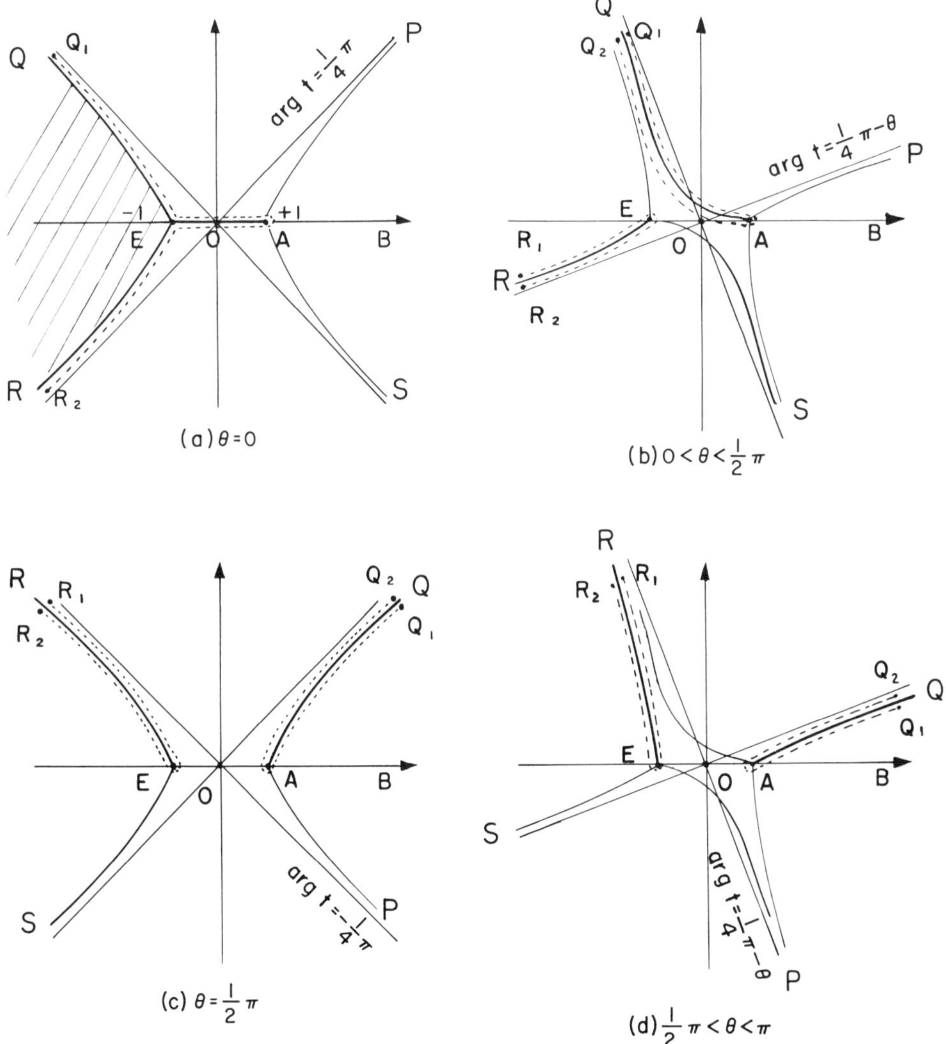

FIGURE 2. *t-plane: principal curves and domains* $\mathbf{S}(\theta)$.

When $-\pi<\theta\leq 0$ the cuts are taken to be the conjugates of those for $-\theta$, and the branch is fixed by

$$\xi(0)=-\frac{1}{4}i\pi \quad (-\pi<\theta<0), \qquad \xi(0+0i)=-\frac{1}{4}i\pi \quad (\theta=0). \tag{3.14}$$

It is evident from (3.6) that (3.13) and (3.14) are consistent when $\theta=0$. *Henceforth we understand ξ and $\xi(t)$ to denote this branch.* For a given value of θ $(-\pi<\theta<\pi)$, $\xi(t)$ is a single-valued function of t defined everywhere in the t-plane except when $\theta=0$ and t lies in the shaded region to the left of ER+EQ in figure 2(a). When $t>1$, $\xi(t)$ is positive provided that $|\theta|<\frac{3}{4}\pi$; this result is false if $\frac{3}{4}\pi<|\theta|<\pi$ for the cut AQ passes across the real axis as θ passes through $\frac{3}{4}\pi$.

Let $x\equiv x(t)$ now denote the branch $e^{2i\theta}\xi(t)$. We then take the x-domain $\mathbf{D}(\theta)$ of [9], section 2 to be the map of the function $x(t)$ from which circles surrounding the points A and E and of radius $\frac{1}{2}\delta$ have been removed, δ being a fixed small positive number.

The map of $x(t)$ has two sheets when $\theta=0$ and three sheets when $0<|\theta|<\pi$. It is illustrated in figure 3 for $\theta=0$ and $0<\theta<\frac{1}{2}\pi$. The passage from one sheet to another takes place across the positive real axis and, when $0<\theta<\frac{1}{2}\pi$, the join AE. The heavy lines correspond to the cuts in the t-plane and are boundaries of $\mathbf{D}(\theta)$.

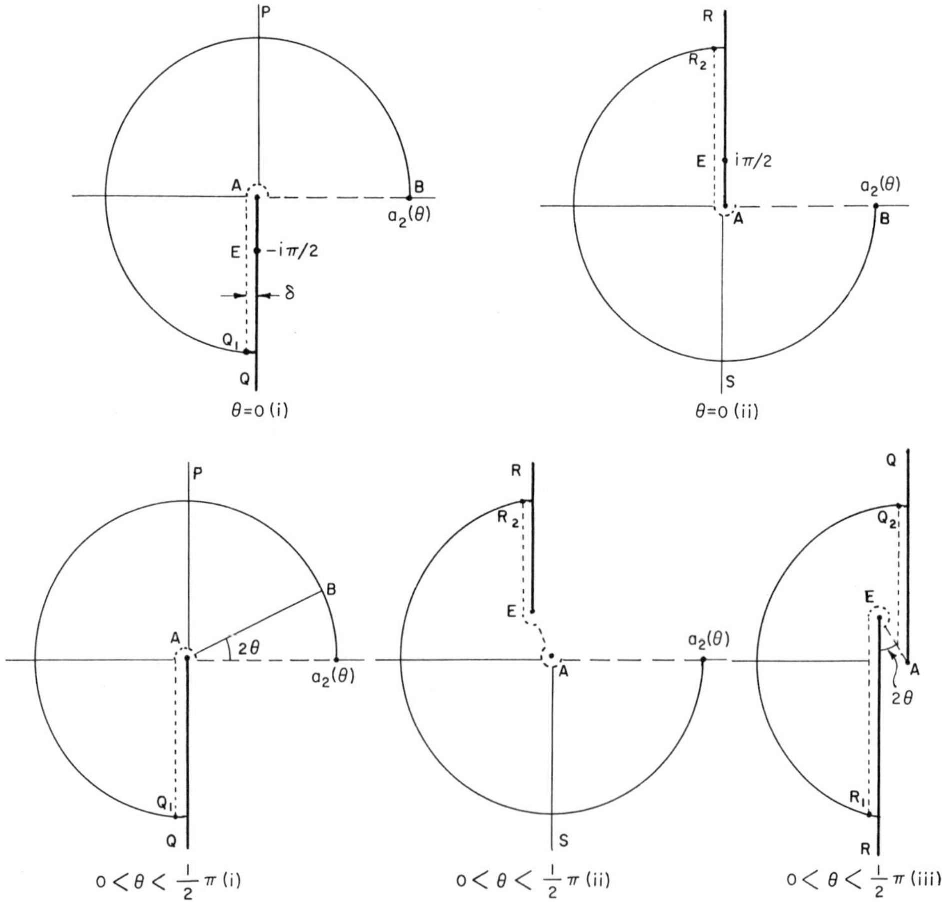

FIGURE 3. *x-plane: domains $\mathbf{D}(\theta)$ and $\mathbf{G}(\theta)$.*

The map of $x(t)$ is clearly a single unbounded domain. In order to ensure that $\mathbf{D}(\theta)$ is a single domain we have to impose the restriction $|\theta| \leq \pi - \epsilon$, omit sheet (iii) when $|\theta| < \epsilon$ and make $\delta = \delta(\epsilon)$ small enough. Otherwise, for example, the circle center E radius $\frac{1}{2}\delta$ would intersect the boundary AQ.

Next, we define the x-domain $\mathbf{G}(\theta)$ required by [9], section 2 to be that part of the map of $x(t)$ which remains after removing all points at a distance less than δ from its boundaries. Sheet (iii) is again to be omitted when $|\theta| < \epsilon$. The boundaries of $\mathbf{G}(\theta)$ are the dotted lines and circular arcs of figure 3; their t-maps are the dotted curves of figure 2. Clearly when $|\theta| \leq \pi - \epsilon$, $\mathbf{G}(\theta)$ is a single domain if $\delta = \delta(\epsilon)$ is small enough. We readily show that conditions (i) and (ii) of [9], section 2 are fulfilled (cf. [9], sec. 3(i)).

Following [9], section 2 further, we take $a_2(\theta)$ to be the point at infinity on the positive real axis. We then find that $\mathbf{H}_2(\theta) \equiv \mathbf{G}(\theta)$. From theorem A it now follows that there exists a solution W_2 of eq (3.8) such that for large positive values of u

$$W_2 \sim e^{-ux} \sum_{s=0}^{\infty} (-)^s \frac{\mathscr{A}_s^*(\theta, x)}{u^s} \qquad (x \in \mathbf{G}(\theta), \quad |\theta| \leq \pi - \epsilon), \tag{3.15}$$

uniformly with respect to x and θ. The coefficients are given by the recurrence relation

$$\mathscr{A}_{s+1}^*(\theta, x) = -\frac{1}{2} \frac{d}{dx} \mathscr{A}_s^*(\theta, x) + \frac{1}{2} \int e^{-4i\theta} \mathscr{F}(e^{-2i\theta} x) \mathscr{A}_s^*(\theta, x) dx, \tag{3.16}$$

the path of integration lying in $\mathbf{D}(\theta)$.

If we now restore the variables μ and ξ by means of eqs (3.7), and let

$$\mathscr{A}_s(\xi) \equiv (-)^s e^{2si\theta} \mathscr{A}_s^*(\theta, x), \tag{3.17}$$

the relations (3.15) and (3.16) become

$$W_2 \sim e^{-\mu^2 \xi} \sum_{s=0}^{\infty} \frac{\mathscr{A}_s(\xi)}{\mu^{2s}} \qquad (t \in \mathbf{S}(\arg \mu), \quad |\arg \mu| \leq \pi - \epsilon) \tag{3.18}$$

as $|\mu| \to \infty$, where

$$\mathscr{A}_{s+1}(\xi) = \frac{1}{2} \mathscr{A}_s'(\xi) - \frac{1}{2} \int \mathscr{F}(\xi) \mathscr{A}_s(\xi) d\xi. \tag{3.19}$$

Here $\mathbf{S}(\arg \mu) \equiv \mathbf{S}(\theta)$ denotes the t-map of $\mathbf{G}(\theta)$; its boundaries are the dotted curves of figure 2.

It is of some interest to determine the asymptotic behavior of the boundaries of $\mathbf{S}(\theta)$. Consider, for example, the dotted curve AQ_1. Its equation may be written

$$\xi = (y - i\delta) e^{(\frac{3}{2}\pi - 2\theta)i}, \tag{3.20}$$

where the real parameter y ranges from 0 to $+\infty$. Reversion of (3.5) yields

$$t^2 = 2\xi + \frac{1}{2} \ln 8\xi + \frac{1}{2} + O(\xi^{-1} \ln \xi) \qquad \text{as } |\xi| \to \infty. \tag{3.21}$$

When y is large, we may substitute (3.20) in (3.21) and thence obtain

$$\arg t = \frac{3}{4}\pi - \theta + \frac{\cos 2\theta}{8} \frac{\ln 8y}{y} + \frac{\cos 2\theta + (2\theta - \frac{3}{2}\pi) \sin 2\theta - 4\delta}{8y} + O\left\{\left(\frac{\ln y}{y}\right)^2\right\}. \tag{3.22}$$

The corresponding equation of the principal curve AQ is of course obtained on setting $\delta = 0$ in this result. Clearly the principal curve AQ and the dotted curve AQ_1 ultimately always lie on the same side of their common asymptote, except when $\theta = \frac{3}{4}\pi$, in which event AQ is its own asymptote.

4. Identification of Solutions

The substitutions

$$a=-\frac{1}{2}\mu^2, \qquad z=\mu t\sqrt{2}, \tag{4.1}$$

transform eq (2.1) into (1.1). This result and the second of eqs (3.1) show that

$$W=(t^2-1)^{\frac{1}{4}}U\left(-\frac{1}{2}\mu^2,\mu t\sqrt{2}\right) \tag{4.2}$$

is a solution of (3.2). Equation (2.2) shows that when μ is fixed and $t\to\infty e^{-i\arg\mu}$ this solution becomes exponentially small. The same is true of the solution W_2 introduced in section 3, provided that $|\mu|$ is sufficiently large and $|\arg\mu|\leq\pi-\epsilon$; this can be seen from (3.18) and (3.5). Since all solutions which are linearly independent of (4.2) are exponentially large in the same circumstances, we see immediately that the solution (4.2) is a multiple of W_2.

Hence we have

$$U\left(-\frac{1}{2}\mu^2,\mu t\sqrt{2}\right)\sim g(\mu)\frac{e^{-\mu^2\xi}}{(t^2-1)^{\frac{1}{4}}}\sum_{s=0}^{\infty}\frac{\mathscr{A}_s(\xi)}{\mu^{2s}} \qquad (t\in \mathbf{S}(\arg\mu),\ |\arg\mu|\leq\pi-\epsilon) \tag{4.3}$$

as $|\mu|\to\infty$, uniformly with respect to t and $\arg\mu$. The branch of $(t^2-1)^{\frac{1}{4}}$ here and elsewhere is understood to be continuous in $\mathbf{S}(\arg\mu)$ and to be asymptotic to the principal value of $\sqrt{t}$ as $t\to\infty e^{-i\arg\mu}$. Similarly for the branch of $(t^2-1)^{\frac{1}{2}}$ used later. The function $g(\mu)$ is independent of t. Its asymptotic expansion for large $|\mu|$ can be found from the limiting form of the ratio of the two sides of the relation (4.3) as $t\to\infty e^{-i\arg\mu}$. Thus with the aid of (2.2) and (3.5) we find that

$$\frac{1}{g(\mu)}\sim 2^{\frac{1}{4}\mu^2+\frac{1}{4}}e^{\frac{1}{4}\mu^2}\mu^{-\frac{1}{2}\mu^2+\frac{1}{2}}\sum_{s=0}^{\infty}\frac{g_s}{\mu^{2s}} \qquad (|\arg\mu|\leq\pi-\epsilon) \tag{4.4}$$

as $|\mu|\to\infty$, uniformly with respect to $\arg\mu$. Here

$$g_s=\lim\mathscr{A}_s(\xi) \qquad \text{as } \xi\to\infty e^{-2i\arg\mu}. \tag{4.5}$$

That this limit exists and is independent of $\arg\mu$ can be seen from the lemma of [9], section 5. The g_s are not determined however until the arbitrary constants in the recurrence relation (3.19) have been fixed, and we now consider how to do this in the most advantageous manner.

From (3.19), (3.3) and the first of (3.1), we obtain

$$\mathscr{A}_{s+1}=\frac{1}{2}\frac{1}{(t^2-1)^{\frac{1}{2}}}\frac{d\mathscr{A}_s}{dt}+\frac{1}{8}\int\frac{3t^2+2}{(t^2-1)^{\frac{5}{2}}}\mathscr{A}_s\,dt. \tag{4.6}$$

Taking $\mathscr{A}_0=1$, we find

$$\mathscr{A}_1=\frac{t^3-6t}{24(t^2-1)^{\frac{3}{2}}}+\text{constant}. \tag{4.7}$$

A convenient value for this constant is zero, then $\mathscr{A}_1$ has the form of a polynomial in t divided by $(t^2-1)^{\frac{3}{2}}$.

The eq (4.7) suggests the substitution

$$\mathscr{A}_s=u_s(t)/(t^2-1)^{\frac{3s}{2}} \tag{4.8}$$

in (4.6). This leads to the recurrence relation

$$(t^2-1)u_s'(t)-3stu_s(t)=r_{s-1}(t), \tag{4.9}$$

where

$$8r_s(t)=(3t^2+2)u_s(t)-12(s+1)tr_{s-1}(t)+4(t^2-1)r_{s-1}'(t). \tag{4.10}$$

We can satisfy this relation by taking $u_s(t)$ to be a polynomial in t of degree $3s$ which is an odd or even function of t according as s is odd or even. All the coefficients are determined automatically if s is odd, but there is a degree of freedom if s is even, since the left of eq (4.9) vanishes when u_s is a multiple of $(t^2-1)^{\frac{3}{2}s}$. We remove this by making the coefficient of the highest power of t vanish.

Thus we prescribe that $u_s(t)$ is a polynomial in t of degree $3s$ (s odd), $3s-2$ (s even, $s\geq 2$). With $u_0(t)=1$ this condition determines the coefficients completely. Numerical calculation based on (4.9) and (4.10) yields

$$\left.\begin{aligned} &u_0(t)=1, \quad u_1(t)=(t^3-6t)/24, \quad u_2(t)=(-9t^4+249t^2+145)/1152, \\ &u_3(t)=(-4042t^9+18189t^7-28287t^5-1\,51995t^3-2\,59290t)/4\,14720. \end{aligned}\right\} \tag{4.11}$$

Hence

$$g_{2s}=0 \quad (s=1,2,\ldots), \qquad g_0=1, \quad g_1=\frac{1}{24}, \quad g_3=-\frac{2021}{2\,07360}. \tag{4.12}$$

An independent method of calculating the asymptotic expansion of $g(\mu)$ for large μ will be given in section 5.

Expansion for the derivative. Theorem A of [9] shows that the asymptotic series (4.3) may be differentiated term by term with respect to t. With the aid of the first of eqs (3.1) we deduce that

$$U'\left(-\frac{1}{2}\mu^2,\mu t\sqrt{2}\right)\sim-\frac{\mu}{\sqrt{2}}\,g(\mu)\,(t^2-1)^{\frac{1}{4}}e^{-\mu^2\xi}\sum_{s=0}^{\infty}\frac{\mathscr{B}_s(\xi)}{\mu^{2s}} \qquad (t\in \mathbf{S}(\arg\mu), \quad |\arg\mu|\leq\pi-\epsilon) \tag{4.13}$$

as $|\mu|\to\infty$, uniformly with respect to t and $\arg\mu$. Here, in the notation of Miller,

$$U'(a,z)\equiv\partial U(a,z)/\partial z.$$

The coefficients are given by

$$\mathscr{B}_s(\xi)=v_s(t)/(t^2-1)^{\frac{3}{2}s}, \tag{4.14}$$

where

$$v_s(t)=u_s(t)+\frac{1}{2}\,tu_{s-1}(t)-r_{s-2}(t). \tag{4.15}$$

The first four $v_s(t)$ are

$$\left.\begin{aligned} &v_0(t)=1, \quad v_1(t)=(t^3+6t)/24, \quad v_2(t)=(15t^4-327t^2-143)/1152, \\ &v_3(t)=(-4042t^9+18189t^7-36387t^5+2\,38425t^3+2\,59290t)/4\,14720. \end{aligned}\right\} \tag{4.16}$$

5. Use of Connection Formulas

In this section we seek expansions which hold in the complement of the domain $\mathbf{S}(\arg\mu)$, Since $U(a,z)$ is an integral function of a it will clearly suffice to obtain a set of expansions which, together with (4.3), cover the whole t-plane (except of course the neighborhoods of the points $t=\pm1$) when $|\arg\mu|\leq\frac{1}{2}\pi$.

For convenience we introduce the notation $\mathbf{U}_j\equiv\mathbf{U}_j(\theta)$ ($j=0$, 1, 2, 3, 4) for the closed domains into which the t-plane is divided by the principal curves defined in section 3. Figure 4 shows the enumeration of these regions when $0\leq\theta<\pi$. When $-\pi<\theta\leq0$ we define

$$\mathbf{U}_j(\theta)=\mathbf{U}_j^*(-\theta) \quad (j=0,2,4), \qquad \mathbf{U}_1(\theta)=\mathbf{U}_3^*(-\theta), \quad \mathbf{U}_3(\theta)=\mathbf{U}_1^*(-\theta), \tag{5.1}$$

where the star denotes the conjugate region. Then the boundaries of each region $\mathbf{U}_j(\theta)$ vary continuously with θ in the interval $(-\pi,\pi)$.

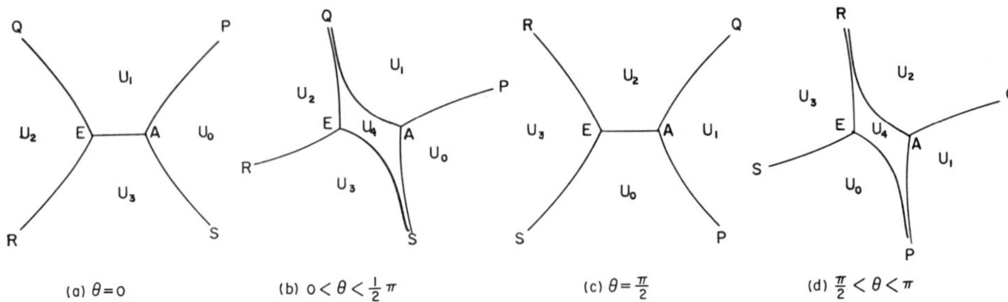

FIGURE 4. *t-plane: domains* $\mathbf{U}_j(\theta)$.

The following relations, suggested by figure 4, are easily verified:

$$\left.\begin{aligned}\mathbf{U}_0(\theta+\tfrac{1}{2}\pi)=\mathbf{U}_3(\theta),\quad \mathbf{U}_1(\theta+\tfrac{1}{2}\pi)=\mathbf{U}_0(\theta),\quad \mathbf{U}_2(\theta+\tfrac{1}{2}\pi)=\mathbf{U}_1(\theta),\\ \mathbf{U}_3(\theta+\tfrac{1}{2}\pi)=\mathbf{U}_2(\theta),\quad \mathbf{U}_4(\theta+\tfrac{1}{2}\pi)=\mathbf{U}_4(\theta),\end{aligned}\right\} \tag{5.2}$$

$$\left.\begin{aligned}\mathbf{U}_0(\theta+\pi)=\mathbf{U}_2(\theta),\quad \mathbf{U}_1(\theta+\pi)=\mathbf{U}_3(\theta),\quad \mathbf{U}_2(\theta+\pi)=\mathbf{U}_0(\theta),\\ \mathbf{U}_3(\theta+\pi)=\mathbf{U}_1(\theta),\quad \mathbf{U}_4(\theta+\pi)=\mathbf{U}_4(\theta).\end{aligned}\right\} \tag{5.3}$$

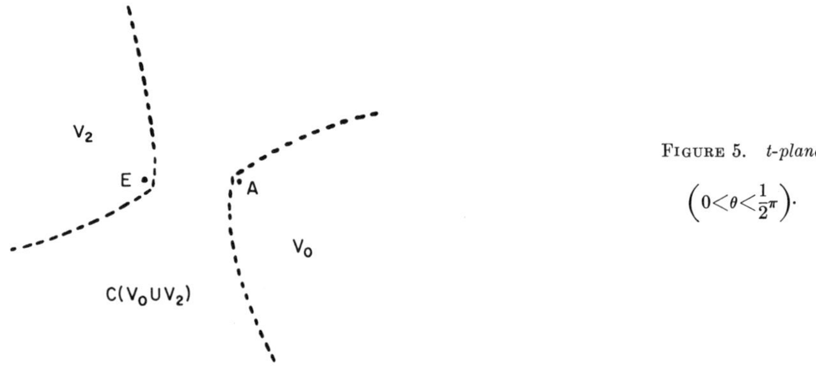

FIGURE 5. *t-plane* $\left(0<\theta<\frac{1}{2}\pi\right)$.

The region formed by adding to $\mathbf{U}_j(\theta)$ points whose x-maps (cf. fig. 3) are at a distance less than δ from the x-map of the boundary of $\mathbf{U}_j(\theta)$ we shall denote by $\mathbf{V}_j(\theta)$ or $\mathbf{V}_j$. Figure 5 illustrates for example $\mathbf{V}_0(\theta)$, $\mathbf{V}_2(\theta)$, and $C\{\mathbf{V}_0(\theta)\cup\mathbf{V}_2(\theta)\}$, the complement of the union of $\mathbf{V}_0(\theta)$ and $\mathbf{V}_2(\theta)$, when $0<\theta<\frac{1}{2}\pi$. The boundaries are of the same character as the dotted curves of figure 2.

The first connection formula we use is

$$U\left(-\frac{1}{2}\mu^2,\mu t\sqrt{2}\right)=(2\pi)^{-\frac{1}{2}}\Gamma\left(\frac{1}{2}+\frac{1}{2}\mu^2\right)\left\{e^{i\pi(-\frac{1}{4}\mu^2+\frac{1}{4})}U\left(\frac{1}{2}\mu^2,-i\mu t\sqrt{2}\right)+e^{i\pi(\frac{1}{4}\mu^2-\frac{1}{4})}U\left(\frac{1}{2}\mu^2,i\mu t\sqrt{2}\right)\right\}, \tag{5.4}$$

obtained from (2.5). Replacing μ by $\mp i\mu$ in (4.3), we have

$$U\left(\frac{1}{2}\mu^2,-i\mu t\sqrt{2}\right)\sim g(-i\mu)\,\frac{e^{\mu^2\xi}}{(t^2-1)^{\frac{1}{4}}}\sum_{s=0}^{\infty}(-)^s\,\frac{\mathscr{A}_s(\xi)}{\mu^{2s}}\qquad\left(t\in\mathbf{S}\left(\theta-\frac{1}{2}\pi\right),\ -\frac{1}{2}\pi+\epsilon\leq\theta\leq\frac{3}{2}\pi-\epsilon\right), \tag{5.5}$$

$$U\left(\frac{1}{2}\mu^2, i\mu t\sqrt{2}\right) \sim g(i\mu) \frac{e^{\mu^2\xi}}{(t^2-1)^{\frac{1}{4}}} \sum_{s=0}^{\infty} (-)^s \frac{\mathscr{A}_s(\xi)}{\mu^{2s}} \quad \left(t \in \mathbf{S}\left(\theta+\frac{1}{2}\pi\right),\ -\frac{3}{2}\pi+\epsilon \leq \theta \leq \frac{1}{2}\pi-\epsilon\right), \tag{5.6}$$

and from (4.4) and (4.12),

$$g(\mp i\mu) \sim e^{\pm i\pi(\frac{1}{4}\mu^2+\frac{1}{4})} 2^{\frac{1}{4}\mu^2-\frac{1}{4}} e^{\frac{1}{4}\mu^2} \mu^{-\frac{1}{2}\mu^2-\frac{1}{2}} \left(1-\sum_{s=0}^{\infty} \frac{g_{2s+1}}{\mu^{4s+2}}\right)^{-1} \quad \left\{\begin{matrix} -\frac{1}{2}\pi+\epsilon \leq \theta \leq \frac{3}{2}\pi-\epsilon \\ -\frac{3}{2}\pi+\epsilon \leq \theta \leq \frac{1}{2}\pi-\epsilon \end{matrix}\right\}. \tag{5.7}$$

When $|\theta| \leq \frac{1}{2}\pi-\epsilon$, we can substitute (5.5) and (5.6) in (5.4). Care is required, however, because at any given point in the t-plane the branches of $\xi=\xi(t)$ and $(t^2-1)^{\frac{1}{4}}$ may not be the same in the two formulas.

When $0 \leq \theta \leq \frac{1}{2}\pi-\epsilon$, $\mathbf{S}(\theta-\frac{1}{2}\pi)$ is the complement of the region enclosed by "dotted curves" surrounding the principal curves AS and ER of figures 2(a) and 2(b); $\mathbf{S}(\theta+\frac{1}{2}\pi)$ is the complement of the region enclosed by dotted curves surrounding AP and EQ. To the right of AP+AS, i.e., in $\mathbf{U}_0(\theta)$, the branches of $\xi(t)$ and $(t^2-1)^{\frac{1}{4}}$ are the same in (5.5) and (5.6). Substitution of these formulas in (5.4) leads to a nugatory result, however, as we would expect, since $e^{\mu^2\xi}$ is exponentially large here whereas $U(-\frac{1}{2}\mu^2, \mu t\sqrt{2})$ is exponentially small.

In the central region between AP+AS and EQ+ER, i.e., in $C(\mathbf{U}_0 \cup \mathbf{U}_2)$, the two branches of $\xi(t)$ have opposite signs. Let

$$\xi=-i\eta \text{ in (5.5)}, \qquad \xi=i\eta \text{ in (5.6)}. \tag{5.8}$$

Then η is real and positive when $-1<t<1$; from (3.4) we find

$$\eta=\int_t^1 (1-t^2)^{\frac{1}{2}} dt = \frac{1}{2}\cos^{-1}t - \frac{1}{2}\, t(1-t^2)^{\frac{1}{2}}. \tag{5.9}$$

The appropriate branches of $(t^2-1)^{\frac{1}{4}}$ and $\mathscr{A}_s(\xi)$ in (5.5) are given by

$$(t^2-1)^{\frac{1}{4}} = e^{\frac{1}{4}i\pi}(1-t^2)^{\frac{1}{4}}, \qquad \mathscr{A}_s(\xi)=e^{-\frac{3}{2}is\pi} u_s(t)/(1-t^2)^{\frac{3}{2}s} \tag{5.10}$$

(cf. (4.8)), in which $(1-t^2)^{\frac{1}{4}}$ and $(1-t^2)^{\frac{3}{2}s}$ have their principal values when $-1<t<1$. For (5.6) we change the sign of i.

Carrying out the substitution, we obtain

$$U(-\tfrac{1}{2}\mu^2, \mu t\sqrt{2}) \sim \frac{2\tilde{g}(\mu)}{(1-t^2)^{\frac{1}{4}}} \left[\cos(\mu^2\eta-\tfrac{1}{4}\pi) \sum_{s=0}^{\infty} \frac{(-)^s u_{2s}(t)}{(1-t^2)^{3s}\mu^{4s}} - \sin(\mu^2\eta-\tfrac{1}{4}\pi) \sum_{s=0}^{\infty} \frac{(-)^s u_{2s+1}(t)}{(1-t^2)^{3s+\frac{3}{2}}\mu^{4s+2}}\right] \tag{5.11}$$

as $|\mu|\rightarrow\infty$, uniformly with respect to $t \in C(\mathbf{V}_0 \cup \mathbf{V}_2)$ and θ in the interval $(-\frac{1}{2}\pi+\epsilon, \frac{1}{2}\pi-\epsilon)$. Here

$$\begin{aligned} \tilde{g}(\mu) &\sim (2\pi)^{-\frac{1}{2}} \Gamma(\tfrac{1}{2}+\tfrac{1}{2}\mu^2) 2^{\frac{1}{4}\mu^2-\frac{1}{4}} e^{\frac{1}{4}\mu^2} \mu^{-\frac{1}{2}\mu^2-\frac{1}{2}} \left\{1-\sum_{s=0}^{\infty} \frac{g_{2s+1}}{\mu^{4s+2}}\right\}^{-1} \\ &\sim 2^{-\frac{1}{4}\mu^2-\frac{1}{4}} e^{-\frac{1}{4}\mu^2} \mu^{\frac{1}{2}\mu^2-\frac{1}{2}} \left\{\sum_{s=0}^{\infty} \frac{\gamma_s}{(\frac{1}{2}\mu^2)^s}\right\} \left\{1-\sum_{s=0}^{\infty} \frac{g_{2s+1}}{\mu^{4s+2}}\right\}^{-1} \qquad (|\theta| \leq \tfrac{1}{2}\pi-\epsilon), \end{aligned} \tag{5.12}$$

on substituting for $\Gamma(\frac{1}{2}+\frac{1}{2}\mu^2)$ by means of (2.22).

The expansion (5.11) is one of the required formulas. Before leaving it we shall show that $\tilde{g}(\mu)=g(\mu)$, and in the process obtain a new derivation of the coefficients in the asymptotic expansion (4.4) for $1/g(\mu)$.

When $\epsilon \leq \theta \leq \frac{1}{2}\pi-\epsilon$ the principal curve AQ is included in the region of validity of (5.11); this is easily proved by considering the x-map. On this curve $\mu^2\eta$ is positive. Let us set

$$\eta=\eta_m \equiv \left(m+\frac{1}{4}\right)\frac{\pi}{\mu^2}, \tag{5.13}$$

where m is a positive integer. Then (5.11) reduces to

$$U\left(-\frac{1}{2}\mu^2, \mu t_m\sqrt{2}\right)\sim(-)^m\frac{2\tilde{g}(\mu)}{(1-t_m^2)^{\frac{1}{4}}}\sum_{s=0}^{\infty}\frac{(-)^s u_{2s}(t_m)}{(1-t_m^2)^{3s}\mu^{4s}}, \tag{5.14}$$

where t_m is the value of t corresponding to $\eta=\eta_m$.

Now keep μ fixed and let $m\to\infty$. Then $t_m\to\infty e^{i(\frac{3}{4}\pi-\theta)}$, and so from (2.7)

$$U\left(-\frac{1}{2}\mu^2, \mu t_m\sqrt{2}\right)\sim e^{-\frac{1}{2}\mu^2 t_m^2}(\mu t_m\sqrt{2})^{\frac{1}{2}\mu^2-\frac{1}{2}}+\frac{(2\pi)^{\frac{1}{2}}e^{i\pi(\frac{1}{2}\mu^2+\frac{1}{2})}}{\Gamma(\frac{1}{2}-\frac{1}{2}\mu^2)}e^{\frac{1}{2}\mu^2 t_m^2}(\mu t_m\sqrt{2})^{-\frac{1}{2}\mu^2-\frac{1}{2}}. \tag{5.15}$$

From (3.5) and the first of (5.8)

$$\frac{1}{2}t_m^2-\frac{1}{2}\ln 2t_m=-i\eta_m+\frac{1}{4}+O(|t_m|^{-2}). \tag{5.16}$$

Hence

$$e^{-\frac{1}{2}\mu^2 t_m^2}(\mu t_m\sqrt{2})^{\frac{1}{2}\mu^2-\frac{1}{2}}\sim(-)^m e^{\frac{1}{4}\pi i-\frac{1}{4}\mu^2}2^{-\frac{1}{4}\mu^2-\frac{1}{4}}\mu^{\frac{1}{2}\mu^2-\frac{1}{2}}t_m^{-\frac{1}{2}},$$

$$e^{\frac{1}{2}\mu^2 t_m^2}(\mu t_m\sqrt{2})^{-\frac{1}{2}\mu^2-\frac{1}{2}}\sim(-)^m e^{-\frac{1}{4}\pi i+\frac{1}{4}\mu^2}2^{\frac{1}{4}\mu^2-\frac{1}{4}}\mu^{-\frac{1}{2}\mu^2-\frac{1}{2}}t_m^{-\frac{1}{2}},$$

and substituting these results in (5.15), we obtain

$$(-)^m e^{\frac{1}{4}\pi i-\frac{1}{4}\mu^2}2^{-\frac{1}{4}\mu^2-\frac{1}{4}}\mu^{\frac{1}{2}\mu^2-\frac{1}{2}}\left\{1+\frac{(2\pi)^{\frac{1}{2}}e^{\frac{1}{2}i\pi\mu^2+\frac{1}{2}\mu^2}2^{\frac{1}{2}\mu^2}\mu^{-\mu^2}}{\Gamma(\frac{1}{2}-\frac{1}{2}\mu^2)}\right\}t_m^{-\frac{1}{2}}, \tag{5.17}$$

as the limiting form of the left-hand side of (5.14).

On the right of (5.14) we have

$$(1-t_m^2)^{\frac{1}{4}}\sim e^{-\frac{1}{4}\pi i}t_m^{\frac{1}{2}}, \qquad u_{2s}(t_m)/(1-t_m^2)^{3s}\to 0 \quad (s\geq 1)$$

(cf. (5.10), (4.5) and (4.12)). Comparing the two sides we see that

$$\tilde{g}(\mu)\simeq 2^{-\frac{1}{4}\mu^2-\frac{1}{4}}e^{-\frac{1}{4}\mu^2}\mu^{\frac{1}{2}\mu^2-\frac{1}{2}}\left\{\frac{1}{2}+\frac{(2\pi)^{\frac{1}{2}}e^{\frac{1}{2}i\pi\mu^2+\frac{1}{2}\mu^2}2^{\frac{1}{2}\mu^2}\mu^{-\mu^2}}{2\Gamma(\frac{1}{2}-\frac{1}{2}\mu^2)}\right\} \qquad \left(\epsilon\leq\theta\leq\frac{1}{2}\pi-\epsilon\right), \tag{5.18}$$

where we use the symbol $\simeq$ to mean that the two sides of this relation have the same asymptotic expansion for large $|\mu|$.

We now substitute (2.27), with $z=-\frac{1}{2}\mu^2$, in (5.18) and obtain

$$\tilde{g}(\mu)\sim 2^{-\frac{1}{4}\mu^2-\frac{1}{4}}e^{-\frac{1}{4}\mu^2}\mu^{\frac{1}{2}\mu^2-\frac{1}{2}}\left\{\frac{1}{2}+\frac{1}{2}\sum_{s=0}^{\infty}\frac{\gamma_s}{(\frac{1}{2}\mu^2)^s}\right\}. \tag{5.19}$$

Although this result has been proved on the assumption that $\epsilon\leq\theta\leq\frac{1}{2}\pi-\epsilon$, comparison with (5.12) shows immediately that it holds in the range $|\theta|\leq\frac{1}{2}\pi-\epsilon$, and that

$$\frac{1}{2}+\frac{1}{2}\sum_{s=0}^{\infty}\frac{\gamma_s}{(\frac{1}{2}\mu^2)^s}\sim\left\{\sum_{s=0}^{\infty}\frac{\gamma_s}{(\frac{1}{2}\mu^2)^s}\right\}\left\{1-\sum_{s=0}^{\infty}\frac{g_{2s+1}}{\mu^{4s+2}}\right\}^{-1}, \tag{5.20}$$

identically. Thence using (2.26) and changing the sign of μ^2, we find

$$\left\{1+\sum_{s=0}^{\infty}\frac{g_{2s+1}}{\mu^{4s+2}}\right\}\left\{\frac{1}{2}+\frac{1}{2}\sum_{s=0}^{\infty}\frac{\gamma_s}{(\frac{1}{2}\mu^2)^s}\right\}\sim 1. \tag{5.21}$$

Multiplying this identity with (5.19), and comparing the result with (4.4) (remembering that $g_{2s}=0$ if $s\geq 1$), we see that we can take

$$\tilde{g}(\mu)=g(\mu) \tag{5.22}$$

in the expansion (5.11), as stated earlier.

We can express the coefficients g_{2s+1} in terms of γ_s, by equating coefficients in (5.21); thus

$$g_1=-\gamma_1=\frac{1}{24}, \qquad g_3=-2g_1\gamma_2-4\gamma_3=-\frac{2021}{2\,07360},$$

on using (2.25). These values agree with those given in (4.12).

The expansion (5.11) is the first of our expansions for $U(-\frac{1}{2}\mu^2, \mu t\sqrt{2})$ in regions complementary to $\mathbf{S}(\theta)$. The corresponding expansion for the derivative is

$$U'(-\tfrac{1}{2}\mu^2, \mu t\sqrt{2}) \sim \mu\sqrt{2}g(\mu)(1-t^2)^{\frac{1}{4}}\left[\sin(\mu^2\eta-\tfrac{1}{4}\pi)\sum_{s=0}^{\infty}\frac{(-)^s v_{2s}(t)}{(1-t^2)^{3s}\mu^{4s}}\right.$$
$$\left.+\cos(\mu^2\eta-\tfrac{1}{4}\pi)\sum_{s=0}^{\infty}\frac{(-)^s v_{2s+1}(t)}{(1-t^2)^{3s+\frac{3}{2}}\mu^{4s+2}}\right], \tag{5.23}$$

valid under the same conditions as (5.11); the coefficients $v_s(t)$ are given by (4.15) and (4.16).

A further expansion can be found by substituting (5.5) and (5.6) in (5.4) when t lies in the region ($\mathbf{U}_2$) to the left of EQ+ER. We shall not carry out this substitution however, because the expansion so obtained is contained in a more general result (5.28), which we shall now derive. As a preliminary we record the following relations in which the symbol $\simeq$ has the same meaning as in (5.18).

$$g(\mu e^{i\pi}) \simeq e^{i\pi(\frac{1}{2}\mu^2-\frac{1}{2})}g(\mu) \qquad (-\pi+\epsilon\leq\arg\mu\leq-\epsilon), \tag{5.24}$$

$$\left.\begin{aligned}\frac{(2\pi)^{\frac{1}{2}}e^{-i\pi(\frac{1}{4}\mu^2+\frac{1}{4})}}{\Gamma(\frac{1}{2}-\frac{1}{2}\mu^2)}g(\mu e^{-\frac{1}{2}i\pi}) &\simeq 2\cos\frac{1}{2}\pi\mu^2\cdot g(\mu) \qquad (|\arg\mu|\leq\frac{1}{2}\pi-\epsilon),\\ &\simeq e^{-\frac{1}{2}i\pi\mu^2}g(\mu) \qquad (\epsilon\leq\arg\mu\leq\pi-\epsilon).\end{aligned}\right\} \tag{5.25}$$

We use the connection formula

$$U\left(-\frac{1}{2}\mu^2,-\mu t\sqrt{2}\right)=e^{i\pi(-\frac{1}{2}\mu^2+\frac{1}{2})}U\left(-\frac{1}{2}\mu^2,\mu t\sqrt{2}\right)+\frac{(2\pi)^{\frac{1}{2}}e^{-i\pi(\frac{1}{4}\mu^2+\frac{1}{4})}}{\Gamma(\frac{1}{2}-\frac{1}{2}\mu^2)}U\left(\frac{1}{2}\mu^2,-i\mu t\sqrt{2}\right), \tag{5.26}$$

obtained from (2.4). We may substitute the expansions (4.3) and (5.5) when

$$-\tfrac{1}{2}\pi+\epsilon\leq\theta\leq\pi-\epsilon.$$

The branches of $\xi(t)$ are the same in both expansions when

$$\left.\begin{aligned}&t\in C(\mathbf{V}_2\cup\mathbf{V}_3\cup\mathbf{V}_4) && \left(0\leq\theta\leq\frac{1}{2}\pi\right),\\ &t\in C(\mathbf{V}_2\cup\mathbf{V}_3) && \left(-\frac{1}{2}\pi+\epsilon\leq\theta\leq0,\ \ \frac{1}{2}\pi\leq\theta\leq\pi-\epsilon\right).\end{aligned}\right\} \tag{5.27}$$

Using (5.25) we obtain immediately

$$\left.\begin{aligned}U\left(-\frac{1}{2}\mu^2,-\mu t\sqrt{2}\right) \sim \frac{g(\mu)}{(t^2-1)^{\frac{1}{4}}}&\left[\left(\sin\frac{1}{2}\pi\mu^2+i\cos\frac{1}{2}\pi\mu^2\right)e^{-\mu^2\xi}\sum_{s=0}^{\infty}\frac{\mathscr{A}_s(\xi)}{\mu^{2s}}\right.\\ &\left.+2\cos\frac{1}{2}\pi\mu^2\cdot e^{\mu^2\xi}\sum_{s=0}^{\infty}(-)^s\frac{\mathscr{A}_s(\xi)}{\mu^{2s}}\right]\\ t\in C(\mathbf{V}_2\cup\mathbf{V}_3) \qquad &\left(-\frac{1}{2}\pi+\epsilon\leq\theta\leq0\right),\\ t\in C(\mathbf{V}_2\cup\mathbf{V}_3\cup\mathbf{V}_4) \qquad &\left(0\leq\theta\leq\frac{1}{2}\pi-\epsilon\right),\end{aligned}\right\} \tag{5.28}$$

and

$$\left.\begin{aligned}U\left(-\frac{1}{2}\mu^2,-\mu t\sqrt{2}\right)\sim e^{-\frac{1}{2}i\pi\mu^2}\frac{g(\mu)}{(t^2-1)^{\frac{1}{4}}}\left[ie^{-\mu^2\xi}\sum_{s=0}^{\infty}\frac{\mathscr{A}_s(\xi)}{\mu^{2s}}+e^{\mu^2\xi}\sum_{s=0}^{\infty}(-)^s\frac{\mathscr{A}_s(\xi)}{\mu^{2s}}\right]\\ t\in C(\mathbf{V}_2\cup\mathbf{V}_3\cup\mathbf{V}_4)\qquad\left(\epsilon\leq\theta\leq\frac{1}{2}\pi\right),\\ t\in C(\mathbf{V}_2\cup\mathbf{V}_3)\qquad\left(\frac{1}{2}\pi\leq\theta\leq\pi-\epsilon\right).\end{aligned}\right\}\tag{5.29}$$

The last result may be expressed in a form more pertinent to the range $|\theta|\leq\frac{1}{2}\pi$ by replacing μ by $\mu e^{i\pi}$ and using (5.24). This gives

$$\left.\begin{aligned}U\left(-\frac{1}{2}\mu^2,\mu t\sqrt{2}\right)\sim\frac{g(\mu)}{(t^2-1)^{\frac{1}{4}}}\left[e^{-\mu^2\xi}\sum_{s=0}^{\infty}\frac{\mathscr{A}_s(\xi)}{\mu^{2s}}-ie^{\mu^2\xi}\sum_{s=0}^{\infty}(-)^s\frac{\mathscr{A}_s(\xi)}{\mu^{2s}}\right]\\ t\in C(\mathbf{V}_0\cup\mathbf{V}_1)\qquad\left(-\frac{1}{2}\pi\leq\theta\leq-\epsilon\right).\end{aligned}\right\}\tag{5.30}$$

Three further formulas, (5.28*), (5.29*), and (5.30*), can be derived immediately from (5.28), (5.29), and (5.30), respectively, merely by changing the sign of i. They are valid with the conditions

$$\left.\begin{aligned}t\in C(\mathbf{V}_1\cup\mathbf{V}_2\cup\mathbf{V}_4)\qquad\left(-\frac{1}{2}\pi+\epsilon\leq\theta\leq 0\right),\\ t\in C(\mathbf{V}_1\cup\mathbf{V}_2)\qquad\left(0\leq\theta\leq\frac{1}{2}\pi-\epsilon\right),\end{aligned}\right\}\text{for (5.28*)},$$

$$\left.\begin{aligned}t\in C(\mathbf{V}_1\cup\mathbf{V}_2)\qquad\left(-\pi+\epsilon\leq\theta\leq-\frac{1}{2}\pi\right),\\ t\in C(\mathbf{V}_1\cup\mathbf{V}_2\cup\mathbf{V}_4)\qquad\left(-\frac{1}{2}\pi\leq\theta\leq-\epsilon\right),\end{aligned}\right\}\text{for (5.29*)},$$

$$t\in C(\mathbf{V}_0\cup\mathbf{V}_3)\qquad\left(\epsilon\leq\theta\leq\frac{1}{2}\pi\right),\qquad\text{for (5.30*)}.$$

6. Summary of Expansions in Terms of Elementary Functions

The principal expansion of this kind is (4.3). It is valid for large $|\mu|$ uniformly with respect to arg $\mu\equiv\theta$ and t when $-\pi+\epsilon\leq\theta\leq\pi-\epsilon$ and t lies in the unbounded domain $\mathbf{S}(\theta)$, the boundaries of which are the dotted curves illustrated in figure 2. The function $\xi(t)$ is given by (3.4), that branch being chosen which is continuous in $\mathbf{S}(\theta)$ and takes the value (3.13) or (3.14) at the origin. The branch of $(t^2-1)^{\frac{1}{4}}$ in (4.3) is continuous in $\mathbf{S}(\theta)$ and is asymptotic to the principal value of $\sqrt{t}$ as $t\to\infty e^{-i\theta}$ Both $\xi(t)$ and $(t^2-1)^{\frac{1}{4}}$ are positive when $t>1$ and $|\theta|<\frac{3}{4}\pi$. The coefficients $\mathscr{A}_s(\xi)$ are given by (4.8) and (4.11), the branch of $(t^2-1)^{\frac{3}{2}s}$ being defined similarly. The function $g(\mu)$ is calculable from either of the asymptotic expansions

$$\frac{1}{g(\mu)}\sim 2^{\frac{1}{4}\mu^2+\frac{1}{4}}e^{\frac{1}{4}\mu^2}\mu^{-\frac{1}{2}\mu^2+\frac{1}{2}}\left\{1+\sum_{s=0}^{\infty}\frac{g_{2s+1}}{\mu^{4s+2}}\right\},\tag{6.1}$$

$$g(\mu)\sim 2^{-\frac{1}{4}\mu^2-\frac{1}{4}}e^{-\frac{1}{4}\mu^2}\mu^{\frac{1}{2}\mu^2-\frac{1}{2}}\left\{1+\frac{1}{2}\sum_{s=1}^{\infty}\frac{\gamma_s}{(\frac{1}{2}\mu^2)^s}\right\},\tag{6.2}$$

both of which are uniformly valid with respect to arg μ in the interval $(-\pi+\epsilon,\ \pi-\epsilon)$. Here g_1, g_3 are given by (4.12); γ_1, γ_2, γ_3, γ_4 by (2.25).

Expansions valid in regions complementary to $\mathbf{S}(\theta)$ are (5.11), (5.28), (5.29), (5.30), (5.28*), (5.29*), and (5.30*). When $0\leq\theta\leq\frac{1}{2}\pi$ the expansions (5.28) and (5.29) between them cover the principal curve ER of figure 2; expansions (5.28*) and (5.30*) cover the principal curve AQ, except when $0\leq\theta\leq\epsilon$ and $\operatorname{Re} t\geq-1-\delta'$. The region of validity of (5.11) however includes that part of AQ to the right of $\operatorname{Re} t\geq-1+\delta'$ when $0\leq\theta\leq\epsilon$. Hence the expansions we have obtained cover the whole t-plane (except of course the neighborhoods of the points $t=\pm1$) when $0\leq\theta\leq\frac{1}{2}\pi$, and similarly also when $-\frac{1}{2}\pi\leq\theta\leq0$.

The branches of the functions ξ, $(t^2-1)^{\frac{1}{4}}$ and $\mathscr{A}_s(\xi)$ in (5.28), (5.29), (5.30), and the conjugate results are the same as in (4.3), and the function $g(\mu)$ again has the asymptotic expansions (6.1) and (6.2).

In (5.11) $\tilde{g}(\mu)$ is identical with $g(\mu)$, and $\eta=\eta(t)$ is given by (5.9). The branch of the latter function is continuous in the region of validity of (5.11) and positive when $-1<t<1$; the same is true of the branches of $(1-t^2)^{\frac{1}{4}}$ and $(1-t^2)^{\frac{3}{2}}$.

The expansions for the derivative $U'\,(-\frac{1}{2}\mu^2,\,\mu t\sqrt{2})$ corresponding to (4.3) and (5.11) are (4.13) and (5.23) respectively, the coefficients $\mathscr{B}_s(\xi)$ and $v_s(t)$ being given by (4.14) to (4.16). The derivative expansions corresponding to (5.28) to (5.30*) have not been recorded, but they can be deduced immediately by use of the formal identities

$$\frac{d}{dt}\frac{e^{\pm\mu^2\xi}}{(t^2-1)^{\frac{1}{4}}}\sum_{s=0}^{\infty}(\mp)^s\frac{\mathscr{A}_s(\xi)}{\mu^{2s}}=\pm\mu^2(t^2-1)^{\frac{1}{4}}e^{\pm\mu^2\xi}\sum_{s=0}^{\infty}(\mp)^s\frac{\mathscr{B}_s(\xi)}{\mu^{2s}}, \tag{6.3}$$

and are valid under the same conditions as the parent formulas.

Uniform asymptotic expansions of solutions of eq (1.1) for large $|\mu|$ and unbounded t have previously been given by Watson [12], Schwid [10], and Darwin [1].

Watson applies the method of steepest descents to integral representations of

$$D_n(z)\equiv U\left(-n-\frac{1}{2},\,z\right)$$

and obtains a set of asymptotic expansions for the function $D_n(2\zeta\sqrt{n})$ when $|n|$ is large and $|\arg\,n|\leq\frac{1}{2}\,\pi-\epsilon$, which cover the whole ζ-plane other than the neighborhoods of the points $\zeta=\pm1$. For each expansion the ζ-region of validity is fixed, unlike the regions obtained in the present paper which are permitted to depend on arg n. For example, the region of validity of Watson's expansion analogous to our principal expansion (4.3) is

$$\mathbf{U}_0\left(\frac{1}{4}\pi\right)\cup\mathbf{U}_1\left(\frac{1}{4}\pi\right)\quad(|\zeta-1|\geq\delta').$$

This is, as we would expect, the region common to $\mathbf{S}(\arg\mu)$ as $\arg\mu$ varies over the interval $(-\frac{1}{4}\pi+\epsilon,\,\frac{1}{4}\pi-\epsilon)$.

Schwid applies the methods of Langer for the asymptotic solution of differential equations and his results resemble those we have obtained. The main differences are that Schwid confines attention to the regions $\operatorname{Re} t\geq0$ and $|\arg\,\mu|<\frac{3}{4}\pi$, gives only the leading terms in each expansion and identifies the asymptotic solutions by considering their behavior at $t=0$ rather than $t=\infty$.

Schwid's results apply to the even and odd solutions of the differential equation. An effect of carrying out the identification at $t=0$ is that the error term in the asymptotic representation of each of these solutions is obtained essentially in the form

$$e^{-\mu^2\xi}O(|\mu|^{-2})+e^{\mu^2\xi}O(|\mu|^{-2}) \tag{6.4}$$

in the present notation. If we try to deduce from these results the asymptotic form of the function $U\,(-\frac{1}{2}\mu^2,\,\mu t\sqrt{2})$ by means of connection formulas, the error term will necessarily also be of the form (6.4). Accordingly, a nugatory result will be obtained in the region $(\mathbf{U}_0(\arg\mu))$ in which $U\,(-\frac{1}{2}\mu^2,\,\mu t\sqrt{2})$ is exponentially small for large $|\mu|$. Indeed, the proof given on page

358 of Schwid's paper of the fact that the error term in the asymptotic expression for the Hermite polynomials is of the form $e^{-\mu^2\xi}O(|\mu|^{-2})$ instead of (6.4) is false. The mistake is apparently due to overlooking the fact that the O's are functions of the independent variable as well as of the parameter μ.

Darwin develops series solutions of the eq (2.10), and in [4] Miller applies the same methods to eq (2.1). The results are typified by the series

$$U(a,z)=\frac{\{\Gamma(\frac{1}{2}-a)\}^{\frac{1}{2}}}{(2\pi)^{\frac{1}{4}}}\exp\left\{-\theta-\frac{1}{2}\ln X+\sum_{s=1}^{\infty}(-)^s\frac{d_{3s}}{X^{3s}}\right\} \tag{6.5}$$

for negative a and real positive z, in which

$$X=\sqrt{(z^2+4a)}, \qquad \theta=\frac{1}{2}\int_{2\sqrt{(-a)}}^{z} Xdz,$$

and d_{3s} is a polynomial in z of degree $3s$ if s is odd, or s if s is even. Explicit expressions for $d_3, d_6, \ldots, d_{24}$ are given on page 68 of [4].

The series (6.5) is derived by purely formal methods from the differential equation without any investigation of its asymptotic nature. Applying the substitutions $a=-\frac{1}{2}\mu^2$ and $z=\mu t\sqrt{2}$ we find that $X=\mu(t^2-1)^{\frac{1}{2}}\sqrt{2}$ and $\theta=\mu^2\xi(t)$, in the notation of sections 3 and 4. Next, replacing $\{\Gamma(\frac{1}{2}-a)\}^{\frac{1}{2}}\equiv\{\Gamma(\frac{1}{2}+\frac{1}{2}\mu^2)\}^{\frac{1}{2}}$ by its asymptotic expansion for large positive μ (cf. (2.22)) and expanding $\exp\{\sum(-)^s d_{3s}X^{-3s}\}$ in descending powers of X^3, we find that the series (6.5) reduces to the form (4.3). Hence (6.5) must be a uniform asymptotic expansion for large $|a|$ of the same character as (4.3).

Darwin remarks that the explicit expressions for the coefficients d_{3s} are simpler in form and easier to calculate than the expressions for the coefficients in series of the form (4.3). This is indeed true, but we cannot however regard the form (6.5) as being entirely superior to (4.3); it does not, for example, lead to convenient forms of expansion when we differentiate with respect to z.

Part 3. Expansions in Terms of Airy Functions

7. Asymptotic Solutions of the Differential Equation

In this section we seek to apply theorem B of [9] to the differential eq (1.1). The preliminary step is to take new variables $\zeta=\zeta(t)$ and W, defined by

$$\zeta\left(\frac{d\zeta}{dt}\right)^2=t^2-1, \qquad W=\left(\frac{d\zeta}{dt}\right)^{\frac{1}{2}}w=\left(\frac{t^2-1}{\zeta}\right)^{\frac{1}{4}}w \tag{7.1}$$

(cf. [7], sec. 2). Equation (1.1) then becomes

$$\frac{d^2W}{d\zeta^2}=\{\mu^4\zeta+F(\zeta)\}W, \tag{7.2}$$

where

$$F(\zeta)\equiv\left(\frac{dt}{d\zeta}\right)^{\frac{1}{2}}\frac{d^2}{d\zeta^2}\left\{\left(\frac{dt}{d\zeta}\right)^{-\frac{1}{2}}\right\}=\frac{5}{16\zeta^2}+\zeta\mathscr{F}(\xi)=\frac{5}{16\zeta^2}-\frac{(3t^2+2)}{4(t^2-1)^3}\zeta \tag{7.3}$$

on reduction (cf. (3.3)).

The first of eqs (7.1) may be integrated to give

$$\frac{2}{3}\zeta^{\frac{3}{2}}=\int_1^t(t^2-1)^{\frac{1}{2}}dt=\frac{1}{2}t(t^2-1)^{\frac{1}{2}}-\frac{1}{2}\ln\{t+(t^2-1)^{\frac{1}{2}}\}, \tag{7.4}$$

on specifying that $\zeta=0$ when $t=1$. In terms of the variable ξ of section 3,

$$\frac{2}{3}\zeta^{\frac{3}{2}}=\xi. \tag{7.5}$$

An essential difference between the functions $\xi(t)$ and $\zeta(t)$ is that the former has a branch point at $t=1$, whereas the latter is analytic at this point. Both functions have branch points at $t=-1$. The branch of $\zeta(t)$ which will be the most convenient for our purposes depends on $\theta\equiv\arg\mu$. We define it to be positive when $t>1$ and continuous in the t-plane cut along the principal curve ER of figure 2 when $0<\theta<\pi$, along the curve conjugate to ER when $-\pi<\theta<0$, or along EQ and ER when $\theta=0$. This determines $\zeta(t)$ uniquely everywhere except in the region ($\mathbf{U}_2(0)$) to the left of EQ+ER when $\theta=0$.

Following [9], section 1, we make the further substitutions

$$u=|\mu|^2, \qquad \theta=\arg\mu, \qquad \zeta=e^{-\frac{2}{3}i\theta}z. \tag{7.6}$$

Equation (7.2) then becomes

$$\frac{d^2W}{dz^2}=\{u^2z+e^{-\frac{4}{3}i\theta}F(e^{-\frac{2}{3}i\theta}z)\}W. \tag{7.7}$$

The mapping of the t-plane on the z-plane is conveniently carried out by using $x(t)$ (defined in sec. 3) as an intermediate variable; in terms of z

$$x=\frac{2}{3}z^{\frac{3}{2}}. \tag{7.8}$$

In our first application of theorem B to eq (7.7) we make a cut in the z-plane along the part of the level curve of the function exp $(\frac{2}{3}z^{\frac{3}{2}})$ which joins the point E of affix$-(\frac{3}{4}\pi)^{\frac{2}{3}}e^{\frac{2}{3}i\theta}$ and corresponding to $t=-1$, to the point at infinity on the negative real z-axis. Such a join can be made when $|\theta|\leq\frac{1}{2}\pi$ (cf. [7], fig. 2).

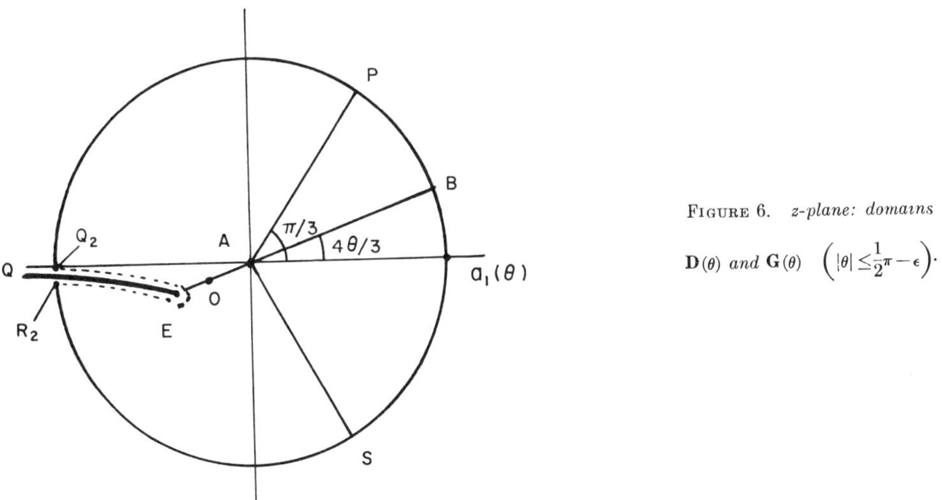

FIGURE 6. *z-plane: domains* $\mathbf{D}(\theta)$ *and* $\mathbf{G}(\theta)$ $\left(|\theta|\leq\frac{1}{2}\pi-\epsilon\right)$.

Figure 6 illustrates the mapping, the lettering corresponding to figures 2 and 3. The cut is the heavy curve EQ. The principal curves through A are mapped on the rays $\arg z=\pm\frac{1}{3}\pi$ and π; the principal curves through E are mapped on the two sides of the cut EQ and the remaining part of the level curve of exp $(\frac{2}{3}z^{\frac{3}{2}})$ passing through E. Cut in this way, the z-plane is mapped on the complement of the domain $\mathbf{U}_2(\theta)$ (see fig. 4) in the t-plane.

When $|\theta|\leq\frac{1}{2}\pi-\epsilon$, we take the z-domain $\mathbf{D}(\theta)$ of [9], section 9, to be the cut plane with the circle center E and radius $\frac{1}{2}\delta$ removed, δ being the fixed small positive number introduced in section 3. Then $F(e^{-\frac{4}{3}i\theta}z)$ is regular in $\mathbf{D}(\theta)$ (including $z=0$), and the condition (9.2) of [9] is satisfied with $\sigma=\frac{3}{2}$ because

$$e^{-\frac{8}{3}i\theta}F(e^{-\frac{4}{3}i\theta}z)\sim-\frac{7}{64z^2} \tag{7.9}$$

as $|z|\to\infty$, uniformly with respect to bounded arg z and bounded θ; this result follows from (7.3), (7.5), and (3.5).

We take the z-domain $\mathbf{G}(\theta)$ required by [9], section 9, to be the complement of the map of the t-domain $\mathbf{V}_2(\theta)$, defined in section 5. The boundaries of $\mathbf{G}(\theta)$ are the dotted curves surrounding the cut EQ in figure 6. All the preliminary conditions of [9], section 9, are then satisfied if $\delta=\delta(\epsilon)$ is taken to be small enough.

Taking the point $a_1(\theta)$ of [9], section 9 to be at infinity on the positive real axis, we find $\mathbf{H}_1(\theta)\equiv\mathbf{G}(\theta)$. Hence from theorem B, a solution W_1 of (7.7) exists, such that for large positive values of u

$$W_1\sim\mathrm{Ai}(u^{\frac{2}{3}}z)\sum_{s=0}^{\infty}\frac{A_s(\theta,z)}{u^{2s}}+\frac{\mathrm{Ai}'(u^{\frac{2}{3}}z)}{u^{\frac{4}{3}}}\sum_{s=0}^{\infty}\frac{B_s(\theta,z)}{u^{2s}} \qquad \left(z\in\mathbf{G}(\theta),\ |\theta|\leq\frac{1}{2}\pi-\epsilon\right)$$

uniformly with respect to z and θ, the coefficients $A_s(\theta,z)$ and $B_s(\theta,z)$ being given by recurrence relations derived from eqs (9.12) and (9.13) of [9]. When the variables μ and ζ are restored, this result becomes

$$W_1\sim\mathrm{Ai}(\mu^{\frac{4}{3}}\zeta)\sum_{s=0}^{\infty}\frac{A_s(\zeta)}{\mu^{4s}}+\frac{\mathrm{Ai}'(\mu^{\frac{4}{3}}\zeta)}{\mu^{\frac{8}{3}}}\sum_{s=0}^{\infty}\frac{B_s(\zeta)}{\mu^{4s}} \tag{7.10}$$

as $|\mu|\to\infty$, when

$$t\in C\mathbf{V}_2(\theta)\text{ and }-\frac{1}{2}\pi+\epsilon\leq\theta\leq\frac{1}{2}\pi-\epsilon, \tag{7.11}$$

where $A_0(\zeta)=$constant and

$$\left.\begin{aligned}B_s(\zeta)&=\frac{1}{2}\zeta^{-\frac{1}{2}}\int_0^{\zeta}\zeta^{-\frac{1}{2}}\{-A_s''(\zeta)+F(\zeta)A_s(\zeta)\}d\zeta\\ A_{s+1}(\zeta)&=-\frac{1}{2}B_s'(\zeta)+\frac{1}{2}\int^{\zeta}F(\zeta)B_s(\zeta)d\zeta\end{aligned}\right\}\quad(s=0,1,2,\ldots), \tag{7.12}$$

the paths of integration lying in the map of $C\mathbf{U}_2(\theta)$.

A second application of theorem B can be made to the differential eq (7.7) by taking the cut in the z-plane for the function $F(e^{-\frac{4}{3}i\theta}z)$ to be the part of the level curve of exp $(\frac{2}{3}z^{\frac{3}{2}})$ joining E to S, the point at infinity on arg $z=-\frac{1}{3}\pi$. This join is possible if $0\leq\theta\leq\pi$; the corresponding t-domain is then $C\mathbf{U}_3(\theta)$.

With similar analysis, again taking $a_1(\theta)$ to be the point $z=+\infty$, we may show that a solution of eq (7.2) exists having the right-hand side of (7.10) as its asymptotic expansion for large $|\mu|$, uniformly valid with respect to t and θ when

$$t\in C\mathbf{V}_3(\theta)\text{ and }\epsilon\leq\theta\leq\pi-\epsilon. \tag{7.13}$$

The coefficients $A_s(\zeta)$ and $B_s(\zeta)$ are again given by (7.12), the paths of integration now lying in the map of $C\mathbf{U}_3(\theta)$.

By considering the limiting behavior as $\zeta\to\infty e^{-\frac{4}{3}i\theta}$, we see that when $\epsilon\leq\theta\leq\frac{1}{2}\pi-\epsilon$, the ratio of the new solution to W_1 is independent of ζ and has an asymptotic expansion for large $|\mu|$ identically equal to unity. Hence the expansion (7.10) for W_1 holds with the conditions (7.13). Similar analysis, or an appeal to symmetry, shows that this expansion also holds with the further conditions

$$t\in C\mathbf{V}_1(\theta)\text{ and }-\pi+\epsilon\leq\theta\leq-\epsilon. \tag{7.14}$$

We can combine the three t-regions of validity into a single region $\mathbf{T}(\theta)$, given by

$$\left.\begin{array}{l} CV_1, \quad CV_1 \cup CV_2, \quad CV_2, \quad CV_2 \cup CV_3, \quad CV_3, \\ \text{according as } \theta \text{ lies in the respective intervals} \\ (-\pi+\epsilon, -\frac{1}{2}\pi+\epsilon), \quad (-\frac{1}{2}\pi+\epsilon, -\epsilon), \quad (-\epsilon, \epsilon), \quad (\epsilon, \frac{1}{2}\pi-\epsilon), \quad (\frac{1}{2}\pi-\epsilon, \pi-\epsilon). \end{array}\right\} \tag{7.15}$$

$\mathbf{T}(\theta)$ is illustrated in figure 7 for $0 \leq \theta \leq \pi-\epsilon$; $\mathbf{T}(-\theta)$ is the conjugate of $\mathbf{T}(\theta)$. The dotted curves are of the same character as the dotted curves of figure 2; their x-maps parallel the x-maps of the principal curves through E at a distance δ away. $\mathbf{T}(\theta)$ is the unshaded domain bounded by the dotted curves.

Collecting together the results of this section, we have shown that there exists a solution W_1 of the differential eq (7.2) whose asymptotic expansion for large $|\mu|$ is given by (7.10) and is uniformly valid with respect to $t \in \mathbf{T}(\theta)$ and θ in the interval $(-\pi+\epsilon, \pi-\epsilon)$. The functions $\zeta(t)$, $F(\zeta)$, $A_s(\zeta)$ and $B_s(\zeta)$ are given by (7.4), (7.3), and (7.12), and are regular when $t \in \mathbf{T}(\theta)$.

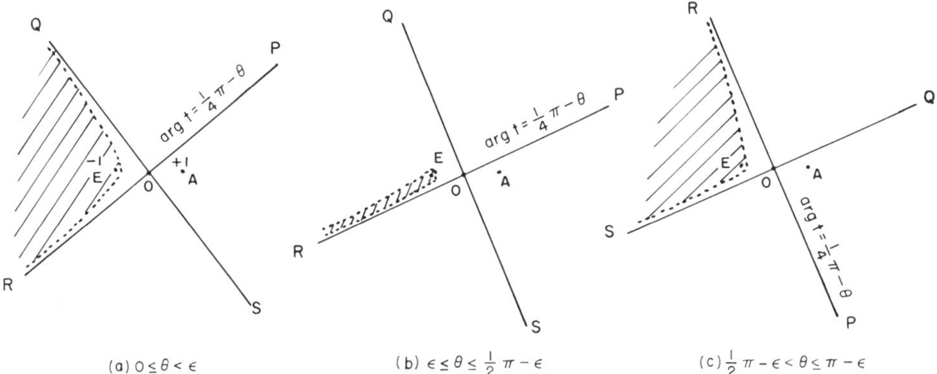

FIGURE 7. *t-plane: domains* $\mathbf{T}(\theta)$.

8. Identification of Solutions

From (4.1) and the second of eqs (7.1), we see that

$$W=\left(\frac{t^2-1}{\zeta}\right) U\left(-\frac{1}{2}\mu^2, \mu t\sqrt{2}\right), \tag{8.1}$$

satisfies the differential eq (7.2). When μ is fixed and $t \to \infty e^{-i\arg\mu}$ this solution becomes exponentially small. If $|\mu|$ is sufficiently large and $|\arg\mu| \leq \pi-\epsilon$, the solution W_1 of section 7 is also exponentially small in the same circumstances, for from (7.4) we have $\zeta \sim (\frac{3}{4})^{\frac{2}{3}} t^{\frac{4}{3}}$, and so $\mu^{\frac{4}{3}}\zeta \to +\infty$. Hence the two solutions are linearly dependent. Accordingly,

$$U\left(-\frac{1}{2}\mu^2, \mu t\sqrt{2}\right) \sim h(\mu)\left(\frac{\zeta}{t^2-1}\right)^{\frac{1}{4}}\left[\mathrm{Ai}\,(\mu^{\frac{4}{3}}\zeta)\sum_{s=0}^{\infty}\frac{A_s(\zeta)}{\mu^{4s}}+\frac{\mathrm{Ai}'(\mu^{\frac{4}{3}}\zeta)}{\mu^{\frac{8}{3}}}\sum_{s=0}^{\infty}\frac{B_s(\zeta)}{\mu^{4s}}\right] \tag{8.2}$$

as $|\mu| \to \infty$, uniformly with respect to $t \in \mathbf{T}(\arg\mu)$ and $\arg\mu$ in the interval $(-\pi+\epsilon, \pi-\epsilon)$.

This is the fundamental expansion in terms of Airy functions. In order to determine the function $h(\mu)$, we first fix the arbitrary constants of integration in the recurrence relations (7.12) by specifying the conditions

$$A_0(\zeta)=1, \qquad \lim_{|\zeta|\to\infty} A_s(\zeta)=0 \quad (s=1,2,\ldots). \tag{8.3}$$

We now examine the limiting form of the ratio of the two sides of the relation (8.2) as $t \to \infty e^{-i \arg \mu}$. With the aid of (2.2), (3.5), (7.5), (8.3), and the asymptotic forms

$$\mathrm{Ai}\,(\mu^{\frac{4}{3}}\zeta) \sim \tfrac{1}{2}\,\pi^{-\frac{1}{2}}(\mu^{\frac{4}{3}}\zeta)^{-\frac{1}{4}}e^{-\mu^2\xi}, \qquad \mathrm{Ai}'\,(\mu^{\frac{4}{3}}\zeta) \sim -\tfrac{1}{2}\,\pi^{-\frac{1}{2}}(\mu^{\frac{4}{3}}\zeta)^{\frac{1}{4}}e^{-\mu^2\xi}$$

(cf. [8], appendix), we find that $1/h(\mu)$ has the asymptotic expansion

$$\frac{1}{h(\mu)} \sim \pi^{-\frac{1}{2}}\, 2^{\frac{1}{4}\mu^2-\frac{3}{4}}\, \mu^{-\frac{1}{2}\mu^2+\frac{1}{6}}\, e^{\frac{1}{4}\mu^2} \left(1 - \sum_{s=0}^{\infty} \frac{h_s}{\mu^{4s+2}}\right) \qquad (|\arg \mu| \leq \pi - \epsilon) \tag{8.4}$$

as $|\mu| \to \infty$, uniformly valid with respect to $\arg \mu$, where

$$h_s = \lim_{|\zeta| \to \infty} \{\zeta^{\frac{1}{2}} B_s(\zeta)\}. \tag{8.5}$$

An alternative form of expansion, not involving the quantities h_s, may be found as follows. Suppose that $\epsilon \leq \theta \leq \frac{1}{2}\pi - \epsilon$. The principal curve AQ is then included in the domain $\mathbf{T}(\theta)$. On this curve $\mu^{\frac{4}{3}}\zeta$ is negative (cf. fig. 6). Let us set

$$\zeta = \zeta_m \equiv \mu^{-\frac{4}{3}} a_m', \tag{8.6}$$

where a_m' is the mth negative zero of the function Ai′. Then (8.2) becomes

$$U\left(-\frac{1}{2}\mu^2, \mu t_m \sqrt{2}\right) \sim h(\mu) \left(\frac{\zeta_m}{t_m^2 - 1}\right)^{\frac{1}{4}} \mathrm{Ai}\,(a_m') \sum_{s=0}^{\infty} \frac{A_s(\zeta_m)}{\mu^{4s}}, \tag{8.7}$$

in which t_m denotes the value of t corresponding to $\zeta = \zeta_m$. Now keep μ fixed and let $m \to \infty$, so that $t \to \infty e^{i(\frac{3}{4}\pi - \theta)}$. Then from [8], appendix, we have

$$a_m' = -\left\{\frac{3}{8}\pi(4m-3)\right\}^{\frac{2}{3}} \{1 + O(m^{-2})\}, \qquad \mathrm{Ai}\,(a_m') = (-)^{m-1} \pi^{-\frac{1}{2}} (-a_m')^{-\frac{1}{4}} \{1 + O(m^{-2})\}, \tag{8.8}$$

and from (7.5) and (3.5)

$$t_m^2 - \ln 2t_m = \frac{4}{3}\zeta_m^{\frac{3}{2}} + \frac{1}{2} + O(|t_m|^{-2}). \tag{8.9}$$

Substituting in (2.7) by means of these equations, we find that the limiting form of the left-hand side of (8.7) is minus the expression (5.17), with t_m defined of course as in the present section. On the other hand, the limiting form of the right of (8.7) is

$$h(\mu) \left(\frac{\mu^{-\frac{4}{3}} a_m'}{t_m^2 - 1}\right)^{\frac{1}{4}} (-)^{m-1} \pi^{-\frac{1}{2}} (-a_m')^{-\frac{1}{4}} \sim (-)^{m-1} \pi^{-\frac{1}{2}} e^{\frac{i\pi}{4}} \mu^{-\frac{1}{3}}\, h(\mu)\, t_m^{-\frac{1}{2}}$$

(cf. (8.3) and (8.8)). Comparing the two sides we find that $\frac{1}{2}\pi^{-\frac{1}{2}}\mu^{-\frac{1}{3}}h(\mu)$ has the same asymptotic expansion for large $|\mu|$ as the right-hand side of the relation (5.18). Hence we obtain

$$h(\mu) \simeq 2\pi^{\frac{1}{2}} \mu^{\frac{1}{3}} \tilde{g}(\mu) \simeq 2\pi^{\frac{1}{2}} \mu^{\frac{1}{3}} g(\mu), \tag{8.10}$$

where $\tilde{g}(\mu)$, $g(\mu)$ are defined in sections 4 and 5; $g(\mu)$ has the asymptotic expansions (6.1) and (6.2) for large $|\mu|$. Although (8.10) has been proved on the assumption that $\epsilon \leq \arg \mu \leq \frac{1}{2}\pi - \epsilon$, (8.4) shows that $1/h(\mu)$, and hence also $h(\mu)$, possesses a single asymptotic expansion over the range $|\arg \mu| \leq \pi - \epsilon$. Therefore (8.10) must also hold over this wider range of $\arg \mu$.

Substituting (8.10) in (8.2), we obtain the final form of the fundamental expansion, given by

$$U\left(-\frac{1}{2}\mu^2, \mu t\sqrt{2}\right) \sim 2\pi^{\frac{1}{2}} \mu^{\frac{1}{3}} g(\mu) \left(\frac{\zeta}{t^2 - 1}\right)^{\frac{1}{4}} \left[\mathrm{Ai}\,(\mu^{\frac{4}{3}}\zeta) \sum_{s=0}^{\infty} \frac{A_s(\zeta)}{\mu^{4s}} + \frac{\mathrm{Ai}'(\mu^{\frac{4}{3}}\zeta)}{\mu^{\frac{8}{3}}} \sum_{s=0}^{\infty} \frac{B_s(\zeta)}{\mu^{4s}}\right] \tag{8.11}$$

as $|\mu| \to \infty$, uniformly with respect to $t \in \mathbf{T}(\arg \mu)$, $\arg \mu$ in the interval $(-\pi + \epsilon, \pi - \epsilon)$.

The coefficients $A_s(\zeta)$ and $B_s(\zeta)$ appearing in (8.11) are determined by the recurrence relations (7.12) and the conditions (8.3). Explicit expressions can be found by the method used in [8], section 6 for the determination of the coefficients in the uniform expansions of Bessel functions of large order. We take ζ to be fixed and positive, and let $\mu\to+\infty$. The Airy functions appearing in (8.11) can then be expanded in their asymptotic series for large positive arguments and the result compared with (4.3). Thus we derive the expressions

$$A_s(\zeta)=\sum_{m=0}^{2s} b_m\zeta^{-\frac{3}{2}m}\mathscr{A}_{2s-m}(\xi),\qquad \zeta^{\frac{1}{2}}B_s(\zeta)=-\sum_{m=0}^{2s+1} a_m\zeta^{-\frac{3}{2}m}\mathscr{A}_{2s-m+1}(\xi) \tag{8.12}$$

(cf. [8], (6.6)), where $\mathscr{A}_m(\xi)$ is defined in section 4, $a_0=1$ and

$$a_m=\frac{(2m+1)\,(2m+3)\,.\,.\,.\,(6m-1)}{m!(144)^m},\qquad b_m=-\frac{6m+1}{6m-1}\,a_m. \tag{8.13}$$

Analytic continuation shows that eqs (8.12) hold for all values of ζ.

We note from (8.10), (8.4), and (6.1) that

$$h_s\equiv\lim_{|\zeta|\to\infty}\{\zeta^{\frac{1}{2}}B_s(\zeta)\}=-g_{2s+1}\qquad (s=0,1,\,.\,.\,.), \tag{8.14}$$

where g_s is given by (4.12) and (5.21).

Expansion for the derivative. The expansion (8.11) may be differentiated term by term with respect to t; this is a consequence of theorem B. With the aid of the first of eqs (7.1), we find that

$$U'\left(-\frac{1}{2}\mu^2,\,\mu t\sqrt{2}\right)\sim\frac{(2\pi)^{\frac{1}{2}}\mu^{\frac{2}{3}}g(\mu)}{\phi(\zeta)}\left[\frac{\mathrm{Ai}\,(\mu^{\frac{4}{3}}\zeta)}{\mu^{\frac{4}{3}}}\sum_{s=0}^{\infty}\frac{C_s(\zeta)}{\mu^{4s}}+\mathrm{Ai}'\,(\mu^{\frac{4}{3}}\zeta)\sum_{s=0}^{\infty}\frac{D_s(\zeta)}{\mu^{4s}}\right] \tag{8.15}$$

as $|\mu|\to\infty$, uniformly with respect to $t\in\mathbf{T}(\arg\mu)$ and $\arg\mu$ in the interval $(-\pi+\epsilon,\pi-\epsilon)$.

Here we have used the notation

$$\phi(\zeta)\equiv\left(\frac{\zeta}{t^2-1}\right)^{\frac{1}{4}}=\left(\frac{dt}{d\zeta}\right)^{\frac{1}{2}}. \tag{8.16}$$

The coefficients $C_s(\zeta)$ and $D_s(\zeta)$ are expressed in terms of $A_s(\zeta)$ and $B_s(\zeta)$ by the equations

$$C_s(\zeta)=\chi(\zeta)A_s(\zeta)+A_s'(\zeta)+\zeta B_s(\zeta),\qquad D_s(\zeta)=A_s(\zeta)+\chi(\zeta)B_{s-1}(\zeta)+B_{s-1}'(\zeta), \tag{8.17}$$

(cf. [8], (6.10)), where

$$\chi(\zeta)\equiv\frac{\phi'(\zeta)}{\phi(\zeta)}=\frac{1}{4\zeta}-\frac{t}{2}\frac{\zeta^{\frac{1}{2}}}{(t^2-1)^{\frac{3}{2}}}=\frac{1-2t\{\phi(\zeta)\}^6}{4\zeta}. \tag{8.18}$$

Explicitly,

$$\zeta^{-\frac{1}{2}}C_s(\zeta)=-\sum_{m=0}^{2s+1} b_m\zeta^{-\frac{3}{2}m}\mathscr{B}_{2s-m+1}(\xi),\qquad D_s(\zeta)=\sum_{m=0}^{2s} a_m\zeta^{-\frac{3}{2}m}\mathscr{B}_{2s-m}(\xi), \tag{8.19}$$

(cf. [8], (6.12)), where $\mathscr{B}_m(\xi)$ is given by (4.14) to (4.16), and a_m, b_m by (8.13).

9. Use of Connection Formulas

In this section we seek expansions which hold in t-regions complementary to $\mathbf{T}(\arg\mu)$ when $|\arg\mu|\le\frac{1}{2}\pi$ (cf. sec 5).

In (8.11) replace μ by $\mu e^{i\pi}$. Then using (5.24) we obtain

$$U\left(-\frac{1}{2}\mu^2,-\mu t\sqrt{2}\right)\sim 2\pi^{\frac{1}{2}}e^{i\pi(\frac{1}{2}\mu^2-\frac{1}{6})}\mu^{\frac{1}{3}}g(\mu)\left(\frac{\zeta}{t^2-1}\right)^{\frac{1}{4}}$$
$$\times\left[\mathrm{Ai}\,(e^{-\frac{2}{3}\pi i}\mu^{\frac{4}{3}}\zeta)\sum_{s=0}^{\infty}\frac{A_s(\zeta)}{\mu^{4s}}+\frac{\mathrm{Ai}'(e^{-\frac{2}{3}\pi i}\mu^{\frac{4}{3}}\zeta)}{e^{\frac{2}{3}\pi i}\mu^{\frac{8}{3}}}\sum_{s=0}^{\infty}\frac{B_s(\zeta)}{\mu^{4s}}\right], \tag{9.1}$$

when

$$-\pi+\epsilon\leq\theta\leq-\epsilon, \qquad t\in \mathbf{T}(\theta+\pi). \tag{9.2}$$

In particular, we see from (7.15) and (5.3) that (9.1) holds when

$$-\frac{1}{2}\pi\leq\theta\leq-\epsilon, \qquad t\in C\mathbf{V}_3(\theta+\pi)=C\mathbf{V}_1(\theta). \tag{9.3}$$

Similarly,

$$U\left(-\frac{1}{2}\mu^2,-\mu t\sqrt{2}\right)\sim 2\pi^{\frac{1}{2}}e^{-i\pi(\frac{1}{2}\mu^2-\frac{1}{6})}\mu^{\frac{1}{3}}g(\mu)\left(\frac{\zeta}{t^2-1}\right)^{\frac{1}{4}}$$

$$\times\left[\mathrm{Ai}\,(e^{\frac{2}{3}\pi i}\mu^{\frac{4}{3}}\zeta)\sum_{s=0}^{\infty}\frac{A_s(\zeta)}{\mu^{4s}}+\frac{\mathrm{Ai}'(e^{\frac{2}{3}\pi i}\mu^{\frac{4}{3}}\zeta)}{e^{-\frac{2}{3}\pi i}\mu^{\frac{8}{3}}}\sum_{s=0}^{\infty}\frac{B_s(\zeta)}{\mu^{4s}}\right], \tag{9.4}$$

when

$$\epsilon\leq\theta\leq\frac{1}{2}\pi, \qquad t\in C\mathbf{V}_3(\theta). \tag{9.5}$$

The expansions (9.1) and (9.4) are of course the expansions for $U(-\frac{1}{2}\mu^2, \mu t\sqrt{2})$ in terms of Airy functions valid at the turning point $t=-1$. Combined with (8.11) they cover the whole of the t-plane when $\epsilon\leq|\theta|\leq\frac{1}{2}\pi$.

When $|\theta|\leq\epsilon$ we cannot expect to be able to represent the function $U(-\frac{1}{2}\mu^2, \mu t\sqrt{2})$ in $\mathbf{V}_2(\theta)=C\mathbf{T}(\theta)$ by means of a single Airy function, because as $|t|\to\infty$ in this region $U(-\frac{1}{2}\mu^2, \mu t\sqrt{2})$ becomes exponentially small or exponentially large according as μ^2 is or is not an odd positive integer (cf. (2.7)).

To deal with this remaining case we use the connection formula

$$U\left(-\frac{1}{2}\mu^2, -\mu t\sqrt{2}\right)-\sin\frac{1}{2}\pi\mu^2\cdot U\left(-\frac{1}{2}\mu^2, \mu t\sqrt{2}\right)=\left(\frac{\pi}{2}\right)^{\frac{1}{2}}\frac{e^{-i\pi(\frac{1}{4}\mu^2+\frac{1}{4})}}{\Gamma(\frac{1}{2}-\frac{1}{2}\mu^2)}U\left(\frac{1}{2}\mu^2,-i\mu t\sqrt{2}\right)$$

$$+\left(\frac{\pi}{2}\right)^{\frac{1}{2}}\frac{e^{i\pi(\frac{1}{4}\mu^2+\frac{1}{4})}}{\Gamma(\frac{1}{2}-\frac{1}{2}\mu^2)}U\left(\frac{1}{2}\mu^2, i\mu t\sqrt{2}\right), \tag{9.6}$$

obtained in an obvious manner from (5.26). Replacing μ by $\mu e^{-\frac{1}{2}i\pi}$ in (8.11) and using the first of eqs (5.25), we find

$$\left(\frac{\pi}{2}\right)^{\frac{1}{2}}\frac{e^{-i\pi(\frac{1}{4}\mu^2+\frac{1}{4})}}{\Gamma(\frac{1}{2}-\frac{1}{2}\mu^2)}U\left(\frac{1}{2}\mu^2, -i\mu t\sqrt{2}\right)\sim 2\pi^{\frac{1}{2}}e^{-\frac{1}{6}\pi i}\mu^{\frac{1}{3}}\cos\frac{1}{2}\pi\mu^2\cdot g(\mu)\left(\frac{\zeta}{t^2-1}\right)^{\frac{1}{4}}$$

$$\times\left[\mathrm{Ai}\,(e^{-\frac{2}{3}\pi i}\mu^{\frac{4}{3}}\zeta)\sum_{s=0}^{\infty}\frac{A_s(\zeta)}{\mu^{4s}}+\frac{\mathrm{Ai}'(e^{-\frac{2}{3}\pi i}\mu^{\frac{4}{3}}\zeta)}{e^{\frac{2}{3}\pi i}\mu^{\frac{8}{3}}}\sum_{s=0}^{\infty}\frac{B_s(\zeta)}{\mu^{4s}}\right].$$

We substitute this result and its conjugate form on the right of (9.6), and on the left we substitute (8.11). Using the eq [3]

$$\mathrm{Bi}\,(\mu^{\frac{4}{3}}\zeta)=e^{-\frac{1}{6}\pi i}\mathrm{Ai}\,(e^{-\frac{2}{3}\pi i}\mu^{\frac{4}{3}}\zeta)+e^{\frac{1}{6}\pi i}\mathrm{Ai}\,(e^{\frac{2}{3}\pi i}\mu^{\frac{4}{3}}\zeta),$$

and its differentiated form, we obtain

$$U\left(-\frac{1}{2}\mu^2,-\mu t\sqrt{2}\right)\sim 2\pi^{\frac{1}{2}}\mu^{\frac{1}{3}}g(\mu)\left(\frac{\zeta}{t^2-1}\right)^{\frac{1}{4}}\left[\sin\frac{1}{2}\pi\mu^2\left\{\mathrm{Ai}\,(\mu^{\frac{4}{3}}\zeta)\sum_{s=0}^{\infty}\frac{A_s(\zeta)}{\mu^{4s}}+\frac{\mathrm{Ai}'(\mu^{\frac{4}{3}}\zeta)}{\mu^{\frac{8}{3}}}\sum_{s=0}^{\infty}\frac{B_s(\zeta)}{\mu^{4s}}\right\}\right.$$

$$\left.+\cos\frac{1}{2}\pi\mu^2\left\{\mathrm{Bi}\,(\mu^{\frac{4}{3}}\zeta)\sum_{s=0}^{\infty}\frac{A_s(\zeta)}{\mu^{4s}}+\frac{\mathrm{Bi}'(\mu^{\frac{4}{3}}\zeta)}{\mu^{\frac{8}{3}}}\sum_{s=0}^{\infty}\frac{B_s(\zeta)}{\mu^{4s}}\right\}\right]. \tag{9.7}$$

This is valid when $|\theta|\leq\frac{1}{2}\pi-\epsilon$ in the t-region common to the domains $\mathbf{T}(\theta)$, $\mathbf{T}(\theta-\frac{1}{2}\pi)$, and

$\mathbf{T}(\theta+\frac{1}{2}\pi)$. Using (7.15) and (5.2), we see that this means (9.7) is valid when

$$|\theta|\leq\frac{1}{2}\pi-\epsilon, \qquad t\,\epsilon\, C\mathbf{V}_2(\theta). \tag{9.8}$$

This is the required result. We note that when μ^2 is an odd integer the terms in Bi, Bi′ in (9.7) vanish, as we require from a consideration of the behavior of $U(-\frac{1}{2}\mu^2, \mu t\sqrt{2})$ as $|t|\to\infty$ in the sector $\mathbf{V}_2(\theta)$. We may also verify without difficulty that (9.7) agrees with (9.1) and (9.4) in their common t-regions of validity when $-\frac{1}{2}\pi+\epsilon\leq\theta\leq-\epsilon$ and $\epsilon\leq\theta\leq\frac{1}{2}\pi-\epsilon$, respectively.

10. Summary of Expansions in Terms of Airy Functions

The fundamental expansion of this kind is (8.11). It is valid for large $|\mu|$ uniformly with respect to $\theta\equiv\arg\mu$ and t, when $-\pi+\epsilon\leq\theta\leq\pi-\epsilon$ and t lies in the domain $\mathbf{T}(\theta)$, defined by (7.15) and illustrated in figure 7. The function $g(\mu)$ is calculable from either of the asymptotic expansions (6.1) and (6.2). The function $\zeta(t)$ is given by (7.4), that branch being chosen which is continuous in $\mathbf{T}(\theta)$ and positive when $t>1$; similarly the branch of $\{\zeta/(t^2-1)\}^{\frac{1}{4}}$ is continuous in $\mathbf{T}(\theta)$ and positive when $t>1$. The coefficients $A_s(\zeta)$ and $B_s(\zeta)$ are given by eqs (8.12); they satisfy the recurrence relations (7.12) and the eqs (8.3). They are regular functions of ζ in the ζ-map of $\mathbf{T}(\theta)$ and real when ζ lies on the real axis to the right of the point of affix $-(\frac{3}{4}\pi)^{\frac{2}{3}}$.

Other asymptotic expansions for $U(-\frac{1}{2}\mu^2, \mu t\sqrt{2})$ are (9.1), (9.4), and (9.7); they are valid with the conditions (9.3), (9.5), and (9.8), respectively. The combination of these expansions and (8.11) covers the whole t-plane when $|\theta|\leq\frac{1}{2}\pi$.

The expansion for the derivative $U'(-\frac{1}{2}\mu^2, \mu t\sqrt{2})$ corresponding to (8.11) is (8.15), and holds under the same conditions. The coefficients $C_s(\zeta)$ and $D_s(\zeta)$ are given by (8.17) and (8.19). The derivative expansions corresponding to (9.1), (9.4), and (9.7) are not recorded, but they can be obtained immediately by analogy with (8.15). They are valid under the same conditions as the parent formulas.

Approximations for parabolic cylinder functions of large orders which hold in regions containing one of the turning points have been given by Watson ([12], sec. 17), Schwid ([10], sec. 6), Erdélyi, Kennedy, and McGregor [2], and Kazarinoff [15].

Watson derives asymptotic expansions in terms of elementary functions for the functions $D_n(z\pm2\sqrt{n})\equiv U(-n-\frac{1}{2}, z\pm2\sqrt{n})$ as $|n|\to\infty$, with the conditions $|\arg n|<\frac{1}{2}\pi$ and $z=o(|n|^{-\frac{1}{6}})$. In the present notation these conditions correspond to $|\arg\mu|<\frac{1}{4}\pi$ and $t\mp1=o(|\mu|^{-\frac{4}{3}})$. Watson's expansions are analogous to Meissel's expansions for Bessel functions of large order and may be obtained from the present uniform expansions by suitable re-expansion (cf. [8], sec. 6).

Schwid gives approximations in terms of Bessel functions of order one-third (Airy functions) which are valid in the neighborhood $t-1=O(|\mu|^{-2})$ as $|\mu|\to\infty$, using the present notation. This neighborhood is complementary to the region of validity of Schwid's approximations in terms of elementary functions, described in section 6. No investigation is made of the validity of the Bessel function approximations outside the neighborhood.

The approach of Erdélyi, Kennedy, and McGregor is similar to the one we have used. These writers obtain a set of asymptotic approximations in terms of Airy functions, to solutions of eq (1.1) when μ and t are both complex (using the present notation). The results are contained in those given in the present paper; the principal ways in which they have been extended and improved upon are as follows.

First, the approximations apply to fixed values of arg μ, no investigation is made of their uniform validity with respect to arg μ.

Secondly, the approximations are established with the condition $|\arg\mu|\leq\frac{1}{4}\pi$; outside this interval connection formulas are to be used. This leads to complexity in the number of results; eleven are given compared with the four we have found to be sufficient.

Finally, only the leading terms of the asymptotic expansions are found; no terms involving the first derivative of the Airy functions appear in the approximations. An effect of this is to exclude regions in the t-plane containing the zeros of the Airy functions. No special treatment of such regions is required in the present theory.

Kazarinoff's paper (which has appeared since the present paper was written) contains a brief investigation of the extension of the results of Erdélyi, Kennedy, and McGregor to include uniformity with respect to arg μ and higher terms in the asymptotic expansions. The results are rather more complicated and less complete than those given here.

Part 4. Results for Real Variables

In this part we suppose μ, t, a, and x to denote real variables except where otherwise indicated; μ is always large and positive. The variable x is not to be confused with the complex function $x(t)$ introduced in parts 2 and 3.

11. The Equation $w''=(\frac{1}{4}x^2+a)w$

The principal solutions of this equation are $U(a,x)$, $U(a,-x)$, $V(a,x)$, and $\overline{U}(a,x)$, defined in section 2.

(i) *a positive*. When $a \to +\infty$ the differential equation has no turning points on the real x-axis. The expansions of greatest interest are therefore those in terms of elementary functions, and we shall not consider the forms of the Airy expansions.

In (4.3) replace μ,t by $i\mu$, $-it$, respectively. Then using (4.8), we obtain

$$U\left(\frac{1}{2}\mu^2, \mu t\sqrt{2}\right) \sim g(\mu e^{\frac{1}{2}i\pi}) \frac{e^{\mu^2\xi(-it)}}{(-t^2-1)^{\frac{1}{4}}} \sum_{s=0}^{\infty} (-)^s \frac{u_s(-it)}{(-t^2-1)^{\frac{3}{2}s}} \frac{1}{\mu^{2s}} \tag{11.1}$$

as $\mu \to +\infty$. In the complex plane the region of validity of this expansion is $e^{\frac{1}{2}i\pi}\mathrm{S}(\frac{1}{2}\pi)$, and from figure 2(c) it is clear that this region includes the whole of the real t-axis.

We have now to interpret the branches of the various functions.

From (6.1) and (6.2) we derive

$$g(\mu e^{\frac{1}{2}i\pi}) = e^{-i\pi(\frac{1}{4}+\frac{1}{4}\mu^2)}\overline{g}(\mu), \tag{11.2}$$

where the function $\overline{g}(\mu)$ is real and has the expansions

$$\frac{1}{\overline{g}(\mu)} \sim 2^{-\frac{1}{4}\mu^2+\frac{1}{4}} e^{-\frac{1}{4}\mu^2} \mu^{\frac{1}{2}\mu^2+\frac{1}{2}} \left\{1 - \sum_{s=0}^{\infty} \frac{g_{2s+1}}{\mu^{4s+2}}\right\}, \tag{11.3}$$

$$\overline{g}(\mu) \sim 2^{\frac{1}{4}\mu^2-\frac{1}{4}} e^{\frac{1}{4}\mu^2} \mu^{-\frac{1}{2}\mu^2-\frac{1}{2}} \left\{1 + \frac{1}{2} \sum_{s=1}^{\infty} (-)^s \frac{\gamma_s}{(\frac{1}{2}\mu^2)^s}\right\}, \tag{11.4}$$

as $\mu \to +\infty$. Comparison of (11.4) with (2.27) shows incidentally that

$$\overline{g}(\mu) \simeq 2^{\frac{1}{4}\mu^2-\frac{5}{4}} e^{\frac{1}{4}\mu^2} \mu^{-\frac{1}{2}\mu^2-\frac{1}{2}} + \frac{\pi^{\frac{1}{2}} 2^{-\frac{1}{4}\mu^2-\frac{3}{4}} e^{-\frac{1}{4}\mu^2} \mu^{\frac{1}{2}\mu^2-\frac{1}{2}}}{\Gamma(\frac{1}{2}+\frac{1}{2}\mu^2)}, \tag{11.5}$$

where the symbol $\simeq$ is defined in section 5.

Next, when t lies on the negative imaginary axis of figure 2(c), $\xi(t)$ lies on the straight line O′C′ shown in the third diagram of figure 1. Hence

$$\xi(-it) = \frac{1}{4} i\pi - \overline{\xi}(t), \tag{11.6}$$

where $\overline{\xi}(t)$ is positive when $t>0$. From (3.4) we find

$$\overline{\xi}(t) = \int_0^t (t^2+1)^{\frac{1}{2}} dt = \frac{1}{2} t(t^2+1)^{\frac{1}{2}} + \frac{1}{2} \ln \{t+(t^2+1)^{\frac{1}{2}}\}, \tag{11.7}$$

the branches here having their principal values.

Lastly, we have

$$(-t^2-1)^{\frac{1}{4}}=e^{-\frac{1}{4}i\pi}(t^2+1)^{\frac{1}{4}}, \qquad (-t^2-1)^{\frac{3}{2}s}=e^{-\frac{3}{2}si\pi}(t^2+1)^{\frac{3}{2}s}, \tag{11.8}$$

where the functions on the right have their principal values.

Substituting in (11.1) by means of (11.2), (11.6), (11.8), and the relation

$$\bar{u}_s(t)\equiv i^s u_s(-it), \tag{11.9}$$

we obtain the required result in real form, given by

$$U\left(\frac{1}{2}\mu^2,\mu t\sqrt{2}\right)\sim\bar{g}(\mu)\frac{e^{-\mu^2\bar{\xi}(t)}}{(t^2+1)^{\frac{1}{4}}}\sum_{s=0}^{\infty}\frac{\bar{u}_s(t)}{(t^2+1)^{\frac{3}{2}s}}\frac{1}{\mu^{2s}} \qquad (-\infty<t<\infty), \tag{11.10}$$

uniformly with respect to t. Explicit expressions for the coefficients $\bar{u}_s(t)$, obtained from (4.11), are

$$\left.\begin{aligned}&\bar{u}_0(t)=1, \qquad \bar{u}_1(t)=(-t^3-6t)/24, \qquad \bar{u}_2(t)=(9t^4+249t^2-145)/1152,\\&\bar{u}_3(t)=(4042t^9+18189t^7+28287t^5-1\,51995t^3+2\,59290t)/4\,14720.\end{aligned}\right\} \tag{11.11}$$

The corresponding result for the derivative is

$$U'\left(\frac{1}{2}\mu^2,\mu t\sqrt{2}\right)\sim-\frac{\mu}{\sqrt{2}}\,\bar{g}(\mu)(t^2+1)^{\frac{1}{4}}e^{-\mu^2\bar{\xi}(t)}\sum_{s=0}^{\infty}\frac{\bar{v}_s(t)}{(t^2+1)^{\frac{3}{2}s}}\frac{1}{\mu^{2s}}, \tag{11.12}$$

valid with the same conditions. Here

$$\bar{v}_s(t)\equiv i^s v_s(-it). \tag{11.13}$$

Explicitly,

$$\left.\begin{aligned}&\bar{v}_0(t)=1, \qquad \bar{v}_1(t)=(-t^3+6t)/24, \qquad \bar{v}_2(t)=(-15t^4-327t^2+143)/1152,\\&\bar{v}_3(t)=(4042t^9+18189t^7+36387t^5+2\,38425t^3-2\,59290t)/4\,14720,\end{aligned}\right\} \tag{11.14}$$

(cf. (4.13) to (4.16)).

Since $U(a,x)$ and $U(a,-x)$ comprise a satisfactory pair of solutions of the differential equation when a is positive, it is unnecessary to give the corresponding expansions for the solutions $V(a,x)$ and $\overline{U}(a,x)$.

(ii) *a negative; expansions in terms of elementary functions.* For $t\geq 1+\delta$, the expansion of $U(-\frac{1}{2}\mu^2,\ \mu t\sqrt{2})$ is given in real form by (4.3)[3]. In this expansion $g(\mu)$ is calculable from (6.1), (6.2), or the relation

$$g(\mu)\simeq 2^{-\frac{1}{4}\mu^2-\frac{5}{4}}e^{-\frac{1}{4}\mu^2}\mu^{\frac{1}{2}\mu^2-\frac{1}{2}}+\pi^{-\frac{1}{2}}2^{\frac{1}{4}\mu^2-\frac{7}{4}}e^{\frac{1}{4}\mu^2}\mu^{-\frac{1}{2}\mu^2-\frac{1}{2}}\Gamma\left(\frac{1}{2}+\frac{1}{2}\mu^2\right), \tag{11.15}$$

and ξ, $\mathscr{A}_s(\xi)$, by use of (3.4), (4.8), and (4.11), all fractional powers now taking their principal values.

When $-1+\delta\leq t\leq 1-\delta$, the expansion of $U(-\frac{1}{2}\mu^2,\ \mu t\sqrt{2})$ is given by (5.11). Here $\tilde{g}(\mu)$ is identical with $g(\mu)$ and η is given by (5.9).

For the remaining part of the real axis we have the expansion (5.28). In the present circumstances the function $\xi(t)$ is positive and bounded away from zero. Hence the contribution of the series $i\cos\frac{1}{2}\pi\mu^2\cdot e^{-\mu^2\xi}\sum\mathscr{A}_s(\xi)\mu^{-2s}$ is always asymptotically negligible compared with that of $2\cos\frac{1}{2}\pi\mu^2\cdot e^{\mu^2\xi}\sum(-)^s\mathscr{A}_s(\xi)\mu^{-2s}$. Thus (5.28) reduces to

$$U\left(-\frac{1}{2}\mu^2,-\mu t\sqrt{2}\right)\sim\frac{g(\mu)}{(t^2-1)^{\frac{1}{4}}}\left[2\cos\frac{1}{2}\pi\mu^2\cdot e^{\mu^2\xi}\sum_{s=0}^{\infty}(-)^s\frac{\mathscr{A}_s(\xi)}{\mu^{2s}}+\sin\frac{1}{2}\pi\mu^2\cdot e^{-\mu^2\xi}\sum_{s=0}^{\infty}\frac{\mathscr{A}_s(\xi)}{\mu^{2s}}\right], \tag{11.16}$$

[3] For convenience the symbol δ' of parts 2 and 3, denoting an arbitrary fixed positive number, is replaced here by δ.

valid when $t \geq 1+\delta$. The functions $g(\mu)$, ξ and $\mathscr{A}_s(\xi)$ are the same as in (4.3), discussed above.

The corresponding expansions for the second solution $\overline{U}(a, x)$ are of interest, particularly when $x \geq 0$. It is not possible to obtain them trivially by substituting in the connection formula (2.12) the expansions we have just obtained for $U(a, x)$ and $U(a, -x)$, because although cancellation of terms in the asymptotic series takes place, we cannot assert immediately that the remainder terms implied in these series also cancel. To put the results on a sound basis we substitute the expansions (5.5) and (5.6) in the formula

$$\overline{U}\left(-\frac{1}{2}\mu^2, \mu t\sqrt{2}\right) = (2\pi)^{-\frac{1}{2}}\Gamma\left(\frac{1}{2}+\frac{1}{2}\mu^2\right)\left[e^{-i\pi(\frac{1}{4}+\frac{1}{4}\mu^2)}U\left(\frac{1}{2}\mu^2, -i\mu t\sqrt{2}\right)+e^{i\pi(\frac{1}{4}+\frac{1}{4}\mu^2)}U\left(\frac{1}{2}\mu^2, i\mu t\sqrt{2}\right)\right], \tag{11.17}$$

obtained from (2.5) and (2.12).

When $t \geq 1+\delta$ the branches of $\xi(t)$ and $\mathscr{A}_s(\xi)$ in both expansions are the principal ones. With the aid of the relation

$$(2\pi)^{-\frac{1}{2}}\Gamma\left(\frac{1}{2}+\frac{1}{2}\mu^2\right)e^{\mp i\pi(\frac{1}{4}+\frac{1}{4}\mu^2)}g(\mu e^{\mp\frac{1}{2}i\pi}) \simeq g(\mu), \tag{11.18}$$

proved by use of (2.22), (5.7), (5.20), and (6.2), we deduce that

$$\overline{U}\left(-\frac{1}{2}\mu^2, \mu t\sqrt{2}\right) \sim 2g(\mu)\frac{e^{\mu^2\xi}}{(t^2-1)^{\frac{1}{4}}}\sum_{s=0}^{\infty}(-)^s\frac{u_s(t)}{(t^2-1)^{\frac{3}{2}s}}\frac{1}{\mu^{2s}} \qquad (t \geq 1+\delta). \tag{11.19}$$

When $-1+\delta \leq t \leq 1-\delta$, we interpret the branches of the many-valued functions by use of (5.8) and (5.10). With the aid of (11.18) we obtain on reduction

$$\overline{U}\left(-\frac{1}{2}\mu^2, \mu t\sqrt{2}\right) \sim \frac{2g(\mu)}{(1-t^2)^{\frac{1}{4}}}\left[\cos(\mu^2\eta+\tfrac{1}{4}\pi)\sum_{s=0}^{\infty}\frac{(-)^s u_{2s}(t)}{(1-t^2)^{3s}\mu^{4s}} - \sin(\mu^2\eta+\tfrac{1}{4}\pi)\sum_{s=0}^{\infty}\frac{(-)^s u_{2s+1}(t)}{(1-t^2)^{3s+\frac{3}{2}}\mu^{4s+2}}\right], \tag{11.20}$$

valid when $-1+\delta \leq t \leq 1-\delta$.

The expansions for the derivatives corresponding to the expansions obtained in this subsection can be written down immediately by use of (6.3), or by analogy with (5.23).

(iii) *a negative; expansions in terms of Airy functions.* For $t \geq -1+\delta$ the expansions of $U(-\frac{1}{2}\mu^2, \mu t\sqrt{2})$ and $U(-\frac{1}{2}\mu^2, -\mu t\sqrt{2})$ are given by (8.11) and (9.7), respectively. In these expansions $\zeta(t)$ is given by (7.4) and is real when $t > -1$; a more convenient form of this relation when $-1 < t < 1$ is

$$\frac{2}{3}(-\zeta)^{\frac{3}{2}} = \int_t^1 (1-t^2)^{\frac{1}{2}}dt = \frac{1}{2}\cos^{-1}t - \frac{1}{2}t(1-t^2)^{\frac{1}{2}} = \eta(t) \tag{11.21}$$

(cf. (5.9)). The coefficients $A_s(\zeta)$ and $B_s(\zeta)$ are real and given by (8.12), (8.13), and (4.8)[4].

An expansion for the solution $\overline{U}(a,x)$ may be derived by use of (11.17), as in the case of the exponential-type expansions. The result is

$$\overline{U}\left(-\frac{1}{2}\mu^2, \mu t\sqrt{2}\right) \sim 2\pi^{\frac{1}{2}}\mu^{\frac{1}{3}}g(\mu)\left(\frac{\zeta}{t^2-1}\right)^{\frac{1}{4}}\left[\mathrm{Bi}(\mu^{\frac{4}{3}}\zeta)\sum_{s=0}^{\infty}\frac{A_s(\zeta)}{\mu^{4s}} + \frac{\mathrm{Bi}'(\mu^{\frac{4}{3}}\zeta)}{\mu^{\frac{8}{3}}}\sum_{s=0}^{\infty}\frac{B_s(\zeta)}{\mu^{4s}}\right], \tag{11.22}$$

valid when $t \geq -1+\delta$.

The corresponding expansion for $\overline{U}'(-\frac{1}{2}\mu^2, \mu t\sqrt{2})$ may be written down by analogy with (8.15).

(iv) *Hermite polynomials of large degree.* Expansions for the Hermite polynomial $H_n(x)$

[4] More convenient expressions when $-1<t<1$ are given by (13.4) and (13.5).

when n is large can be obtained from the results in subsections (ii) and (iii) above by use of the equation

$$H_{\frac{1}{2}\mu^2-\frac{1}{2}}(\mu t)=2^{\frac{1}{4}\mu^2-\frac{1}{4}}e^{\frac{1}{2}\mu^2t^2}U\left(-\frac{1}{2}\mu^2,\mu t\sqrt{2}\right). \tag{11.23}$$

Thus (4.3) and (5.11) yield expansions which hold uniformly in the ranges $x\geq(1+\delta)\ (2n+1)^{\frac{1}{2}}$ and $-(1-\delta)(2n+1)^{\frac{1}{2}}\leq x\leq(1-\delta)(2n+1)^{\frac{1}{2}}$, respectively. An expansion uniformly valid throughout the range $x\geq-(1-\delta)(2n+1)^{\frac{1}{2}}$ is furnished by the Airy expansion (8.11).

12. The Equation $w''=(a-\frac{1}{4}x^2)w$

(i) *a positive; expansions in terms of elementary functions.* From (2.16), (2.17) and (2.18), we have

$$k^{-\frac{1}{2}}W\left(\frac{1}{2}\mu^2,\mu t\sqrt{2}\right)+ik^{\frac{1}{2}}W\left(\frac{1}{2}\mu^2,-\mu t\sqrt{2}\right)=2^{\frac{1}{2}}e^{\frac{1}{8}\pi\mu^2}e^{i(\frac{1}{2}\phi_2+\frac{1}{8}\pi)}U\left(\frac{1}{2}i\mu^2,e^{-\frac{1}{4}\pi i}\mu t\sqrt{2}\right), \tag{12.1}$$

where

$$k=(1+e^{\pi\mu^2})^{\frac{1}{2}}-e^{\frac{1}{2}\pi\mu^2},\qquad e^{\frac{1}{2}i\phi_2}=\pm\left\{\Gamma\left(\frac{1}{2}+\frac{1}{2}i\mu^2\right)\Big/\left|\Gamma\left(\frac{1}{2}+\frac{1}{2}i\mu^2\right)\right|\right\}^{\frac{1}{2}}. \tag{12.2}$$

For large μ,

$$k=\frac{1}{2}e^{-\frac{1}{2}\pi\mu^2}+O(e^{-\frac{3}{2}\pi\mu^2}),\qquad e^{\frac{1}{2}i\phi_2}=e^{-\frac{1}{4}i\mu^2}\left(\frac{1}{2}\mu^2\right)^{\frac{1}{4}i\mu^2}\{1+O(\mu^{-2})\}. \tag{12.3}$$

In (4.3) replace μ by $\mu e^{-\frac{1}{4}\pi i}$. Then we obtain

$$U\left(\frac{1}{2}i\mu^2,e^{-\frac{1}{4}\pi i}\mu t\sqrt{2}\right)\sim g(\mu e^{-\frac{1}{4}\pi i})\ \frac{e^{i\mu^2\xi}}{(t^2-1)^{\frac{1}{4}}}\sum_{s=0}^{\infty}i^s\ \frac{u_s(t)}{(t^2-1)^{\frac{3}{2}s}}\frac{1}{\mu^{2s}} \tag{12.4}$$

as $\mu\to+\infty$. The t-region of validity of this expansion, $\mathbf{S}(-\frac{1}{4}\pi)$, includes the segments $t\geq1+\delta$ and $-1+\delta\leq t\leq1-\delta$ of the real axis (cf. fig. 2(b)).

When $\mu\to+\infty$, we have from (12.2) and (2.22)

$$e^{\frac{1}{2}i\phi_2}\sim e^{-\frac{1}{4}i\mu^2}\left(\frac{1}{2}\mu^2\right)^{\frac{1}{4}i\mu^2}\left[\sum_{s=0}^{\infty}\frac{\gamma_s}{(\frac{1}{2}i\mu^2)^s}\right]^{\frac{1}{2}}\left|\sum_{s=0}^{\infty}\frac{\gamma_s}{(\frac{1}{2}i\mu^2)^s}\right|^{-\frac{1}{2}}, \tag{12.5}$$

the choice of ambiguous sign on taking the square root being resolved by use of (12.3). From (6.1)

$$g(\mu e^{-\frac{1}{4}\pi i})\sim2^{\frac{1}{4}i\mu^2-\frac{1}{4}}e^{\frac{1}{4}i\mu^2}e^{\frac{1}{8}\pi i-\frac{1}{8}\pi\mu^2}\mu^{-\frac{1}{2}i\mu^2-\frac{1}{2}}(1+iL)^{-1}, \tag{12.6}$$

where the symbol L denotes the asymptotic series

$$L\equiv\sum_{s=0}^{\infty}(-)^s\ \frac{g_{2s+1}}{\mu^{4s+2}}. \tag{12.7}$$

Further from (5.21), with μ replaced by $\mu e^{\frac{1}{4}\pi i}$, we have

$$\sum_{s=0}^{\infty}\frac{\gamma_s}{(\frac{1}{2}i\mu^2)^s}\sim\frac{2}{1-iL}-1=\frac{1+iL}{1-iL}. \tag{12.8}$$

Introducing a function $l(\mu)$, defined by

$$l(\mu)\equiv2^{\frac{1}{2}}e^{\frac{1}{8}\pi\mu^2}e^{i(\frac{1}{2}\phi_2-\frac{1}{8}\pi)}g(\mu e^{-\frac{1}{4}\pi i}), \tag{12.9}$$

and using (12.5), (12.6) and (12.8), we find that

$$l(\mu)\ \sim\ \frac{2^{\frac{1}{4}}}{\mu^{\frac{1}{2}}}\left(\frac{1+iL}{1-iL}\right)^{\frac{1}{2}}\left|\frac{1+iL}{1-iL}\right|^{-\frac{1}{2}}\ \frac{1}{1+iL}\ \sim\ \frac{2^{\frac{1}{4}}}{\mu^{\frac{1}{2}}}\ \frac{1}{(1+L^2)^{\frac{1}{2}}}. \tag{12.10}$$

Thus $l(\mu)$ is a real function, having an asymptotic expansion of the form

$$l(\mu)\sim\frac{2^{\frac{1}{4}}}{\mu^{\frac{1}{2}}}\sum_{s=0}^{\infty}\frac{l_s}{\mu^{4s}}\quad\text{as }\mu\to+\infty. \tag{12.11}$$

From (12.10), (12.7), and (4.12), we find

$$l_0=1,\qquad l_1=-\frac{1}{2}g_1^2=-\frac{1}{1152},\qquad l_2=g_1g_3+\frac{3}{8}g_1^4=-\frac{16123}{398\,13120}. \tag{12.12}$$

We are now in a position to combine (12.1) with (12.4). When $t\geq 1+\delta$ each of the functions $\xi(t)$, $(t^2-1)^{\frac{1}{4}}$ and $(t^2-1)^{\frac{3}{2}s}$ is positive. Substituting by means of (12.9) and the first of (12.3), and equating real and imaginary parts, we find

$$W(\tfrac{1}{2}\mu^2,\mu t\sqrt{2})\sim\frac{2^{-\frac{1}{2}}e^{-\frac{1}{4}\pi\mu^2}l(\mu)}{(t^2-1)^{\frac{1}{4}}}\left[\cos(\mu^2\xi+\tfrac{1}{4}\pi)\sum_{s=0}^{\infty}\frac{(-)^s u_{2s}(t)}{(t^2-1)^{3s}\mu^{4s}}-\sin(\mu^2\xi+\tfrac{1}{4}\pi)\sum_{s=0}^{\infty}\frac{(-)^s u_{2s+1}(t)}{(t^2-1)^{3s+\frac{3}{2}}\mu^{4s+2}}\right], \tag{12.13}$$

$$W(\tfrac{1}{2}\mu^2,-\mu t\sqrt{2})\sim\frac{2^{\frac{1}{2}}e^{\frac{1}{4}\pi\mu^2}l(\mu)}{(t^2-1)^{\frac{1}{4}}}\left[\sin(\mu^2\xi+\tfrac{1}{4}\pi)\sum_{s=0}^{\infty}\frac{(-)^s u_{2s}(t)}{(t^2-1)^{3s}\mu^{4s}}+\cos(\mu^2\xi+\tfrac{1}{4}\pi)\sum_{s=0}^{\infty}\frac{(-)^s u_{2s+1}(t)}{(t^2-1)^{3s+\frac{3}{2}}\mu^{4s+2}}\right], \tag{12.14}$$

uniformly with respect to t when $t\geq 1+\delta$. The function $l(\mu)$ is calculable from (12.11).

When $-1+\delta\leq t\leq 1-\delta$ the appropriate branches of ξ and $(t^2-1)^{\frac{1}{4}}$ are given by $\xi=-i\eta$ and $(t^2-1)^{\frac{1}{4}}=e^{\frac{1}{4}\pi i}(1-t^2)^{\frac{1}{4}}$, where η is the real function (5.9) and $(1-t^2)^{\frac{1}{4}}$ is positive. Hence we derive

$$W\left(\frac{1}{2}\mu^2,\mu t\sqrt{2}\right)\sim\frac{l(\mu)}{2^{\frac{1}{2}}e^{\frac{1}{4}\pi\mu^2}}\frac{e^{\mu^2\eta}}{(1-t^2)^{\frac{1}{4}}}\sum_{s=0}^{\infty}\frac{(-)^s u_s(t)}{(1-t^2)^{\frac{3}{2}s}}\frac{1}{\mu^{2s}}\qquad(-1+\delta\leq t\leq 1-\delta). \tag{12.15}$$

The expansions for the derivatives corresponding to (12.13), (12.14), and (12.15) may be obtained by term-by-term differentiation with respect to t (cf. (5.23) and (4.13)).

It may be noted that (12.13), (12.14), and (12.15) are respectively equivalent to the formal series (317), (318) and (324) given on pages 84 and 85 of [4] (cf. sec. 6).

(ii) *a positive; expansions in terms of Airy functions.* In (8.11) replace μ by $\mu e^{-\frac{1}{4}\pi i}$. Then we obtain

$$U\left(\frac{1}{2}i\mu^2,e^{-\frac{1}{4}\pi i}\mu t\sqrt{2}\right)\sim\frac{2\pi^{\frac{1}{2}}\mu^{\frac{1}{3}}}{e^{\frac{1}{12}\pi i}}g(\mu e^{-\frac{1}{4}\pi i})\left(\frac{\zeta}{t^2-1}\right)^{\frac{1}{4}}$$
$$\times\left[\mathrm{Ai}(e^{-\frac{1}{3}\pi i}\mu^{\frac{4}{3}}\zeta)\sum_{s=0}^{\infty}(-)^s\frac{A_s(\zeta)}{\mu^{4s}}+\frac{\mathrm{Ai}'(e^{-\frac{1}{3}\pi i}\mu^{\frac{4}{3}}\zeta)}{e^{-\frac{2}{3}\pi i}\mu^{\frac{8}{3}}}\sum_{s=0}^{\infty}(-)^s\frac{B_s(\zeta)}{\mu^{4s}}\right]. \tag{12.16}$$

The t-region of validity of this expansion, $\mathbf{T}(-\frac{1}{4}\pi)$, includes the segment $t\geq -1+\delta$ (cf. fig. 7(b)). Now from [3]

$$2e^{\frac{1}{6}\pi i}\,\mathrm{Ai}(e^{-\frac{1}{3}\pi i}\mu^{\frac{4}{3}}\zeta)=\mathrm{Bi}(-\mu^{\frac{4}{3}}\zeta)+i\,\mathrm{Ai}(-\mu^{\frac{4}{3}}\zeta). \tag{12.17}$$

Hence substituting (12.16) in (12.1) and equating real parts, we find, with the aid of (12.9) and the first of (12.3),

$$W\left(\frac{1}{2}\mu^2,\mu t\sqrt{2}\right)\sim\frac{\pi^{\frac{1}{2}}\mu^{\frac{1}{3}}l(\mu)}{2^{\frac{1}{2}}e^{\frac{1}{4}\pi\mu^2}}\left(\frac{\zeta}{t^2-1}\right)^{\frac{1}{4}}\left[\mathrm{Bi}(-\mu^{\frac{4}{3}}\zeta)\sum_{s=0}^{\infty}(-)^s\frac{A_s(\zeta)}{\mu^{4s}}+\frac{\mathrm{Bi}'(-\mu^{\frac{4}{3}}\zeta)}{\mu^{\frac{8}{3}}}\sum_{s=0}^{\infty}(-)^s\frac{B_s(\zeta)}{\mu^{4s}}\right] \tag{12.18}$$

valid when $t\geq -1+\delta$. The corresponding expansion for the derivative is

$$W'\left(\frac{1}{2}\mu^2,\mu t\sqrt{2}\right)\sim -\frac{\pi^{\frac{1}{2}}\mu^{\frac{2}{3}}l(\mu)}{2e^{\frac{1}{4}\pi\mu^2}}\left(\frac{t^2-1}{\zeta}\right)^{\frac{1}{4}}$$

$$\times\left[-\frac{\mathrm{Bi}(-\mu^{\frac{4}{3}}\zeta)}{\mu^{\frac{4}{3}}}\sum_{s=0}^{\infty}(-)^s\frac{C_s(\zeta)}{\mu^{4s}}+\mathrm{Bi}'(-\mu^{\frac{4}{3}}\zeta)\sum_{s=0}^{\infty}(-)^s\frac{D_s(\zeta)}{\mu^{4s}}\right] \quad (12.19)$$

(cf. (8.15)).

The expansion for the second solution may be obtained by equating imaginary parts. This gives

$$W\left(\frac{1}{2}\mu^2,-\mu t\sqrt{2}\right)\sim\frac{\pi^{\frac{1}{2}}\mu^{\frac{1}{3}}l(\mu)}{2^{-\frac{1}{2}}e^{-\frac{1}{4}\pi\mu^2}}\left(\frac{\zeta}{t^2-1}\right)^{\frac{1}{4}}\left[\mathrm{Ai}(-\mu^{\frac{4}{3}}\zeta)\sum_{s=0}^{\infty}(-)^s\frac{A_s(\zeta)}{\mu^{4s}}+\frac{\mathrm{Ai}'(-\mu^{\frac{4}{3}}\zeta)}{\mu^{\frac{8}{3}}}\sum_{s=0}^{\infty}(-)^s\frac{B_s(\zeta)}{\mu^{4s}}\right]. \quad (12.20)$$

However, this result is only proved in this way for $t\geq 1$, i.e., non-negative values of ζ. This is because the error term on truncating the series inside the square brackets at the mth terms is

$$\{\mathrm{Bi}(-\mu^{\frac{4}{3}}\zeta)+i\,\mathrm{Ai}(-\mu^{\frac{4}{3}}\zeta)\}O(\mu^{-4m})+\{\mathrm{Bi}'(-\mu^{\frac{4}{3}}\zeta)+i\,\mathrm{Ai}'(-\mu^{\frac{4}{3}}\zeta)\}(1+|\zeta|^{\frac{1}{2}})^{-1}O(\mu^{-4m-\frac{8}{3}}).$$

When ζ is negative the terms in Bi and Bi′ cannot be absorbed in the terms in Ai and Ai′ because they are exponentially larger.

That (12.20) *is* valid in the wider range $t\geq -1+\delta$ can be proved as follows. In (8.11) replace μ by $\mu e^{\frac{3}{4}\pi i}$. Then we obtain

$$U\left(\frac{1}{2}i\mu^2,e^{\frac{3}{4}\pi i}\mu t\sqrt{2}\right)\sim 2\pi^{\frac{1}{2}}e^{\frac{1}{4}\pi i}\mu^{\frac{1}{3}}g(\mu e^{\frac{3}{4}\pi i})\left(\frac{\zeta}{t^2-1}\right)^{\frac{1}{4}}$$

$$\times\left[\mathrm{Ai}(-\mu^{\frac{4}{3}}\zeta)\sum_{s=0}^{\infty}(-)^s\frac{A_s(\zeta)}{\mu^{4s}}+\frac{\mathrm{Ai}'(-\mu^{\frac{4}{3}}\zeta)}{\mu^{\frac{8}{3}}}\sum_{s=0}^{\infty}(-)^s\frac{B_s(\zeta)}{\mu^{4s}}\right] \quad (12.21)$$

valid when $t\in\mathbf{T}(\frac{3}{4}\pi)$and so including $t\geq -1+\delta$ (cf. fig. 7). Equation (12.1), with t replaced by $-t$, shows that the left-hand side equals

$$2^{-\frac{1}{2}}e^{-\frac{1}{8}\pi\mu^2}e^{-i(\frac{1}{2}\phi_2+\frac{1}{8}\pi)}\left[k^{-\frac{1}{2}}W\left(\frac{1}{2}\mu^2,-\mu t\sqrt{2}\right)+ik^{\frac{1}{2}}W\left(\frac{1}{2}\mu^2,\mu t\sqrt{2}\right)\right]. \quad (12.22)$$

Also from (6.1) and (12.7) we derive

$$g(\mu e^{\frac{3}{4}\pi i})\sim 2^{\frac{1}{4}i\mu^2-\frac{1}{4}}e^{\frac{1}{4}i\mu^2}e^{-\frac{3}{8}\pi i+\frac{3}{8}\pi\mu^2}\mu^{-\frac{1}{2}i\mu^2-\frac{1}{2}}(1+iL)^{-1}$$

$$\simeq e^{\frac{1}{2}\pi\mu^2-\frac{1}{2}\pi i}g(\mu e^{-\frac{1}{4}\pi i})$$

$$=2^{-\frac{1}{2}}e^{\frac{3}{8}\pi\mu^2}e^{-i(\frac{1}{2}\phi_2+\frac{3}{8}\pi)}l(\mu) \quad (12.23)$$

(cf. (12.6) and (12.9)). The expansion (12.20) may now be obtained by substituting (12.22) and (12.23) in (12.21), equating real parts and using the first of (12.3).

(iii) *a negative*. When a is negative the differential equation has no turning points on the real axis and we may confine our attention to the exponential-type expansions.

From (2.16), (2.17) and (2.18), we have

$$k^{-\frac{1}{2}}W\left(-\frac{1}{2}\mu^2,\mu t\sqrt{2}\right)+ik^{\frac{1}{2}}W\left(-\frac{1}{2}\mu^2,-\mu t\sqrt{2}\right)=2^{\frac{1}{2}}e^{-\frac{1}{8}\pi\mu^2}e^{i(\frac{1}{2}\phi_2+\frac{1}{8}\pi)}U\left(-\frac{1}{2}i\mu^2,e^{-\frac{1}{4}\pi i}\mu t\sqrt{2}\right), \quad (12.24)$$

where

$$k=(1+e^{-\pi\mu^2})^{\frac{1}{2}}-e^{-\frac{1}{2}\pi\mu^2}=1+O(e^{-\frac{1}{2}\pi\mu^2}), \quad (12.25)$$

and

$$e^{\frac{1}{2}i\phi_2}=\pm\left\{\Gamma\left(\frac{1}{2}-\frac{1}{2}i\mu^2\right)\Big/\left|\Gamma\left(\frac{1}{2}-\frac{1}{2}i\mu^2\right)\right|\right\}^{\frac{1}{2}}=e^{\frac{1}{4}i\mu^2}\left(\frac{1}{2}\mu^2\right)^{-\frac{1}{4}i\mu^2}\{1+O(\mu^{-2})\} \tag{12.26}$$

(cf. (12.2)and(12.3)).

In (4.3) replace μ, t by $\mu e^{\frac{1}{4}\pi i}$, $te^{-\frac{1}{2}\pi i}$, respectively. Then we obtain

$$U\left(-\frac{1}{2}i\mu^2,e^{-\frac{1}{4}\pi i}\mu t\sqrt{2}\right)\sim g(\mu e^{\frac{1}{4}\pi i})\frac{e^{-i\mu^2\xi(-it)}}{(-t^2-1)^{\frac{1}{4}}}\sum_{s=0}^{\infty}(-i)^s\frac{u_s(-it)}{(-t^2-1)^{\frac{3}{2}s}}\frac{1}{\mu^{2s}}. \tag{12.27}$$

The region of validity of this expansion, $e^{\frac{1}{2}\pi i}\mathbf{S}(\frac{1}{4}\pi)$, includes the whole of the real t-axis; this is clear from figures 2(b), 3(iii) and 1.

Substituting (12.25), (12.26), and (12.27) in (12.24) in the manner of (i) and (ii) and determining the branches of the many-valued functions, we obtain finally

$$W(-\tfrac{1}{2}\mu^2,\mu t\sqrt{2})\sim\frac{l(\mu)}{(t^2+1)^{\frac{1}{4}}}\left[\cos(\mu^2\bar{\xi}+\tfrac{1}{4}\pi)\sum_{s=0}^{\infty}\frac{(-)^s\bar{u}_{2s}(t)}{(t^2+1)^{3s}\mu^{4s}}\right.$$
$$\left.-\sin(\mu^2\bar{\xi}+\tfrac{1}{4}\pi)\sum_{s=0}^{\infty}\frac{(-)^s\bar{u}_{2s+1}(t)}{(t^2+1)^{3s+\frac{3}{2}}\mu^{4s+2}}\right], \tag{12.28}$$

uniformly when $-\infty<t<\infty$. Here $\bar{\xi}\equiv\bar{\xi}(t)$ and $\bar{u}_s(t)$ are given by (11.7), (11.9), and (11.11), and $l(\mu)$ is calculable from (12.11).

The corresponding expansion for the derivative is

$$W'(-\tfrac{1}{2}\mu^2,\mu t\sqrt{2})\sim-\frac{\mu}{\sqrt{2}}l(\mu)(t^2+1)^{\frac{1}{4}}\left[\sin(\mu^2\bar{\xi}+\tfrac{1}{4}\pi)\sum_{s=0}^{\infty}\frac{(-)^s\bar{v}_{2s}(t)}{(t^2+1)^{3s}\mu^{4s}}\right.$$
$$\left.+\cos(\mu^2\bar{\xi}+\tfrac{1}{4}\pi)\sum_{s=0}^{\infty}\frac{(-)^s\bar{v}_{2s+1}(t)}{(t^2+1)^{3s+\frac{3}{2}}\mu^{4s+2}}\right], \tag{12.29}$$

$\bar{v}_s(t)$ being given by (11.13) and (11.14).

13. Numerical Examples

Here we illustrate the use of the Airy-type expansions by two examples. It may be remarked that applications would be greatly facilitated by the preparation of suitable numerical tables of the functions $g(\mu)$, $l(\mu)$, $\zeta(t)$, $\{\zeta/(t^2-1)\}^{\frac{1}{4}}$ and $A_s(\zeta)$, $B_s(\zeta)$, $C_s(\zeta)$, $D_s(\zeta)$ $(s=0,1,\ldots)$. In the present examples these functions are calculated directly from formulas given in this paper.

It will appear from the examples that the early coefficients in the expansions decrease in magnitude as s increases. This property, shared with uniform asymptotic expansions for Bessel functions of large order ([8], sec. 6), makes the expansions well suited to numerical work.

Example 1. To evaluate $W(a,\pm x)$, $W'(a,\pm x)$ for $a=8$, $x=7$.

We use the expansions (12.18), (12.19), and (12.20). Here

$$\mu=4, \qquad t=7/(4\sqrt{2})=1.23743\ 6867.$$

Hence

$$2^{\frac{1}{2}}e^{\frac{1}{4}\pi\mu^2}=4.05527\ 596\times10^5, \tag{13.1}$$

and from (12.11), (12.12)

$$l(\mu)\sim(2^{\frac{1}{4}}/2)\ (1-0.00000\ 3391-0.00000\ 0006+\ldots)=0.59460\ 1538,$$

so that

$$\pi^{\frac{1}{2}}\mu^{\frac{1}{3}}l(\mu)=1.67296\ 798. \tag{13.2}$$

Next, from (7.4) we have

$$\zeta^{\frac{3}{2}}=\frac{3}{4}t(t^2-1)^{\frac{1}{2}}-\frac{3}{4}\ln\{t+(t^2-1)^{\frac{1}{2}}\}=0.16932\ 9581.$$

Hence

$$\zeta=0.30607\,0793, \quad \zeta^{\frac{1}{2}}=0.55323\,6652, \quad \{\zeta/(t^2-1)\}^{\frac{1}{4}}=0.87122\,5817. \tag{13.3}$$

From (4.8) and (4.11) we compute

$$\mathscr{A}_0=1, \quad \mathscr{A}_1=-0.59504\,451, \quad \mathscr{A}_2=2.92479\,70, \quad \mathscr{A}_3=-26.48264\,7,$$

and from (8.13)

$$a_0=1, \quad a_1=\frac{5}{48}, \quad a_2=\frac{385}{4608}, \quad a_3=\frac{85085}{6\,63552},$$

$$b_0=1, \quad b_1=-\frac{7}{48}, \quad b_2=-\frac{455}{4608}, \quad b_3=-\frac{95095}{6\,63552}.$$

Substitution in (8.12) yields

$$A_0(\zeta)=1, \quad A_1(\zeta)=-0.00649\,0, \quad B_0(\zeta)=-0.03637\,97, \quad B_1(\zeta)=0.01202.$$

Next, we have $\mu^{\frac{4}{3}}\zeta=1.94342\,840$, and by interpolation in table VII of [3]

$$\mathrm{Bi}(-\mu^{\frac{4}{3}}\zeta)=-0.39524\,090, \qquad \mathrm{Bi}'(-\mu^{\frac{4}{3}}\zeta)=0.32386\,134,$$

$$\mathrm{Ai}(-\mu^{\frac{4}{3}}\zeta)=0.26162\,602, \qquad \mathrm{Ai}'(-\mu^{\frac{4}{3}}\zeta)=0.59097\,966.$$

The content of the square brackets on the right of (12.18) is accordingly

$$-0.39524\,090(1+0.00002\,535+\ldots)+4^{-\frac{8}{3}}\times0.32386\,134(-0.03637\,97-0.00004\,70+\ldots)$$
$$=-0.39554\,353.$$

Substituting this value and (13.1), (13.2), (13.3) in (12.18), we obtain

$$W(8,7)=-\frac{1.67296\,798\times0.87122\,5817\times0.39554\,353}{4.05527\,596\times10^5}=\mathbf{-0.14216\,485\times10^{-5}}.$$

Similarly,

$$W(8,-7)=1.67296\,798\times0.87122\,5817\times0.26109\,871\times4.05527\,596\times10^5=\mathbf{1.54327\,57\times10^5}.$$

For the derivatives, we have first

$$2^{-\frac{1}{2}}\pi^{\frac{1}{2}}\mu^{\frac{2}{3}}l(\mu)=1.87784\,306.$$

From (4.14) and (4.16), we compute

$$\mathscr{B}_0=1, \quad \mathscr{B}_1=1.00283\,725, \quad \mathscr{B}_2=-3.52326\,59, \quad \mathscr{B}_3=29.91942\,3,$$

and thence from (8.19)

$$C_0(\zeta)=-0.07833\,70, \quad C_1(\zeta)=0.00971, \quad D_0(\zeta)=1, \quad D_1(\zeta)=0.00760\,4.$$

Hence

$$W'(8,7)=-\frac{1.87784\,306\times0.31897\,315}{4.05527\,596\times10^5\times0.87122\,5817}=\mathbf{-0.16953\,613\times10^{-5}},$$

$$W'(8,-7)=\frac{1.87784\,306\times0.59419\,143\times4.05527\,596\times10^5}{0.87122\,5817}=\mathbf{5.19368\,20\times10^5}.$$

The results of this example agree adequately with the entries given in [4], table I.

Example 2. To evaluate the Hermite polynomial $H_n(x)$ for $n=10$, $x=4$ by means of the expansion (8.11).

Here (cf. (11.23))

$$\mu=\sqrt{21}=4.58257\ 569, \qquad t=4/\sqrt{21}=0.87287\ 1561.$$

From (6.1) and (4.12)

$$1/g(\mu)\sim 2^{5.5}e^{5.25}(21)^{-5}(1+0.00198\ 4127-0.00000\ 1052+\ .\ .\ .)=2.11580\ 149\times 10^{-3},$$

and hence

$$2\pi^{\frac{1}{2}}\mu^{\frac{1}{3}}g(\mu)=2.78291\ 471\times 10^{3}.$$

Next, from (11.21)

$$(-\zeta)^{\frac{3}{2}}=\frac{3}{4}\cos^{-1}t-\frac{3}{4}t\,(1-t^2)^{\frac{1}{2}}=0.06286\ 6477.$$

Hence

$$\zeta=-0.15810\ 5174, \quad (-\zeta)^{\frac{1}{2}}=0.39762\ 4413, \quad \{\zeta/(t^2-1)\}^{\frac{1}{4}}=0.90271\ 123.$$

When ζ is negative, real forms of equations (8.12) and (4.8) are

$$A_s(\zeta)=(-)^s\sum_{m=0}^{2s} b_m(-\zeta)^{-\frac{3}{2}m}\overline{\mathscr{A}}_{2s-m}, \qquad (-\zeta)^{\frac{1}{2}}B_s(\zeta)=(-)^{s-1}\sum_{m=0}^{2s+1} a_m(-\zeta)^{-\frac{3}{2}m}\overline{\mathscr{A}}_{2s-m+1}, \tag{13.4}$$

where

$$\overline{\mathscr{A}}_s=u_s(t)/(1-t^2)^{\frac{3}{2}s}. \tag{13.5}$$

In the present example we find

$$\overline{\mathscr{A}}_0=1, \quad \overline{\mathscr{A}}_1=-1.63978\ 319, \quad \overline{\mathscr{A}}_2=21.19031\ 25, \quad \overline{\mathscr{A}}_3=-516.52363,$$

and thence

$$A_0(\zeta)=1, \quad A_1(\zeta)=-0.01024\ 3, \quad B_0(\zeta)=-0.04317\ 57, \quad B_1(\zeta)=0.0166.$$

Next, $\mu^{\frac{4}{3}}\zeta=-1.20344\ 324$. Interpolation in table I of [3], yields

$$\mathrm{Ai}(\mu^{\frac{4}{3}}\zeta)=0.52582\ 209, \qquad \mathrm{Ai}'(\mu^{\frac{4}{3}}\zeta)=0.10920\ 810.$$

Substituting these results in (8.11), we obtain

$$U(-\tfrac{21}{2}, 4\sqrt{2})=2.78291\ 471\times 10^3\times 0.90271\ 123\times 0.52572\ 856=1.32071\ 865\times 10^3,$$

and thence using (11.23)

$$H_{10}(4)=2^5e^8U\left(-\frac{21}{2}, 4\sqrt{2}\right)=\mathbf{1.25984\ 22\times 10^8}.$$

The exact value of this polynomial, computed from explicit expressions given on page 75 of [4], is 1259 84224.

Part 5. Zeros and Associated Values

14. Zeros of $U(a, z)$, $U'(a, z)$, $\overline{U}(a, z)$, and $\overline{U}'(a, z)$

(i) *a negative;* $-2(1-\delta)\sqrt{(-a)}\leq z\leq 2\sqrt{(-a)}$. Let the sth real zeros, counted in descending order away from the point $z=2\sqrt{(-a)}$, of the functions $U(a,z)$, $U'(a,z)$, $\overline{U}(a,z)$, and $\overline{U}'(a,z)$ be denoted by $u_{a,s}$, $u'_{a,s}$, $\bar{u}_{a,s}$, and $\bar{u}'_{a,s}$, respectively.

The expansion (8.11) is of exactly the same form as the expansion (4.24) of [8]. The analysis of section 7 of that paper is therefore immediately applicable, and we find

$$u_{a,s}\sim 2^{\frac{1}{2}}\mu\left[p_0(\alpha)+\frac{p_1(\alpha)}{\mu^4}+\frac{p_2(\alpha)}{\mu^8}+\ .\ .\ .\right] \tag{14.1}$$

as $\mu \equiv \sqrt{(-2a)} \to +\infty$, s fixed. Here

$$\alpha = \mu^{-\frac{4}{3}} a_s, \tag{14.2}$$

a_s denoting the sth negative zero of the function Ai and not to be confused with the a_m of (8.13). The coefficients $p_r(\zeta)$ are given by

$$p_0(\zeta) = t(\zeta), \qquad p_1(\zeta) = \alpha_1(\zeta) t'(\zeta), \qquad p_2(\zeta) = \alpha_2(\zeta) t'(\zeta) + \frac{1}{2} \{\alpha_1(\zeta)\}^2 t''(\zeta), \ldots, \tag{14.3}$$

where $t(\zeta)$ is the function inverse to $\zeta(t)$, defined by (11.21), and the $\alpha_r(\zeta)$ are given by relations of the form of (7.13) of [8]. Explicitly,

$$p_1(\zeta) = \frac{t^3 - 6t}{24(t^2-1)^2} + \frac{5}{48\{(t^2-1)\zeta^3\}^{\frac{1}{2}}}. \tag{14.4}$$

The corresponding expansion for the associated value $U'(a, u_{a,s})$ can be found by substituting (14.1) in (8.15). More convenient expressions for the coefficients are obtained however, if we substitute instead in the equation

$$U'(a, u_{a,s}) = \pm (2\pi)^{\frac{1}{4}} \left\{ \Gamma\left(\frac{1}{2} - a\right) \right\}^{\frac{1}{2}} \left(-\frac{du_{a,s}}{ds} \right)^{-\frac{1}{2}} \tag{14.5}$$

obtained from eq (2.15) of [6] and the second of (2.13) in the present paper. This yields the required result in the form

$$U'(a, u_{a,s}) \sim \pi^{\frac{1}{4}} \mu^{\frac{1}{6}} \left\{ \Gamma\left(\frac{1}{2} + \frac{1}{2}\mu^2\right) \right\}^{\frac{1}{2}} \mathrm{Ai}'(a_s) \psi(\alpha) \left[1 + \sum_{r=1}^{\infty} \frac{P_r(\alpha)}{\mu^{4r}} \right], \tag{14.6}$$

where

$$\psi(\zeta) \equiv \frac{1}{\phi(\zeta)} = \left(\frac{t^2-1}{\zeta} \right)^{\frac{1}{4}}, \tag{14.7}$$

and the coefficients $P_r(\zeta)$ satisfy the asymptotic identity

$$\left[1 + \sum_{r=1}^{\infty} \frac{P_r(\zeta)}{\mu^{4r}} \right]^{-2} \sim 1 + \{\psi(\zeta)\}^2 \sum_{r=1}^{\infty} \frac{p_r'(\zeta)}{\mu^{4r}}. \tag{14.8}$$

Thus

$$P_1(\zeta) = -\frac{1}{2} \{\psi(\zeta)\}^2 p_1'(\zeta), \qquad P_2(\zeta) = \frac{3}{2} \{P_1(\zeta)\}^2 - \frac{1}{2} \{\psi(\zeta)\}^2 p_2'(\zeta). \tag{14.9}$$

Explicitly,

$$P_1(\zeta) = \frac{t^4 - 15t^2 - 6}{48(t^2-1)^3} + \frac{5t}{96(t^2-1)^{\frac{3}{2}} \zeta^{\frac{3}{2}}} + \frac{5}{64\zeta^3}. \tag{14.10}$$

For the zeros of the derivatives of the parabolic cylinder functions, we may show by similar analysis that

$$u_{a,s}' \sim 2^{\frac{1}{2}} \mu \left[q_0(\beta) + \frac{q_1(\beta)}{\mu^4} + \frac{q_2(\beta)}{\mu^8} + \ldots \right], \tag{14.11}$$

where

$$\beta = \mu^{-\frac{4}{3}} a_s', \tag{14.12}$$

$$q_0(\zeta) = t(\zeta), \qquad q_1(\zeta) = \beta_1(\zeta) t'(\zeta), \qquad q_2(\zeta) = \beta_2(\zeta) t'(\zeta) + \frac{1}{2} \{\beta_1(\zeta)\}^2 t''(\zeta), \ldots, \tag{14.13}$$

and the $\beta_r(\zeta)$ are given by eq (7.22) of [8].

The equation corresponding to (14.5) is

$$U(a, u_{a,s}') = \pm (2\pi)^{\frac{1}{4}} \left\{ \Gamma\left(\frac{1}{2} - a\right) \right\}^{\frac{1}{2}} \left(-a - \frac{1}{4} u_{a,s}'^2 \right)^{-\frac{1}{2}} \left(-\frac{du_{a,s}'}{ds} \right)^{-\frac{1}{2}}, \tag{14.14}$$

and from it we derive

$$U(a,u'_{a,s})\sim\frac{(4\pi)^{\frac{1}{4}}}{\mu^{\frac{1}{6}}}\left\{\Gamma\left(\frac{1}{2}+\frac{1}{2}\mu^2\right)\right\}^{\frac{1}{2}}\mathrm{Ai}\,(a'_s)\phi(\beta)\left[1+\sum_{r=1}^{\infty}\frac{Q_r(\beta)}{\mu^{4r}}\right], \tag{14.15}$$

in which the coefficients satisfy the identity

$$\left[1+\sum_{r=1}^{\infty}\frac{Q_r(\zeta)}{\mu^{4r}}\right]^{-2}\sim\frac{\{\phi(\zeta)\}^2}{\zeta}\left[\left\{\sum_{r=0}^{\infty}\frac{q_r(\zeta)}{\mu^{4r}}\right\}^2-1\right]\left[\sum_{r=0}^{\infty}\frac{q'_r(\zeta)}{\mu^{4r}}\right]. \tag{14.16}$$

For the function $\overline{U}(a,z)$, the expansions corresponding to (14.1), (14.6), (14.11), and (14.15) are obtained on replacing the symbols U, u, Ai, and a_s by $\overline{U}$, $\bar{u}$, Bi, and b_s, respectively.

(ii) *a negative; other zeros.* Since the zeros of the function Ai are all real and negative, it follows from (8.11) that for all sufficiently large positive values of the parameter μ, the zeros of $U(-\frac{1}{2}\mu^2,\ \mu t\sqrt{2})$ in the t-domain $\mathbf{T}(0)$ (the unshaded region of fig. 7(a)), all lie in the interval $-1<t<1$. This is also true of the zeros of the derived function $U'(-\frac{1}{2}\mu^2,\mu t\sqrt{2})$. The function $\overline{U}(-\frac{1}{2}\mu^2,\mu t\sqrt{2})$ and its derivative however, have complex zeros in this region corresponding to the complex zeros of Bi and Bi′ (cf. [8], appendix).

In the complementary region $C\mathbf{T}(0)$ the distribution of the zeros of $U(-\frac{1}{2}\mu^2,\mu t\sqrt{2})$ can be investigated by reversion of the expansion (9.7). The situation is more complicated; the pattern of the zeros depends largely upon the non-integer part of $\frac{1}{2}\mu^2$. It should be noted that expansions of zeros in the interval $-1<t<1$ of fixed enumeration, counted away from $t=-1$, come within this category.

(iii) *a complex.* The extension of the results of this section to general values of a presents no difficulty. We readily see, for example, that when $\epsilon\leq\arg\mu\leq\pi-\epsilon$ the zeros of $U(-\frac{1}{2}\mu^2,\ \mu t\sqrt{2})$ in the t-domain $\mathbf{T}(\arg\mu)$ are asymptotically close to the principal curve AQ, illustrated in figure 4.

15. Zeros of $W(a,x)$ and $W'(a,x)$

We denote the sth positive zeros of the functions $W(a,x)$, $W'(a,x)$, $W(a,-x)$ and $W'(a,-x)$, by $w_{a,s}$, $w'_{a,s}$, $\overline{w}_{a,s}$ and $\overline{w}'_{a,s}$, respectively.

(i) *a positive.* Writing $\mu=\sqrt{(2a)}$ and

$$\alpha=\mu^{-\frac{4}{3}}b_s,\qquad \beta=\mu^{-\frac{4}{3}}b'_s, \tag{15.1}$$

where b_s and b'_s are the sth negative zeros of Bi and Bi′ respectively, we find by reversion of (12.18) and (12.20)

$$w_{a,s}\sim2^{\frac{1}{2}}\mu\left[p_0(-\alpha)-\frac{p_1(-\alpha)}{\mu^4}+\frac{p_2(-\alpha)}{\mu^8}-\ldots\right], \tag{15.2}$$

$$2^{\frac{1}{2}}e^{\frac{1}{2}\pi a}W'(a,w_{a,s})\sim-2^{-\frac{1}{4}}\pi^{\frac{1}{2}}\mu^{\frac{1}{6}}\mathrm{Bi}'(b_s)\psi(-\alpha)\left[1+\sum_{r=1}^{\infty}(-)^r\frac{P_r(-\alpha)}{\mu^{4r}}\right], \tag{15.3}$$

$$w'_{a,s}\sim2^{\frac{1}{2}}\mu\left[q_0(-\beta)-\frac{q_1(-\beta)}{\mu^4}+\frac{q_2(-\beta)}{\mu^8}-\ldots\right], \tag{15.4}$$

$$2^{\frac{1}{2}}e^{\frac{1}{2}\pi a}W(a,w'_{a,s})\sim2^{\frac{1}{4}}\pi^{\frac{1}{2}}\mu^{-\frac{1}{6}}\,\mathrm{Bi}\,(b'_s)\phi(-\beta)\left[1+\sum_{r=1}^{\infty}(-)^r\frac{Q_r(-\beta)}{\mu^{4r}}\right], \tag{15.5}$$

as $a\to+\infty$, uniformly with respect to s. Here $\psi(\zeta)$, $\phi(\zeta)$, $p_r(\zeta)$, $P_r(\zeta)$, $q_r(\zeta)$ and $Q_r(\zeta)$ are the functions defined in section 14(i).

With α, β given by (14.2) and (14.12) instead of (15.1) and Bi, b replaced by Ai, a, the right-hand sides of (15.2), (15.3), (15.4), and (15.5) are the corresponding expansions of $\overline{w}_{a,s}$, $-2^{-\frac{1}{2}}e^{-\frac{1}{2}\pi a}W'(a,-\overline{w}_{a,s})$, $\overline{w}'_{a,s}$ and $2^{-\frac{1}{2}}e^{-\frac{1}{2}\pi a}W(a,-\overline{w}'_{a,s})$, respectively.

(ii) *a negative.* Let $\mu=\sqrt{(-2a)}$, and

$$t=\tau(\bar{\xi}) \tag{15.6}$$

be the function inverse to $\bar{\xi}(t)$, defined by (11.7). Then by reversion of the expansion (12.28), we find

$$w_{a,s}\sim 2^{\frac{1}{2}}\mu\left[\tau+\frac{e_1(\tau)}{\mu^4}+\frac{e_2(\tau)}{\mu^8}+\cdots\right] \tag{15.7}$$

as $a\to-\infty$, uniformly with respect to s. Here

$$\tau=\tau\left\{\frac{(s-\frac{3}{4})\pi}{\mu^2}\right\}, \tag{15.8}$$

and the leading coefficients are given by the formulas

$$e_1(\tau)=\frac{\tau^3+6\tau}{24(\tau^2+1)^2},\qquad e_2(\tau)=\frac{56\tau^9+247\tau^7+291\tau^5-2520\tau^3+3780\tau}{5760(\tau^2+1)^5}. \tag{15.9}$$

The expansion of the associated value of the derivative can be obtained by substitution of (15.7) in the equation

$$W'(a,w_{a,s})=\pm\pi^{\frac{1}{2}}\left(\frac{dw_{a\ s}}{ds}\right)^{-\frac{1}{2}} \tag{15.10}$$

obtained from eq (2.15) of [6] and (2.19). Thence we derive

$$W'(a,w_{a,s})\sim(-)^s 2^{-\frac{1}{4}}\mu^{\frac{1}{2}}(\tau^2+1)^{\frac{1}{4}}\left[1+\sum_{r=1}^{\infty}\frac{E_r(\tau)}{\mu^{4r}}\right], \tag{15.11}$$

where

$$E_1(\tau)=\frac{\tau^4+15\tau^2-6}{48(\tau^2+1)^3},\qquad E_2(\tau)=\frac{112\tau^{10}+489\tau^8-98\tau^6-34995\tau^4+80460\tau^2-7020}{23040(\tau^2+1)^6}. \tag{15.12}$$

In a similar manner we obtain from (12.29)

$$w'_{a,s}\sim 2^{\frac{1}{2}}\mu\left[\tau+\frac{f_1(\tau)}{\mu^4}+\frac{f_2(\tau)}{\mu^8}+\ldots\right], \tag{15.13}$$

where now

$$\tau=\tau\left\{\frac{(s-\frac{1}{4})\pi}{\mu^2}\right\}, \tag{15.14}$$

and

$$f_1(\tau)=\frac{\tau^3-6\tau}{24(\tau^2+1)^2},\qquad f_2(\tau)=\frac{56\tau^9+247\tau^7+651\tau^5+2640\tau^3-3420\tau}{5760(\tau^2+1)^5}. \tag{15.15}$$

Finally, from

$$W(a,w'_{a,s})=\pm\pi^{\frac{1}{2}}\left(\frac{1}{4}w'^{2}_{a,s}-a\right)^{-\frac{1}{4}}\left(\frac{dw'_{a,s}}{ds}\right)^{-\frac{1}{2}}, \tag{15.16}$$

we find

$$W(a,w'_{a,s})\sim(-)^s 2^{\frac{1}{4}}\mu^{-\frac{1}{2}}(\tau^2+1)^{-\frac{1}{4}}\left[1+\sum_{r=1}^{\infty}\frac{F_r(\tau)}{\mu^{4r}}\right], \tag{15.17}$$

where τ is again given by (15.14), and

$$F_1(\tau)=\frac{-\tau^4-9\tau^2+6}{48(\tau^2+1)^3},\qquad F_2(\tau)=\frac{-112\tau^{10}-479\tau^8-142\tau^6+25725\tau^4-67500\tau^2+7380}{23040(\tau^2+1)^6}. \tag{15.18}$$

If in the right-hand sides of (15.7) and (15.11) we take the value (15.14) for τ instead of (15.8), we obtain the asymptotic expansions of $\overline{w}_{a,s}$ and $-W'(a,-\overline{w}_{a,s})$, respectively. Similarly the right-hand sides of (15.13) and (15.17) with τ given by (15.8) are the expansions of $\overline{w}'_{a,s}$ and $-W(a,-\overline{w}'_{a,s})$.

It may be noticed that some of the numerical coefficients occurring in the expressions (14.4) and (14.10) for the coefficients in the reverted Airy-type expansions are the same as those in the expressions (15.9) and (15.12) pertaining to the reverted exponential-type expansions. This is no mere coincidence; we may expect that by suitable re-expansion of the reverted Airy forms, and subsequent comparison with exponential-type expansions obtained by reversion of (5.11), we shall obtain explicit expressions for the coefficients p_r and P_r in terms of the e_r and E_r respectively, analogous to the formulas (8.12). Similarly for q_r and Q_r.

16. Numerical Examples

Example 1. To compute the smallest positive zero of $W(10,x)$ and the corresponding $W'(10,x)$.

We use (15.2) and (15.3) with $\mu=\sqrt{20}=4.47213\,595$, $s=1$. Entering table V of [3], we find

$$b_1=-1.17371\,322, \qquad \mathrm{Bi}'(b_1)=0.60195\,789.$$

Hence

$$\zeta=-\alpha=-\mu^{-\frac{4}{3}}b_1=0.15929\,7392.$$

The value of t is given by

$$t(t^2-1)^{\frac{1}{2}}-\ln\{t+(t^2-1)^{\frac{1}{2}}\}=\frac{4}{3}\zeta^{\frac{3}{2}}=0.08477\,1864.$$

Solving this equation by successive approximation, using Newton's rule, we find

$$t=1.12489\,6229=p_0(-\alpha),$$

and thence from (14.4)

$$p_1(-\alpha)=0.02960\,04.$$

Hence

$$w_{10,1}\sim 2^{\frac{1}{2}}\times 4.47213\,595(1.12489\,6229-0.00007\,4001+\ .\ .\ .\ )=\mathbf{7.11400\,040}.$$

For the derivative, we compute from (14.7) and (14.10)

$$1/\psi(-\alpha)=\phi(-\alpha)=0.88019\,8053, \qquad P_1(-\alpha)=0.00942\,2.$$

Substitution in (15.3) then yields

$$W'(10,w_{10,1})\sim-\frac{2^{-\frac{3}{4}}\pi^{\frac{1}{2}}(20)^{\frac{1}{12}}\times 0.60195\,789}{e^{5\pi}\times 0.88019\,8053}(1-0.00002\,356+\ .\ .\ .\ )=\mathbf{-1.39416\,865\times 10^{-7}}.$$

By inverse and direct interpolation in table III of [4], we find

$$w_{10,1}=7.11400\,000, \qquad W'(10,w_{10,1})=-1.39416\,86\times 10^{-7}.$$

The discrepancy in the values of $w_{10,1}$ is undoubtedly due to neglect of the third term of the asymptotic series.

Example 2. To compute the fifth positive zero of $W'(-10,x)$ and the corresponding $W(-10,x)$.

We use (15.13) and (15.17). From (15.14), (15.6), and (11.7) we see that the appropriate value of τ satisfies

$$\tau(\tau^2+1)^{\frac{1}{2}}+\ln\{\tau+(\tau^2+1)^{\frac{1}{2}}\}=2\bar{\xi}=2(5-\tfrac{1}{4})\pi/20=1.49225\,6510.$$

Solving by successive approximation, we find

$$\tau=0.69390\,0720, \qquad (\tau^2+1)^{\frac{1}{2}}=1.21716\,8110,$$

and thence from (15.15)

$$f_1(\tau) = -0.07269\ 508, \qquad f_2(\tau) = -0.03321.$$

Thus

$$w'_{-10,5} \sim 2^{\frac{1}{2}} \times 4.47213\ 595(0.69390\ 0720 - 0.00018\ 1738 - 0.00000\ 0208 + \ldots) = \mathbf{4.38746\ 276}.$$

From (15.18) we compute

$$F_1(\tau) = 0.00919\ 198, \qquad F_2(\tau) = -0.07882,$$

and thence

$$W(-10, w'_{-10,5}) \sim -2^{\frac{1}{4}}(20)^{-\frac{1}{4}}(1.21716\ 8110)^{-\frac{1}{2}}(1 + 0.00002\ 2980 - 0.00000\ 0493 + \ldots)$$
$$= -\mathbf{0.50972\ 3297}.$$

The results of this example agree with values obtained by interpolation in table III of [4].

References

[1] C. G. Darwin, On Weber's function, Quart. J. Mech. Appl. Math. **2,** 311 (1949).
[2] A. Erdélyi, M. Kennedy, and J. L. McGregor, Parabolic cylinder functions of large order, J. Rat. Mech. Anal. **3,** 459 (1954).
[3] J. C. P. Miller, The Airy integral, British Assoc. Math. Tables, Part-Volume B (Cambridge University Press, Cambridge, Eng. 1946).
[4] J. C. P. Miller, Tables of Weber parabolic cylinder functions, London: H. M. Stationery Office (1955).
[5] L. M. Milne-Thomson, The calculus of finite differences, London: Macmillan (1933).
[6] F. W. J. Olver, A new method for the evaluation of zeros of Bessel functions and of other solutions of second-order differential equations, Proc. Cambridge Philos. Soc. **46,** 570 (1950).
[7] F. W. J. Olver, The asymptotic solution of linear differential equations of the second order for large values of a parameter, Phil. Trans. Roy. Soc. London, Ser. A **247,** 307 (1954).
[8] F. W. J. Olver, The asymptotic expansion of Bessel functions of large order, Phil. Trans. Roy. Soc. London, Ser. A **247,** 328 (1954).
[9] F. W. J. Olver, Uniform asymptotic expansions of solutions of linear second-order differential equations for large values of a parameter, Phil. Trans. Roy. Soc. London, Ser. A **250,** 479 (1958).
[10] N. Schwid, The asymptotic forms of the Hermite and Weber functions, Trans. Amer. Math. Soc. **37,** 339 (1935).
[11] G. N. Watson, A theory of asymptotic series, Phil. Trans. Roy. Soc. London, Ser. A **211,** 279 (1911).
[12] G. N. Watson, The harmonic functions associated with the parabolic cylinder, Proc. London Math. Soc. **17,** 116 (1918).
[13] E. T. Whittaker, On the functions associated with the parabolic cylinder in harmonic analysis, Proc. London Math. Soc. **35,** 417 (1903).
[14] E. T. Whittaker and G. N. Watson, A course of modern analysis (Cambridge University Press, Cambridge, Eng., 1952).
[15] N. D. Kazarinoff, Asymptotic theory of second order differential equations with two simple turning points, Arch. Rat. Mech. Anal. **2,** 129 (1958).

WASHINGTON, D. C.

(Paper 63B2–14)

[790]

ERROR BOUNDS FOR THE LIOUVILLE–GREEN (OR WKB) APPROXIMATION

BY F. W. J. OLVER

Communicated by E. T. GOODWIN

Received 18 January 1961

ABSTRACT. Error bounds are derived and examined for approximate solutions in terms of elementary functions of the differential equations

$$\frac{d^2w}{dx^2} = \{\pm u^2 + f(u,x)\}w, \quad \frac{d^2w}{dx^2} = \pm u^2 p(u,x)w,$$

in which u is a positive parameter, the functions f and p are free from singularities and p does not vanish. Bounds are also obtained for the remainder terms in the asymptotic expansions of the solutions in descending powers of u. The variable x ranges over a real interval, finite or infinite, or over a region of the complex plane, bounded or unbounded.

Applications are made to parabolic cylinder functions of large orders, and modified Bessel functions of large orders.

1. *Introduction.* An approximation frequently made in applied mathematics is the representation of the solution of the differential equation

$$\frac{d^2w}{dx^2} = q(x)\,w, \tag{1·1}$$

in the form

$$w \doteqdot Aq^{-\frac{1}{4}}\exp\left(\int q^{\frac{1}{2}}dx\right) + Bq^{-\frac{1}{4}}\exp\left(-\int q^{\frac{1}{2}}dx\right), \tag{1·2}$$

where A, B are arbitrary constants. The validity of this approximation can be examined by constructing the second-order differential equation satisfied by the right of (1·2), that is,

$$\frac{d^2w}{dx^2} = q(x)\{1 - f(x)\}\,w, \tag{1·3}$$

where

$$f(x) = \frac{4qq'' - 5q'^2}{16q^3} = -\frac{1}{q^{\frac{3}{4}}}\frac{d^2}{dx^2}\left(\frac{1}{q^{\frac{1}{4}}}\right). \tag{1·4}$$

An essential condition, therefore, is that $|f(x)|$ be small throughout the x-interval under consideration. This is satisfied if $\{q(x)\}^{-\frac{1}{4}}$ is a bounded, slowly-varying function of x.

The naming of this type of approximation has a varied history. Apparently the formula (1·2) was first derived by Liouville (1) and Green (2) in 1837. Since 1926, however, many writers in theoretical physics have referred to it as the 'WKB' approximation in consequence of publications in that year by Wentzel (3), Kramers (4), and Brillouin (5). Other writers have restricted the name WKB to an approximate method for determining the arbitrary constants in the formulae which connect the approximate solutions (1·2) valid on either side of a zero of $q(x)$; see (6). Often in the latter usage the initial 'J' is added to WKB to acknowledge the fact that Jeffreys (7) had previously given this method of connexion in 1924, and Jeffreys (8) has recently pointed out that he himself had been anticipated by Gans (9).

Frequently in applications the function $q(x)$ in (1·1) contains a large positive parameter, which we denote by u, in the form $q(x) = u^2 p(x)$. In this event (1·1), (1·2) take the forms

$$\frac{d^2w}{dx^2} = u^2 p(x)\, w, \tag{1·5}$$

$$w \doteq Ap^{-\frac{1}{4}} \exp\left(u \int p^{\frac{1}{2}}\, dx\right) + Bp^{-\frac{1}{4}} \exp\left(-u \int p^{\frac{1}{2}}\, dx\right), \tag{1·6}$$

and (1·4) becomes

$$f(x) = \frac{4pp'' - 5p'^2}{16u^2p^3} = -\frac{1}{u^2 p^{\frac{3}{4}}} \frac{d^2}{dx^2}\left(\frac{1}{p^{\frac{1}{4}}}\right). \tag{1·7}$$

Therefore if, in the interval considered, the function $p(x)$ is free from zeros and its only singularities are double poles, then $|f|$ becomes uniformly small as $u \to \infty$. The approximation (1·6) must therefore then be expected to provide an asymptotic representation of the solution for large u.

The literature establishing precise conditions for the existence of asymptotic solutions of this form is already extensive. What does not appear to have been generally realized is that it is possible to deduce from the existence proofs strict upper bounds for the errors in the approximate solutions, and in their derivatives, *which are both realistic and easy to evaluate*. Such bounds were in fact obtained in 1912 by Blumenthal (10), but the significance of his results seems to have been overlooked. The first step in the present paper is a re-derivation of Blumenthal's results, but we improve upon them by permitting the x-interval under consideration to be infinite and by sharpening the bounds.

We consider a more general form of differential equation than (1·5), allowing the coefficient p to depend on u as well as on x. Thus as standard forms of equation we take

$$\frac{d^2w}{dx^2} = \pm\, u^2 p(u,x)\, w \tag{1·8}$$

in which $p(u,x)$ is positive. In § 2 we establish, by analysis similar to that of (8), § 3, two theorems (1 and 2), corresponding to Blumenthal's results, which give error bounds for the approximate solutions of this equation when $p(u,x)$ has the form

$$p(u,x) = 1 + u^{-2} f(u,x). \tag{1·9}$$

The function $f(u,x)$ is a continuous function of x, and no extra complication occurs in these theorems by assuming it to be a complex function of the real variables u and x. The x-interval may be finite or infinite.

In § 3 we examine these bounds, and among other results we show that if f is real, of constant sign and independent of u, and the upper sign is taken in (1·8), the bounds overestimate the actual relative errors merely by a quantity which is of order u^{-2} for large u. Also included in this section is a simple illustrative example in which $f =$ constant.

In § 4 we derive the corresponding theorems (3 and 4) for equation (1·8) by use of a transformation of this equation which converts $p(u,x)$ into the form (1·9); these theorems may be applied to equation (1·1) by setting $u = 1$. An application to the differential equations of parabolic cylinder functions follows in § 5.

The expression (1·6) can be regarded as a first approximation to the solution of (1·5) and formal solutions can be constructed in infinite series of descending powers of u. Conditions under which these series are asymptotic expansions of solutions have been established, for example, in (11) and (12). In § 6 we prove, by an extension of the methods of § 2, two further theorems (5 and 6) which give sharp bounds for the remainder terms in these series and include Theorems 1 and 2 as special cases. Blumenthal (10), also, has obtained bounds for the remainder terms in the series. His method of derivation, which is quite different and yields less satisfactory results, is briefly described at the end of this section.

The essential form we obtain for each bound appears to be new. It is expressed as an explicit multiple (which tends either to 1 or to 2 as $u \to \infty$) of the *total variation*, over the x-interval, of the first neglected term in the series. We use the expression 'total variation', or simply 'variation', of an arbitrary real or complex function $\phi(x)$ over an interval $a \leqslant x \leqslant b$ in the same sense as in the classical theory of functions of a real variable, that is, to denote the upper bound of

$$\sum_{s=0}^{n-1} |\phi(x_{s+1}) - \phi(x_s)| \tag{1·10}$$

as $n \to \infty$, for all possible modes of subdivision

$$a = x_0 < x_1 < x_2 < \ldots < x_{n-1} < x_n = b$$

of the interval (a, b) (see, for example, (13), § 11·4). We denote this upper bound by $\mathscr{V}_{x=a,b}(\phi)$, or $\mathscr{V}_{a,b}(\phi)$, or even $\mathscr{V}(\phi)$, when there is no ambiguity. If $\phi(x)$ has a continuous (or piecewise continuous) derivative, then

$$\mathscr{V}_{x=a,b}(\phi) = \int_a^b |\phi'(x)|\, dx. \tag{1·11}$$

In practice, the variation of a real continuous function is easily found from its maxima and minima and values at the ends of the interval. In the example shown in the accompanying diagram, we have

$$\begin{aligned}\mathscr{V}_{x=a,b}(\phi) &= \{\phi(a) - \phi(x_1)\} + \{\phi(x_2) - \phi(x_1)\} + \{\phi(x_2) - \phi(x_3)\} + \{\phi(b) - \phi(x_3)\} \\ &= \phi(a) - 2\phi(x_1) + 2\phi(x_2) - 2\phi(x_3) + \phi(b).\end{aligned} \tag{1·12}$$

In the next section of the paper, § 7, we illustrate Theorem 5 by applying it to a transformed form of the differential equation for the modified Bessel functions $I_n(x)$ and $K_n(x)$. Asymptotic expansions of these functions and their derivatives which are uniformly valid in a region which includes the infinite interval $0 \leqslant x \leqslant \infty$ have previously been given by the present writer in (14); we are now able to assess precisely and easily the remainder terms in these expansions.

Throughout §§ 2–7 the variables u and x are assumed to be real. This restriction is inessential, and in the concluding section, § 8, we state the extension of the principal results to complex values of the variables (Theorem 7).

As far as possible, the analysis presented below is self-contained and independent of results of the earlier papers, so that the reader need not refer elsewhere except in the later sections.

The methods of this paper are capable of development for other general types of approximate solutions of linear differential equations, and in (15) error bounds are established for Airy-function representations which are valid in intervals or complex regions containing a turning point of the differential equation, that is, a simple zero of $p(u, x)$ in (1·8). It is hoped that the present paper and (15) will serve as starting points for the development of more exact theories in this branch of mathematical analysis.

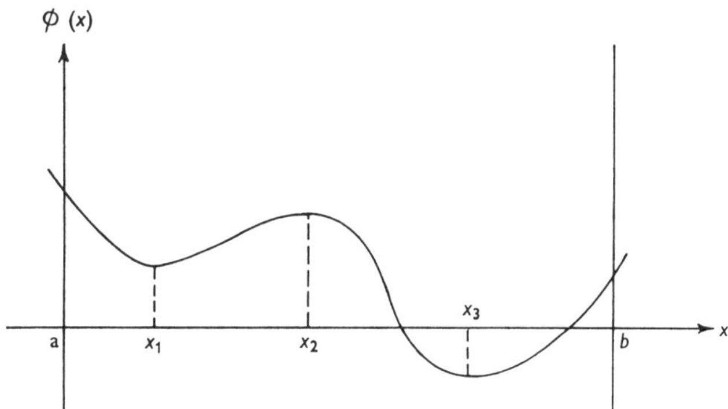

Fig. 1

2. *First approximation theorems.*

THEOREM 1. *Let u be a positive parameter, and $f(u,x)$ be a continuous real or complex function of x in the interval $a \leqslant x \leqslant b$. Then in this interval the differential equation*

$$\frac{d^2w}{dx^2} = \{u^2 + f(u,x)\}\, w \tag{2·1}$$

has solutions $w_1(u,x)$, $w_2(u,x)$, such that

$$w_1(u,x) = e^{ux}\{1+\epsilon_1(u,x)\}, \quad w_2(u,x) = e^{-ux}\{1+\epsilon_2(u,x)\}, \tag{2·2}$$

$$\frac{d}{dx}w_1(u,x) = u\, e^{ux}\{1+2\eta_1(u,x)\}, \quad \frac{d}{dx}w_2(u,x) = -u\, e^{-ux}\{1+2\eta_2(u,x)\}, \tag{2·3}$$

where

$$|\epsilon_1(u,x)|,\ |\eta_1(u,x)| \leqslant \exp\left\{\frac{F_1(u,x)}{2u}\right\} - 1, \tag{2·4}$$

$$|\epsilon_2(u,x)|,\ |\eta_2(u,x)| \leqslant \exp\left\{\frac{F_2(u,x)}{2u}\right\} - 1, \tag{2·5}$$

and

$$F_1(u,x) = \int_a^x |f(u,t)|\, dt, \quad F_2(u,x) = \int_x^b |f(u,t)|\, dt. \tag{2·6}$$

The interval (a,b) may be infinite, provided that the integrals (2·6) converge.

THEOREM 2. *With the conditions of Theorem 1, the differential equation*

$$\frac{d^2w}{dx^2} = \{-u^2 + f(u,x)\}\, w \tag{2·7}$$

has solutions $w_1(u,x)$, $w_2(u,x)$, *such that*

$$w_1(u,x) = e^{iux} + \epsilon_1(u,x), \quad w_2(u,x) = e^{-iux} + \epsilon_2(u,x), \tag{2·8}$$

$$\frac{d}{dx} w_1(u,x) = iu\, e^{iux} + u\eta_1(u,x), \quad \frac{d}{dx} w_2(u,x) = -iu\, e^{-iux} + u\eta_2(u,x), \tag{2·9}$$

where
$$|\epsilon_1(u,x)|,\ |\eta_1(u,x)|,\ |\epsilon_2(u,x)|,\ |\eta_2(u,x)| \leqslant \exp\left\{\frac{F(u,x)}{u}\right\} - 1, \tag{2·10}$$

and
$$F(u,x) = \left| \int_c^x |f(u,t)|\, dt \right|, \tag{2·11}$$

c being an arbitrary point such that $a \leqslant c \leqslant b$. *The interval* (a,b) *and the value of c may be infinite provided that the integral* (2·11) *converges.*

Proof of Theorem 1. We first suppose that a and b are finite. We introduce a sequence of functions $h_n(u,x)$, $n = 0, 1, \ldots$, defined by $h_0(u,x) = 0$ and

$$h_n(u,x) = \frac{1}{2u} \int_a^x \{e^{u(x-t)} - e^{u(t-x)}\} \{h_{n-1}(u,t) + e^{ut}\} f(u,t)\, dt \quad (n \geqslant 1). \tag{2·12}$$

Thus
$$h_1(u,x) = \frac{1}{2u} \int_a^x \{e^{ux} - e^{u(2t-x)}\} f(u,t)\, dt. \tag{2·13}$$

Since $t \leqslant x$, we have $e^{u(2t-x)} \leqslant e^{ux}$, and hence

$$0 \leqslant e^{ux} - e^{u(2t-x)} < e^{ux}. \tag{2·14}$$

Therefore
$$|h_1(u,x)| \leqslant \frac{e^{ux}}{2u} \int_a^x |f(u,t)|\, dt = e^{ux} \frac{F_1(u,x)}{2u}. \tag{2·15}$$

Next, we have

$$\begin{aligned} |h_2(u,x) - h_1(u,x)| &= \left| \frac{1}{2u} \int_a^x \{e^{ux} - e^{u(2t-x)}\} h_1(u,t)\, e^{-ut} f(u,t)\, dt \right| \\ &\leqslant \frac{1}{2u} \int_a^x e^{ux} \frac{F_1(u,t)}{2u} |f(u,t)|\, dt \leqslant \frac{e^{ux}}{2} \left\{\frac{F_1(u,x)}{2u}\right\}^2, \end{aligned} \tag{2·16}$$

since $\partial F_1(u,t)/\partial t = |f(u,t)|$. Continuing the argument, we establish that†

$$|h_{n+1}(u,x) - h_n(u,x)| \leqslant \frac{e^{ux}}{(n+1)!} \left\{\frac{F_1(u,x)}{2u}\right\}^{n+1} \quad (n = 0, 1, \ldots). \tag{2·17}$$

Therefore for any given positive value of u the series

$$h(u,t) = \sum_{n=0}^{\infty} \{h_{n+1}(u,t) - h_n(u,t)\} \tag{2·18}$$

converges uniformly with respect to t in the interval $a \leqslant t \leqslant x$. Term-by-term differentiation of this series is permissible, provided that the resulting series is uniformly convergent ((13), § 1·72). Using primes to denote differentiations with respect to x (or t), we have

$$h_1'(u,x) = \frac{1}{2} \int_a^x \{e^{ux} + e^{u(2t-x)}\} f(u,t)\, dt, \tag{2·19}$$

$$h_{n+1}'(u,x) - h_n'(u,x) = \frac{1}{2} \int_a^x \{e^{ux} + e^{u(2t-x)}\} \{h_n(u,t) - h_{n-1}(u,t)\} e^{-ut} f(u,t)\, dt \quad (n \geqslant 1). \tag{2·20}$$

† The introduction of the factor $1/(n+1)!$ on the right of (2·17) was suggested by a study of (**16**). See also Remark (i) of § 3.

With the aid of (2·17), we deduce that

$$|h'_{n+1}(u,x)-h'_n(u,x)| \leqslant \frac{e^{ux}}{n!}\int_a^x \left\{\frac{F_1(u,t)}{2u}\right\}^n |f(u,t)|\,dt \leqslant \frac{2u\,e^{ux}}{(n+1)!}\left\{\frac{F_1(u,x)}{2u}\right\}^{n+1}, \quad (2{\cdot}21)$$

when $n = 0, 1, \ldots$. Therefore term-by-term differentiation of (2·18) *is* permissible.

Next, we have

$$h''_1(u,x) = u^2 h_1(u,x) + e^{ux} f(u,x), \quad (2{\cdot}22)$$

$$h''_{n+1}(u,x)-h''_n(u,x) = u^2\{h_{n+1}(u,x)-h_n(u,x)\}+\{h_n(u,x)-h_{n-1}(u,x)\}f(u,x) \quad (n \geqslant 1). \quad (2{\cdot}23)$$

Therefore a second differentiation of (2·18) is permissible; moreover

$$h''(u,t) = \{u^2+f(u,t)\}\,h(u,t)+e^{ut}f(u,t). \quad (2{\cdot}24)$$

We now define

$$w_1(u,x) = e^{ux}+h(u,x). \quad (2{\cdot}25)$$

From (2·24), with $t = x$, it follows that $w_1(u,x)$ satisfies (2·1). Also from (2·2), (2·3), (2·17) and (2·21), we have

$$|\epsilon_1(u,x)| = |e^{-ux}\,h(u,x)| \leqslant \sum_{n=0}^{\infty}\frac{1}{(n+1)!}\left\{\frac{F_1(u,x)}{2u}\right\}^{n+1} = \exp\left\{\frac{F_1(u,x)}{2u}\right\}-1,$$

$$|\eta_1(u,x)| = \left|\frac{1}{2u}e^{-ux}\,h'(u,x)\right| \leqslant \sum_{n=0}^{\infty}\frac{1}{(n+1)!}\left\{\frac{F_1(u,x)}{2u}\right\}^{n+1} = \exp\left\{\frac{F_1(u,x)}{2u}\right\}-1.$$

These are the stated inequalities (2·4).†

When $a = -\infty$ no essential modification to the proof is necessary. The definition (2·12) is now in terms of an infinite integral, the convergence of which is a consequence of the inequalities (2·17).

An exactly similar proof holds for the second solution $w_2(u,x)$.

Proof of Theorem 2. This is similar to the proof of Theorem 1. The sequence $h_n(u,x)$ is now defined by $h_0(u,x) = 0$ and

$$h_n(u,x) = \frac{1}{u}\int_c^x \sin\{u(x-t)\}\{h_{n-1}(u,t)+e^{iut}\}f(u,t)\,dt \quad (n \geqslant 1). \quad (2{\cdot}26)$$

Corresponding to (2·14), we have

$$|\sin\{u(x-t)\}| \leqslant 1, \quad (2{\cdot}27)$$

and this leads to

$$|h_{n+1}(u,x)-h_n(u,x)| \leqslant \frac{1}{(n+1)!}\left\{\frac{F(u,x)}{u}\right\}^{n+1} \quad (2{\cdot}28)$$

in place of the key inequality (2·17). Other changes are quite straightforward.

3. *Remarks on Theorems* 1 *and* 2.

(i) *Simplification of the bounds.* If in (2·17) and (2·21) we replace the factor $1/(n+1)!$ by 2^{-n}, we obtain

$$|\epsilon_1(u,x)|,\ |\eta_1(u,x)| \leqslant \frac{\frac{1}{2}F_1(u,x)}{u-\frac{1}{4}F_1(u,x)} \quad (3{\cdot}1)$$

† It may be noticed in passing, as a consequence of (2·14), that unless $f(u,t)$ vanishes identically in the interval (a,x), equality can hold in (2·17) and (2·21) only when $x = a$. The same is therefore true of (2·4), and similarly of (2·5) with b taking the place of a.

in place of (2·4), provided that $u > \frac{1}{4}F_1(u,x)$. Although (3·1) is less sharp than (2·4) the difference is unlikely to be important in applications, and this simplification is often worthwhile. Similarly for (2·5) and (2·10).

(ii) *Blumenthal's results.* Previous proofs of results similar to Theorems 1 and 2 have been given by Blumenthal (10). His bounds are of the simplified kind just described; in place of (3·1) he obtains the weaker inequalities

$$|\epsilon_1(u,x)| \leqslant \frac{F_1(u,x)}{u-F_1(u,x)}, \quad |\eta_1(u,x)| \leqslant \frac{\frac{1}{2}F_1(u,x)}{u-F_1(u,x)}.$$

Blumenthal also restricts a and b to be finite.

(iii) *Uniformity of the bounds.* Let

$$I = \int_a^b |f(u,t)|\,dt.$$

Then clearly the error terms $\epsilon_j(u,x)$, $\eta_j(u,x)$, $j=1,2$, in (2·2) and (2·3) are subject to

$$|\epsilon_j(u,x)|,\ |\eta_j(u,x)| \leqslant \exp\left(\frac{I}{2u}\right)-1 \leqslant \frac{\frac{1}{2}I}{u-\frac{1}{4}I}, \tag{3·2}$$

which are *uniform* bounds for all x in (a,b). Similarly for the error terms in (2·8) and (2·9).

(iv) *Boundary conditions for the solutions.* When $x \to a$, the function $F_1(u,x)$ tends to zero, and therefore also the right of (2·4) tends to zero. Hence the $w_1(u,x)$ of Theorem 1 is the solution of (2·1) satisfying the conditions

$$w_1(u,a) = e^{ua}, \quad w_1'(u,a) = u\,e^{ua}, \quad \text{if } a \text{ is finite}, \tag{3·3}$$

or

$$w_1(u,x) \sim e^{ux}, \quad w_1'(u,x) \sim u\,e^{ux}, \quad \text{as} \quad x \to -\infty, \quad \text{if } a = -\infty. \tag{3·4}$$

Similarly for $w_2(u,x)$ as $x \to b$.

Actually, for infinite intervals the bounds provide information on the rate at which the errors tend to zero as $|x| \to \infty$. For suppose that $f(u,x) = O(|x|^{-\alpha})$ as $x \to -\infty$, so that

$$|f(u,x)| \leqslant k\,|x|^{-\alpha} \quad (x < x_0 < 0), \tag{3·5}$$

where $\alpha\ (>1)$, k and x_0 are constants. Then

$$F_1(u,x) \leqslant \int_{-\infty}^{x} k\,|t|^{-\alpha}\,dt = \frac{k}{(\alpha-1)\,|x|^{\alpha-1}}, \tag{3·6}$$

and so from (3·1) we have

$$|\epsilon_1(u,x)|,\ |\eta_1(u,x)| \leqslant \frac{k}{2u(\alpha-1)\,|x|^{\alpha-1}-\frac{1}{2}k}, \tag{3·7}$$

which is $O(|x|^{-(\alpha-1)})$ as $x \to -\infty$.

Similar remarks apply to Theorem 2. In this case boundary conditions are expressed by

$$\lim_{x\to c}\{w_1(u,x)-e^{iux}\} = 0, \quad \lim_{x\to c}\{w_1'(u,x)-iu\,e^{iux}\} = 0, \tag{3·8}$$

$$\lim_{x\to c}\{w_2(u,x)-e^{-iux}\} = 0, \quad \lim_{x\to c}\{w_2'(u,x)+iu\,e^{-iux}\} = 0. \tag{3·9}$$

The solutions $w_1(u,x)$ and $w_2(u,x)$ therefore depend on the chosen value of c. When $f(u,x)$ is real they are of course complex conjugates.

(v) *Other solutions.* Approximations and bounds for other solutions of the differential equations can be found by taking linear combinations of those for the given solutions. Important special cases occur with Theorem 2; we thus derive solutions $w_3(u,x)$, $w_4(u,x)$ of equation (2·7) having the real approximations

$$w_3(u,x) = \cos ux + \epsilon_3(u,x), \quad w_3'(u,x) = -u\sin ux + u\eta_3(u,x), \tag{3·10}$$

$$w_4(u,x) = \sin ux + \epsilon_4(u,x), \quad w_4'(u,x) = u\cos ux + u\eta_4(u,x), \tag{3·11}$$

where ϵ_3, η_3, ϵ_4 and η_4 are all subject to the bound (2·10).

(vi) *The conditions on* $f(u,x)$. The assumed condition that $f(u,x)$ is a continuous function of x is not essential. The same proofs are applicable if, for example, $f(u,x)$ has a finite number of discontinuities in (a,b). The solutions $w_j(u,x)$ and their first derivatives are then continuous, but their second derivatives are discontinuous at the points of discontinuity of $f(u,x)$.

(vii) *Alternative bounds.* From (2·13), (2·18) and (2·25), we see that the error term in the first of (2·2) can be expressed in the form

$$\epsilon_1(u,x) = \frac{1}{2u}\int_a^x f(u,t)\,dt + \theta_1(u,x) + \theta_2(u,x), \tag{3·12}$$

where
$$\theta_1(u,x) = -\frac{1}{2u}\int_a^x e^{2u(t-x)} f(u,t)\,dt, \tag{3·13}$$

$$\theta_2(u,x) = e^{-ux}\sum_{n=1}^{\infty}\{h_{n+1}(u,x) - h_n(u,x)\}. \tag{3·14}$$

Denoting the upper bound of $|f(u,t)|$ in $a \leqslant t \leqslant x$ by $M_1(u,x)$, we have

$$|\theta_1(u,x)| \leqslant \frac{M_1(u,x)}{2u}\int_a^x e^{2u(t-x)}\,dt \leqslant \frac{M_1(u,x)}{4u^2}, \tag{3·15}$$

and from (2·17) we find that

$$|\theta_2(u,x)| \leqslant \exp\left\{\frac{F_1(u,x)}{2u}\right\} - 1 - \frac{F_1(u,x)}{2u}. \tag{3·16}$$

If, for example, f is independent of u, then M_1 and F_1 are independent of u, and therefore $\theta_1 + \theta_2$ is $O(u^{-2})$. If, also, f is real and of constant sign throughout the interval (a,x), then

$$F_1(u,x) = \pm\int_a^x f(u,t)\,dt,$$

and therefore the bounds given by (2·4) and (3·1) both overestimate the actual value of $|\epsilon_1(u,x)|$ by an amount which is $O(u^{-2})$ for large u.

The corresponding results for Theorem 2 are less satisfactory. The expression corresponding to (3·12) is given by

$$\epsilon_1(u,x) = \frac{e^{iux}}{2iu}\int_c^x f(u,t)\,dt + \theta_1(u,x) + \theta_2(u,x), \tag{3·17}$$

where
$$\theta_1(u,x) = -\frac{e^{-iux}}{2iu}\int_c^x e^{2iut} f(u,t)\,dt, \tag{3·18}$$

and $$|\theta_2(u,x)| = \left|\sum_{n=1}^{\infty}\{h_{n+1}(u,x)-h_n(u,x)\}\right| \leqslant \exp\left\{\frac{F(u,x)}{u}\right\}-1-\frac{F(u,x)}{u}; \tag{3·19}$$
compare (2·28).

For many functions $f(u,x)$, the values of $\theta_1(u,x)$ and $\theta_2(u,x)$ for large values of u are negligible compared with $F(u,x)/u$; this is true, for example, when $f(u,x)$ is independent of u, in consequence of the Riemann–Lebesgue lemma ((13), § 13·21). In these circumstances, the bound (2·10) for $|\epsilon_1(u,x)|$ overestimates the actual value of this quantity by a factor which is at least 2.

An exception to this general conclusion is provided by the following example

$$\left.\begin{aligned} f(u,x) &= 1 \quad \left(x_0-\frac{\delta}{2u}\leqslant x\leqslant x_0+\frac{\delta}{2u}\right),\\ &= 0 \quad \text{(otherwise)}, \end{aligned}\right\} \tag{3·20}$$

where x_0 is any fixed point of the interval (a,b) and δ a small, but arbitrary, positive number. Then taking $c = a$, we find that

$$\begin{aligned}\theta_1\left(u, x_0+\frac{\pi}{2u}\right) &= \frac{e^{-iux_0}}{2u}\int_a^{x_0+(\pi/2u)} e^{2iut}f(u,t)\,dt\\ &= \frac{e^{-iux_0}}{2u}\int_{x_0-(\delta/2u)}^{x_0+(\delta/2u)} e^{2iut}\,dt = \frac{e^{iux_0}\sin\delta}{2u^2},\end{aligned} \tag{3·21}$$

and for $x = x_0+(\pi/2u)$ we have

$$\frac{e^{iux}}{2iu}\int_a^x f(u,t)\,dt = \frac{e^{iux_0}\,\delta}{2u^2}. \tag{3·22}$$

Since δ is small, the modulus of the sum of (3·21) and (3·22) is approximately δ/u^2, that is, $F(u,x)/u$; the remaining term on the right of (3·17), namely $\theta_2(u,x)$, is clearly negligible in consequence of (3·19).

For alternative forms of error bound in the expressions (2·3) and (2·9) for the derivatives, and also in (2·8), see § 6.

(viii) *Example.* For illustrative purposes, let $f(x) = \lambda$, a constant, and take $a = 0$. Then (2·1) becomes

$$\frac{d^2w}{dx^2} = (u^2+\lambda)\,w. \tag{3·23}$$

The solution of this equation which satisfies (3·3) is given by

$$w_1 = A\exp\{(u^2+\lambda)^{\frac{1}{2}}x\}+B\exp\{-(u^2+\lambda)^{\frac{1}{2}}x\},$$
where $$A+B = 1, \quad A-B = u(u^2+\lambda)^{-\frac{1}{2}}. \tag{3·24}$$

Solving these equations for A, B, we find that

$$\begin{aligned}\epsilon_1(u,x) = e^{-ux}w_1-1 &= \tfrac{1}{2}\{1+u(u^2+\lambda)^{-\frac{1}{2}}\}\exp\{(u^2+\lambda)^{\frac{1}{2}}x-ux\}\\ &\quad+\tfrac{1}{2}\{1-u(u^2+\lambda)^{-\frac{1}{2}}\}\exp\{-(u^2+\lambda)^{\frac{1}{2}}x-ux\}-1.\end{aligned} \tag{3·25}$$

Since $F_1(u,x) = |\lambda|\,x$, the bound (2·4) for $\epsilon_1(u,x)$ is $\exp(\frac{1}{2}|\lambda|\,x/u)-1$. It is readily verified that for large u this bound and the right of (3·25) have the same asymptotic modulus, namely $|\lambda|\,x/(2u)$, as we expect from Remark (vii) above.

In the case of Theorem 2, the corresponding formula for $\epsilon_1(u,x)$ is, with $c = 0$,

$$\epsilon_1(u,x) = \tfrac{1}{2}\{1+u(u^2-\lambda)^{-\frac{1}{2}}\}\exp\{i(u^2-\lambda)^{\frac{1}{2}}x\} + \tfrac{1}{2}\{1-u(u^2-\lambda)^{-\frac{1}{2}}\}\exp\{-i(u^2-\lambda)^{\frac{1}{2}}x\} - e^{iux}. \tag{3·26}$$

For large u this is asymptotic to $-i\lambda x\, e^{iux}/(2u)$, the modulus of which is one half of the asymptotic form of the bound $\exp(|\lambda x|/u)-1$ obtained from (2·10) and (2·11), again in accordance with Remark (vii).

(ix) *Evaluation of the bounds.* In this example, and in examples given later in §§ 5 and 7, it is possible to evaluate the integrals (2·6) and (2·11) in closed form. In other applications this may not be so. The theorems are still useful, however, in effect they reduce the problem of estimating the error in an approximate solution of a differential equation to the much easier problem of estimating the value of a definite integral.

4. *Approximation theorems for the equations*

$$\frac{d^2w}{dx^2} = \pm u^2 p(u,x)\, w. \tag{4·1}$$

Let $p(u,x)$ be real and positive in an interval $a \leqslant x \leqslant b$, which again can be infinite. We take a new variable ξ, defined by

$$\xi = \int \{p(u,x)\}^{\frac{1}{2}}\, dx. \tag{4·2}$$

Clearly, ξ has a one-to-one correspondence with x. Let α, β be the values of ξ corresponding respectively to a, b. Also, let

$$W = p^{\frac{1}{4}} w. \tag{4·3}$$

Then W satisfies the differential equation

$$\frac{d^2W}{d\xi^2} = \{\pm u^2 + f(u,\xi)\}\, W, \tag{4·4}$$

where
$$f(u,\xi) = \frac{4pp'' - 5p'^2}{16p^3} = -\frac{1}{p^{\frac{3}{4}}}\frac{d^2}{dx^2}\left(\frac{1}{p^{\frac{1}{4}}}\right), \tag{4·5}$$

primes denoting x-derivatives of p; compare (1·4).

We now suppose f to be a continuous function of ξ. This is clearly so if p'' is a continuous function of x,† since by hypothesis p does not vanish. Consider first (4·4) with the upper sign, and apply Theorem 1 of § 2. A solution $W_1(u,\xi)$ exists such that

$$W_1(u,\xi) = e^{u\xi}(1+\epsilon_1), \quad \frac{d}{d\xi}W_1(u,\xi) = u\, e^{u\xi}(1+2\eta_1), \tag{4·6}$$

where
$$|\epsilon_1|,\ |\eta_1| \leqslant \exp\left(\frac{F_1}{2u}\right) - 1; \quad F_1 = \int_\alpha^\xi |f(u,\tau)|\, d\tau. \tag{4·7}$$

This result may be transformed back to the original variables by defining

$$w_1(u,x) = p^{-\frac{1}{4}} W_1(u,\xi), \tag{4·8}$$

† This condition is not essential, however, for if $p = (x+\text{constant})^{-4}$ then f vanishes identically and (4·4) has the *exact* solutions $e^{\pm u\xi}$ or $e^{\pm iu\xi}$.

the derivative of which is given by

$$\frac{d}{dx} w_1(u,x) = p^{\frac{1}{4}} \left\{ \frac{d}{d\xi} W_1(u,\xi) - \frac{p'}{4p^{\frac{3}{2}}} W_1(u,\xi) \right\}. \tag{4·9}$$

A second solution of (4·1) can be constructed in the same way. Thus we have proved

THEOREM 3. *Let u be a positive parameter, and $p(u,x)$ be positive and $p''(u,x)$ be a continuous function of x in the interval $a \leqslant x \leqslant b$. Then the differential equation*

$$\frac{d^2w}{dx^2} = u^2 p(u,x)\, w \tag{4·10}$$

has solutions $w_1(u,x)$, $w_2(u,x)$, such that

$$w_1(u,x) = p^{-\frac{1}{4}} \exp\left(u \int p^{\frac{1}{2}} dx\right)(1+\epsilon_1), \tag{4·11}$$

$$w_2(u,x) = p^{-\frac{1}{4}} \exp\left(-u \int p^{\frac{1}{2}} dx\right)(1+\epsilon_2), \tag{4·12}$$

$$w_1'(u,x) = up^{\frac{1}{4}} \exp\left(u \int p^{\frac{1}{2}} dx\right)\left\{1 + 2\eta_1 - \frac{p'}{4up^{\frac{3}{2}}}(1+\epsilon_1)\right\}, \tag{4·13}$$

$$w_2'(u,x) = -up^{\frac{1}{4}} \exp\left(-u \int p^{\frac{1}{2}} dx\right)\left\{1 + 2\eta_2 + \frac{p'}{4up^{\frac{3}{2}}}(1+\epsilon_2)\right\}, \tag{4·14}$$

where

$$|\epsilon_j|, |\eta_j| \leqslant \exp\left\{\frac{F_j(u,x)}{2u}\right\} - 1 \leqslant \frac{\frac{1}{2}F_j(u,x)}{u - \frac{1}{4}F_j(u,x)}, \tag{4·15}$$

and

$$F_1(u,x) = \int_a^x \frac{1}{p^{\frac{1}{4}}} \left| \frac{d^2}{dx^2}\left(\frac{1}{p^{\frac{1}{4}}}\right) \right| dx, \quad F_2(u,x) = \int_x^b \frac{1}{p^{\frac{1}{4}}} \left| \frac{d^2}{dx^2}\left(\frac{1}{p^{\frac{1}{4}}}\right) \right| dx. \tag{4·16}$$

The interval (a,b) may be infinite, provided that the integrals (4·16) *converge.*

The corresponding result for (4·1) with the lower sign, obtained by application of Theorem 2, is expressed by

THEOREM 4. *With the conditions of Theorem* 3, *the differential equation*

$$\frac{d^2w}{dx^2} = -u^2 p(u,x)\, w \tag{4·17}$$

has conjugate solutions $w(u,x)$, $w^*(u,x)$ such that

$$w(u,x) = p^{-\frac{1}{4}} \left\{ \exp\left(iu \int p^{\frac{1}{2}} dx\right) + \epsilon \right\}, \tag{4·18}$$

$$w'(u,x) = iup^{\frac{1}{4}} \left\{ \exp\left(iu \int p^{\frac{1}{2}} dx\right) - i\eta + \frac{ip'}{4up^{\frac{3}{2}}} \left[\exp\left(iu \int p^{\frac{1}{2}} dx\right) + \epsilon \right] \right\}, \tag{4·19}$$

where

$$|\epsilon|, |\eta| \leqslant \exp\left\{\frac{F(u,x)}{u}\right\} - 1 \leqslant \frac{F(u,x)}{u - \frac{1}{2}F(u,x)}, \tag{4·20}$$

and

$$F(u,x) = \left| \int_c^x \frac{1}{p^{\frac{1}{4}}} \left| \frac{d^2}{dx^2}\left(\frac{1}{p^{\frac{1}{4}}}\right) \right| dx \right|, \tag{4·21}$$

c being an arbitrary point such that $a \leqslant c \leqslant b$. The interval (a,b) and the value of c may be infinite provided that (4·21) *converges.*

Remarks on Theorems 3 *and* 4. Remarks (iii) to (vii) on Theorems 1 and 2 made in §3 carry over to their transformed forms Theorems 3 and 4. Thus if, for example, $F_j(u, x)$ in (4·15) is replaced by

$$I = \int_a^b \frac{1}{p^{\frac{1}{4}}} \left| \frac{d^2}{dx^2}\left(\frac{1}{p^{\frac{1}{4}}}\right) \right| dx, \tag{4·22}$$

then uniform bounds are obtained for the errors ϵ_j, η_j.

Next, if $x \to a$ then ϵ_1 and η_1 tend to zero, and we obtain boundary values for w_1 and its derivative; similarly for w_2 as $x \to b$. The particular case of the corresponding result for Theorem 4 in which $b = c = \infty$ embraces a recent theorem of Wintner (17) concerning equation (4·17) with $u = 1$.

Finally, alternative forms of the error terms in the expressions (4·13), (4·14) and (4·19) for the derivatives may be obtained by applying the transformations (4·2) and (4·3) to the forms of Theorems 5 and 6 (given below in §6) which result on taking the integer m of these theorems to be zero.

Additional observations are the following.

(i) If the second derivative of $p^{-\frac{1}{4}}$ is of constant sign, (4·16) may be integrated by parts, yielding the alternative form

$$\int \frac{1}{p^{\frac{1}{4}}} \frac{d^2}{dx^2}\left(\frac{1}{p^{\frac{1}{4}}}\right) dx = \frac{1}{p^{\frac{1}{4}}} \frac{d}{dx}\left(\frac{1}{p^{\frac{1}{4}}}\right) - \int \left\{\frac{d}{dx}\left(\frac{1}{p^{\frac{1}{4}}}\right)\right\}^2 dx. \tag{4·23}$$

(ii) Generalizations of Theorems 3 and 4 are obtained if we apply Theorems 1 and 2 to the transformed forms of the equations

$$\frac{d^2w}{dx^2} = \{\pm u^2 p(u, x) + q(u, x)\} w, \tag{4·24}$$

in which $q(u, x)$ is assumed to be a continuous function of x, which are obtained on application of (4·2) and (4·3). We find that relations (4·11) to (4·15) again hold, provided that now

$$F_j(u, x) = \int \frac{1}{p^{\frac{1}{4}}} \left| \frac{d^2}{dx^2}\left(\frac{1}{p^{\frac{1}{4}}}\right) - \frac{q}{p^{\frac{1}{4}}} \right| dx, \tag{4·25}$$

with the integration limits a, x for $j = 1$, and x, b for $j = 2$. Similarly for Theorem 4.

5. *Parabolic cylinder functions of large orders.* As an example of the application of Theorem 3, we consider the equation

$$\frac{d^2w}{dt^2} = \mu^4(t^2+1) w \tag{5·1}$$

over the infinite interval $-\infty \leqslant t \leqslant \infty$. Here $u = \mu^2$,

$$p = t^2+1, \quad \frac{d^2}{dt^2}(p^{-\frac{1}{4}}) = \frac{3t^2-2}{4(t^2+1)^{\frac{9}{4}}}, \tag{5·2}$$

and

$$F_2(\mu^2, t) = F_2(t) = \int_t^\infty \frac{|3t^2-2|}{4(t^2+1)^{\frac{5}{2}}} dt. \tag{5·3}$$

On evaluating this integral we obtain the following expressions for $F_2(t)$

$$\frac{t^3+6t}{12(t^2+1)^{\frac{3}{2}}}-\frac{1}{12}, \quad -\frac{t^3+6t}{12(t^2+1)^{\frac{3}{2}}}+\frac{2}{3}\sqrt{\frac{2}{5}}-\frac{1}{12}, \quad \frac{t^3+6t}{12(t^2+1)^{\frac{3}{2}}}+\frac{4}{3}\sqrt{\frac{2}{5}}-\frac{1}{12}, \tag{5·4}$$

valid respectively when

$$t \geqslant \sqrt{\tfrac{2}{3}}, \quad -\sqrt{\tfrac{2}{3}} \leqslant t \leqslant \sqrt{\tfrac{2}{3}}, \quad t \leqslant -\sqrt{\tfrac{2}{3}}. \tag{5·5}$$

For all values of t, $$F_2(t) \leqslant F_2(-\infty) = \tfrac{4}{3}\sqrt{\tfrac{2}{5}}-\tfrac{1}{6} = 0{\cdot}6766\ldots. \tag{5·6}$$

Defining $$\xi = \int_0^t (t^2+1)^{\frac{1}{2}}\,dt = \tfrac{1}{2}t(t^2+1)^{\frac{1}{2}}+\tfrac{1}{2}\ln\{t+(t^2+1)^{\frac{1}{2}}\}, \tag{5·7}$$

we see from Theorem 3 that (5·1) has a solution w_2, given by

$$w_2 = \frac{e^{-\mu^2\xi}}{(t^2+1)^{\frac{1}{4}}}(1+\epsilon_2), \quad |\epsilon_2| \leqslant \frac{\frac{1}{2}F_2(t)}{\mu^2-\frac{1}{4}F_2(t)}. \tag{5·8}$$

A uniform bound is provided by (5·6); thus

$$|\epsilon_2| \leqslant \frac{(0{\cdot}3383\ldots)}{\mu^2-(0{\cdot}1692\ldots)}, \tag{5·9}$$

showing, for example, that for all values of t the error term in (5·8) cannot exceed 10 % when $\mu^2 > 3{\cdot}552\ldots$. Incidentally, had we used the sharper bound

$$\exp\{(0{\cdot}3383\ldots)/\mu^2\}-1$$

for $|\epsilon_2|$, the last criterion would only be changed to $\mu^2 > 3{\cdot}550\ldots$, in confirmation of Remark (i) of § 3.

In a similar way, we can construct the other solution w_1.

It is of interest to compare (5·8) with the uniform asymptotic expansion of w_2 for large μ. Since $\epsilon_2 \to 0$ as $t \to \infty$, w_2 is determined by the condition $w_2 \sim (t^2+1)^{-\frac{1}{4}}e^{-\mu^2\xi}$. Accordingly, from (18), (11·10) and (11·11), we find that

$$w_2 = \frac{e^{-\mu^2\xi}}{(t^2+1)^{\frac{1}{4}}}\left[1+\left\{\frac{1}{24}-\frac{t^3+6t}{24(t^2+1)^{\frac{3}{2}}}\right\}\frac{1}{\mu^2}+O\left(\frac{1}{\mu^4}\right)\right]. \tag{5·10}$$

When $t \geqslant \sqrt{\frac{2}{3}}$, the second term in the square brackets is $-\frac{1}{2}\mu^{-2}F_2(t)$, and the bound (5·8) overestimates the actual error by a factor which tends to unity as $\mu \to \infty$, as we expect from Remark (vii) of § 3. In the next section, we shall see that the error is intimately connected with the first neglected term of the expansion, even when the condition typified here by $t \geqslant \sqrt{\frac{2}{3}}$ is not satisfied.

A similar application of Theorem 4, with $c = \infty$, to the equation

$$\frac{d^2w}{dt^2} = -\mu^4(t^2+1)\,w, \tag{5·11}$$

shows that a solution w exists of the form

$$w = \frac{e^{i\mu^2\xi}}{(t^2+1)^{\frac{1}{4}}}(1+\epsilon), \quad |\epsilon| \leqslant \frac{F_2(t)}{\mu^2-\frac{1}{2}F_2(t)}, \tag{5·12}$$

$F_2(t)$ again being given by (5·4). This is to be compared with

$$w = \frac{e^{i\mu^2\xi}}{(t^2+1)^{\frac{1}{4}}}\left[1+\left\{\frac{1}{24}-\frac{t^3+6t}{24(t^2+1)^{\frac{3}{2}}}\right\}\frac{i}{\mu^2}+O\left(\frac{1}{\mu^4}\right)\right], \tag{5·13}$$

obtained by replacing μ^2 in (5·10) by $-i\mu^2$ (compare (18), (12·27)).

6. *Higher approximation theorems.* Formal series solutions of (2·1), that is, of the equation

$$\frac{d^2w}{dx^2} = \{u^2+f(u,x)\}\,w, \tag{6·1}$$

can be constructed in the form

$$w = e^{ux}\sum_{s=0}^{\infty}\frac{A_s(u,x)}{u^s}. \tag{6·2}$$

Differentiating with respect to x, we find that

$$\frac{dw}{dx} = u\,e^{ux}\sum_{s=0}^{\infty}\frac{B_s(u,x)}{u^s}, \quad \frac{d^2w}{dx^2} = u^2\,e^{ux}\sum_{s=0}^{\infty}\frac{C_s(u,x)}{u^s}, \tag{6·3}$$

where
$$B_s = A_s + A'_{s-1}, \quad C_s = B_s + B'_{s-1} = A_s + 2A'_{s-1} + A''_{s-2}, \tag{6·4}$$

primes denoting x-derivatives. Accordingly, (6·1) is formally satisfied if

$$C_{s+2} = A_{s+2} + fA_s, \tag{6·5}$$

that is,
$$2A'_{s+1} = -A''_s + fA_s, \tag{6·6}$$

or, on integration,
$$A_{s+1} = -\tfrac{1}{2}A'_s + \tfrac{1}{2}\int fA_s\,dx. \tag{6·7}$$

Thus A_0 = constant, which we shall take to be unity, and the higher coefficients are determined recursively, apart from arbitrary integration constants.

THEOREM 5. *Let u be a positive parameter, and $f(u,x)$ and at least its first m x-derivatives be continuous real or complex functions of x in the interval $a \leqslant x \leqslant b$. Then in this interval the differential equation* (6·1) *has solutions $w_{m,j}(u,x)$ $(j = 1, 2)$, which depend on m and are such that*

$$w_{m,1}(u,x) = e^{ux}\left\{\sum_{s=0}^{m}\frac{A_s(u,x)}{u^s} + \epsilon_{m,1}(u,x)\right\}, \tag{6·8}$$

$$w_{m,2}(u,x) = e^{-ux}\left\{\sum_{s=0}^{m}(-)^s\frac{A_s(u,x)}{u^s} + \epsilon_{m,2}(u,x)\right\}, \tag{6·9}$$

$$w'_{m,1}(u,x) = u\,e^{ux}\left\{\sum_{s=0}^{m}\frac{B_s(u,x)}{u^s} + \frac{B_{m+1}(u,x)-A_{m+1}(u,a)}{u^{m+1}} + \eta_{m,1}(u,x)\right\}, \tag{6·10}$$

$$w'_{m,2}(u,x) = -u\,e^{-ux}\left\{\sum_{s=0}^{m}(-)^s\frac{B_s(u,x)}{u^s} + (-)^{m+1}\frac{B_{m+1}(u,x)-A_{m+1}(u,b)}{u^{m+1}} + \eta_{m,2}(u,x)\right\}, \tag{6·11}$$

where
$$|\epsilon_{m,j}(u,x)| \leqslant e^{\mathscr{V}(A_1)/u}\frac{\mathscr{V}(A_{m+1})}{u^{m+1}} \leqslant \frac{u}{u-\mathscr{V}(A_1)}\,\frac{\mathscr{V}(A_{m+1})}{u^{m+1}}, \tag{6·12}$$

$$|\eta_{m,j}(u,x)| \leqslant \{2e^{\mathscr{V}(A_1)/u}-1\}\frac{\mathscr{V}(A_{m+1})}{u^{m+1}} \leqslant \frac{u+\mathscr{V}(A_1)}{u-\mathscr{V}(A_1)}\,\frac{\mathscr{V}(A_{m+1})}{u^{m+1}}. \tag{6·13}$$

The variations† $\mathscr{V}$ *are taken over the intervals* (a,x) *for* $j=1$ *and* (x,b) *for* $j=2$. *The interval* (a,b) *may be infinite provided that* $\mathscr{V}(A_1)$ *and* $\mathscr{V}(A_{m+1})$ *are finite.*

Since
$$\mathscr{V}(A_1) = \int |A_1'(u,t)|\,dt = \frac{1}{2}\int |f(u,t)|\,dt, \tag{6·14}$$

Theorem 5 reduces to Theorem 1 when $m=0$, except for slight differences in the error terms.

We use the same method of proof as for Theorem 1. Let

$$L_m(u,x) = e^{ux}\sum_{s=0}^{m}\frac{A_s(u,x)}{u^s}. \tag{6·15}$$

Then with the aid of (6·4) and (6·6), we find

$$L_m'(u,x) = u\,e^{ux}\left\{\sum_{s=0}^{m}\frac{B_s(u,x)}{u^s}+\frac{A_m'(u,x)}{u^{m+1}}\right\}, \tag{6·16}$$

and
$$L_m''(u,x)-(u^2+f)\,L_m(u,x) = -\frac{2e^{ux}\,A_{m+1}(u,x)}{u^m}. \tag{6·17}$$

Combining this equation with (6·1), we obtain

$$\frac{d^2}{dx^2}\{w-L_m(u,x)\}-(u^2+f)\{w-L_m(u,x)\} = \frac{2e^{ux}\,A_{m+1}'(u,x)}{u^m}. \tag{6·18}$$

In order to solve the last equation for $w-L_m(u,x)$, we introduce a sequence $h_{m,n}(u,x)$, $n=0,1,\ldots$, defined by $h_{m,0}(u,x)=0$ and

$$h_{m,n}(u,x) = \frac{1}{2u}\int_a^x\{e^{u(x-t)}-e^{u(t-x)}\}\left\{f(u,t)\,h_{m,n-1}(u,t)+\frac{2e^{ut}\,A_{m+1}'(u,t)}{u^m}\right\}dt \quad (n\geqslant 1) \tag{6·19}$$

(compare (2·12)), the continuity, and therefore integrability, of $A_{m+1}'(u,t)$ being a consequence of the assumed differentiability conditions on $f(u,x)$. With $n=1$, we have

$$|h_{m,1}(u,x)| = \left|\frac{1}{u^{m+1}}\int_a^x\{e^{ux}-e^{u(2t-x)}\}\,A_{m+1}'(u,t)\,dt\right| \leqslant \frac{e^{ux}\mathscr{V}_{a,x}(A_{m+1})}{u^{m+1}} \tag{6·20}$$

(compare (2·14) and (2·15)). Corresponding to (2·17) and (2·21), we find

$$|h_{m,n+1}(u,x)-h_{m,n}(u,x)| \leqslant \frac{1}{n!}\left\{\frac{\mathscr{V}_{a,x}(A_1)}{u}\right\}^n\frac{e^{ux}\,\mathscr{V}_{a,x}(A_{m+1})}{u^{m+1}}, \tag{6·21}$$

$$|h_{m,n+1}'(u,x)-h_{m,n}'(u,x)| \leqslant \frac{2u}{n!}\left\{\frac{\mathscr{V}_{a,x}(A_1)}{u}\right\}^n\frac{e^{ux}\,\mathscr{V}_{a,x}(A_{m+1})}{u^{m+1}}. \tag{6·22}$$

Using these results we can show, as in § 2, that

$$h(u,x) \equiv \sum_{n=0}^{\infty}\{h_{m,n+1}(u,x)-h_{m,n}(u,x)\} \tag{6·23}$$

is a solution of (6·18), and therefore that

$$w_{m,1}(u,x) \equiv L_m(u,x)+h(u,x) \tag{6·24}$$

satisfies (6·1). The desired inequality (6·12) follows immediately by summing (6·21). If we sum (6·22) directly we obtain a bound for the error term in the expression for

† Defined in § 1.

$w'_{m,1}(u,x)$ in a form which is similar to that of Theorem 1. To obtain (6·10), we split the expression for $h'_{m,1}(u,x)$; thus

$$h'_{m,1}(u,x) = \frac{1}{u^m}\int_a^x \{e^{ux}+e^{u(2t-x)}\} A'_{m+1}(u,t)\,dt$$

$$= \frac{e^{ux}}{u^m}\left\{A_{m+1}(u,x)-A_{m+1}(u,a)+\int_a^x e^{2u(t-x)}A'_{m+1}(u,t)\,dt\right\}. \qquad (6\cdot25)$$

It then follows from (6·4), (6·10), (6·16), and the differentiated forms of (6·23) and (6·24) that

$$|\eta_{m,1}(u,x)| = \left|\frac{1}{u^{m+1}}\int_a^x e^{2u(t-x)}A'_{m+1}(u,t)\,dt+\frac{e^{-ux}}{u}\sum_{n=1}^{\infty}\{h'_{m,n+1}(u,x)-h'_{m,n}(u,x)\}\right|$$

$$\leqslant \frac{\mathscr{V}(A_{m+1})}{u^{m+1}}+2\sum_{n=1}^{\infty}\frac{1}{n!}\left\{\frac{\mathscr{V}(A_1)}{u}\right\}^n\frac{\mathscr{V}(A_{m+1})}{u^{m+1}} = \{2e^{\mathscr{V}(A_1)/u}-1\}\frac{\mathscr{V}(A_{m+1})}{u^{m+1}},$$

as stated in (6·13).

Similarly for the second solution $w_{m,2}(u,x)$ and its derivative.

The corresponding generalization of Theorem 2 may be achieved in the same way, and is expressed by

THEOREM 6. *With the conditions of Theorem 5, the differential equation*

$$\frac{d^2w}{dx^2} = \{-u^2+f(u,x)\}\,w \qquad (6\cdot26)$$

has solutions $w_{m,j}(u,x)$ $(j=1,2)$, which depend on m and are such that

$$w_{m,j}(u,x) = e^{\pm iux}\left\{\sum_{s=0}^{m}\frac{A_s(u,x)}{(\pm iu)^s}+\frac{A_{m+1}(u,x)-A_{m+1}(u,c)}{(\pm iu)^{m+1}}\right\}+\epsilon_{m,j}(u,x), \qquad (6\cdot27)$$

$$w'_{m,j}(u,x) = \pm iu\,e^{\pm iux}\left\{\sum_{s=0}^{m}\frac{B_s(u,x)}{(\pm iu)^s}+\frac{B_{m+1}(u,x)-A_{m+1}(u,c)}{(\pm iu)^{m+1}}\right\}+u\eta_{m,j}(u,x), \qquad (6\cdot28)$$

the plus signs being taken throughout for $j=1$ and the minus signs for $j=2$. The error terms are bounded by

$$|\epsilon_{m,j}(u,x)|,\ |\eta_{m,j}(u,x)| \leqslant \{2e^{2\mathscr{V}(A_1)/u}-1\}\frac{\mathscr{V}(A_{m+1})}{u^{m+1}} \leqslant \frac{u+2\mathscr{V}(A_1)}{u-2\mathscr{V}(A_1)}\frac{\mathscr{V}(A_{m+1})}{u^{m+1}}, \qquad (6\cdot29)$$

c being an arbitrary point in the interval $a \leqslant c \leqslant b$, and all variations $\mathscr{V}$ being taken over the interval (c,x). The values of a, b and c may be infinite provided that $\mathscr{V}_{c,x}(A_1)$ and $\mathscr{V}_{c,x}(A_{m+1})$ are finite.

Remarks (iii) to (vii) in § 3 may be extended to apply to Theorems 5 and 6.

Blumenthal (**10**) has also proved results which give bounds for the error terms in the case when a and b are finite, but they are more complicated and less sharp than (6·12), (6·13) and (6·29). The difference arises in the way in which the solutions of the differential equation (6·18) are constructed. We have solved (6·18) by successive approximation, taking $e^{\pm ux}$ as complementary functions; see (6·19). Blumenthal solves (6·18) directly, thus

$$w_{m,1}(u,x) - L_m(u,x) = \frac{1}{D}\int_a^x \{w_2(u,x)\,w_1(u,t) - w_1(u,x)\,w_2(u,t)\}\,\frac{2e^{ut}A'_{m+1}(u,t)}{u^m}\,dt, \qquad (6·30)$$

where $w_1(u,x)$ and $w_2(u,x)$ are the solutions defined by Theorem 1, and D is their Wronskian.

7. *Modified Bessel functions of large orders.* We illustrate the powerful nature of the results of the previous section by applying Theorem 5 to derive error bounds for the remainder terms in the uniform asymptotic expansions of $I_n(nz)$ and $K_n(nz)$ given by the writer in (14), § 2.

In the notation of that paper, the functions $(1+z^2)^{\frac{1}{4}} I_n(nz)$ and $(1+z^2)^{\frac{1}{4}} K_n(nz)$ satisfy the equation

$$\frac{d^2W}{d\zeta^2} = \{n^2 + f(\zeta)\}\,W, \qquad (7·1)$$

where
$$\zeta = \int \frac{(1+z^2)^{\frac{1}{2}}}{z}\,dz = (1+z^2)^{\frac{1}{2}} + \ln\left\{\frac{z}{1+(1+z^2)^{\frac{1}{2}}}\right\}, \quad f(\zeta) = \frac{z^2(4-z^2)}{4(1+z^2)^3}. \qquad (7·2)$$

The coefficients A_s of Theorem 5 are denoted by U_s, and are polynomials of degree $3s$ in the variable

$$t = (1+z^2)^{-\frac{1}{2}}; \qquad (7·3)$$

see† (14), (2·20). In particular,

$$U_1 = \tfrac{1}{8}t - \tfrac{5}{24}t^3. \qquad (7·4)$$

The z-interval we consider is $0 \leqslant z \leqslant \infty$; this corresponds to $-\infty \leqslant \zeta \leqslant \infty$ and $1 \geqslant t \geqslant 0$. The function U_1 takes the values $-\frac{1}{12}$ and 0 at the ends of the interval, and has just the one stationary value $1/(12\sqrt{5})$ at $t = 1/\sqrt{5}$. Therefore (compare (1·12))

$$\mathscr{V}_{\zeta=-\infty,\infty}(U_1) = \mathscr{V}_{t=0,1}(U_1) = \frac{1}{6\sqrt{5}} + \frac{1}{12} = 0·158\ldots = n_0, \qquad (7·5)$$

say.

Theorem 5 shows that solutions $W_{m,1}(n,\zeta)$, $W_{m,2}(n,\zeta)$ of (7·1) exist such that

$$W_{m,1}(n,\zeta) = e^{n\zeta}\left\{\sum_{s=0}^{m}\frac{U_s}{n^s} + \epsilon_{m,1}(n,t)\right\}, \quad W_{m,2}(n,\zeta) = e^{-n\zeta}\left\{\sum_{s=0}^{m}(-)^s\frac{U_s}{n^s} + \epsilon_{m,2}(n,t)\right\}, \qquad (7·6)$$

where‡
$$|\epsilon_{m,1}(n,t)| \leqslant \frac{n}{n-n_0}\,\frac{\mathscr{V}_{t,1}(U_{m+1})}{n^{m+1}}, \quad |\epsilon_{m,2}(n,t)| \leqslant \frac{n}{n-n_0}\,\frac{\mathscr{V}_{0,t}(U_{m+1})}{n^{m+1}}. \qquad (7·7)$$

Therefore
$$(1+z^2)^{\frac{1}{4}} K_n(nz) = \lambda_{m,1} W_{m,1}(n,\zeta) + \lambda_{m,2} W_{m,2}(n,\zeta), \qquad (7·8)$$

where $\lambda_{m,1}$, $\lambda_{m,2}$ are independent of z. If we multiply both sides of this equation by $e^{-n\zeta}$ and let $z \to \infty$, so that $\zeta \to \infty$ and $t \to 0$, the left-hand side and the second term on the right both vanish, and since $U_1, U_2, \ldots$ also vanish we are left with

$$0 = \lambda_{m,1}\{1 + \epsilon_{m,1}(n,0)\}. \qquad (7·9)$$

† In this reference $(1+z^2)^{-\frac{1}{2}}$ is denoted by u.

‡ For simplicity, we have replaced $\mathscr{V}_{t,1}(U_1)$ and $\mathscr{V}_{0,t}(U_1)$ by their common upper bound n_0 in the denominators of (7·7), at the expense of weakening these inequalities slightly.

From (7·7) we see that the factor $1+\epsilon_{m,1}(n,0)$ cannot vanish if n is large enough to ensure that

$$\frac{n}{n-n_0}\,\frac{\mathscr{V}_{t=0,1}(U_{m+1})}{n^{m+1}} < 1; \tag{7·10}$$

therefore in these circumstances $\lambda_{m,1}=0$. Hence

$$\lambda_{m,2} = (1+z^2)^{\frac{1}{4}}\,K_n(nz)/W_{m,2}(n,\zeta), \tag{7·11}$$

and evaluating this expression at $z=\infty$, we find that $\lambda_{m,2}=(\pi/2n)^{\frac{1}{2}}$, exactly, since $\epsilon_{m,2}(n,0)=0$.

We have thus achieved the first of the desired results

$$K_n(nz) = \left(\frac{\pi}{2n}\right)^{\frac{1}{2}}\frac{e^{-n\zeta}}{(1+z^2)^{\frac{1}{4}}}\left\{\sum_{s=0}^{m}(-)^s\frac{U_s}{n^s}+\epsilon_{m,2}(n,t)\right\}, \tag{7·12}$$

where $\epsilon_{m,2}(n,t)$ is subject to (7·7). Equation (7·12) is valid under the very mild restriction (7·10).

The corresponding result for the derivative may be obtained with the aid of (6·11), and is found to be

$$K_n'(nz) = -\left(\frac{\pi}{2n}\right)^{\frac{1}{2}}\frac{(1+z^2)^{\frac{1}{4}}}{z}\,e^{-n\zeta}\left\{\sum_{s=0}^{m+1}(-)^s\frac{V_s}{n^s}+\eta_{m,2}(n,t)+\frac{t(1-t^2)}{2n}\,\epsilon_{m,2}(n,t)\right\}, \tag{7·13}$$

where the V_s are polynomials in t defined in (**14**), § 2, and

$$|\eta_{m,2}(n,t)| \leqslant \frac{n+n_0}{n-n_0}\,\frac{\mathscr{V}_{0,t}(U_{m+1})}{n^{m+1}}. \tag{7·14}$$

For the other modified Bessel function, we have

$$(1+z^2)^{\frac{1}{4}}\,I_n(nz) = \mu_{m,1}W_{m,1}(n,\zeta)+\mu_{m,2}W_{m,2}(n,\zeta). \tag{7·15}$$

Multiplying both sides by $e^{n\zeta}$ and letting $z\to 0$, we find that the left-hand side and the first term on the right both vanish; therefore $\mu_{m,2}=0$, provided that

$$\frac{n}{n-n_0}\,\frac{\mathscr{V}_{t=0,1}(U_{m+1})}{n^{m+1}} < \left|1+\sum_{s=1}^{m}(-)^s\frac{\gamma_s}{n^s}\right|, \tag{7·16}$$

where γ_s is the value of U_s at $t=1$; see (**14**), (2·16) and (2·17). In this event, we find that

$$\mu_{m,1} = \frac{(1+z^2)^{\frac{1}{4}}\,I_n(nz)}{W_{m,1}(n,\zeta)} = \frac{1}{(2\pi n)^{\frac{1}{2}}}\,\frac{1}{1+\epsilon_{m,1}(n,0)}, \tag{7·17}$$

on evaluation at $z=\infty$.

Thus we have

$$I_n(nz) = \frac{1}{1+\epsilon_{m,1}(n,0)}\,\frac{e^{n\zeta}}{(2\pi n)^{\frac{1}{2}}(1+z^2)^{\frac{1}{4}}}\left\{\sum_{s=0}^{m}\frac{U_s}{n^s}+\epsilon_{m,1}(n,t)\right\}, \tag{7·18}$$

valid when (7·16) holds. Similarly,

$$I_n'(nz) = \frac{1}{1+\epsilon_{m,1}(n,0)}\,\frac{(1+z^2)^{\frac{1}{4}}\,e^{n\zeta}}{(2\pi n)^{\frac{1}{2}}\,z}\left\{\sum_{s=0}^{m+1}\frac{V_s}{n^s}-\frac{\gamma_{m+1}}{n^{m+1}}+\eta_{m,1}(n,t)-\frac{t(1-t^2)}{2n}\,\epsilon_{m,1}(n,t)\right\}, \tag{7·19}$$

where

$$|\eta_{m,1}(n,t)| \leqslant \frac{n+n_0}{n-n_0}\,\frac{\mathscr{V}_{t,1}(U_{m+1})}{n^{m+1}}. \tag{7·20}$$

Uniform bounds for the errors are obtained on replacing $\mathscr{V}_{t,1}(U_{m+1})$ and $\mathscr{V}_{0,t}(U_{m+1})$ in (7·7), (7·14) and (7·20) by $\mathscr{V}_{0,1}(U_{m+1})$. By numerical evaluation, using tables of U_s given in (19), we find, for example, that $\mathscr{V}_{0,1}(U_4) = 0{\cdot}065\ldots$. Hence on truncation of the asymptotic series at the terms in U_3 and V_3, the highest ones given explicitly in (14), we find that the errors are bounded by

$$\left.\begin{aligned} |\epsilon_{3,j}| &\leqslant \frac{n}{n-(0{\cdot}158\ldots)}\,\frac{(0{\cdot}065\ldots)}{n^4} \\ |\eta_{3,j}| &\leqslant \frac{n+(0{\cdot}158\ldots)}{n-(0{\cdot}158\ldots)}\,\frac{(0{\cdot}065\ldots)}{n^4} \end{aligned}\right\} \quad (0 \leqslant z \leqslant \infty). \tag{7·21}$$

As a matter of interest, the maximum value of $|U_4|$ in $0 \leqslant t \leqslant 1$ is $0{\cdot}020\ldots$, so that the uniform bound exceeds the maximum value of the first neglected terms in (7·12) and (7·18) by a factor of about 3.

Further practical details concerning the error bounds in this example are given in (19).

8. *Extension to complex variables.* We consider the equation

$$\frac{d^2w}{dz^2} = \{u^2 + f(u,z)\}\,w, \tag{8·1}$$

in which u is a positive parameter and $f(u,z)$ a regular (holomorphic) function of the complex variable z when z lies in a simply-connected domain $\mathbf{D}(u)$, bounded or otherwise. Complex values of u may be included in this problem by application of the transformation

$$z' = z\,e^{i\arg u}; \tag{8·2}$$

in particular, this accommodates the case when u^2 is replaced by $-u^2$ in (8·1).

Formal series solutions are again given by (6·2) to (6·7), with x now replaced by z. The analysis of § 6 may be extended in a straightforward manner, and the most important restriction which emerges is that the path of integration for the integrals (6·19) must be such that

$$\Re t \leqslant \Re z, \tag{8·3}$$

otherwise the factor $1 - e^{2u(t-z)}$ would be unbounded (compare (2·14)). We give here no further details of the proof; after introducing preliminary definitions similar to those of (12), § 2, we state the resulting theorem below.

Let $a_j = a_j(u)$, $j = 1, 2$, be any prescribed point of $\mathbf{D}(u)$ or the point at infinity on a straight line $\mathscr{L}_j$ lying in $\mathbf{D}(u)$; in the latter event we suppose that $|\arg(-a_1)| \leqslant \frac{1}{2}\pi$, $|\arg a_2| \leqslant \frac{1}{2}\pi$. Let $\mathbf{H}_j(u)$ comprise those points z of $\mathbf{D}(u)$ which can be joined to a_j by a path $\mathscr{P}_j$ having the following properties, t being a typical point of $\mathscr{P}_j$.

(i) $\mathscr{P}_j$ lies in $\mathbf{D}(u)$.

(ii) $\mathscr{P}_j$ comprises a finite number of Jordan arcs, each with a parametric equation of the form $t = t(\tau)$, where τ is the real parameter of the arc; $t''(\tau)$ is continuous and $t'(\tau)$ does not vanish. If a_j is at infinity, $\mathscr{P}_j$ coincides with $\mathscr{L}_j$ for all sufficiently large $|t|$.

(iii) As t traverses $\mathscr{P}_j$ from a_j to z, $\Re t$ is monotonic non-decreasing if $j = 1$, and monotonic non-increasing if $j = 2$.

THEOREM 7. *The differential equation* (8·1) *has solutions* $w_{m,j}(u,z)$ $(j = 1, 2)$, *which depend on an arbitrary integer m and are such that*

$$w_{m,j}(u,z) = e^{\pm uz}\left\{\sum_{s=0}^{m}(\pm)^s\frac{A_s(u,z)}{u^s} + (\pm)^{m+1}\frac{A_{m+1}(u,z) - A_{m+1}(u,a_j)}{u^{m+1}} + \epsilon_{m,j}(u,z)\right\}, \quad (8\cdot4)$$

$$w'_{m,j}(u,z) = \pm u\,e^{\pm uz}\left\{\sum_{s=0}^{m}(\pm)^s\frac{B_s(u,z)}{u^s} + (\pm)^{m+1}\frac{B_{m+1}(u,z) - A_{m+1}(u,a_j)}{u^{m+1}} + \eta_{m,j}(u,z)\right\}, \quad (8\cdot5)$$

the plus signs being taken throughout for $j = 1$, *and the minus signs for* $j = 2$. *When z lies in* $\mathbf{H}_j(u)$ *the error terms are bounded by*

$$|\epsilon_{m,j}(u,z)|,\ |\eta_{m,j}(u,z)| \leqslant \{2e^{2\mathscr{V}(A_1)/u} - 1\}\frac{\mathscr{V}(A_{m+1})}{u^{m+1}} \leqslant \frac{u + 2\mathscr{V}(A_1)}{u - 2\mathscr{V}(A_1)}\,\frac{\mathscr{V}(A_{m+1})}{u^{m+1}}. \quad (8\cdot6)$$

In this result, the variation of the complex function A_1 is taken along the path $\mathscr{P}_j$, and is defined by

$$\mathscr{V}(A_1) = \int_{\mathscr{P}_j} |A'_1(u,t)\,dt|; \quad (8\cdot7)$$

compare (1·11). Similarly for $\mathscr{V}(A_{m+1})$.

It should be noted that in circumstances in which Theorems 5 and 7 are both applicable, the former is not included in the latter since it has sharper results for the error terms. Theorem 6, however, is included in Theorem 7 when both of these theorems are applicable.

The author wishes to acknowledge useful suggestions by Dr J. Heading and Sir Harold Jeffreys, and by his colleagues, Mr A. R. Curtis and Mr G. F. Miller, and the assistance of Mr Curtis in checking the manuscript.

The work described above has been carried out as part of the research programme of the National Physical Laboratory and this paper is published by permission of the Director of the Laboratory.

REFERENCES

(1) LIOUVILLE, J. Sur le développement des fonctions ou parties de fonctions en séries.... *J. Math. Pures Appl.* 2 (1837), 16–35.

(2) GREEN, G. On the motion of waves in a variable canal of small depth and width. *Trans. Camb. Phil. Soc.* 6 (1837), 457–62.

(3) WENTZEL, G. Eine Verallgemeinerung der Quantenbedingungen für die Zwecke der Wellenmechanik. *Z. Phys.* 38 (1926), 518–29.

(4) KRAMERS, H. A. Wellenmechanik und halbzahlige Quantisierung. *Z. Phys.* 39 (1926), 828–40.

(5) BRILLOUIN, L. Remarques sur la Méchanique Ondulatoire. *J. Phys. Radium*, 7 (1926), 353–68.

(6) ERDÉLYI, A. Asymptotic solution of differential equations with transition points. *Proc. Int. Congr. Math. Amsterdam*, 3 (1954), 92–101.

(7) JEFFREYS, H. On certain approximate solutions of linear differential equations of the second order. *Proc. Lond. Math. Soc.* 23 (1924–5), 428–36.

(8) JEFFREYS, H. On approximate solutions of linear differential equations. *Proc. Camb. Phil. Soc.* 49 (1953), 601–11.

(9) GANS, R. Fortpflanzung des Lichts durch ein inhomogenes Medium. *Ann. Phys., Lpz.*, 47 (1915), 709–36.

(10) BLUMENTHAL, O. Uber asymptotische Integration linearer Differentialgleichungen, mit Anwendung auf eine asymptotische Theorie der Kugelfunktionen. *Arch. Math. Phys., Lpz.*, 19 (1912), 136–74.

(11) HORN, J. Ueber eine lineare Differentialgleichung zweiter Ordnung mit einem willkürlichen Parameter. *Math. Ann.* 52 (1899), 271–92.

(12) OLVER, F. W. J. Uniform asymptotic expansions of solutions of linear second-order differential equations for large values of a parameter. *Phil. Trans.* A, 250 (1958), 479–517.

(13) TITCHMARSH, E. C. *The theory of functions*, 2nd ed. (Oxford, 1939).

(14) OLVER, F. W. J. The asymptotic expansion of Bessel functions of large order. *Phil. Trans.* A, 247 (1954), 328–68.

(15) OLVER, F. W. J. Error bounds for Airy-function expansions in turning-point problems. (In preparation.)

(16) ERDÉLYI, A. Singular Volterra integral equations and their use in asymptotic expansions. *U.S. Army Mathematics Research Center Technical Summary Report, No.* 194 (Madison, Wisconsin, 1960).

(17) WINTNER, A. The Schwarzian derivative and the approximation method of Brillouin. *Quart. Appl. Math.* 16 (1958), 82–6.

(18) OLVER, F. W. J. Uniform asymptotic expansions for Weber parabolic cylinder functions of large orders. *J. Res. Nat. Bur. Stand.* 63B (1959), 131–69.

(19) OLVER, F. W. J. Tables for Bessel functions of moderate or large orders. *Math. Tab. Nat. Phys. Lab.* 6 (London: H.M. Stationery Office, In the Press).

NATIONAL PHYSICAL LABORATORY
TEDDINGTON
MIDDLESEX

J. Soc. Indust. Appl. Math.
Vol. 11, No. 3, September, 1963

ERROR BOUNDS FOR FIRST APPROXIMATIONS IN TURNING-POINT PROBLEMS*

F. W. J. OLVER†

1. Introduction and summary. In this paper we consider approximate solutions of the differential equation

$$\frac{d^2w}{dx^2} = \{u^2p(u, x) + q(u, x)\}w, \tag{1.01}$$

in which u is a large parameter. We suppose that $p(u, x)$ and $q(u, x)$ are free from singularities in the x-region considered, and that $p(u, x)$ has there a simple zero, a so-called *turning-point* or *transition-point* of the differential equation. It is well known that in these circumstances approximate solutions can be expressed in terms of Airy functions. The asymptotic nature of these approximations as $|u| \to \infty$ has been investigated by many writers, particularly Langer [1, 2], Cherry [3], Jeffreys [4], Erdélyi [5, 6] and Olver [7, 8]. In the present investigation, we shall establish, for the first time, explicit strict bounds for the errors of the Airy-function approximations. Similar bounds have been given by the writer in [9] for the 'Liouville-Green' approximation which holds in regions in which the differential equation is free from singularities and turning points, and the present investigation follows similar lines.

In this paper, we consider in detail the case of the first approximation, and real (though not necessarily bounded) values of x. The theory then takes its simplest form, but nevertheless should prove quite adequate for many physical problems involving a second-order differential equation with a single turning point. In a second paper [10], corresponding theorems for complete asymptotic expansions and complex variables are established.

The paper is arranged as follows. The next section, §2, is a preliminary in which a number of relevant properties of the functions Ai and Bi are stated. We also introduce certain auxiliary functions and constants associated with Ai and Bi which are used in the following sections, and give some numerical tables. In §3 we establish our main theorem. This gives error bounds for the approximate solutions Ai(x) and Bi(x) of the differential equation

$$\frac{d^2w}{dx^2} = \{x + f(x)\}w, \tag{1.02}$$

* Received by the editors May 16, 1962, and in final revised form February 18, 1963.

† National Bureau of Standards, Washington, D. C.

in which $f(x)$ is an arbitrary continuous function, which may be real or complex.

In §4 the corresponding result for equation (1.01) (Theorem 2) is established by application of the Langer modification of the Liouville transformation which converts (1.01) into the form (1.02).

In §5 the theorems of the previous two sections are discussed. We derive alternative expressions for the error bounds which are sometimes more readily computable. We also show that the known asymptotic nature of the approximations for large values of the parameter u of (1.01) is an immediate consequence of the nature of the error bounds. Lastly, in this section, the bounds are compared with those given in [9] for the Liouville-Green approximation. We find that the present bounds are a little less sharp in character, and need to be reduced by factors ranging from 1 to about 5 to bring them into line with [9]. The question naturally arises whether this is possible. This is answered in the affirmative in §§6 and 7 by Theorem 3 which is a sharper form of Theorem 1. It is, however, a more complicated result and its proof is less straightforward. Although the improvement is undoubtedly a desirable attainment on purely mathematical grounds, it is probably true that there are few physical applications in which the improvement is likely to be important. In consequence, the writer believes §§6 and 7 can be safely ignored by most readers who are more interested in applications than in an intrinsic study of the error analysis.

In the concluding section, §8, we illustrate both the simple and the more refined bounds by applying Theorems 2 and 3 to a differential equation satisfied by the Hermite polynomials. We derive error bounds, which are new, for the known [11, 12] Airy asymptotic forms of these polynomials when the degree is large, and illustrate the results by numerical examples.

2. Relevant properties of Airy functions. The Airy functions are the solutions of the differential equation

$$\frac{d^2y}{dx^2} = xy. \tag{2.01}$$

For real values of x the standard solutions are denoted by $\mathrm{Ai}(x)$ and $\mathrm{Bi}(x)$. They have the initial values [13]

$$\begin{aligned} \mathrm{Ai}(0) &= \frac{1}{\sqrt{3}}\,\mathrm{Bi}(0) = \frac{1}{3^{2/3}\Gamma(\frac{2}{3})}, \\ \mathrm{Ai}'(0) &= -\frac{1}{\sqrt{3}}\,\mathrm{Bi}'(0) = -\frac{1}{3^{1/3}\Gamma(\frac{1}{3})}, \end{aligned} \tag{2.02}$$

and satisfy the Wronskian relation

$$\mathrm{Ai}(x)\mathrm{Bi}'(x) - \mathrm{Ai}'(x)\mathrm{Bi}(x) = 1/\pi. \tag{2.03}$$

When x is positive $\mathrm{Ai}(x)$, $\mathrm{Bi}(x)$, $-\mathrm{Ai}'(x)$ and $\mathrm{Bi}'(x)$ are all positive and monotonic; when x is negative these functions are oscillatory, with diminishing period as $x \to -\infty$. Their precise asymptotic behavior is given by [13, 14]

$$(2.04)\quad \begin{aligned} \mathrm{Ai}(x) &= \tfrac{1}{2}\pi^{-1/2}x^{-1/4}e^{-(2/3)x^{3/2}}\{1 + O(x^{-3/2})\}, \\ \mathrm{Ai}'(x) &= -\tfrac{1}{2}\pi^{-1/2}x^{1/4}e^{-(2/3)x^{3/2}}\{1 + O(x^{-3/2})\}, \\ \mathrm{Bi}(x) &= \pi^{-1/2}x^{-1/4}e^{(2/3)x^{3/2}}\{1 + O(x^{-3/2})\}, \\ \mathrm{Bi}'(x) &= \pi^{-1/2}x^{1/4}e^{(2/3)x^{3/2}}\{1 + O(x^{-3/2})\}, \end{aligned}$$

as $x \to +\infty$, and

$$(2.05)\quad \begin{aligned} \mathrm{Ai}(x) &= \pi^{-1/2}\,|\,x\,|^{-1/4}\{\cos\theta + O(|\,x\,|^{-3/2})\}, \\ \mathrm{Ai}'(x) &= \pi^{-1/2}\,|\,x\,|^{1/4}\{\sin\theta + O(|\,x\,|^{-3/2})\}, \\ \mathrm{Bi}(x) &= -\pi^{-1/2}\,|\,x\,|^{-1/4}\{\sin\theta + O(|\,x\,|^{-3/2})\}, \\ \mathrm{Bi}'(x) &= \pi^{-1/2}\,|\,x\,|^{1/4}\{\cos\theta + O(|\,x\,|^{-3/2})\}, \end{aligned}$$

as $x \to -\infty$, where $\theta = \frac{2}{3}\,|\,x\,|^{3/2} - \frac{1}{4}\pi$.

In order to have a simple way of assessing the magnitudes of the Airy functions, we define

$$(2.06)\quad \begin{gathered} E(x) = \exp\left(\tfrac{2}{3}x^{3/2}\right), \quad (x > 0); \\ E(x) = 1, \quad (x \leqq 0); \qquad E^{-1}(x) = 1/E(x), \end{gathered}$$

and introduce four auxiliary functions $M(x)$, $\chi(x)$, $N(x)$ and $\psi(x)$, defined for all real values of x by the equations

$$(2.07)\quad \begin{aligned} \mathrm{Ai}(x) &= E^{-1}(x)M(x)\sin\chi(x), \\ \mathrm{Bi}(x) &= E(x)M(x)\cos\chi(x), \\ \mathrm{Ai}'(x) &= E^{-1}(x)N(x)\sin\psi(x), \\ \mathrm{Bi}'(x) &= E(x)N(x)\cos\psi(x). \end{aligned}$$

Thus

$$(2.08)\quad \begin{aligned} M^2(x) &= E^2(x)\mathrm{Ai}^2(x) + E^{-2}(x)\mathrm{Bi}^2(x), \\ N^2(x) &= E^2(x)\mathrm{Ai}'^2(x) + E^{-2}(x)\mathrm{Bi}'^2(x). \end{aligned}$$

Similar auxiliary functions have been defined by Miller ([13], page B48). For negative x his $F(x)$, $\chi(x)$, $G(x)$ and $\psi(x)$ are identical with our $M(x)$, $\chi(x)$, $N(x)$ and $\psi(x)$, respectively. When x is positive, however, this is no longer true: Miller's functions, in fact, are analytic at $x = 0$, whereas ours have discontinuities in their second derivatives at this point.

From (2.04) and (2.05), we deduce that

$$M(x) \sim \left(\frac{5}{4\pi}\right)^{1/2} \frac{1}{x^{1/4}}, \qquad N(x) \sim \left(\frac{5}{4\pi}\right)^{1/2} x^{1/4}, \qquad x \to +\infty, \tag{2.09}$$

and

$$M(x) \sim \frac{1}{\pi^{1/2}|x|^{1/4}}, \qquad N(x) \sim \frac{|x|^{1/4}}{\pi^{1/2}}, \qquad x \to -\infty. \tag{2.10}$$

To facilitate the application of results of later sections, we give, in Tables 1 and 2 below, two-decimal values of the four auxiliary functions and also of $|x|^{1/4}M(x)$, $|x|^{-1/4}N(x)$, computed with the aid of the tables in [13]. The values of $\chi(x)$ and $\psi(x)$ are in radians. The slowly-varying nature of the functions $M(x)$ and $N(x)$ should be noticed.

The following constants occur in the subsequent analysis.

$$\begin{aligned} \lambda_1 &= \max_{(-\infty,\infty)} \{\pi |x|^{1/2} M^2(x)\} = 1.430 \cdots, \\ \lambda_2 &= \max_{(-\infty,\infty)} \{\pi |x^{1/2}\mathrm{Bi}(x)| E^{-1}(x) M(x)\} = 1.315 \cdots, \end{aligned} \tag{2.11}$$

$$\begin{aligned} \mu_1 &= \max_{(-\infty,0)} \{\pi |x|^{1/2} M^2(x)\} = 1, \\ \mu_2 &= \max_{(-\infty,\infty)} \{\pi |x^{1/2}\mathrm{Ai}(x)| E(x) M(x)\} = 1. \end{aligned} \tag{2.12}$$

The values of λ_1 and λ_2 were computed with the aid of the tables in [13]; the maxima occur at $x = 1.12 \cdots, 1.10 \cdots$, respectively.

For μ_1 we use the identity

$$\pi |x|^{1/2} M^2(x) = \tfrac{1}{2}\pi X[J_{1/3}^2(X) + Y_{1/3}^2(X)], \qquad x \leqq 0, \tag{2.13}$$

in which $X = \frac{2}{3}|x|^{3/2}$ (compare [13], page B17). Watson ([15], §13.74) has proved that the right of the last equation is an increasing function of X. That $\mu_1 = 1$ now follows from (2.10).

For the remaining constant, μ_2, we observe that

$$\pi |x^{1/2}\mathrm{Ai}(x)| E(x) M(x) = \pi |x|^{1/2} M^2(x) |\sin \chi(x)|. \tag{2.14}$$

Since $\sin \chi(x)$ takes the values ± 1 at a sequence of values of x tending to $-\infty$, it follows that the maximum value of the function (2.14) for negative x is unity. That $\mu_2 = 1$ follows from this result and the numerical fact that the maximum value of (2.14) for positive x is $0.567 \cdots$ (occurring at $x = 1.46 \cdots$).

3. First approximation theorem.

Theorem 1. *Let $f(x)$ be a continuous (or piecewise continuous) real or complex function of x in the interval $a \leqq x \leqq b$. Then in this interval the*

TABLE 1

Auxiliary functions for negative x

x	$M(x)$	$N(x)$	$\lvert x\rvert^{1/4}M(x)$	$\lvert x\rvert^{-1/4}N(x)$	$\chi(x)$	$\psi(x)$
0.0	0.71	0.52	0.00	∞	0.52	−0.52
−0.2	.66	.52	.44	0.78	.66	−.50
−0.4	.63	.54	.50	.68	.81	−.43
−0.6	.59	.55	.52	.63	0.98	−.33
−0.8	.57	.57	.54	.61	1.17	−.19
−1.0	0.55	0.59	0.55	0.59	1.38	−0.02
−1.2	.53	.61	.55	.58	1.60	.18
−1.4	.51	.63	.55	.58	1.84	.39
−1.6	.50	.65	.56	.57	2.09	.63
−1.8	.48	.66	.56	.57	2.36	0.88
−2	0.47	0.68	0.56	0.57	2.64	1.15
−3	.43	.75	.56	.57	4.23	2.71
−4	.40	.80	.56	.57	6.11	4.57
−5	0.38	0.84	0.56	0.56	8.23	6.68
−6	.36	.88	.56	.56	10.58	9.02
−7	.35	.92	.56	.56	13.13	11.57
−8	.34	.95	.56	.56	15.87	14.31
−9	.33	0.98	.56	.56	18.78	17.22
−10	0.32	1.00	0.56	0.56	21.86	20.30
$-\infty$	0.00	∞	0.56	0.56	∞	∞

differential equation

$$\frac{d^2w}{dx^2} = \{x + f(x)\}w \tag{3.01}$$

has solutions $w_1(x)$, $w_2(x)$, *such that*

$$w_1(x) = \mathrm{Ai}(x) + \epsilon_1(x), \qquad w_1'(x) = \mathrm{Ai}'(x) + \eta_1(x), \tag{3.02}$$

$$w_2(x) = \mathrm{Bi}(x) + \epsilon_2(x), \qquad w_2'(x) = \mathrm{Bi}'(x) + \eta_2(x), \tag{3.03}$$

where

$$\begin{aligned} \lvert \epsilon_1(x) \rvert &\leqq \lambda_1^{-1}\{e^{\lambda_1 F_1(x)} - 1\}E^{-1}(x)M(x), \\ \lvert \eta_1(x) \rvert &\leqq \lambda_1^{-1}\{e^{\lambda_1 F_1(x)} - 1\}E^{-1}(x)N(x), \end{aligned} \tag{3.04}$$

$$\begin{aligned} \lvert \epsilon_2(x) \rvert &\leqq (\lambda_2/\lambda_1)\{e^{\lambda_1 F_2(x)} - 1\}E(x)M(x), \\ \lvert \eta_2(x) \rvert &\leqq (\lambda_2/\lambda_1)\{e^{\lambda_1 F_2(x)} - 1\}E(x)N(x); \end{aligned} \tag{3.05}$$

TABLE 2

Auxiliary functions for positive x

x	$M(x)$	$N(x)$	$x^{1/4}M(x)$	$x^{-1/4}N(x)$	$\chi(x)$	$\psi(x)$
0.0	0.71	0.52	0.00	∞	0.52	−0.52
0.2	.74	.51	.49	0.76	.45	−.55
0.4	.74	.51	.59	.64	.42	−.58
0.6	.73	.52	.64	.59	.40	−.59
0.8	.70	.54	.66	.57	.40	−.59
1.0	0.67	0.57	0.67	0.57	0.40	−0.57
1.2	.64	.60	.67	.57	.41	−.56
1.4	.62	.63	.67	.58	.41	−.55
1.6	.59	.66	.67	.59	.42	−.53
1.8	.57	.69	.66	.59	.43	−.52
2	0.55	0.71	0.66	0.60	0.43	−0.51
3	.49	.81	.64	.62	.45	−.49
4	.45	.88	.64	.62	.45	−.48
5	0.42	0.94	0.63	0.63	0.46	−0.47
6	.40	0.98	.63	.63	.46	−.47
7	.39	1.02	.63	.63	.46	−.47
8	.38	1.06	.63	.63	.46	−.47
9	.37	1.09	.63	.63	.46	−.47
10	0.36	1.12	0.63	0.63	0.46	−0.47
∞	0.00	∞	0.63	0.63	0.46	−0.46

$\lambda_1 = 1.430 \cdots$ *and* $\lambda_2 = 1.315 \cdots$ *are defined in* §2, *and*

$$(3.06) \qquad F_1(x) = \int_x^b | t^{-1/2} f(t) | \, dt, \qquad F_2(x) = \int_a^x | t^{-1/2} f(t) | \, dt.$$

The interval (a, b) *may be infinite provided that the integrals* (3.06) *converge.*

The following proof is similar to that of Theorem 1 of [9]. From (3.01) and (3.02) we obtain

$$(3.07) \qquad \epsilon_1''(x) - x\epsilon_1(x) = f(x)\{\epsilon_1(x) + \mathrm{Ai}(x)\}.$$

Applying the principle of variation of parameters to this inhomogeneous differential equation for $\epsilon_1(x)$, we obtain an equivalent integral equation, given by

$$(3.08) \quad \epsilon_1(x) = \pi \int_x^b \{\mathrm{Ai}(x)\mathrm{Bi}(t) - \mathrm{Bi}(x)\mathrm{Ai}(t)\} f(t) \{\epsilon_1(t) + \mathrm{Ai}(t)\} \, dt.$$

We solve this equation by the standard procedure of successive approximation†, but first we require an upper bound for the kernel.

From (2.07) we deduce that

$$\begin{aligned}|\operatorname{Ai}(x)\operatorname{Bi}(t) - \operatorname{Bi}(x)\operatorname{Ai}(t)| &= M(x)M(t)\,|E^{-1}(x)E(t)\sin\chi(x)\cos\chi(t)\\ &\qquad - E(x)E^{-1}(t)\cos\chi(x)\sin\chi(t)|\\ &\leqq E^{-1}(x)E(t)M(x)M(t)\{|\sin\chi(x)\cos\chi(t)|\\ &\qquad + |\cos\chi(x)\sin\chi(t)|\},\end{aligned}$$

if $t \geqq x$, for then $E(x)E^{-1}(t) \leqq E^{-1}(x)E(t)$. The content of the braces cannot exceed unity; hence we have

$$|\operatorname{Ai}(x)\operatorname{Bi}(t) - \operatorname{Bi}(x)\operatorname{Ai}(t)| \leqq E^{-1}(x)E(t)M(x)M(t), \qquad t \geqq x. \tag{3.09}$$

Returning to the integral equation (3.08), we suppose temporarily that b is finite. We introduce a sequence of functions $h_n(x)$, defined by $h_0(x) = 0$, and

$$\begin{aligned}h_n(x) = \pi \int_x^b \{\operatorname{Ai}(x)\operatorname{Bi}(t) - \operatorname{Bi}(x)\operatorname{Ai}(t)\}&\\ \times f(t)\{h_{n-1}(t) + \operatorname{Ai}(t)\}\,dt,& \qquad n \geqq 1.\end{aligned} \tag{3.10}$$

Taking $n = 1$ and applying (3.09), we find that

$$\begin{aligned}|h_1(x)| &\leqq \pi E^{-1}(x)M(x)\int_x^b E(t)M(t)\,|f(t)\operatorname{Ai}(t)|\,dt\\ &\leqq E^{-1}(x)M(x)\int_x^b |t^{-1/2}f(t)|\,dt = E^{-1}(x)M(x)F_1(x),\end{aligned} \tag{3.11}$$

in consequence of the second of (2.12). Continuing the argument as in [9], §2, we establish that

$$\begin{aligned}|h_{n+1}(x) - h_n(x)| \leqq \lambda_1^n E^{-1}(x)M(x)F_1^{n+1}(x)/(n+1)!,&\\ n = 0, 1, \cdots.&\end{aligned} \tag{3.12}$$

The corresponding bound for the derivative may be established with the aid of the inequality

$$|\operatorname{Ai}'(x)\operatorname{Bi}(t) - \operatorname{Bi}'(x)\operatorname{Ai}(t)| \leqq E^{-1}(x)E(t)N(x)M(t), \qquad t \geqq x, \tag{3.13}$$

(compare (3.09)), and is given by

† A general theory of Volterra integral equations having special reference to the asymptotic solution of ordinary differential equations is given in [16].

$$(3.14)\qquad | h'_{n+1}(x) - h_n'(x) | \leqq \lambda_1{}^n E^{-1}(x) N(x) F_1{}^{n+1}(x)/(n+1)!, \qquad n = 0, 1, \cdots .$$

Summation of (3.12) and (3.14) leads to (3.04), as in [9], §2.

In the case $b = \infty$, we again define $h_n(x)$ by the integral (3.10), the convergence of which is now a consequence of the inequalities (3.11) and (3.12).

It may be observed that the particular solution $w_1(x)$ which we have constructed of the differential equation, satisfies the conditions

$$(3.15)\qquad w_1(b) = \mathrm{Ai}(b), \qquad w_1'(b) = \mathrm{Ai}'(b),$$

if b is finite, and

$$(3.16)\qquad \lim_{x\to\infty} [E(x)\{w_1(x) - \mathrm{Ai}(x)\}/M(x)] = \lim_{x\to\infty} [E(x)\{w_1'(x) - \mathrm{Ai}'(x)\}/N(x)] = 0,$$

when $b = \infty$.

A similar proof holds for the second solution $w_2(x)$. In this case the bounds for the kernel of the integral equation are furnished by (3.09) with x and t interchanged, and

$$(3.17)\qquad | \mathrm{Bi}'(x)\mathrm{Ai}(t) - \mathrm{Ai}'(x)\mathrm{Bi}(t) | \leqq E(x)E^{-1}(t)N(x)M(t), \qquad t \leqq x.$$

4. Approximation theorem for the equation

$$(4.01)\qquad \frac{d^2w}{dx^2} = \{p(x) + q(x)\}w.$$

In a given interval (a, b), which may be infinite, let $p(x)$ be a continuous real function of x with one zero, a simple zero x_0, and $q(x)$ be a continuous real or complex function. Without loss of generality we may suppose that $p(x)$ has the same sign as $x - x_0$. We take a new variable $\zeta = \zeta(x)$ satisfying the equation

$$(4.02)\qquad \zeta \left(\frac{d\zeta}{dx}\right)^2 = p(x)$$

with the conditions $\zeta(x_0) = 0$, $\zeta'(x_0) > 0$. Hence

$$(4.03)\qquad \frac{2}{3}\zeta^{3/2} = \int_{x_0}^{x} \{p(t)\}^{1/2}\, dt, \qquad x \geqq x_0 .$$

and

$$(4.04)\qquad \frac{2}{3}(-\zeta)^{3/2} = \int_{x}^{x_0} \{-p(t)\}^{1/2}\, dt, \qquad x \leqq x_0 .$$

Clearly ζ has a one-to-one correspondence with x; let $\zeta = \alpha, \beta$, correspond respectively to $x = a, b$. If we transform (4.01) from the variable x to the variable ζ, we introduce a term in the first derivative. This term is eliminated if we take a new dependent variable W defined by

$$W = \left(\frac{d\zeta}{dx}\right)^{1/2} w = \left(\frac{p}{\zeta}\right)^{1/4} w. \tag{4.05}$$

We find that W satisfies the equation

$$\frac{d^2W}{d\zeta^2} = \left\{ \zeta + f(\zeta) + \frac{\zeta q(x)}{p(x)} \right\} W, \tag{4.06}$$

where

$$f(\zeta) = \left(\frac{dx}{d\zeta}\right)^{1/2} \frac{d^2}{d\zeta^2} \left(\frac{d\zeta}{dx}\right)^{1/2} = \left(\frac{\zeta}{p}\right)^{1/4} \frac{d^2}{d\zeta^2} \left(\frac{p}{\zeta}\right)^{1/4}. \tag{4.07}$$

By considering the inverse transformation from (4.06) back to (4.01), or by direct calculation, we may verify that

$$f(\zeta) = -\left(\frac{\zeta}{p}\right)^{3/4} \frac{d^2}{dx^2} \left(\frac{\zeta}{p}\right)^{1/4}, \tag{4.08}$$

and from this derive the further expression

$$f(\zeta) = \frac{5}{16\zeta^2} + [4p(x)p''(x) - 5\{p'(x)\}^2] \frac{\zeta}{16\{p(x)\}^3}. \tag{4.09}$$

In order to be able to apply Theorem 1 of §3 to the differential equation (4.06), we require $f(\zeta)$ to be continuous. Although the foregoing transformation is well known ([1], §2; [7], §2), the following lemma appears to be new.

LEMMA. *Let $(x - x_0)^{-1}p(x)$ be a positive, twice-continuously differentiable function of x in the interval (a, b). Then $(x - x_0)^{-1}\zeta(x)$ is a positive, twice-continuously differentiable function of x in (a, b), and $f(\zeta)$ is continuous in the corresponding ζ-interval.*

To prove this result, consider first $x > x_0$ and let

$$r(x) = \left\{\frac{p(x)}{x - x_0}\right\}^{1/2}, \qquad s(x) = (x - x_0)^{-3/2} \int_{x_0}^{x} (t - x_0)^{1/2} r(t)\, dt. \tag{4.10}$$

Then $r(x)$ is clearly positive and twice-continuously differentiable in (a, b). By use of the mean-value theorem for integrals, we readily show that $s(x)$ is right-continuous at x_0, and takes the (nonzero) value $\frac{2}{3}r(x_0)$ there.

Next, by use of integration by parts we find that

$$s'(x) = (x - x_0)^{-5/2} \int_{x_0}^{x} (t - x_0)^{3/2} r'(t)\, dt, \qquad x > x_0. \tag{4.11}$$

Thus $s'(x)$ is continuous in the open interval (x_0, b), and by another application of the mean-value theorem for integrals it is seen that $s'(x) \to \frac{2}{5}r'(x_0)$ as $x \to x_0$. The existence and continuity of the right derivative $s'(x)$ at $x = x_0$ now follow from the mean-value theorem for derivatives. In a similar way we can show that $s''(x)$ also has these properties. Since

$$(x - x_0)^{-1}\zeta(x) = \left(\frac{3}{2}\right)^{2/3}\{s(x)\}^{2/3}, \qquad x > x_0, \tag{4.12}$$

it follows that $(x - x_0)^{-1}\zeta(x)$ is non-vanishing and twice-continuously differentiable on the right at x_0. In an exactly similar manner we can show that the same properties hold on the left at x_0, and moreover that the corresponding left and right derivatives are equal.

Finally, the continuity of $f(\zeta)$ follows from (4.08) and the relation

$$\left(\frac{\zeta}{p}\right)^{1/4} = \left(\frac{3}{2}\right)^{1/6}\{s(x)\}^{1/6}\{r(x)\}^{-1/2}. \tag{4.13}$$

It is of interest to obtain the value of $f(0)$ in terms of the derivatives of p at x_0. We find that

$$f(0) = \frac{10p'(x_0)p'''(x_0) - 9\{p''(x_0)\}^2}{140\{p'(x_0)\}^{8/3}}, \tag{4.14}$$

the existence of $p'''(x_0)$ being a consequence of the assumed conditions of the lemma.

Having established the lemma, we return to equation (4.06). Applying Theorem 1 of §3, and subsequently restoring our original dependent variable w, we arrive at the following theorem, in which we have used the notation

$$\phi(x) = (\zeta/p)^{1/4}. \tag{4.15}$$

THEOREM 2. *Let $(x - x_0)^{-1}p(x)$ be a positive, twice-continuously differentiable function of x in the interval $a \leqq x \leqq b$, and $q(x)$ be a continuous real or complex function in the same interval. Then the differential equation*

$$\frac{d^2w}{dx^2} = \{p(x) + q(x)\}w \tag{4.16}$$

has solutions $w_1(x)$, $w_2(x)$, such that

$$\begin{aligned} w_1(x) &= \phi(x)\{\mathrm{Ai}(\zeta) + \epsilon_1\}, \\ w_1'(x) &= \{\phi(x)\}^{-1}\{\mathrm{Ai}'(\zeta) + \eta_1 + \phi'(x)w_1(x)\}, \end{aligned} \tag{4.17}$$

$$\begin{aligned} w_2(x) &= \phi(x)\{\mathrm{Bi}(\zeta) + \epsilon_2\}, \\ w_2'(x) &= \{\phi(x)\}^{-1}\{\mathrm{Bi}'(\zeta) + \eta_2 + \phi'(x)w_2(x)\}, \end{aligned} \tag{4.18}$$

where

$$\frac{E(\zeta)}{M(\zeta)} \mid \epsilon_1 \mid, \qquad \frac{E(\zeta)}{N(\zeta)} \mid \eta_1 \mid \leqq \frac{e^{\lambda_1 F_1} - 1}{\lambda_1}, \tag{4.19}$$

and

$$\frac{E^{-1}(\zeta)}{M(\zeta)} \mid \epsilon_2 \mid, \qquad \frac{E^{-1}(\zeta)}{N(\zeta)} \mid \eta_2 \mid \leqq \frac{\lambda_2}{\lambda_1} (e^{\lambda_1 F_2} - 1). \tag{4.20}$$

In this result, the functions F_1, F_2 are given by

$$\begin{aligned} F_1 &= \int_x^b \left| \frac{\phi''(x)}{\phi(x)} - q(x) \right| \frac{dx}{\mid p(x) \mid^{1/2}}, \\ F_2 &= \int_a^x \left| \frac{\phi''(x)}{\phi(x)} - q(x) \right| \frac{dx}{\mid p(x) \mid^{1/2}}, \end{aligned} \tag{4.21}$$

or equivalently,

$$\begin{aligned} F_1 &= \int_\zeta^\beta \left| f(\zeta) + \frac{\zeta q(x)}{p(x)} \right| \frac{d\zeta}{\mid \zeta \mid^{1/2}}, \\ F_2 &= \int_\alpha^\zeta \left| f(\zeta) + \frac{\zeta q(x)}{p(x)} \right| \frac{d\zeta}{\mid \zeta \mid^{1/2}}. \end{aligned} \tag{4.22}$$

The interval (a, b) may be infinite, provided that these integrals converge.

5. Remarks on Theorems 1 and 2. Remarks (i), (iii)–(vi) and (ix) made in §3 of [9] are also applicable to the present theorems. Other observations are as follows.

(i) *The vanishing of* $p(x)$. It will be observed that the conditions of Theorem 1 do not require that the interval (a, b) essentially contains the point $x = 0$. In consequence, the assumption of §4 that $p(x)$ has a zero x_0 in the interval (a, b) is not essential. Theorem 2 still holds if $p(x)$ is nonvanishing, provided that $p''(x)$ is continuous. The x_0 occurring in the definition (4.03) or (4.04) of ζ is then taken to be an arbitrary number less than a or greater than b, depending whether $p(x)$ is positive or negative in (a, b). Of course, in these circumstances uniform approximations to the solutions also exist in terms of elementary functions ([9], Theorems 3 and 4).

(ii) *Alternative expressions for the integrals* F_1 *and* F_2. Let

$$H(x) = \mp \int_0^{\zeta(x)} \mid \zeta \mid^{-1/2} f(\zeta) \, d\zeta, \tag{5.01}$$

where the upper sign is taken if $\zeta > 0$ and the lower sign if $\zeta < 0$. Then

$H(x)$ is continuous at $x = x_0$, although its derivative here is infinite, in general. By definition ([9], §1) the variation of H is given by

$$\mathcal{V}(H) = \int \left| \frac{dH}{d\zeta} \right| d\zeta = \int | \zeta^{-1/2} f(\zeta) | \, d\zeta. \tag{5.02}$$

Hence when $q(x) = 0$, we have from (4.22)

$$F_1 = \mathcal{V}_{x,b}(H), \qquad F_2 = \mathcal{V}_{a,x}(H). \tag{5.03}$$

From (4.08) and (4.02) we find that

$$f(\zeta) = \frac{5}{16\zeta^2} - \frac{|\zeta|}{|p|^{3/4}} \frac{d^2}{dx^2}\left(\frac{1}{|p|^{1/4}}\right), \qquad x \neq x_0. \tag{5.04}$$

Equations (5.01) and (5.04) yield

$$H(x) = \frac{5}{24|\zeta|^{3/2}} \pm \int^x \frac{1}{|p|^{1/4}} \frac{d^2}{dx^2}\left(\frac{1}{|p|^{1/4}}\right) dx + \text{constant}, \tag{5.05}$$

where the upper or lower sign is taken according as $x > x_0$ or $x < x_0$, and the lower limits of integration and the integration constants are chosen in such a way that

$$H(x_0) = 0. \tag{5.06}$$

The expression (5.05) should be compared with (4.16) of [9].

In the case when $q(x)$ is not zero, we may again use (5.03), provided that (5.05) is replaced by

$$\begin{aligned} H(x) &= \frac{5}{24|\zeta|^{3/2}} \\ &\pm \int^x \left\{ \frac{1}{|p|^{1/4}} \frac{d^2}{dx^2}\left(\frac{1}{|p|^{1/4}}\right) - \frac{q}{|p|^{1/2}} \right\} dx + \text{constant}. \end{aligned} \tag{5.07}$$

(iii) *Asymptotic nature of the bounds.* Let us apply Theorem 2 to the equation

$$\frac{d^2w}{dx^2} = \{u^2 p(x) + q(x)\} w, \tag{5.08}$$

in which $p(x)$ and $q(x)$ again satisfy the conditions of the theorem, and u is a positive parameter. Again defining ζ by (4.03) and (4.04) and $\phi(x)$ by (4.15), we find that equation (5.08) has a solution $w_1(u, x)$ given by

$$\begin{aligned} w_1(u, x) &= \phi(x)\{\mathrm{Ai}(u^{2/3}\zeta) + \epsilon_1\}, \\ \frac{dw_1(u, x)}{dx} &= \frac{u^{2/3}}{\phi(x)} \{\mathrm{Ai}'(u^{2/3}\zeta) + \eta_1 + u^{-2/3}\phi'(x) w_1(u, x)\}, \end{aligned} \tag{5.09}$$

where

$$(5.10)\qquad \frac{E(u^{2/3}\zeta)}{M(u^{2/3}\zeta)}\,|\,\epsilon_1\,|,\qquad \frac{E(u^{2/3}\zeta)}{N(u^{2/3}\zeta)}\,|\,\eta_1\,|\leqq\frac{1}{\lambda_1}\left\{\exp\left(\frac{\lambda_1F_1}{u}\right)-1\right\},$$

and F_1 is given by (4.21) or (4.22). Similar expressions hold for a second solution $w_2(u, x)$.

If u is large then the right side of (5.10) is clearly $O(u^{-1})$ uniformly with respect to x. Hence we have

$$(5.11)\qquad |\,\epsilon_1\,| = E^{-1}(u^{2/3}\zeta)M(u^{2/3}\zeta)O(u^{-1}).$$

Now $E^{-1}(u^{2/3}\zeta)M(u^{2/3}\zeta)$ is of the same order of magnitude as $\mathrm{Ai}(u^{2/3}\zeta)$; therefore the approximation $\phi(x)\mathrm{Ai}(u^{2/3}\zeta)$ is a uniform asymptotic representation of $w_1(u, x)$ for large u. This result is well-known [2, 4, 5, 6]; the importance of the present theorem is that it supplies numerical values for the constants implied in the O's. It is instructive to probe this point more deeply. We can regard $\mathrm{Ai}(u^{2/3}\zeta)$ as the leading term in an asymptotic expansion of $w_1(u, x)/\phi(x)$ in descending powers of u. We therefore compare the bound (5.10) for ϵ_1 with the next term in the expansion.

To simplify the comparison, we assume that $p(x)$ is an analytic function, $q(x) = 0$ and $b = \infty$. The complete asymptotic expansion is then given by Theorem B of [8], and applying the known boundary conditions on $w_1(u, x)$ at $x = \infty$ (compare (3.16)), we find that the second term is

$$(5.12)\qquad u^{-1}\beta_0\mathrm{Ai}(u^{2/3}\zeta) + u^{-4/3}B_0(\zeta)\mathrm{Ai}'(u^{2/3}\zeta),$$

where $B_0(\zeta)$ is given by

$$\frac{1}{2}\zeta^{-1/2}\int_0^\zeta t^{-1/2}f(t)\,dt,\quad (\zeta\geqq 0);\qquad \frac{1}{2}(-\zeta)^{-1/2}\int_\zeta^0(-t)^{-1/2}f(t)\,dt,\quad (\zeta\leqq 0),$$

and

$$(5.13)\qquad \beta_0 = \lim_{\zeta\to\infty}\{\zeta^{1/2}B_0(\zeta)\}.$$

From (5.01) we see that

$$(5.14)\qquad H(x) = -2\,|\,\zeta\,|^{1/2}B_0(\zeta).$$

Hence if $(\epsilon_1)_B$ denotes the *bound* for $|\,\epsilon_1\,|$ given by (5.10), we find that for large u

$$(5.15)\qquad \begin{aligned}(\epsilon_1)_B &\sim E^{-1}(u^{2/3}\zeta)M(u^{2/3}\zeta)u^{-1}F_1\\ &= u^{-1}E^{-1}(u^{2/3}\zeta)M(u^{2/3}\zeta)\times 2\mathcal{V}_{x,\infty}\{|\,\zeta\,|^{1/2}B_0(\zeta)\},\end{aligned}$$

in consequence of (5.03).

The connection between (5.12) and (5.15) may be exhibited by writing the former in the form

$$u^{-1}E^{-1}(u^{2/3}\zeta)M(u^{2/3}\zeta)\{\beta_0 \sin \chi(u^{2/3}\zeta) + |\zeta|^{1/2}B_0(\zeta)\Lambda(u^{2/3}\zeta) \sin \psi(u^{2/3}\zeta)\}, \tag{5.16}$$

where $\Lambda(z) \equiv |z|^{-1/2}N(z)/M(z)$ and is approximately unity unless $|z|$ is small (see (2.09) and (2.10)). In the case when ζ is positive and bounded away from zero, we find from (2.04) and (2.07)

$$\sin \chi(u^{2/3}\zeta) \doteqdot -\sin \psi(u^{2/3}\zeta) \doteqdot \frac{1}{\sqrt{5}}. \tag{5.17}$$

Therefore in these circumstances (5.16) reduces to

$$5^{-1/2}u^{-1}E^{-1}(u^{2/3}\zeta)M(u^{2/3}\zeta)[\zeta^{1/2}B_0(\zeta)]_\zeta^\infty \tag{5.18}$$

approximately, showing that the ratio of the bound $(\epsilon_1)_B$ to the variation of the first neglected term of the series is about $2\sqrt{5} = 4.47 \cdots$. It will be recalled from [9] that the corresponding overestimation ratios for the asymptotic expansions in terms of elementary functions are 1 for exponential-type solutions, 2 for oscillatory solutions and 2 for the derivatives of both types of solution. The object of our next two sections is to derive bounds for ϵ_1, ϵ_2, η_1 and η_2 which have overestimation ratios close to these "ideal" values of 1 or 2 on the exponential side ($\zeta > 0$) and 2 on the oscillatory side ($\zeta < 0$).

6. Further constants associated with the Airy functions. The following constants appear in the analysis of the next section. We enumerate them in the order in which they occur there.

$$\lambda_3 = \max_{(0,\infty)} [\pi x^{1/2}\text{Ai}(x)\{\text{Bi}(x) - \sqrt{3}\text{Ai}(x)\}] = 0.504 \cdots, \tag{6.01}$$

$$\mu_3 = \max_{(0,\infty)} \{x^{1/2}\text{Ai}(x)/|\text{Ai}'(x)|\} = 1, \tag{6.02}$$

$$\lambda_4 = \max_{(0,\infty)} [\pi x^{1/2}\text{Ai}(x)\{\text{Ai}^2(x) + \text{Bi}^2(x)\}^{1/2}] = 0.521 \cdots, \tag{6.03}$$

$$\lambda_5 = \max_{(0,\infty)} \{\pi x^{1/2}\text{Ai}(x)\ \text{Bi}(x)\} = 0.520 \cdots, \tag{6.04}$$

$$\lambda_6 = \max_{(0,\infty)} \{x^{1/2}\text{Bi}(x)/\text{Bi}'(x)\} = 1.295 \cdots. \tag{6.05}$$

The given numerical values were computed with the aid of the tables in [13]; approximate values of the corresponding x's are respectively 2.4, ∞, 1.28, 1.33 and 0.97. Only μ_3 admits of analytical derivation. Using

(2.01), we find that

$$\frac{d}{dx}\{x\,\mathrm{Ai}^2(x) - \mathrm{Ai}'^2(x)\} = \mathrm{Ai}^2(x) \geqq 0. \tag{6.06}$$

From (2.04) we see that the value of $x\mathrm{Ai}^2(x) - \mathrm{Ai}'^2(x)$ at infinity is zero. Hence $x\,\mathrm{Ai}^2(x) - \mathrm{Ai}'^2(x) \leqq 0$, and from this it follows that $\mu_3 = 1$.

7. Sharper bounds. We return to the analysis of §3 and apply inequalities for the Airy functions which are more refined than (3.09), (3.13) and (3.17).

(i) Ai-*type solution:* $x \geqq 0$. When x is positive, $\mathrm{Ai}(x)$ is a decreasing function and $\mathrm{Bi}(x)$ an increasing function. Hence

$$0 \leqq \mathrm{Ai}(x)\mathrm{Bi}(t) - \mathrm{Bi}(x)\mathrm{Ai}(t) < \mathrm{Ai}(x)\mathrm{Bi}(t), \qquad t \geqq x \geqq 0. \tag{7.01}$$

This result corresponds closely to (2.14) of [9], and is sharper than (3.09). A slight further sharpening can be achieved by using the inequality

$$\frac{\mathrm{Ai}(x)}{\mathrm{Bi}(x)} \leqq \frac{\mathrm{Ai}(0)}{\mathrm{Bi}(0)} = \frac{1}{\sqrt{3}}; \tag{7.02}$$

(see (2.02)). Thus we derive

$$0 \leqq \mathrm{Ai}(x)\mathrm{Bi}(t) - \mathrm{Bi}(x)\mathrm{Ai}(t) \leqq \mathrm{Ai}(x)\{\mathrm{Bi}(t) - \sqrt{3}\mathrm{Ai}(t)\}, \qquad t \geqq x \geqq 0. \tag{7.03}$$

Applying this bound instead of (3.09) to (3.10), we deduce that

$$|h_{n+1}(x) - h_n(x)| \leqq \lambda_3^{n+1}\mathrm{Ai}(x)F_1^{n+1}(x)/(n+1)!, \qquad x \geqq 0, \tag{7.04}$$

in place of (3.12), where λ_3 is given by (6.01).

For the corresponding formula for the derivative, we have to consider the kernel $\mathrm{Ai}'(x)\mathrm{Bi}(t) - \mathrm{Bi}'(x)\mathrm{Ai}(t)$. The quantities $\mathrm{Ai}'(x)\mathrm{Bi}(t)$ and $\mathrm{Bi}'(x)\mathrm{Ai}(t)$ are of opposite sign, hence bounds as powerful as (7.01) and (7.03) are not available; the same situation in fact obtains for the derivatives of exponential-type approximations ([9], (2.21)). We use instead the fact that $\mathrm{Bi}'(x)/|\mathrm{Ai}'(x)|$ is an increasing function when x is positive. Thus when $t \geqq x \geqq 0$,

$$\begin{aligned} &|\mathrm{Ai}'(x)\mathrm{Bi}(t) - \mathrm{Bi}'(x)\mathrm{Ai}(t)| \\ &\qquad \leqq |\mathrm{Ai}'(x)|\,\mathrm{Bi}(t) + \frac{|\mathrm{Ai}'(x)|\,\mathrm{Bi}'(t)\mathrm{Ai}(t)}{|\mathrm{Ai}'(t)|} = \frac{|\mathrm{Ai}'(x)|}{\pi\,|\mathrm{Ai}'(t)|} \end{aligned} \tag{7.05}$$

(compare (2.03)). Hence we deduce that

$$|h'_{n+1}(x) - h_n'(x)| \leqq \lambda_3^n\,|\mathrm{Ai}'(x)|\,F_1^{n+1}(x)/(n+1)!, \qquad x \geqq 0. \tag{7.06}$$

(ii) Ai-*type solution:* $x \leqq 0$. We treat this case by splitting the range of integration in the expression for $h_n(x)$ at $t = 0$. With the notation of §2, we have

$$(7.07)\qquad \begin{aligned} &| \operatorname{Ai}(x)\operatorname{Bi}(t) - \operatorname{Bi}(x)\operatorname{Ai}(t) | \leqq M(x)\{\operatorname{Ai}^2(t) + \operatorname{Bi}^2(t)\}^{1/2}, \\ &\qquad\qquad\qquad\qquad\qquad\qquad\qquad x \leqq 0, \quad t \geqq 0, \\ &| \operatorname{Ai}(x)\operatorname{Bi}(t) - \operatorname{Bi}(x)\operatorname{Ai}(t) | \leqq M(x)M(t), \qquad x \leqq 0, \quad t \leqq 0. \end{aligned}$$

Using these bounds, we obtain from (3.10), with $n = 1$, and (6.03),

$$(7.08)\qquad | h_1(x) | \leqq M(x)G_1(x), \qquad x \leqq 0,$$

where

$$(7.09)\qquad G_1(x) = \int_x^0 | t^{-1/2}f(t) | \, dt + \lambda_4 \int_0^b | t^{-1/2}f(t) | \, dt.$$

The corresponding bound for $| h_2(x) - h_1(x) |$ is obtained from (3.10) with $n = 2$; we substitute for $h_1(t)$ by means of (7.04) when $t \geqq 0$, and (7.08) when $t \leqq 0$. Using the fact that $\lambda_3 < \lambda_4$ (an immediate consequence of the definitions (6.01) and (6.03)), we find that

$$| h_2(x) - h_1(x) | \leqq M(x)G_1^2(x)/2!, \qquad x \leqq 0,$$

and more generally

$$(7.10)\qquad \begin{aligned} &| h_{n+1}(x) - h_n(x) | \leqq M(x)G_1^{n+1}(x)/(n + 1)!, \\ &\qquad\qquad\qquad\qquad x \leqq 0, \quad n = 0, 1, \cdots. \end{aligned}$$

Similarly

$$(7.11)\qquad | h'_{n+1}(x) - h_n'(x) | \leqq N(x)G_1^{n+1}(x)/(n + 1)!, \qquad x \leqq 0.$$

(iii) Bi-*type solution:* $x \leqq 0$. For $w_2(x)$ the sequence $h_n(x)$ is defined by $h_0(x) = 0$ and

$$(7.12)\qquad \begin{aligned} h_n(x) = \pi \int_a^x &\{\operatorname{Bi}(x)\operatorname{Ai}(t) - \operatorname{Ai}(x)\operatorname{Bi}(t)\} \\ &\times f(t)\{h_{n-1}(t) + \operatorname{Bi}(t)\} \, dt, \qquad n \geqq 1 \end{aligned}$$

(compare (3.10)). When $x \leqq 0$ a simple improvement can be effected in the proof of Theorem 1. We replace the constants λ_1 and λ_2 by the corresponding maxima of the expressions (2.11) in the range $(-\infty, 0)$. This means that we replace λ_1 by μ_1, that is, by unity, and λ_2 by

$$\max_{(-\infty,0)} \{\pi \, | x|^{1/2} \operatorname{Bi}(x) \, | \, M(x)\} = \max_{(-\infty,0)} \{\pi \, | x|^{1/2} \cos \chi(x) \, | \, M^2(x)\},$$

which also is unity. Hence when $x \leqq 0$

$$(7.13)\qquad \begin{aligned} |h_{n+1}(x) - h_n(x)| &\leqq M(x)F_2^{n+1}(x)/(n+1)!, \\ |h'_{n+1}(x) - h_n'(x)| &\leqq N(x)F_2^{n+1}(x)/(n+1)!. \end{aligned}$$

(iv) Bi-*type solution:* $x \geqq 0$. As in subsection (ii) above, we split the range of integration in (7.12) at $t = 0$. Bounds on the kernels are supplied by

$$(7.14)\qquad \begin{aligned} |\mathrm{Bi}(x)\,\mathrm{Ai}(t) - \mathrm{Ai}(x)\,\mathrm{Bi}(t)| &\leqq M(t)\,\{\mathrm{Ai}^2(x) + \mathrm{Bi}^2(x)\}^{1/2} \\ &\leqq (2/\sqrt{3})\,\mathrm{Bi}(x)M(t), \\ |\mathrm{Bi}'(x)\,\mathrm{Ai}(t) - \mathrm{Ai}'(x)\,\mathrm{Bi}(t)| &\leqq M(t)\,\{\mathrm{Ai}'^2(x) + \mathrm{Bi}'^2(x)\}^{1/2} \\ &\leqq (2/\sqrt{3})\,\mathrm{Bi}'(x)M(t), \end{aligned}$$

when $t \leqq 0$ and $x \geqq 0$, and by

$$(7.15)\qquad \begin{aligned} 0 \leqq \mathrm{Bi}(x)\mathrm{Ai}(t) - \mathrm{Ai}(x)\mathrm{Bi}(t) &< \mathrm{Bi}(x)\mathrm{Ai}(t), \\ |\mathrm{Bi}'(x)\mathrm{Ai}(t) - \mathrm{Ai}'(x)\mathrm{Bi}(t)| &\leqq \mathrm{Bi}'(x)/\{\pi \mathrm{Bi}'(t)\}, \end{aligned}$$

when $x \geqq t \geqq 0$; compare (7.07), (7.01), (7.02) and (7.05). Thence we derive

$$(7.16)\qquad \begin{aligned} |h_{n+1}(x) - h_n(x)| &\leqq \mathrm{Bi}(x)G_2^{n+1}(x)/(n+1)!, \\ |h'_{n+1}(x) - h_n'(x)| &\leqq \mathrm{Bi}'(x)G_3^{n+1}(x)/(n+1)!, \end{aligned}$$

when $x \geqq 0$, where

$$(7.17)\qquad G_2(x) = \frac{2}{\sqrt{3}} \int_a^0 |t^{-1/2}f(t)|\,dt + \lambda_5 \int_0^x |t^{-1/2}f(t)|\,dt,$$

$$(7.18)\qquad G_3(x) = \frac{2}{\sqrt{3}} \int_a^0 |t^{-1/2}f(t)|\,dt + \lambda_6 \int_0^x |t^{-1/2}f(t)|\,dt,$$

and λ_5, λ_6 are defined in §6.

Collecting together the results of the four subsections we have the following.

THEOREM 3. *The error terms* $\epsilon_1(x)$, $\eta_1(x)$, $\epsilon_2(x)$ *and* $\eta_2(x)$ *of Theorem* 1 *satisfy the inequalities*

$$(7.19)\qquad \begin{aligned} |\epsilon_1(x)| &\leqq (e^{\lambda_3 F_1(x)} - 1)\mathrm{Ai}(x), \\ |\eta_1(x)| &\leqq \lambda_3^{-1}(e^{\lambda_3 F_1(x)} - 1)\,|\mathrm{Ai}'(x)|, \end{aligned} \qquad x \geqq 0,$$

$$(7.20)\qquad \begin{aligned} |\epsilon_1(x)| &\leqq (e^{G_1(x)} - 1)M(x), \\ |\eta_1(x)| &\leqq (e^{G_1(x)} - 1)N(x), \end{aligned} \qquad x \leqq 0,$$

$$(7.21)\qquad \begin{aligned} |\epsilon_2(x)| &\leqq (e^{G_2(x)} - 1)\mathrm{Bi}(x), \\ |\eta_2(x)| &\leqq (e^{G_3(x)} - 1)\mathrm{Bi}'(x), \end{aligned} \qquad x \geqq 0,$$

$$(7.22)\qquad \begin{aligned} |\epsilon_2(x)| &\leqq (e^{F_2(x)} - 1)M(x), \\ |\eta_2(x)| &\leqq (e^{F_2(x)} - 1)N(x), \end{aligned} \qquad x \leqq 0.$$

Here $F_1(x)$, $F_2(x)$ *are defined by* (3.06), *and* $G_1(x)$, $G_2(x)$, $G_3(x)$ *by* (7.09), (7.17), (7.18).

The corresponding sharper form of Theorem 2 follows immediately from this theorem and the transformation of §4, and we do not need to record the result.

The bounds given by Theorem 3 are close to the "ideal" bounds suggested by the discussion in §5(iii) above. These ideal bounds would be given by replacing λ_3, λ_4 and λ_5 all by $\frac{1}{2}$, and λ_6 and $2/\sqrt{3}$ $(= 1.155 \cdots)$ by 1. Hence although further improvements in the error bounds may be possible†, the resulting gains cannot be substantial.

8. Hermite polynomials of large degree. The Hermite polynomial of degree n

$$(8.01)\qquad H_n(x) = (-)^n e^{x^2} \frac{d^n}{dx^n} e^{-x^2}$$

has the associated differential equation

$$(8.02)\qquad \frac{d^2w}{dx^2} = (x^2 - \mu^2)w,$$

in which

$$(8.03)\qquad w = e^{-x^2/2} H_n(x), \qquad \mu = (2n+1)^{1/2}.$$

We apply Theorem 2 of §4 to (8.02), taking

$$p(x) = x^2 - \mu^2, \qquad q(x) = 0, \qquad x_0 = \mu, \qquad b = \infty,$$

but leaving a to be an arbitrary number in the open interval $(-\mu, \infty)$ (compare §5(i)). From (4.03) and (4.04), we have

† For example, if a and b are finite, we could substitute their values for the infinite end-points of the maximization intervals in the definitions of λ_s and μ_s.

$$(8.04)\quad \begin{aligned}(2/3)\zeta^{3/2} &= \int_\mu^x (t^2 - \mu^2)^{1/2}\,dt \\ &= (1/2)x(x^2 - \mu^2)^{1/2} - (1/2)\mu^2 \cosh^{-1}(x/\mu), \quad x \geqq \mu,\end{aligned}$$

$$(8.05)\quad \begin{aligned}(2/3)(-\zeta)^{3/2} &= \int_x^\mu (\mu^2 - t^2)^{1/2}\,dt \\ &= (1/2)\mu^2 \cos^{-1}(x/\mu) - (1/2)x(\mu^2 - x^2)^{1/2}, \quad -\mu < x \leqq \mu,\end{aligned}$$

and from (4.09) and (4.15),

$$(8.06)\quad f(\zeta) = \frac{5}{16\zeta^2} + \frac{(3x^2 + 2\mu^2)}{(\mu^2 - x^2)^3}\frac{\zeta}{4}, \qquad \phi(x) = \left(\frac{\zeta}{x^2 - \mu^2}\right)^{1/4}.$$

In terms of the solutions given by (4.17) and (4.18), we can express

$$(8.07)\quad e^{-x^2/2}H_n(x) = Aw_1(x) + Bw_2(x), \qquad a \leqq x \leqq \infty,$$

where A, B are independent of x. In order to determine these constants, we consider the behavior of each term in this equation as $x \to \infty$, or, equivalently, $\zeta \to \infty$. We readily verify that

$$(8.08)\quad x \sim \frac{2}{\sqrt{3}}\zeta^{3/4}, \qquad f(\zeta) \sim -\frac{7}{64\zeta^2}, \qquad \phi(x) \sim \left(\frac{3}{4}\right)^{1/4}\frac{1}{\zeta^{1/8}}.$$

Hence $\int^\infty |\zeta^{-1/2}f(\zeta)|\,d\zeta$ converges, confirming that ∞ is a legitimate value for b (see (4.22)), and showing also that $\epsilon_1 \to 0$ as $\zeta \to \infty$. It now follows from (4.17) that $w_1(x) \to 0$. On the other hand, we have

$$(8.09)\quad \begin{aligned}&\mathrm{Bi}(\zeta) \sim (2/\sqrt{5})E(\zeta)M(\zeta), \\ &|\epsilon_2| \leqq (\lambda_2/\lambda_1)(e^{\lambda_1 F_2(\infty)} - 1)E(\zeta)M(\zeta);\end{aligned}$$

see (2.04), (2.09) and (4.20). Therefore if we take a to be positive and sufficiently large to ensure that

$$(8.10)\quad e^{\lambda_1 F_2(\infty)} - 1 < \lambda_1/(\lambda_2\sqrt{5}),$$

(and this is always possible since $F_2(\infty) \to 0$ as $a \to \infty$), we can be certain that $w_2(x) \to \infty$. Since $e^{-x^2/2}H_n(x)$ obviously vanishes, it follows that the B of (8.07) is zero. Therefore

$$(8.11)\quad e^{-x^2/2}H_n(x) = Aw_1(x).$$

This result is proved for $a \leqq x \leqq \infty$, with a condition on a implied by (8.10). Since, however, $w_1(x)$ is independent of a and both sides of (8.11) are continuous solutions of the differential equation (8.02) when $-\mu < x \leqq \infty$, it follows that (8.11) holds in this larger interval†.

† A similar argument shows that in the derivation of error bounds for the asymptotic expansions of modified Bessel functions given in [9], the restrictions (7.10) and (7.16) given there are in fact unnecessary.

The value of A can be found by again letting $x \to \infty$ and determining the limiting value of the ratio of the two sides of (8.11). In this way we obtain our major result in this section:

$$H_n(x) = (2\pi)^{1/2} e^{-\mu^2/4} \mu^{\mu^2/2} e^{x^2/2} \left(\frac{\zeta}{x^2 - \mu^2}\right)^{1/4} \{\mathrm{Ai}(\zeta) + \epsilon_1\}, \qquad x > -\mu = -\sqrt{2n+1}, \tag{8.12}$$

where ϵ_1 is subject to (4.19).

In order to evaluate this bound for ϵ_1, we use the formulas of §5(ii). With the aid of (5.03) and (5.05), we find that

$$| \epsilon_1 | \leqq \lambda_1^{-1} [\exp \{\lambda_1 \mathcal{V}_{x,\infty}(H)\} - 1] E^{-1}(\zeta) M(\zeta), \tag{8.13}$$

where

$$H(x) = \frac{5}{24 | \zeta |^{3/2}} + \frac{x^3 - 6\mu^2 x}{12\mu^2 | x^2 - \mu^2 |^{3/2}}, \tag{8.14}$$

the undetermined constants in (5.05) necessarily vanishing here as a consequence of (5.06), since the expansion of the right of (8.14) in powers of $| x - \mu |^{1/2}$ has no constant term.

The determination of the variation of $H(x)$ requires a knowledge of the stationary values of this function. These occur at the zeros of $f(\zeta)$, and we may show that none occur in the interval $-\mu < x < \infty$, as follows. Consider first the interval (μ, ∞). From the first of (8.06), we see that the condition for a zero can be expressed as

$$\frac{2}{3} \zeta^{3/2} = \frac{\sqrt{5}}{3} \frac{(x^2 - \mu^2)^{3/2}}{(3x^2 + 2\mu^2)^{1/2}}. \tag{8.15}$$

The ratio of the x-derivatives of the two sides of this equation is found to be $5^{-1/2} x^{-1} (2x^2 + 3\mu^2)^{-1} (3x^2 + 2\mu^2)^{3/2}$, and it is straightforward to show that this quantity exceeds unity for all $x > \mu$, thereby justifying our assertion. Similarly for the interval $-\mu < x < \mu$.

The graph of $H(x)$ is shown in Fig. 1. From (8.14) and (8.05), we find that

$$H(\infty) = \frac{1}{12\mu^2}, \qquad H(0) = \frac{5}{9\pi\mu^2}, \qquad H(x) \sim \frac{5}{24(2\mu)^{1/2}(x + \mu)^{3/2}} \quad \text{as } x \to -\mu, \tag{8.16}$$

and with the aid of (4.14), we obtain

$$f(0) = -\tfrac{9}{35}(2\mu)^{-8/3}; \qquad H(x) \sim -2 | \zeta |^{1/2} f(0) \sim \tfrac{18}{35}(2\mu)^{-5/2} | x - \mu |^{1/2} \quad \text{as } x \to \mu. \tag{8.17}$$

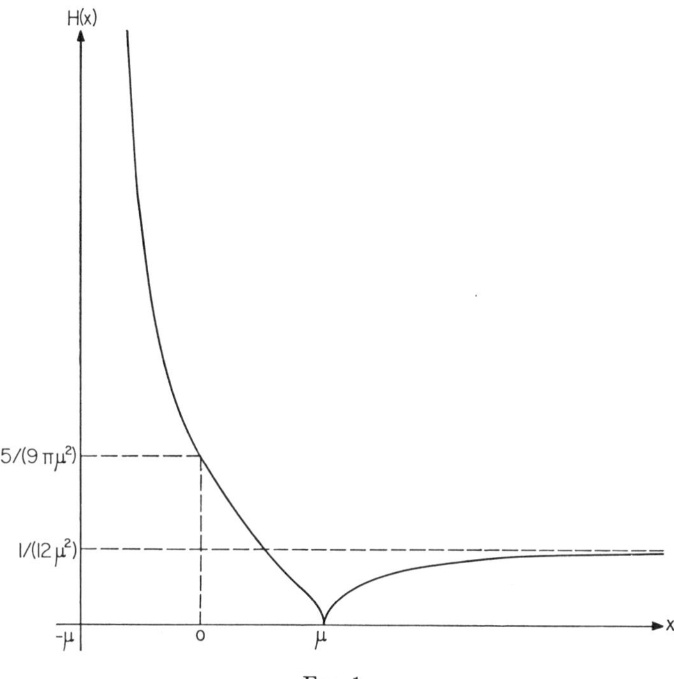

FIG. 1

The variation of $H(x)$ is given by (compare [9], (1.12))

$$\mathcal{V}_{x,\infty}(H) = \frac{1}{12\mu^2} - H, \qquad x \geqq \mu;$$
(8.18)
$$\mathcal{V}_{x,\infty}(H) = \frac{1}{12\mu^2} + H, \qquad -\mu < x \leqq \mu.$$

We observe from (8.16) that this variation tends to infinity as $x \to -\mu$, resulting in large values for the error bound (8.13) in the neighborhood of this point. This is only to be expected, because the differential equation (8.02) has another turning point at $x = -\mu$, causing the approximation (8.12) to break down there. In applications, a more accurate procedure is to confine the use of (8.12) to positive values of x, obtaining $H_n(x)$ for negative x by use of the reflection formula

(8.19) $$H_n(-x) = (-)^n H_n(x).$$

From (8.16) and (8.18), we have

(8.20) $$\mathcal{V}_{\mu,\infty}(H) = \frac{1}{12\mu^2}, \qquad \mathcal{V}_{0,\infty}(H) = \left(\frac{5}{9\pi} + \frac{1}{12}\right)\frac{1}{\mu^2} = \frac{0.260\cdots}{\mu^2}.$$

Thus uniform bounds on the error term ϵ_1 of the approximation (8.12) are supplied by

$$|\epsilon_1| \leqq \frac{1}{\lambda_1}\left\{\exp\left(\frac{\lambda_1}{12\mu^2}\right) - 1\right\} E^{-1}(\zeta)M(\zeta) \leqq \frac{E^{-1}(\zeta)M(\zeta)}{12\mu^2 - 0.715}, \quad x \geqq \mu, \tag{8.21}$$

$$|\epsilon_1| \leqq \frac{1}{\lambda_1}\left\{\exp\left(\frac{0.260\lambda_1}{\mu^2}\right) - 1\right\} M(\zeta) \leqq \frac{M(\zeta)}{3.84\mu^2 - 0.715}, \quad 0 \leqq x \leqq \mu, \tag{8.22}$$

(compare Remark (i) of [9], §3).

Sharper bounds are given by Theorem 3 of §7. We find that

$$|\epsilon_1| \leqq [\exp\{\lambda_3 \mathcal{V}_{x,\infty}(H)\} - 1]\mathrm{Ai}(\zeta), \qquad x \geqq \mu, \tag{8.23}$$

$$|\epsilon_1| \leqq [\exp\{\mathcal{V}^{(1)}_{x,\infty}(H)\} - 1]M(\zeta), \qquad 0 \leqq x \leqq \mu, \tag{8.24}$$

where

$$\mathcal{V}^{(1)}_{x,\infty}(H) = \mathcal{V}_{x,\mu}(H) + \lambda_4 \mathcal{V}_{\mu,\infty}(H) = H + \frac{(0.521 \cdots)}{12\mu^2}. \tag{8.25}$$

Uniform bounds sharper than (8.21) and (8.22) are thus provided by

$$|\epsilon_1| \leqq \left\{\exp\left(\frac{\lambda_3}{12\mu^2}\right) - 1\right\}\mathrm{Ai}(\zeta) \leqq \frac{\lambda_3 \mathrm{Ai}(\zeta)}{12\mu^2 - (\lambda_3/2)} = \frac{\mathrm{Ai}(\zeta)}{23.8\mu^2 - (1/2)}, \quad x \geqq \mu, \tag{8.26}$$

$$|\epsilon_1| \leqq \left[\exp\left\{\left(\frac{5}{9\pi} + \frac{\lambda_4}{12}\right)\frac{1}{\mu^2}\right\} - 1\right] M(\zeta) \leqq \frac{M(\zeta)}{4.54\mu^2 - (1/2)}, \quad 0 \leqq x \leqq \mu. \tag{8.27}$$

The corresponding approximation and error bounds for the derivative $H_n'(x)$ can be obtained without difficulty with the aid of the formulas for $w_1'(x)$ and η_1 given by Theorems 2 and 3, but we shall not derive these results here.

It is to be noticed that the bounds for ϵ_1 are $O(\mu^{-2})$ when μ is large, as anticipated in §5(iii) above. The approximation (8.12) is, in fact, the leading term of a uniform asymptotic expansion of $H_n(x)$ in descending powers of μ. The leading term of this expansion has been given before, but without error bounds; see for example [17], pages 69 and 74, [12] and [6]. The full expansion, again without error bounds, is supplied by the relations (11.23) and (8.11) of [11]; the notation of this reference is similar to that used here except that ζ is replaced by $\mu^{4/3}\zeta$. Bounds for the remainder term on truncation of this uniform asymptotic expansion at a term beyond the first cannot be obtained by application of the theorems of the present paper, but they can be obtained quite easily by the extensions of the present theory given in [10].

Example 1. To evaluate ϵ_1 *for* $n = 10$, $x = 6$.

Here $\mu = \sqrt{21} = 4.5825757$. From (8.04), we obtain

$$\tfrac{2}{3}\zeta^{3/2} = 3\sqrt{15} - 10.5\cosh^{-1}(1.3093073) = 3.559771,$$

$$\zeta = 3.054984,$$

and hence $(2\pi)^{1/2}e^{-\mu^2/4}\mu^{\mu^2/2}e^{x^2/2}\{\zeta/(x^2-\mu^2)\}^{1/4} = 5.07255 \times 10^{12}$.

Interpolating in Table II of [13], we find $\mathrm{Ai}(\zeta) = 5.96521 \times 10^{-3}$, and multiplying together the last two results, we obtain 3.02588×10^{10} as our approximation to $H_{10}(6)$. By direct computation from the explicit polynomial expression given on page 75 of [17], we derive the exact value $H_{10}(6) = 3\,02320\,00224$. Hence (see (8.12))

$$\epsilon_1 = -0.00268 \times 10^{10}/(5.07255 \times 10^{12}) = -\mathbf{5.3} \times \mathbf{10^{-6}}. \tag{8.28}$$

How do our bounds compare with this value? The sharpest one is provided by (8.23). From (8.14) and (8.18), we find

$$H(x) = 0.00213, \qquad \mathcal{V}_{x,\infty}(H) = 0.00184, \qquad \text{RHS of (8.23)} = 5.5 \times 10^{-6}.$$

This is very close to the actual value of $|\epsilon_1|$, as we expect for the Ai-type solution on the exponential side of the turning point; compare Remark (iii) of §5 and the last paragraph of §7.

The uniform bounds are also of interest. We have $E^{-1}(\zeta) = 0.0284$ and $M(\zeta) = 0.49$, the latter value being extracted from Table 2 of §2. Therefore

$$\text{RHS of (8.21)} = 55 \times 10^{-6}; \quad \text{RHS of (8.26)} = 12 \times 10^{-6}.$$

As a check on the values of ϵ_1 and $H(x)$, we may compute the next term in the asymptotic expansion. This is given by

$$-\tfrac{1}{24}\mu^{-2}\mathrm{Ai}(\zeta) - \tfrac{1}{2}|\zeta|^{-1/2}H(x)\mathrm{Ai}'(\zeta); \tag{8.29}$$

see (5.12), with $u = 1$, and (5.14). On evaluation, we obtain -5.2×10^{-6}, in agreement with (8.28).

Example 2. To evaluate ϵ_1 *for* $n = 10$, $x = 4$.

(These values of n and x were used in Example 2 on page 164 of [11]). Following the previous example, we have from (8.05)

$$\tfrac{2}{3}(-\zeta)^{3/2} = 10.5\cos^{-1}(0.8728716) - 2\sqrt{5} = 0.880129,$$

$$\zeta = -1.203442,$$

and thence

$$(2\pi)^{1/2} e^{-\mu^2/4} \mu^{\mu^2/2} e^{x^2/2} \{\zeta/(x^2 - \mu^2)\}^{1/4} = 2.40113 \times 10^8,$$

$$\mathrm{Ai}\,(\zeta) = 0.525819.$$

Also [11] $H_{10}(4) = 1259\ 84224$. From these results and (8.12), we compute $\epsilon_1 = -$ **0.00113.**

We now evaluate the bounds. From (8.14), (8.25) and Table 1 of §2, we find

$$H(x) = 0.00164, \qquad \mathcal{V}_{x,\infty}^{(1)}(H) = 0.00371, \qquad M(\zeta) = 0.53.$$

Hence for the sharpest bound, we have RHS of (8.24) = 0.0020. This result is not quite so striking as in the previous example, but this is consistent with our expectations since we are now on the oscillatory side of the turning point. For the uniform bounds in this example, we find

$$\text{RHS of } (8.22) = 0.0066; \qquad \text{RHS of } (8.27) = 0.0056.$$

Again, as a check we obtain -0.00112 on evaluation of (8.29), in satisfactory agreement with the above value of ϵ_1.

Acknowledgments. This work was begun whilst the author was at the National Physical Laboratory, Teddington. Acknowledgment is made of the assistance of Miss Jean Wark (N.P.L.) and Miss Theresa Phillips (N.B.S.) in the computation of the tables and constants given in §§2 and 6, and of helpful criticisms of the manuscript by Dr. A. Ghaffari, Mr. G. F. Miller and Dr. E. R. Pike.

REFERENCES

[1] R. E. Langer, *On the asymptotic solutions of ordinary differential equations, with an application to the Bessel functions of large order*, Trans. Amer. Math. Soc., 33 (1931), pp. 23–64.

[2] ———, *The asymptotic solutions of ordinary linear differential equations of the second order, with special reference to a turning point*, Ibid., 67 (1949), pp. 461–490.

[3] T. M. Cherry, *Uniform asymptotic formulae for functions with transition points*, Ibid., 68 (1950), pp. 224–257.

[4] H. Jeffreys, *On approximate solutions of linear differential equations*, Proc. Cambridge Philos. Soc., 49 (1953), pp. 601–611.

[5] A. Erdélyi, *Differential equations with transition points. I. The first approximation*, Tech. Report 6, Contract Nonr-220 (11), Department of Mathematics, California Institute of Technology, 1955.

[6] ———, *Asymptotic solutions of differential equations with transition points or singularities*, J. Mathematical Phys., 1 (1960), pp. 16–26.

[7] F. W. J. Olver, *The asymptotic solution of linear differential equations of the*

second order for large values of a parameter, Philos. Trans. Roy. Soc. London. Ser. A, 247 (1954), pp. 307–327.

[8] ———, *Uniform asymptotic expansions of solutions of linear second-order differential equations for large values of a parameter*, Ibid., 250 (1958), pp. 479–517.

[9] ———, *Error bounds for the Liouville-Green (or WKB) approximation*, Proc. Cambridge Philos. Soc., 57 (1961), pp. 790–810.

[10] ———, *Error bounds for asymptotic expansions in turning-point problems*, (To appear).

[11] ———, *Uniform asymptotic expansions for Weber parabolic cylinder functions of large orders*, J. Res. Nat. Bur. Standards, 63B (1959), pp. 131–169.

[12] H. Skovgaard, *Asymptotic forms of Hermite polynomials*, Tech. Report 18, Contract Nonr-220 (11), Department of Mathematics California Institute of Technology, 1959.

[13] J. C. P. Miller, *The Airy integral*, British Association Math. Tables, Part-vol. B, Cambridge University Press, 1946.

[14] F. W. J. Olver, *The asymptotic expansion of Bessel functions of large order*, Philos. Trans. Roy. Soc. London. Ser. A, 247 (1954), pp. 328–368.

[15] G. N. Watson, *Theory of Bessel functions*, Second edition, Cambridge University Press, 1944.

[16] A. Erdélyi, *Singular Volterra integral equations and their use in asymptotic expansions*, U. S. Army Mathematics Research Center Technical Summary Report No. 194, Madison, Wisconsin, 1960.

[17] J. C. P. Miller, *Tables of Weber parabolic cylinder functions*, Her Majesty's Stationery Office, London, 1955.

J. Soc. Indust. Appl. Math.
Vol. 12, No. 1, March, 1964

ERROR BOUNDS FOR ASYMPTOTIC EXPANSIONS IN TURNING-POINT PROBLEMS*

F. W. J. OLVER†

1. Introduction and summary. In a previous paper [1], error bounds were established for approximate solutions in terms of Airy functions of the differential equation

$$\frac{d^2w}{dx^2} = \{u^2\, p(u, x) + q(u, x)\}\, w, \tag{1.01}$$

in which u is a large positive parameter, and x ranges over a real interval. This interval contains no singularities of the functions $p(u, x)$ and $q(u, x)$ and just one zero of $p(u, x)$; it need not be bounded. In the present paper, we extend these results in two ways. First, we consider higher approximations which are obtained by truncating certain formal (generally divergent) series solutions of the differential equation involving the Airy functions and their first derivatives. Second, we consider complex values of x and u.

The series we are investigating are not themselves new; it is already known [2, 3, 4, 5] that they represent uniform asymptotic expansions of solutions for large values of the parameter u, subject to certain restrictions on $p(u, x)$ and $q(u, x)$. The new feature of the results given below is that they supply precise, and realistic, bounds for the remainder terms associated with the truncated series.

The paper is arranged as follows. In the next section, §2, we derive the formal series solutions of the special form

$$\frac{d^2w}{dx^2} = \{x + f(x)\}\, w \tag{1.02}$$

of (1.01), and we establish sufficient conditions on $f(x)$ to ensure the continuity and boundedness of the terms in the series.

In §3 we state and prove two theorems which give error bounds, when x is real, for the approximate solutions obtained by truncation of the formal series of §2; Theorem 1 relates to an odd number of terms of the series, and Theorem 2 to an even number. The two theorems are generalizations of Theorem 1 of [1], which gives error bounds for the leading term of the series. We also discuss in this section the application of these theorems to the more general equation (1.01).

* Received by the editors March 28, 1963, and in revised form September 12, 1963.
† National Bureau of Standards, Washington, D. C.

In §4 we illustrate the theory developed so far by applying Theorem 1 to Bessel's differential equation, thence deriving precise error bounds, which are new, for the uniform asymptotic expansions of $J_n(x)$ and $Y_n(x)$ for large n which were given by the writer in [6].

In the concluding section, §5, we develop the general theory corresponding to §3 in the case when x ranges over a complex domain.

2. The formal series. Equation (1.02) is formally satisfied by the series ([4, §3])

$$w = \mathrm{Ci}(x)\sum_{s=0}^{\infty} A_s(x) + \mathrm{Ci}'(x)\sum_{s=0}^{\infty} B_s(x), \tag{2.01}$$

in which $\mathrm{Ci}(x)$ denotes any Airy function, and the coefficients are defined recursively by $A_0(x) = 1$ and

$$\begin{aligned} B_s(x) &= \frac{1}{2}x^{-1/2}\int_0^x t^{-1/2}\{f(t)A_s(t) - A_s''(t)\}\,dt, & x > 0, \\ B_s(x) &= \frac{1}{2}(-x)^{-1/2}\int_x^0 (-t)^{-1/2}\{f(t)A_s(t) - A_s''(t)\}\,dt, & x < 0, \end{aligned} \tag{2.02}$$

$$A_{s+1}(x) = -\frac{1}{2}B_s'(x) + \frac{1}{2}\int f(x)B_s(x)\,dx. \tag{2.03}$$

The derivative of w is given by

$$w' = \mathrm{Ci}(x)\sum_{s=0}^{\infty} C_s(x) + \mathrm{Ci}'(x)\sum_{s=0}^{\infty} D_s(x), \tag{2.04}$$

where

$$C_s(x) = A_s'(x) + xB_s(x), \qquad D_s(x) = A_s(x) + B_{s-1}'(x). \tag{2.05}$$

In order that the series (2.01) may meaningfully represent a solution of the differential equation (1.02), the coefficients $A_s(x)$ and $B_s(x)$ must be bounded. For finite intervals it suffices that they be continuous, and we first examine this condition. Then, at the end of this section, we consider the boundedness of the coefficients in infinite intervals.

LEMMA 1. *Let $g(x)$ and its first m derivatives be continuous real or complex functions of x in the interval (a, b), and let $G(x)$ be defined by*

$$G(x) = \begin{cases} x^{-1/2}\displaystyle\int_0^x t^{-1/2}g(t)\,dt, & x > 0, \\ 2g(0), & x = 0, \\ (-x)^{-1/2}\displaystyle\int_x^0 (-t)^{-1/2}g(t)\,dt, & x < 0. \end{cases} \tag{2.06}$$

Then $G(x)$ and its first m derivatives are continuous in (a, b). Also, $G^{(m+1)}(x)$ is continuous, except possibly at $x = 0$.

This result may be proved with the aid of the mean-value theorems for integrals and derivatives, in a similar manner to the Lemma of [1, §4], and it is unnecessary to give details. The interval (a, b) need not contain $x = 0$; the result then becomes trivial.

Let us suppose that for some arbitrary integer j, $f^{(j)}(x)$ is continuous in (a, b). With $s = 0$, equations (2.02) become

$$(2.07)\qquad \begin{aligned} B_0(x) &= \frac{1}{2}x^{-1/2}\int_0^x t^{-1/2}f(t)\,dt, && x > 0,\\ B_0(x) &= \frac{1}{2}(-x)^{-1/2}\int_x^0 (-t)^{-1/2}f(t)\,dt, && x < 0. \end{aligned}$$

Hence defining $B_0(0) = f(0)$, we see from Lemma 1 that $B_0^{(j)}$ is continuous in (a, b). From (2.03) we now see that $A_1^{(j-1)}$ is continuous, and therefore that $(fA_1 - A_1'')^{(j-3)}$ is continuous. A further application of Lemma 1 shows that $B_1^{(j-3)}$ is continuous. Repeating the argument we establish the continuity of the following derivatives: $B_0^{(j)}$, $B_1^{(j-3)}$, $B_2^{(j-6)}$, $\cdots$, and $A_1^{(j-1)}$, $A_2^{(j-4)}$, $A_3^{(j-7)}$, $\cdots$.

In a similar way, we can show that if $f^{(j)}(x)$ is continuous in a closed interval not containing $x = 0$, then $B_0^{(j+1)}$, $B_1^{(j-1)}$, $B_2^{(j-3)}$, $\cdots$, and $A_1^{(j)}$, $A_2^{(j-2)}$, $A_3^{(j-4)}$, $\cdots$, are all continuous in this interval.

Of particular interest in the next section is the continuity of the functions A_m' and $2xB_m' + B_m$, that is, $fA_m - A_m''$, for given m. Taking $j = 3m - 1$, $3m$ and also $2m - 1$, $2m$, in the results of the preceding two paragraphs we have the following:

LEMMA 2. (i) *If $f^{(2m-1)}(x)$ is continuous in the interval (a, b) and $f^{(3m-1)}(x)$ is continuous in a neighborhood of $x = 0$, then $A_m'(x)$ is continuous in (a, b).*

(ii) *If $f^{(2m)}(x)$ is continuous in (a, b) and $f^{(3m)}(x)$ is continuous in a neighborhood of $x = 0$, then $2xB_m'(x) + B_m(x)$ is continuous in (a, b).*

We turn now to the behavior of the coefficients as $|x| \to \infty$. This has been investigated previously in [4, Lemma 1] and [5, §11] by means of complex variable theory. Corresponding results for real variables can be established in an analogous manner and are expressed in the final lemma of this section.

LEMMA 3. (i) *Let $f^{(2m-1)}(x)$ be continuous and*

$$(2.08)\qquad f^{(r)}(x) = O(|x|^{-r-(1/2)-\sigma}), \qquad r = 0, 1, \cdots, 2m - 1,$$

as $|x| \to \infty$, where $\sigma > 0$ is a constant. If $s \leqq m$, then $A_s(x)$, $D_s(x)$, and $\mathcal{V}(A_s)$ tend to finite limits as $|x| \to \infty$, and

$$(2.09) \qquad A_s^{(r)}(x),\, D_s^{(r)}(x) = O(|\,x\,|^{-r-\sigma_1}), \quad 1 \leqq r \leqq 2m - 2s + 1.$$

$$(2.10) \qquad |\,x\,|^{1/2}B_s^{(r)}(x),\, |\,x\,|^{-1/2}C_s^{(r)}(x) = O(|\,x\,|^{-r}), \quad 0 \leqq r \leqq 2m - 2s,$$

where $\sigma_1 = \min(\sigma, \frac{3}{2})$.

(ii) *Let* $f^{(2m)}(x)$ *be continuous and* (2.08) *hold when* $r = 0, 1, \cdots, 2m$. *If* $s \leqq m$, *then* $A_s(x)$, $D_s(x)$ *and* $\mathcal{V}(|\,x\,|^{1/2}B_s)$ *tend to finite limits as* $|\,x\,| \to \infty$, *and* (2.09) *and* (2.10) *hold with the conditions* $1 \leqq r \leqq 2m - 2s + 2$ *and* $0 \leqq r \leqq 2m - 2s + 1$, *respectively*.

In this result the variations $\mathcal{V}$ are defined ([7, §1]) by

$$(2.11) \qquad \mathcal{V}(A_s) = \int |\,A_s'(x)\,|\,dx, \qquad \mathcal{V}(|\,x\,|^{1/2}B_s) = \int |\,\{x^{1/2}B_s(x)\}'\,|\,dx.$$

The conditions of the Lemma are sufficient but not necessary for the convergence of the variations. For this reason we shall not incorporate them in the approximation theorems of the next section; instead, we regard Lemma 3 as a test for convergence. In most applications in which the variations converge, however, the conditions of Lemma 3 undoubtedly will be satisfied.

3. Approximation theorems for real variables. To introduce these theorems, we first recall from [1, §§2, 3] the following definitions associated with the Airy functions $\mathrm{Ai}(x)$ and $\mathrm{Bi}(x)$.

$$(3.01) \qquad E(x) = \exp(\tfrac{2}{3}x^{3/2}), \quad x > 0;$$
$$E(x) = 1, \quad x \leqq 0; \qquad E^{-1}(x) = 1/E(x),$$

$$(3.02) \qquad \mathrm{Ai}(x) = E^{-1}(x)M(x)\sin\chi(x), \qquad \mathrm{Bi}(x) = E(x)M(x)\cos\chi(x),$$

$$(3.03) \qquad \mathrm{Ai}'(x) = E^{-1}(x)N(x)\sin\psi(x), \qquad \mathrm{Bi}'(x) = E(x)N(x)\cos\psi(x).$$

$$(3.04) \qquad \lambda_1 = \max_{(-\infty,\infty)} \{\pi\,|\,x\,|^{1/2}M^2(x)\} = 1.430\cdots,$$

$$(3.05) \qquad \lambda_2 = \max_{(-\infty,\infty)} \{\pi E^{-1}(x)M(x)|\,x^{1/2}\mathrm{Bi}(x)|\} = 1.315\cdots,$$

$$(3.06) \qquad \mu_2 = \max_{(-\infty,\infty)} \{\pi E(x)M(x)|\,x^{1/2}\mathrm{Ai}(x)|\} = 1.$$

In this paper we do not need the constants λ_3 to λ_6 and μ_3 introduced in [1, §6], but we shall require the following new constants:

$$(3.07) \qquad \mu_4 = \max_{(-\infty,\infty)} \{\pi E^{-1}(x)M(x)|\,\mathrm{Bi}'(x)|\} = \tfrac{1}{2}\sqrt{5} = 1.118\cdots,$$

$$(3.08) \qquad \lambda_7 = \max_{(-\infty,\infty)} \{\pi E(x)M(x)|\,\mathrm{Ai}'(x)|\} = 1.002\cdots.$$

These maxima are attained at $x = +\infty$ and $x = -2.29\cdots$, respectively.

THEOREM 1. *Let $f(x)$ and its first $2m$ derivatives be continuous real or complex functions of x in the interval (a, b), and let $f^{(3m)}(x)$ be continuous in a neighborhood of $x = 0$. Then in the interval (a, b) the differential equation*

$$\frac{d^2w}{dx^2} = \{x + f(x)\}w \tag{3.09}$$

has solutions $w_{2m+1,1}(x)$ and $w_{2m+1,2}(x)$, which depend on m and are such that

$$w_{2m+1,1}(x) = \mathrm{Ai}(x) \sum_{s=0}^{m} A_s(x) + \mathrm{Ai}'(x) \sum_{s=0}^{m-1} B_s(x) + \epsilon_{2m+1,1}(x), \tag{3.10}$$

$$\begin{aligned} w'_{2m+1,1}(x) = \mathrm{Ai}(x) &\left\{\sum_{s=0}^{m-1} C_s(x) + A_m'(x)\right\} \\ &+ \mathrm{Ai}'(x) \sum_{s=0}^{m} D_s(x) + \eta_{2m+1,1}(x), \end{aligned} \tag{3.11}$$

$$w_{2m+1,2}(x) = \mathrm{Bi}(x) \sum_{s=0}^{m} A_s(x) + \mathrm{Bi}'(x) \sum_{s=0}^{m-1} B_s(x) + \epsilon_{2m+1,2}(x), \tag{3.12}$$

$$\begin{aligned} w'_{2m+1,2}(x) = \mathrm{Bi}(x) &\left\{\sum_{s=0}^{m-1} C_s(x) + A_m'(x)\right\} \\ &+ \mathrm{Bi}'(x) \sum_{s=0}^{m} D_s(x) + \eta_{2m+1,2}(x), \end{aligned} \tag{3.13}$$

where

$$\begin{aligned} &\frac{E(x)\,|\,\epsilon_{2m+1,1}(x)\,|}{M(x)}, \frac{E(x)\,|\,\eta_{2m+1,1}(x)\,|}{N(x)} \\ &\qquad \leqq 2 \exp\{2\lambda_1 \mathcal{V}_{x,b}(|\,x\,|^{1/2}B_0)\}\mathcal{V}_{x,b}(|\,x\,|^{1/2}B_m), \end{aligned} \tag{3.14}$$

$$\begin{aligned} &\frac{|\,\epsilon_{2m+1,2}(x)\,|}{E(x)M(x)}, \frac{|\,\eta_{2m+1,2}(x)\,|}{E(x)N(x)} \\ &\qquad \leqq 2\lambda_2 \exp\{2\lambda_1 \mathcal{V}_{a,x}(|\,x\,|^{1/2}B_0)\}\mathcal{V}_{a,x}(|\,x\,|^{1/2}B_m). \end{aligned} \tag{3.15}$$

The interval (a, b) may be infinite, provided that the variations in (3.14) and (3.15) are finite.

THEOREM 2. *Let $f(x)$ and its first $2m - 1$ derivatives be continuous real or complex functions of x in the interval (a, b), and let $f^{(3m-1)}(x)$ be continuous in a neighborhood of $x = 0$. Then in the interval (a, b) the differential equation (3.09) has solutions $w_{2m,1}(x)$ and $w_{2m,2}(x)$, which depend on m and are such that*

$$w_{2m,1}(x) = \mathrm{Ai}(x) \sum_{s=0}^{m-1} A_s(x) + \mathrm{Ai}'(x) \sum_{s=0}^{m-1} B_s(x) + \epsilon_{2m,1}(x), \tag{3.16}$$

$$w'_{2m,1}(x) = \mathrm{Ai}(x) \sum_{s=0}^{m-1} C_s(x) + \mathrm{Ai}'(x)\left\{\sum_{s=0}^{m-1} D_s(x) + B'_{m-1}(x)\right\} + \eta_{2m,1}(x), \tag{3.17}$$

$$w_{2m,2}(x) = \mathrm{Bi}(x) \sum_{s=0}^{m-1} A_s(x) + \mathrm{Bi}'(x) \sum_{s=0}^{m-1} B_s(x) + \epsilon_{2m,2}(x), \tag{3.18}$$

$$w'_{2m,2}(x) = \mathrm{Bi}(x) \sum_{s=0}^{m-1} C_s(x) + \mathrm{Bi}'(x)\left\{\sum_{s=0}^{m-1} D_s(x) + B'_{m-1}(x)\right\} + \eta_{2m,2}(x), \tag{3.19}$$

where

$$\frac{E(x)\mid \epsilon_{2m,1}(x)\mid}{M(x)}, \frac{E(x)\mid \eta_{2m,1}(x)\mid}{N(x)} \leqq 2\lambda_7 \exp\{2\lambda_1 \mathcal{V}_{x,b}(\mid x\mid^{1/2}B_0)\}\mathcal{V}_{x,b}(A_m), \tag{3.20}$$

$$\frac{\mid \epsilon_{2m,2}(x)\mid}{E(x)M(x)}, \frac{\mid \eta_{2m,2}(x)\mid}{E(x)N(x)} \leqq \sqrt{5} \exp\{2\lambda_1 \mathcal{V}_{a,x}(\mid x\mid^{1/2}B_0)\}\mathcal{V}_{a,x}(A_m). \tag{3.21}$$

The interval (a, b) *may be infinite, provided that the variations in* (3.20) *and* (3.21) *are finite.*

The proofs of these theorems are similar to that of Theorem 1 of [1]. Substituting (3.10) in (3.09), for example, we find that

$$\epsilon''_{2m+1,1}(x) - \{x + f(x)\}\epsilon_{2m+1,1}(x) = -R_{2m+1,1}(x), \tag{3.22}$$

where

$$R_{2m+1,1}(x) = -\{2xB_m'(x) + B_m(x)\}\mathrm{Ai}(x) = \mp 2\mid x\mid^{1/2} \frac{d\{\mid x\mid^{1/2}B_m(x)\}}{dx}\mathrm{Ai}(x), \tag{3.23}$$

the upper or lower sign being taken according as $x > 0$ or $x < 0$. From the conditions on $f(x)$ and Lemma 2(ii) we see that $R_{2m+1,1}(x)$ is continuous in (a, b).

We construct a convergent series solution of the inhomogeneous differential equation (3.22) for $\epsilon_{2m+1,1}(x)$ in the form

$$\epsilon_{2m+1,1}(x) = \sum_{n=0}^{\infty}\{h_{n+1}(x) - h_n(x)\}, \tag{3.24}$$

where $h_0(x) = 0$ and

$$h_n(x) = \pi \int_x^b \{\mathrm{Ai}(x)\mathrm{Bi}(t) - \mathrm{Bi}(x)\mathrm{Ai}(t)\} \times \{f(t)h_{n-1}(t) - R_{2m+1,1}(t)\}\, dt, \quad n \geqq 1. \tag{3.25}$$

With the aid of (3.09) of [1], (3.04) and (3.06) above, and the relation

$$\int | t^{-1/2} f(t) | \, dt = 2\mathcal{V}(| t |^{1/2} B_0), \tag{3.26}$$

obtained from (2.07), we find that

$$| h_{n+1}(x) - h_n(x) | \leqq 2^{n+1} \lambda_1^n E^{-1}(x) M(x) \mathcal{V}_{x,b}(| x |^{1/2} B_m) \frac{\{\mathcal{V}_{x,b}(| x |^{1/2} B_0)\}^n}{n!}. \tag{3.27}$$

On summation we obtain the inequality for $| \epsilon_{2m+1,1}(x) |$ stated in (3.14). The remaining inequalities may be established similarly.

Remarks on Theorems 1 *and* 2. Many of the Remarks made in [7, §3] are immediately applicable to the present theorems. We also comment as follows.

(i) More refined, but more complicated, bounds for the error terms can be established by the methods used in [1, §7] for the first approximations.

(ii) In order to apply Theorems 1 and 2 to the differential equation (1.01) we use the transformation to new variables ζ and W given by (4.02) to (4.09) of [1]. The differentiability conditions required by the theorems can be investigated by means of the following result, which is an extension of the Lemma of [1, §4] and can be proved in a similar way.

LEMMA 4. *In the interval* $a \leqq x \leqq b$, *let* $q(x)$ *be an* m *times continuously differentiable real or complex function, and let* $(x - x_0)^{-1} p(x)$ *be an* $m + 2$ *times continuously differentiable real function. Then* $(x - x_0)^{-1} \zeta(x)$ *is an* $m + 2$ *times continuously differentiable real function of* x *in* $a \leqq x \leqq b$, *and* $F(\zeta) \equiv f(\zeta) + \zeta\{q(x)/p(x)\}$ *is an* m *times continuously differentiable function of* ζ *in the corresponding* ζ*-interval.*

In the next section we give an example of the application of the theorems.

4. Bessel functions of large orders. The functions $z^{1/2} J_n(nz)$ and $z^{1/2} Y_n(nz)$ satisfy the differential equation

$$\frac{d^2 w}{dz^2} = \left\{ n^2 \left(\frac{1 - z^2}{z^2} \right) - \frac{1}{4z^2} \right\} w. \tag{4.01}$$

In this paper we restrict n and z to real positive values. Extensions to complex variables can be achieved by means of the theorems of §5.

Equation (4.01) is of the form (1.01). Applying the transformations of [1, §5(iii)] with

$$x = z, \qquad u = n, \qquad p(z) = z^{-2}(1 - z^2), \qquad q(z) = -\tfrac{1}{4}z^{-2}, \qquad x_0 = 1,$$

we obtain the equation (compare [6, §4])

$$\frac{d^2W}{d\zeta^2} = \{n^2\zeta + F(\zeta)\}W, \tag{4.02}$$

in which

$$\begin{aligned} \tfrac{2}{3}(-\zeta)^{3/2} &= (z^2 - 1)^{1/2} - \sec^{-1} z, & z \geqq 1, \\ \tfrac{2}{3}\,\zeta^{3/2} &= \ln \frac{1 + (1 - z^2)^{1/2}}{z} - (1 - z^2)^{1/2}, & 0 < z \leqq 1, \end{aligned} \tag{4.03}$$

$$W = \{(1 - z^2)/\zeta\}^{1/4} J_n(nz) \quad \text{or} \quad \{(1 - z^2)/\zeta\}^{1/4} Y_n(nz), \tag{4.04}$$

and

$$F(\zeta) = \frac{5}{16\zeta^2} + \frac{\zeta}{4}\,\frac{z^2(z^2 + 4)}{(z^2 - 1)^3}. \tag{4.05}$$

The interval $0 < z < \infty$ transforms into $\infty > \zeta > -\infty$, with $\zeta = 0$ corresponding to $z = 1$. Lemma 4 shows that $F(\zeta)$ has derivatives of all orders in the transformed interval. We readily verify that the conditions of Lemma 3 (§2) are satisfied by $F(\zeta)$ with $\sigma = \frac{3}{2}$, for all values of the integer m; compare [6, (4.14) and (4.15)].

With the substitution $x = n^{2/3}\zeta$, (4.02) takes exactly the form (3.09) with $f(x) = n^{-4/3}F(n^{-2/3}x)$. Applying Theorem 1, taking $a = -\infty$ and $b = \infty$, and subsequently restoring the variable ζ, we find that (4.02) has a solution $W_{2m+1,1}$, with m ($\geqq 0$) arbitrary, given by

$$W_{2m+1,1} = \mathrm{Ai}(n^{2/3}\zeta) \sum_{s=0}^{m} \frac{A_s(\zeta)}{n^{2s}} + \frac{\mathrm{Ai}'(n^{2/3}\zeta)}{n^{4/3}} \sum_{s=0}^{m-1} \frac{B_s(\zeta)}{n^{2s}} + \epsilon_{2m+1,1}\,, \tag{4.06}$$

where the coefficients $A_s(\zeta)$ and $B_s(\zeta)$ are defined by (4.17) and (4.23) of [6] (see also [8, §7]), and

$$\begin{aligned} |\,\epsilon_{2m+1,1}\,| \leqq 2n^{-2m-1} &\exp\{2\lambda_1 n^{-1} \mathcal{V}_{\zeta,\infty}(|\,\zeta\,|^{1/2} B_0)\} \\ &\times \mathcal{V}_{\zeta,\infty}(|\,\zeta\,|^{1/2} B_m) E^{-1}(n^{2/3}\zeta) M(n^{2/3}\zeta), \end{aligned} \tag{4.07}$$

when $-\infty < \zeta < \infty$. To simplify this bound and also subsequent similar bounds (at negligible practical expense) we replace $2\lambda_1 \mathcal{V}_{\zeta,\infty}(|\,\zeta\,|^{1/2} B_0)$ by

$$n_0 = 2\lambda_1 \mathcal{V}_{-\infty,\infty}(|\,\zeta\,|^{1/2} B_0). \tag{4.08}$$

(The numerical value $n_0 = 0.30 \cdots$ is computed in [8, §5].) Then (4.07) becomes

$$|\,\epsilon_{2m+1,1}\,| \leqq 2e^{n_0/n} n^{-2m-1} \mathcal{V}_{\zeta,\infty}(|\,\zeta\,|^{1/2} B_m) E^{-1}(n^{2/3}\zeta) M(n^{2/3}\zeta). \tag{4.09}$$

From (4.04), we see that

$$(4.10)\quad W_{2m+1,1} = c_1\{(1 - z^2)/\zeta\}^{1/4} J_n(nz) + c_2\{(1 - z^2)/\zeta\}^{1/4} Y_n(nz),$$

where c_1, c_2 depend on m and n, but are independent of ζ. Multiplying both sides of this equation by $\zeta^{1/4}$ and letting $\zeta \to +\infty$, we readily verify, with the aid of Lemma 3, that the left side and the first term on the right both vanish, but the second term on the right becomes infinite. Hence $c_2 = 0$. To evaluate c_1, we set $\zeta = n^{-2/3} a_r'$, where a_r' is the rth negative zero of the function Ai$'$, and take r to be large, so that ζ is large and negative. Then using relations given in [6, §4] we arrive at the first of our desired results:

$$(4.11)\quad \begin{aligned} J_n(nz) = \frac{1}{1+\delta_{2m+1}} \frac{1}{n^{1/3}} \left(\frac{4\zeta}{1-z^2}\right)^{1/4} \Big\{ &\mathrm{Ai}(n^{2/3}\zeta) \sum_{s=0}^{m} \frac{A_s(\zeta)}{n^{2s}} \\ &+ \frac{\mathrm{Ai}'(n^{2/3}\zeta)}{n^{4/3}} \sum_{s=0}^{m-1} \frac{B_s(\zeta)}{n^{2s}} + \epsilon_{2m+1,1} \Big\}, \end{aligned}$$

in which $\epsilon_{2m+1,1}$ is bounded by (4.09), and δ_{2m+1} is independent of z and bounded by

$$(4.12)\quad |\delta_{2m+1}| \leqq 2e^{n_0/n} n^{-2m-1} \mathcal{V}_{-\infty,\infty}(|\zeta|^{1/2} B_m).$$

These bounds are valid when $n > 0$ and $-\infty < \zeta < \infty$, that is, $0 < z < \infty$. Equation (4.11) is the complete form of the asymptotic expansion (4.24) of [6], for these ranges of the variables.

The corresponding result for the derivative, valid with the same conditions, may be verified to be

$$(4.13)\quad \begin{aligned} J_n'(nz) = -\frac{1}{1+\delta_{2m+1}} \frac{2}{n^{2/3} z \varphi(\zeta)} &\times \left[\frac{\mathrm{Ai}(n^{2/3}\zeta)}{n^{2/3}} \left\{ \sum_{s=0}^{m-1} \frac{C_s(\zeta)}{n^{2s}} + \frac{C_m(\zeta) - \zeta B_m(\zeta)}{n^{2m}} \right\} \right. \\ &\left. + \mathrm{Ai}'(n^{2/3}\zeta) \sum_{s=0}^{m} \frac{D_s(\zeta)}{n^{2s}} + \eta_{2m+1,1} + \frac{\varphi'(\zeta)}{\varphi(\zeta)} \frac{\epsilon_{2m+1,1}}{n^{2/3}} \right], \end{aligned}$$

where $\varphi(\zeta)$, $C_s(\zeta)$ and $D_s(\zeta)$ are defined as in [6, §6], and

$$(4.14)\quad |\eta_{2m+1,1}| \leqq 2e^{n_0/n} n^{-2m-1} \mathcal{V}_{\zeta,\infty}(|\zeta|^{1/2} B_m) E^{-1}(n^{2/3}\zeta) N(n^{2/3}\zeta).$$

For the Bessel function of the second kind, we observe from Theorem 1 that (4.02) has a second solution, given by

$$(4.15)\quad W_{2m+1,2} = \mathrm{Bi}(n^{2/3}\zeta) \sum_{s=0}^{m} \frac{A_s(\zeta)}{n^{2s}} + \frac{\mathrm{Bi}'(n^{2/3}\zeta)}{n^{4/3}} \sum_{s=0}^{m-1} \frac{B_s(\zeta)}{n^{2s}} + \epsilon_{2m+1,2},$$

and

$$\frac{dW_{2m+1,2}}{d\zeta} = \mathrm{Bi}(n^{2/3}\zeta)\left\{\sum_{s=0}^{m-1}\frac{A_s'(\zeta)+\zeta B_s(\zeta)}{n^{2s}}+\frac{A_m'(\zeta)}{n^{2m}}\right\} \tag{4.16}$$
$$+ n^{2/3}\,\mathrm{Bi}'(n^{2/3}\zeta)\sum_{s=0}^{m}\frac{A_s(\zeta)+B_{s-1}'(\zeta)}{n^{2s}} + n^{2/3}\,\eta_{2m+1,2},$$

where

$$\frac{|\epsilon_{2m+1,2}|}{M(n^{2/3}\zeta)}, \frac{|\eta_{2m+1,2}|}{N(n^{2/3}\zeta)} \leqq \frac{2\lambda_2\, e^{\nu_0/n}}{n^{2m+1}}\,\mathcal{V}_{-\infty,\zeta}(|\zeta|^{1/2}B_m)E(n^{2/3}\zeta). \tag{4.17}$$

From (4.04), we see that

$$W_{2m+1,2} = c_3 J + c_4 Y, \tag{4.18}$$

in which we have used the notation

$$J = \{(1-z^2)/\zeta\}^{1/4}J_n(nz), \qquad Y = \{(1-z^2)/\zeta\}^{1/4}Y_n(nz). \tag{4.19}$$

The values of c_3 and c_4 are given by

$$c_3 = \frac{YW_{2m+1,2}' - Y'W_{2m+1,2}}{J'Y-JY'}, \qquad c_4 = \frac{J'W_{2m+1,2}-JW_{2m+1,2}'}{J'Y-JY'}. \tag{4.20}$$

When $\zeta = -\infty$, the right side of (4.17) vanishes. Evaluating the Wronskians in the expressions (4.20) at this point, we find without difficulty that $c_3 = 0$ and $c_4 = -2^{-1/2}n^{1/3}$. Thus we have

$$Y_n(nz) = -\frac{1}{n^{1/3}}\left(\frac{4\zeta}{1-z^2}\right)^{1/4}\left\{\mathrm{Bi}(n^{2/3}\zeta)\sum_{s=0}^{m}\frac{A_s(\zeta)}{n^{2s}}\right. \tag{4.21}$$
$$\left. + \frac{\mathrm{Bi}'(n^{2/3}\zeta)}{n^{4/3}}\sum_{s=0}^{m-1}\frac{B_s(\zeta)}{n^{2s}} + \epsilon_{2m+1,2}\right\},$$

where $\epsilon_{2m+1,2}$ is bounded by (4.17) when $n > 0$ and $z > 0$. The corresponding result for the derivative, valid with the same conditions, is given by

$$Y_n'(nz) = \frac{2}{n^{2/3}z\varphi(\zeta)}\left[\frac{\mathrm{Bi}(n^{2/3}\zeta)}{n^{2/3}}\left\{\sum_{s=0}^{m-1}\frac{C_s(\zeta)}{n^{2s}}+\frac{C_m(\zeta)-\zeta B_m(\zeta)}{n^{2m}}\right\}\right. \tag{4.22}$$
$$\left. + \mathrm{Bi}'(n^{2/3}\zeta)\sum_{s=0}^{m}\frac{D_s(\zeta)}{n^{2s}} + \eta_{2m+1,2} + \frac{\varphi'(\zeta)}{\varphi(\zeta)}\frac{\epsilon_{2m+1,2}}{n^{2/3}}\right].$$

Corresponding bounds for truncation of the asymptotic expansions of $J_n(nz)$, $Y_n(nz)$ and their derivatives at an even number of terms can be obtained by application of Theorem 2, but we shall not give these results here.

Further practical details concerning the bounds of this section are given in [8, §5]. In this reference $e^{n_0/n}$ is replaced by the slightly weaker, but generally quite adequate, bound $n/(n - n_0)$ when $n > n_0$, and the suffix $2m + 1$ on ϵ, η and δ is changed to $2m$.

5. Approximation theorems for complex variables. We consider now the solution of the differential equation

$$\frac{d^2w}{dz^2} = \{z + f(z)\}\, w \tag{5.01}$$

for values of z lying in a given simply-connected domain **D**, bounded or unbounded, in which $f(z)$ is regular (holomorphic). The formal series solutions of §2 above are again applicable, z now replacing x. By considering the Taylor expansions at $z = 0$ we readily see that (2.02) and (2.03) define an infinite set of regular functions $A_s(z)$ and $B_s(z)$. Moreover, since **D** is simply-connected these functions are single-valued.

Let the sectors

$$|\arg z| \leqq \tfrac{1}{3}\pi, \qquad -\pi \leqq \arg z \leqq -\tfrac{1}{3}\pi, \qquad \tfrac{1}{3}\pi \leqq \arg z \leqq \pi,$$

be denoted by $\mathbf{S}_1$, $\mathbf{S}_2$, $\mathbf{S}_3$, respectively, as in [4]. The appropriate pair of complementary functions for constructing the solution of (3.22) is now $\mathrm{Ai}(z)$ and $\mathrm{Ai}(ze^{2\pi i/3})$ in $\mathbf{S}_1 \cup \mathbf{S}_2$, and $\mathrm{Ai}(z)$ and $\mathrm{Ai}(ze^{-2\pi i/3})$ in $\mathbf{S}_1 \cup \mathbf{S}_3$. Thus as integral equations representing (3.22), we use

$$\begin{aligned}\epsilon_{2m+1}(z) = 2\pi e^{\mp\pi i/6} \int_z^b \{\mathrm{Ai}(z)\mathrm{Ai}(te^{\mp 2\pi i/3}) - \mathrm{Ai}(ze^{\mp 2\pi i/3})\mathrm{Ai}(t)\}& \\ \times \{f(t)\epsilon_{2m+1}(t) - R_{2m+1,1}(t)\}\, dt,&\end{aligned} \tag{5.02}$$

where b is an arbitrary point of **D** (b corresponds to the right endpoint of the interval in the real case).

To derive bounds for the kernels of these equations, we define

$$E(z) = |\exp(\tfrac{2}{3}z^{3/2})|, \qquad E^{-1}(z) = |\exp(-\tfrac{2}{3}z^{3/2})|, \tag{5.03}$$

where $z^{3/2}$ has its principal value. (It may be noticed that this is consistent with (3.01).) We also define

$$\begin{aligned} E(z)|\mathrm{Ai}(z)| &= M_2(z)\sin\chi_2(z), \\ E^{-1}(z)|\mathrm{Ai}(ze^{-2\pi i/3})| &= M_2(z)\cos\chi_2(z), \\ E(z)|\mathrm{Ai}'(z)| &= N_2(z)\sin\psi_2(z), \\ E^{-1}(z)|\mathrm{Ai}'(ze^{-2\pi i/3})| &= N_2(z)\cos\psi_2(z), \end{aligned} \tag{5.04}$$

when $z \in \mathbf{S}_1 \cup \mathbf{S}_3$, where $M_2(z)$, $N_2(z)$, $\chi_2(z)$ and $\psi_2(z)$ are real nonnegative functions. Similarly,

$$
(5.05)\qquad
\begin{aligned}
E(z)|\,\mathrm{Ai}(z)\,| &= M_3(z)\sin\chi_3(z),\\
E^{-1}(z)|\,\mathrm{Ai}(ze^{2\pi i/3})\,| &= M_3(z)\cos\chi_3(z),\\
E(z)|\,\mathrm{Ai}'(z)\,| &= N_3(z)\sin\psi_3(z),\\
E^{-1}(z)|\,\mathrm{Ai}'(ze^{2\pi i/3})\,| &= N_3(z)\cos\psi_3(z),
\end{aligned}
$$

when $z \in \mathbf{S}_1 \cup \mathbf{S}_2$. Thus

$$
(5.06)\qquad
\left.
\begin{aligned}
M_2^2(z) &= E^2(z)|\,\mathrm{Ai}^2(z)\,| + E^{-2}(z)|\,\mathrm{Ai}^2(ze^{-2\pi i/3})\,|\\
N_2^2(z) &= E^2(z)|\,\mathrm{Ai}'^2(z)\,| + E^{-2}(z)|\,\mathrm{Ai}'^2(ze^{-2\pi i/3})\,|
\end{aligned}
\right\},\ z \in \mathbf{S}_1 \cup \mathbf{S}_3,
$$

with similar relations holding for the auxiliary functions $M_3^2(z)$ and $N_3^2(z)$ in $\mathbf{S}_1 \cup \mathbf{S}_2$.

Let us suppose that t and z are complex numbers lying in $\mathbf{S}_1 \cup \mathbf{S}_3$, such that

$$
(5.07)\qquad E(t) \geqq E(z).
$$

We then readily verify that

$$
(5.08)\qquad
\begin{aligned}
&|\,\mathrm{Ai}(z)\,\mathrm{Ai}(te^{-2\pi i/3}) - \mathrm{Ai}(ze^{-2\pi i/3})\,\mathrm{Ai}(t)\,|\\
&\qquad\qquad \leqq E^{-1}(z)E(t)M_2(z)M_2(t),
\end{aligned}
$$

and

$$
(5.09)\qquad
\begin{aligned}
&|\,\mathrm{Ai}'(z)\,\mathrm{Ai}(te^{-2\pi i/3}) - e^{-2\pi i/3}\,\mathrm{Ai}'(ze^{-2\pi i/3})\,\mathrm{Ai}(t)\,|\\
&\qquad\qquad \leqq E^{-1}(z)E(t)N_2(z)M_2(t).
\end{aligned}
$$

Corresponding results when $t, z \in \mathbf{S}_1 \cup \mathbf{S}_2$ are obtained by changing the sign of i and replacing M_2, N_2 by M_3, N_3, respectively.

The relations (5.08) and (5.09) correspond to the inequalities in the real case given by (3.09) and (3.13) of [1]. The role of the constants λ_s and μ_s is played here by

$$
(5.10)\qquad \nu_1 = \max_{z\in \mathbf{S}_1\cup \mathbf{S}_3}\{\pi|\,z\,|^{1/2}M_2^2(z)\} = \max_{z\in \mathbf{S}_1\cup \mathbf{S}_2}\{\pi|\,z\,|^{1/2}M_3^2(z)\},
$$

$$
(5.11)\qquad
\begin{aligned}
\nu_2 &= \max_{z\in \mathbf{S}_1\cup \mathbf{S}_3}\{\pi E(z)M_2(z)|\,z^{1/2}\,\mathrm{Ai}(z)\,|\}\\
&= \max_{z\in \mathbf{S}_1\cup \mathbf{S}_2}\{\pi E(z)M_3(z)|\,z^{1/2}\,\mathrm{Ai}(z)\,|\},
\end{aligned}
$$

and

$$
(5.12)\qquad
\begin{aligned}
\nu_3 &= \max_{z\in \mathbf{S}_1\cup \mathbf{S}_3}\{\pi E(z)M_2(z)|\,\mathrm{Ai}'(z)\,|\}\\
&= \max_{z\in \mathbf{S}_1\cup \mathbf{S}_2}\{\pi E(z)M_3(z)|\,\mathrm{Ai}'(z)\,|\}.
\end{aligned}
$$

The equality of the two maxima in (5.10) follows from symmetry; similarly for (5.11) and (5.12).

The analysis of §3 may now be extended in a straightforward manner (compare also [5, §§7, 13] and [7, §8]), and the resulting theorems are as follows.

THEOREM 3. *Let $f(z)$ be regular in a simply-connected domain* **D**, *and let* $\mathbf{H}(b)$ *comprise those points z of* **D** *which can be joined to any prescribed point b of* **D** *by a path $\mathcal{P}$ satisfying the conditions*:

(i) *$\mathcal{P}$ lies in* **D**.

(ii) *$\mathcal{P}$ comprises a finite number of Jordan arcs, each with a parametric equation of the form $t = t(\tau)$, where τ is the real parameter of the arc; $t''(\tau)$ is continuous and $t'(\tau)$ does not vanish. If b is the point at infinity on a straight line $\mathcal{L}$, then $\mathcal{P}$ coincides with $\mathcal{L}$ for all sufficiently large $|t|$.*

(iii) *As t traverses $\mathcal{P}$ from z to b then $\mathrm{Re}(\frac{2}{3}t^{3/2})$ is monotonic nondecreasing.*

Then (5.01) *has a regular solution $w_{2m+1}(z)$, which depends on b and on an arbitrary nonnegative integer m, and is given by*

$$w_{2m+1}(z) = \mathrm{Ai}(z) \sum_{s=0}^{m} A_s(z) + \mathrm{Ai}'(z) \sum_{s=0}^{m-1} B_s(z) + \epsilon_{2m+1}(z), \tag{5.13}$$

$$\begin{aligned} w'_{2m+1}(z) = \mathrm{Ai}(z) &\left\{ \sum_{s=0}^{m-1} C_s(z) + A_m'(z) \right\} \\ &+ \mathrm{Ai}'(z) \sum_{s=0}^{m} D_s(z) + \eta_{2m+1}(z), \end{aligned} \tag{5.14}$$

where

$$\frac{|\epsilon_{2m+1}(z)|}{M_j(z)}, \frac{|\eta_{2m+1}(z)|}{N_j(z)} \leqq 4\nu_2 E^{-1}(z) \exp\{4\nu_1 \mathcal{V}_{\mathcal{P}}(t^{1/2}B_0)\}\mathcal{V}_{\mathcal{P}}(t^{1/2}B_m), \tag{5.15}$$

when $z \in \mathbf{H}(b)$. Here $j = 2$ when $z \in \mathbf{S}_1 \cup \mathbf{S}_3$, and $j = 3$ when $z \in \mathbf{S}_1 \cup \mathbf{S}_2$.

THEOREM 4. *With the conditions of Theorem* 3, (5.01) *has a regular solution $w_{2m}(z)$, which depends on b and on an arbitrary nonnegative integer m, and is given by*

$$w_{2m}(z) = \mathrm{Ai}(z) \sum_{s=0}^{m-1} A_s(z) + \mathrm{Ai}'(z) \sum_{s=0}^{m-1} B_s(z) + \epsilon_{2m}(z), \tag{5.16}$$

$$\begin{aligned} w'_{2m}(z) = \mathrm{Ai}(z) &\sum_{s=0}^{m-1} C_s(z) \\ &+ \mathrm{Ai}'(z) \left\{ \sum_{s=0}^{m-1} D_s(z) + B'_{m-1}(z) \right\} + \eta_{2m}(z), \end{aligned} \tag{5.17}$$

where

$$(5.18)\qquad \frac{|\epsilon_{2m}(z)|}{M_j(z)},\ \frac{|\eta_{2m}(z)|}{N_j(z)} \leqq 4\nu_3 E^{-1}(z)\exp\{4\nu_1 \mathcal{V}_{\mathcal{P}}(t^{1/2}B_0)\}\, \mathcal{V}_{\mathcal{P}}(A_m),$$

when $z \in \mathbf{H}(b)$. *Again*, $j = 2$ *when* $z \in \mathbf{S}_1 \cup \mathbf{S}_3$, *and* $j = 3$ *when* $z \in \mathbf{S}_1 \cup \mathbf{S}_2$.

Remarks on Theorems 3 *and* 4. (i) When $z \in \mathbf{S}_1$ each theorem gives two bounds for ϵ and η, since we may take $j = 2$ and $j = 3$ in (5.15) and (5.18).

(ii) For prescribed b and m, each theorem gives only one solution of the differential equation. However, corresponding expressions for two further solutions can be obtained by replacing z by $ze^{\pm 2\pi i/3}$ in (5.01), applying Theorems 3 and 4 to the resulting equation and then restoring the original variable z. If the new solutions constructed in this way are compared with (5.13), (5.14), (5.16) and (5.17), we find that $\mathrm{Ai}(z)$ and $\mathrm{Ai}'(z)$ are replaced by $\mathrm{Ai}(ze^{\mp 2\pi i/3})$ and $e^{\mp 2\pi i/3}\,\mathrm{Ai}'(ze^{\mp 2\pi i/3})$, respectively, but that the rest of these expressions is unaffected. The modifications to (5.15) and (5.18) and to the definition of the regions of validity are quite straightforward.

(iii) Numerical values of the constants ν_1, ν_2, and ν_3 have not been computed. We may show, however, by considering the asymptotic forms of the functions in (5.10) to (5.12) for large negative z, that

$$(5.19)\qquad \nu_1 \geqq \tfrac{5}{4},\qquad \nu_2 \geqq \sqrt{\tfrac{5}{4}},\qquad \nu_3 \geqq \tfrac{5}{8}.$$

We observe also from (5.06) that $\nu_1 \geqq \nu_2$.

Actually, in (5.15) and (5.18), ν_1, ν_2, and ν_3 may be replaced by the maxima of the respective functions in (5.10) to (5.12) *on the path* $\mathcal{P}$. This will yield slightly sharper bounds in some regions.

REFERENCES

[1] F. W. J. Olver, *Error bounds for first approximations in turning-point problems*, this Journal, 11 (1963), pp. 748–772.

[2] R. E. Langer, *The asymptotic solutions of ordinary linear differential equations of the second order, with special reference to a turning point*, Trans. Amer. Math. Soc., 67 (1949), pp. 461–490.

[3] T. M. Cherry, *Uniform asymptotic formulae for functions with transition points*, Ibid., 68 (1950), pp. 224–257.

[4] F. W. J. Olver, *The asymptotic solution of linear differential equations of the second order for large values of a parameter*, Philos. Trans. Roy. Soc. London. Ser. A, 247 (1954), pp. 307–327.

[5] ———, *Uniform asymptotic expansions of solutions of linear second-order differential equations for large values of a parameter*, Ibid., 250 (1958), pp. 479–517.

[6] ———, *The asymptotic expansion of Bessel functions of large order*, Ibid., 247 (1954), pp. 328–368.

[7] ———, *Error bounds for the Liouville-Green (or WKB) approximation*, Proc. Cambridge Philos. Soc., 57 (1961), pp. 790–810.

[8] ———, *Tables for Bessel functions of moderate or large orders*, Math. Tables Nat. Phys. Lab., vol. 6, Her Majesty's Stationery Office, London, 1962.

J. SIAM NUMER. ANAL.
Ser. B, Vol. 2, No. 2

ON THE ASYMPTOTIC SOLUTIONS OF SECOND-ORDER DIFFERENTIAL EQUATIONS HAVING AN IRREGULAR SINGULARITY OF RANK ONE, WITH AN APPLICATION TO WHITTAKER FUNCTIONS*

F. W. J. OLVER†

1. Introduction. Recently [1], I discussed in a general way the computational use of asymptotic expansions, and in particular drew attention to their limitations near the boundaries of their regions of validity in the complex plane. The first purpose of the present paper is to consider the well-known asymptotic solutions of second-order ordinary differential equations in a region containing an irregular singularity of rank one, and derive strict error bounds of comparatively simple form which accurately reflect the behavior of the actual errors of the asymptotic solutions throughout the regions of validity.

The second purpose of this paper is to apply the new theory to the Whittaker form of the confluent hypergeometric equation, and thence derive explicit error bounds for the asymptotic expansions of the Whittaker functions. These functions have many important specializations, including Bessel functions, parabolic cylinder functions, the exponential integral, sine and cosine integrals, Fresnel integrals, and the error function. With the aid of the new bounds automatic computing routines can be constructed for generating these functions for large complex arguments, which are completely free from any vagueness concerning the error.

Without loss of generality, the irregular singularity of the differential equation can be taken to be the point at infinity. The general second-order equation of rank one is then given by ([2, §3.1])

$$\frac{d^2w}{dz^2} + f(z)\frac{dw}{dz} + g(z)w = 0, \tag{1.01}$$

in which the coefficients $f(z)$ and $g(z)$ can be expanded in series of powers of $1/z$:

$$f(z) = \sum_{s=0}^{\infty} \frac{f_s}{z^s}, \qquad g(z) = \sum_{s=0}^{\infty} \frac{g_s}{z^s}, \tag{1.02}$$

convergent in a neighborhood of infinity. Not all of f_0, g_0 and g_1 vanish, otherwise the singularity would be regular and the solutions capable of expansion in convergent power series. For our purposes it is unnecessary

* Received by the editors June 24, 1964, and in revised form November 12, 1964.

† National Bureau of Standards, Washington, D. C. This work was supported by the U. S. Army Research Office, Durham, North Carolina, under Project 4238-M.

for the series (1.02) to converge; no extra complication results in assuming that they are merely asymptotic. More precisely, we suppose that $f(z)$ and $g(z)$ are either continuous on a given real interval **I** or regular (holomorphic) in a given complex domain **D**, and in both cases

$$f(z) = \sum_{s=0}^{n-1} \frac{f_s}{z^s} + \frac{F_n(z)}{z^n}, \qquad g(z) = \sum_{s=0}^{n-1} \frac{g_s}{z^s} + \frac{G_n(z)}{z^n}, \tag{1.03}$$

where $F_n(z)$ and $G_n(z)$, $n = 0, 1, \cdots$, are bounded functions of z. Although interest naturally is focussed on cases in which **I** and **D** are unbounded, we do not need to restrict the analysis to them.

2. The formal series. Formal expansions satisfying (1.01) are provided by the so-called normal series of Thomé, of the form

$$w = z^{\mu} e^{\lambda z} \sum_{s=0}^{\infty} \frac{a_s}{z^s}. \tag{2.01}$$

Substituting this in the differential equation and equating to zero successive powers of $1/z$, we obtain the equations

$$\lambda^2 + f_0\lambda + g_0 = 0, \quad (f_0 + 2\lambda)\mu = -(f_1\lambda + g_1), \tag{2.02}$$

and

$$\begin{aligned}(f_0 + 2\lambda)sa_s &= (s - \mu)(s - 1 - \mu)a_{s-1} \\ &\quad + \{\lambda f_2 + g_2 - (s - 1 - \mu)f_1\}a_{s-1} \\ &\quad + \{\lambda f_3 + g_3 - (s - 2 - \mu)f_2\}a_{s-2} \\ &\quad + \cdots + \{\lambda f_{s+1} + g_{s+1} + \mu f_s\}a_0 .\end{aligned} \tag{2.03}$$

The first of these equations, called by Horn [3] the *characteristic equation*, yields two possible values λ_1, λ_2 for λ. The second equation determines corresponding values μ_1, μ_2 of μ, and (2.03) determines recursively corresponding sets of coefficients $a_{s,1}$, $a_{s,2}$, $s = 1, 2, \cdots$, in terms of arbitrary multiplicative constants $a_{0,1}$, $a_{0,2}$, respectively.

The process obviously fails only if $f_0 + 2\lambda = 0$, that is, if $\lambda_1 = \lambda_2$, since

$$f_0 + 2\lambda_1 = \lambda_1 - \lambda_2 = -(f_0 + 2\lambda_2). \tag{2.04}$$

This exceptional case can be handled by the transformation of Fabry ([2, §3.2], [4, §17.53]), by means of which (1.01) is converted into a similar differential equation having either an irregular singularity of rank one with distinct values of λ_j, or a regular singularity.

The numbers μ_j, $j = 1, 2$, or their negatives, are known as the *exponents* of the singularity, and $\exp(-2\pi i\mu_j)$ as the *latent roots of the circuit matrix* or *circuit roots* ([2, §3.1], [4, §17.5], [5, §3.2]). The factors $e^{\lambda_j z}$ are sometimes called the *determining factors* ([4, §17.5]); we shall call the λ_j themselves the *leading exponents* to distinguish them from the μ_j.

That the series (2.01) generally diverges can be seen immediately from (2.03). Neglecting all terms on the right beyond the first, we obtain

$$\frac{a_s}{a_{s-1}} \sim \frac{s}{f_0 + 2\lambda}, \quad \text{as} \quad s \to \infty.$$

Hence, only in cases in which the f_s and g_s are such that the first term is largely cancelled by the contribution of the others could there be any possibility of convergence.

The series (2.01) provides asymptotic expansions of solutions of (1.01), in the sense of Poincaré, in restricted regions of the complex plane. Several proofs of this are available (see [2]–[5] and, especially, [6, §3.2]), but none lends itself satisfactorily to the problem of determining error bounds. We shall use a new proof, which resembles earlier proofs in that it depends on the solution of a Volterra integral equation by successive approximation. It differs by using a new form of kernel, and by basing the integral equation on an inhomogeneous form of differential equation for the error term.

3. Construction of an integral equation for the error term. Let

$$L_n(z) = z^\mu e^{\lambda z} \sum_{s=0}^{n-1} \frac{a_s}{z^s}, \qquad n \geqq 1, \tag{3.01}$$

and $L_0(z) = 0$. By differentiation of this expression and use of (2.02) and (2.03), we may verify that

$$L_n''(z) + f(z)L_n'(z) + g(z)L_n(z) = z^\mu e^{\lambda z} R_n(z), \tag{3.02}$$

where

$$R_n(z) = \frac{(f_0 + 2\lambda)na_n}{z^{n+1}} + \frac{\tilde{R}_{n+1}(z)}{z^{n+2}}, \tag{3.03}$$

and

$$\tilde{R}_{n+1}(z) = \sum_{s=0}^{n-1} a_s\{(\mu - s)F_{n+1-s}(z) + \lambda F_{n+2-s}(z) + G_{n+2-s}(z)\}. \tag{3.04}$$

Hence, from (2.04) we have

$$| R_n(z) | \leqq | \lambda_1 - \lambda_2 | \left\{ \frac{n \,| a_n |}{| z |^{n+1}} + \frac{(n+1)r_{n+1}}{| z |^{n+2}} \right\}, \tag{3.05}$$

where

$$r_{n+1} = \frac{\max | \tilde{R}_{n+1}(z) |}{(n+1) | \lambda_1 - \lambda_2 |}, \tag{3.06}$$

and is finite in consequence of the assumptions of §1.

We now write the solution of (1.01) in the form

$$w(z) = L_n(z) + \epsilon_n(z). \tag{3.07}$$

Then from (3.02) we obtain

$$\epsilon_n''(z) + f(z)\epsilon_n'(z) + g(z)\epsilon_n(z) = -z^\mu e^{\lambda z} R_n(z). \tag{3.08}$$

We formulate an integral equation for the error term $\epsilon_n(z)$ by a standard procedure in the asymptotic theory of differential equations: (3.08) is compared with an asymptotically similar homogeneous equation having known solutions. The precise way in which this is done, however, is of decisive importance in the theory of error bounds. We approach from the following result, obtained from the Liouville transformation [2, §4.1] and easily verifiable directly.

LEMMA. *The differential equation*

$$Y'' + qY' + \left\{\frac{1}{2}q' + \frac{1}{4}q^2 - p - p^{1/4}\frac{d^2}{dz^2}(p^{-1/4})\right\} Y = 0, \tag{3.09}$$

in which p and q are arbitrary functions of the independent variable z, is satisfied by

$$Y = p^{-1/4}\exp\left(\pm\int p^{1/2}\,dz - \frac{1}{2}\int q\,dz\right). \tag{3.10}$$

In applying this lemma, we select p and q in such a way that the expansions of the coefficients of Y' and Y in (3.09) match those of the corresponding coefficients in (3.08) through the terms in $1/z$. A convenient choice is given by

$$q = f_0 + \frac{f_1}{z}, \qquad p = \frac{1}{2}q' + \frac{1}{4}q^2 - g_0 - \frac{g_1}{z} + \frac{\text{const.}}{z^2}, \tag{3.11}$$

and selecting the constant in the second of these equations to "complete the square", we obtain

$$p = \frac{1}{4}(f_0^2 - 4g_0)\left(1 + \frac{\rho}{z}\right)^2, \quad \text{where} \quad \rho = \frac{f_0 f_1 - 2g_1}{f_0^2 - 4g_0}. \tag{3.12}$$

With these substitutions, (3.09) becomes

$$Y'' + \left(f_0 + \frac{f_1}{z}\right) Y' + \left\{g_0 + \frac{g_1}{z} + \frac{\tilde{g}_2}{z^2} + \chi(z)\right\} Y = 0, \tag{3.13}$$

where

$$\begin{aligned} 4\tilde{g}_2 &= f_1(f_1 - 2) - \rho^2(f_0^2 - 4g_0), \\ \chi(z) &= -p^{1/4}(p^{-1/4})'' = \frac{\rho}{z^3}\left(1 + \frac{\rho}{4z}\right)\left(1 + \frac{\rho}{z}\right)^{-2}. \end{aligned} \tag{3.14}$$

Equation (3.13) clearly has the desired matching with (3.08).

The quantities f_0, f_1, g_0, and g_1 can be expressed in terms of λ_1, λ_2, μ_1, and μ_2 by means of (2.02). We find that

$$f_0 = -(\lambda_1 + \lambda_2), \quad g_0 = \lambda_1\lambda_2, \tag{3.15}$$
$$f_1 = -(\mu_1 + \mu_2), \quad g_1 = \lambda_1\mu_2 + \lambda_2\mu_1,$$

and thence,

$$\rho = \frac{\mu_1 - \mu_2}{\lambda_1 - \lambda_2}, \qquad \tilde{g}_2 = \mu_1\mu_2 + \frac{1}{2}(\mu_1 + \mu_2). \tag{3.16}$$

We also find from the expression (3.10) that solutions of (3.13) are given by

$$Y_1(z) = \left(1 + \frac{\rho}{z}\right)^{-1/2} z^{\mu_1} e^{\lambda_1 z}, \qquad Y_2(z) = \left(1 + \frac{\rho}{z}\right)^{-1/2} z^{\mu_2} e^{\lambda_2 z}. \tag{3.17}$$

Using (1.03) with $n = 2$, we now rearrange (3.08) in the form

$$\epsilon_n''(z) + \left(f_0 + \frac{f_1}{z}\right)\epsilon_n'(z) + \left\{g_0 + \frac{g_1}{z} + \frac{\tilde{g}_2}{z^2} + \chi(z)\right\}\epsilon_n(z)$$
$$= -\frac{F_2(z)}{z^2}\epsilon_n'(z) - \frac{\tilde{G}_2(z)}{z^2}\epsilon_n(z) - z^{\mu}e^{\lambda z}R_n(z), \tag{3.18}$$

where

$$\tilde{G}_2(z) = G_2(z) - \tilde{g}_2 - z^2\chi(z). \tag{3.19}$$

The desired integral equation is obtained on solving (3.18) by variation of parameters. Thus we have

$$\epsilon_n(z) = \int_z^{z_j} K(z, t)\left\{\frac{F_2(t)}{t^2}\epsilon_n'(t) + \frac{\tilde{G}_2(t)}{t^2}\epsilon_n(t) + t^{\mu}e^{\lambda t}R_n(t)\right\} dt, \tag{3.20}$$

where

$$K(z, t) = \frac{Y_1(z)Y_2(t) - Y_2(z)Y_1(t)}{(\lambda_1 - \lambda_2)t^{\mu_1+\mu_2}e^{(\lambda_1+\lambda_2)t}}, \tag{3.21}$$

and z_j, $j = 1, 2$, is an arbitrary fixed point which can be taken at infinity; indeed, this is usually done (see Remark (i) of §5).

4. Solution of the integral equation. Let us take the case $j = 1$. We construct the solution of (3.20) in the usual manner (see, e.g., [7]) as the sum of a convergent series

$$\epsilon_n(z) = \sum_{s=0}^{\infty}\{h_{s+1}(z) - h_s(z)\}, \tag{4.01}$$

in which the sequence $\{h_s(z)\}$ is defined by $h_0(z) = 0$ and

$$h_{s+1}(z) = \int_z^{z_1} K(z, t) \left\{ \frac{F_2(t)}{t^2} h_s'(t) + \frac{\tilde{G}_2(t)}{t^2} h_s(t) + t^{\mu_1} e^{\lambda_1 t} R_n(t) \right\} dt, \tag{4.02}$$

when $s \geqq 0$. Clearly

$$h_{s+1}'(z) = \int_z^{z_1} \frac{\partial K(z, t)}{\partial z} \left\{ \frac{F_2(t)}{t^2} h_s'(t) + \frac{\tilde{G}_2(t)}{t^2} h_s(t) + t^{\mu_1} e^{\lambda_1 t} R_n(t) \right\} dt. \tag{4.03}$$

Bounds are needed for the kernel $K(z, t)$ and its z-derivative. We introduce the notation

$$\xi_1(t) = \lambda_1 t + \mu_1 \log t, \quad \xi_2(t) = \lambda_2 t + \mu_2 \log t, \tag{4.04}$$

in which the branch of $\log t$ is continuous on the path of integration. Then from (3.17) and (3.21) we have

$$\begin{aligned} K(z, t) = {} & \frac{1}{\lambda_1 - \lambda_2} \left(1 + \frac{\rho}{z}\right)^{-1/2} \left(1 + \frac{\rho}{t}\right)^{-1/2} \\ & \times \{ e^{\xi_1(z) - \xi_1(t)} - e^{\xi_2(z) - \xi_2(t)} \}. \end{aligned} \tag{4.05}$$

Let the path of integration be denoted by $\mathcal{P}$, and impose the restriction that $\operatorname{Re}\{\xi_2(t) - \xi_1(t)\}$ *is monotonic nondecreasing as* t *traverses* $\mathcal{P}$ *from* z *to* z_1 (cf. [5, §11.3]). Then

$$| e^{\xi_2(z) - \xi_2(t)} | \leqq | e^{\xi_1(z) - \xi_1(t)} |, \tag{4.06}$$

and therefore‡

$$| K(z, t) | \leqq 2\alpha \, | \lambda_1 - \lambda_2 |^{-1} \, | e^{\xi_1(z) - \xi_1(t)} |, \tag{4.07}$$

where

$$\alpha = \max_{t \in \mathcal{P}} \left| 1 + \frac{\rho}{t} \right|^{-1}. \tag{4.08}$$

For the derivative, we readily show that

$$\left| \frac{\partial K(z, t)}{\partial z} \right| \leqq 2\alpha\beta \, | \lambda_1 - \lambda_2 |^{-1} \, | e^{\xi_1(z) - \xi_1(t)} |, \tag{4.09}$$

where

$$\begin{aligned} 2\beta = \max_{t \in \mathcal{P}} \Bigg[& \left| \lambda_1 + \frac{\mu_1}{t} + \frac{\rho}{2t^2} \left(1 + \frac{\rho}{t}\right)^{-1} \right| \\ & + \left| \lambda_2 + \frac{\mu_2}{t} + \frac{\rho}{2t^2} \left(1 + \frac{\rho}{t}\right)^{-1} \right| \Bigg]. \end{aligned} \tag{4.10}$$

Inequalities (3.05), (4.07), and (4.09) are now applied to (4.02) and

‡ In the case in which the variables are real, the factor 2 can be omitted from the right of (4.07) with a consequent improvement in the final bounds (cf. [7, (2.14)]).

(4.03), and using induction we readily establish that

$$\begin{aligned}(4.11)\qquad &|h_{s+1}(z) - h_s(z)|,\ \beta^{-1}|h'_{s+1}(z) - h_s'(z)| \\ &\leqq \frac{(2\alpha)^{s+1}(\beta\gamma+\delta)^s}{|\lambda_2-\lambda_1|^s s!}\mathcal{V}_{\mathcal{P}}{}^s\left(\frac{1}{t}\right)|e^{\xi_1(z)}|\left\{\mathcal{V}_{\mathcal{P}}\left(\frac{a_n}{t^n}\right)+\mathcal{V}_{\mathcal{P}}\left(\frac{r_{n+1}}{t^{n+1}}\right)\right\},\end{aligned}$$

for $s = 0, 1, \cdots$, where

$$(4.12)\qquad \gamma = \max_{t\in\mathcal{P}}|F_2(t)|, \quad \delta = \max_{t\in\mathcal{P}}|\tilde{G}_2(t)|,$$

and $\mathcal{V}_{\mathcal{P}}$ denotes the variational operator on the path $\mathcal{P}$, that is, the integral over $\mathcal{P}$ of the modulus of the derivative [7, §8].

5. Main results for the general differential equation. One further step in the analysis is required before we can assert that the sum of the series $\sum\{h_{s+1}(z) - h_s(z)\}$ constructed in §4 is in fact a solution of the differential equation (3.08). We need to justify that this series may be twice differentiated term by term. The effect of this requirement is to introduce further conditions on the nature of the path of integration in the complex case; those adopted in a similar proof in [8, pp. 491–492] are adequate here.

A similar analysis can be carried out for the case $j = 2$, and the final result is expressed as follows:

THEOREM. *With the conditions of* §1, *the differential equation* (1.01) *has solutions* $w_{n,j}(z)$, $j = 1, 2$, *depending on an arbitrary positive integer* n *and arbitrary fixed points* z_j, *given by*

$$(5.01)\qquad w_{n,j}(z) = z^{\mu_j}e^{\lambda_j z}\left(\sum_{s=0}^{n-1}\frac{a_{s,j}}{z^s}\right) + \epsilon_{n,j}(z),$$

in which λ_j, μ_j, *and* $a_{s,j}$ *are defined in* §2, *and*

$$\begin{aligned}(5.02)\qquad |\epsilon_{n,j}(z)|, \frac{|\epsilon'_{n,j}(z)|}{\beta_j} &\leqq 2\alpha_j|z^{\mu_j}e^{\lambda_j z}|\left\{\mathcal{V}_{\mathcal{P}_j}\left(\frac{a_{n,j}}{t^n}\right)+\mathcal{V}_{\mathcal{P}_j}\left(\frac{r_{n+1,j}}{t^{n+1}}\right)\right\} \\ &\times\exp\left\{\frac{2\alpha_j(\beta_j\gamma_j+\delta_j)}{|\lambda_2-\lambda_1|}\mathcal{V}_{\mathcal{P}_j}\left(\frac{1}{t}\right)\right\}, z\in\mathbf{H}_j(z_j).\end{aligned}$$

In this theorem the region $\mathbf{H}_j(z_j)$ is a subset of **I** or **D** comprising those points for which there exists a path $\mathcal{P}_j$ joining z and z_j such that

(i) $\mathcal{P}_j$ lies entirely in **I** or **D**;

(ii) $\mathcal{P}_j$ consists of a finite number of Jordan arcs, each having a parametric equation of the form $t = t(\tau)$ with $t''(\tau)$ continuous and $t'(\tau)$ nonvanishing;

(iii) As t traverses $\mathcal{P}_j$ from z to z_j, the quantity

$$\operatorname{Re}\{(\lambda_2-\lambda_1)t + (\mu_2-\mu_1)\log t\}$$

is monotonic nondecreasing if $j = 1$, or monotonic nonincreasing if $j = 2$.

The point z_j may be at infinity in **I**, or the point at infinity on a straight line $\mathcal{L}_j$ lying in **D**. In the latter case we require that $\mathcal{P}_j$ coincide with $\mathcal{L}_j$ for all sufficiently large $|t|$.

The quantities appearing in the bounds (5.02) are defined as follows:

$$\alpha_j = \max\left|1 + \frac{\rho}{t}\right|^{-1}, \qquad \gamma_j = \max|F_2(t)|, \tag{5.03}$$

$$\beta_j = \frac{1}{2}\max\left[\left|\lambda_1 + \frac{\mu_1}{t} + \frac{\rho}{2t^2}\left(1 + \frac{\rho}{t}\right)^{-1}\right| + \left|\lambda_2 + \frac{\mu_2}{t} + \frac{\rho}{2t^2}\left(1 + \frac{\rho}{t}\right)^{-1}\right|\right], \tag{5.04}$$

$$\delta_j = \max\left|G_2(t) - \mu_1\mu_2 - \frac{1}{2}(\mu_1 + \mu_2) - \frac{\rho}{t}\left(1 + \frac{\rho}{4t}\right)\left(1 + \frac{\rho}{t}\right)^{-2}\right|, \tag{5.05}$$

$$r_{n+1,j} = \frac{\max\left|\sum_{s=0}^{n-1} a_{s,j}\{(\mu_j - s)F_{n+1-s}(t) + \lambda_j F_{n+2-s}(t) + G_{n+2-s}(t)\}\right|}{(n+1)|\lambda_2 - \lambda_1|}, \tag{5.06}$$

where all maxima are taken over $\mathcal{P}_j$, and

$$\rho = \frac{\mu_1 - \mu_2}{\lambda_1 - \lambda_2}. \tag{5.07}$$

Remarks on the theorem. (i) *Choice of* z_j. From (5.02) it is seen that the solutions $w_{n,j}(z)$ satisfy the boundary conditions

$$z^{-\mu_j}e^{-\lambda_j z}\epsilon_{n,j}(z) \to 0, \qquad z^{-\mu_j}e^{-\lambda_j z}\epsilon'_{n,j}(z) \to 0 \qquad \text{as} \qquad z \to z_j.$$

Different choices of z_j will thus result in different solutions of the differential equation. Except when a pair of points z_1, z_2 can be found for which both $\mathbf{H}_1(z_1)$ and $\mathbf{H}_2(z_2)$ embrace the whole of **I** or **D**, more than one choice of z_1 and z_2, in fact, will be necessary to cover **I** or **D** completely with overlapping regions of validity.

Generally, to achieve maximal regions of validity the z_j would be taken at infinity. This choice may also ease the problem of identifying other solutions of the differential equation in terms of the asymptotic solutions. A finite choice of z_j is sometimes unavoidable, however. For example, if **I** is the infinite interval $(-\infty, \infty)$, the choice $z_1 = +\infty$ and $z_2 = -\infty$ yields at most one solution for positive z and another solution for negative z, because the variations in the error bounds (5.02) become infinite when $\mathcal{P}_j$ passes through the origin.

(ii) *Choice of path.* Condition (ii) on $\mathcal{P}_j$ is quite weak and is fulfilled, for example, if the path consists of a finite chain of straight lines.

Condition (iii) is of more consequence. For large $|t|$ it means roughly that $\mathrm{Re}\{(\lambda_2 - \lambda_1)t\}$ is monotonic. Because of this, the total range of $\arg z$ in $\mathbf{H}_j(z_j)$ cannot exceed 3π when $|z|$ is large. This result, incidentally, is already known [3], [6], in the proofs of the asymptotic nature of the formal expansion (2.01).

For a given choice of z_j, the sharpness of the bounds can be increased by selecting the path which minimizes the variations appearing in (5.02). In general terms this means keeping the $\mathcal{P}_j$ as far away from the origin as possible, subject to conditions (i) to (iii). In the following sections we show how to effect this choice in the case of the Whittaker form of the confluent hypergeometric equation; much of the analysis is applicable, however, to all second-order equations having an irregular singularity of rank one.

(iii) *Nature of the bound.* If $|z|$ is large enough to ensure that successive terms in the series (5.01) are decreasing reasonably rapidly at the point of truncation (a necessary condition for the expansion to be useful), both t^{-1} and its variation on the path will often be sufficiently small to cause both α_j and the exponential factor containing $\mathcal{V}_{\mathcal{P}_j}(t^{-1})$ in (5.02) to approximate to unity. The bound for $z^{-\mu_j}e^{-\lambda_j z}\epsilon_{n,j}(z)$ then reduces effectively to twice the variation of the first neglected term $a_{n,j}t^{-n}$; cf. [1, §4]. An exception occurs in the special case when $a_{n,j}$ vanishes, or is exceptionally small compared with preceding coefficients: the contribution of the variation of $r_{n+1,j}t^{-n-1}$ is then significant.

(iv) *Other bounds.* In the case in which $f(z) = 0$ and $g_0 \neq 0$, Bochenek [9] has derived bounds for $\epsilon_{n,j}(z)$ and $\epsilon'_{n,j}(z)$ by following through the asymptotic proof given in [2] with an explicit treatment of the error term. The results involve considerable calculation and are applicable only if z is real. They do not reveal the intimate connection of the error term with the variation of the first neglected term in the expansion.

6. Whittaker functions of large argument. As an important example of the foregoing theory, because of the many special functions of mathematical physics it embraces, we consider the solution of the Whittaker equation

$$\frac{d^2w}{dz^2} = \left(\frac{1}{4} - \frac{k}{z} + \frac{m^2 - \frac{1}{4}}{z^2}\right) w, \tag{6.01}$$

in which the variable z and parameters k, m are real or complex. Application of the theorem of §5 immediately yields the solutions

$$w_{n,j}(z) = z^{\pm k} e^{\mp\frac{1}{2}z} \left(\sum_{s=0}^{n-1} \frac{a_{s,j}}{z^s}\right) + \epsilon_{n,j}(z), \qquad j = 1, 2. \tag{6.02}$$

The upper signs are taken if $j = 1$, the lower signs if $j = 2$, and the coefficients are given by

$$(6.03)\quad \begin{aligned} a_{s,1} &= \{m^2 - (k - \tfrac{1}{2})^2\}\{m^2 - (k - \tfrac{3}{2})^2\} \\ &\qquad \cdots \{m^2 - (k - s + \tfrac{1}{2})^2\}/s!, \\ a_{s,2} &= \{(k + \tfrac{1}{2})^2 - m^2\}\{(k + \tfrac{3}{2})^2 - m^2\} \\ &\qquad \cdots \{(k + s - \tfrac{1}{2})^2 - m^2\}/s!. \end{aligned}$$

The functions $F_s(z)$ and $G_s(z)$, defined in §1, vanish identically in the present case when $s \geqq 0$ and $s \geqq 3$, respectively; in consequence the numbers $r_{n+1,j}$ defined by (5.06) also vanish. Thus we find that

$$(6.04)\quad \begin{aligned} &| \epsilon_{n,j}(z)|, \beta_j^{-1} | \epsilon'_{n,j}(z)| \\ &\qquad \leqq 2\alpha_j | z^{\pm k} e^{\mp\frac{1}{2}z} | \mathcal{V}_{\mathcal{P}_j}(a_{n,j}t^{-n}) \exp\{2\alpha_j\delta_j\mathcal{V}_{\mathcal{P}_j}(t^{-1})\}, \end{aligned}$$

where

$$(6.05)\quad \begin{aligned} \alpha_j &= \max_{t \in \mathcal{P}_j} \left| 1 - \frac{2k}{t} \right|^{-1}, \\ \beta_j &= \max_{t \in \mathcal{P}_j} \left[\left| \frac{1}{4} - \frac{k}{2t} + \frac{k}{2t^2}\left(1 - \frac{2k}{t}\right)^{-1} \right| \right. \\ &\qquad \left. + \left| \frac{1}{4} - \frac{k}{2t} - \frac{k}{2t^2}\left(1 - \frac{2k}{t}\right)^{-1} \right| \right], \\ \delta_j &= \max_{t \in \mathcal{P}_j} \left| k^2 - m^2 + \frac{1}{4} + \frac{2k}{t}\left(1 - \frac{k}{2t}\right)\left(1 - \frac{2k}{t}\right)^{-2} \right|, \end{aligned}$$

and $\mathcal{P}_j$ denotes a path linking z to an arbitrary fixed point $z_j(\neq 0)$ of the complex plane. The regions of validity $\mathbf{H}_j(z_j)$ of the inequalities (6.04) are determined by the condition that $\mathcal{P}_j$ must be such that if t traverses it from z to z_j, then $\operatorname{Re}(t - 2k \log t)$ is monotonic nondecreasing ($j = 1$) or monotonic nonincreasing ($j = 2$). We now complete the definition of the solutions $w_{n,j}(z)$ by choosing $z_1 = +\infty$, $z_2 = -\infty$. A precise investigation of the extent of $\mathbf{H}_1(\infty)$ is given in the next section. At this stage we merely observe that since $t - 2k \log t \sim t$ as $|t| \to \infty$, it follows that in the neighborhood of infinity the regions $\mathbf{H}_1(\infty)$ and $\mathbf{H}_2(-\infty)$ coincide roughly with the sectors $|\arg z| < 3\pi/2$ and $|\arg(-z)| < 3\pi/2$, respectively.

The fundamental solution of (6.01) for large $|z|$ is the Whittaker function $W_{k,m}(z)$, defined by the condition [10, Chap. 16]

$$(6.06)\quad W_{k,m}(z) \sim z^k e^{-\frac{1}{2}z} \quad \text{as} \quad z \to +\infty.$$

A full investigation of this single solution will suffice for the complete description of the asymptotic solutions of (6.01), because all other solutions can be expressed satisfactorily either in terms of the linearly independent pair $W_{k,m}(z)$ and $W_{-k,m}(ze^{\pi i})$, or in terms of the pair $W_{k,m}(z)$ and $W_{-k,m}(ze^{-\pi i})$. From (6.02), (6.04), and (6.06) we deduce that

$$W_{k,m}(z) = w_{n,1}(z); \tag{6.07}$$

cf. [11, p. 766].

The problem we now face is the evaluation of the variations of the functions t^{-1} and t^{-n} occurring in (6.04). For each admissible z there exists an infinity of possible paths $\mathcal{P}_1$ linking z to $+\infty$ and fulfilling the condition that Re $(t - 2k \log t)$ is monotonic: we naturally try to select one which minimizes the variations. When $k = 0$, (6.01) reduces to the differential equation satisfied by $z^{\frac{1}{2}}\mathcal{C}_m(\frac{1}{2}iz)$, where $\mathcal{C}_m$ is the general cylinder function of order m. For this case the minimization of the variations has been fully investigated in [1]. The extension to nonzero k given in the next two sections follows similar lines, but is more difficult.

7. Evaluation of the variations. We may now discard the suffix j from $\mathcal{P}_j$ and other symbols. Let τ be the arc-parameter of $\mathcal{P}$ measured from z. Then

$$\frac{d\,(\text{Re}\, t)}{d\tau} = \cos\varphi, \tag{7.01}$$

where φ is the angle of slope of the path at t (Fig. 1). We also have

$$\left|\frac{d}{d\tau}\text{Re}\,(2k\log t)\right| \leqq \left|\frac{d}{d\tau}(2k\log t)\right| = \left|\frac{2k}{t}\right|. \tag{7.02}$$

Hence the condition that Re $(t - 2k \log t)$ be monotonic is fulfilled if

$$\cos\varphi \geqq \left|\frac{2k}{t}\right| \tag{7.03}$$

at all points of the path.

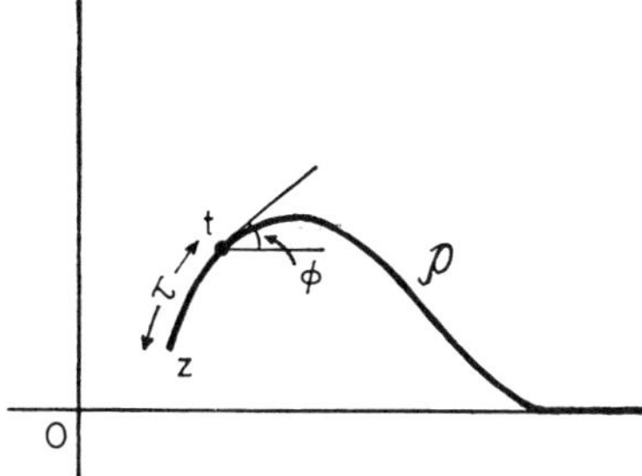

FIG. 1. *t-plane*

Condition (7.03) is sufficient but not necessary; nevertheless on account of its simplicity we shall confine our choice to paths on which it is fulfilled.

We consider in turn various subregions of the sector $|\arg z| < 3\pi/2$, and for convenience introduce the notations

$$\theta = \arg z, \qquad \kappa = 2\,|\,k\,|\,. \tag{7.04}$$

Region $\mathbf{R}_1$. This is the halfplane $\operatorname{Re} z \geqq \kappa$ (see Fig. 2).

As in [1, §6(i)] let $\mathcal{P}$ be taken to be the outward ray through z. Then since $|\,2k/t\,| \leqq \kappa/|\,z\,| \leqq \cos\theta = \cos\varphi$, it follows that (7.03) is satisfied on this path. Again, as in [1], we see that

$$\mathcal{V}_{\mathcal{P}}(t^{-n}) = |\,z\,|^{-n}, \qquad z \in \mathbf{R}_1, \tag{7.05}$$

and that no other path can yield a smaller variation.

Regions $\mathbf{R}_2$, $\bar{\mathbf{R}}_2$. These comprise the parts of the halfplane $\operatorname{Re} z < \kappa$ for which $|\,z\,| > \kappa$ when $\operatorname{Re} z \geqq 0$, and $|\operatorname{Im} z\,| \geqq \kappa$ when $\operatorname{Re} z < 0$. $\mathbf{R}_2$ is

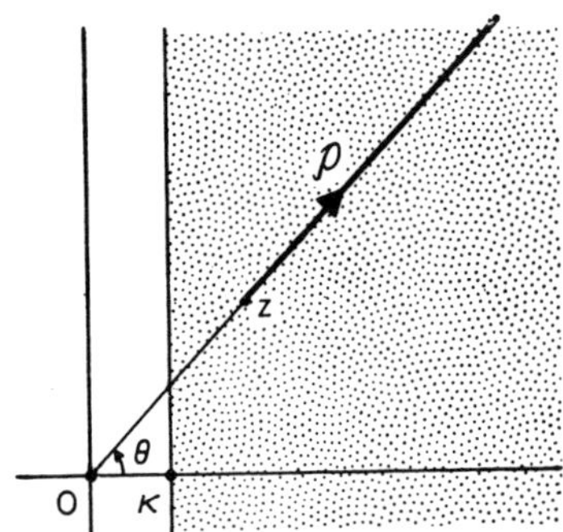

FIG. 2. *Region* $\mathbf{R}_1$

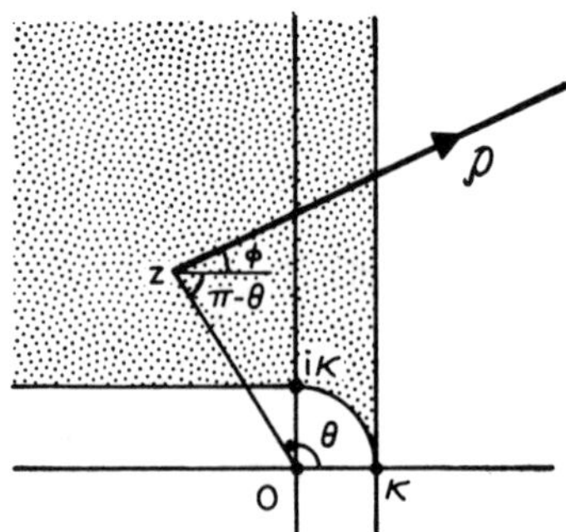

FIG. 3. *Region* $\mathbf{R}_2$

the part which lies above the real axis and is the shaded region of Fig. 3; $\bar{\mathbf{R}}_2$ is the complex conjugate of $\mathbf{R}_2$.

Consider the path of constant angle of slope φ, given by

$$\cos \varphi = \frac{\kappa}{|z|}. \tag{7.06}$$

This path clearly fulfils (7.03), provided that $|z|$ is the shortest distance from the origin to the path. Obviously this is the case when $0 \leqq \theta \leqq \pi/2$. When $\pi/2 < \theta \leqq \pi$ we require $(\pi - \theta) + \varphi \geqq \pi/2$, giving $\sin \theta \geqq \cos \varphi$, and therefore $|z| \sin \theta \geqq \kappa$, which in fact determines the left part of the boundary of $\mathbf{R}_2$.

With this choice we have

$$\begin{aligned} \mathcal{V}_{\mathcal{P}}(t^{-n}) &= \int_0^\infty \frac{n\, d\tau}{|z + \tau e^{i\varphi}|^{n+1}} \\ &= \int_0^\infty \frac{n\, d\tau}{||z| e^{i(\theta-\varphi)} + \tau|^{n+1}} \leqq \int_0^\infty \frac{n\, d\tau}{(|z|^2 + \tau^2)^{\frac{1}{2}n+\frac{1}{2}}}, \end{aligned} \tag{7.07}$$

since, as we have just seen, $|\theta - \varphi| \leqq \pi/2$. Hence

$$\mathcal{V}_{\mathcal{P}}(t^{-n}) \leqq \frac{\chi(n)}{|z|^n}, \qquad z \in \mathbf{R}_2, \tag{7.08}$$

where

$$\chi(n) = \frac{\sqrt{\pi}\,\Gamma(\frac{1}{2}n + 1)}{\Gamma(\frac{1}{2}n + \frac{1}{2})}. \tag{7.09}$$

Clearly by symmetry (7.08) also holds in $\bar{\mathbf{R}}_2$.

That the right of (7.08) is satisfactorily close to the minimum possible variation can be seen from the fact that for the case $k = 0$ treated in [1, §6(ii)], the path of minimum variation *was* used and led to exactly the same bound $\chi(n)/|z|^n$.

Regions $\mathbf{R}_3$, $\bar{\mathbf{R}}_3$. Consider now the remaining part of the sector

$$\pi/2 < \arg z < 3\pi/2.$$

We again select a path of constant angle of slope φ ($\varphi < \pi/2$), which is clearly permissible provided that $\cos \varphi \geqq \kappa/|z_0|$, where z_0 is the foot of the perpendicular from O (Fig. 4). Thus we require $\operatorname{Im} z_0 \geqq \kappa$. Since z_0 is located on the semicircle on Oz as diameter, it follows that an admissible path of this kind exists only when the line $\operatorname{Im} t = \kappa$ intersects the semicircle, that is, when

$$|\tfrac{1}{2}z| + \operatorname{Im}(\tfrac{1}{2}z) \geqq \kappa, \tag{7.10}$$

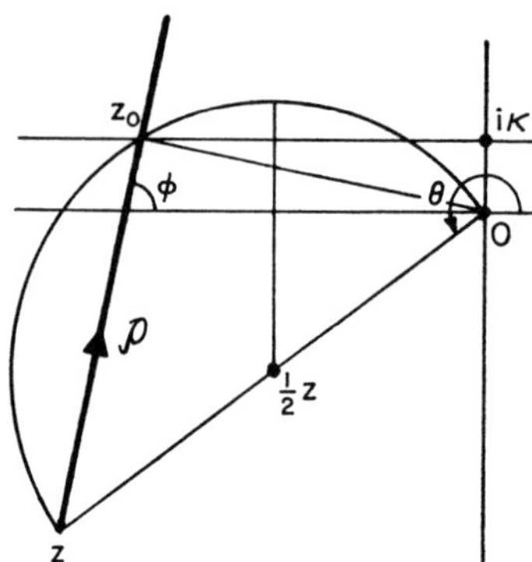

FIG. 4

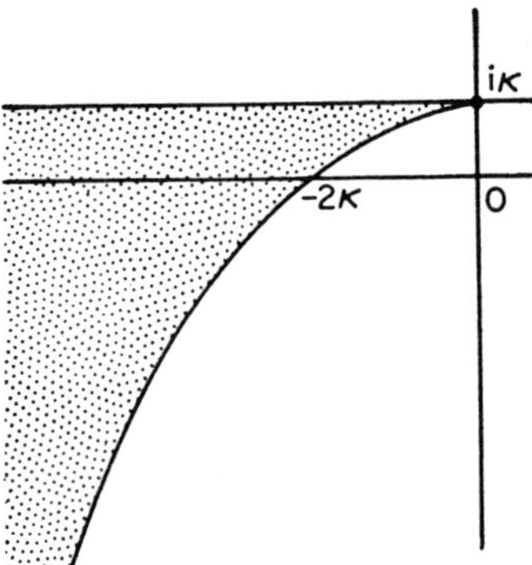

FIG. 5. *Region* $\mathbf{R}_3$

or in terms of Cartesian coordinates $z = x + iy$,

$$x^2 + 4\kappa y \geqq 4\kappa^2. \tag{7.11}$$

This determines the outside of a parabola, and we now define $\mathbf{R}_3$ to be the region in the left halfplane lying between this parabola and the line $\operatorname{Im} t = \kappa$. $\mathbf{R}_3$ is the shaded region of Fig. 5; $\bar{\mathbf{R}}_3$ is its conjugate.

The variation over this path is given by

$$\mathcal{V}_{\mathcal{P}}(t^{-n}) = \int_z^{z_0} \left| \frac{n\, dt}{t^{n+1}} \right| + \int_{z_0}^{z_0 + \infty e^{i\varphi}} \left| \frac{n\, dt}{t^{n+1}} \right|. \tag{7.12}$$

The second integral equals

$$\int_0^\infty \frac{n\, d\tau}{|z_0 + \tau e^{i\varphi}|^{n+1}} = \int_0^\infty \frac{n\, d\tau}{|i|z_0| + \tau|^{n+1}} = \frac{\chi(n)}{|z_0|^n}, \tag{7.13}$$

and the first integral is bounded by the same quantity. Therefore

$$\mathcal{V}_{\mathcal{P}}(t^{-n}) < \frac{2\chi(n)}{|z_0|^n}, \qquad z \in \mathbf{R}_3. \tag{7.14}$$

For a given z in $\mathbf{R}_3$, the point z_0 can lie anywhere on the arc of the semicircle of Fig. 4 which is cut off by the line $\operatorname{Im} t = \kappa$. To minimize the right side of (7.14), we take z_0 to be the leftmost point of this arc, as shown in Fig. 4. This yields

$$|z_0|^2 = \kappa y + \tfrac{1}{2}x^2 - x(\tfrac{1}{4}x^2 - \kappa^2 + \kappa y)^{1/2}, \tag{7.15}$$

the coordinate x being, of course, negative here.

By symmetry, (7.14) holds when $z \in \bar{\mathbf{R}}_3$, with κ replaced by $-\kappa$ on the right side of (7.15).

8. Expressions for the bounds; use of connection formulae. The bounds (7.05), (7.08), and (7.14) may be substituted on the right of (6.04) for $\mathcal{V}_{\mathcal{P}}(t^{-n})$ and $\mathcal{V}_{\mathcal{P}}(t^{-1})$. The other quantities α, β, and δ appearing in (6.04) can be evaluated without difficulty directly from their definitions (6.05). Little is lost, however, on replacing them by slightly weaker, but simpler, expressions, as follows.

Let

$$\sigma = \frac{\kappa}{|z|} = \left|\frac{2k}{z}\right|, \quad \sigma_0 = \frac{\kappa}{|z_0|} = \left|\frac{2k}{z_0}\right|, \tag{8.01}$$

where $|z_0|$ is given by (7.15). Clearly in the regions considered in §7, we have σ, $\sigma_0 \leqq 1$; indeed, whenever the early terms in the expansion (6.02) are decreasing reasonably rapidly σ, and often also σ_0, will be quite small compared with unity. On the selected paths $\mathcal{P}$ of §7, we have

$$\left|\frac{2k}{t}\right| \leqq \sigma, \quad z \in \mathbf{R}_1 \cup \mathbf{R}_2 \cup \bar{\mathbf{R}}_2; \qquad \left|\frac{2k}{t}\right| \leqq \sigma_0, \quad z \in \mathbf{R}_3, \bar{\mathbf{R}}_3. \tag{8.02}$$

Therefore, in $\mathbf{R}_1$, $\mathbf{R}_2$, and $\bar{\mathbf{R}}_2$,

$$\begin{aligned} &\alpha \leqq (1-\sigma)^{-1}, \\ &\beta \leqq \tfrac{1}{2} + \tfrac{1}{2}\sigma + \tfrac{1}{2}\sigma(1-\sigma)^{-1}|z|^{-1}, \\ &\delta \leqq |k^2 - m^2 + \tfrac{1}{4}| + \sigma(1 + \tfrac{1}{4}\sigma)(1-\sigma)^{-2}. \end{aligned} \tag{8.03}$$

For $\mathbf{R}_3$ and $\bar{\mathbf{R}}_3$ we replace σ by σ_0 and $|z|^{-1}$ by $|z_0|^{-1}$.

We have now achieved a complete evaluation of the bounds (6.04) when $|\arg z| < 3\pi/2$ and z lies in the union of the regions $\mathbf{R}_1$, $\mathbf{R}_2$, $\bar{\mathbf{R}}_2$, $\mathbf{R}_3$, and $\bar{\mathbf{R}}_3$. In other phase ranges $W_{k,m}(z)$ may be computed by use of the connection formulae ([12, p. 27])

$$\begin{aligned} (-)^s W_{k,m}(ze^{\pm 2s\pi i}) = &-\frac{e^{\pm 2k\pi i}\sin(2sm\pi) + \sin\{(2s-2)m\pi\}}{\sin(2m\pi)} W_{k,m}(z) \\ &\mp \frac{\sin(2sm\pi)}{\sin(2m\pi)} \frac{2i\pi e^{\pm k\pi i}}{\Gamma(\frac{1}{2}+m-k)\Gamma(\frac{1}{2}-m-k)} W_{-k,m}(ze^{\pm\pi i}), \end{aligned} \tag{8.04}$$

in which s is an integer, positive or negative.

We may further observe that by using (8.04) we can avoid the direct use of (6.04) in those parts of $\mathbf{R}_3$ and $\bar{\mathbf{R}}_3$ which lie outside the phase range $|\arg z| \leqq \pi$. Indeed, it is desirable that we should do this, for the bound (7.14) can be quite large here compared with those applicable when $|\arg z| \leqq \pi$. This is merely another example of the gradual failure of an asymptotic expansion as the boundaries of its regions of validity are approached, and the restoration of the "lost" accuracy by the inclusion of exponentially small contributions obtained from connection formulae [1, §§2, 7].

The parts of $\mathbf{R}_3$ and $\bar{\mathbf{R}}_3$ of greatest interest accordingly belong to the sector $|\arg z| \leqq \pi$. In these subregions it is possible to improve upon the bound (7.14), and at the same time avoid the computation of (7.15), as follows.

Let $\mathbf{R}_4$ denote that part of the strip $0 \leqq \operatorname{Im} z \leqq \kappa$ which lies outside and to the left of the circle $|z| = 2\kappa$, and $\bar{\mathbf{R}}_4$ be the conjugate of $\mathbf{R}_4$ (Fig. 6). With a given value of $|z|$ and arbitrary choice of $\arg z$ in $\mathbf{R}_4$, the minimum value of $|z_0|$ occurs when z lies on the negative real axis (Fig. 7): this can be established by considering the intersections of the semicircles on Oz as diameter with the line $\operatorname{Im} z = \kappa$. From (7.15) we find that this minimum is given by

$$\begin{aligned} |z_0|^2 &= \tfrac{1}{2}|z|^2 + |z|(\tfrac{1}{4}|z|^2 - \kappa^2)^{1/2} \\ &= |z|^2\{\tfrac{1}{2} + \tfrac{1}{2}(1 - 4\sigma^2)^{1/2}\}, \end{aligned} \tag{8.05}$$

where σ is defined by (8.01). From this result and (7.13) we obtain the bound

$$\chi(n)\{\tfrac{1}{2} + \tfrac{1}{2}(1 - 4\sigma^2)^{1/2}\}^{-n/2}|z|^{-n} \tag{8.06}$$

for the second integral on the right of (7.12). For the other integral, we

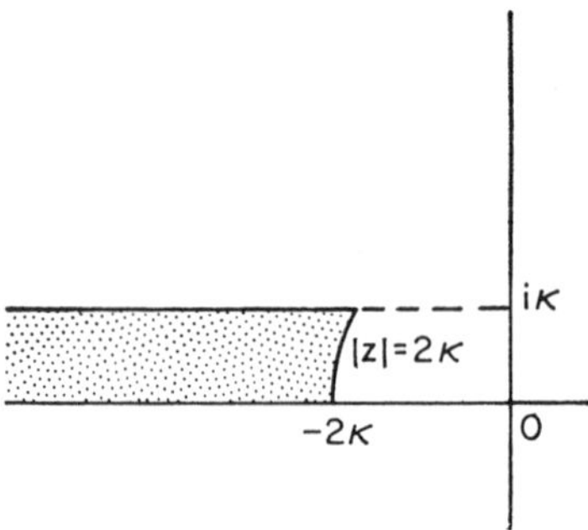

FIG. 6. *Region* $\mathbf{R}_4$

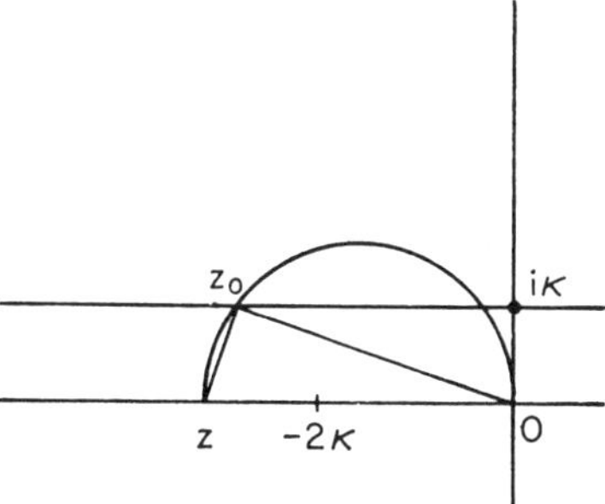

FIG. 7. *Minimum* $|z_0|$ *position*

have clearly

$$(8.07) \qquad \mathcal{V}_{z,z_0}(t^{-n}) \leqq \frac{n\,|z - z_0|}{|z_0|^{n+1}}.$$

Again, for a given value of $|z|$ the maximum of $|z - z_0|$ occurs when z lies on the negative real axis (Fig. 7), and in this case

$$(8.08) \qquad |z - z_0|^2 = |z|^2 - |z_0|^2 = |z|^2\{\tfrac{1}{2} - \tfrac{1}{2}(1 - 4\sigma^2)^{1/2}\}.$$

Substituting the last result in (8.07) and combining with (8.06), we obtain the desired bound

$$(8.09) \quad \mathcal{V}_{\mathcal{P}}(t^{-n}) \leqq \left[\frac{\chi(n)}{\{\frac{1}{2} + \frac{1}{2}(1 - 4\sigma^2)^{\frac{1}{2}}\}^{\frac{1}{2}n}} + \frac{n\sigma}{\{\frac{1}{2} + \frac{1}{2}(1 - 4\sigma^2)^{\frac{1}{2}}\}^{\frac{1}{2}n+1}}\right]\frac{1}{|z|^n}, \; z \in \mathbf{R}_4.$$

This result obviously also holds in $\bar{\mathbf{R}}_4$.

A cruder, but simpler, result is obtained on setting σ equal to its maximum value of $\frac{1}{2}$ in $\mathbf{R}_4$ and $\bar{\mathbf{R}}_4$:

$$(8.10) \qquad \mathcal{V}_{\mathcal{P}}(t^{-n}) \leqq \{\chi(n) + n\}2^{n/2}\,|z|^{-n}.$$

Often, however, σ will be quite small, in which case the right of (8.09) becomes approximately $\chi(n)|z|^{-n}$, in agreement with the bound applicable in $\mathbf{R}_2$ and $\bar{\mathbf{R}}_2$ (see (7.08)).

9. Collected results for Whittaker functions. Let k, m be any numbers, real or complex. Then

$$(9.01) \qquad W_{k,m}(z) = z^k e^{-\frac{1}{2}z}\left(\sum_{s=0}^{n-1}\frac{a_s}{z^s}\right) + \epsilon_n(z),$$

where n is an arbitrary positive integer, $a_s \equiv a_{s,1}$ is given by the first of (6.03), and

$$(9.02) \qquad |\epsilon_n(z)|,\, \beta^{-1}\,|\epsilon_n'(z)| \leqq 2\alpha\,|z^k e^{-\frac{1}{2}z}a_n|\,\mathcal{V}_{\mathcal{P}}(t^{-n})\exp\{2\alpha\delta\mathcal{V}_{\mathcal{P}}(t^{-1})\}.$$

These inequalities hold when $|\arg z| \leqq \pi$ and z lies in the union of the regions $\mathbf{R}_1$, $\mathbf{R}_2$, $\bar{\mathbf{R}}_2$, $\mathbf{R}_4$, and $\bar{\mathbf{R}}_4$, where $\mathbf{R}_1$, $\mathbf{R}_2$, and $\mathbf{R}_4$ are respectively the shaded regions of Figs. 2, 3 and 6, and $\bar{\mathbf{R}}_2$, $\bar{\mathbf{R}}_4$ are the conjugates of $\mathbf{R}_2$, $\mathbf{R}_4$. In these diagrams $\kappa = 2|k|$.

Upper bounds for the variation $\mathcal{V}_{\mathcal{P}}(t^{-n})$ appearing in (9.02) are

$$(9.03)\qquad \begin{aligned} &|z|^{-n}, && z \in \mathbf{R}_1, \\ &\chi(n)|z|^{-n}, && z \in \mathbf{R}_2, \bar{\mathbf{R}}_2, \\ &\{\chi(n) + \sigma v^2 n\} v^n |z|^{-n}, && z \in \mathbf{R}_4, \bar{\mathbf{R}}_4, \end{aligned}$$

where $\chi(n) = \sqrt{\pi}\,\Gamma(\frac{1}{2}n + 1)/\Gamma(\frac{1}{2}n + \frac{1}{2})$ and is tabulated in [1], and

$$(9.04)\qquad \sigma = \left|\frac{2k}{z}\right|, \qquad v = \{\tfrac{1}{2} + \tfrac{1}{2}(1 - 4\sigma^2)^{1/2}\}^{-1/2}.$$

For $\mathcal{V}_{\mathcal{P}}(t^{-1})$, set $n = 1$ in (9.03). Bounds for α, β, and δ are supplied by (8.03) when $z \in \mathbf{R}_1 \cup \mathbf{R}_2 \cup \bar{\mathbf{R}}_2$. For $z \in \mathbf{R}_4$ or $\bar{\mathbf{R}}_4$, replace σ by $v\sigma$ and $|z|^{-1}$ by $v|z|^{-1}$ in (8.03).

For other ranges of $\arg z$, appropriate asymptotic expansions, complete with error bounds, can be obtained by substituting for $W_{k,m}(z)$ and $W_{-k,m}(ze^{\pm\pi i})$ in (8.04) by means of (9.01) and (9.02). In general, these expansions involve both positive and negative exponentials.

Bounds for the error term $\epsilon_n(z)$ of (9.01) and its derivative have also been evaluated in §§7, 8 in the case when z lies in regions $\mathbf{R}_3$ and $\bar{\mathbf{R}}_3$ which extend outside $|\arg z| \leqq \pi$ and include $\mathbf{R}_4$ and $\bar{\mathbf{R}}_4$ as subregions (Fig. 5). For computational purposes, however, these extensions should be avoided and the connection formula (8.04) used instead.

10. Summary. In §§1–5 strict error bounds have been established for the well-known asymptotic expansions of solutions of the general linear second-order ordinary differential equation in the neighborhood of an irregular singularity of rank one. These bounds are closely related to the variation of the first neglected term of the truncated series, and in this respect resemble bounds obtained previously for asymptotic solutions of second-order differential equations containing a large parameter.

In §§6–9 the theory is applied to the Whittaker form of the differential equation satisfied by the confluent hypergeometric functions. Error bounds are obtained, for the first time, for the asymptotic expansions of these functions for large complex values of the independent variable z, and unrestricted values of the parameters k and m. The bounds are in a form immediately suitable for numerical evaluation. By incorporating them in automatic computing routines for evaluating the Whittaker functions, any uncertainty concerning the errors associated with the use of the asymptotic expansions can be eliminated completely. The bounds also provide theoretical insight into the accuracy of the expansions in various circumstances.

Much of the analysis of §§6–9 is applicable to all second-order differential equations having an irregular singularity of rank one, and not merely the Whittaker equation.

REFERENCES

[1] F. W. J. OLVER, *Error bounds for asymptotic expansions, with an application to cylinder functions of large argument*, Asymptotic Solutions of Differential Equations and their Applications, C. H. Wilcox, ed., John Wiley, New York, 1964.

[2] A. ERDÉLYI, *Asymptotic Expansions*, Dover, New York, 1956.

[3] J. HORN, *Untersuchung der Integrale einer linearen Differentialgleichung in der Umgebung einer Unbestimmtheitsstelle vermittelst successiver Annäherungen*, Arch. Math. Phys., 4 (1903), pp. 213–230.

[4] E. L. INCE, *Ordinary Differential Equations*, Dover, New York, 1944.

[5] W. WASOW, *Introduction to the asymptotic theory of ordinary linear differential equations*, Working paper, National Bureau of Standards, Washington, 1953.

[6] H. JEFFREYS, *Asymptotic Approximations*, Oxford University Press, Oxford, 1962.

[7] F. W. J. OLVER, *Error bounds for the Liouville-Green (or WKB) approximation*, Proc. Cambridge Philos. Soc., 57 (1961), pp. 790–810.

[8] ——, *Uniform asymptotic expansions of solutions of linear second-order differential equations for large values of a parameter*, Philos. Trans. Roy. Soc. London Ser. A., 250 (1958), pp. 479–517.

[9] K. BOCHENEK, *Les solutions asymptotiques des équations différentielles et l'évaluation des restes*, Proceedings of a Symposium on the Numerical Treatment of Ordinary Differential Equations, Integral and Integro-differential Equations, Birkhäuser, Basel, 1960.

[10] E. T. WHITTAKER AND G. N. WATSON, *Modern Analysis*, 4th ed., Cambridge University Press, Cambridge, 1927.

[11] F. W. J. OLVER, *Error bounds for first approximations in turning-point problems*, J. Soc. Indust. Appl. Math., 11 (1963), pp. 748–772.

[12] H. BUCHHOLZ, *Die konfluente hypergeometrische Funktion*, Springer, Berlin, 1953.

J. SIAM Numer. Anal.
Ser. B, Vol. 2, No. 2

ERROR BOUNDS FOR ASYMPTOTIC SOLUTIONS OF SECOND-ORDER DIFFERENTIAL EQUATIONS HAVING AN IRREGULAR SINGULARITY OF ARBITRARY RANK*

F. W. J. OLVER† AND F. STENGER‡

1. Introduction. Recently [1], error bounds have been derived for asymptotic expansions of solutions of the differential equation

$$\frac{d^2w}{dz^2} + f(z)\frac{dw}{dz} + g(z)w = 0 \tag{1.01}$$

in a region containing an irregular singularity of rank one. The purpose of the present paper is to give similar bounds for the corresponding expansions in a region containing a singularity of higher rank. We again assume that $f(z)$ and $g(z)$ are either continuous** on a given real interval **I**, or regular (holomorphic) in a given complex domain **D**, and we now suppose that

$$f(z) = z^k\left\{\sum_{s=0}^{n-1}\frac{f_s}{z^s} + \frac{F_n(z)}{z^n}\right\}, \quad g(z) = z^{2k}\left\{\sum_{s=0}^{n-1}\frac{g_s}{z^s} + \frac{G_n(z)}{z^n}\right\}, \tag{1.02}$$

where $F_n(z)$ and $G_n(z)$ are bounded functions of z on **I** or in **D**, for each $n = 0, 1, \cdots$. Here k is an arbitrary nonnegative integer, and at least one of the coefficients f_0, g_0 and g_1 is nonzero (otherwise k could be replaced by $k - 1$). The number $k + 1$ is called the *rank* of the singularity at infinity [2, §3.1].

2. Formal series solutions. Generalized asymptotic expansions of the solutions of (1.01), complete with error bounds and valid for arbitrary values of k (including nonintegral k), can be obtained by combining the theorem of [3, §5], with the Liouville transformation [2, §4.1]. The precise form of the power-series type of expansion with which we are now concerned may be inferred by re-expanding these more general series. Except in this respect however, our investigation does not depend on the theory of generalized expansions.

Setting

$$y = \exp\left\{\frac{1}{2}\int f(z)\,dz\right\}w,$$

* Received by the editors August 21, 1964, and in revised form November 12, 1964. This work was supported by the U. S. Army Research Office, Durham, North Carolina, under Project 4238-M, and by a graduate fellowship from the Province of Alberta.

† National Bureau of Standards, Washington, D. C.
‡ University of Alberta, Edmonton, Alberta.

** Although $f'(z)$ appears in the next section, the analysis there is purely formal, and the existence of $f'(z)$ is not required in the results of §3.

we find that (1.01) becomes

$$\frac{d^2y}{dz^2} = \varphi(z)y, \tag{2.01}$$

where

$$\varphi(z) = \tfrac{1}{4}f^2(z) + \tfrac{1}{2}f'(z) - g(z). \tag{2.02}$$

Hence for large $|z|$, we have

$$w = \{\varphi(z)\}^{-1/4} \exp\left[\pm \int \{\varphi(z)\}^{1/2}\, dz - \frac{1}{2}\int f(z)\, dz\right]\{1 + o(1)\}. \tag{2.03}$$

Let $\varphi_0, \varphi_1, \cdots$ denote the coefficients in the expansion

$$\{\varphi(z)\}^{1/2} = z^k \sum_{s=0}^{\infty} \frac{\varphi_s}{z^s} \tag{2.04}$$

obtained on formally substituting (1.02) in (2.02). For example,

$$\varphi_0 = (\tfrac{1}{4}f_0^2 - g_0)^{1/2}, \tag{2.05}$$

where for definiteness we take the principal branch of the right side, and assume for the moment that $\varphi_0 \neq 0$. Then (2.03) indicates that solutions of (1.01) exist of the form

$$w_j = e^{\xi_j(z)} \sum_{s=0}^{\infty} \frac{a_{s,j}}{z^s}, \qquad j = 1, 2, \tag{2.06}$$

where

$$\xi_j(z) = z^{k+1} \sum_{s=0}^{k} \frac{(\pm\varphi_s - \frac{1}{2}f_s)}{(k + 1 - s)z^s} + (\pm\varphi_{k+1} - \tfrac{1}{2}f_{k+1} - \tfrac{1}{2}k) \log z, \tag{2.07}$$

the upper signs being taken here (and elsewhere) when $j = 1$, and the lower signs when $j = 2$.

We determine the coefficients in (2.06) directly by substitution in (1.01). This yields immediately the formal identity

$$u_j(z) \sum_{s=0}^{\infty} \frac{a_{s,j}}{z^s} - \frac{2v_j(z)}{z} \sum_{s=0}^{\infty} \frac{sa_{s,j}}{z^s} + \frac{1}{z^2} \sum_{s=0}^{\infty} \frac{s(s + 1)a_{s,j}}{z^s} = 0, \tag{2.08}$$

in which

$$\begin{aligned} u_j(z) &= \xi_j''(z) + \xi_j'^2(z) + f(z)\xi_j'(z) + g(z), \\ v_j(z) &= \xi_j'(z) + \tfrac{1}{2}f(z). \end{aligned} \tag{2.09}$$

These two functions can be expanded in descending powers of z by means of (1.02), and we find that the coefficients of $z^{2k}, z^{2k-1}, \cdots, z^{k-1}$ in the expansion of $u_j(z)$ vanish as a consequence of the identity

$$(2.10)\quad \left(\sum_{s=0}^{\infty} \frac{\varphi_s - \frac{1}{2}f_s}{z^s}\right)\left(\sum_{s=0}^{\infty} \frac{\varphi_s + \frac{1}{2}f_s}{z^s}\right) + \sum_{s=0}^{\infty} \frac{g_s}{z^s} = \frac{1}{2z^{k+1}} \sum_{s=0}^{\infty} \frac{(k-s)f_s}{z^s}$$

obtained from (1.02), (2.02), and (2.04). Thus we have

$$(2.11)\qquad u_j(z) \sim z^{k-1} \sum_{s=1}^{\infty} \frac{u_{s,j}}{z^s}, \quad v_j(z) \sim z^k \sum_{s=0}^{\infty} \frac{v_{s,j}}{z^s},$$

where the coefficients $u_{s,j}$ and $v_{s,j}$ are calculable in terms of f_s and g_s; in particular,

$$(2.12)\qquad \begin{aligned} v_{s,j} &= \pm\varphi_s \quad \text{if} \quad s < k+1, \\ v_{k+1,j} &= \pm\varphi_{k+1} - \tfrac{1}{2}k, \\ v_{s,j} &= \tfrac{1}{2}f_s \quad \text{if} \quad s > k+1. \end{aligned}$$

Equating coefficients in (2.08), we find that

$$(2.13)\qquad \begin{aligned} 2sv_{0,j}a_{s,j} = \sum_{m=1}^{s} \{u_{m,j} &- 2(s-m)v_{m,j}\}a_{s-m,j} \\ &+ (s-k-1)(s-k)a_{s-k-1,j}, \end{aligned}$$

the last term on the right being absent when $s < k + 1$. For arbitrary given values of $a_{0,1}$ and $a_{0,2}$, (2.13) determines recurrently two sets of coefficients $a_{0,j}, a_{1,j}, a_{2,j}, \cdots, j = 1, 2$. The formal solutions (2.06) are now completely determined.

When $\varphi_0 = 0$, that is, when $f_0^2 = 4g_0$, the above construction fails and solutions of the form (2.06) cannot be found in general: this can be seen directly from (2.03). This exceptional case is characterized by the fact that as $\varphi_0 \to 0$ the *leading exponent*, that is, the coefficient of the highest power of z in the polynomial part of $\xi_j(z)$, tends to the same value for $j = 1$ and $j = 2$. The difficulty can be overcome by the following transformation of Fabry's type:

$$(2.14)\qquad W = \exp\left\{\frac{1}{2}\int\left(\sum_{s=0}^{k} f_s z^{k-s}\right) dz\right\} w, \quad t = z^{1/2}.$$

It is readily verified that at $t = \infty$ the transformed differential equation has either (i) a regular singularity, or (ii) an irregular singularity of rank not exceeding $2k + 1$ with unequal leading exponents, and therefore is amenable to the present analysis.

3. Theorem and proof. To discuss specific solutions of (1.01) we suppose that z_j is an arbitrary fixed point of **I** or **D**. Any branch may be selected at $z = z_j$ for the function $\log z$ occurring in the definition (2.07) of $\xi_j(z)$; then elsewhere the branch is determined by continuity.

THEOREM. *The differential equation* (1.01) *has solutions* $w_{n,j}(z)$, $j = 1, 2$, *depending on* z_j *and on an arbitrary positive integer* n, *such that*

$$w_{n,j}(z) = e^{\xi_j(z)}\left(\sum_{s=0}^{n-1}\frac{a_{s,j}}{z^s}\right) + \epsilon_{n,j}(z), \tag{3.01}$$

where

$$\begin{aligned}&|\epsilon_{n,j}(z)|,\ \beta_j^{-1}|z^{-k}\epsilon_{n,j}'(z)|\\ &\leqq 2\alpha_j|e^{\xi_j(z)}|\{\mathcal{V}_{\mathcal{P}_j}(a_{n,j}t^{-n}) + \mathcal{V}_{\mathcal{P}_j}(r_{n+1,j}t^{-n-1})\}\\ &\qquad\times \exp\{\alpha_j|\varphi_0|^{-1}(\beta_j\gamma_j + \delta_j)\mathcal{V}_{\mathcal{P}_j}(t^{-1})\}\end{aligned} \tag{3.02}$$

when $z \in \mathbf{H}_j(z_j)$.

The regions of validity $\mathbf{H}_j(z_j)$ of this result are defined as in [1, §5], except that the quantity $\operatorname{Re}\{(\lambda_2 - \lambda_1)t + (\mu_2 - \mu_1)\log t\}$ appearing in the monotonicity condition (iii) is replaced by $\operatorname{Re}\{\xi_2(t) - \xi_1(t)\}$. (As a consequence, the total range of $\arg z$ in $\mathbf{H}_j(z_j)$ cannot exceed $3\pi/(k+1)$ when $|z|$ is large.) Other quantities in (3.02) are defined by the following expressions, in which all maxima are taken with respect to $t \in \mathcal{P}_j$. Denote

$$\psi(z) \equiv \sum_{s=0}^{k+1}\frac{\varphi_s}{z^s}. \tag{3.03}$$

Then

$$\alpha_j = |\varphi_0|\max|\psi(t)|^{-1}, \qquad \gamma_j = \max|F_{k+2}(t)|, \tag{3.04}$$

$$\beta_j = \frac{1}{2}\max\left[|t|^{-k}\left|\xi_1'(t) - \frac{1}{2}\left\{\frac{\psi'(t)}{\psi(t)}\right\}\right| + |t|^{-k}\left|\xi_2'(t) - \frac{1}{2}\left\{\frac{\psi'(t)}{\psi(t)}\right\}\right|\right], \tag{3.05}$$

$$\begin{aligned}\delta_j = \max\Bigg|&t^{-k}\xi_j'(t)F_{k+2}(t) - t^{2-k}u_j(t) \pm t^2\psi'(t) - \frac{1}{2}kt^{1-k}\left\{\frac{\psi'(t)}{\psi(t)}\right\}\\ &+ t^{2-k}\{\psi(t)\}^{-2}\left\{\frac{1}{2}\psi(t)\psi''(t) - \frac{3}{4}\psi'^2(t)\right\}\Bigg|,\end{aligned} \tag{3.06}$$

the upper or lower sign again being taken according as $j = 1$ or 2. Lastly,

$$\begin{aligned}&2(n+1)|\varphi_0|r_{n+1,j}\\ &= \max\left|\sum_{s=0}^{n-1}a_{s,j}\{U_{n+1-s,j}(t) - 2sV_{n+1-s,j}(t)\} + \sum_{s=n-k}^{n-1}\frac{s(s+1)a_{s,j}}{t^{s+k-n}}\right|,\end{aligned} \tag{3.07}$$

where $U_{n,j}(z)$ and $V_{n,j}(z)$ are given by

$$\frac{u_j(z)}{z^{k-1}} = \sum_{s=1}^{n-1}\frac{u_{s,j}}{z^s} + \frac{U_{n,j}(z)}{z^n}, \qquad \frac{v_j(z)}{z^k} = \sum_{s=0}^{n-1}\frac{v_{s,j}}{z^s} + \frac{V_{n,j}(z)}{z^n}, \tag{3.08}$$

and are bounded functions of z in consequence of the assumptions of §1.

The proof of this theorem is similar to that of the theorem of [1], and only the principal steps need be mentioned. The inhomogeneous differential equation satisfied by $\epsilon_{n,j}(z)$ is found to be

$$\epsilon''_{n,j}(z) + f(z)\epsilon'_{n,j}(z) + g(z)\epsilon_{n,j}(z) = -z^k e^{\xi_j(z)} R_{n,j}(z), \tag{3.09}$$

where

$$|R_{n,j}(z)| \leqq 2|\varphi_0|\left\{\frac{n|a_{n,j}|}{|z|^{n+1}} + \frac{(n+1)r_{n+1,j}}{|z|^{n+2}}\right\}. \tag{3.10}$$

The coefficients on the left side of (3.09) are matched, through the first $k+2$ terms of their expansions in powers of $1/z$, with the corresponding coefficients in [1, (3.09)] by choosing

$$q(z) = z^k \sum_{s=0}^{k+1} \frac{f_s}{z^s}, \qquad \{p(z)\}^{1/2} = z^k \psi(z), \tag{3.11}$$

where $\psi(z)$ is given by (3.03). Equation (3.09) is rearranged as

$$\begin{aligned} &\epsilon''_{n,j}(z) + q(z)\epsilon'_{n,j}(z) + l(z)\epsilon_{n,j}(z) \\ &\quad = \{q(z) - f(z)\}\epsilon'_{n,j}(z) + \{l(z) - g(z)\}\epsilon_{n,j}(z) - z^k e^{\xi_j(z)} R_{n,j}(z), \end{aligned} \tag{3.12}$$

where

$$l = \tfrac{1}{2}q' + \tfrac{1}{4}q^2 - p - p^{1/4}(p^{-1/4})''. \tag{3.13}$$

The complementary functions of the left side of (3.12) are

$$Y_1(z) = \{\psi(z)\}^{-1/2} e^{\xi_1(z)} \quad \text{and} \quad Y_2(z) = \{\psi(z)\}^{-1/2} e^{\xi_2(z)}. \tag{3.14}$$

Accordingly, an integral equation equivalent to (3.12) is

$$\begin{aligned} \epsilon_{n,j}(z) = \int_{z_j}^{z} K(z,t)[&\{q(t) - f(t)\}\epsilon'_{n,j}(t) \\ &+ \{l(t) - g(t)\}\epsilon_{n,j}(t) - t^k e^{\xi_j(t)} R_{n,j}(t)]\, dt, \end{aligned} \tag{3.15}$$

where

$$K(z,t) = \frac{Y_1(z)Y_2(t) - Y_2(z)Y_1(t)}{Y_1'(t)Y_2(t) - Y_2'(t)Y_1(t)}. \tag{3.16}$$

The following inequalities hold on $\mathcal{P}_j$:

$$|K(z,t)| \leqq \alpha_j \left|\frac{e^{\xi_j(z)-\xi_j(t)}}{\varphi_0 t^k}\right|, \qquad \left|\frac{\partial K(z,t)}{\partial z}\right| \leqq \alpha_j \beta_j \left|\frac{z^k e^{\xi_j(z)-\xi_j(t)}}{\varphi_0 t^k}\right|, \tag{3.17}$$

$$|q(t) - f(t)| \leqq \gamma_j |t|^{-2}, \qquad |l(t) - g(t)| \leqq \delta_j |t^{k-2}|. \tag{3.18}$$

The bounds (3.02) may now be established by solving (3.15) by the standard procedure of successive approximation.

REFERENCES

[1] F. W. J. Olver, *On the asymptotic solutions of second-order differential equations having an irregular singularity of rank one, with an application to Whittaker functions*, this Journal, 2 (1965), pp. 225–243.

[2] A. Erdélyi, *Asymptotic Expansions*, Dover, New York, 1956.

[3] F. W. J. Olver, *Error bounds for asymptotic expansions, with an application to cylinder functions of large argument*, Asymptotic Solutions of Differential Equations and their Applications, C. H. Wilcox, ed., John Wiley, New York, 1964.

JOURNAL OF RESEARCH of the National Bureau of Standards—B. Mathematics and Mathematical Physics
Vol. 69B, No. 4, October–December 1965

Error Analysis of Phase-Integral Methods. I. General Theory for Simple Turning Points

F. W. J. Olver

(July 1, 1965)

An approximate general solution of the differential equation

$$d^2w/dz^2=f(z)w$$

is

$$w\approx Af^{-1/4}\exp\left(\int f^{1/2}dz\right)+Bf^{-1/4}\exp\left(-\int f^{1/2}dz\right).$$

To represent a particular solution throughout a given complex domain, differing pairs of values of the arbitrary constants A and B are generally needed in various subdomains. Strict error bounds are established in this paper for the coefficients in the linear equations connecting the pairs of values of A and B, in the case in which the z-domain under consideration is unbounded and contains a simple zero and no singularities of $f(z)$.

The results are used to place the "WKBJ" method on a firm mathematical foundation.

Key Words: Differential equations, asymptotics, phase-integral, WKBJ method, turning point, connection formulas, error bounds.

1. Introduction

A form of solution of the differential equation

$$\frac{d^2w}{dz^2}=f(z)w, \tag{1.01}$$

which is frequently used in applied mathematical problems, is given by

$$w\approx Af^{-1/4}\exp\left(\int f^{1/2}dz\right)+Bf^{-1/4}\exp\left(-\int f^{1/2}dz\right). \tag{1.02}$$

in which A, B are arbitrary constants. This is commonly referred to as the *Liouville-Green approximation*, the *WKB* or *WKBJ approximation*, or the *phase-integral solution*. A necessary condition for the applicability of the approximation is that $f^{-1/4}$ be a bounded, slowly varying function of z [1].[1] This condition automatically excludes from the region of validity most kinds of singularity of f, and also zeros of f, the so-called turning or transition points of the differential equation.

A further condition concerns the monotonicity of the function Re $\int f^{1/2}dz$ (see [1], sec. 8, and sec. 4, below). This requirement has the effect of dividing the z-region under consideration, **D**, say, into a number of subregions, in each of which a different combination of values of A and B in (1.02) has to be used in order to represent the same solution of (1.01). This is, of course, an example of the Stokes phenomenon. The linear equations linking the values of A and B in one subregion with those in another are called the *connection formulas* or *Stokes equations*. The purpose of this paper is to determine the connection formulas in the case in which **D** contains no singularity of $f(z)$, and just one simple turning point, z_0, say.

[1] Figures in brackets indicate the literature references at the end of this paper.

Two principal methods are available for attacking this problem. The first, applicable when $f(z)$ is real on the real axis, starts with the Liouville-Green approximation in real exponential form on one side of z_0, then traces its behavior along a semicircle in the complex plane to the real axis on the other side of z_0. The condition that the solution must be real when z is real suffices to determine the main features of the appropriate connection formula. This method was originated by Zwaan [2], and has been developed further by Kemble [3] and Furry [4].

The second method consists of using approximations to the solutions in terms of Airy functions (or equivalently Bessel functions of order one-third), the region of validity of which includes z_0. Away from z_0 the Airy functions are approximated by their asymptotic forms for large argument, leading immediately to the Liouville-Green forms. In the original version [5]–[7], the Airy-function representations were constructed by replacing $f(z)$ by a linear approximation in the neighborhood of z_0; this is, perhaps, the true "WKBJ method." In more recent work [8]–[13] more powerful uniform approximations in terms of Airy functions have been used to derive the connection formulas.

Hitherto, the method of Zwaan has not been put on a sound mathematical footing: writing in 1934 Langer [14] commented ". . . an estimation of the relative error involved in the result is almost impossible . . .". The treatment of the Airy-function procedure has been more satisfactory. By supposing that $f(z)$ contains as multiplicative factor (the square of) a large parameter u, say, it is shown in the references cited that the connection formulas are asymptotically correct as $|u| \to \infty$.

In the present paper we give rigorous analytical derivations of both methods, and supply explicit bounds for the errors in the connection formulas (thus overcoming one of Langer's objections to Zwaan's method). Since we are dealing with bounds there is no need for the explicit introduction of a parameter u in the theory: asymptotic properties in applications can be regarded as immediate consequences of the form of error bound.[2] We also dispense with the requirement, in Zwaan's method, that $f(z)$ be real on the axis.

The starting points of our investigation are the theories of error bounds for the Liouville-Green and uniform Airy approximations which have been developed by the writer in [1] and [15].

Since the present paper and its sequel [19] were written, there has appeared a monograph on the same subject by N. Fröman and P. O. Fröman [20]. Although the objective of these authors is essentially the same as that of the present writer, the methods used and results achieved are sufficiently different to warrant publication of this paper and [19] in substantially their original forms. Some comparisons with the results of [20] are given in section 11.

2. First Error Theorem

We begin with the differential equation

$$\frac{d^2w}{dz^2} = \{1 + f(z)\}w, \tag{2.01}$$

in which $f(z)$ is a regular (holomorphic) function in a simply-connected complex z-domain $\mathbf{D}$.

THEOREM 1. *Equation (2.01) has solutions $w_j(z)$ ($j = 1, 2$) regular in $\mathbf{D}$, and depending on arbitrary fixed points a_j of $\mathbf{D}$, such that*

$$w_j(z) = e^{\pm z}\{1 + \epsilon_j(z)\}, \tag{2.02}$$

where

$$|\epsilon_j(z)|, \ |\epsilon_j'(z)|, \ |\epsilon_j(z) \pm \epsilon_j'(z)| \leqslant \exp\{\mathscr{V}_{\mathscr{P}_j}(G)\} - 1 \tag{2.03}$$

when $z \in \mathbf{H}_j$.

[2] In the course of a valuable historical introduction to phase-integral methods, Heading ([13], p. 8) critizes other writers, particularly Budden [12], for failing to standardize a large parameter in their analysis. The theory of error bounds presented in this paper provides a broader basis for justifying the work of these writers.

In this result, the upper signs are to be taken when $j=1$, and the lower signs when $j=2$. The regions of validity $\mathbf{H}_j$ ($j=1, 2$) comprise the aggregate of those points z for which there exists a path $\mathscr{P}_j$ joining z with a_j and lying entirely in $\mathbf{D}$, with the properties:

(i) $\mathscr{P}_j$ consists of a finite number of Jordan arcs, each having a parametric equation of the form $t=t(\tau)$ with $t''(\tau)$ continuous and $t'(\tau)$ nonvanishing.

(ii) Re t is monotonic nonincreasing ($j=1$) or monotonic nondecreasing ($j=2$), as t traverses $\mathscr{P}_j$ from z to a_j.

Lastly, $\mathscr{V}_{\mathscr{P}_j}(G)$ denotes the variation along $\mathscr{P}_j$ of the function

$$G(z)=\int f(z)dz; \tag{2.04}$$

thus

$$\mathscr{V}_{\mathscr{P}_j}(G)=\int_z^{a_j} |f(t)dt|. \tag{2.05}$$

Theorem 1 is essentially a special case of Theorem 7 of [1]. We record the main steps of the proof, however, since they are needed in the derivation of further properties of the error terms in the next section.

For illustration, we take $j=1$. From (2.01) and (2.02) we find that

$$\epsilon_1''(z)+2\epsilon_1'(z)=f(z)\{1+\epsilon_1(z)\}. \tag{2.06}$$

Therefore $\epsilon_1(z)$ satisfies the integral equation

$$\epsilon_1(z)=\tfrac{1}{2}\int_{a_1}^{z}\{1-e^{2(t-z)}\}\{1+\epsilon_1(t)\}f(t)dt, \tag{2.07}$$

in which the path of integration is taken to be $\mathscr{P}_1$. The solution of this equation is constructed as a convergent series

$$\epsilon_1(z)=\sum_{n=1}^{\infty} l_{1,n}(z), \tag{2.08}$$

in which $l_{1,0}(z)=1$, and

$$l_{1,n}(z)=\tfrac{1}{2}\int_{a_1}^{z}\{1-e^{2(t-z)}\}f(t)l_{1,n-1}(t)dt \qquad (n\geqslant 1). \tag{2.09}$$

In consequence of Condition (ii) on $\mathscr{P}_1$ we have

$$|e^{2(t-z)}|\leqslant 1, \tag{2.10}$$

and by substituting this bound in (2.09) we readily establish by induction that

$$|l_{1,n}(z)|\leqslant\{\mathscr{V}_{a_1,z}(G)\}^n/n! \qquad (n\geqslant 0), \tag{2.11}$$

showing that the series (2.08) is uniformly convergent. In order that solutions of the integral equation (2.07) also represent solutions of the differential equation (2.06), some restrictions are needed on the path $\mathscr{P}_1$. Those given in (i) above suffice (compare [16], pp. 491–492).

The first of the inequalities (2.03) now follows by summation of (2.11). The other two are derived similarly from the differentiated form of (2.09).

3. Further Properties of the Error Terms

For conciseness, we shall refer to a path $\mathscr{P}_j$ which fulfills Condition (i) of section 2 and is such that Re t is monotonic as t traverses $\mathscr{P}_j$ as a *progressive* path. If the monotonicity is strict everywhere, we say that the path is *strictly progressive*. It may be noted, incidentally, that Condition (i) is quite weak, and can be fulfilled, for example, by any finite chain of straight lines and circular arcs.

The condition that $\mathbf{D}$ be simply connected, mentioned in the opening paragraph of section 2, merely ensures that each $w_j(z)$ is single-valued in $\mathbf{D}$. For a multiply connected domain, each branch of $w_j(z)$ satisfies (2.02) and (2.03) whenever a progressive path can be found in $\mathbf{D}$ joining z with a_j.

In section 2 we tacitly assumed that the points a_j are finite. Theorem 1 still holds, however, if $\mathbf{D}$ is unbounded and the a_j are points at infinity on progressive paths $\mathscr{L}_j$, provided that $\mathscr{P}_j$ coincides with $\mathscr{L}_j$ as a_j is approached and that the variations (2.05) converge. The convergence of the infinite integrals (2.09) is then an immediate consequence of the inequality (2.11).

A situation of importance in the later sections occurs when a_1 and a_2 are distinct points at infinity, and can be joined by a progressive path $\mathscr{P}$ lying in $\mathbf{D}$. From (2.03) it is seen that $\epsilon_1(z)$ and $\epsilon_1'(z)$ both vanish as $z \to a_1$ along $\mathscr{P}$; indeed we may regard these properties as the boundary conditions defining $w_1(z)$. What happens, however, to $\epsilon_1(z)$ and $\epsilon_1'(z)$ as $z \to a_2$?

THEOREM 2. *If* $|\text{Re}\, z| \to \infty$ *as* $z \to a_1$ *and as* $z \to a_2$, *then*

$$\left.\begin{aligned} &\epsilon_1(z) \to \textit{a constant } \epsilon_1(a_2), \textit{ say, and } \epsilon_1'(z) \to 0 \textit{ as } z \to a_2, \\ &\epsilon_2(z) \to \textit{a constant } \epsilon_2(a_1), \textit{ say, and } \epsilon_2'(z) \to 0 \textit{ as } z \to a_1, \end{aligned}\right\} \tag{3.01}$$

$$\epsilon_1(a_2) = \epsilon_2(a_1), \tag{3.02}$$

and

$$|\epsilon_1(a_2)| \leqslant \tfrac{1}{2}\,[\exp\{\mathscr{V}_{\mathscr{P}}(G)\} - 1]. \tag{3.03}$$

To prove these results, let

$$\alpha_{1,n}(z) = \int_{a_1}^{z} f(t) l_{1,n-1}(t)\,dt, \qquad \beta_{1,n}(z) = \int_{a_1}^{z} e^{2(t-z)} f(t) l_{1,n-1}(t)\,dt. \tag{3.04}$$

Then from (2.08) and (2.09), we have

$$\epsilon_1(z) = \tfrac{1}{2}\,\Sigma\alpha_{1,n}(z) - \tfrac{1}{2}\,\Sigma\beta_{1,n}(z), \qquad \epsilon_1'(z) = \Sigma\beta_{1,n}(z), \tag{3.05}$$

the summations extending in each case from $n = 1$ to $n = \infty$. Since by hypothesis $\int_{a_1}^{a_2} |f(t)dt|$ converges, it follows that for any given positive number δ which is less than the value of this doubly infinite integral there exists a point z' on $\mathscr{P}$ such that

$$\int_{z'}^{a_2} |f(t)dt| = \delta. \tag{3.06}$$

Let z be any point of $\mathscr{P}$ between z' and a_2 (fig. 1). Then by splitting the integration range in the second of (3.04) at z', and applying (2.10), (2.11), we find that

$$|\beta_{1,n}(z)| \leqslant |e^{2(z'-z)}|\,\frac{\{\mathscr{V}_{a_1,z'}(G)\}^n}{n!} + \frac{\{\mathscr{V}_{a_1,a_2}(G)\}^{n-1}}{(n-1)!}\,\delta. \tag{3.07}$$

FIGURE 1.

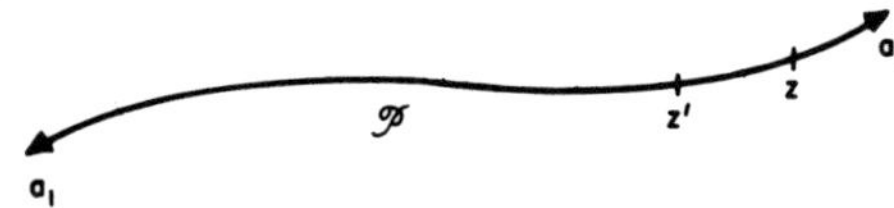

Therefore

$$|\Sigma\beta_{1,n}(z)| \leqslant |e^{2(z'-z)}| \left[\exp\{\mathscr{V}_{a_1,z'}(G)\}-1\right]+\delta\exp\{\mathscr{V}_{a_1,a_2}(G)\}. \tag{3.08}$$

With the assumed conditions, $e^{2(z'-z)}$ vanishes as $z\to a_2$. Hence $\Sigma\beta_{1,n}(z)$ also vanishes, that is, $\epsilon_1'(z)\to 0$.

Next, we have

$$|\alpha_{1,n}(z)-\alpha_{1,n}(z')| = \left|\int_{z'}^{z} f(t)l_{1,n-1}(t)dt\right| \leqslant \frac{\{\mathscr{V}_{a_1,z}(G)\}^n-\{\mathscr{V}_{a_1,z'}(G)\}^n}{n!}, \tag{3.09}$$

and so

$$|\Sigma\alpha_{1,n}(z)-\Sigma\alpha_{1,n}(z')| \leqslant \exp\{\mathscr{V}_{a_1,z}(G)\}-\exp\{\mathscr{V}_{a_1,z'}(G)\}. \tag{3.10}$$

The right side of this inequality vanishes as z and z' tend independently to a_2, therefore $\Sigma\alpha_{1,n}(z)$ tends to a constant limiting value. The proof of the first line of (3.01) is now complete. The second line follows by symmetry.

To prove (3.02), we construct the Wronskian from (2.02):

$$\mathscr{W}(w_2,w_1)=\{1+\epsilon_2(z)\}\{1+\epsilon_1(z)+\epsilon_1'(z)\}+\{1+\epsilon_1(z)\}\{1+\epsilon_2(z)-\epsilon_2'(z)\}. \tag{3.11}$$

From (3.01) we see that this reduces to $2\{1+\epsilon_2(a_1)\}$ at $z=a_1$, and $2\{1+\epsilon_1(a_2)\}$ at $z=a_2$. Since the Wronskian is constant, (3.02) follows.

Lastly, we derive from (2.11) and (3.04)

$$|\alpha_{1,n}(a_2)| \leqslant \{\mathscr{V}_{a_1,a_2}(G)\}^n/n!. \tag{3.12}$$

The improved form (3.03) of the main result (2.03) for the special value $z=a_2$ now follows immediately from (3.05).

4. Basic Error Theorem for the Liouville-Green Approximation

We now proceed to apply the results of the preceding two sections to the differential equation (1.01). A useful generalization is achieved on replacing the coefficient of w in this equation by the sum of two functions $f(z)$ and $h(z)$, say. In applications the second function $h(z)$ will either be zero, or small, in some sense, compared with $f(z)$ (cf. [1], sec. 4(ii)). From the standpoint of the error analysis, however, the relative magnitudes of $f(z)$ and $h(z)$ are immaterial.

Thus we consider the differential equation

$$\frac{d^2w}{dz^2}=\{f(z)+h(z)\}w, \tag{4.01}$$

in which $f(z)$ and $h(z)$ are regular in an unbounded domain $\mathbf{D}$, not necessarily simply connected, and $f(z)$ is nonvanishing. The Liouville transformation

$$\xi=\xi(z)=\int\{f(z)\}^{1/2}dz, \qquad W=\{f(z)\}^{1/4}w, \tag{4.02}$$

throws (4.01) into the form

$$d^2W/d\xi^2=\{1+\varphi(\xi)\}W, \tag{4.03}$$

in which

$$\varphi(\xi)=\frac{h(z)}{f(z)}+\frac{4f(z)f''(z)-5\{f'(z)\}^2}{16\{f(z)\}^3}=\frac{h}{f}-\frac{1}{f^{3/4}}\frac{d^2}{dz^2}\left(\frac{1}{f^{1/4}}\right). \tag{4.04}$$

Equation (4.03) is of the form (2.01). The z-domain $\mathbf{D}$ is mapped on a ξ-domain $\mathbf{\Delta}$, say, and $\varphi(\xi)$ is regular in $\mathbf{\Delta}$. By applying Theorem 1 and then restoring the original variables we establish the following:

THEOREM 3. *Let $z=c$ be an arbitrary point at infinity in $\mathbf{D}$ such that $|\mathrm{Re}\ \xi(c)|=\infty$, and let the branch of $\xi(z)$ be chosen in such a way that* Re $\xi(z)$ *is positive as c is approached. Then the differential equation* (4.01) *has a regular solution, given by*

$$w(z)=\{f(z)\}^{-1/4}e^{-\xi(z)}\{1+\epsilon(z)\}, \tag{4.05}$$

where

$$|\epsilon(z)|,\ |f^{-1/2}\epsilon'(z)|,\ |\epsilon(z)-\{\epsilon'(z)/\xi'(z)\}|\leqslant\exp\{\mathscr{V}_{z,c}(F)\}-1 \quad (z\in\mathbf{H}(c)), \tag{4.06}$$

and

$$F(z)=\int\varphi(\xi)d\xi=\int\left\{\frac{h}{f^{1/2}}-\frac{1}{f^{1/4}}\frac{d^2}{dz^2}\left(\frac{1}{f^{1/4}}\right)\right\}dz. \tag{4.07}$$

The region of validity $\mathbf{H}(c)$ comprises those points z which can be joined in $\mathbf{D}$ to c by a "ξ-progressive" path $\mathscr{P}$, that is, one whose map in the ξ-plane is progressive (sec. 3). All branches in (4.05) to (4.07) are continuous along $\mathscr{P}$, that of $f^{1/2}$ being the square of $f^{1/4}$. The variation in (4.06) is taken over $\mathscr{P}$; naturally we assume that it converges at c, otherwise (4.06) is meaningless.

5. Fundamental Connection Formula for a Simple Turning Point

In this section we suppose that $f(z)$ and $h(z)$ are regular in an unbounded, simply connected domain $\mathbf{D}$, and that $f(z)$ has just one zero in $\mathbf{D}$, a simple zero at an interior point z_0, say. We define $\xi(z)$ by (4.02), fixing the arbitrary lower integration limit at z_0:

$$\xi(z)=\int_{z_0}^{z}\{f(t)\}^{1/2}dt. \tag{5.01}$$

Clearly $\xi(z)$ is an analytic function in $\mathbf{D}$ whose only singularity is a branch point at z_0. If the Taylor expansion of $f(z)$ in the neighborhood of z_0 is denoted by

$$f(z)=\sum_{n=1}^{\infty}f_n(z-z_0)^n, \tag{5.02}$$

then

$$\xi(z)=\frac{2}{3}f_1^{1/2}(z-z_0)^{3/2}\{1+O(z-z_0)\}. \tag{5.03}$$

The set of points in the z-plane for which

$$\mathrm{Re}\ \xi(z)=0 \tag{5.04}$$

we call the *principal curves* [3] *associated with the turning point* z_0. Using the theory of conformal mapping, we see that each principal curve is a regular Jordan arc which can terminate only at z_0 or at the boundary of **D**. Moreover, no principal curve can intersect itself or any other principal curve except at z_0. Equation (5.03) shows that three principal curves emerge from z_0, inclined there at angle $2\pi/3$ with each other; we denote them by $\mathscr{C}_1, \mathscr{C}_2, \mathscr{C}_3$, enumerated in the positive rotational sense. These and other principal curves divide **D** into a number of subdomains. Those with z_0 on their boundary we call the *principal subdomains*, and denote them by $\mathbf{D}_1$, $\mathbf{D}_2$, $\mathbf{D}_3$, as indicated on figure 2. Clearly each $\mathbf{D}_j$ is itself simply connected. For terminological convenience in what follows we shall enumerate j with modulo 3; thus $\mathbf{D}_0\equiv\mathbf{D}_3$, $\mathbf{D}_4\equiv\mathbf{D}_1$.

We define $\xi_j(z)$ $(j=1, 2, 3)$ to be that branch of $\xi(z)$ which is continuous in **D** cut along the corresponding $\mathscr{C}_j$, and whose real part is positive in $\mathbf{D}_j$ and negative in the other two principal subdomains. Analytic continuation across the cut $\mathscr{C}_j$ is expressed by

$$\xi_j\{(z-z_0)e^{\pm 2\pi i}\}=-\xi_j(z-z_0); \tag{5.05}$$

compare (5.03). We define $g(z)$ to be the four-valued function

$$g(z)=\{f(z)\}^{-1/4}. \tag{5.06}$$

Its analytic continuation is expressed by

$$g\{(z-z_0)e^{\pm 2\pi i}\}=\mp ig(z-z_0). \tag{5.07}$$

We denote by $g_j(z)$ $(j=1, 2, 3)$ an arbitrary constant multiple of any branch of $g(z)$ which is continuous in **D** cut along the corresponding $\mathscr{C}_j$. In order to achieve a symmetric form of final result, we stipulate that these multiples must be chosen in such a way that

$$g_j(z)=e^{\pi i/6}g_{j-1}(z) \qquad (z\in\mathbf{D}_{j-1}\cup\mathbf{D}_j). \tag{5.08}$$

[3] Also called anti-Stokes lines [12], [13].

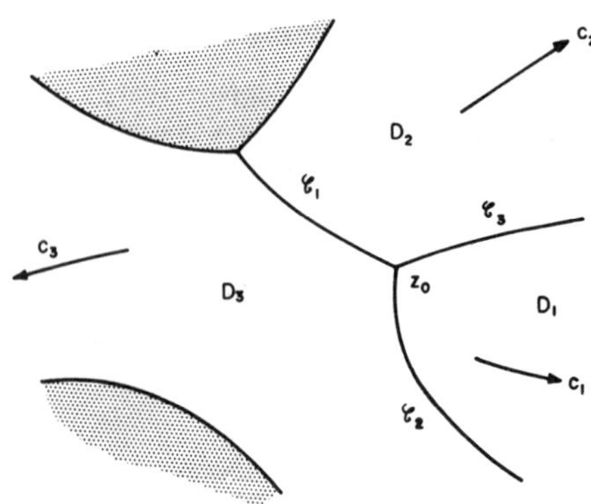

FIGURE 2. *Principal subdomains* $\mathbf{D}_j$.

The self-consistency of this equation for $j=1, 2, 3$ follows from (5.07).

Let us suppose that each $\mathbf{D}_j$ is unbounded, and that c_j is an arbitrary point at infinity in $\mathbf{D}_j$ such that $|\operatorname{Re}\,\xi(c_j)|=\infty$. Excluding the point z_0 from $\mathbf{D}$, and then applying Theorem 3 of section 4, we find that eq (4.01) has solutions $w_j(z)$ $(j=1, 2, 3)$, given by

$$w_j(z)=g_j(z)e^{-\xi_j(z)}\{1+\epsilon_j(z)\}, \tag{5.09}$$

where

$$|\epsilon_j(z)|,|f^{-1/2}\epsilon_j'(z)|,|\epsilon_j(z)-f^{-1/2}\epsilon_j'(z)|\leq \exp\{\mathscr{V}_{z,c_j}(F)\}-1 \qquad (z\in \mathbf{H}(c_j)) \tag{5.10}$$

and F is defined by (4.07). In (5.10) the branch of $f^{-1/2}$ is $1/\xi_j'(z)$, and $\mathbf{H}(c_j)$ is the set of z-points which can be joined to c_j by a ξ-progressive path $\mathscr{P}_j$ lying in $\mathbf{D}$. (We observe, in passing, that the corresponding principal curve $\mathscr{C}_j$ is excluded from $\mathbf{H}(c_j)$, since passage of $\mathscr{P}_j$ through z_0 is not permitted.) $\mathscr{V}_{z,c_j}(F)$ denotes the variation over $\mathscr{P}_j$ of F.

Although each solution $w_j(z)$ satisfies (5.09) and (5.10) in only part of $\mathbf{D}$, the theory of linear differential equations shows that each $w_j(z)$ exists and is regular throughout the whole of $\mathbf{D}$. Further, there is an identity of the form

$$A_1w_1(z)+A_2w_2(z)+A_3w_3(z)=0, \tag{5.11}$$

in which the A_j are independent of z. Equation (5.11) is the connection formula for the turning point z_0, and the values of the A_j constitute our primary quest. In order to obtain them, we assume that ξ-progressive paths exist in $\mathbf{D}$ linking each pair of reference points c_j, c_k.

Consider (5.11) in $\mathbf{D}_1$. Substituting by means of (5.09) and (5.08), and using the relations $\xi_2(z)=\xi_3(z)=-\xi_1(z)$, we derive

$$A_1e^{-\xi_1(z)}\{1+\epsilon_1(z)\}+A_2e^{(\pi i/6)+\xi_1(z)}\{1+\epsilon_2(z)\}+A_3e^{-(\pi i/6)+\xi_1(z)}\{1+\epsilon_3(z)\}=0. \tag{5.12}$$

Now let $z\to c_1$. Then $\epsilon_1(z)$ vanishes and, by Theorem 2, $\epsilon_2(z)$ and $\epsilon_3(z)$ tend to constant values $\epsilon_2(c_1)$ and $\epsilon_3(c_1)$. Also by hypothesis $\operatorname{Re}\,\xi_1(z)\to+\infty$. Hence we have

$$e^{\pi i/6}\{1+\epsilon_2(c_1)\}A_2+e^{-\pi i/6}\{1+\epsilon_3(c_1)\}A_3=0. \tag{5.13}$$

Similarly,

$$e^{-\pi i/6}\{1+\epsilon_1(c_2)\}A_1+e^{\pi i/6}\{1+\epsilon_3(c_2)\}A_3=0, \tag{5.14}$$

$$e^{\pi i/6}\{1+\epsilon_1(c_3)\}A_1+e^{-\pi i/6}\{1+\epsilon_2(c_3)\}A_2=0. \tag{5.15}$$

The consistency of this set of homogeneous equations for the A_j immediately follows from the relation $\epsilon_j(c_k)=\epsilon_k(c_j)$ obtained from eq (3.02). Solving for the A_j and substituting back in (5.11), we arrive at the desired form of the connection formula:

$$\{1+\epsilon_2(c_3)\}w_1(z)+e^{-2\pi i/3}\{1+\epsilon_3(c_1)\}w_2(z)+e^{2\pi i/3}\{1+\epsilon_1(c_2)\}w_3(z)=0. \tag{5.16}$$

The error terms are bounded by (compare (3.03))

$$|\epsilon_j(c_k)| \leq \tfrac{1}{2}[\exp\{\mathcal{V}_{c_j,c_k}(F)\}-1],$$

that is,

$$|\epsilon_j(c_k)| \leq \frac{1}{2}\exp\left\{\int_{c_j}^{c_k}\left|\frac{h}{f^{1/2}}-\frac{1}{f^{1/4}}\frac{d^2}{dz^2}\left(\frac{1}{f^{1/4}}\right)\right||dz|\right\}-\frac{1}{2}, \tag{5.17}$$

in which the variation (or integral) is evaluated along a ξ-progressive path linking c_j with c_k.

We summarize this section so far. The assumed conditions on $f(z)$ and $h(z)$ are stated in the opening sentence. The principal subdomains $\mathbf{D}_j$ associated with the turning point z_0 are defined in the second paragraph. Three solutions $w_j(z)$ $(j=1, 2, 3)$ of the differential equation (4.01) are then constructed, given by (5.09) and (5.10). Here $\xi_j(z)$ is the branch of the integral (5.01) which has positive real part in $\mathbf{D}_j$; $g_j(z)$ $(j=1, 2, 3)$ is any set of constant multiples of branches of $\{f(z)\}^{-1/4}$ satisfying (5.08); c_j is any point at infinity in $\mathbf{D}_j$ such that $|\text{Re}\ \xi(c_j)|=\infty$. The error bounds (5.10) are valid at all points z which can be joined to c_j by a ξ-progressive path lying in $\mathbf{D}$. Finally, provided that the c_j can be joined in pairs by paths of this kind, the fundamental connection formula between the $w_j(z)$ is expressed by (5.16) and (5.17).

Remark. It will be observed that an outstanding problem in the application of the connection formula (5.16) will be the determination of suitable ξ-progressive paths linking together the points at infinity c_j. This appears to be an unavoidable feature of this method,[4] and its solution will generally necessitate a study of the actual conformal transformation between the ξ and z-planes.

Usually it will not be difficult to demonstrate the existence of suitable ξ-progressive paths. These paths are not unique, however, and the most effective use of the connection formula clearly calls for the determination of those paths which minimize the total variation of the function F along their length. Roughly speaking, this means that we need to keep the ξ-progressive paths as far away as possible from all the singularities of $f(z)$ and $h(z)$ and zeros of $f(z)$ (including z_0), subject to fulfillment of the monotonicity condition.

The construction of ξ-progressive paths in the neighborhood of infinity is discussed in section 7 below, and an example of the determination of ξ-progressive paths which minimize the variation of an analytic function along their length is given in [17], section 7.

6. Airy's Equation

The simplest example to which the theory of section 5 is applicable is provided by the differential equation

$$d^2w/dz^2=zw. \tag{6.01}$$

In the notation of sections 4 and 5, we take $f(z)=z$ and $h(z)=0$. Then $z_0=0$ and

$$\xi(z)=\frac{2}{3}z^{3/2}, \qquad g(z)=z^{-1/4}. \tag{6.02}$$

$\mathbf{D}$ comprises the whole z-plane, and the principal subdomains $\mathbf{D}_j$ are the sectors of angle $2\pi/3$ indicated in figure 3.

We define $g_1(z)$ to be the principal branch of $z^{-1/4}$. Then from (5.08)

$$g_2(z)=e^{\pi i/6}z^{-1/4} \quad (z\in\mathbf{D}_1\cup\mathbf{D}_2), \qquad g_3(z)=e^{-\pi i/6}z^{-1/4} \quad (z\in\mathbf{D}_1\cup\mathbf{D}_3). \tag{6.03}$$

[4] Compare [20], chapter 4.

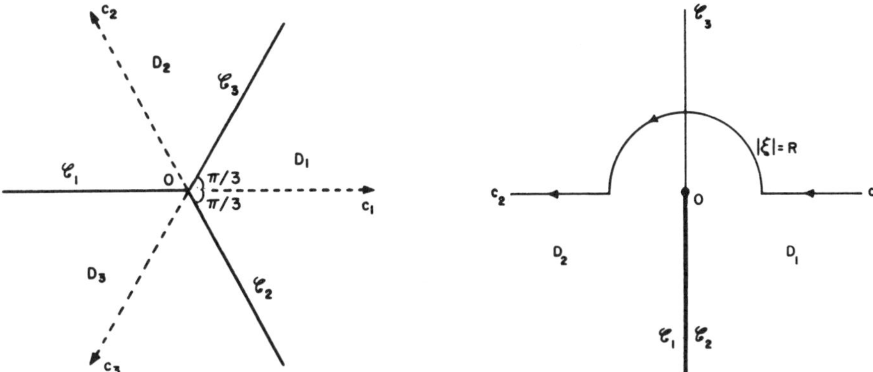

FIGURE 3. *Principal subdomains for Airy's equation.*

FIGURE 4. *ξ-plane.*

We take $c_1=+\infty$, $c_2=\infty\, e^{2\pi i/3}$, $c_3=\infty\, e^{-2\pi i/3}$. By considering the ξ-map we see that each region $\mathbf{H}(c_j)$ comprises the part of the z-plane remaining after the removal of the corresponding principal curve $\mathscr{C}_j$. From (4.07) we have

$$F(z)=\frac{5}{24z^{3/2}}=\frac{5}{36\xi}. \tag{6.04}$$

The ξ-map of a suitable ξ-progressive path linking c_1 with c_2 is shown in figure 4. As the radius R of the semicircular part tends to infinity, the variation of F along the whole path vanishes. Therefore from (5.17) $\epsilon_1(c_2)=0$. Similarly $\epsilon_2(c_3)$ and $\epsilon_3(c_1)$ both vanish. Accordingly (5.16) reduces to

$$w_1(z)+e^{-2\pi i/3}w_2(z)+e^{2\pi i/3}w_3(z)=0. \tag{6.05}$$

By comparing (5.09) with the asymptotic forms of the Airy functions [18], we identify

$$w_1(z)=2\pi^{1/2}\,\mathrm{Ai}\,(z),\quad w_2(z)=2\pi^{1/2}\,\mathrm{Ai}\,(ze^{-2\pi i/3}),\qquad w_3(z)=2\pi^{1/2}\,\mathrm{Ai}\,(ze^{2\pi i/3}). \tag{6.06}$$

Equation (6.05) is the well-known connection formula for these functions. It is interesting to observe that our theory yields it *exactly*.

The vanishing of the error terms in this example is essentially a consequence of there being no boundary of $\mathbf{D}$ intervening between c_j and c_k in the neighborhood of infinity. The path of figure 4 generally cannot be used to establish this in other cases. Using different paths, we formulate sufficient conditions for the vanishing of $\epsilon_j(c_k)$ in the next section.

7. Sufficient Conditions for the Vanishing of the Error Terms

We again use the notation of section 5, and we suppose, without loss of generality, that the turning point z_0 is at the origin.

THEOREM 4. *Let $f(z)$ and $h(z)$ be regular in the sector* $\mathbf{S}$: $\gamma_1<\arg z<\gamma_2$ *for all sufficiently large $|z|$, and*

$$f(z)\sim Az^r,\qquad h(z)=O(z^{(r/2)-1-\delta}),\ \text{as}\ |z|\to\infty, \tag{7.01}$$

uniformly with respect to $\arg z$ *in* $\mathbf{S}$. *Here A is a real or complex non-zero constant, and r, δ real* *constants such that*[5] $r > -2$, $\delta > 0$. *Also, within* **S**, *let* $\mathscr{L}_1$, $\mathscr{L}_2$ *be outward-drawn rays from the origin which lie in adjacent principal subdomains* $\mathbf{D}_1$, $\mathbf{D}_2$, *respectively. Then provided that* $\mathscr{L}_1$, $\mathscr{L}_2$ *are not parallel to the asymptotes of the neighboring boundaries of* $\mathbf{D}_1$, $\mathbf{D}_2$, *they are strictly* ξ*-progressive paths in the neighborhood of infinity, and if* c_1, c_2 *denote the points at infinity on them,* $\epsilon_1(c_2) = 0$.

To prove this result, we first observe that

$$g(z) \equiv \{f(z)\}^{-1/4} \sim A^{-1/4} z^{-r/4} \tag{7.02}$$

as $|z| \to \infty$ in **S**. Hence ([10], section 1.36)

$$g''(z) \sim \frac{r}{4}\left(\frac{r}{4}+1\right) A^{-1/4} z^{-(r/4)-2} \quad (r \neq 0) \tag{7.03}$$

as $|z| \to \infty$ in any sector $\mathbf{S}': \gamma_1' < \arg z < \gamma_2'$ which is properly interior to **S**. When $r = 0$, the result corresponding to (7.03) is $g''(z) = o(z^{-2})$. Therefore, whether or not r is zero, we have

$$g(z) g''(z) = O(z^{-s-1}) \tag{7.04}$$

as $|z| \to \infty$ in $\mathbf{S}'$, uniformly with respect to arg z, where

$$s = \tfrac{1}{2} r + 1,$$

and is positive, by hypothesis. From (4.07), (7.01), and (7.04) we derive

$$F'(z) = O(z^{-1-\delta_1}), \tag{7.05}$$

where $\delta_1 = \min(\delta, s)$. Hence the integral in (5.17) converges. The essential problem is to show how to construct a ξ-progressive path connecting c_1 with c_2 in the neighborhood of infinity.

From (5.01), we have

$$\xi = \frac{A^{1/2} z^s}{s} + o(z^s) \qquad (z \in \mathbf{S}). \tag{7.06}$$

Therefore the principal curves approach infinity in directions parallel to the rays

$$\arg z = \{(n + \tfrac{1}{2})\pi - \tfrac{1}{2}\alpha\}/s, \tag{7.07}$$

where $\alpha = \arg A$, and $n = 0, \pm 1, \pm 2, \ldots$. We call these the *principal directions.* Consider the behavior of Re ξ along the ray with parametric equation $z = \tau e^{i\gamma}$, γ being real and fixed, and $0 < \tau < \infty$. We have

$$d\xi/d\tau = \{f(z)\}^{1/2}(dz/d\tau) = A^{1/2}\tau^{r/2} e^{is\gamma}\{1 + o(1)\} \tag{7.08}$$

as $\tau \to \infty$. Hence

$$d\,(\mathrm{Re}\,\xi)/d\tau = |A|^{1/2}\tau^{r/2}\{\cos(s\gamma + \tfrac{1}{2}\alpha) + o(1)\}. \tag{7.09}$$

The constant $\cos(s\gamma + \frac{1}{2}\alpha)$ vanishes only when γ has one of the values given by the right of (7.07). *Therefore any outward ray in* **S** *not in a principal direction is a strictly* ξ*-progressive path for all sufficiently large* $|z|$. In particular this includes $\mathscr{L}_1$ and $\mathscr{L}_2$.

[5] Thus the differential equation (4.01) has an irregular singularity at infinity.

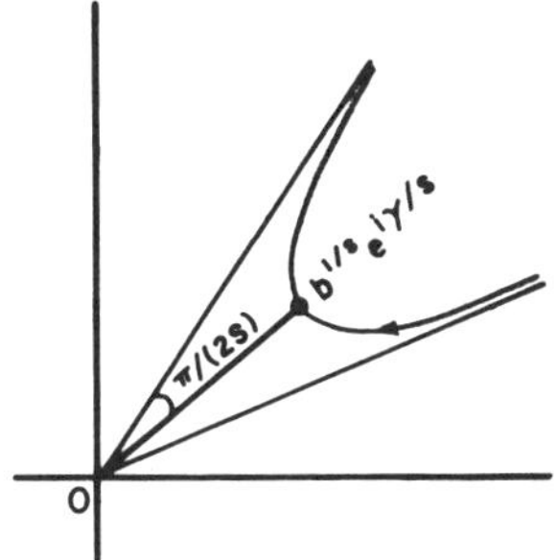

FIGURE 5. *z-plane.*

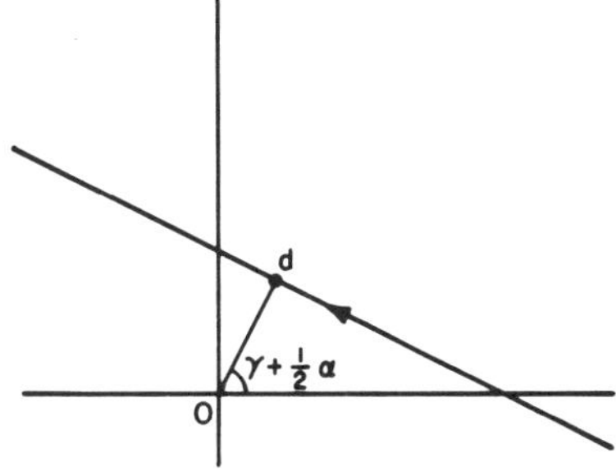

FIGURE 6. *Approximate path in ξ-plane:* $|d|=|A|^{1/2}b/s$.

Next, consider the path with parametric equation

$$z=\{(b+i\tau)e^{i\gamma}\}^{1/s} \qquad (-\infty<\tau<\infty), \tag{7.10}$$

where γ and b are real constants, b being large and positive.

As indicated in figure 5 this path begins at infinity, sweeps back to its nearest approach to the origin at $z=b^{1/s}e^{i\gamma/s}$ and then returns to infinity, the angle between its asymptotes being π/s. When b is large, we have from (7.06)

$$\xi\sim s^{-1}|A|^{1/2}(b+i\tau)e^{i\{\gamma+(\alpha/2)\}},$$

showing that the ξ-map of (7.10) is approximately the straight line indicated in figure 6. Corresponding to (7.09), we find that

$$\frac{d}{d\tau}(\mathrm{Re}\,\xi)=-\frac{|A|^{1/2}}{s}\left\{\sin\left(\gamma+\frac{1}{2}\alpha\right)+o(1)\right\}, \tag{7.11}$$

as $b\to\infty$, uniformly with respect to τ. Therefore *the curve (7.10) is a strictly ξ-progressive path for all sufficiently large b, except when its asymptotes are in principal directions.*

Now consider c_1 and c_2. By taking γ_1' and γ_2' sufficiently close to γ_1 and γ_2, respectively, we can ensure that $\mathbf{S}'$ contains $\mathscr{L}_1$ and $\mathscr{L}_2$. From (7.06) and the hypotheses on $\mathscr{L}_1$ and $\mathscr{L}_2$, it follows that

$$|\arg c_2-\arg c_1|<2\pi/s. \tag{7.12}$$

Figure 7 indicates how c_1 and c_2 can be connected by a strictly ξ-progressive path in $\mathbf{S}'$, consisting of $\mathscr{L}_1$, $\mathscr{L}_2$ and either one or two arcs of the type (7.10), depending whether or not $|\arg c_2-\arg c_1|<\pi/s$. The corresponding path in the ξ-plane is approximately that indicated in figure 8. On letting the b-parameters of the arcs tend to infinity, we see immediately from (7.05) that the variations of F along $\mathscr{L}_1$ and $\mathscr{L}_2$ tend to zero. For the contribution to the variation from the arcs (7.10), we observe that the corresponding values of $\int|z^{-1-\delta_1}dz|$ are bounded by

$$\frac{1}{s}\int_{-\infty}^{\infty}\frac{d\tau}{|b+i\tau|^{1+(\delta_1/s)}}$$

and accordingly vanish as $b\to\infty$. The proof of Theorem 4 is now complete.

An alternative set of sufficient conditions for the vanishing of the error terms is provided by the next theorem, which can be proved in a similar way. In this result m, A denote real or complex nonzero constants, and δ is a positive constant.

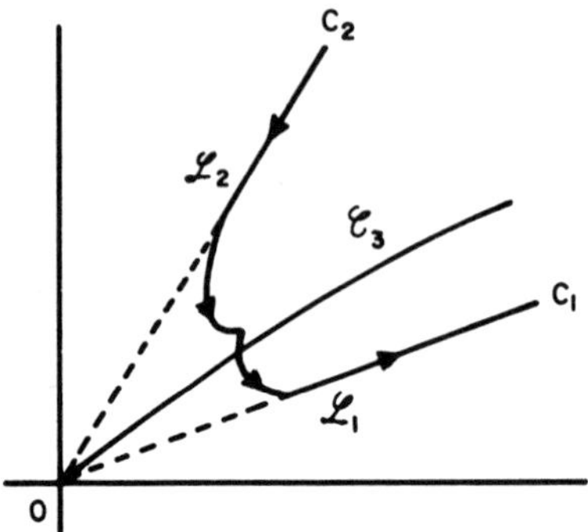

FIGURE 7. ξ-progressive path in z-plane.

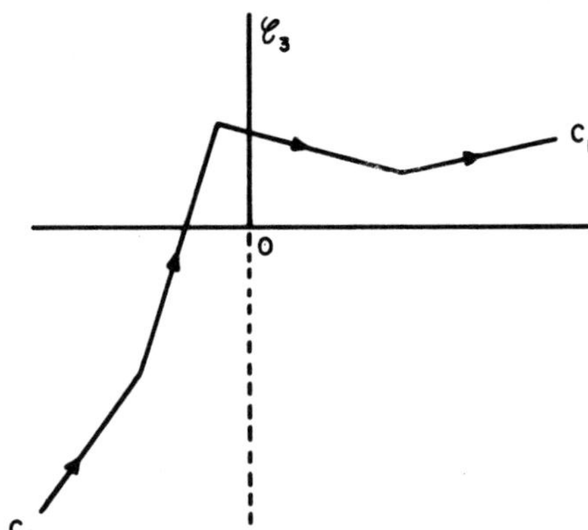

FIGURE 8. *Approximate path in ξ-plane.*

THEOREM 5. *Let $f(z)$ and $h(z)$ be regular in the strip* **T**: $\gamma_1 < \mathrm{Im}(mz) < \gamma_2$ *for all sufficiently large positive values of* Re (mz), *and*

$$f(z) \sim Ae^{mz}, \qquad h(z) = O(e^{mz(1-\delta)/2}), \text{ as } \mathrm{Re}\ (mz) \to +\infty, \tag{7.13}$$

uniformly with respect to Im (mz) *in* **T**. *Also, within* **T**, *let* $\mathscr{L}_1$, $\mathscr{L}_2$ *be straight lines which are parallel to the boundaries of* **T**, *and lie in adjacent principal subdomains* $\mathbf{D}_1$, $\mathbf{D}_2$, *respectively. Then provided that* $\mathscr{L}_1$, $\mathscr{L}_2$ *do not coincide with the asymptotes of the boundaries of* $\mathbf{D}_1$, $\mathbf{D}_2$, *they are strictly ξ-progressive paths in the neighborhood of infinity, and if* c_1, c_2 *are the points at infinity on them,* $\epsilon_1(c_2) = 0$.

8. The Gans-Jeffreys Connection Formulas (i)

The first of these two formulas may be regarded as a special case of the fundamental connection formula of section 5. We again consider the differential equation (4.01) in which $f(z)$ and $h(z)$ are regular in an unbounded, simply connected domain **D** and $f(z)$ has just one zero in **D**, a simple zero at $z=0$. This time we suppose that **D** includes the whole of the real axis, and that $f(z)$ and $h(z)$ are real when z is real. Without loss of generality, we take $f(z)/z$ to be positive on the real axis.

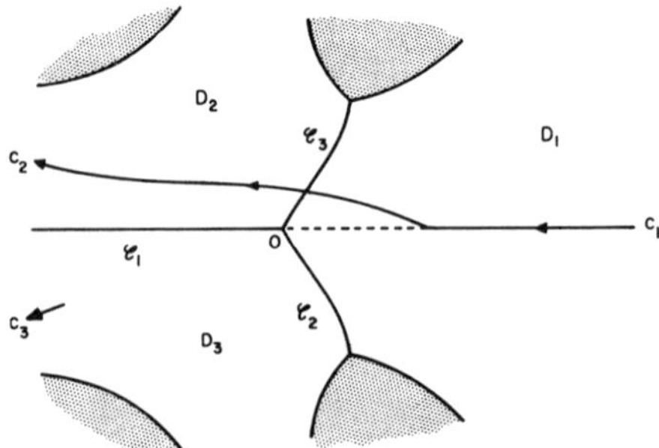

FIGURE 9. *Typical principal subdomains* $\mathbf{D}_j$ *for the Gans-Jeffreys formulas.*

One of the three principal curves, say $\mathscr{C}_1$, which emerges from the origin is the negative real axis. The other two emerge at directions $\pm\pi/3$ and are symmetric with respect to the real axis. As in section 5 we denote by $\mathbf{D}_1$, $\mathbf{D}_2$, $\mathbf{D}_3$ the three subdomains bounded by the principal curves which have the origin on their boundary.

We take c_1 to be the point at infinity on the positive real axis, and c_2, c_3 to be points at infinity on conjugate ξ-progressive curves $\mathscr{L}_2$, $\mathscr{L}_3$. These curves must lie within $\mathbf{D}_2$, $\mathbf{D}_3$, respectively, and be such that no boundary of $\mathbf{D}$ intervenes between them in the neighborhood of infinity. We assume that $|\mathrm{Re}\ \xi(z)| \to \infty$ when $z \to c_j$ $(j=1, 2, 3)$ (the "dominancy condition") and that the variation of the function F, defined by (4.07), converges when $z \to c_j$. The other assumptions we need are

(i) ξ-progressive paths can be found in $\mathbf{D}$ which link c_2 or c_3 with any point of the real axis, other than the origin, and which do not pass through the origin.

(ii) $\epsilon_2(c_3)$, $\epsilon_2(-\infty)$ and $\epsilon_3(-\infty)$ all vanish.

In applications, Condition (i) can be tested by considering the ξ-map. Sufficient sets of conditions for (ii) are to be found in section 7.

The connection formula (5.16) for the solutions of (4.01) takes the form

$$w_1(z) = e^{\pi i/3}\{1+\epsilon_3(c_1)\}w_2(z) + e^{-\pi i/3}\{1+\epsilon_2(c_1)\}w_3(z). \tag{8.01}$$

To fix the solutions completely, we define $g_1(z)$ to be the branch of $\{f(z)\}^{-1/4}$ which takes its principal value on the positive real axis (compare (5.08)). The branches of the $w_j(z)$ are then interpreted in real form as follows.

When z is positive, we have immediately from (5.09) and (5.10)

$$w_1(z) = f^{-1/4} \exp\left(-\int_0^z f^{1/2}dz\right)\{1+\epsilon_1(z)\} \qquad (z>0), \tag{8.02}$$

where

$$|\epsilon_1(z)| \leqslant \exp\{\mathcal{V}_{z,\infty}(F)\} - 1, \tag{8.03}$$

the variation being taken along the real axis.

When z is negative, we find that

$$\xi_2(z) = i\int_z^0 |f|^{1/2}dz, \qquad g_2(z) = e^{-\pi i/12}|f|^{-1/4}. \tag{8.04}$$

Substitution of (5.09) in (8.01) accordingly yields

$$w_1(z) = \{1+\epsilon_3(c_1)\}e^{\pi i/4}|f|^{-1/4} \exp\left(-i\int_z^0 |f|^{1/2}dz\right)\{1+\epsilon_2(z)\} + \text{conjugate}. \tag{8.05}$$

Let

$$1+\epsilon_2(z) = \{1+\mu(z)\}e^{-i\eta(z)}, \qquad 1+\epsilon_3(z) = \{1+\mu(z)\}e^{i\eta(z)}, \tag{8.06}$$

where $\mu(z)$ and $\eta(z)$ are real and $\mu(z) \geqslant -1$. Then by elementary considerations

$$|\mu(z)| \leqslant |\epsilon_2(z)|, \qquad |\eta(z)| \leqslant \sin^{-1}|\epsilon_2(z)| \leqslant \pi|\epsilon_2(z)|/2, \tag{8.07}$$

provided that $|\epsilon_2(z)| \leqslant 1$. Substitution in (8.05) yields

$$w_1(z) = 2\{1+\mu(c_1)\}\{1+\mu(z)\}|f|^{-1/4} \sin\left\{\int_z^0 |f|^{1/2}dz + \frac{1}{4}\pi + \eta(z) - \eta(c_1)\right\} \qquad (z<0). \tag{8.08}$$

From (8.07), (5.17) and (5.10), we deduce that

$$|\mu(c_1)|,\ |2\eta(c_1)/\pi| \leqslant \frac{1}{2}[\exp\{\mathcal{V}_{c_1, c_2}(F)\}-1], \tag{8.09}$$

and

$$|\mu(z)|,\ |2\eta(z)/\pi| \leqslant \exp\{\mathcal{V}_{z, c_2}(F)\}-1, \tag{8.10}$$

the variations in both cases being evaluated along ξ-progressive curves.

The aggregate of (8.02), (8.03), (8.08), (8.09), and (8.10), may be regarded as the full form of the first Gans-Jeffreys connection formula [5], [6], [9]. It expresses the subdominant exponential solution on one side of the turning point in terms of the oscillatory solutions on the other side. Boundary conditions satisfied by $w_1(z)$ are clearly

$$\epsilon_1(z)=o(1) \quad (z\to+\infty); \qquad \mu(z),\ \eta(z)=o(1) \quad (z\to-\infty). \tag{8.11}$$

We note that the bounds for the error terms $\mu(c_1)$, $\eta(c_1)$ depend solely on the variation of F along the ξ-progressive path linking c_1 and c_2, and can be minimized by proper choice of path.

The second connection formula concerns the dominant exponential solution. Unlike the subdominant solution, this is not unique. Consider first $w_2(z)$. For negative z, we have from (8.04) and (5.09)

$$w_2(z)=e^{-\pi i/12}|f|^{-1/4}\exp\left(-i\int_z^0|f|^{1/2}dz\right)\{1+\epsilon_2(z)\} \qquad (z<0). \tag{8.12}$$

On the other side of the turning point

$$w_2(z)=e^{\pi i/6}f^{-1/4}\exp\left(\int_0^z f^{1/2}dz\right)\{1+\epsilon_2(z)\} \qquad (z>0). \tag{8.13}$$

Instead of $w_2(z)$, however, we prefer the real solution $w_4(z)=\mathrm{Re}\{e^{-\pi i/6}w_2(z)\}$. Then with the aid of (8.06), we derive

$$w_4(z)=[1+\mathrm{Re}\{\epsilon_2(z)\}]f^{-1/4}\exp\left(\int_0^z f^{1/2}dz\right) \qquad (z>0), \tag{8.14}$$

and

$$w_4(z)=\{1+\mu(z)\}|f|^{-1/4}\cos\left\{\int_z^0|f|^{1/2}dz+\frac{1}{4}\pi+\eta(z)\right\} \qquad (z<0), \tag{8.15}$$

where $|\epsilon_2(z)|$, $|\mu(z)|$ and $|2\eta(z)/\pi|$ are all bounded by the right of (8.10). These relations constitute the second Gans-Jeffreys formula. We note that $w_4(z)$ satisfies the boundary conditions

$$\epsilon_2(z)=\epsilon_2(c_1)+o(1) \quad (z\to+\infty); \qquad \mu(z),\ \eta(z)=o(1) \quad (z\to-\infty); \tag{8.16}$$

where $|\epsilon_2(c_1)|$ is bounded by the right of (8.09).

9. The Gans-Jeffreys Connection Formulas (ii)

In this section we attack the problem of section 8 with real-variable theory. Our hypotheses are

(i) The function $f(z)/z$ is positive and twice-continuously differentiable throughout the interval $-\infty<z<\infty$, and $h(z)$ is a continuous real function in the same interval.

(ii) As $z \to \pm\infty$, $\left|\int_0^z f^{1/2}dz\right|$ is unbounded and the variation of the function F (defined by (4.07)) converges.

In these circumstances it is shown in [15], sections 4, 5 that (4.01) has a solution [6]

$$w_1(z) = (\zeta/f)^{1/4}\{\mathrm{Ai}\,(\zeta) + \epsilon_1(z)\}, \tag{9.01}$$

where

$$\zeta \equiv \zeta(z) = \pm\left|\frac{3}{2}\int_0^z \{f(t)\}^{1/2}dt\right|^{2/3} \tag{9.02}$$

$$|\epsilon_1(z)| \leqslant \lambda_1^{-1}E^{-1}(\zeta)M(\zeta)\,[\exp\{\lambda_1\mathscr{V}_{z,\,\infty}(H)\} - 1] \quad (-\infty < z < \infty), \tag{9.03}$$

and

$$H(z) = \frac{5}{24|\zeta|^{3/2}} \pm \int^z \left\{\frac{1}{|f|^{1/4}}\frac{d^2}{dz^2}\left(\frac{1}{|f|^{1/4}}\right) - \frac{h}{|f|^{1/2}}\right\}dz. \tag{9.04}$$

The upper or lower signs are taken in (9.02) and (9.04) according as z is positive or negative, and the integration limits in (9.04) are determined by the condition $H(0) = 0$. The auxiliary functions $E(\zeta)$, $M(\zeta)$ and the constant $\lambda_1 = 1.430\ \ldots$ are defined in [15], section 2.

On substituting for Ai (ζ) by means of its appropriate asymptotic forms, we obtain

$$w_1(z) = \tfrac{1}{2}\pi^{-1/2}f^{-1/4}\{1 + o(1)\}\exp\left(-\int_0^z f^{1/2}dz\right) \quad (z \to +\infty), \tag{9.05}$$

and

$$w_1(z) = \pi^{-1/2}|f|^{-1/4}\left[\sin\left(\int_z^0 |f|^{1/2}dz + \frac{1}{4}\pi\right) + \rho(z) + o(1)\right] \quad (z \to -\infty), \tag{9.06}$$

where

$$\rho(z) = \pi^{1/2}|\zeta|^{1/4}\epsilon_1(z), \tag{9.07}$$

and is bounded.

In order to throw (9.06) fully into "modulus-phase" form, we appeal to Theorem 4 of [1]. This shows that solutions of (4.01) exist of the form

$$|f|^{-1/4}\{1 + o(1)\}\exp\left\{\pm i\int_z^0 |f|^{1/2}dz\right\} \text{ as } z \to -\infty. \tag{9.08}$$

Therefore we can express

$$w_1(z) = \pi^{-1/2}|f|^{-1/4}\{1 + \nu_1 + o(1)\}\sin\left\{\int_z^0 |f|^{1/2}dz + \frac{1}{4}\pi + \nu_2 + o(1)\right\} \quad (z \to -\infty), \tag{9.09}$$

where ν_1, ν_2 are real constants to be determined. Comparison of (9.06) with (9.09) shows that

$$\rho(z) = (1 + \nu_1)\sin\{\theta(z) + \nu_2\} - \sin\theta(z) + o(1), \tag{9.10}$$

[6] Not to be confused with the $w_1(z)$ of section 8.

where

$$\theta(z) \equiv \int_z^0 |f|^{1/2} dz + \frac{1}{4}\pi. \tag{9.11}$$

Hence $\rho(z)$ has the form

$$\rho(z) = \rho_L \sin\{\theta(z) + \nu_3\} + o(1), \tag{9.12}$$

where ρ_L and ν_3 are real constants ($\rho_L \geqslant 0$) related to ν_1 and ν_2. Clearly

$$\rho_L = \overline{\lim_{z\to-\infty}} |\rho(z)|. \tag{9.13}$$

Accordingly, from (9.03), (9.07) and the fact that $|\zeta|^{1/4}M(\zeta) \to \pi^{-1/2}$ as $\zeta \to -\infty$, we have

$$\rho_L \leqslant \lambda_1^{-1}[\exp\{\lambda_1 \mathscr{V}_{-\infty,\infty}(H)\} - 1]. \tag{9.14}$$

In (9.10) and (9.12) let $z \to -\infty$ through a sequence of values for which $\theta(z)$ is an integral multiple of π (compare hypothesis (ii) above). This gives

$$(1+\nu_1)\sin\nu_2 = \rho_L \sin\nu_3. \tag{9.15}$$

Alternatively, by using a sequence of odd integral multiples of $\frac{1}{2}\pi$ we derive

$$(1+\nu_1)\cos\nu_2 - 1 = \rho_L \cos\nu_3. \tag{9.16}$$

Combination of (9.15) and (9.16) yields

$$(1+\nu_1)e^{i\nu_2} = 1 + \rho_L e^{i\nu_3}. \tag{9.17}$$

$$|\nu_1| \leqslant \rho_L, \qquad |\nu_2| \leqslant \sin^{-1}\rho_L \leqslant \pi\rho_L/2, \tag{9.18}$$

provided that $\rho_L \leqslant 1$ (compare (8.07)).

The first Gans-Jeffreys formula is represented by the aggregate of (9.05), (9.09), (9.14), and (9.18).

A second solution of (4.01) is [15]

$$w_2(z) = (\zeta/f)^{1/4}\{\mathrm{Bi}(\zeta) + \epsilon_2(z)\}, \tag{9.19}$$

where

$$|\epsilon_2(z)| \leqslant (\lambda_2/\lambda_1)E(\zeta)M(\zeta)[\exp\{\lambda_1 \mathscr{V}_{-\infty,z}(H)\} - 1], \tag{9.20}$$

and $\lambda_2 = 1.315\ldots.$ Substitution of the asymptotic forms for $\mathrm{Bi}(\zeta)$ yields

$$w_2(z) = \pi^{-1/2} f^{-1/4}\{1 + \sigma(z) + o(1)\}\exp\left(\frac{2}{3}\zeta^{3/2}\right) \qquad (z \to +\infty), \tag{9.21}$$

and

$$w_2(z) = \pi^{-1/2}|f|^{-1/4}\left\{\cos\left(\frac{2}{3}|\zeta|^{3/2}+\frac{1}{4}\pi\right)+o(1)\right\} \qquad (z\to-\infty), \tag{9.22}$$

where

$$\sigma(z) = \pi^{1/2}\zeta^{1/4}\exp\left(-\frac{2}{3}\zeta^{3/2}\right)\epsilon_2(z). \tag{9.23}$$

An obvious extension of Theorem 2 of the present paper shows that $\sigma(z)$ tends to a finite limit, $\sigma(\infty)$, as $z\to+\infty$. Accordingly

$$w_2(z) = \pi^{-1/2}f^{-1/4}\{1+\sigma(\infty)+o(1)\}\exp\left(\int_0^z f^{1/2}\,dz\right) \qquad (z\to+\infty). \tag{9.24}$$

From (9.20) and the asymptotic form of $M(\zeta)$, we obtain

$$|\sigma(\infty)| \leq \frac{1}{2}\sqrt{5}\,\lambda_2\lambda_1^{-1}\,[\exp\{\lambda_1\mathcal{V}_{-\infty,\infty}(H)\}-1]. \tag{9.25}$$

And with the aid of analysis similar to that used for throwing (9.06) into the form (9.09), we readily show that

$$w_2(z) = \pi^{-1/2}|f|^{-1/4}\{1+o(1)\}\cos\left\{\int_z^0 |f|^{1/2}\,dz+\frac{1}{4}\pi+o(1)\right\} \qquad (z\to-\infty). \tag{9.26}$$

Relations (9.24), (9.25), and (9.26) consititute the second Gans-Jeffreys connection formula.

Sharper bounds. Somewhat improved bounds for the error terms of this section can be achieved by application of Theorem 3 of [15], and, in the case of the Bi-type solution, of an inequality of the same character as (3.03) above. The results are expressed by

$$\rho_L \leq \exp\{\mathcal{V}_{-\infty,0}(H)+\lambda_4\mathcal{V}_{0,\infty}(H)\}-1, \tag{9.27}$$

and

$$|\sigma(\infty)| \leq \exp\left\{\frac{2}{\sqrt{3}}\mathcal{V}_{-\infty,0}(H)+\lambda_5\mathcal{V}_{0,\infty}(H)\right\}-1-\left(\frac{2}{\sqrt{3}}-\frac{1}{2}\right)\mathcal{V}_{-\infty,0}(H), \tag{9.28}$$

where $\lambda_4 = 0.521\ldots$ and $\lambda_5 = 0.520\ldots.$

10. Summary

The primary purpose of this paper has been to investigate approximate solutions of the differential equations (1.01) and, more generally, (4.01) in a complex domain containing no singularities of the functions $f(z)$ and $h(z)$, and a simple zero of $f(z)$. Some preliminary results concerning the Liouville-Green approximation are given in sections 2 through 4. The connection formula for the three fundamental approximations of this kind is then established in section 5, complete with strict error bounds, by considering limiting behavior at the point at infinity. An illustrative application to Airy's differential equation is made in section 6, and some general circumstances in which the error terms vanish are investigated in section 7.

An important special case concerns real variables: the problem here is to connect the exponential-type solutions on one side of the turning point with the oscillatory solutions on the other.

This is solved in section 8 by use of the general formula of section 5. An alternative solution is constructed in section 9 solely from real-variable theory, by re-expansion of known uniform Airy-function approximations. Naturally, different hypotheses on $f(z)$ and $h(z)$ are made in the two methods, but the principal difference lies in the forms of the error bounds. In the first method the bounds depend on the variation of a certain function along an infinite path of specified type in the complex plane. In the second method the bounds depend on the variation of a more complicated function along the real axis.

The error bounds given in this paper have computational applications. As shown in the paper which follows [19], they may also be used to provide theoretical insight into the accuracy of the connection formulas in various circumstances, including cases when the differential equation contains a large parameter. Hitherto, in certain applications it has often been difficult to assess the validity of assumptions which have been made.

11. Addendum

The problems studied in the present paper and its sequel [19] have also been considered in a recent monograph by Fröman and Fröman [20]. These authors have the same principal objective, that is, the determination of strict error bounds for the coefficients in the connection formulas. The method they use resembles the first method of the present paper in that it is a rigorous formulation of the procedure of Zwaan. The difference in the approaches is as follows.

We constructed (in sec. 5) the connection formula for the three fundamental Liouville-Green solutions associated with a simple turning point of the differential equation (4.01) by examining the asymptotic behavior of these solutions in the neighborhood of the point at infinity. Fröman and Fröman, following Kemble [3], express the general solution of (4.01) in the form

$$w(z) = a_1(z) f^{-1/4} \exp\left(\int f^{1/2}\, dz\right) + a_2(z) f^{-1/4} \exp\left(-\int f^{1/2}\, dz\right),$$

and derive a pair of first-order simultaneous differential equations for the functions $a_1(z)$ and $a_2(z)$. These equations are then used to determine the changes in these functions as z passes from one principal subdomain to another. Error bounds for the changes are expressed in terms of contour integrals which are equivalent to the variations of our function $F(z)$.

Generally the path for the contour integrals consists of the real z-axis with indentations bypassing the turning points. Fröman and Fröman show that the appropriate form of monotonicity condition is fulfilled when the indentations are semicircles of sufficiently small radius ([20], p. 39). Owing to the singularities of $F(z)$ at the turning points however, the error bounds tend to infinity as the indentations shrink to zero. In consequence Fröman and Fröman assume that the indentations can actually be taken sufficiently large to yield small error bounds, without violating the monotonicity condition. In the present paper we allow greater flexibility in the choice of paths, with a view to achieving minimum bounds in applications.

Other comparisons with the first method of the present paper are as follows:

(i) There is greater emphasis in [20] on the case in which the coefficients $f(z)$ and $h(z)$ in the differential equation (4.01) are real on the real axis.

(ii) Approximations and expansions are frequently used in [20] to simplify the error bounds. There is greater emphasis on strict bounds in the present paper and [19].

(iii) In the special limiting case in which z is at infinity the error bounds can be sharpened by a factor of 2 (see Theorem 2 of sec. 3 above). This refinement is freely used in the present paper and [19], but is not in evidence in [20].

The author acknowledges helpful criticisms of the draft of this paper by Drs. J. Heading, L. Maximon, and F. Stenger.

The work described above has been supported by the U.S. Army Research Office, Durham, North Carolina (Project No. 4238–M).

12. References

[1] Olver, F. W. J., Error bounds for the Liouville-Green (or WKB) approximation, Proc. Camb. Phil. Soc. **57**, 790–810 (1961).

[2] Zwaan, A., Intensitäten im *Ca*-Funkenspektrum, Arch. Néerlandaises des Sciences Exactes, Ser. 3A **12**, 1–76 (1929).

[3] Kemble, E. C., A contribution to the theory of the BWK method, Phys. Rev. **48**, 549–561 (1935).

[4] Furry, W. H., Two notes on phase-integral methods, Phys. Rev. **71**, 360–371 (1947).

[5] Gans, R., Fortpflanzung des Lichts durch ein inhomogenes Medium, Ann. Phys. Lpz. **47**, 709–736 (1915).

[6] Jeffreys, H., On certain approximate solutions of linear differential equations of the second order, Proc. Lond. Math. Soc. (2) **23**, 428–436 (1924–5).

[7] Kramers, H. A., Wellenmechanik und halbzahlige Quantisierung, Z. Phys. **39**, 828–840 (1926).

[8] Langer, R. E., On the connection formulas and the solutions of the wave equation, Phys. Rev. **51**, 669–676 (1937).

[9] Jeffreys, H., On approximate solutions of linear differential equations, Proc. Camb. Phil. Soc. **49**, 601–611 (1953).

[10] Jeffreys, H., Asymptotic approximations (Oxford University Press, 1962).

[11] Morse, P. M. and Feshbach, H., Methods of theoretical physics, Pt 2 (McGraw-Hill Book Co., Inc., New York, N.Y. 1953).

[12] Budden, K. G., Radio waves in the ionosphere. (Cambridge University Press, 1961).

[13] Heading, J., An introduction to phase-integral methods, (John Wiley & Sons, Inc., New York, N.Y., 1962).

[14] Langer, R. E., The asymptotic solutions of ordinary linear differential equations of the second order, with special reference to the Stokes phenomenon, Bull. Amer. Math. Soc. **40**, 545–582 (1934).

[15] Olver, F. W. J., Error bounds for first approximations in turning-point problems, J. Soc. Indust. Appl. Math. **11**, 748–772 (1963).

[16] Olver, F. W. J., Uniform asymptotic expansions of solutions of linear second-order differential equations for large values of a parameter, Phil. Trans. Roy. Soc. London, Ser. A **250**, 479–517 (1958).

[17] Olver, F. W. J., On the asymptotic solutions of second-order differential equations having an irregular singularity of rank one, with an application to Whittaker functions, J. Soc. Indust. Appl. Math. Ser. B **2**, 225–243 (1965).

[18] Miller, J. C. P., The Airy Integral, Brit. Ass. Math. Tab. Part – Vol. B. (Cambridge University Press, 1946).

[19] Olver, F. W. J., Error analysis of phase-integral methods. II. Application to wave-penetration problems, J. Res. NBS **69B**, No. 4, 291–300 (1965).

[20] Fröman, N. and Fröman, P. O., JWKB approximation-Contributions to the theory (North-Holland Publishing Co., Amsterdam, 1965).

(Paper 69B4–157)

JOURNAL OF RESEARCH of the National Bureau of Standards—B. Mathematics and Mathematical Physics
Vol. 69B, No. 4, October–December 1965

Error Analysis of Phase-Integral Methods.

II. Application to Wave-Penetration Problems

F. W. J. Olver

(July 1, 1965)

A study is made of the differential equation

$$d^2w/dz^2 = \{f(z) + h(z)\}w,$$

in which $f(z)$ and $h(z)$ are real and regular on the real axis, and $f(z)$ has exactly two zeros there. Strict error bounds are derived for the coefficients in the formulas which connect the asymptotic solutions at $z = +\infty$ with the asymptotic solutions at $z = -\infty$. Applications are made to two physical problems.

Key Words: Wave penetration, connection formulas, error bounds, potential barrier, transmission coefficient, harmonic oscillator.

1. Introduction and Summary

The purpose of this paper is to illustrate the theory of the preceding paper [1][1] with two problems arising from the one-dimensional wave equation

$$\frac{d^2w}{dz^2} = \{f(z) + h(z)\}w. \tag{1.01}$$

In both problems the functions $f(z)$ and $h(z)$ are real when z is real, and in the region considered $f(z)$ has no singularities and just two zeros, both of which are real and simple. The function $h(z)$ is to be regarded in some sense as being small compared with $f(z)$, but we do not need to formulate this assumption precisely at the outset.

In the first problem (secs. 2, 3) $f(z)$ is positive between the zeros. The object is to determine the ratio of the amplitude of an oscillatory solution at $z = +\infty$ to the amplitudes of its components at $z = -\infty$.

In the second problem (secs. 4, 5) $f(z)$ is negative between the zeros. The object here is to determine eigenconditions admitting a solution having subdominant (exponentially small) character at both $-\infty$ and $+\infty$.

Both problems have been treated frequently in the literature; see the references cited in [1], and [2]–[7] given at the end of this paper. Generally, the viewpoint adopted has been that $f(z)$ contains a multiplicative factor u^2, $h(z)$ is independent of u, and approximate solutions are sought which are asymptotically correct for large values of the parameter u. In the present paper, we establish explicit strict error bounds for the approximate solutions. Asymptotic properties with respect to parameters are immediate consequences of the forms of the bounds.

Two methods for solution were given in [1]. We shall concentrate on the first (Zwaan's method). As we shall see, from the standpoint of deriving error bounds this affords the simpler approach to the present problems.

2. The Overdense Potential Barrier

In this section we suppose that $f(z)$ and $h(z)$ are regular in an unbounded, simply connected complex domain **D**, which includes the whole of the real axis, and that the only zeros of $f(z)$ in **D** are simple zeros at $z = \pm a$ $(a > 0)$. When $-a < z < a$, $f(z)$ is positive.

[1] Figures in brackets indicate the literature references at the end of this paper.

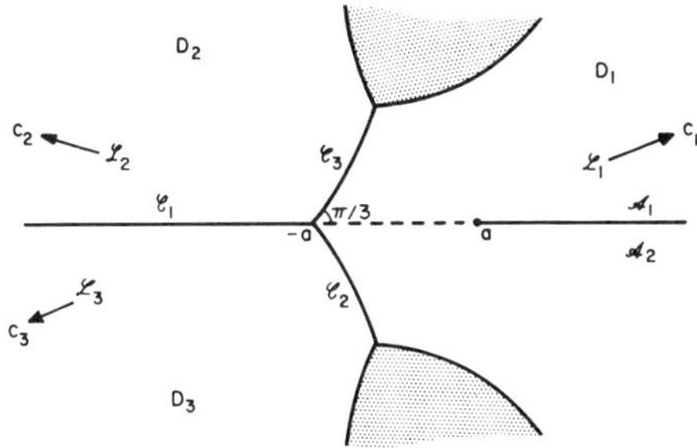

FIGURE 1. *Typical principal subdomains for the over-dense potential barrier.*

Let us consider the turning point at $z=-a$. In order to apply the theory of [1], section 5, we have to exclude the other zero $z=a$ from the domain of consideration, and we do this by introducing a cut $\mathscr{A}$ along the real axis from $z=a$ to $z=+\infty$. We shall refer to the upper and lower sides of this cut as $\mathscr{A}_1$ and $\mathscr{A}_2$, respectively. As in [1], section 5, we define

$$\xi(z)=\int_{-a}^{z}\{f(t)\}^{1/2}dt. \tag{2.01}$$

The associated principal curves $\mathscr{C}_j$ and subdomains $\mathbf{D}_j$ are indicated in figure 1. $\mathscr{C}_1$ is the part of the real axis which extends from $-a$ to $-\infty$. $\mathscr{C}_2, \mathscr{C}_3$ are conjugate curves emerging from $-a$ at directions $\mp\pi/3$. They cannot intersect the real axis at any other point. We define $\xi_j(z)$ to be the branch of $\xi(z)$ which is continuous in $\mathbf{D}$ cut along $\mathscr{A}$ and $\mathscr{C}_j$, and has positive real part in $\mathbf{D}_j$. Similarly, $g_1(z)$, $g_2(z)$, $g_3(z)$ are multiples of any branches of $g(z)\equiv\{f(z)\}^{-1/4}$ which are continuous in $\mathbf{D}$ cut along $\mathscr{A}$ and the corresponding $\mathscr{C}_j$, and satisfy eq (5.08) of [1].

As in [1], let c_j be points at infinity on arbitrary ξ-progressive curves $\mathscr{L}_j$ lying within $\mathbf{D}_j$ such that $|\mathrm{Re}\ \xi(c_j)|=\infty$. $\mathscr{L}_1$ is taken in the upper half-plane, and $\mathscr{L}_2$, $\mathscr{L}_3$ are complex conjugates, but no boundary of $\mathbf{D}$ may intervene between $\mathscr{L}_2$ and $\mathscr{L}_3$ in the neighborhood of infinity. The other assumptions we make are

(i) The variation of the function

$$F\equiv\int\left\{\frac{h}{f^{1/2}}-\frac{1}{f^{1/4}}\frac{d^2}{dz^2}\left(\frac{1}{f^{1/4}}\right)\right\}dz \tag{2.02}$$

converges as $z\to c_j$ along $\mathscr{L}_j$, and also as $z\to\pm\infty$ along the real axis.

(ii) ξ-progressive curves can be found in $\mathbf{D}$ linking: (a) c_1 with c_2; (b) c_1 with c_3; (c) c_2 or c_3 with any point of the interval $(-\infty, -a)$; (d) c_1 with any point of $\mathscr{A}_1$.

(iii) $\epsilon_2(c_3)=0$; $\epsilon_2(z)$ and $\epsilon_3(z)\to 0$ as $z\to-\infty$; $\epsilon_1(z)\to 0$ as $z\to+\infty$ on $\mathscr{A}_1$.

(A brief discussion of the determination of ξ-progressive paths is given in the remark at the end of [1], sec. 5, and sufficient conditions for the fulfillment of (iii) are given in sec. 7 of the same paper.)

With these conditions, the fundamental connection formula ([1], sec. 5) for the solutions

$$w_j(z)=g_j(z)e^{-\xi_j(z)}\{1+\epsilon_j(z)\} \tag{2.03}$$

of the differential eq (1.01), becomes

$$w_1(z)=e^{\pi i/3}\{1+\epsilon_1(c_3)\}w_2(z)+e^{-\pi i/3}\{1+\epsilon_1(c_2)\}w_3(z). \tag{2.04}$$

We fix the $w_j(z)$ by specifying that $g_1(z)$ is the branch of $\{f(z)\}^{-1/4}$ which takes its principal value on the join of $-a$ and $+a$, and interpret the branches of the two sides of (2.04) in real form, as follows.

When $z \in \mathscr{A}_1$, we have

$$\xi_1(z) = \alpha - i\int_a^z |f|^{1/2}dt, \qquad g_1(z) = e^{\pi i/4}|f|^{-1/4}, \tag{2.05}$$

where

$$\alpha = \int_{-a}^{a} |f|^{1/2}dt \tag{2.06}$$

and is positive. Therefore from (2.03) and hypothesis (iii)

$$w_1(z) = e^{-\alpha}|f|^{-1/4}\{1+o(1)\}\ \exp\left(i\int_a^z |f|^{1/2}dt + \frac{1}{4}\pi i\right) \qquad (z \to +\infty). \tag{2.07}$$

This expression represents a *transmitted wave* ([8], p. 1100; [6]).

When $-\infty < z < -a$, we have

$$\xi_2(z) = i\int_z^{-a} |f|^{1/2}dt, \qquad g_2(z) = e^{-\pi i/12}|f|^{-1/4}. \tag{2.08}$$

Hence

$$e^{\pi i/3}\{1+\epsilon_1(c_3)\}w_2(z) = |f|^{-1/4}\{1+\epsilon_1(c_3)+o(1)\}\ \exp\left(-i\int_z^{-a} |f|^{1/2}dt + \frac{1}{4}\pi i\right) \qquad (z \to -\infty). \tag{2.09}$$

This is the *incident wave*, moving in the same direction as the transmitted wave. Similarly

$$e^{-\pi i/3}\{1+\epsilon_1(c_2)\}w_3(z) = |f|^{-1/4}\{1+\epsilon_1(c_2)+o(1)\}\ \exp\left(i\int_z^{-a} |f|^{1/2}dt - \frac{1}{4}\pi i\right) \qquad (z \to -\infty). \tag{2.10}$$

This is the *reflected wave*, moving in the opposite direction to the transmitted wave.

The *transmission coefficient* [2] T and the *reflection coefficient* R are the squares of the limiting ratios of the wave amplitudes. Thus

$$T = e^{-2\alpha}|1+\epsilon_1(c_3)|^{-2}, \tag{2.11}$$

$$R = |1+\epsilon_1(c_2)|^2|1+\epsilon_1(c_3)|^{-2}, \tag{2.12}$$

where ([1], (5.17))

$$|\epsilon_1(c_j)| \leqslant \tfrac{1}{2}\exp\{\mathscr{V}_{c_1, c_j}(F)\} - \tfrac{1}{2} \qquad (j = 2, 3), \tag{2.13}$$

the variations being evaluated along ξ-progressive curves.

Alternative, and sometimes more effective, formulas for T and, especially, R, can be derived with the aid of the energy-conservation eq ([8], p. 1066; [9], sec. 4.4)

$$T + R = 1. \tag{2.14}$$

[2] Or "tunnelling probability" [7].

(This relation may be verified analytically by means of Theorem 4 of [10] and the Wronskian property of the solutions of the differential equation.) Substitution of (2.11) and (2.12) in (2.14) yields

$$|1+\epsilon_1(c_3)|^2-|1+\epsilon_1(c_2)|^2=e^{-2\alpha}. \tag{2.15}$$

Hence

$$T=\frac{e^{-2\alpha}}{|1+\epsilon_1(c_2)|^2+e^{-2\alpha}}, \tag{2.16}$$

$$R=1-\frac{e^{-2\alpha}}{|1+\epsilon_1(c_3)|^2}=\frac{|1+\epsilon_1(c_2)|^2}{|1+\epsilon_1(c_2)|^2+e^{-2\alpha}}. \tag{2.17}$$

Remark. It is interesting to observe that we have not used the connection formula for the turning point at $z=+a$. In contrast, if the problem of this section were attacked by real-variable theory ([1], sec. 9) connection formulas at both $-a$ and $+a$ would have to be used, and the resulting approximate solutions of the differential equation then matched in the interval $-a<z<a$; see, for example, [5, 6].[3] The error analysis obviously would be lengthier and lead to more complicated bounds.

3. Example [4]

Consider the modified Weber equation

$$d^2w/dz^2=(a^2-z^2)w \qquad (a>0). \tag{3.01}$$

We take $f(z)=a^2-z^2$ and $h(z)=0$. Then

$$\xi(z)=\int_{-a}^{z}(a^2-t^2)^{1/2}dt=\tfrac{1}{2}\pi a^2-\tfrac{1}{2}a^2\cos^{-1}(z/a)+\tfrac{1}{2}z(a^2-z^2)^{1/2}. \tag{3.02}$$

The principal subdomains in the z-plane are shown in figure 2 (compare [12], sec. 3). The

[3] A real-variable method which avoids this matching has been suggested by Miller and Good [4] and by Pike [7], but it has yet to be placed on a firm mathematical foundation.

[4] For an account of some direct numerical calculations of transmission coefficients to high precision see [11].

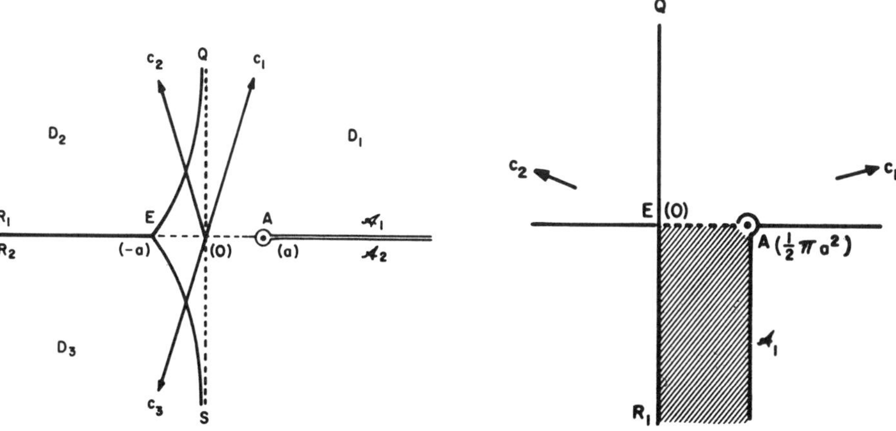

FIGURE 2. *z-plane: principal subdomains.*

FIGURE 3. *ξ_1-plane.*

corresponding map of the upper half of the z-plane on the ξ_1-plane is indicated in figure 3; the map of the lower half is of course the image of figure 3 in the real axis.

The reference points c_1, c_2, and c_3 are taken at infinity on outward-drawn rays from the origin in the first, second, and third quadrants of the z-plane, respectively. With the aid of the ξ_1-map, we see immediately that hypotheses (i) and (ii) of section 2 are satisfied. Theorem 4 of [1] shows that hypothesis (iii) is satisfied, and also that $\epsilon_1(c_2)=0$. From the definition (2.06) we have $\alpha=\frac{1}{2}\pi a^2$. Hence from eqs (2.16) and (2.17), we derive

$$T=\frac{e^{-\pi a^2}}{1+e^{-\pi a^2}}, \qquad R=\frac{1}{1+e^{-\pi a^2}}. \tag{3.03}$$

These results may be verified analytically with the aid of asymptotic expansions and connection formulas for the Weber functions given, for example, in [13]. The actual expression for $w_1(z)$ is found to be, in Miller's notation,

$$w_1(z)=2^{-(ia^2+1)/4}a^{ia^2/2}\exp\left(-\frac{1}{2}\pi a^2-\frac{1}{4}ia^2-\frac{1}{2}i\varphi_2\right)\left\{k^{-1/2}W\left(\frac{1}{2}a^2,\sqrt{2}z\right)+ik^{1/2}W\left(\frac{1}{2}a^2,-\sqrt{2}z\right)\right\}. \tag{3.04}$$

It is noteworthy that our theory [5] proves sufficiently powerful to yield exact expressions for T and R. This example, however, is unique in this respect.

4. The Approximate Harmonic Oscillator

The conditions assumed in this section are those stated in the opening paragraph of section 2, except that $f(z)$ is negative when $-a<z<a$.

We distinguish the auxiliary functions and principal subdomains associated with the turning point at $z=-a$ from those associated with $z=+a$ by the addition of tildes. Thus

$$\xi(z)=\int_a^z \{f(t)\}^{1/2}dt, \qquad \tilde{\xi}(z)=\int_{-a}^z \{f(t)\}^{1/2}dt. \tag{4.01}$$

Three principal curves emerge from $-a$ and three from $+a$, one of which in each case is the join of $-a$ and $+a$; see figure 4. Let $\mathbf{D}_1$, $\tilde{\mathbf{D}}_1$ be the principal subdomains which include the points

[5] This is also true of the theory of [14], section 9.1.

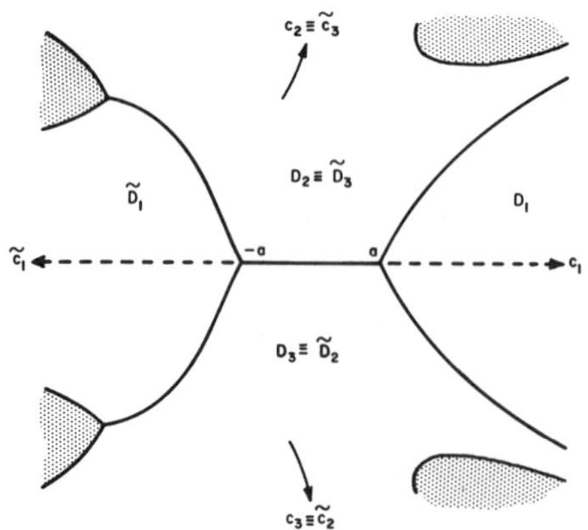

FIGURE 4. *Typical principal subdomains for the approximate harmonic oscillator.*

$c_1 \equiv +\infty$ and $\tilde{c}_1 \equiv -\infty$, respectively. Then clearly $\tilde{\mathbf{D}}_2 \equiv \mathbf{D}_3$ and $\tilde{\mathbf{D}}_3 \equiv \mathbf{D}_2$. We take c_2, c_3 to be points at infinity on conjugate ξ-progressive curves lying within $\mathbf{D}_2$, $\mathbf{D}_3$, respectively, and $\tilde{c}_2 = c_3$, $\tilde{c}_3 = c_2$. We assume that ξ-progressive curves can be found in $\mathbf{D}$ linking c_2 and c_3 with $+\infty$ and $-\infty$, and also with each other, and that $|\text{Re}\ \xi(z)| \to \infty$ as z tends to any of the c_j or $\tilde{c}_j$.

The fundamental connection formulas for the two turning points are given by

$$\{1+\epsilon_2(c_3)\}w_1(z) + e^{-2\pi i/3}\{1+\epsilon_3(c_1)\}w_2(z) + e^{2\pi i/3}\{1+\epsilon_1(c_2)\}w_3(z) = 0, \tag{4.02}$$

and

$$\{1+\tilde{\epsilon}_2(\tilde{c}_3)\}\tilde{w}_1(z) + e^{-2\pi i/3}\{1+\tilde{\epsilon}_3(\tilde{c}_1)\}\tilde{w}_2(z) + e^{2\pi i/3}\{1+\tilde{\epsilon}_1(\tilde{c}_2)\}\tilde{w}_3(z) = 0. \tag{4.03}$$

Clearly $\tilde{w}_3(z)$ is a multiple of $w_2(z)$, and $\tilde{w}_2(z)$ a multiple of $w_3(z)$. To fix these relationships we suppose that $g_1(z)$ and $\tilde{g}_1(z)$ are the branches of the function $\{f(z)\}^{-1/4}$ which take their principal values when $z > a$ and $z < -a$, respectively. We then find that if $z \in \mathbf{D}_2$

$$\xi_2(z) = i\alpha + \tilde{\xi}_3(z), \qquad g_2(z) = e^{-\pi i/6}\tilde{g}_3(z), \tag{4.04}$$

where α is again defined by (2.06). Therefore

$$\tilde{w}_3(z) = e^{(\pi i/6)+i\alpha}\, w_2(z), \qquad \tilde{\epsilon}_3(z) = \epsilon_2(z). \tag{4.05}$$

Substituting these and the corresponding results for $\tilde{w}_2(z)$ and $\tilde{\epsilon}_2(z)$ in (4.03), we obtain the connection formula for $\tilde{w}_1(z)$, $w_2(z)$ and $w_3(z)$. Elimination of $w_3(z)$ by means of (4.02) then yields

$$w_1(z) = \left[\frac{1+\epsilon_1(c_3)}{1+\epsilon_2(c_3)} + \frac{\{1+\epsilon_1(c_2)\}\{1+\tilde{\epsilon}_1(c_3)\}}{\{1+\epsilon_2(c_3)\}\{1+\tilde{\epsilon}_1(c_2)\}}e^{2i\alpha}\right]e^{\pi i/3}w_2(z) - \frac{1+\epsilon_1(c_2)}{1+\tilde{\epsilon}_1(c_2)}\, ie^{i\alpha}\tilde{w}_1(z). \tag{4.06}$$

The eigencondition or "quantization condition" is that $w_1(z)$ be a multiple of $\tilde{w}_1(z)$. Accordingly, this is expressed by

$$e^{2i\alpha} = -\frac{\{1+\epsilon_1(c_3)\}\{1+\tilde{\epsilon}_1(c_2)\}}{\{1+\epsilon_1(c_2)\}\{1+\tilde{\epsilon}_1(c_3)\}}. \tag{4.07}$$

Neglecting the error terms, we obtain at once the well-known approximations $\alpha = \frac{1}{2}\pi, \frac{3}{2}\pi, \frac{5}{2}\pi, \ldots$. To assess the errors in these approximations, we write

$$\epsilon_1(c_3) = pe^{iq}, \qquad \epsilon_1(c_2) = pe^{-iq}, \qquad \tilde{\epsilon}_1(c_2) = \tilde{p}e^{i\tilde{q}}, \qquad \tilde{\epsilon}_1(c_3) = \tilde{p}e^{-i\tilde{q}}, \tag{4.08}$$

where p, $\tilde{p} > 0$ and q, $\tilde{q}$ are real. Then

$$\ln\left[\frac{\{1+\epsilon_1(c_3)\}\{1+\tilde{\epsilon}_1(c_2)\}}{\{1+\epsilon_1(c_2)\}\{1+\tilde{\epsilon}_1(c_3)\}}\right] = 2i(\omega + \tilde{\omega}), \tag{4.09}$$

where ω, $\tilde{\omega}$ are real, and

$$|\omega| \leq \sin^{-1} p \leq \tfrac{1}{2}\pi p, \qquad |\tilde{\omega}| \leq \sin^{-1}\tilde{p} \leq \tfrac{1}{2}\pi\tilde{p}, \tag{4.10}$$

provided that p, $\tilde{p} \leq 1$. Hence the eigencondition is given by

$$\alpha = (n + \tfrac{1}{2} + \beta)\pi \qquad (n = 0, 1, 2, \ldots), \tag{4.11}$$

where $\beta=(\omega+\bar{\omega})/\pi$ and is real. From (4.08), (4.10) and [1], (5.17), we obtain the desired bound

$$|\beta| \leqslant \frac{1}{4} \exp \{\mathscr{V}_{\infty, c_2}(F)\} + \frac{1}{4} \exp \{\mathscr{V}_{-\infty, c_2}(F)\} - \frac{1}{2}, \tag{4.12}$$

where F is any continuous branch of the integral defined by (2.02) above, and the variations are taken along ξ-progressive paths.

When the eigencondition is satisfied, eq (4.06) reduces to

$$w_1(z) = (-)^n \frac{1+\delta}{1+\bar{\delta}} \bar{w}_1(z), \tag{4.13}$$

where

$$|\delta| \leqslant \tfrac{1}{2} \exp \{\mathscr{V}_{\infty, c_2}(F)\} - \tfrac{1}{2}, \qquad |\bar{\delta}| \leqslant \tfrac{1}{2} \exp \{\mathscr{V}_{-\infty, c_2}(F)\} - \tfrac{1}{2}. \tag{4.14}$$

Thus on the real axis the eigensolution is given by

$$w_1(z) = \{1+\epsilon_1(z)\} f^{-1/4} \exp \left(-\int_a^z f^{1/2} dt\right), \tag{4.15}$$

where

$$|\epsilon_1(z)| \leqslant \exp \{\mathscr{V}_{\infty, z}(F)\} - 1 \qquad (z > a), \tag{4.16}$$

and

$$w_1(z) = (-)^n (1+\delta)(1+\bar{\delta})^{-1} \{1+\bar{\epsilon}_1(z)\} f^{-1/4} \exp \left(-\int_z^{-a} f^{1/2} dt\right), \tag{4.17}$$

where

$$|\bar{\epsilon}_1(z)| \leqslant \exp \{\mathscr{V}_{-\infty, z}(F)\} - 1 \qquad (z < -a). \tag{4.18}$$

The fractional powers in (4.15) and (4.17) are positive, and the variations in (4.16) and (4.18) are taken along the real axis.

Remark. Let $f(z)$ and $h(z)$ be even functions of z, and assume that c_2, c_3 can be taken as the points $\pm i\infty$. Then the bound (4.12) and eq (4.13) reduce to

$$|\beta| \leqslant \tfrac{1}{2} \exp \{\mathscr{V}_{\infty, i\infty}(F)\} - \tfrac{1}{2}, \tag{4.19}$$

and

$$w_1(z) = (-)^n \bar{w}_1(z) = (-)^n w_1(-z). \tag{4.20}$$

5. Examples

(i) *Weber equation.*

$$d^2w/dz^2 = (z^2 - a^2)w \qquad (a > 0). \tag{5.01}$$

We take $f(z) = z^2 - a^2$ and $h(z) = 0$. Then

$$\xi(z) = \int_a^z (t^2 - a^2)^{1/2} dt = \tfrac{1}{2} z(z^2 - a^2)^{1/2} - \tfrac{1}{2} a^2 \ln \{z + (z^2 - a^2)^{1/2}\} + \tfrac{1}{2} a^2 \ln a. \tag{5.02}$$

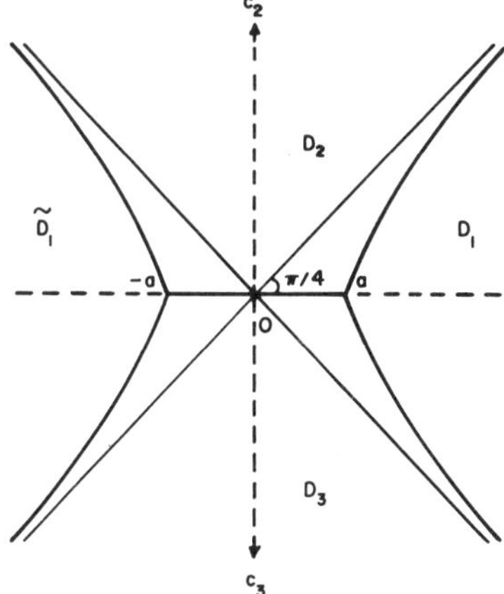

FIGURE 5. *Principal subdomains for the Weber equation.*

Figure 5 illustrates the principal subdomains. We take c_2, c_3 to be the points $\pm i\infty$. From [1], theorem 4, we see that $\epsilon_1(c_2)=\bar{\epsilon}_1(c_2)=0$. Hence in (4.08) $p=\bar{p}=0$. Therefore β vanishes and $\alpha=(n+\frac{1}{2})\pi$, *exactly*.[6] Using (2.06), we see that this implies $a^2=2n+1$. The eigensolution of course obeys (4.20). These results accord with the known analytical properties of the Weber functions [13].

(ii) *Large parameter.* Let $f(z)=u^2f_0(z)$, where u is a large positive parameter, and the functions $f_0(z)$ and $h(z)$ are independent of u (though this restriction could be eased without significantly affecting the following analysis). Then

$$\alpha=u\alpha_0, \qquad F(z)=F_0(z)/u, \tag{5.03}$$

where

$$\alpha_0=\int_{-a}^{a}|f_0|^{1/2}dt, \qquad F_0(z)=\int\left\{\frac{h}{f_0^{1/2}}-\frac{1}{f_0^{1/4}}\frac{d^2}{dz^2}\left(\frac{1}{f_0^{1/4}}\right)\right\}dz, \tag{5.04}$$

and are independent of u. Write

$$\chi(u)\equiv\frac{1}{4}\exp\{u^{-1}\mathscr{V}_{\infty,c_2}(F_0)\}+\frac{1}{4}\exp\{u^{-1}\mathscr{V}_{-\infty,c_2}(F_0)\}-\frac{1}{2}. \tag{5.05}$$

Then the eigencondition (4.11) is expressed by the following equation for u:

$$\psi(u)\equiv u\alpha_0-(n+\tfrac{1}{2})\pi-\pi\beta(u)=0, \tag{5.06}$$

in which

$$|\beta(u)|\leq\chi(u). \tag{5.07}$$

For large u, $\chi(u)=O(u^{-1})$. Therefore for large values of the positive integer n

$$u=\alpha_0^{-1}(n+\tfrac{1}{2})\pi+O(n^{-1}). \tag{5.08}$$

[6] Compare also [14], page 109.

In order to determine precise bounds for the O term in the last equation, consider the function $\chi(u)$. This tends to infinity as $u\to 0$, and to zero as $u\to\infty$, and is monotonic strictly decreasing when u is positive. Let u_0 be the root of $\chi(u_0)=\frac{1}{2}$. Then provided that $n\pi/\alpha_0>u_0$, we have from (5.06) and (5.07)

$$\psi(n\pi/\alpha_0)=-\tfrac{1}{2}\pi-\pi\beta(n\pi/\alpha_0)<0, \tag{5.09}$$

and

$$\psi\{(n+1)\pi/\alpha_0\}=\tfrac{1}{2}\pi-\pi\beta\{(n+1)\pi/\alpha_0\}>0. \tag{5.10}$$

Therefore at least one eigenvalue exists in the interval $n\pi/\alpha_0<u<(n+1)\pi/\alpha_0$. To delimit it in a shorter interval, let

$$u=(n+\tfrac{1}{2}+v)\pi/\alpha_0, \tag{5.11}$$

where $|v|<\frac{1}{2}$. Then substituting in (5.06), we obtain

$$v=\beta\{(n+\tfrac{1}{2}+v)\pi/\alpha_0\}. \tag{5.12}$$

Hence $|v|<\chi(n\pi/\alpha_0)$, that is,

$$|v|<\frac{1}{4}\exp\left\{\frac{\alpha_0}{n\pi}\mathscr{V}_{\infty,c_2}(F_0)\right\}+\frac{1}{4}\exp\left\{\frac{\alpha_0}{n\pi}\mathscr{V}_{-\infty,c_2}(F_0)\right\}-\frac{1}{2}. \tag{5.13}$$

Relations (5.11) and (5.13) constitute the precise form of the eigencondition (5.08), and are valid when $n>\alpha_0u_0/\pi$. The eigensolution satisfies (4.13), where

$$|\delta|\leqslant\frac{1}{2}\exp\left\{\frac{\alpha_0}{n\pi}\mathscr{V}_{\infty,c_2}(F_0)\right\}-\frac{1}{2},\qquad |\tilde{\delta}|\leqslant\frac{1}{2}\exp\left\{\frac{\alpha_0}{n\pi}\mathscr{V}_{-\infty,c_2}(F_0)\right\}-\frac{1}{2}. \tag{5.14}$$

The above proof shows that there is at least one eigenvalue (5.11) which satisfies (5.13). In order to establish that there is *exactly* one eigenvalue of this form, we need to investigate the sign of the derivative $\beta'(u)$ in the interval considered.[7] This takes us back to fundamentals. We have to consider the Liouville-Green approximate solutions of the differential equation

$$d^2w/dz^2=\{u^2+f(z)\}w \tag{5.15}$$

(compare [1], sec. 2) as functions of u, and seek bounds for the u-derivatives of their error terms. The analysis is somewhat lengthy, and we record only the result:

Let

$$X(u)=\frac{le^l}{3-e^l}\,\frac{1+\kappa}{u}+\frac{\tilde{l}e^{\tilde{l}}}{3-e^{\tilde{l}}}\,\frac{1+\tilde{\kappa}}{u}-\alpha_0, \tag{5.16}$$

where

$$l=u^{-1}\mathscr{V}_{\infty,c_2}(F_0),\qquad \tilde{l}=u^{-1}\mathscr{V}_{-\infty,c_2}(F_0), \tag{5.17}$$

$$\kappa=(2e\cos\gamma)^{-1},\qquad \tilde{\kappa}=(2e\cos\tilde{\gamma})^{-1}, \tag{5.18}$$

[7] An alternative, but apparently no easier, approach would be to consider complex values of u and apply Rouché's theorem.

and γ, $\tilde{\gamma}$ are the maximum angles of slope with the real ξ-axis of the ξ-maps of the progressive paths linking ∞ to c_2, and $-\infty$ to c_2, respectively ($|\gamma|$, $|\tilde{\gamma}| < \frac{1}{2}\pi$). Then exactly one eigenvalue satisfies (5.11) and (5.13), provided that $n \geqslant \alpha_0 u_1/\pi$ where u_1 is the largest positive zero of $X(u)$.

Some simplification can be effected by using upper bounds for the zeros of $\chi(u) - \frac{1}{2}$ and $X(u)$. In this way we may verify that both of the conditions $n > \alpha_0 u_0/\pi$ and $n \geqslant \alpha_0 u_1/\pi$ are certainly fulfilled if n exceeds each of the three quantities

$$\frac{\alpha_0 \mathcal{V}_{\infty, c_2}(F_0)}{\pi \ln 2}, \quad \frac{\alpha_0 \mathcal{V}_{-\infty, c_2}(F_0)}{\pi \ln 2}, \quad \frac{2 \ln 2}{\pi}(2 + \kappa + \tilde{\kappa}). \tag{5.19}$$

The author acknowledges helpful criticisms of the draft of this paper by Drs. J. Heading and L. Maximon.

The work described above has been supported by the U.S. Army Research Office, Durham, N. Carolina (Project No. 4238–M).

6. References

[1] Olver F. W. J., Error analysis of phase-integral methods. I. General theory for simple turning points, J. Res. NBS **69B**, No. 4, 271–290 (1965).
[2] Dunham, J. L., The Wentzel-Brillouin-Kramers method of solving the wave equation, Phys. Rev. **41**, 713–720 (1932).
[3] Birkhoff, G. D., Quantum mechanics and asymptotic series. Bull. Amer. Math. Soc. **39**, 681–700 (1933).
[4] Miller, S. C., and Good, R. H., A WKB-type approximation to the Schrödinger equation, Phys. Rev. **91**, 174–179 (1953).
[5] Jeffreys, H., On the use of asymptotic approximations of Green's type when the coefficient has zeros, Proc. Camb. Phil. Soc. **52**, 61–66 (1956).
[6] McLeod, J. B., The determination of the transmission coefficient. Quart. J. Math. Oxford (2) **12**, 153–158 (1961).
[7] Pike, E. R., On the related-equation method of asymptotic approximation II. Direct solutions of wave-penetration problems, Quart. J. Mech. Appl. Math. **17**, 369–379 (1964).
[8] Morse, P. M., and Feshbach, H., Methods of Theoretical Physics, Pt 2 (McGraw-Hill Book Co., Inc., New York, N.Y., 1953).
[9] Heading, J., An Introduction to Phase-Integral Methods (John Wiley & Sons, Inc., New York, N.Y., 1962).
[10] Olver, F. W. J., Error bounds for the Liouville-Green (or WKB) approximation. Proc. Camb. Phil. Soc. **57**, 790–810 (1961).
[11] Jamet, P., Calcul exact du coefficient de transmission d'une barrière de potentiel: étude de l'erreur de l'approximation BKW (in press).
[12] Olver, F. W. J., Uniform asymptotic expansions for Weber parabolic cylinder functions of large orders, J. Res. NBS **63B** (Math. and Math. Phys.), No. 2, 131–169 (1959).
[13] Miller, J. C. P., Tables of Weber parabolic cylinder functions (Her Majesty's Stationery Office, London, 1955).
[14] Fröman, N., and Fröman, P. O., JWKB approximation—Contributions to the theory (North-Holland Publishing Co., Amsterdam, 1965).

(Paper 69B4–158)

JOURNAL OF RESEARCH of the National Bureau of Standards — B. Mathematics and Mathematical Physics
Vol. 71B, Nos. 2 and 3, April-September 1967

Numerical Solution of Second-Order Linear Difference Equations

F. W. J. Olver

Institute for Basic Standards, National Bureau of Standards, Washington, D.C. 20234

(May 4, 1967)

A new algorithm is given for computing the solution of any second-order linear difference equation which is applicable when simple recurrence procedures cannot be used because of instability. Compared with the well-known Miller algorithm the new method has the advantages of (i) automatically determining the correct number of recurrence steps, (ii) applying to inhomogeneous difference equations, (iii) enabling more powerful error analyses to be constructed.

The method is illustrated by numerical computations, including error analyses, of Anger-Weber, Struve, and Bessel functions, and the solution of a differential equation in Chebyshev series.

Key Words: Chebyshev series, difference equations, error analysis, Miller algorithm, recurrence methods, special functions.

1. Introduction

A powerful computational algorithm for evaluating the most rapidly decreasing solution of a second-order homogeneous linear difference equation was published in 1952 by J. C. P. Miller ([1],[1] page xvii) in connection with the tabulation of modified Bessel functions. Since then, various writers have applied the algorithm to other special functions, and similar computational processes have been used by Clenshaw [2] for the numerical solution of ordinary differential equations in series of Chebyshev polynomials. Error analyses of the algorithm have been supplied by the present writer [3] and Oliver [12] and quite recently Gautschi [4] has examined the relation of the algorithm to classical results in the theory of continued fractions.

The present investigation stems from the observation that Miller's algorithm can be regarded as a procedure for solving a tridiagonal set of simultaneous linear algebraic equations. Adopting this more general standpoint, we shall show how to recast the algorithm into a new form which enables the correct number of recurrence steps to be determined automatically without appeal to an asymptotic or other analytical formula. In this respect it resembles an algorithm proposed recently by Shintani [5].

The new formulation has the further advantages of (i) being applicable to inhomogeneous difference equations, (ii) lending itself readily to powerful error analyses. There seems to be no alternative method of comparable power available at present for computing solutions of inhomogeneous equations in the case when forward recurrence and backward recurrence are both unstable.

2. Statement of the Problem

Let the given difference equation be denoted by

$$a_r y_{r-1} - b_r y_r + c_r y_{r+1} = d_r, \tag{2.01}$$

[1]Figures in brackets indicate the literature references at the end of this paper.

where a_r, b_r, c_r, and d_r are given functions of the nonnegative integer variable r. We assume that the general solution of (2.01) has the form

$$y_r = Af_r + Bg_r + h_r, \tag{2.02}$$

in which A and B are arbitrary constants, and the complementary functions f_r, g_r, and the particular solution h_r have the properties $f_0 \neq 0$, $g_r \neq 0$ for all sufficiently large r, and

$$f_r/g_r \to 0, \qquad h_r/g_r \to 0, \qquad (r \to \infty). \tag{2.03}$$

(It may be noted that we do not require either f_r or h_r to tend to zero as $r \to \infty$.)

The first problem we investigate is the computation of the solution of (2.01) which has the property

$$y_r/g_r \to 0 \qquad (r \to \infty), \tag{2.04}$$

and satisfies the *normalizing condition*

$$y_0 = k \tag{2.05}$$

for an arbitrarily assigned value of the constant k. Later (secs. 9–11) we allow for a more general form of normalizing condition and also drop the restriction $f_0 \neq 0$.

The given conditions ensure that y_r exists and is unique. For, from (2.03) and (2.04) the B of (2.02) is seen to be zero, and from (2.05) we derive $A = (k - h_0)/f_0$. Therefore

$$y_r = \frac{k - h_0}{f_0} f_r + h_r. \tag{2.06}$$

It is well known that direct use of (2.01) as a recurrence relation for generating $y_2, y_3, \ldots$ from given values of y_0 and y_1 (if available) is an unstable procedure. Essentially, each computational rounding error introduces into the numerical solution a small multiple of f_r and a small multiple of g_r, and in consequence of (2.04) the latter ultimately grows faster than the wanted solution.

It may also happen [2] in the inhomogeneous case that f_r grows more rapidly than y_r in the direction of decreasing r. In this event recurrence by use of (2.01) is unstable in this direction too.

3. Approach

Analogous work in the numerical solution of linear differential equations [3] suggests that a stable way of solving the present problem is to treat it directly as a boundary-value problem rather than use initial-value techniques. We are already given the value of y_0. Suppose that for some large integer N, the value of y_N can be obtained from an asymptotic formula or by other means. Then eqs (2.01) with $r = 1, 2, \ldots, N-1$ comprise a set of simultaneous linear algebraic equations for the unknowns $y_1, y_2, \ldots, y_{N-1}$, which are solvable by standard matrix computational processes.

This possibility has already been noted by Gautschi ([4], Introduction) following a suggestion by M. E. Rose, but Gautschi did not pursue the idea because of the difficulty of obtaining the value of y_N, in general. Following Miller's approach in the homogeneous case [1], we solve the algebraic equations with the value of y_N arbitrarily set equal to zero. It transpires that for large N the great majority of the y_r produced in this way are generally excellent approximations to the true values; only in the neighborhood of $r = N$ can substantial errors occur.

We first establish the convergence of the process.

[2] See Example 1 of section 6.
[3] See, for example, [6].

Theorem 1. *With the conditions of section 2, suppose that for all sufficiently large* N *the system of equations*

$$a_r y_{r-1}^{(N)} - b_r y_r^{(N)} + c_r y_{r+1}^{(N)} = d_r, \quad (r = 1, 2, \ldots, N-1), \tag{3.01}$$

and

$$y_0^{(N)} = k, \qquad y_N^{(N)} = 0, \tag{3.02}$$

has a solution $y_0^{(N)}, y_1^{(N)}, \ldots, y_N^{(N)}$. *Then if* r *is fixed and* $N \to \infty$, $y_r^{(N)} \to y_r$.

To establish this result, we observe that since y_r is a particular solution of (2.01), we can express

$$y_r^{(N)} = A_N f_r + B_N g_r + y_r,$$

where A_N and B_N are independent of r. Setting $r=0$, N in turn and using (2.05) and (3.02), we derive

$$A_N = \frac{g_0 y_N}{f_0 g_N - g_0 f_N}, \qquad B_N = -\frac{f_0 y_N}{f_0 g_N - g_0 f_N},$$

the denominator $f_0 g_N - g_0 f_N$ being nonvanishing for all sufficiently large N in consequence of the assumed conditions. In fact, from (2.03) we have

$$f_0 g_N - g_0 f_N \sim f_0 g_N \qquad (N \to \infty).$$

Hence from (2.04) it follows that $A_N \to 0$ and $B_N \to 0$. This completes the proof.

Thus for any given value of r, or for any finite range of values, y_r can be calculated to prescribed accuracy by solving the system of equations (3.01) and (3.02) with a sufficiently high value of N. Naturally, we now enquire what exactly constitutes a "sufficiently high" value? Or, to put the question another way, given N, to what accuracy does $y_r^{(N)}$ approximate y_r for $r < N$?

A simple practical way of providing an answer is to solve eqs (3.01) and (3.02) for increasing values of N until the results are in satisfactory numerical agreement. This procedure has two drawbacks. First, it is wasteful of computing time if the originally guessed values of N are either too low or much too high. Second, there is no absolute guarantee that values of y_r computed with two (or more) different values of N must be correct when they agree.

The optimum value of N, that is, the minimum value necessary to achieve specified accuracy in y_r for a given range of values of r, can be determined automatically when a suitable method is used to solve the algebraic equations. Accordingly, we consider this process next.

4. Solution of the Algebraic Equations

We shall solve the tridiagonal system of equations

$$\left.\begin{aligned}
-b_1 y_1^{(N)} + c_1 y_2^{(N)} \qquad\qquad\qquad &= d_1 - a_1 k,\\
a_2 y_1^{(N)} - b_2 y_2^{(N)} + c_2 y_3^{(N)} \qquad &= d_2,\\
a_3 y_2^{(N)} - b_3 y_3^{(N)} + c_3 y_4^{(N)} &= d_3,\\
\cdots\cdots\cdots\cdots &,\\
a_{N-2} y_{N-3}^{(N)} - b_{N-2} y_{N-2}^{(N)} + c_{N-2} y_{N-1}^{(N)} &= d_{N-2},\\
a_{N-1} y_{N-2}^{(N)} - b_{N-1} y_{N-1}^{(N)} &= d_{N-1},
\end{aligned}\right\} \tag{4.01}$$

by simple elimination followed by back-substitution. To begin with we suppose that none of the c_r vanish.

Let the first of (4.01) be rewritten in the form

$$p_2 y_1^{(N)} - p_1 y_2^{(N)} = e_1, \tag{4.02}$$

where

$$p_1 = 1, \qquad p_2 = \frac{b_1}{c_1}, \qquad e_1 = \frac{a_1 k - d_1}{c_1}. \tag{4.03}$$

The result of eliminating $y_1^{(N)}$ from (4.02) and the second of (4.01) can be expressed as

$$p_3 y_2^{(N)} - p_2 y_3^{(N)} = e_2,$$

where

$$p_3 = \frac{b_2 p_2 - a_2 p_1}{c_2}, \qquad e_2 = \frac{a_2 e_1 - d_2 p_2}{c_2}.$$

Continuing the elimination, we obtain

$$p_{r+1} y_r^{(N)} - p_r y_{r+1}^{(N)} = e_r \qquad (r = 1, 2, \ldots, N-1), \tag{4.04}$$

where

$$p_{r+1} = \frac{b_r p_r - a_r p_{r-1}}{c_r}, \qquad e_r = \frac{a_r e_{r-1} - d_r p_r}{c_r}. \tag{4.05}$$

Thus p_r is the solution of the homogeneous form of the difference eq (2.01), with the initial conditions $p_0 = 0$ and $p_1 = 1$. We also observe that the second of (4.05) holds for $r = 1$ if we define

$$e_0 = k.$$

The final equation of the form (4.04) is used to begin the back-substitution. On substituting the second of (3.02), we derive

$$y_{N-1}^{(N)} = e_{N-1}/p_N; \tag{4.06}$$

thence $y_{N-2}^{(N)}, y_{N-3}^{(N)}, \ldots, y_1^{(N)}$ may be computed by use of (4.04) with descending values of r. The process fails if, and only if, one of the numbers $p_2, p_3, \ldots, p_N$ vanishes. In this event the set of eqs (3.01) and (3.02) has either no solution or an infinity of solutions, and the algorithm breaks down.

When one or more of the coefficients c_r vanishes the set of eqs (4.01) becomes uncoupled. A simple modification takes care of the situation. Suppose, for example, that $c_s = 0$ but all other c_r are nonzero. Then the first s equations of (4.01) determine $y_1^{(N)}, y_2^{(N)}, \ldots, y_s^{(N)}$ completely: they can be solved by application of the recurrence relations (4.05) for $r = 1, 2, \ldots, s-1$ and use of the back-substitution relation (4.04), beginning with

$$y_s^{(N)} = \frac{a_s e_{s-1} - d_s p_s}{b_s p_s - a_s p_{s-1}}. \tag{4.07}$$

The remaining $N-s-1$ equations are solvable for $y_{s+1}^{(N)}, y_{s+2}^{(N)}, \ldots, y_{N-1}^{(N)}$ by the method already described: eq (4.07) takes the place of the first of (3.02).

To ease the presentation we shall suppose in the remainder of the paper that none of the c_r vanish.

Applying ourselves to the problem of determining the optimum value of N, we observe that the effect of replacing N by $N+1$ is to prolong the elimination process by one step, beginning the back-substitution with $y_{N+1}^{(N+1)}=0$ instead of $y_N^{(N)}=0$. Thus we have

$$p_{r+1}y_r^{(N+1)}-p_r y_{r+1}^{(N+1)}=e_r \qquad (r=1, 2, \ldots, N). \tag{4.08}$$

Subtraction of (4.04) from (4.08) gives

$$y_r^{(N+1)}-y_r^{(N)}=\frac{p_r}{p_{r+1}}\left(y_{r+1}^{(N+1)}-y_{r+1}^{(N)}\right) \qquad (r\leq N-1), \tag{4.09}$$

and repeated application of this result leads to

$$y_r^{(N+1)}-y_r^{(N)}=\frac{p_r}{p_{r+1}}\frac{p_{r+1}}{p_{r+2}}\cdots\frac{p_{N-1}}{p_N}\left(y_N^{(N+1)}-y_N^{(N)}\right),$$

that is,

$$y_r^{(N+1)}-y_r^{(N)}=\frac{p_r e_N}{p_N p_{N+1}} \qquad (r=1, 2, \ldots, N). \tag{4.10}$$

By use of this formula we can predict the effect of changing N into $N+1$ *before* any back-substitution is carried out.

Suppose, for example, that we wish to compute y_L to D decimal places for given values of the integers L and D. Then the recurrence relations (4.05) are applied from $r=1$ past $r=L$ until a value of r is reached for which

$$\left|\frac{p_L e_r}{p_r p_{r+1}}\right|<\tfrac{1}{2}\times 10^{-D}. \tag{4.11}$$

If this value of r is taken as N, then we can be sure that the approximation $y_L^{(N)}$ yielded by the back-substitution agrees to D decimal places with the value $y_L^{(N+1)}$ that would be obtained from the next higher approximation. (Whether this value of N *is* adequate is considered in the next section.)

If, as is more usual, accurate values of y_r are required for a whole range of values of r, then the criterion (4.11) is used with $|p_L|$ denoting the greatest value of $|p_r|$ in the given range. We might, for example, desire the computation to D decimal places of *all* values of y_r that exceed $1/2\times 10^{-D}$ in absolute value—it being assumed, of course, in this case that $y_r\to 0$ as $r\to\infty$. Then N is determined by the condition that

$$|e_N/p_{N+1}|<\tfrac{1}{2}\times 10^{-D}, \tag{4.12}$$

provided that $|p_N|\geq|p_r|$ when $r<N$.

5. Expansions for the Solution and the Truncation Error

The method suggested in the last section for determining N is based on the criterion that the values of $y_r^{(N)}$ and $y_r^{(N+1)}$ must agree to within the prescribed tolerance for y_r. This does not guarantee, however, that their common value is y_r. To resolve this doubt we consider higher approximations.

Replacing N by $N+1$ in (4.10), and adding the result to (4.10) itself, we obtain

$$y_r^{(N+2)} - y_r^{(N)} = p_r \left(\frac{e_N}{p_N p_{N+1}} + \frac{e_{N+1}}{p_{N+1} p_{N+2}} \right).$$

Continuation of this process yields

$$y_r^{(N+s)} - y_r^{(N)} = p_r \left(\frac{e_N}{p_N p_{N+1}} + \frac{e_{N+1}}{p_{N+1} p_{N+2}} + \ . \ . \ . + \frac{e_{N+s-1}}{p_{N+s-1} p_{N+s}} \right),$$

where s is an arbitrary positive integer. Letting $s \to \infty$ and using Theorem 1, we derive the following expression for the *truncation error*

$$\epsilon_r^{(N)} \equiv y_r - y_r^{(N)} = E_N p_r, \tag{5.01}$$

where E_N is the sum of the (necessarily convergent) series

$$E_N = \sum_{s=N}^{\infty} \frac{e_s}{p_s p_{s+1}}. \tag{5.02}$$

Thus the precise criterion for determining N is that $|E_N p_r|$ must not exceed the specified tolerance in y_r for each wanted value of r.

Once the value of N has been decided, the actual value of the truncation error can be found by continuing the computation of p_r and e_{r-1} beyond $r=N$ and using (5.01) and (5.02). Later [7], we shall show how to use these expansions to determine strict bounds for $\epsilon_r^{(N)}$ directly from the properties of the coefficients a_r, b_r, c_r, and d_r.

As a special case of (5.01) we have the expansion

$$y_r = p_r \sum_{s=r}^{\infty} \frac{e_s}{p_s p_{s+1}}. \tag{5.03}$$

Subtraction of (5.01) from (5.03) yields

$$y_r^{(N)} = p_r \sum_{s=r}^{N-1} \frac{e_s}{p_s p_{s+1}} \qquad (r < N), \tag{5.04}$$

a result which is obtainable more directly by repeated use of the back-substitution relation (4.04). Thus *the whole of our computing scheme is equivalent to approximating the convergent infinite series (5.03) by the partial sum (5.04).*

6. Examples

EXAMPLE 1. *Anger-Weber functions.*

For integer values of r the function $\mathbf{E}_r(x)$ satisfies eq (2.01) with

$$a_r = c_r = 1, \qquad b_r = \frac{2r}{x}, \qquad d_r = -\frac{2\{1-(-1)^r\}}{\pi x}.$$

We restrict ourselves here to positive values of the argument x. The principal properties of $\mathbf{E}_r(x)$ are established in [8], chapter 10. In particular, we have

$$\mathbf{E}_{2r}(x) = \frac{2x}{(4r^2-1)\pi} \sum_{s=0}^{\infty} \alpha_s(r) x^{2s}, \qquad \mathbf{E}_{2r+1}(x) = \frac{2}{(2r+1)\pi} \sum_{s=0}^{\infty} \frac{2r-2s-1}{2r-1} \alpha_s(r) x^{2s}, \tag{6.01}$$

where $\alpha_0(r)=1$, and

$$\alpha_s(r)=\frac{1}{(4r^2-3^2)(4r^2-5^2)\ldots\{4r^2-(2s+1)^2\}} \qquad (s>0). \tag{6.02}$$

Using the inequality

$$\left|\frac{\alpha_s(r)}{\alpha_{s-1}(r)}\right|\leq\frac{1}{|4r-1|},$$

we deduce that if x is fixed and $r\to\infty$, then

$$\mathbf{E}_{2r}(x)\sim\frac{2x}{(4r^2-1)\pi}, \qquad \mathbf{E}_{2r+1}(x)\sim\frac{2}{(2r+1)\pi}. \tag{6.03}$$

The corresponding homogeneous form of (2.01) has the Bessel functions $J_r(x)$ and $Y_r(x)$ as solutions. For fixed x and large r, we have

$$J_r(x)\sim\frac{1}{(2\pi r)^{1/2}}\left(\frac{ex}{2r}\right)^r, \qquad Y_r(x)\sim-\left(\frac{2}{\pi r}\right)^{1/2}\left(\frac{2r}{ex}\right)^r. \tag{6.04}$$

Thus ultimately $J_r(x)$ decays more rapidly than $\mathbf{E}_r(x)$, and $|Y_r(x)|$ grows rapidly. In consequence, both simple forward recurrence and simple backward recurrence are unstable methods for generating $\mathbf{E}_r(x)$ from (2.01) when $r>x$.

With

$$f_r=J_r(x), \qquad g_r=Y_r(x), \qquad h_r=\mathbf{E}_r(x),$$

the conditions of section 2 are satisfied, provided that $J_0(x)\neq 0$. Let us apply the method of section 4 to a specific example, say the computation of $\mathbf{E}_r(x)$ for $x=1$, $r=1(1)10$, correct to within 2 units of the eighth decimal place. We suppose that we are given $\mathbf{E}_0(1)=-0.56865\ 6627$, this value having been extracted from [9] and confirmed by evaluation of the first of (6.01).

Beginning with $p_0=0$, $p_1=1$, and $e_0=\mathbf{E}_0(1)$, values of p_r and e_r were generated by use of (4.05). They are recorded in the upper part of table 1, correct to 9 significant figures. After passing the last of the given values of r, namely 10, the "test function" $p_{10}e_r/(p_rp_{r+1})$ was computed. This falls below the value 2×10^{-8} for the first time when $r=14$. In accordance with the criterion of section 4 this is the value [4] to be assigned to N. The column of values $y_r^{(14)}$ was then generated by backward use of (4.04), beginning with $y_{14}^{(14)}=0$. For $r=1(1)10$ these are the wanted approximations to $\mathbf{E}_r(1)$.

To test the accuracy of the results, the computations were repeated for $N=32$ and $N=34$, using a time-sharing automatic computer and working to 36 floating binary figures, with an exponent of 12 binary figures. As further checks the values for $r=1(1)5$ were compared with the 10-decimal values given in [9], and the values for $r=10$ and 11 computed from the expansions (6.01). The full results of these computations are not included here, but the digits in $y_r^{(14)}$ which differ from those in the more accurate values of $\mathbf{E}_r(1)$ are printed in italic type, and the difference $\epsilon_r^{(14)}$ between the two values is recorded, in units of the 9th decimal place, in the penultimate column of the upper part of table 1. As expected, this error does not exceed 2×10^{-8} in absolute value within the wanted range $r=1(1)10$.

[4] The more precise criterion of section 5 requires that the sum of $p_{10}e_r/(p_rp_{r+1})$ and all subsequent values in this column be less than 2×10^{-8}. Inspection of table 1 indicates that this also is satisfied for $N=14$. This would not be the case, however, if we reduced our error tolerance in y_r from 2×10^{-8} to 1×10^{-8}.

TABLE 1. *Anger-Weber function* $\mathbf{E}_r(1)$

r	p_r	e_r	$\dfrac{p_{10}e_r}{p_r p_{r+1}}$	$y_r^{(14)}$	$10^9\epsilon_r^{(14)}$	$10^9 E_{14} p_r$
0	0	−0.56865 6627				
1	1	0.70458 2918		0.43816 2436		
2	2	0.70458 2918		.17174 1955		
3	7	9.61725 973		.24880 5382		
4	40	9.61725 973		.04785 0795		
5	313	$4.08141\ 237\times10^{2}$		.13400 0978		
6	3090	$4.08141\ 237\times10^{2}$		.01891 9443		
7	36767	$4.72213\ 396\times10^{4}$		.09303 2343		
8	5 11648	$4.72213\ 396\times10^{4}$		.01029 3811		
9	81 49601	$1.04236\ 156\times10^{7}$		.07166 8637	1	1
10	1461 81170	$1.04236\ 156\times10^{7}$	3.6×10^{-3}	.00650 2117	12	12
11	$2.91547\ 380\times10^{9}$	$3.72252\ 015\times10^{9}$	2.9×10^{-3}	.05837 3706	240	240
12	$6.39942\ 424\times10^{10}$	$3.72252\ 015\times10^{9}$	5.5×10^{-6}	.00447 9865	5279	5279
13	$1.53294\ 634\times10^{12}$	$1.95553\ 042\times10^{12}$	4.7×10^{-6}	.04914 3054	12 6445	12 6444
14	$3.97926\ 106\times10^{13}$	$1.95553\ 042\times10^{12}$	6.5×10^{-9}	.00000 0000		
15	$1.11266\ 015\times10^{15}$	$1.41863\ 843\times10^{15}$	5.6×10^{-9}			
16	$3.33400\ 119\times10^{16}$					

r	p_r	e_r	$\dfrac{e_r}{p_r p_{r+1}}$	
14	$3.97926\ 106\times10^{13}$	$1.95553\ 042\times10^{12}$	4.41672×10^{-17}	$E_{14}=8.24845\times10^{-17}$
15	$1.11266\ 015\times10^{15}$	$1.41863\ 843\times10^{15}$	3.82422×10^{-17}	
16	$3.33400\ 119\times10^{16}$	$1.41863\ 843\times10^{15}$	0.00399×10^{-17}	
17	$1.06576\ 772\times10^{18}$	$1.35839\ 625\times10^{18}$	$.00352\times10^{-17}$	
18	$3.62027\ 625\times10^{19}$	$1.35839\ 625\times10^{18}$	$.00000\times10^{-17}$	
19	$1.30223\ 368\times10^{21}$			

It is of interest to apply the expansions of section 5 to this example. The necessary computations for evaluating the expansion (5.02) are given in the lower part of table 1, and the values of $10^9E_{14}p_r$ appear in the final column of the upper part of the same table. They agree with $10^9\epsilon_r^{(14)}$ to within a unit.

EXAMPLE 2. *Struve functions.*

The function $\mathbf{H}_r(x)$ satisfies eq (2.01) with

$$a_r=c_r=1,\qquad b_r=\frac{2r}{x},\qquad d_r=\frac{(\frac{1}{2}x)^r}{\sqrt{\pi}\,\Gamma\left(r+\frac{3}{2}\right)}.$$

For fixed x and large r we have ([8], sec. 10.4)

$$\mathbf{H}_r(x)\sim\frac{x}{\sqrt{2}\,\pi r}\left(\frac{ex}{2r}\right)^r. \tag{6.05}$$

Since the complementary functions of the difference equation are again the Bessel functions $J_r(x)$ and $Y_r(x)$, the conditions of section 2 are satisfied, provided that $J_0(x)\neq 0$.

Let us evaluate $\mathbf{H}_r(0.1)$ to 8 significant figures for all positive integer values of r such that $|\mathbf{H}_r(0.1)|$ exceeds $\frac{1}{2}\times10^{-30}$.

The computations are shown in table 2. The value

$$e_0=\mathbf{H}_0(0.1)=0.06359\ 12700$$

TABLE 2. *Struve function* $\mathbf{H}_r(0.1)$

r	p_r	d_r	e_r	$\dfrac{e_r}{p_r p_{r+1}}$	$y_r^{(15)}$	$\dfrac{10^9 \epsilon_r^{(15)}}{y_r^{(15)}}$
0	0	0.63661 9772	0.06359 12700			
1	1	2.12206 591 × 10^-2^	.04237 06109		2.12065 160 × 10^-3^	0
2	20	4.24413 182 × 10^-4^	.03388 23473		4.24211 125 × 10^-5^	0
3	799	6.06304 546 × 10^-6^	.02903 79740		6.06080 029 × 10^-7^	−1
4	47920	6.73671 718 × 10^-8^	.02580 97391		6.73467 605 × 10^-9^	0
5	38 32801	6.12428 835 × 10^-10^	.02346 24212		6.12271 820 × 10^-11^	2
6	3832 32180	4.71099 104 × 10^-12^	.02165 70178		4.70994 424 × 10^-13^	4
7	4.59840 288 × 10^10^	3.14066 069 × 10^-14^	.02021 28155		3.14004 492 × 10^-15^	4
8	6.43738 080 × 10^12^	1.84744 746 × 10^-16^	.01902 35432		1.84712 338 × 10^-17^	−1
9	1.02993 494 × 10^15^	9.72340 768 × 10^-19^	.01802 20955		9.72186 442 × 10^-20^	0
10	1.85381 852 × 10^17^	4.63019 413 × 10^-21^	.01716 37415		4.62952 313 × 10^-22^	4
11	3.70753 405 × 10^19^	2.01312 788 × 10^-23^	.01641 73675		2.01285 948 × 10^-24^	4
12	8.15638 953 × 10^21^	8.05251 152 × 10^-26^	.01576 05733		8.05151 746 × 10^-27^	2
13	1.95749 641 × 10^24^	2.98241 167 × 10^-28^	.01517 67673	1.5 × 10^-53^	2.98206 890 × 10^-29^	−1
14	5.08940 910 × 10^26^	1.02841 782 × 10^-30^	.01465 33634	2.0 × 10^-58^	1.02829 540 × 10^-31^	1 1520
15	1.42501 497 × 10^29^	3.31747 684 × 10^-33^	.01418 06180	2.3 × 10^-63^	0	
16	4.27499 402 × 10^31^					

was extracted from [9], and confirmed by evaluation of the expansion

$$\mathbf{H}_r(x) = (\tfrac{1}{2}x)^{r+1} \sum_{s=0}^{\infty} \frac{(-)^s (\tfrac{1}{2}x)^{2s}}{\Gamma\left(s+\dfrac{3}{2}\right) \Gamma\left(s+r+\dfrac{3}{2}\right)}. \tag{6.06}$$

The largest of the wanted values of r was determined by the criterion

$$|e_r/p_{r+1}| > \tfrac{1}{2} \times 10^{-30} \quad \text{and} \quad |e_{r+1}/p_{r+2}| \leq \tfrac{1}{2} \times 10^{-30};$$

compare (5.03). This gave $r=13$. Next, we have from (5.01), (5.02), and (5.03),

$$\frac{\epsilon_r^{(N)}}{y_r} = \frac{p_r p_{r+1}}{e_r} \frac{e_N}{p_N p_{N+1}}.$$

From table 2 we see that the right of this relation is an increasing function of r, hence N is the least value for which

$$\frac{e_N}{p_N p_{N+1}} \leq (\tfrac{1}{2} \times 10^{-8}) \frac{e_{13}}{p_{13} p_{14}}.$$

From the entries in the column headed $e_r/(p_r p_{r+1})$ we see immediately that this gives $N=15$.

The values of $y_r^{(15)}$, computed from (4.04), appear in the penultimate column of the table. For $r \leq 13$ they are the required approximations to $\mathbf{H}_r(0.1)$. Again, more accurate values were obtained by automatic computation with a higher value of $N(26)$, and also by evaluation of the expansion (6.06) for $r=1(1)15$. In the final column the relative error $\epsilon_r^{(15)}/y_r^{(15)}$ is given in units of the 9th decimal place. As expected, it lies within the stipulated limit $\frac{1}{2} \times 10^{-8}$ in the required range.

7. Propagation of Rounding Errors

In addition to the truncation error $\epsilon_r^{(N)}$ which has been analyzed in sections 4 and 5, the other possible sources of error in the final solution are the rounding errors introduced during the calculations. Since the computing process is essentially the solution of a finite system of linear algebraic equations, the nature of the transmission of these errors is available from general theory [10] chapter 4; [6], chapter 9. However, because of special features of the present problem, including the fact that in our form of elimination the absolute values of the multipliers are not bounded by unity, some comments on the effects of rounding errors may be helpful.

Consider first the computation of the sequence p_r. From the conditions $p_0=0$, $p_1=1$, we see that in terms of the fundamental solutions f_r and g_r of section 2

$$p_r=(f_0g_r-g_0f_r)/(f_0g_1-g_0f_1), \tag{7.01}$$

the denominator here necessarily being nonzero since f_r and g_r are independent solutions of the difference equation. By hypothesis, $f_0 \neq 0$; therefore p_r always contains a multiple of g_r. And since $f_r/g_r \to 0$ as $r \to \infty$, p_r ultimately becomes proportional to g_r when a fixed number of significant figures is maintained in the computations.

Each rounding error in the formation of the p_r can be regarded as introducing unwanted small multiples of f_r and g_r. Ultimately, the former dies out in comparison with the latter; the error is then propagated at the same rate as p_r itself. Before this stage is attained, however, some loss of accuracy is possible. If the value of $|f_0|$ is unduly small compared with $|g_0f_1/g_1|$, then from (7.01) we see that initially p_r behaves like a multiple of f_r. But the rounding errors are still propagated in proportion to g_r, and this generally causes a steady loss of significant figures. The loss ceases when the term f_0g_r in (7.01) overtakes g_0f_r in magnitude, at which stage the computation becomes completely stable.

It should be realized that this loss of accuracy is not attributable to the method of computation, but to the fact that, as a rule, the whole problem is ill-posed when $|f_0|$ is small compared with $|g_0f_r/g_r|$ for at least one value of r. For from (2.06) we see that

$$\left|\frac{\delta y_r}{y_r}\right|=\left|\frac{f_r\delta k}{f_0y_r}\right| \gg \left|\frac{g_r\delta k}{g_0y_r}\right|, \tag{7.02}$$

where δy_r is the change in y_r consequent upon an arbitrary change δk in the value of k. Since $|g_r/g_0|$ is generally large compared with $|y_r|$ (see (2.04)), the relative error in y_r is very sensitive to rounding errors in the given value of k.

Examples 1 and 2 of section 6 would be ill-posed in this way if the chosen value of x were close to a zero of $J_0(x)$, say $x=5.52$. This would become apparent at the beginning of the computations: the early p_r would diminish in size, in contrast to the behavior they exhibit in tables 1 and 2.

The difficulty could be overcome in these and other examples by carrying out the computation of k and p_r to higher precision, and making the necessary prolongation of the recurrences until the criteria of sections 4 and 5 for terminating them are met.

If the value of y_1 can be found, however, a preferable alternative is to apply the algorithm of sections 3 and 4 with the given y_1 as normalizing value, instead of $y_0=k$. In effect, this means that the recurrences (4.05) are begun with $p_1=0$, $p_2=1$, and $e_1=y_1$. Subsequently the value of y_0 can be computed from y_1 and y_2 by a single backward application of (2.01).

The other part of the elimination process is the computation of the right-hand sides e_r. From (4.05) we see that in the inhomogeneous case instability could arise from this source if there were persistent heavy cancellation between a_re_{r-1} and d_rp_r. No naturally occurring examples of this phenomenon have been encountered so far, however.

Lastly, we see from (4.04) that a rounding error introduced in $y_s^{(N)}$ during the back-substitution is multiplied by the factor p_r/p_s when it is transmitted to $y_r^{(N)}$ $(r<s)$. Except when the problem is ill-posed, this factor decays with diminishing r at a faster rate than $y_r^{(N)}$ itself, because p_r contains a substantial multiple of g_r, and y_r contains no multiple of this function.

Summarizing this section, we have shown that unstable transmission of rounding errors can occur only when the original problem is ill-posed or when heavy cancellation takes place during the calculation of e_r from the second of (4.05).

8. Comparison With the Algorithms of Miller and Shintani

In section 4 we solved the set of eqs (3.01) and (3.02) by eliminating the variables in the order $y_1^{(N)}, y_2^{(N)}, \ldots, y_{N-2}^{(N)}$: we may call this *forward elimination.* Suppose now that these variables are eliminated in the reverse order: *backward elimination.* The resulting set of pivotal equations can be expressed in the form

$$u_{r+1}^{(N)} y_r^{(N)} - u_r^{(N)} y_{r+1}^{(N)} = v_r^{(N)} \quad (r = N-2, N-3, \ldots, 0), \tag{8.01}$$

where the quantities $u_r^{(N)}$ and $v_r^{(N)}$ are defined by

$$u_{N-1}^{(N)} = 1, \qquad u_{N-2}^{(N)} = b_{N-1}/a_{N-1}, \qquad v_{N-2}^{(N)} = d_{N-1}/a_{N-1}, \tag{8.02}$$

and

$$u_{r-1}^{(N)} = \frac{b_r u_r^{(N)} - c_r u_{r+1}^{(N)}}{a_r}, \qquad v_{r-1}^{(N)} = \frac{c_r v_r^{(N)} + d_r u_r^{(N)}}{a_r}, \qquad (r < N-1). \tag{8.03}$$

(It should be observed that $u_r^{(N)}$ and $v_r^{(N)}$ depend on N as well as r, unlike the p_r and e_r of section 4.)

The last of eqs (8.01) is used to begin the back-substitution. It yields

$$y_1^{(N)} = (u_1^{(N)} k - v_0^{(N)})/u_0^{(N)},$$

where k is again the given value of y_0. Then $y_2^{(N)}, y_3^{(N)}, \ldots, y_{N-1}^{(N)}$ may be computed by successive application of (8.01) with $r = 1, 2, \ldots, N-2$.

Thus the elimination process consists of constructing a sequence $u_r^{(N)}$ which satisfies the homogeneous form of the given difference equation (2.01) and the conditions $u_N^{(N)} = 0$, $u_{N-1}^{(N)} = 1$. This is exactly the first stage of Miller's algorithm [1], [3]: the $u_r^{(N)}$ are the so-called trial values. And in the homogeneous case, given by $d_r = 0$, all the quantities $v_r^{(N)}$ vanish, causing the formulas (8.01) for back-substitution to reduce to

$$y_r^{(N)} = \frac{u_r^{(N)}}{u_{r-1}^{(N)}} y_{r-1}^{(N)} = \frac{u_r^{(N)}}{u_{r-1}^{(N)}} \frac{u_{r-1}^{(N)}}{u_{r-2}^{(N)}} \cdots \frac{u_1^{(N)}}{u_0^{(N)}} y_0^{(N)} = \frac{k}{u_0^{(N)}} u_r^{(N)}.$$

This is the second stage of the Miller algorithm: $k/u_0^{(N)}$ is the normalizing factor.

Accordingly, *in the homogeneous case the Miller recurrence algorithm can be regarded as the solution of the set of equations (3.01) and (3.02), with* $\mathrm{d}_r = 0$, *by backward elimination.* In the inhomogeneous case the solution by backward elimination, described above, can be regarded as a generalization of the Miller algorithm.

Compared with the forward elimination process of section 4, the Miller algorithm suffers from the disadvantages that it does not determine automatically the correct value of N, and if a second value of N is used as a check on the adequacy of the original value, then the computations must begin afresh. The advantage of the Miller algorithm is that the process of back-substitution is less laborious; this advantage is restricted to the homogeneous case, however, and is offset if more than one trial value of N has to be used.

The method Shintani [5] [5] has developed for solving second-order linear difference equations in the homogeneous case consists of the use of the Miller algorithm preceded by two forward recurrence processes to determine the optimum value of N. In the present notation, Shintani takes $a_r = 1$ and $d_r = 0$. His formulas for forward recurrence are given by ([5], Theorem 1)

[5] See also [4], section 4.

$$P_{r+1}(\nu)=b_{r+1}P_r(\nu)-c_rP_{r-1}(\nu), \tag{8.04}$$

where $\nu=0$ or 1, and

$$P_{-1}(0)=0, \qquad P_0(0)=1; \qquad P_0(1)=0, \qquad P_1(1)=1. \tag{8.05}$$

It is easily verified, for example, that the quantities $P_r(0)$ appear when the forward elimination procedure is applied to eqs (4.01) with $d_r=0$ and the multipliers chosen in such a way that the constant value $-k$ is preserved on the right-hand sides. The resulting pivotal equations are in fact

$$-P_r(0)y_r^{(N)}+c_rP_{r-1}(0)y_{r+1}^{(N)}=-k \qquad (r=1,2,\ldots,N-1). \tag{8.06}$$

In our notation

$$P_r(0)=c_1c_2\ldots c_rp_{r+1}. \tag{8.07}$$

From the computational standpoint, the evaluation of Shintani's sequence $P_r(0)$ may be compared with the evaluation of our sequence p_r, the evaluation of his $P_r(1)$ with our e_r, and the application of the Miller algorithm with our process of back-substitution. In the first stage the computing effort is identical, but in the second and third stages our method requires considerably less effort.

9. More General Form of Normalizing Condition

Let us consider now the solution of the difference eq (2.01) when (2.05) is replaced by the more general normalizing condition

$$m_0y_0+m_1y_1+m_2y_2+\ldots=k, \tag{9.01}$$

in which m_0, m_1, . . ., and k are given constants. We again suppose that the general solution of (2.01) has the form (2.02), but instead of the conditions imposed on f_r, g_r, and h_r in section 2, we assume that

$$\left|\sum_{r=0}^{N} m_rg_r\right|\to\infty \text{ as } N\to\infty, \tag{9.02}$$

and

$$\sum_{r=0}^{\infty} m_rf_r=F, \qquad \sum_{r=0}^{\infty} m_rh_r=H, \tag{9.03}$$

where F and H are finite, and $F\neq 0$. Then (2.01) has a unique solution fulfilling (9.01). It is given by

$$y_r=\frac{k-H}{F}f_r+h_r; \tag{9.04}$$

compare (2.06).

The obvious extension of the approach of section 3 is to solve the system of linear algebraic equations given by

$$a_ry_{r-1}^{(N)}-b_ry_r^{(N)}+c_ry_{r+1}^{(N)}=d_r \qquad (r=1,2,\ldots,N-1), \tag{9.05}$$

$$\sum_{r=0}^{N} m_ry_r^{(N)}=k, \tag{9.06}$$

and

$$y_N^{(N)}=0. \tag{9.07}$$

THEOREM 2. *In addition to the other conditions of this section, assume that for all sufficiently large* N *the system of equations* (9.05), (9.06), *and* (9.07) *has a solution, that* $g_N \neq 0$, *and that*

$$\frac{f_N}{g_N}\sum_{r=0}^{N} m_r g_r \to 0, \qquad \frac{h_N}{g_N}\sum_{r=0}^{N} m_r g_r \to 0, \qquad (N\to\infty). \tag{9.08}$$

Then if r *is fixed and* $N\to\infty$, $y_r^{(N)}\to y_r$.

This result may be established by expressing $y_r^{(N)}$ in the form

$$y_r^{(N)} = A_N f_r + B_N g_r + h_r. \tag{9.09}$$

Using (9.06) and (9.07), we find that

$$A_N = \frac{h_N \sum_{r=0}^{N} m_r g_r - g_N\left(\sum_{r=0}^{N} m_r h_r - k\right)}{g_N \sum_{r=0}^{N} m_r f_r - f_N \sum_{r=0}^{N} m_r g_r}, \qquad B_N = \frac{f_N\left(\sum_{r=0}^{N} m_r h_r - k\right) - h_N \sum_{r=0}^{N} m_r f_r}{g_N \sum_{r=0}^{N} m_r f_r - f_N \sum_{r=0}^{N} m_r g_r}.$$

In consequence of the assumed conditions, the denominators are asymptotic to Fg_N as $N\to\infty$. Hence $A_N\to(k-H)/F$. Next, the assumed conditions imply that f_N/g_N and h_N/g_N both tend to zero. Hence $B_N\to 0$. Comparison of (9.04) and (9.09) completes the proof.

When the forward elimination process of section 4 is applied to eqs (9.05), (9.06), and (9.07), the following pivotal equations are obtained:

$$p_{r+1}y_r^{(N)} - p_r y_{r+1}^{(N)} + q_r\left(\sum_{s=r+1}^{N-1} m_s y_s^{(N)}\right) = e_r \qquad (r=0, 1, \ldots, N-1), \tag{9.10}$$

(compare (4.04)), where

$$q_0 = 1, \qquad q_r = \frac{a_1 a_2 \ldots a_r}{c_1 c_2 \ldots c_r} \qquad (r \geqslant 1), \tag{9.11}$$

$$p_0 = 0, \qquad p_1 = m_0, \qquad e_0 = k, \tag{9.12}$$

and, if $r \geqslant 1$,

$$p_{r+1} = \frac{b_r p_r - a_r p_{r-1}}{c_r} + q_r m_r, \qquad e_r = \frac{a_r e_{r-1} - d_r p_r}{c_r}. \tag{9.13}$$

In consequence of (9.07), the final equation of the form (9.10) reduces to

$$p_N y_{N-1}^{(N)} = e_{N-1}. \tag{9.14}$$

This yields the value of $y_{N-1}^{(N)}$; thence $y_{N-2}^{(N)}, y_{N-3}^{(N)}, \ldots, y_0^{(N)}$ may be computed from (9.10) by back-substitution.

The value of N may be determined in a similar way to that suggested in section 4. Suppose, for example, that all nonvanishing values of y_r are needed to a fixed number of decimal places,

D, say—a common form of requirement with the present type of normalizing condition. Then N is determined by the condition

$$|e_N/p_{N+1}| < \tfrac{1}{2} \times 10^{-D}, \tag{9.15}$$

provided that $|p_r| \leqslant |p_N|$ when $r \leqslant N$, and also

$$|p_r/q_{r-1}| > |m_r|, \ |m_{r+1}|, \ . \ . \ ., \ |m_N|. \tag{9.16}$$

EXAMPLE 3. *Bessel functions.*

Let us evaluate $J_0(x), J_1(x), . . .,$ for $x=5$ to 5 decimal places, by use of the relations

$$J_{r-1}(x) - (2r/x)\ J_r(x) + J_{r+1}(x) = 0, \tag{9.17}$$

and

$$J_0(x) + 2J_2(x) + 2J_4(x) + . . . = 1. \tag{9.18}$$

In the present notation, we have

$$a_r = c_r = 1, \qquad b_r = 2r/x, \qquad d_r = 0,$$

$$m_0 = 1, \qquad m_1 = m_3 = . . . = 0, \qquad m_2 = m_4 = . . . = 2, \qquad k = 1.$$

Accordingly, eqs (9.11) through (9.13) yield

$$q_r = 1, \qquad e_r = 1, \qquad p_0 = 0, \qquad p_1 = 1,$$

and

$$p_{r+1} = b_r p_r - p_{r-1} + m_r. \tag{9.19}$$

Table 3 gives the values of p_r correct to 6 significant figures. The criterion (9.15) suggests that N be taken as the least value of r for which $|p_{r+1}| > 2 \times 10^5$. This gives[6] $N = 14$. The column of values of $y_r^{(14)}$ is then generated upwards by use of (9.10), starting with $y_{14}^{(14)} = 0$. These are the required approximations to $J_n(5)$: their differences, $\epsilon_r^{(14)}$, from the true values are recorded in the final column in units of the 5th decimal place. The agreement is satisfactory.

TABLE 3. *Bessel function* $J_r(5)$

r	b_r	m_r	p_r	$y_r^{(14)}$	$\sum_r^{13} m_s y_s^{(14)}$	$10^5 \epsilon_r^{(14)}$
0	0.0	1	0	−0.17758		−2
1	0.4	0	1	−.32758	1.17758	0
2	0.8	2	0.4	.04655	1.17758	2
3	1.2	0	1.32	.36482	1.08448	1
4	1.6	2	1.184	.39123	1.08448	0
5	2.0	0	2.5744	.26114	0.30202	0
6	2.4	2	3.9648	.13105	.30202	0
7	2.8	0	8.94112	.05338	.03992	0
8	3.2	2	21.0703	.01841	.03992	0
9	3.6	0	60.4838	.00552	.00310	0
10	4.0	2	196.671	.00147	.00310	0
11	4.4	0	728.200	.00035	.00016	0
12	4.8	2	3007.41	.00008	.00016	0
13	5.2	0	13709.4	.00001	.00000	1
14	5.6	2	68281.5	.00000	.00000	0
15			368669.			

[6] When the condition (9.16) is violated—as it is in this example near the beginning of the range—it would be safer in practice to take N slightly higher than the value predicted by (9.15).

It may be noted that estimates of the optimum value of N for generating Bessel functions from (9.17) and (9.18) by Miller's algorithm have been computed by Makinouchi [11] for $x=0.01(.01)$ $0.1(.1)1(1)10(10)100$ and precisions of 9, 10, 18, 20, and 30 significant figures. These values were obtained by use of the asymptotic approximations (6.04) above. In constructing a program for generating the $J_r(x)$ for arbitrary x and arbitrary precision, however, it would be simpler to determine the optimum N by use of (9.15) (or (4.11)). The resulting gain would tend to offset the extra effort needed in applying the back-substitution relation (9.10) compared with the normalizing of the trial values in the Miller algorithm.

10. Bounds for the Truncation Error

In order to obtain strict bounds for the truncation error associated with the algorithm of section 9, we proceed as in section 5. Write, temporarily,

$$\eta_r = y_r^{(N+1)} - y_r^{(N)}. \tag{10.01}$$

Then from (9.10) we obtain

$$p_{r+1}\eta_r = p_r\eta_{r+1} - q_r(m_{r+1}\eta_{r+1} + \ldots + m_N\eta_N) \qquad (r<N). \tag{10.02}$$

Therefore

$$|\eta_r| \leq \rho_r(|\eta_{r+1}| + |\eta_{r+2}| + \ldots + |\eta_N|) \qquad (r<N), \tag{10.03}$$

where ρ_r is the greater of

$$\left|\frac{p_r - q_r\, m_{r+1}}{p_{r+1}}\right| \text{ and } \left|\frac{q_r}{p_{r+1}}\right| \sup_{2\leq s\leq\infty} |m_{r+s}|. \tag{10.04}$$

Equations (9.07), (9.14), and (10.01) yield $\eta_N = e_N/p_{N+1}$. From this result and (10.03) we may verify that

$$|\eta_{N-1}| \leq \rho_{N-1}|e_N/p_{N+1}|, \tag{10.05}$$

and thence by induction that

$$|\eta_r| \leq \rho_r(1+\rho_{r+1})(1+\rho_{r+2}) \ldots (1+\rho_{N-1})\,|e_N/p_{N+1}| \qquad (r\leq N-2). \tag{10.06}$$

The left-hand side of the last relation is $|y_r^{(N+1)} - y_r^{(N)}|$. Replacing N by $N+1$, $N+2$, . . ., in turn and summing, and applying Theorem 2, we find that

$$|\epsilon_r^{(N)}| \leq \rho_r(1+\rho_{r+1})(1+\rho_{r+2}) \ldots (1+\rho_{N-1})E_N \qquad (r\leq N-2), \tag{10.07}$$

where

$$\epsilon_r^{(N)} = y_r - y_r^{(N)}, \tag{10.08}$$

and

$$E_N = \left|\frac{e_N}{p_{N+1}}\right| + (1+\rho_N)\left|\frac{e_{N+1}}{p_{N+2}}\right| + (1+\rho_N)(1+\rho_{N+1})\left|\frac{e_{N+2}}{p_{N+3}}\right| + \ldots, \tag{10.09}$$

provided that the last series converges. Similarly

$$|\epsilon_{N-1}^{(N)}| \leq \rho_{N-1}E_N, \qquad |y_N| \leq E_N. \tag{10.10}$$

The results (10.07) and (10.10) are strict bounds for the truncation error, in contrast to the expansion of section 5 which is exact (for the algorithm of sections 3 and 4). Often the bound (10.07) is a considerable overestimate.[7] Thus in Example 3, the right-hand side of (10.07) or (10.10) has the following values for $N=14$, in units of the 5th decimal place:

$$568,\ 237,\ 29,\ 12,\ 3,\ 1,\ 1,\ \text{then zero for } r=7,\ 8,\ \ldots,\ 14.$$

In consequence, if N is determined by the criterion that for each required value of r the right-hand sides of (10.07) and (10.10) must not exceed the specified tolerance in y_r, then the resulting value is perfectly safe but often unnecessarily high. Applied to Example 3, this criterion yields $N=18$, compared with the value 14 which we used and found to be quite adequate.

In the next section we give an alternative formulation of the algorithm of section 9. Although perhaps less elegant, it generally yields a sharper assessment of the truncation error than that of this section.

11. Alternative Method for the General Normalizing Condition

The algorithm of sections 3 and 4 can be applied to the problem of section 9 in the following way. First, we construct a solution f_r of the homogeneous form of the given equation (2.01). The choice of this solution is arbitrary, provided that the first of (2.03) is satisfied. Then by means of an additional back-substitution we construct an arbitrary solution h_r of (2.01) itself. The required solution y_r may then be computed from (9.04), in which k is defined by (9.01), and F, H by (9.03). In the case when the given difference eq (2.01) is itself homogeneous, only the solution f_r need be computed, and (9.04) reduces to

$$y_r=(k/F)f_r. \tag{11.01}$$

The simplest choice of the normalizing conditions needed for constructing f_r and h_r is given by

$$f_0=1, \qquad h_0=0. \tag{11.02}$$

The first of these may be an inconvenient or even impossible condition, however; in this event we may follow the suggestion given in section 7 and use instead

$$f_1=1, \qquad h_1=0. \tag{11.03}$$

To assess the truncation error in the final solution y_r, let $\varphi_r^{(N)}$ and $\theta_r^{(N)}$ be the truncation errors in the approximations $f_r^{(N)}$ and $h_r^{(N)}$ to f_r and h_r; thus

$$f_r=f_r^{(N)}+\varphi_r^{(N)}, \qquad h_r=h_r^{(N)}+\theta_r^{(N)}. \tag{11.04}$$

Bounds for $\varphi_r^{(N)}$ and $\theta_r^{(N)}$ are computable from the expansions of section 5. From (9.04) we have

$$y_r=\frac{k-H_N-\tau_N}{F_N+\sigma_N}\left(f_r^{(N)}+\varphi_r^{(N)}\right)+h_r^{(N)}+\theta_r^{(N)} \qquad (r=0,\ 1,\ \ldots,\ N), \tag{11.05}$$

where F_N, H_N are the computed quantities

$$F_N=\sum_{r=0}^{N-1} m_r f_r^{(N)}, \qquad H_N=\sum_{r=0}^{N-1} m_r h_r^{(N)}, \tag{11.06}$$

[7] This can be traced to the fact that over most of the range the second of the two quantities (10.04) is usually very much smaller than the first.

and σ_N, τ_N assessable errors [8]

$$\sigma_N = \sum_{r=0}^{N} m_r \varphi_r^{(N)} + \sum_{r=N+1}^{\infty} m_r f_r, \tag{11.07}$$

$$\tau_N = \sum_{r=0}^{N} m_r \theta_r^{(N)} + \sum_{r=N+1}^{\infty} m_r h_r. \tag{11.08}$$

If $F_N \neq 0$, then to the first order of small quantities the truncation error in the formula

$$y_r \doteqdot \frac{k - H_N}{F_N} f_r^{(N)} + h_r^{(N)} \tag{11.09}$$

is composed of three parts:

$$\frac{k-H_N}{F_N}\varphi_r^{(N)}, \qquad -\left\{\tau_N + \frac{\sigma_N(k-H_N)}{F_N}\right\}\frac{f_r^{(N)}}{F_N}, \qquad \theta_r^{(N)}, \qquad (r \leq N). \tag{11.10}$$

In the homogeneous case they reduce to two:

$$\frac{k}{F_N}\varphi_r^{(N)}, \qquad -\frac{k\sigma_N}{F_N^2} f_r^{(N)}, \qquad (r \leq N). \tag{11.11}$$

The first of (11.11) is the normalized multiple of the truncation error in the formula $f_r \doteqdot f_r^{(N)}$; the second of (11.11) is a fixed relative error arising from the approximate representation of the normalizing factor k/F by k/F_N.

The equivalence of the method of this section to the algorithm of section 9 can be seen from the fact that the function on the right of (11.09) is exactly the solution of the set of eqs (9.05), (9.06), and (9.07).

Example 4.[9]

Let us compute to 5 decimal places the solution of the homogeneous equation

$$(2r-1)y_{r-1} - 12ry_r + (2r+1)y_{r+1} = 0, \tag{11.12}$$

satisfying the condition

$$\tfrac{1}{2}y_0 + y_1 + y_2 + y_3 + \ldots = 1. \tag{11.13}$$

In the notation of sections 2 and 9 we have

$$a_r = 2r-1, \quad b_r = 12r, \quad c_r = 2r+1, \quad d_r = 0, \quad m_0 = \tfrac{1}{2}, \quad m_r = 1 \quad (r>0), \quad k=1.$$

The computations are shown in table 4. Values of p_r were generated from $p_0 = 0$, $p_1 = 1$, and (11.12) when $r > 1$, correct to 6 significant figures. With $e_0 = 1$ (compare the first of (11.02)), we find from the second of (4.05) that $e_r = 1/(2r+1)$. The least value of r for which $e_r/p_{r+1} < \frac{1}{2} \times 10^{-5}$ is 7; in

[8] See also [7].
[9] [3], section 5.
[10] An alternative way of estimating N in this particular example is to observe that, except for small r, the wanted solution of (11.12) behaves roughly like $A\lambda^{-r}$, where A is a constant and $\lambda^2 - 6\lambda + 1 = 0$, giving $\lambda = 3 + \sqrt{8} \doteqdot 5.8$. Assuming that A is of order unity, as is reasonable in view of the condition (11.13), we find that $N \doteqdot 5/\log_{10}\lambda \doteqdot 7$. This method of estimation is somewhat less certain than the one we have used, and it is not universally applicable.

accordance with (4.12) this is the value[10] to ascribe to N. The back-substitution process for the determination of $f_r^{(7)}$ is given by $f_7^{(7)}=0$, and

$$p_{r+1}\,f_r^{(7)}=p_r\,f_{r+1}^{(7)}+e_r \qquad (r=6,5,\ldots,0);$$

compare (4.04). Division of $f_r^{(7)}$ by $F_7=0.599069$, computed from the first of (11.06), yields the wanted approximations $y_r^{(7)}$ to y_r.

TABLE 4

r	p_r	e_r	$f_r^{(7)}$	$y_r^{(7)}$
0	0	1.000000	1.000000	1.66926
1	1	0.333333	0.086107	0.14373
2	4	.200000	.011094	.01852
3	18.6	.142857	.001587	.00265
4	92.8	.111111	.000238	.00040
5	480.467	.090909	.000037	.00006
6	2544.80	.076923	.000006	.00001
7	13687.7	.066667	.000000	.00000
8	74445.6			

The example is now complete, but it is of interest to illustrate the error analysis of this section. Accordingly, the whole calculation was repeated twice, keeping four extra significant figures throughout. In the first repetition the same value $N=7$ was used. In the second repetition a new N was determined by the condition $|e_N/p_{N+1}|<\frac{1}{2}\times 10^{-9}$; this gave $N=12$.

The results appear in table 5. The column headed $10^9\epsilon_r^{(7)}$ gives the difference of $10^9y_r^{(7)}$ from the more accurate values $10^9y_r^{(12)}$. The next columns give $10^9\varphi_r^{(7)}/F_7$ and $-10^9\sigma_7f_r^{(7)}/F_7^2$; the value of $\varphi_r^{(7)}$ was obtained by subtracting $f_r^{(7)}$ from $f_r^{(12)}$, and σ_7 computed from (11.07), using the values of $f_r^{(12)}$ for f_r when $r\geqslant 8$. As expected, the values of $10^9\epsilon_r^{(7)}$ are in good agreement with the sum of the entries on the same row in the following two columns.

TABLE 5

r	$f_r^{(12)}$	$y_r^{(12)}$	$f_r^{(7)}$	$y_r^{(7)}$	$10^9\epsilon_r^{(7)}$	$10^9\dfrac{\phi_r^{(7)}}{F_7}$	$-10^9\dfrac{\sigma_7f_r^{(7)}}{F_7^2}$
0	1.00000 00000	1.66925 3684	1.00000 00000	1.66925 7339	−3655	0	−3655
1	0.08610 68379	0.14373 4156	0.08610 68378	0.14373 4471	−315	0	−315
2	.01109 40183	.01851 8731	.01109 40180	.01851 8771	−40	1	−41
3	.00158 71852	.00264 9415	.00158 71839	.00264 9418	−3	2	−6
4	.00023 83677	.00039 7896	.00023 83614	.00039 7887	9	11	−1
5	.00003 68169	.00006 1457	.00003 67845	.00006 1403	54	54	0
6	.00000 57914	.00000 9667	.00000 56199	.00000 9381	286	286	0
7	.00000 09228	.00000 1540	.00000 00000	.00000 0000	1540	1540	0
8	.00000 01485	.00000 0248					
9	.00000 00241	.00000 0040					
10	.00000 00039	.00000 0007	$F_7=0.59906\ 88055$		$\sigma_7=0.00000\ 13118$		
11	.00000 00006	.00000 0001	$F_{12}=0.59907\ 01173$				
12	.00000 00000	.00000 0000					

12. Summary

In this paper we have described a new algorithm for computing the solution y_r of any second-order linear difference equation, homogeneous or inhomogeneous, which is applicable when simple forward recurrence (and possibly also backward recurrence) cannot be used because of instability.

In the first part (secs. 2–8) we considered the case in which the wanted solution y_r has a specified value at the beginning of the range $r=0$, and an appropriate convergence condition as $r\to\infty$. In this case the algorithm is based on the solution of a finite number, N, of simultaneous linear algebraic equations of tridiagonal form by forward elimination. As $N\to\infty$ the solution $y_r^{(N)}$ of these equations converges to y_r (sec. 3). In sections 4 and 5 it was shown that during the process of computing $y_r^{(N)}$ the minimum value of N necessary to achieve specified tolerance in $|y_r-y_r^{(N)}|$ emerges automatically. Analyses of the truncation error and of the propagation of rounding errors were made in sections 5 and 7. The former leads to a convergent series expansion for y_r; the latter shows that the method of computation is quite stable, unless the problem itself is ill-posed. Numerical examples (sec. 6) illustrated the algorithm and confirmed the error analyses.

In section 8 it was shown that the well-known algorithm of J. C. P. Miller for the homogeneous case can be regarded as the computation of $y_r^{(N)}$ by backward elimination, taking a guessed value of N. It was also shown that the recent extension of Miller's algorithm by Shintani is related to the process of forward elimination.

In the second part of the paper (secs. 9–11) a more general form of normalizing condition for y_r was considered. An extended form of the algorithm was developed in section 9 and applied to a numerical example in the same section. In section 10 bounds for the truncation error were given and discussed. In the concluding section (sec. 11) it was shown that the more general problem can also be solved by application of the original algorithm of sections 3 and 4.

It is hoped that the results of this paper will prove to be of considerable usefulness in the computation of special functions from recurrence relations, in the solution of ordinary differential equations in Chebyshev series by Clenshaw's method, and in the solution of the discretized form of boundary-value problems in ordinary differential equations when one boundary is at infinity. In the last two connections, it may be possible to extend the present approach to difference equations of order higher than the second.

The writer thanks C. W. Clenshaw, G. F. Miller, and D. L. Yarmush for valuable comments on this work, and also his wife, Mrs. G. E. Olver, for carrying out the desk and automatic computations of the numerical examples, only a few of which are included in the paper. The automatic computation was supported by National Institutes of Health Grant No. NB05613.01.

13. References

[1] British Association for the Advancement of Science, Bessel functions—Part II, Mathematical Tables, v. **10**. (Cambridge University Press, 1952.)

[2] Clenshaw, C. W., The numerical solution of linear differential equations in Chebyshev series, Proc. Camb. Philos. Soc. **53**, 134–149 (1957).

[3] Olver, F. W. J., Error analysis of Miller's recurrence algorithm, Math. Comp. **18**, 65–74 (1964).

[4] Gautschi, W., Computational aspects of three-term recurrence relations, S.I.A.M. Rev. **9**, 24–82 (1967).

[5] Shintani, H., Note on Miller's recurrence algorithm, J. Sci. Hiroshima Univ. Ser. A-I **29**, 121–133 (1965).

[6] Fox, L., The numerical solution of two-point boundary problems in ordinary differential equations. (Oxford University Press, 1957.)

[7] Olver, F. W. J., Bounds for the solutions of second-order linear difference equations, J. Res. NBS.

[8] Watson, G. N., Theory of Bessel functions, Second edition. (Cambridge University Press, 1944.)

[9] Robinson, C., A numerical and analytical investigation of Struve's function. Thesis, London Univ. (1948).

[10] Wilkinson, J. H., The algebraic eigenvalue problem. (Oxford University Press, 1965.)

[11] Makinouchi, S., Note on the recurrence techniques for the calculation of Bessel functions $J_\nu(x)$, Tech. rep. Osaka Univ. **15**, 185–201 (1965).

[12] Oliver, J., Relative error propagation in the recursive solution of linear recurrence relations, Num. Math. **9**, 323–340 (1967).

(Paper 71B2&3–206)

JOURNAL OF RESEARCH of the National Bureau of Standards — B. Mathematics and Mathematical Physics
Vol. 71B, No. 4, October–December 1967

Bounds for the Solutions of Second-Order Linear Difference Equations

F. W. J. Olver

Institute for Basic Standards, National Bureau of Standards, Washington, D.C. 20234

(August 22, 1967)

Simple bounds are established for the solutions of second-order homogeneous linear difference equations in ranges in which the solutions are exponential in character. The results are applied to a recent algorithm for the computation of subdominant solutions of second-order linear difference equations, homogeneous or otherwise. Strict and extremely realistic bounds are obtained for the truncation error associated with the algorithm in a number of examples, including Anger-Weber functions, Struve functions, and the solution of a differential equation in Chebyshev series.

Key Words: Chebyshev series, difference equations, error bounds, Miller algorithm, recurrence relations, special functions.

1. Introduction

Consider the difference equation

$$p_{r+1}-bp_r+p_{r-1}=0, \qquad (1.01)$$

in which b is a real constant exceeding 2. The solution may be expressed as

$$p_r=\frac{p_1-\lambda_2 p_0}{\lambda_1-\lambda_2}\lambda_1^r+\frac{\lambda_1 p_0-p_1}{\lambda_1-\lambda_2}\lambda_2^r,$$

where p_0 and p_1 are at our disposal, and

$$\lambda_1=\frac{1}{2}b+\sqrt{\frac{1}{4}b^2-1}, \qquad \lambda_2=\frac{1}{2}b-\sqrt{\frac{1}{4}b^2-1}.$$

Eventually $|p_r|$ grows in proportion to λ_1^r as $r\to\infty$, unless $p_1=\lambda_2 p_0$. The larger the value of b, the faster is the rate of growth.

Now consider the equation

$$p_{r+1}-b_r p_r+p_{r-1}=0, \qquad (1.02)$$

in which $b_1, b_2, b_3, \ldots$ is a given sequence of real numbers, not necessarily monotonic, each of which exceeds the b of eq (1.01). If the same values of p_0 and p_1 are used for the two difference equations, then it is reasonable to expect—and not difficult to prove—that the solution of (1.02) generally increases at a faster rate than that of (1.01). Similarly, if, for every r, $b_r<B$, where B is another constant, then we expect the solution of (1.02) to grow at a slower rate than the solution of

$$p_{r+1}-Bp_r+p_{r-1}=0 \qquad (1.03)$$

having the same initial values.

The first purpose of the present paper is to generalize these intuitive results and present them as precise theorems. We shall allow the coefficient of p_{r-1} in (1.02) to take values other than unity, and we shall not insist that the solutions of difference equations under comparison agree at both $r=0$ and $r=1$. Some related results for the tails of continued fractions have been given by Blanch [1] [1] and Merkes [2].

The second purpose of this paper is to apply these theorems to a recent algorithm for computing subdominant solutions of homogeneous or inhomogeneous second-order difference equations [3], with a view to obtaining bounds for the truncation error.

2. Comparison Theorem

THEOREM 1. *Let*

$$\mathrm{q}_{r+1}=\beta_r \mathrm{q}_r-\alpha_r \mathrm{q}_{r-1}, \qquad (2.01)$$

$$\mathrm{Q}_{r+1}=\mathrm{B}_r\mathrm{Q}_r-\mathrm{A}_r\mathrm{Q}_{r-1}, \qquad (2.02)$$

where $\{\alpha_r\}$, $\{\beta_r\}$, $\{\mathrm{A}_r\}$, $\{\mathrm{B}_r\}$, $r=1, 2, \ldots$, *are sequences of numbers such that*

[1] Figures in brackets indicate the literature references at the end of this paper.

$$A_r \geqq \alpha_r \geqq 0, \qquad B_r - A_r \geqq \beta_r - \alpha_r \geqq 1. \tag{2.03}$$

If

$$Q_1 - q_1 \geqq Q_0 - q_0 \geqq 0, \text{ and } q_1 \geqq \max(q_0, 0), \tag{2.04}$$

then q_r *and* Q_r *are nondecreasing functions of* r, *and* $q_r \leqq Q_r$, $r=0, 1, \ldots$.

To establish this result, we begin with the identity

$$q_{r+1} - q_r = (\beta_r - \alpha_r - 1)q_r + \alpha_r(q_r - q_{r-1}), \tag{2.05}$$

and observe that both $\beta_r - \alpha_r - 1$ and α_r are nonnegative. From (2.04) we see that $q_1 \geqq 0$ and $q_1 - q_0 \geqq 0$. Hence from (2.05), with $r=1$, it follows that $q_2 - q_1 \geqq 0$. Continuing this argument by induction, we see that $q_{r+1} \geqq q_r$, $r=0, 1, \ldots$. Similarly from the identity

$$Q_{r+1} - Q_r = (B_r - A_r - 1)Q_r + A_r(Q_r - Q_{r-1}), \tag{2.06}$$

we deduce that $\{Q_r\}$ is a nondecreasing sequence.

Next, subtraction of (2.05) and (2.06) yields

$$\begin{aligned}(Q_{r+1} - q_{r+1}) - (Q_r - q_r) &= (\beta_r - \alpha_r - 1)(Q_r - q_r) \\ &+ \{(B_r - A_r) - (\beta_r - \alpha_r)\}Q_r \\ &+ \alpha_r\{(Q_r - q_r) - (Q_{r-1} - q_{r-1})\} \\ &+ (A_r - \alpha_r)(Q_r - Q_{r-1}).\end{aligned}$$

Assume that $r \geqq 1$ and $Q_r - q_r \geqq Q_{r-1} - q_{r-1} \geqq 0$—as is certainly the case when $r=1$ in consequence of (2.04). Then using (2.03) and the facts that $Q_r \geqq 0$ and $Q_r - Q_{r-1} \geqq 0$, we deduce that $(Q_{r+1} - q_{r+1}) - (Q_r - q_r) \geqq 0$. Hence by induction $Q_r \geqq q_r$. This completes the proof.

3. Bounds for the Solutions When the a_r Are Nonnegative

THEOREM 2. *Let*

$$p_{r+1} = b_r p_r - a_r p_{r-1}, \tag{3.01}$$

where $\{a_r\}$ *and* $\{b_r\}$, $r=1, 2, \ldots$, *are sequences of numbers such that*

$$b_r \geqq 1 + a_r \geqq 1, \tag{3.02}$$

and suppose that

$$p_0 > 0, \qquad \kappa \equiv p_1/p_0 \geqq 1. \tag{3.03}$$

Then p_r *is a nondecreasing function of* r, *and*

$$p_0\{\min(\kappa, \lambda)\}^r \leqq p_r \leqq p_0\{\max(\kappa, \Lambda)\}^r \quad (r=0, 1, \ldots). \tag{3.04}$$

Here λ, Λ *are the largest roots of the equations*

$$\lambda^2 - b\lambda + a = 0, \qquad \Lambda^2 - B\Lambda + A = 0, \tag{3.05}$$

respectively, and

$$a = \inf_{r \geqq 1} a_r, \qquad b = a + \inf_{r \geqq 1}(b_r - a_r), \tag{3.06}$$

$$A = \sup_{r \geqq 1} a_r, \qquad B = A + \sup_{r \geqq 1}(b_r - a_r). \tag{3.07}$$

Before entering into the proof of this theorem, we observe that the numbers b and a always exist and are subject to the inequalities

$$b \geqq 1 + a \geqq 1. \tag{3.08}$$

Moreover,

$$\lambda = \frac{1}{2}b + \sqrt{\frac{1}{4}b^2 - a} \geqq \frac{1}{2}a + \frac{1}{2} + \left|\frac{1}{2}a - \frac{1}{2}\right| \geqq \max(a, 1). \tag{3.09}$$

On the other hand, either or both of the sequences $\{a_r\}$ and $\{b_r - a_r\}$ may be unbounded, causing either B or both B and A to be infinite: in this event Λ is undefined and the second of the inequalities (3.04) is inapplicable. Except in these cases, we have [2]

$$A \geqq a, \qquad B \geqq b, \qquad \Lambda \geqq A, \qquad \Lambda \geqq \lambda. \tag{3.10}$$

That p_r is a nondecreasing function of r is an immediate consequence of (3.02), (3.03), and Theorem 1. We shall establish the inequalities (3.04) in turn by further use of this theorem.

Suppose first that $\kappa \geqq \lambda$. Define $q_r = p_0\lambda^r$. Then $q_0 = p_0$, $q_1 = p_0\lambda$ and, in consequence of (3.05),

$$q_{r+1} = bq_r - aq_{r-1} \quad (r \geqq 1). \tag{3.11}$$

We apply Theorem 1, with (3.11) filling the role of (2.01), and (3.01) that of (2.02). Conditions (2.03) are satisfied in consequence of the definitions of a and b, and (2.04) is satisfied since $\kappa \geqq \lambda \geqq 1$. Therefore $p_r \geqq q_r = p_0\lambda^r$, in agreement with the first of (3.04).

Now consider the case when $\kappa < \lambda$. Define $q_r = p_0\kappa^r$. Then $q_0 = p_0$, $q_1 = p_1$, and

$$q_{r+1} = \tilde{b}q_r - \tilde{a}q_{r-1} \quad (r \geqq 1), \tag{3.12}$$

where $\tilde{a} = \min(a, \kappa)$ and $\tilde{b} = \kappa + \tilde{a}\kappa^{-1}$. The desired result $p_r \geqq q_r$ will follow if Theorem 1 can be applied to eqs (3.12) and (3.01). Conditions (2.04) are clearly satisfied; the remaining conditions are

$$a_r \geqq \tilde{a} \geqq 0, \qquad b_r - a_r \geqq \tilde{b} - \tilde{a} \geqq 1.$$

The first two inequalities follow from the definitions of a and $\tilde{a}$. And since

$$\tilde{b} - \tilde{a} - 1 = (\kappa - 1)(1 - \tilde{a}\kappa^{-1}) \geqq 0, \qquad b - a \leqq b_r - a_r,$$

[2] The last of (3.10) may be verified by expressing $2\lambda = b + (b^2 - 4b + 4c)^{1/2}$, where $c \equiv \inf(b_r - a_r) \geqq 1$, showing that $\partial\lambda/\partial b$ and $\partial\lambda/\partial c$ are both positive.

it will suffice to show that

$$b-a \geqq \tilde{b}-\tilde{a}. \tag{3.13}$$

If $\tilde{a}=\kappa$, then $\tilde{b}=\kappa+1$ and (3.13) follows immediately from (3.08). If, on the other hand, $\tilde{a}=a<\kappa$, then

$$(b-a)-(\tilde{b}-\tilde{a})=b-\kappa-a\kappa^{-1}$$

$$=\lambda+\frac{a}{\lambda}-\left(\kappa+\frac{a}{\kappa}\right)$$

$$=(\lambda-\kappa)\left(1-\frac{a}{\lambda\kappa}\right)>0,$$

since $a<\kappa$ and $1 \leqq \kappa<\lambda$. This completes the proof of the left-hand inequality in (3.04).

To establish the right-hand inequality in (3.04), suppose first that $\kappa \leqq \Lambda$. Defining $Q_r=p_0\Lambda^r$, we have $Q_0=p_0$, $Q_1=p_0\Lambda$, and

$$Q_{r+1}=BQ_r-AQ_{r-1} \qquad (r \geqq 1). \tag{3.14}$$

It is readily verified that Theorem 1 is applicable to eqs (3.01) and (3.14). Hence $p_r \leqq p_0\Lambda^r$, in agreement with (3.04).

Now consider the case when $\kappa>\Lambda$. Define $Q_r=p_0\kappa^r$, so that $Q_0=p_0$, $Q_1=p_1$, and

$$Q_{r+1}=\tilde{B}Q_r-AQ_{r-1} \quad (r \geqq 1), \tag{3.15}$$

where $\tilde{B}=\kappa+A\kappa^{-1}$. The desired result $p_r \leqq Q_r$ will follow immediately if Theorem 1 is applicable to (3.01) and (3.15). Conditions (2.04) are obviously satisfied. Conditions (2.03) demand that

$$A \geqq a_r \geqq 0, \qquad \tilde{B}-A \geqq b_r-a_r \geqq 1.$$

In consequence of (3.02) and (3.07), these inequalities will follow if we can show that $\tilde{B} \geqq B$. Now

$$\tilde{B}-B=\kappa+\frac{A}{\kappa}-\left(\Lambda+\frac{A}{\Lambda}\right)=(\kappa-\Lambda)\left(1-\frac{A}{\Lambda\kappa}\right).$$

This is nonnegative because $\kappa>\Lambda$, $\kappa \geqq 1$, and $\Lambda \geqq A$; compare (3.10). The proof of Theorem 2 is now complete.

4. Bounds for the Solutions When the a_r Are Nonpositive

If the quantities a_r are negative then Theorem 2 is inapplicable. The analysis of this case is somewhat easier, however, and Theorem 1 is not needed. For convenience, we replace a_r by $-a_r$.

THEOREM 3. *Let*

$$\mathrm{p}_{r+1}=\mathrm{b}_r\mathrm{p}_r+\mathrm{a}_r\mathrm{p}_{r-1}, \tag{4.01}$$

where

$$\mathrm{a}_r \geqq 0, \qquad \mathrm{b}_r \geqq 0, \qquad \mathrm{p}_0>0, \qquad \kappa \equiv \mathrm{p}_1/\mathrm{p}_0 \geqq 1. \tag{4.02}$$

Then

$$\mathrm{p}_0\{\min(\kappa, \lambda)\}^r \leqq \mathrm{p}_r \leqq \mathrm{p}_0\{\max(\kappa, \Lambda)\}^r \qquad (r=0, 1, \ldots), \tag{4.03}$$

where λ, Λ *are the largest roots of the equations*

$$\lambda^2-b\lambda-a=0, \qquad \Lambda^2-B\Lambda-A=0, \tag{4.04}$$

respectively, and

$$a=\inf_{r \geqq 1} a_r, \qquad b=\inf_{r \geqq 1} b_r, \tag{4.05}$$

$$A=\sup_{r \geqq 1} a_r, \qquad B=\sup_{r \geqq 1} b_r. \tag{4.06}$$

To prove the left-hand inequality in (4.03) when $\kappa \geqq \lambda$, assume that $p_r \geqq p_0\lambda^r$ and $p_{r-1} \geqq p_0\lambda^{r-1}$, — as is the case when $r=1$. Then from (4.01) we have

$$p_{r+1} \geqq bp_0\lambda^r+ap_0\lambda^{r-1}=p_0\lambda^{r+1},$$

as required.

Now consider the case when $\kappa<\lambda$. The roots of the quadratic equation $x^2-bx-a=0$ are λ and $-a/\lambda$. Since $-a/\lambda<\kappa<\lambda$, it follows that

$$\kappa^2-b\kappa-a<0. \tag{4.07}$$

Assume that $p_r \geqq p_0\kappa^r$ and $p_{r-1} \geqq p_0\kappa^{r-1}$ — as is the case when $r=1$. Then from (4.01), (4.05), and (4.07), we derive

$$p_{r+1} \geqq bp_0\kappa^r+ap_0\kappa^{r-1}>p_0\kappa^{r+1}.$$

The right-hand inequality in (4.03) may be established in a similar way.

5. Anger-Weber Functions

In the remaining part of this paper we show how the preceding theorems may be applied to the truncation errors associated with the algorithm of [3].

Consider first the computation of the Anger-Weber function $\mathbf{E}_r(1)$ given in [3], section 6, Example 1. The truncation error of the approximation $y_r^{(14)}$ is given by [3]

[3] [3], section 5.

$$\epsilon_r^{(14)} \equiv \mathbf{E}_r(1) - y_r^{(14)} = E_{14} p_r, \tag{5.01}$$

where

$$E_{14} = \sum_{s=14}^{\infty} \frac{e_s}{p_s p_{s+1}}, \tag{5.02}$$

p_r satisfies (3.01) above with $a_r = 1$, $b_r = 2r$, $p_0 = 0$, $p_1 = 1$, and

$$e_r = e_{r-1} - d_r p_r, \qquad d_r = -(2/\pi)\{1-(-1)^r\}, \tag{5.03}$$

with $e_0 = \mathbf{E}_0(1)$.

We apply Theorem 2 of section 3 to the sequence $\{p_{r+14}\}$, using the numerical entries given in [3], table 1. Working to five significant figures (for the sake of illustration), we have

$$\kappa = p_{15}/p_{14} = 1.1127 \times 10^{15}/(3.9793 \times 10^{13}) = 27.962,$$

$$a = 1, \qquad b = 1 + \inf_{r \geqq 15}(2r-1) = 30.$$

Hence

$$\lambda^2 - 30\lambda + 1 = 0, \qquad \lambda = 29.967 > \kappa.$$

Therefore

$$p_r \geqq p_{14} \kappa^{r-14} \quad (r \geqq 14). \tag{5.04}$$

Next, from (5.03) we have

$$e_s = e_{14} + (4/\pi)(p_{15} + p_{17} + \ldots + p_{s'}) \quad (s > 14), \tag{5.05}$$

where $s' = s$ if s is odd, and $s' = s-1$ if s is even. Since p_s is a nondecreasing function of s (Theorem 2), we have

$$e_s \leqq e_{14} + (4/\pi)(s-14)p_s \quad (s \geqq 14). \tag{5.06}$$

Application of this inequality to (5.02) gives

$$E_{14} \leqq e_{14} \sum_{s=14}^{\infty} \frac{1}{p_s p_{s+1}} + \frac{4}{\pi}\left(\frac{1}{p_{16}} + \frac{2}{p_{17}} + \frac{3}{p_{18}} + \ldots\right). \tag{5.07}$$

Substituting (5.04) on the right-hand side of (5.07), summing the resulting series and using numerical entries given in [3], table 1, we obtain

$$E_{14} \leqq \frac{e_{14}}{p_{14}^2(\kappa - \kappa^{-1})} + \frac{4}{\pi p_{14}(\kappa-1)^2} = 8.824 \times 10^{-17}. \tag{5.08}$$

The actual value of E_{14} was estimated in [3] to be 8.248×10^{-17} by direct numerical summation of the terms in the expansion (5.02). Our strict bound (5.08) exceeds this estimate by about 7 percent, which is very satisfactory in view of the somewhat crude step leading from (5.05) to (5.06).

Strict bounds for the errors $\epsilon_r^{(14)} (r \leqq 14)$ may be computed by substituting the numerical value (5.08) for E_{14} in (5.01) and using the values of p_r given in [3], table 1. Obviously they, too, will overestimate the actual truncation errors by about 7 percent.

6. Struve Functions

As a second example, consider the Struve function $\mathbf{H}_r(0.1)$ computed in [3], section 6, Example 2. The relevant formulas in this case are

$$\epsilon_r^{(15)} = \mathbf{H}_r(0.1) - y_r^{(15)} = E_{15} p_r, \qquad E_{15} = \sum_{s=15}^{\infty} \frac{e_s}{p_s p_{s+1}}, \tag{6.01}$$

and

$$p_{r+1} = 20 r p_r - p_{r-1}, \qquad e_r = e_{r-1} - \frac{(0.05)^r}{\sqrt{\pi}\,\Gamma\left(r+\frac{3}{2}\right)} p_r. \tag{6.02}$$

Applying Theorem 2 to the sequence $\{p_{r+15}\}$ using the numerical entries in [3], table 2, we have

$$\kappa = p_{16}/p_{15} = 299.99643 \ldots,$$

$$a = 1, \qquad b = 1 + \inf_{r \geqq 16}(20r - 1) = 320,$$

$$\lambda = 160 + \sqrt{25599} > \kappa.$$

Hence

$$p_r \geqq p_{15} \kappa^{r-15} \quad (r \geqq 15). \tag{6.03}$$

Inspection of table 2 of [3] suggests that

$$|e_r| \leqq e_{15} \quad (r \geqq 15). \tag{6.04}$$

Deferring the proof of this inequality for the moment and substituting (6.03) and (6.04) in the second of (6.01), we derive

$$|E_{15}| \leqq \frac{e_{15}}{p_{15}^2(\kappa - \kappa^{-1})} = (0.23277\ 966 \ldots) \times 10^{-62}. \tag{6.05}$$

The actual value of E_{15} may be estimated by computation of p_{17}, p_{18}, . . . and direct summation of the expansion in (6.01). This gives $(0.23277\ 943 \ldots) \times 10^{-62}$. Therefore strict bounds for $\epsilon_r^{(15)} (r \leqq 15)$ computed from (6.05) and the first of (6.01) will exceed the estimated errors by less than one part in 10^6, a striking example of the power of Theorem 2.

It remains to establish (6.04). Write

$$u_r = (0.05)^r p_r \bigg/ \left\{ \sqrt{\pi}\,\Gamma\left(r+\frac{3}{2}\right) \right\},$$

so that

$$e_r = e_{15} - u_{16} - u_{17} - \ldots - u_r \quad (r \geqq 16).$$

The p_r and u_r are all positive, in consequence of (6.03). From the first of (6.02) we see that

$$p_{r+1} < 20rp_r.$$

Accordingly, we derive

$$u_{r+1} < 2ru_r/(2r+3),$$

and thence

$$\sum_{r=16}^{\infty} u_r < u_{15}\left(\frac{30}{33} + \frac{30}{33}\cdot\frac{32}{35} + \frac{30}{33}\cdot\frac{32}{35}\cdot\frac{34}{37} + \ldots\right) = 30u_{15},$$

the last relation being a consequence of Gauss' formula for the hypergeometric function F (15, 1; $16\frac{1}{2}$; 1); see [4], eq (15.1.20). Since

$$30u_{15} = 0.01418\,23625\ \ldots$$

the stated result now follows.

7. Clenshaw's Numerical Method for Ordinary Differential Equations

As a final illustration, we consider a recurrence relation arising in the solution of a differential equation by the method of Clenshaw; compare Example 4 of [3], section 11. Suppose first that we are interested in the solution of the equation

$$(2r-1)f_{r-1} - 12rf_r + (2r+1)f_{r+1} = 0 \qquad (7.01)$$

satisfying the conditions $f_0 = 1$ and $f_r \to 0$ as $r \to \infty$. Approximations to f_r are provided by the function $f_r^{(7)}$ given to six decimal places in table 4 of [3] and to ten decimal places in table 5 of the same reference. Let us seek bounds for the magnitude of the truncation error

$$\phi_r^{(7)} = f_r - f_r^{(7)}. \qquad (7.02)$$

As in the other examples, we have

$$\phi_r^{(7)} = E_7 p_r, \qquad E_7 = \sum_{s=7}^{\infty} \frac{e_s}{p_s p_{s+1}}. \qquad (7.03)$$

In the present case p_r satisfies the difference eq (7.01), and $e_r = 1/(2r+1)$. Numerical values of p_r are given in table 4 of [3] for $r = 0, 1, \ldots, 8$. Applying Theorem 2 to the sequence $\{p_{r+7}\}$, we have

$$\kappa = p_8/p_7 = 5.4389, \qquad (7.04)$$

correct to five significant figures, and

$$a = \inf_{r \geqq 8} \frac{2r-1}{2r+1} = \frac{15}{17}, \qquad b = \frac{15}{17} + \inf_{r \geqq 8} \frac{10r+1}{2r+1} = \frac{96}{17}.$$

Accordingly,

$$17\lambda^2 - 96\lambda + 15 = 0, \qquad \lambda = 5.4862 > \kappa.$$

Hence

$$p_r \geqq p_7 \kappa^{r-7} \qquad (r \geqq 7). \qquad (7.05)$$

Substitution of the last inequality in the second of (7.03) yields

$$E_7 \leqq \frac{1}{p_7^2}\left(\frac{1}{15\kappa} + \frac{1}{17\kappa^3} + \frac{1}{19\kappa^5} + \ldots\right)$$

$$< \frac{1}{15p_7^2(\kappa - \kappa^{-1})} = 0.67713 \times 10^{-10}. \qquad (7.06)$$

The corresponding bounds for the errors $\phi_r^{(7)}$ are given in the second column of table A below, in units of the tenth decimal place. They exceed only slightly the estimates for the errors obtained by subtracting the entries in the columns headed $f_r^{(12)}$ and $f_r^{(7)}$ in table 5 of [3], and given in the third column of table A.

TABLE A

r	$0.67713p_r$	$10^{10}\phi_r^{(7)}$ (estimated)	$0.11303p_r$	$-3735\,f_r^{(7)}$	$10^9\epsilon_r^{(7)}$ (estimated)
0	0	0	0	−3735	−3655
1	1	1	0	−322	−315
2	3	3	0	−41	−40
3	13	13	2	−6	−3
4	63	63	10	−1	9
5	325	324	54	0	54
6	1723	1715	288	0	286
7	9268	9228	1547	0	1540

Secondly, suppose that we are interested in the solution of (7.01), y_r, say, which satisfies the condition

$$\tfrac{1}{2}y_0 + y_1 + y_2 + y_3 + \ldots = 1. \qquad (7.07)$$

(This is the form in which the example was originally proposed in [3].) Then y_r is related to f_r, defined above, by

$$y_r = f_r/F, \qquad F = \tfrac{1}{2}f_0 + f_1 + f_2 + \ldots. \qquad (7.08)$$

We seek bounds for the errors

$$\epsilon_r^{(7)} = y_r - y_r^{(7)} \qquad (r = 0, 1, \ldots 7), \qquad (7.09)$$

where

$$y_r^{(7)}=f_r^{(7)}/F_7, \qquad F_7=\tfrac{1}{2}f_0^{(7)}+f_1^{(7)}+f_2^{(7)}+\ldots+f_6^{(7)}. \tag{7.10}$$

Using (7.02) and (7.08), we find that

$$\epsilon_r^{(7)}=\epsilon_r^{(7,1)}-\epsilon_r^{(7,2)}, \tag{7.11}$$

where

$$\epsilon_r^{(7,1)}=\frac{\phi_r^{(7)}}{F_7+\sigma_7}, \qquad \epsilon_r^{(7,2)}=\frac{\sigma_7 f_r^{(7)}}{F_7(F_7+\sigma_7)}, \tag{7.12}$$

and

$$\sigma_7=\tfrac{1}{2}\phi_0^{(7)}+\phi_1^{(7)}+\phi_2^{(7)}+\ldots+\phi_7^{(7)}+ +f_8+f_9+f_{10}+\ldots. \tag{7.13}$$

In order to assess an upper bound for σ_7, we need an upper bound for

$$f_r=p_r\sum_{s=r}^{\infty}\frac{1}{(2s+1)p_s p_{s+1}} \tag{7.14}$$

when $r\geqq 8$; compare [3], eq (5.03). To construct this, we make a second application of Theorem 2 to the sequence $\{p_{r+7}\}$. We have

$$A=\sup_{r\geqq 8}\frac{2r-1}{2r+1}=1, \qquad B=1+\sup_{r\geqq 8}\frac{10r+1}{2r+1}=6.$$

Accordingly

$$\Lambda^2-6\Lambda+1=0, \qquad \Lambda=3+\sqrt{8}=5.8284>\kappa,$$

where κ is defined by (7.04). Hence

$$p_r\leqq p_7\Lambda^{r-7} \qquad (r\geqq 7). \tag{7.15}$$

Substituting this result and (7.05) in (7.14), we find that

$$f_r\leqq\frac{\Lambda^{r-7}}{p_7}\sum_{s=r}^{\infty}\frac{1}{(2s+1)\kappa^{2s-13}}$$

$$<\frac{\Lambda^{r-7}}{p_7(2r+1)(\kappa-\kappa^{-1})\kappa^{2r-14}} \quad (r\geqq 7),$$

and thence

$$\sum_{r=8}^{\infty}f_r<\frac{\Lambda}{17p_7(\kappa-\kappa^{-1})(\kappa^2-\Lambda)}=2007\times 10^{-10},$$

correct to ten decimal places. Adding this value to 10^{-10} times the sum of the entries in the column of table A headed $0.67713p_r$, we derive

$$\sigma_7<13403\times 10^{-10};$$

compare (7.13).

Since F_7 and σ_7 are both positive, we obtain from (7.03), (7.06), (7.12), and the value $F_7=0.59907$ extracted from [3], table 5

$$\epsilon_r^{(7,1)}\leqq\frac{\phi_r^{(7)}}{F_7}=\frac{E_7 p_r}{F_7}\leqq 0.11303\times 10^{-9}p_r,$$

and

$$\epsilon_r^{(7,2)}\leqq\frac{\sigma_7}{F_7^2}f_r^{(7)}\leqq 3735\times 10^{-9}f_r^{(7)},$$

both $\epsilon_r^{(7,1)}$ and $\epsilon_r^{(7,2)}$ obviously being nonnegative. The values of $0.11303p_r$ and $-3735f_r^{(7)}$ are given in the fourth and fifth columns of table A. Comparison with the estimated values of $10^9\epsilon_r^{(7)}$, extracted from table 5 of [3] and given in the final column, is again fully satisfactory.

8. Summary

In the first part of this paper (secs. 1–4) simple lower and upper bounds are established for the solutions of second-order homogeneous linear difference equations, in ranges in which the solutions are exponential in character.

In the second part (secs. 5–7), the results are applied to an algorithm for the computation of subdominant solutions of second-order difference equations which was introduced recently by the writer. It is shown by means of examples how to compute strict and extremely realistic bounds for the truncation error associated with the algorithm. The bounds depend only on simple properties of the coefficients in the difference equation, and are independent of asymptotic theories of the solutions.

The writer is grateful to C. W. Clenshaw and G. F. Miller for valuable criticisms of an earlier form of this paper.

9. References

[1] Blanch, G., Numerical evaluation of continued fractions, Rev. Soc. Indust. Appl. Math. **6**, 383–421 (1964).

[2] Merkes, E. P., On truncation errors for continued fraction computations, J. Numer. Anal. Soc. Indust. Appl. Math. **3**, 486–496 (1966).

[3] Olver, F. W. J., Numerical solution of second-order linear difference equations, J. Res. NBS **71B** (Math. and Math. Phys.) Nos. 2 & 3, 111–129 (1967).

[4] National Bureau of Standards, Handbook of mathematical functions, Appl. Math. Ser. **55** (Government Printing Office, Washington, D.C. 20402, 1964).

(Paper 71B4–238)

JOURNAL OF APPROXIMATION THEORY 1, 293–313 (1968)

Error Bounds for the Laplace Approximation for Definite Integrals

F. W. J. OLVER

National Bureau of Standards, Washington, D.C. 20234

Explicit error bounds are obtained for the well-known asymptotic expansion of integrals of the form

$$\int_a^b e^{-\lambda p(x)} q(x)\, dx,$$

in which λ is a large positive parameter, $p(x)$ and $q(x)$ are real differentiable functions, and $p'(x)$ has a simple zero in the finite or infinite range $[a,b]$. The bounds are expressed in terms of the supremum of a certain function, taken over $[a,b]$, and are asymptotic to the absolute value of the first neglected term in the expansion, as $\lambda \to \infty$. Several illustrative examples are given, including modified Bessel functions and the gamma function.

1. INTRODUCTION

Consider the integral

$$I = \int_a^b e^{-p(x)} q(x)\, dx, \tag{1.1}$$

in which the range of integration is real and may be finite or infinite. Assume that (i) $p(x)$ is real, attains its minimum value at an interior point ξ, say, of $[a,b]$ and increases fairly rapidly as $|x-\xi|$ increases; (ii) $q(x)$ is a relatively slowly-varying function. Then the underlying idea of the Laplace approximation is that almost the whole contribution of the integrand to I arises from the immediate neighborhood of ξ. In consequence, $p(x)$ and $q(x)$ may be replaced by the leading terms in their Taylor-series expansions at $x = \xi$, and the integration limits extended (if necessary) to $-\infty$ and $+\infty$. For example, in the case when $p'(\xi) = 0$, $p''(\xi) \neq 0$, and $q(\xi) \neq 0$, we obtain

$$I \doteqdot \int_a^b \exp\{-p(\xi) - \tfrac{1}{2}(x-\xi)^2 p''(\xi)\} q(\xi)\, dx$$

$$\doteqdot \int_{-\infty}^{\infty} \exp\{-p(\xi) - \tfrac{1}{2}(x-\xi)^2 p''(\xi)\} q(\xi)\, dx = e^{-p(\xi)} q(\xi) \left(\frac{2\pi}{p''(\xi)}\right)^{1/2}. \tag{1.2}$$

This procedure is particularly valuable when the integral contains a parameter: the approximation often possesses an asymptotic property with respect to the parameter. In particular, if $p(x)$ is replaced by $\lambda p(x)$, where λ is positive, then precise circumstances in which

$$\int_a^b e^{-\lambda p(x)} q(x)\, dx \sim e^{-\lambda p(\xi)} q(\xi) \left(\frac{2\pi}{\lambda p''(\xi)}\right)^{1/2} \tag{1.3}$$

as $\lambda \to \infty$ are well known; see, for example, [1]–[5].

The object of this paper is to supply explicit bounds for the error in the Laplace and related approximations, from which (i) asymptotic properties with respect to parameters are directly deducible, (ii) realistic numerical values can be determined. For earlier work on this problem see [6].

2. Formulation

By subdividing the range of integration and changing variables, we can reduce the problem of Section 1 to the approximation of an integral of the form

$$I(\lambda) = \int_0^b e^{-\lambda p(x)} q(x)\, dx, \tag{2.1}$$

in which $p(0) = p'(0) = 0$, and

$$p'(x) > 0 \qquad (0 < x \leqslant b). \tag{2.2}$$

The functions $p(x)$ and $q(x)$ may themselves depend on the positive parameter λ.

We assume that $p(x)$ and $q(x)$ can be expanded in series of ascending powers of x, given by

$$p(x) = p_2 x^2 + p_3 x^3 + p_4 x^4 + \ldots, \tag{2.3}$$

$$q(x) = q_0 + q_1 x + q_2 x^2 + \ldots. \tag{2.4}$$

For our purposes it suffices for these series to be asymptotic expansions in Poincaré's sense, as $x \to 0$. In particular, this includes the common case in which $p(x)$ and $q(x)$ are analytic at $x = 0$ and (2.3) and (2.4) are Maclaurin expansions. In consequence of (2.2), we have $p_2 \geqslant 0$. In this paper we exclude the case $p_2 = 0$; in other words, we restrict $x = 0$ to being a simple saddle-point of the integral (2.1).

Since $p(x)$ is an increasing function in the interval $(0, b)$, we may adopt $p = p(x)$ as a new integration variable. This gives

$$I(\lambda) = \int_0^{p(b)} e^{-\lambda p} f(p) \frac{dp}{p^{1/2}}, \tag{2.5}$$

where

$$f(p) = p^{1/2}(x) q(x) / p'(x). \tag{2.6}$$

Reversion of (2.3) yields an expansion of the form

$$x = c_1 p^{1/2} + c_2 p + c_3 p^{3/2} + \dots, \tag{2.7}$$

this series converging in a neighborhood of $p = 0$ if (2.2) converges, otherwise it is an asymptotic expansion for small p. Substitution of (2.7) in (2.6) yields

$$f(p) = a_0 + a_1 p^{1/2} + a_2 p + a_3 p^{3/2} + \dots, \tag{2.8}$$

the coefficients a_s being expressible in terms of p_s and q_s; in particular,

$$a_0 = q_0/(2p_2^{1/2}).$$

If, in (2.5), we replace the upper limit by infinity, substitute (2.8), and integrate formally term by term, we obtain the series

$$\sum_{s=0}^{\infty} \frac{a_s \Gamma\{(s+1)/2\}}{\lambda^{(s+1)/2}}. \tag{2.9}$$

The first term is one-half the Laplace approximation (1.3).[1] In the case when $p(x)$ and $q(x)$ are independent of λ, the whole series is an asymptotic expansion for $I(\lambda)$ as $\lambda \to \infty$; this is a consequence of Watson's well-known lemma [7], Section 8.3.

We now write

$$I(\lambda) = \sum_{s=0}^{n-1} \frac{a_s \Gamma\{(s+1)/2\}}{\lambda^{(s+1)/2}} + E_n(\lambda), \tag{2.10}$$

where n is an arbitrary positive integer at zero, and seek bounds for the error term $E_n(\lambda)$. In constructing these bounds we shall bear in mind the asymptotic value

$$E_n(\lambda) \sim \frac{a_n \Gamma\{(n+1)/2\}}{\lambda^{(n+1)/2}} + \frac{a_{n+1} \Gamma\{(n+2)/2\}}{\lambda^{(n+2)/2}} + \dots \qquad (\lambda \to \infty), \tag{2.11}$$

applicable when $p(x)$ and $q(x)$ are independent of λ.

We write

$$f(p) = \sum_{s=0}^{n-1} a_s p^{s/2} + f_n(p). \tag{2.12}$$

The remainder term $f_n(p)$ is a continuous function of p, with the property

$$f_n(p) \sim a_n p^{n/2} \qquad (p \to 0), \tag{2.13}$$

provided that $a_n \neq 0$. It is convenient to split $E_n(\lambda)$ in the form

$$E_n(\lambda) = E_n^{(2)}(\lambda) - E_n^{(1)}(\lambda), \tag{2.14}$$

[1] The other half is contributed by the integral over the range $a \leqslant x \leqslant 0$.

where

$$E_n^{(1)}(\lambda) = \sum_{s=0}^{n-1} a_s \int_{p(b)}^{\infty} e^{-\lambda p} p^{(s-1)/2}\, dp, \tag{2.15}$$

$$E_n^{(2)}(\lambda) = \int_0^{p(b)} e^{-\lambda p} f_n(p) \frac{dp}{p^{1/2}}, \tag{2.16}$$

and to consider the two parts separately in succeeding sections. If $p(b) = \infty$, then $E_n^{(1)}(\lambda) = 0$.

3. Bounds for $E_n^{(1)}(\lambda)$

Denoting $p(b)$ by β, and taking a new integration variable $\tau = p - \beta$, we may write

$$E_n^{(1)}(\lambda) = \sum_{s=0}^{n-1} a_s L_s\{p(b), \lambda\}, \tag{3.1}$$

where

$$L_s(\beta, \lambda) = \int_\beta^\infty e^{-\lambda p} p^{(s-1)/2}\, dp = e^{-\lambda\beta} \int_0^\infty e^{-\lambda\tau} (\tau + \beta)^{(s-1)/2}\, d\tau. \tag{3.2}$$

Clearly

$$L_0(\beta, \lambda) \leqslant \frac{e^{-\lambda\beta}}{\lambda\beta^{1/2}}. \tag{3.3}$$

For $s \geqslant 1$, we apply the inequality

$$\tau + \beta \leqslant \beta e^{\tau/\beta}.$$

This yields

$$L_s(\beta, \lambda) \leqslant \frac{\beta^{(s-1)/2} e^{-\lambda\beta}}{\lambda - \frac{1}{2}(s-1)\beta^{-1}}, \tag{3.4}$$

provided that $\lambda > \frac{1}{2}(s-1)\beta^{-1}$. When $s = 1$, the relation (3.4) is an equality. For other values of s and large values of λ, the satisfactory nature of the bound can be seen from the relation

$$L_s(\beta, \lambda) = \frac{\beta^{(s-1)/2} e^{-\lambda\beta}}{\lambda} \left\{1 + \frac{s-1}{2\beta\lambda} + O\left(\frac{1}{\lambda^2}\right)\right\} \qquad (\lambda \to \infty),$$

obtained by applying Watson's lemma to (3.2).

Substitution of (3.3) and (3.4) in (3.1) yields

$$|E_1^{(1)}(\lambda)| \leqslant \frac{e^{-\lambda p(b)} |a_0|}{\lambda p^{1/2}(b)}, \tag{3.5}$$

and

$$|E_n^{(1)}(\lambda)| \leqslant \frac{e^{-\lambda p(b)}}{\lambda - \{(\frac{1}{2}n-1)/p(b)\}} \sum_{s=0}^{n-1} |a_s| \{p(b)\}^{(s-1)/2} \qquad (n \geqslant 2), \tag{3.6}$$

provided that the denominator in the last relation is positive.

4. Bounds for $E_n^{(2)}(\lambda)$

To bound $E_n^{(2)}(\lambda)$, we majorize $f_n(p)$ and then replace the upper integration limit in (2.16) by infinity. One of the simplest results of this kind is given by

$$|E_n^{(2)}(\lambda)| \leqslant A_n \int_0^\infty e^{-\lambda p} p^{(n-1)/2}\, dp = \frac{A_n \Gamma\{(n+1)/2\}}{\lambda^{(n+1)/2}}, \tag{4.1}$$

where

$$A_n = \sup_{0<p<p(b)} |p^{-n/2} f_n(p)|. \tag{4.2}$$

This form of majorant was used by Rosser [6]. However, A_n may not exist when $p(b) = \infty$, and in other cases it may be many times the size of $|a_n|$, causing (4.1) to overestimate grossly the actual error; compare (2.11).

Provided that $a_n \neq 0$, a more realistic bound is obtainable by using a majorant of the form

$$|f_n(p)| \leqslant |a_n|\, p^{n/2} e^{\phi_n \sqrt{p}}, \tag{4.3}$$

where ϕ_n is independent of p. The best value of ϕ_n is obviously

$$\phi_n = \sup_{0<p<p(b)} \left\{ \frac{1}{\sqrt{p}} \ln \left| \frac{f_n(p)}{a_n p^{n/2}} \right| \right\}. \tag{4.4}$$

Combination of (2.16) and (4.3) gives

$$|E_n^{(2)}(\lambda)| \leqslant |a_n| \int_0^\infty \exp(-\lambda p + \phi_n \sqrt{p})\, p^{(n-1)/2}\, dp. \tag{4.5}$$

If $\phi_n \leqslant 0$, we have immediately

$$|E_n^{(2)}(\lambda)| \leqslant \frac{|a_n|\, \Gamma\{(n+1)/2\}}{\lambda^{(n+1)/2}}. \tag{4.6}$$

Now suppose that $\phi_n > 0$. Taking a new integration variable $w = \sqrt{p}$, we find that

$$\int_0^\infty \exp(-\lambda p + \phi_n \sqrt{p}) p^{(n-1)/2}\, dp = 2 \exp\left(\frac{\phi_n^2}{4\lambda}\right) \int_0^\infty \exp\left\{-\lambda \left(w - \frac{\phi_n}{2\lambda}\right)^2\right\} w^n\, dw.$$

Clearly

$$\int_0^{\phi_n/(2\lambda)} \exp\left\{-\lambda\left(w - \frac{\phi_n}{2\lambda}\right)^2\right\} w^n\, dw \leqslant \int_0^{\phi_n/(2\lambda)} w^n\, dw = \frac{1}{n+1} \left(\frac{\phi_n}{2\lambda}\right)^{n+1},$$

and

$$\int_{\phi_n/(2\lambda)}^{\infty} \exp\left\{-\lambda\left(w-\frac{\phi_n}{2\lambda}\right)^2\right\} w^n\,dw = \int_0^{\infty} e^{-\lambda t^2}\left(t+\frac{\phi_n}{2\lambda}\right)^n dt$$

$$= \sum_{s=0}^{n} \binom{n}{s}\left(\frac{\phi_n}{2\lambda}\right)^s \frac{\Gamma\{(n-s+1)/2\}}{2\lambda^{(n-s+1)/2}}.$$

Hence

$$|E_n^{(2)}(\lambda)| \leqslant \frac{|a_n|\,\Gamma\{(n+1)/2\}}{\lambda^{(n+1)/2}} \exp\left(\frac{\phi_n^2}{4\lambda}\right) \sum_{s=0}^{n+1} \alpha_{n,s}\left(\frac{\phi_n}{2\lambda^{1/2}}\right)^s, \tag{4.7}$$

where[2]

$$\alpha_{n,s} = \binom{n}{s}\frac{\Gamma\{(n-s+1)/2\}}{\Gamma\{(n+1)/2\}} \quad (s \leqslant n); \qquad \alpha_{n,n+1} = \frac{1}{\Gamma\{(n+3)/2\}}. \tag{4.8}$$

LEMMA 1.

$$\frac{\Gamma(x)}{\Gamma(x+\frac{1}{2})} \leqslant \left(\frac{\pi}{2x}\right)^{1/2} \qquad (x \geqslant \tfrac{1}{2}). \tag{4.9}$$

Proof. Denoting the logarithmic derivative of $\Gamma(x)$ by $\psi(x)$ and writing $\chi(x) = x^{1/2}\Gamma(x)/\Gamma(x+\frac{1}{2})$, we have

$$\frac{\chi'(x)}{\chi(x)} = \frac{1}{2x} + \psi(x) - \psi(x+\tfrac{1}{2}) = \tfrac{1}{2}\{\psi(x+1) - 2\psi(x+\tfrac{1}{2}) + \psi(x)\} = \tfrac{1}{8}\psi''(\zeta),$$

where $\zeta \in (x, x+1)$. Since

$$\psi''(\zeta) = -\int_0^{\infty} \frac{t^2 e^{-\zeta t}}{1-e^{-t}}\,dt \qquad (\zeta > 0),$$

it follows that $\chi'(x)$ is negative. Hence $\chi(x)$ attains its greatest value in the interval $\frac{1}{2} \leqslant x < \infty$ at $x = \frac{1}{2}$. This agrees with (4.9).

Repeated use of the lemma establishes that

$$\alpha_{n,s} \leqslant (n\pi)^{s/2}/s! \qquad (s = 0, 1, \ldots, n+1, \quad n \geqslant 1),$$

and substitution of this result in (4.7) yields the desired bound

$$|E_n^{(2)}(\lambda)| \leqslant \frac{|a_n|\,\Gamma\{(n+1)/2\}}{\lambda^{(n+1)/2}} \exp\left(\frac{\phi_n^2}{4\lambda} + \frac{\phi_n}{2}\sqrt{\frac{n\pi}{\lambda}}\right) \qquad (\phi_n \geqslant 0, n \geqslant 1), \tag{4.10}$$

[2] Somewhat curiously, as $s \to n+1$, the limiting form of the first of these expressions equals the second expression.

where ϕ_n is defined by (4.4). In terms of the original functions,

$$\phi_n = \sup_{0<x<b} \left\{ \frac{1}{p^{1/2}(x)} \ln \left| \frac{F_n(x)}{a_n p^{n/2}(x)} \right| \right\}, \tag{4.11}$$

where

$$F_n(x) = \frac{p^{1/2}(x) q(x)}{p'(x)} - \sum_{s=0}^{n-1} a_s p^{s/2}(x). \tag{4.12}$$

When a_n vanishes, we may use the same procedure with a suitably chosen constant in place of $|a_n|$ on the right of (4.3). This case is illustrated in Example 5 below.

Comparing the bounds of this and the previous section, we observe that the bound for $|E_n^{(1)}(\lambda)|$ is exponentially smaller than the corresponding bound for $|E_n^{(2)}(\lambda)|$ when λ is large. We also observe that the bound for $|E_n^{(2)}(\lambda)|$ has the same asymptotic form as the modulus of the leading term in (2.11), provided that $a_n \neq 0$.

5. Alternative Bound for $E_n^{(2)}(\lambda)$ When $p(x)$ and $q(x)$ are Even

If the coefficients p_s and q_s of odd suffix vanish, then the a_s of odd suffix also vanish. In this event, in place of (4.3) we may use the majorant

$$|f_{2n}(p)| \leqslant |a_{2n}| p^n e^{\psi_{2n} p}, \tag{5.1}$$

where

$$\psi_{2n} = \sup_{0<p<p(b)} \left\{ \frac{1}{p} \ln \left| \frac{f_{2n}(p)}{a_{2n} p^n} \right| \right\}. \tag{5.2}$$

Substitution of this inequality in (2.16) gives immediately

$$|E_{2n}^{(2)}(\lambda)| \leqslant \frac{|a_{2n}|\, \Gamma(n+\frac{1}{2})}{(\lambda - \psi_{2n})^{n+(1/2)}} \qquad (\lambda > \psi_{2n}). \tag{5.3}$$

This result is applicable to integrals of the form

$$I(\lambda) = \int_a^b e^{-\lambda p(x)} q(x)\, dx \tag{5.4}$$

in the case when $p(x)$ has a minimum at an interior point of $[a,b]$. We may suppose, without loss of generality, that this minimum occurs at $x=0$, that $p(0)=p'(0)=0$, and that $p'(x)$ has the same sign as x throughout $[a,b]$. By truncation of one end of the range of integration, if necessary, we can arrange that

$$p(a) = p(b). \tag{5.5}$$

The contribution to $I(\lambda)$ from the range $0 \leqslant x \leqslant b$ may be transformed, as in Section 2, into the form (2.5). For the range $a \leqslant x \leqslant 0$, we have

$$\int_a^0 e^{-\lambda p(x)} q(x)\, dx = \int_0^{p(a)} e^{-\lambda p} \hat{f}(p) \frac{dp}{p^{1/2}}, \tag{5.6}$$

where

$$\hat{f}(p) = -\frac{p^{1/2}(x) q(x)}{p'(x)}, \tag{5.7}$$

the value of $p^{1/2}(x)$, again, being nonnegative. It is easily verified that the expansion corresponding to (2.8) is given by

$$\hat{f}(p) = a_0 - a_1 p^{1/2} + a_2 p - a_3 p^{3/2} + \dots. \tag{5.8}$$

Using (5.5), we derive

$$I(\lambda) = \int_0^{p(b)} e^{-\lambda p} \{f(p) + \hat{f}(p)\} \frac{dp}{p^{1/2}}. \tag{5.9}$$

Since only integer powers of p occur in the expansion of $f(p) + \hat{f}(p)$, this integral has the desired form.

In terms of the original variables we find that

$$f(p) + \hat{f}(p) = p^{1/2}(x) \left\{ \frac{q(x)}{p'(x)} - \frac{q(\hat{x})}{p'(\hat{x})} \right\}, \tag{5.10}$$

where $x \in [0, b]$ and $\hat{x}$ is the point in the interval $[a, 0]$ such that

$$p(\hat{x}) = p(x). \tag{5.11}$$

Remark. Because of the simplicity of (5.3) compared with (4.10), it is tempting to try using a majorant of the form (5.1) even in the case when the a_s of odd suffix do not vanish. When $a_{2n+1}/a_{2n} \leqslant 0$ this is feasible, but not otherwise. This is because

$$f_{2n}(p) = a_{2n} p^n \{1 + (a_{2n+1}/a_{2n}) p^{1/2} + O(p)\} \qquad (p \to 0).$$

If $a_{2n+1}/a_{2n} > 0$, then regardless of the value of ψ_{2n}, the function $e^{\psi_{2n} p}$ cannot grow quickly enough with p to majorize the quantity in braces. To put this another way, $\psi_{2n} = \infty$ when $a_{2n+1}/a_{2n} > 0$.

6. Evaluation of ϕ_n and ψ_{2n}

The principal difficulty in evaluating the bounds derived above is the determination of the number ϕ_n defined by (4.4) or (4.11), or, alternatively, the number ψ_{2n} defined by (5.2). The required supremum can always be found by numerical computation, but in every example that has been carried out, the supremum has been approached at one of the endpoints of the interval $0 < x < b$. In this section we discuss tests which can sometimes be used to establish this fact without recourse to computation. For convenience, we use the notation

$$w = p^{1/2}, \qquad g(w) = f(p), \qquad g_n(w) = f_n(p).$$

Application of Taylor's theorem to the expansion

$$g(w) = a_0 + a_1 w + a_2 w^2 + \ldots + a_{n-1} w^{n-1} + g_n(w) \tag{6.1}$$

gives

$$g_n(w) = w^n g^{(n)}(\eta)/n!, \tag{6.2}$$

where $0 < \eta < w$. Let $\hat{\phi}_n$ denote the greater of zero and

$$\sup_{0 < w < p^{1/2}(b)} \left\{ \frac{1}{w} \ln \left| \frac{g^{(n)}(w)}{g^{(n)}(0)} \right| \right\}, \tag{6.3}$$

so that

$$|g^{(n)}(w)| \leqslant |g^{(n)}(0)| \exp(\hat{\phi}_n w). \tag{6.4}$$

Then

$$|g_n(w)| \leqslant |a_n| \, w^n \exp(\hat{\phi}_n \eta) \leqslant |a_n| \, w^n \exp(\hat{\phi}_n w). \tag{6.5}$$

Comparison of this result with (4.3) shows that $\phi_n \leqslant \hat{\phi}_n$.

This result furnishes an alternative way of computing ϕ_n, or rather an upper bound for ϕ_n. Usually, however, the evaluation of $\hat{\phi}_n$ is as difficult as the evaluation of ϕ_n. The result is of some value, however, when $g(w)$ is a *completely monotonic* function[3] in the interval $0 < w < p^{1/2}(b)$. By this it is meant that each derivative of $g(w)$ is of constant sign, the sign alternating from one derivative to the next. Obviously, $|g^{(n)}(w)|$ then attains its maximum at $w = 0$; hence $\hat{\phi}_n = 0$ and (4.6) applies. It may be noticed that this particular result is equivalent to the simple error test of Steffensen applied to the integral

$$E_n^{(2)}(\lambda) = 2 \int_0^{p^{1/2}(b)} e^{-\lambda w^2} g_n(w) \, dw. \tag{6.6}$$

This test postulates that if consecutive remainder terms in a series expansion have opposite signs, then the truncation error is numerically less than the first neglected term in the series, and has the same sign [*9*].

In the general case, the essential problem in evaluating ϕ_n, ψ_{2n}, or $\hat{\phi}_n$, is the determination of a supremum of the form

$$\sup_{0 < w < c} \left\{ \frac{\ln H(w)}{w} \right\},$$

[3] Also called an *alternating* function. Tests for complete monotonicity are given in [*1*] and [*8*].

where $(0, c)$ is a finite or infinite interval and $H(w)$ is a nonnegative function such that $H(0) = 1$. We have just considered the trivial situation in which $H'(w)$ is negative throughout $(0, c)$. Let us now suppose that $H'(w)$ is positive.

LEMMA 2. *Let $H(w)$ be a twice-differentiable function in the interval $(0, c)$ such that $H(0) = 1$ and $H'(w) \geqslant 0$.*

(i) *If $H'(w)/H(w)$ is nonincreasing, or equivalently, $H(w)H''(w) \leqslant H'^2(w)$, then $w^{-1}\ln H(w)$ is nonincreasing.*

(ii) *If $H'(w)/H(w)$ is nondecreasing, or equivalently, $H(w)H''(w) \geqslant H'^2(w)$, then $w^{-1}\ln H(w)$ is nondecreasing.*

This result is easily proved by differentiation. It can be applied to a power-series expansion in the following way.

LEMMA 3. *Let*

$$H(w) = h_0 + h_1 w + h_2 w^2 + \dots, \tag{6.7}$$

this expansion converging when $0 \leqslant w < c$. If $h_0 = 1$, $h_s \geqslant 0$ $(s \geqslant 1)$, and

$$(s+1) h_{s+1} h_{s-1} \geqslant s h_s^2 \qquad (s \geqslant 1), \tag{6.8}$$

then $w^{-1}\ln H(w)$ is nondecreasing.

The genesis of this result is that when (6.8) is an equality for all s, $H(w)$ is an exponential function and $w^{-1}\ln H(w)$ is a constant.

When $h_1 = 0$, the condition (6.8) implies that all the higher h_s vanish and the lemma is trivial. Now assume that $h_1 > 0$. We then have

$$H(w) = \sum_{s=0}^{\infty} \frac{\theta_1 \theta_2 \dots \theta_s}{s!} w^s,$$

where $0 < \theta_1 \leqslant \theta_2 \leqslant \theta_3 \leqslant \dots$. Hence

$$H(w) H''(w) - H'^2(w) = \sum_{s=0}^{\infty} \frac{k_s}{s!} w^s,$$

where

$$k_s = s! \sum_{r=0}^{s} \left\{ \frac{\theta_1 \theta_2 \dots \theta_{r+2}}{r!} \frac{\theta_1 \theta_2 \dots \theta_{s-r}}{(s-r)!} - \frac{\theta_1 \theta_2 \dots \theta_{r+1}}{r!} \frac{\theta_1 \theta_2 \dots \theta_{s-r+1}}{(s-r)!} \right\}$$

$$= \sum_{r=0}^{s} \binom{s}{r} \theta_1 \theta_2 \dots \theta_{r+1} \theta_1 \theta_2 \dots \theta_{s-r} (\theta_{r+2} - \theta_{s-r+1}).$$

Addition of this sum to its reverse gives

$$2k_s = \sum_{r=0}^{s} \binom{s}{r} \theta_1 \theta_2 \dots \theta_r \theta_1 \theta_2 \dots \theta_{s-r}\{\theta_{r+1}(\theta_{r+2} - \theta_{s-r+1}) + \theta_{s-r+1}(\theta_{s-r+2} - \theta_{r+1})\}.$$

The quantity in braces is not less than

$$\theta_{r+1}^2 - 2\theta_{r+1}\theta_{s-r+1} + \theta_{s-r+1}^2,$$

and is therefore nonnegative. Hence $k_s \geqslant 0$, from which the stated result follows.

7. Analytical Examples

Example 1[4]

$$I(\lambda) = \int_{-\infty}^{\infty} e^{-\lambda x^2} \ln(1 + x + x^2)\, dx \qquad (\lambda > 0). \tag{7.1}$$

Rewritten in the standard form of Section 2, this integral becomes

$$I(\lambda) = \int_0^{\infty} e^{-\lambda p} f(p) \frac{dp}{p^{1/2}}, \tag{7.2}$$

where $p = x^2$, and

$$f(p) = \tfrac{1}{2}\ln(1 + p^{1/2} + p) + \tfrac{1}{2}\ln(1 - p^{1/2} + p) = \tfrac{1}{2}\ln\left(\frac{1-p^3}{1-p}\right). \tag{7.3}$$

The expansion of $f(p)$ in ascending powers of p is given by (2.8), with $a_0 = 0$, $a_{2s+1} = 0$, and

$$a_{2s} = \frac{1}{2s} \quad (s \equiv 1 \text{ or } 2 \ (\text{mod } 3)), \qquad a_{2s} = -\frac{1}{s} \quad (s \equiv 0 \ (\text{mod } 3)), \qquad (s > 0). \tag{7.4}$$

From (2.10) and (2.14) we derive the desired expansion

$$I(\lambda) = \sum_{s=1}^{n-1} \frac{a_{2s}\Gamma(s + \frac{1}{2})}{\lambda^{s+(1/2)}} + E_{2n}^{(2)}(\lambda), \tag{7.5}$$

where $|E_{2n}^{(2)}(\lambda)|$ is bounded by (5.3). It remains to evaluate the quantity ψ_{2n} defined by (5.2) with $p(b) = \infty$.

In the present case

$$f_{2n}(p) = \tfrac{1}{2}\ln\left(\frac{1-p^3}{1-p}\right) - \sum_{s=1}^{n-1} a_{2s}p^s. \tag{7.6}$$

[4] [3], Section 4.1.

Hence, as $p \to \infty$, we have

$$\frac{1}{p}\ln\left|\frac{f_{2n}(p)}{a_{2n}p^n}\right| \to 0. \tag{7.7}$$

Therefore $\psi_{2n} \geqslant 0$. Also, as $p \to 0$,

$$\frac{1}{p}\ln\left|\frac{f_{2n}(p)}{a_{2n}p^n}\right| \to \frac{a_{2n+2}}{a_{2n}}. \tag{7.8}$$

Since the last quantity is positive when $n \equiv 1 \pmod 3$ and negative otherwise, it is reasonable to conjecture that

$$\psi_{2n} = \frac{a_{2n+2}}{a_{2n}} = \frac{n}{n+1} \quad (n \equiv 1 \pmod 3), \qquad \psi_{2n} = 0 \quad (n \equiv 2 \text{ or } 0 \pmod 3); \tag{7.9}$$

equivalently,

$$\left.\begin{aligned} |f_{6m+2}(p)| &\leqslant \frac{p^{3m+1}}{6m+2}\exp\left\{\frac{(3m+1)p}{3m+2}\right\}, \\ |f_{6m+4}(p)| &\leqslant \frac{p^{3m+2}}{6m+4}, \\ |f_{6m+6}(p)| &\leqslant \frac{p^{3m+3}}{3m+3}, \end{aligned}\right\} \tag{7.10}$$

for any nonnegative integer m. These inequalities may be established as follows.

Differentiation of (7.6) and expansion of the result yields the identity

$$f_{2n}'(p) = \frac{p^{n-1}\{na_{2n}(1+p) + (n+1)a_{2n+2}p\}}{1+p+p^2} \qquad (n \geqslant 1). \tag{7.11}$$

With the aid of (7.4), it is easy to verify that for positive p, the first derivatives of the right sides of the inequalities (7.10) exceed the absolute values of the expression (7.11) for $n = 3m+1$, $3m+2$, $3m+3$, respectively. Each of the inequalities (7.10) is satisfied at $p = 0$; their validity for all positive values of p is now obvious.

Summarizing the results of this example, we have shown that the remainder term $E_{2n}^{(2)}(\lambda)$ in the expansion (7.5) is bounded in absolute value by the first neglected term of the expansion if $n = 0$ or $2 \pmod 3$, and by

$$\{1 - n/(n\lambda + \lambda)\}^{-n-(1/2)}$$

times the first neglected term if $n \equiv 1 \pmod 3$, provided that in the latter event $\lambda > n/(n+1)$.

Example 2. The modified Bessel function I_0 with large argument
Consider the function

$$I_0(\lambda) = \frac{1}{\pi}\int_0^{\pi} e^{\lambda\cos x}\,dx. \tag{7.12}$$

The principal contribution of the integrand comes from the neighborhood of $x = 0$. Owing to the presence of a second saddle-point at $x = \pi$, however, it is necessary to break the range of integration before our analysis is applicable. For simplicity we separate the range at its midpoint, but other subdivisions might lead to slightly better error bounds in certain circumstances.

Thus we write

$$I_0(\lambda) = \frac{e^{\lambda}}{\pi}\{I(\lambda) + E^{(3)}(\lambda)\}, \tag{7.13}$$

where

$$I(\lambda) = \int_0^{\pi/2} e^{-\lambda(1-\cos x)}\,dx, \qquad E^{(3)}(\lambda) = \int_{\pi/2}^{\pi} e^{-\lambda(1-\cos x)}\,dx. \tag{7.14}$$

If $\frac{1}{2}\pi \leqslant x \leqslant \pi$, then $1 - \cos x \geqslant 2x/\pi$. Hence a bound for the second integral is given by

$$E^{(3)}(\lambda) \leqslant \int_{\pi/2}^{\pi} e^{-2\lambda x/\pi}\,dx < \frac{\pi}{2\lambda}e^{-\lambda}. \tag{7.15}$$

For $I(\lambda)$ we have, in the notation of Section 2, $p = 1 - \cos x$ and

$$f(p) = \frac{1}{(2-p)^{1/2}} = \sum_{s=0}^{\infty} a_{2s}p^s \qquad (0 \leqslant p < 2), \tag{7.16}$$

where

$$a_0 = \frac{1}{\sqrt{2}}, \qquad a_{2s} = \frac{1.3.5\ldots(2s-1)}{4^s s!\sqrt{2}} \qquad (s \geqslant 1). \tag{7.17}$$

In the bounds (3.5) and (3.6), we have $p(b) = p(\frac{1}{2}\pi) = 1$. Hence

$$|E_1^{(1)}(\lambda)| \leqslant \frac{e^{-\lambda}}{\lambda\sqrt{2}}, \tag{7.18}$$

and

$$|E_{2n-1}^{(1)}(\lambda)| \leqslant \frac{e^{-\lambda}}{\lambda - n + \frac{3}{2}}\sum_{s=0}^{n-1} a_{2s} < \frac{e^{-\lambda}}{\lambda - n + \frac{3}{2}} \qquad (\lambda > n - \tfrac{3}{2},\ n \geqslant 2), \tag{7.19}$$

since

$$\sum_{s=0}^{\infty} a_{2s} = f(1) = 1.$$

Next, we consider the bound (5.3). The location of the supremum in (5.2) can be found in the present example by use of Lemma 3 of Section 6. In the notation of this lemma

$$w = p, \qquad H(p) = \frac{f_{2n}(p)}{a_{2n}p^n}, \qquad h_s = \frac{a_{2n+2s}}{a_{2n}}.$$

If $N \equiv n + s$ and $s \geqslant 1$, then from (7.17) we see that

$$\frac{(s+1)\,h_{s+1}\,h_{s-1}}{sh_s^2} = \frac{(s+1)(2N+1)\,N}{s(2N-1)(N+1)} = 1 + \frac{2N^2+N+s}{(2N^2+N-1)\,s} > 1.$$

Therefore, the maximum value of $p^{-1}\ln H(p)$ is attained at $p = 1$. This gives

$$\psi_{2n} = \ln\left(\frac{1 - a_0 - a_2 - \ldots - a_{2n-2}}{a_{2n}}\right). \tag{7.20}$$

Numerical values computed from this formula are as follows:

$\psi_0 = 0.35, \quad \psi_2 = 0.50, \quad \psi_4 = 0.56, \quad \psi_6 = 0.59, \quad \psi_8 = 0.61, \quad \psi_{10} = 0.62.$

The aggregate of (7.15), (7.18), (7.19), (7.20), and (5.3) comprises the desired bound for the error term

$$\hat{E}_{2n}(\lambda) = E_{2n}^{(2)}(\lambda) - E_{2n-1}^{(1)}(\lambda) + E^{(3)}(\lambda) \tag{7.21}$$

in the expansion

$$I_0(\lambda) = \frac{e^\lambda}{\pi}\left\{\sum_{s=0}^{n-1}\frac{a_{2s}\,\Gamma(s+\frac{1}{2})}{\lambda^{s+(1/2)}} + \hat{E}_{2n}(\lambda)\right\}. \tag{7.22}$$

For example, taking $n = 1$ and bearing in mind that $E_1^{(1)}(\lambda)$ and $E^{(3)}(\lambda)$ are both positive, we obtain on combination

$$I_0(\lambda) = e^\lambda\left\{\frac{1}{(2\pi\lambda)^{1/2}} + \frac{\hat{E}_2(\lambda)}{\pi}\right\}, \tag{7.23}$$

where

$$\frac{|\hat{E}_2(\lambda)|}{\pi} \leqslant \frac{1}{(2\pi)^{1/2}}\frac{1}{8(\lambda - 0.50)^{3/2}} + \frac{e^{-\lambda}}{2\lambda} \qquad (\lambda > 0.50). \tag{7.24}$$

Other bounds for the remainder terms in the asymptotic expansions of modified Bessel functions for large arguments are included in more general results due to Meijer [*10*], [*11*], and the present writer [*12*], the former being derived from infinite contour integral representations and the latter from differential-equation theory. However, in the particular circumstances of the present example, that is, zero order and real argument, the bounds we have just derived appear to be the only ones whose ratios to the actual errors tend to unity when the argument λ tends to infinity.

Example 3. *Modified Bessel function K_ν with large argument*

Consider the integral

$$e^{\lambda} K_\nu(\lambda) = \int_0^\infty e^{-\lambda(\cosh x - 1)} \cosh \nu x \, dx. \tag{7.25}$$

In the notation of Section 2, we have $p(x) = \cosh x - 1$, $q(x) = \cosh \nu x$, and

$$f(p) = \frac{\cosh \nu x}{\sqrt{2}\cosh \frac{1}{2}x} = \frac{\{1+p+(2p+p^2)^{1/2}\}^\nu + \{1+p-(2p+p^2)^{1/2}\}^\nu}{2\sqrt{2}(1+\frac{1}{2}p)^{1/2}}. \tag{7.26}$$

Thus

$$f(p) = \sum_{s=0}^{\infty} a_{2s} p^s \qquad (0 \leqslant p < 2), \tag{7.27}$$

where $a_0 = 1/\sqrt{2}$, and[5]

$$a_{2s} = \frac{(4\nu^2 - 1^2)(4\nu^2 - 3^2)\ldots\{4\nu^2 - (2s-1)^2\}}{(2s)!\,2^{s+(1/2)}} \qquad (s = 1, 2, \ldots). \tag{7.28}$$

Application of the results of Sections 2 and 5 gives

$$e^{\lambda} K_\nu(\lambda) = \sum_{s=0}^{n-1} \frac{a_{2s}\Gamma(s+\frac{1}{2})}{\lambda^{s+(1/2)}} + E_{2n}^{(2)}(\lambda), \tag{7.29}$$

where $|E_{2n}^{(2)}(\lambda)|$ is bounded by (5.3). In the case $\nu = 0$, we have

$$f(p) = \frac{1}{(2+p)^{1/2}}. \tag{7.30}$$

Clearly $|f^{(n)}(p)|$ attains its maximum value at $p = 0$. Hence by reasoning similar to that of the second paragraph of Section 6, we see that the ψ_{2n} of (5.2) is zero.

This result may be extended to integer values of ν by use of the relation

$$2\cosh 2\nu t = \sum_{s=0}^{\nu} (-)^s \frac{2\nu}{2\nu - s} \binom{2\nu - s}{s} (2\cosh t)^{2\nu - 2s} \qquad (\nu = 1, 2, 3, \ldots). \tag{7.31}$$

Setting $t = \frac{1}{2}x$, we derive

$$f(p) = \sum_{s=0}^{\nu} (-)^s \frac{2\nu}{2\nu - s} \binom{2\nu - s}{s} 2^{\nu - s - 1}(2+p)^{\nu - s - (1/2)}. \tag{7.32}$$

---

[5] Probably the easiest way to derive (7.28) is by way of the differential equation $(2p+p^2)f''(p) + (1+2p)f'(p) + (\frac{1}{4} - \nu^2)f(p) = 0$.

Repeated differentiation yields an expansion of the form

$$f^{(n)}(p) = \sum_{s=0}^{\nu} \frac{A_{n,s}}{(2+p)^{s+n-\nu+(1/2)}}. \tag{7.33}$$

If $n \geqslant \nu$, then it is easily verified that all the coefficients $A_{n,s}$ have the same sign. Therefore, in these circumstances, $|f^{(n)}(p)|$ attains its maximum at $p=0$, and again $\psi_{2n}=0$.

Accordingly, we have shown that when ν is zero or a positive integer, the error term $E_{2n}^{(2)}(\lambda)$ in the asymptotic expansion (7.29) is bounded in magnitude by the first neglected term, provided that $n \geqslant \nu$. This example is purely illustrative, however, because the result has been established in another way; indeed it suffices that ν be real and $n > |\nu| - \frac{1}{2}$ ([7], Section 7.4).

When n does not exceed $|\nu| - \frac{1}{2}$, values of ψ_{2n} may be computed from (5.2). Alternative bounds available from differential-equation theory [12] make no distinction between $n \gtrless |\nu| - \frac{1}{2}$.

8. Numerical Examples

Example 4. *The gamma function*

An integral representation for the gamma function of real positive argument λ is provided by

$$\Gamma(\lambda) = \frac{1}{\lambda}\int_0^\infty e^{-t}\, t^\lambda\, dt = \lambda^\lambda e^{-\lambda}\int_0^\infty (u\, e^{1-u})^\lambda\, du. \tag{8.1}$$

We shift the maximum value of the integrand from $u=1$ to the origin by means of the substitution $u = 1+x$. This gives

$$\Gamma(\lambda) = \lambda^\lambda e^{-\lambda}\int_{-1}^\infty e^{-\lambda p(x)}\, dx, \tag{8.2}$$

where

$$p(x) = x - \ln(1+x). \tag{8.3}$$

We may now either subdivide the range of the new integral at $x=0$ and treat the two parts separately by the theory of Section 4, or apply the theory of Section 5 directly to (8.2). For illustrative purposes let us pursue the latter course. We begin by noting that the preliminary condition (5.5) is satisfied.

Reversion of the expansion

$$p(x) = \frac{x^2}{2} - \frac{x^3}{3} + \frac{x^4}{4} - \cdots \qquad (|x| < 1) \tag{8.4}$$

yields,

$$x = \sqrt{2}p^{1/2} + \frac{2}{3}p + \frac{\sqrt{2}}{18}p^{3/2} - \frac{2}{135}p^2 + \frac{\sqrt{2}}{1080}p^{5/2} + \cdots, \tag{8.5}$$

the reverted expansion converging for all sufficiently small $|p|$. Therefore

$$f(p) = p^{1/2}\frac{dx}{dp} = \sum_{s=0}^{\infty} a_s p^{s/2}, \tag{8.6}$$

where

$$a_0 = \frac{\sqrt{2}}{2}, \qquad a_1 = \frac{2}{3}, \qquad a_2 = \frac{\sqrt{2}}{12}, \qquad a_3 = -\frac{4}{135}, \qquad a_4 = \frac{\sqrt{2}}{432}, \ldots . \tag{8.7}$$

Since $p'(x) = x/(1+x)$, we obtain from (5.10) and (5.8)

$$f(p) + \hat{f}(p) = \left(\frac{1}{x} - \frac{1}{\hat{x}}\right) p^{1/2}(x) = 2\sum_{s=0}^{\infty} a_{2s} p^s, \tag{8.8}$$

where $\hat{x}$ and x are related by

$$p(\hat{x}) = p(x), \tag{8.9}$$

x being positive and $\hat{x}$ negative.

The desired expansion is therefore

$$\Gamma(\lambda) = \lambda^\lambda e^{-\lambda}\left\{\sum_{s=0}^{n-1} \frac{2a_{2s}\,\Gamma(s+\frac{1}{2})}{\lambda^{s+(1/2)}} + E_{2n}^{(2)}(\lambda)\right\}, \tag{8.10}$$

where

$$|E_{2n}^{(2)}(\lambda)| \leqslant \frac{2|a_{2n}|\,\Gamma(n+\frac{1}{2})}{(\lambda - \psi_{2n})^{n+(1/2)}} \qquad (\lambda > \psi_{2n}), \tag{8.11}$$

$$\psi_{2n} = \sup_{0<x<\infty}\left\{\frac{1}{p(x)} \ln\left|\frac{F_{2n}(x)}{2a_{2n}p^n(x)}\right|\right\}, \tag{8.12}$$

and

$$F_{2n}(x) = \left(\frac{1}{x} - \frac{1}{\hat{x}}\right) p^{1/2}(x) - 2\sum_{s=0}^{n-1} a_{2s} p^s(x). \tag{8.13}$$

Numerical values of $F_{2n}(x)$ have been computed for the range $0 \leqslant x < \infty$, the requisite values of $\hat{x}$ being found by solving (8.9) with the aid of Newton's rule. The calculations show that the supremum in (8.12) is approached as $x \to 0$ for $n = 0$, 1, 3, and as $x \to \infty$ for $n = 2$. To four decimal places, the values of the leading ψ_{2n} are given by

$$\psi_0 = 0.1667, \qquad \psi_2 = 0.0278, \qquad \psi_4 = 0.0000, \qquad \psi_6 = 0.0245.$$

This appears to be the first time that error bounds have been evaluated directly for the expansion (8.10). Nevertheless, the example is mainly illustrative. It is well known ([*13*], Section 79) that in the corresponding expansion for the logarithm, given by

$$\ln \Gamma(\lambda) - (\lambda - \tfrac{1}{2}) \ln \lambda + \lambda - \tfrac{1}{2} \ln (2\pi) \sim \sum_{s=0}^{\infty} \frac{B_{2s+2}}{(2s+1)(2s+2)\,\lambda^{2s+1}}, \quad (8.14)$$

in which B_{2s+2} denotes the Bernoulli number of order $2s+2$, the remainder term is numerically less than the first neglected term (and has the same sign), when $\lambda > 0$.

Example 5. Modified Bessel functions K_λ of large order

As our final example we consider the Debye asymptotic expansion for the modified Bessel function

$$K_\lambda(\lambda\rho) = \tfrac{1}{2} \int_{-\infty}^{\infty} e^{\lambda(t - \rho \cosh t)}\, dt, \quad (8.15)$$

for positive ρ and large positive λ. The integrand has a maximum at $t = \sinh^{-1}(1/\rho)$. This is shifted to the origin by the transformation $t = x + \sinh^{-1}(1/\rho)$. Thus we derive

$$K_\lambda(\lambda\rho) = \tfrac{1}{2} \exp\{\lambda \sinh^{-1}(1/\rho) - \lambda(1+\rho^2)^{1/2}\}\, I(\lambda), \quad (8.16)$$

where

$$I(\lambda) = \int_{-\infty}^{\infty} e^{-\lambda p(x)}\, dx, \quad (8.17)$$

and

$$p(x) = (1+\rho^2)^{1/2}(\cosh x - 1) + \sinh x - x. \quad (8.18)$$

Reversion of the Maclaurin expansion for $p(x)$ leads to

$$f(p) \equiv \frac{p^{1/2}(x)}{p'(x)} = \sum_{s=0}^{\infty} a_s p^{s/2}, \quad (8.19)$$

where

$$a_0 = \frac{\sqrt{2}}{2\sigma}, \qquad a_1 = -\frac{1}{3\sigma^4}, \qquad a_2 = -\frac{\sqrt{2}}{8\sigma^3} + \frac{5\sqrt{2}}{24\sigma^7}, \qquad a_3 = \frac{4}{15\sigma^6} - \frac{8}{27\sigma^{10}}, \ldots, \quad (8.20)$$

and $\sigma \equiv (1+\rho^2)^{1/4}$. Applying the theory of Section 4 to the integral (8.17), we find that

$$I(\lambda) = I_1(\lambda) + I_2(\lambda), \quad (8.21)$$

where

$$I_1(\lambda) = \int_0^{\infty} e^{-\lambda p(x)}\, dx = \sum_{s=0}^{n-1} \frac{a_s \Gamma\{(s+1)/2\}}{\lambda^{(s+1)/2}} + E_{n,1}^{(2)}(\lambda), \quad (8.22)$$

$$I_2(\lambda) = \int_0^\infty e^{-\lambda p(-x)}\,dx = \sum_{s=0}^{n-1} (-)^s \frac{a_s \Gamma\{(s+1)/2\}}{\lambda^{(s+1)/2}} + E_{n,2}^{(2)}(\lambda), \tag{8.23}$$

and $E_{n,1}^{(2)}(\lambda)$, $E_{n,2}^{(2)}(\lambda)$ are bounded by expressions of the form (4.10).

Let us evaluate these bounds in the case $n=2$. Combination of (8.22) and (8.23) yields

$$I(\lambda) = \frac{1}{\sigma}\left(\frac{2\pi}{\lambda}\right)^{1/2} + E_2^{(2)}(\lambda), \tag{8.24}$$

where

$$|E_2^{(2)}(\lambda)| \leqslant \frac{2|a_2|\,\Gamma(3/2)}{\lambda^{3/2}} \exp\left\{\frac{\phi_2^2}{4\lambda} + \phi_2\left(\frac{\pi}{2\lambda}\right)^{1/2}\right\}, \tag{8.25}$$

$$\phi_2 = \sup_{-\infty<x<\infty} \left\{\frac{1}{p^{1/2}(x)} \ln\left|\frac{F_2(x)}{a_2 p(x)}\right|\right\}, \tag{8.26}$$

and

$$F_2(x) = \pm\frac{p^{1/2}(x)}{p'(x)} - \frac{\sqrt{2}}{2\sigma} \pm \frac{p^{1/2}(x)}{3\sigma^4}. \tag{8.27}$$

In the last equality the upper or lower signs are taken according as $x \gtrless 0$; $p^{1/2}(x)$ is positive in both cases.

In (8.25), the coefficient a_2 is given by (8.20). As a function of σ, it vanishes at $\sigma = (5/3)^{1/4}$, causing ϕ_2 to become infinite and the bound (8.25) to break down in the neighborhood of this value. This failure can be avoided, if we replace $|a_2|$ in the majorant (4.3) by τ_2/σ, where τ_2 is the supremum of $\sigma|a_2|$ in the range $1 \leqslant \sigma < \infty$.[6] It is easily verified that $\tau_2 = \sqrt{2}/12$; accordingly, we have

$$|E_2^{(2)}(\lambda)| \leqslant \frac{(2\pi)^{1/2}}{12\sigma\lambda^{3/2}} \exp\left\{\frac{\chi_2^2}{4\lambda} + \chi_2\left(\frac{\pi}{2\lambda}\right)^{1/2}\right\} \qquad (\lambda > 0), \tag{8.28}$$

where

$$\chi_2 = \sup\left\{\frac{1}{p^{1/2}(x)} \ln\left|\frac{12\sigma F_2(x)}{\sqrt{2}p(x)}\right|\right\}. \tag{8.29}$$

Furthermore, if this supremum is evaluated with respect to $1 \leqslant \sigma < \infty$ as well as $-\infty < x < \infty$, *then the bound* (8.28) *is uniform with respect to* $\rho \in [0, \infty]$. Numerical calculations yield the value $\chi_2 = 0.251$, this value being approached when $\sigma = 1$ and $x \to 0$ through negative values. Correspondingly,

$$\tfrac{1}{4}\chi_2^2 = 0.016, \qquad \chi_2(\pi/2)^{1/2} = 0.315.$$

---

[6] Instead of τ_2/σ, the supremum of $|a_2|$ could be used, but this would have the disadvantage of causing the ratio of the error bound to the approximation to become infinite as $\sigma \to \infty$.

Error bounds for this example are again available from differential-equation theory [14]. In the present notation, this reference gives the uniform bound

$$|E_2^{(2)}(\lambda)| \leqslant \frac{(2\pi)^{1/2} k}{\sigma\lambda^{3/2}} \exp\left(\frac{k}{\lambda}\right) \qquad (\lambda > 0,\ \rho > 0), \tag{8.30}$$

where

$$k = \frac{1}{6\sqrt{5}} + \frac{1}{12} = 0.158\ldots.$$

As $\lambda \to \infty$, this bound exceeds the right-hand side of (8.28) by the factor $1 + (2/\sqrt{5}) = 1.89\ldots$.

9. Summary and Conclusions

In this paper we have considered the well-known asymptotic expansion for integrals of the form

$$\int_a^b e^{-\lambda p(x)} q(x)\, dx,$$

for large positive values of the parameter λ, in cases where $p(x)$ and $q(x)$ are real differentiable functions and $p'(x)$ has a simple zero in the finite or infinite range $[a, b]$. In the first part (Sections 1–5) we constructed explicit bounds for the error terms associated with the expansion. These bounds are expressed in terms of the supremum of a certain function with respect to the range of integration. The bounds have the desirable property of being asymptotic to the absolute value of the first neglected term in the expansion as $\lambda \to \infty$. Analytical tests for locating the required suprema were given in Section 6 and applied to examples in Section 7. In cases where the tests are inapplicable numerical computation can be used; this was illustrated by examples in Section 8.

The principal difficulty in evaluating the error bounds is the location of the suprema referred to in the preceding paragraph. In every example carried out so far, it has transpired that the suprema occur at the zero of $p'(x)$ or at one of the endpoints of the range of integration. This suggests that it may be worthwhile to attempt the development of further analytical tests for locating the suprema.

Compared with error bounds obtained by application of the asymptotic theory of ordinary differential equations (in cases where such a comparison is possible), the present bounds are slightly sharper but more difficult to evaluate on the whole. Perhaps the last observation is not altogether surprising: it is generally true that higher *terms* in an asymptotic expansion can be obtained more readily from a differential equation than from an integral representation.

Finally, the methods used in this paper appear to be capable of extension to integrals having saddle-points of higher order, that is, points at which $p'(x)$ has a multiple zero.

Acknowledgments

The proof of Lemma 3 (Section 6) is due to M. Newman. The numerical computations in Examples 4 and 5 (Section 8) were carried out by Mrs. G. E. Olver and J. M. Smith. The writer is pleased to acknowledge this assistance, and also the support, in part, of the U.S. Army Research Office, Durham, N.C., under Project No. 4238-M.

References

1. D. V. Widder, "The Laplace Transform." Princeton University Press, Princeton, 1941.
2. A. Erdélyi, "Asymptotic Expansions." Dover, New York, 1956.
3. N. G. De Bruijn, "Asymptotic Methods in Analysis." Interscience, New York, 1958.
4. M. A. Evgrafov, "Asymptotic Estimates and Entire Functions." Gordon and Breach, New York, 1961.
5. E. T. Copson, "Asymptotic Expansions." Cambridge University Press, Cambridge, 1965.
6. J. B. Rosser, Explicit remainder terms for some asymptotic series. *J. Rat. Mech. Anal.* **4** (1955), 595–626.
7. G. N. Watson, "A Treatise on the Theory of Bessel Functions." Second edition. Cambridge University Press, Cambridge, 1944.
8. J. G. van der Corput and J. Franklin, Approximation of integrals by integration by parts. *Koninkl. Nederlandse Akad. Wetensch. Proc.* A **54** (1951), 213–219.
9. J. F. Steffensen, "Interpolation," Second edition. Chelsea, New York, 1950.
10. C. S. Meijer, Asymptotische Entwicklungen von Besselschen, Hankelschen und verwandten Funktionen. I–IV. *Koninkl. Nederlandse Akad. Wetensch. Proc.* **35** (1932), 656–667, 852–866, 948–958, 1079–1090.
11. B. Döring, Über Fehlerschranken zu den Hankelschen asymptotischen Entwicklungen der Besselfunktionen für komplexes Argument und reellen Index. *Z. angew. Math. Mech.* **42** (1962), 62–76.
12. F. W. J. Olver, Error bounds for asymptotic expansions, with an application to cylinder functions of large argument. *In* "Asymptotic Solutions of Differential Equations and Their Applications" (edited by C. H. Wilcox). Wiley, New York, 1964.
13. N. Nielsen, "Handbuch der Theorie der Gammafunktion." Teubner, Leipzig, 1906.
14. F. W. J. Olver, Error bounds for the Liouville-Green (or WKB) approximation. *Proc. Cambridge Philos. Soc.* **57** (1961), 790–810.

SIAM Review
Vol. 12, No. 2, April 1970

WHY STEEPEST DESCENTS?*

F. W. J. OLVER†

1. Introduction. Consider the contour integral

$$I(z) = \int_a^b e^{-zp(t)} q(t)\, dt, \tag{1.01}$$

in which $p(t)$ and $q(t)$ are analytic functions of the complex variable t. A widely used prescription for obtaining the asymptotic expansion of $I(z)$ for large real or complex values of the parameter z is as follows. First, the zeros of $p'(t)$ are determined; these are the so-called *saddle points* or *cols*. Then, if possible, the integration path $\mathscr{P}$, say, is deformed in such a way that:

(a) $\mathscr{P}$ passes through one or more saddle points.

(b) Im $\{p(t)\}$ is constant on $\mathscr{P}$.

On the new path Re $\{p(t)\}$ is taken as integration variable, and the asymptotic expansion of $I(z)$ for $|\text{ph}\, z| < \frac{1}{2}\pi$[1] immediately deduced by application of Watson's lemma [1, § 8.3]. This procedure was sketched in a memoir by Riemann concerning hypergeometric functions with large parameters [2], and developed more fully in a classic paper by Debye on Bessel functions of large orders [3]. For well-known geometrical reasons, paths satisfying condition (b) are called *paths of steepest descent*, or more briefly (when there is no ambiguity) steepest paths.

Frequently it is impractical to construct a path having the desired properties. In consequence, writers have often approximated steepest paths by variously chosen simpler paths. When the method is modified in this way it is sometimes called the *saddle-point method* [4], [5], in contrast to the name *method of steepest descents* which is reserved for the procedure of Riemann, Debye and Watson.

In a recent paper [6] Wyman unified the methods by establishing a more general theorem than Watson's lemma. To describe his result it is convenient to introduce the following notation. Let t_1, t_2 be any two points of $\mathscr{P}$. Then the part of $\mathscr{P}$ lying between t_1 and t_2 is denoted by $(t_1, t_2)_{\mathscr{P}}$ when t_1 and t_2 are both excluded, and by $[t_1, t_2]_{\mathscr{P}}$ when t_1 and t_2 are both included. Similarly for $(t_1, t_2]_{\mathscr{P}}$ and $[t_1, t_2)_{\mathscr{P}}$. Instead of $\mathscr{P}$ having to be a path of steepest descent, the essential conditions imposed by Wyman's theorem are as follows. First, $\mathscr{P}$ lies in the sector

$$|\text{ph}\, \{zp(t) - zp(a)\}| \leqq \tfrac{1}{2}\pi - \delta,$$

* Received by the editors August 12, 1969, and in revised form October 28, 1969. Presented by invitation at the Symposium on Special Functions, sponsored by the Air Force Office of Scientific Research, at the 1969 National Meeting of Society for Industrial and Applied Mathematics, held in Washington, D.C., June 10–12, 1969.

† National Bureau of Standards, Washington, D.C. 20234, and Institute for Fluid Dynamics, University of Maryland, College Park, Maryland 20742. This work was supported in part by the United States Army Research Office, Durham, North Carolina under Project 4238-M.

[1] Throughout this paper we use the notation ph z to denote the phase or argument of the complex number z.

where δ is a fixed positive number. Second, for each point $t_1 \in (a, b]_{\mathscr{P}}$ there exists a positive function $\Delta(t_1)$ such that

$$|p(t) - p(a)| \geqq \Delta(t_1), \qquad t \in [t_1, b]_{\mathscr{P}}.$$

With these conditions fulfilled the asymptotic expansion of $I(z)$ is obtainable in the usual manner: the expansion of $v = p(t) - p(a)$ in ascending powers of $t - a$ is reverted, then the expansion of $e^{-zv}q(t)\,dt/dv$ in powers of v is integrated formally term by term.

We begin the present paper by presenting a theorem which is a minor modification of Wyman's result. The essential condition it imposes on $\mathscr{P}$ is simply that Re $\{zp(t)\}$ attains its minimum at the endpoint a. This is completely analogous to the criterion used for the Laplace approximation for real integrals. The proof of the theorem is similar to Wyman's analysis, but we give it in full because it serves as a preliminary for the second part of the paper, which is the construction of explicit bounds for the error terms in the asymptotic expansion. This is effected by an extension of recently developed work for real variables [7], and in *this* analysis it will be seen that steepest paths regain some of their importance.

2. Conditions. We denote

$$\varpi = \text{angle of slope of } \mathscr{P} \text{ at } a$$

(2.01)

$$= \lim \{\text{ph}\,(t - a)\}, \qquad t \to a \text{ along } \mathscr{P}.$$

Our assumptions are the following:

(i) $p(t)$ and $q(t)$ are independent of z, and single-valued and regular (holomorphic) in an open domain **T**.

(ii) $\mathscr{P}$ is independent of z, a is finite, b is finite or infinite and $(a, b)_{\mathscr{P}}$ lies in **T**.[2]

(iii) In the neighborhood of a, $p(t)$ and $q(t)$ can be expanded in convergent series of the form

$$p(t) = p(a) + \sum_{s=0}^{\infty} p_s(t - a)^{s+\mu}, \qquad q(t) = \sum_{s=0}^{\infty} q_s(t - a)^{s+\lambda-1},$$

where $p_0 \neq 0$, Re $\lambda > 0$ and μ is real and positive. When μ and λ are not integers—and this can only happen when a is a boundary point of **T**—the branches of $(t - a)^\mu$ and $(t - a)^\lambda$ are determined by

$$(t - a)^\mu \sim |t - a|^\mu e^{i\mu\varpi}, \qquad (t - a)^\lambda \sim |t - a|^\lambda e^{i\lambda\varpi},$$

as $t \to a$ along $\mathscr{P}$, and by continuity elsewhere on $\mathscr{P}$.

(iv) z ranges along a ray or over a subsector given by $\theta_1 \leqq \theta \leqq \theta_2$ and $|z| \geqq Z$, where $\theta \equiv \text{ph}\, z$, $\theta_2 - \theta_1 < \pi$ and $Z > 0$. $I(z)$ converges at its upper limit absolutely and uniformly with respect to z.

(v) Re $\{e^{i\theta}p(t) - e^{i\theta}p(a)\}$ is positive when $t \in (a, b)_{\mathscr{P}}$, and is bounded away from zero uniformly with respect to θ as $t \to b$ along $\mathscr{P}$.

Comments. (a) Neither ϖ nor θ need be confined to the principal range $(-\pi, \pi]$, provided that consistency is maintained.

[2] Thus either a or b or both may be boundary points of **T**.

(b) Compared with the conditions of Wyman [6], the forms of expansion permitted in (iii) are less general, but condition (v) on $\mathscr{P}$ is less restrictive; see § 1 above.

3. Asymptotic expansion of $I(z)$. In the whole analysis great care is needed in specifying the branches of the many-valued functions which appear. With this in mind we introduce the following convention: *the value of* $\varpi_0 \equiv \operatorname{ph} p_0$ *is not necessarily the principal one, but is chosen to satisfy*

$$|\varpi_0 + \theta + \mu\varpi| \leqq \tfrac{1}{2}\pi, \tag{3.01}$$

and this "branch" of ph p_0 *is used in constructing all fractional powers of* p_0 *which occur.* For example,

$$p_0^{1/\mu} = \exp\{(\ln|p_0| + i\varpi_0)/\mu\}.$$

Since

$$e^{i\theta}p(t) - e^{i\theta}p(a) \sim e^{i\theta}p_0(t - a)^\mu$$

as $t \to a$ along $\mathscr{P}$ (condition (iii)), and $e^{i\theta}p(t) - e^{i\theta}p(a)$ has nonnegative real part (condition (v)), it is always possible to choose ϖ_0 uniquely in this way. Moreover, because θ is restricted to an interval of length less than π, the value of ϖ_0 which satisfies (3.01) is independent of θ.

We introduce new variables v and w satisfying

$$w^\mu = v = p(t) - p(a). \tag{3.02}$$

The branches of ph v and ph w are determined by

$$\operatorname{ph} v, \quad \mu \operatorname{ph} w \to \varpi_0 + \mu\varpi, \qquad t \to a \text{ along } \mathscr{P}, \tag{3.03}$$

and by continuity elsewhere. Again, it is to be understood that these branches of ph v and ph w are to be used in constructing all fractional powers of v and w. Since v and w cannot vanish on $(a, b)_{\mathscr{P}}$ (condition (v)) this specifies the branches uniquely on $\mathscr{P}$; furthermore ph $v = \mu$ ph w at every point of $\mathscr{P}$. From (3.01), (3.03) and condition (v), it follows that

$$|\theta + \operatorname{ph} v| < \tfrac{1}{2}\pi, \qquad t \in (a, b)_{\mathscr{P}}. \tag{3.04}$$

Accordingly, v is confined to a single Riemann sheet as t traverses $\mathscr{P}$.

For small $|t - a|$ condition (iii) yields

$$w = p_0^{1/\mu}(t - a)\left\{1 + \frac{p_1}{\mu p_0}(t - a) + \cdots\right\}.$$

Thus w is a single-valued regular function of t in the neighborhood of a, and dw/dt does not vanish at a. Application of Landau's theorem on inverse functions [8, § 6.22] shows that for all sufficiently small values of the positive number ρ, the disc $|t - a| < \rho$ is mapped conformally on a domain $\mathbf{W}$ containing $w = 0$. Moreover, if $w \in \mathbf{W}$, then $t - a$ can be expanded in a convergent series

$$t - a = \sum_{s=1}^{\infty} c_s w^s = \sum_{s=1}^{\infty} c_s v^{s/\mu},$$

in which the c_s are expressible in terms of the p_s. In particular,

$$c_1 = \frac{1}{p_0^{1/\mu}}, \quad c_2 = -\frac{p_1}{\mu p_0^{1+(2/\mu)}}, \quad c_3 = \frac{(\mu+3)p_1^2 - 2\mu p_0 p_2}{2\mu^2 p_0^{2+(3/\mu)}}.$$

Let k be a finite point of $(a, b]_{\mathscr{P}}$ chosen independently of z and sufficiently close to a to ensure that the disc $|w| \leqq |p(k) - p(a)|^{1/\mu}$ is contained in $\mathbf{W}$. Then $[a, k]_{\mathscr{P}}$ may be deformed so that its w-map is a straight line. Transformation to the variable v gives

$$\int_a^k e^{-zp(t)} q(t)\, dt = e^{-zp(a)} \int_0^\kappa e^{-zv} f(v)\, dv, \tag{3.05}$$

where

$$\kappa = p(k) - p(a), \qquad f(v) = q(t)\frac{dt}{dv} = \frac{q(t)}{p'(t)}, \tag{3.06}$$

and the path for the integral on the right of (3.05) also is a straight line.

For small $|v|$, $f(v)$ has a convergent expansion of the form

$$f(v) = \sum_{s=0}^{\infty} a_s v^{(s+\lambda-\mu)/\mu}, \tag{3.07}$$

where the coefficients a_s are expressible in terms of p_s and q_s, for example,

$$\begin{aligned} a_0 &= \frac{q_0}{\mu p_0^{\lambda/\mu}}, \qquad a_1 = \left\{\frac{q_1}{\mu} - \frac{(\lambda+1)p_1 q_0}{\mu^2 p_0}\right\}\frac{1}{p_0^{(\lambda+1)/\mu}}, \\ a_2 &= \left[\frac{q_2}{\mu} - \frac{(\lambda+2)p_1 q_1}{\mu^2 p_0} + \{(\lambda+\mu+2)p_1^2 - 2\mu p_0 p_2\}\frac{(\lambda+2)q_0}{2\mu^3 p_0^2}\right]\frac{1}{p_0^{(\lambda+2)/\mu}}. \end{aligned} \tag{3.08}$$

Let $f_n(v)$, $n = 0, 1, 2, \cdots$, be defined by the relations

$$f(v) = \sum_{s=0}^{n-1} a_s v^{(s+\lambda-\mu)/\mu} + v^{(n+\lambda-\mu)/\mu} f_n(v), \qquad f_n(0) = a_n. \tag{3.09}$$

Then $f_n(v)$ is $O(1)$ as $v \to 0$. We rearrange the integral on the right of (3.05) in the form

$$\int_0^\kappa e^{-zv} f(v)\, dv = \sum_{s=0}^{n-1} a_s \int_0^{\infty e^{i \operatorname{ph} \kappa}} e^{-zv} v^{(s+\lambda-\mu)/\mu}\, dv - \varepsilon_{n,2}(z) + \varepsilon_{n,1}(z), \tag{3.10}$$

where ph κ is determined by

$$|\theta + \operatorname{ph} \kappa| < \tfrac{1}{2}\pi \tag{3.11}$$

(compare (3.04)), and

$$\varepsilon_{n,1}(z) = \int_0^\kappa e^{-zv} v^{(n+\lambda-\mu)/\mu} f_n(v)\, dv, \tag{3.12}$$

$$\varepsilon_{n,2}(z) = \sum_{s=0}^{n-1} a_s \int_\kappa^{\infty e^{i \operatorname{ph} \kappa}} e^{-zv} v^{(s+\lambda-\mu)/\mu}\, dv. \tag{3.13}$$

All integration paths are now straight lines.

In consequence of (3.04) and (3.11) we derive

$$\int_0^{\infty e^{i\mathrm{ph}\kappa}} e^{-zv} v^{(s+\lambda-\mu)/\mu}\, dv = \Gamma\left(\frac{s+\lambda}{\mu}\right) \frac{1}{z^{(s+\lambda)/\mu}}, \tag{3.14}$$

and

$$\varepsilon_{n,2}(z) = \sum_{s=0}^{n-1} \Gamma\left(\frac{s+\lambda}{\mu}, z\kappa\right) \frac{a_s}{z^{(s+\lambda)/\mu}}, \tag{3.15}$$

where $\mathrm{ph}\,(z^{(s+\lambda)/\mu}) = (s+\lambda)\theta/\mu$, and $\Gamma(\alpha, \zeta)$ denotes the principal value of the incomplete gamma function

$$\Gamma(\alpha, \zeta) = \int_\zeta^\infty e^{-t} t^{\alpha-1}\, dt. \tag{3.16}$$

For fixed α and large $|\zeta|$ it is known that

$$\Gamma(\alpha, \zeta) = O(e^{-\zeta} \zeta^{\alpha-1}), \tag{3.17}$$

uniformly in a region which includes $|\mathrm{ph}\, \zeta| \leqq \frac{1}{2}\pi$.[3] Hence

$$\varepsilon_{n,2}(z) = O(e^{-z\kappa}/z), \qquad |z| \to \infty, \tag{3.18}$$

uniformly with respect to θ.

For $\varepsilon_{n,1}(z)$ the substitution $v = \kappa\tau$ produces

$$\varepsilon_{n,1}(z) = \int_0^1 e^{-z\kappa\tau} \tau^{(n+\lambda-\mu)/\mu} O(1)\, d\tau. \tag{3.19}$$

In consequence of condition (v) and the fact that θ is restricted to a closed interval, we have

$$\mathrm{Re}\,(z\kappa) = |z|\, \mathrm{Re}\,\{e^{i\theta} p(k) - e^{i\theta} p(a)\} \geqq |z|\eta_k, \tag{3.20}$$

where η_k is independent of z and positive. Hence

$$\varepsilon_{n,1}(z) = O(z^{-(n+\mathrm{Re}\,\lambda)/\mu}) = O(z^{-(n+\lambda)/\mu}), \tag{3.21}$$

uniformly with respect to θ.

Combination of (3.05), (3.10), (3.14), (3.18), (3.20) and (3.21) yields

$$\int_a^k e^{-zp(t)} q(t)\, dt = e^{-zp(a)} \left\{ \sum_{s=0}^{n-1} \Gamma\left(\frac{s+\lambda}{\mu}\right) \frac{a_s}{z^{(s+\lambda)/\mu}} + O\left(\frac{1}{z^{(n+\lambda)/\mu}}\right) \right\}, \tag{3.22}$$

uniformly with respect to θ as $|z| \to \infty$.

It remains to consider the contribution to the original integral from $[k, b]_{\mathscr{P}}$. Let $\hat{b}$ be any finite point of $(k, b)_{\mathscr{P}}$ chosen independently of z. Because

$$\mathrm{Re}\,\{e^{i\theta} p(t) - e^{i\theta} p(a)\}$$

is a continuous function of both t and θ, it attains its lower bound when t ranges over the closed set $[k, \hat{b}]_{\mathscr{P}}$ and θ ranges over its closed interval. Condition (v) ensures that the lower bound is positive. And from the same condition we see that when $\hat{b}$

[3] See [9, Chap. 6] or § 6 below.

is sufficiently close to b and $t \in [\hat{b}, b]_{\mathscr{P}}$, we have

$$\operatorname{Re}\{e^{i\theta}p(t) - e^{i\theta}p(a)\} \geqq \eta > 0, \tag{3.23}$$

where η is independent of θ. Combination of the two results shows that an inequality of the form (3.23) also holds when $t \in [k, b]_{\mathscr{P}}$. Accordingly,

$$\begin{aligned}\operatorname{Re}\{zp(t) - zp(a)\} &= \{(|z| - Z) + Z\} \operatorname{Re}\{e^{i\theta}p(t) - e^{i\theta}p(a)\} \\ &\geqq (|z| - Z)\eta + \operatorname{Re}\{Ze^{i\theta}p(t)\} - \operatorname{Re}\{Ze^{i\theta}p(a)\},\end{aligned}$$

and

$$\begin{aligned}\left|\int_k^b e^{-zp(t)}q(t)\,dt\right| &\leqq |e^{-zp(a)}|\, e^{(Z-|z|)\eta}\, |\exp\{Ze^{i\theta}p(a)\}| \\ &\quad \cdot \int_k^b |\exp\{-Ze^{i\theta}p(t)\}q(t)\,dt|.\end{aligned} \tag{3.24}$$

Condition (iv) shows that the last quantity is $e^{-zp(a)}O(e^{-|z|\eta})$, uniformly with respect to θ. Therefore the asymptotic expansion (3.22) is unaffected by addition of the contribution from $[k, b]_{\mathscr{P}}$.

We have established the following theorem.

THEOREM I. *With the conditions of* § 2,

$$\int_a^b e^{-zp(t)}q(t)\,dt \sim e^{-zp(a)} \sum_{s=0}^{\infty} \Gamma\left(\frac{s+\lambda}{\mu}\right) \frac{a_s}{z^{(s+\lambda)/\mu}}, \qquad |z| \to \infty, \tag{3.25}$$

uniformly with respect to $\theta \equiv \operatorname{ph} z$. *Here the coefficients* a_s *are determined by the procedure given at the beginning of this section, and the branch of* $z^{(s+\lambda)/\mu}$ *has phase* $(s+\lambda)\theta/\mu$.

4. Peak value of integrand at an interior point of path. Consider again the integral (1.01) in the case when $p(t)$ and $q(t)$ are regular in an open domain $\mathbf{T}$ containing $(a, b)_{\mathscr{P}}$, and $\theta \equiv \operatorname{ph} z$ is either fixed or ranges over an interval $[\theta_1, \theta_2]$. But suppose now that the minimum value of $\operatorname{Re}\{e^{i\theta}p(t)\}$ occurs at t_0, where $t_0 \in (a, b)_{\mathscr{P}}$ and is independent of θ.

The path may be subdivided at t_0, giving

$$I(z) = \int_{t_0}^b e^{-zp(t)}q(t)\,dt - \int_{t_0}^a e^{-zp(t)}q(t)\,dt. \tag{4.01}$$

In the neighborhood of t_0, $p(t)$ and $q(t)$ have Taylor series expansions

$$p(t) = p(t_0) + (t - t_0)p'(t_0) + (t - t_0)^2\frac{p''(t_0)}{2!} + \cdots, \tag{4.02}$$

$$q(t) = q(t_0) + (t - t_0)q'(t_0) + (t - t_0)^2\frac{q''(t_0)}{2!} + \cdots. \tag{4.03}$$

For large $|z|$ the asymptotic expansion of each integral on the right of (4.01) is obtainable by application of Theorem I, the role of the series in condition (iii) of § 2 being played by (4.02) and (4.03). However, if $p'(t_0) \neq 0$, then the condition that Re $\{e^{i\theta}p(t)\}$ be a minimum at t_0 gives

$$\cos(\varpi_0 + \theta + \varpi) = 0,$$

where, again, $\varpi_0 = \text{ph}\,\{p'(t_0)\}$ and ϖ is the angle of slope of the path at t_0. Since the two values of ϖ differ by π, $\varpi_0 + \theta + \varpi$ is $\frac{1}{2}\pi$ for one integral and $-\frac{1}{2}\pi$ for the other; compare (3.01). The values of ϖ_0 and the coefficients a_s are exactly the same in the two cases. In consequence, the asymptotic expansions of the two integrals are the same, and all that remains on substitution in (4.01) is an error term $O\{z^{-n}e^{-zp(t_0)}\}$, n being an arbitrary positive integer.[4]

On the other hand, if $p'(t_0) = 0$, that is, if t_0 is a saddle point, then the μ of condition (iii) of § 2 is an integer such that $\mu \geqq 2$. Thus $\mu\varpi$ differs by $\mu\pi$ for the two integrals, causing the values of ϖ_0 which satisfy (3.01) to differ by $\mu\pi$ or $(\mu - 1)\pi$ according as μ is even or odd. In consequence, *different* branches are used for $p_0^{1/\mu}$ in constructing the coefficients a_s, and the asymptotic expansions no longer cancel on substitution in (4.01).

The preceding two paragraphs bring out the importance of saddle points in a clearer way, perhaps, than in previous literature. They also suggest how to handle cases in which $p'(t_0) \neq 0$. Keeping the path of integration within **T**, we try to deform it in such a way that the minimum of Re $\{e^{i\theta}p(t)\}$ occurs either at one of the endpoints, or at a point at which $p'(t)$ vanishes. As we have just seen, the asymptotic expansion of $I(z)$ can then be found by use of Theorem I.

In applications, the construction of a path on which Re $\{e^{i\theta}p(t)\}$ attains its minimum at an endpoint, or at a saddle point, can often be solved by an intelligent guess, especially when the parameter z is confined to the real axis. Failing this, it may be necessary to make a partial study of the conformal mapping between the planes of t and v, where $v = p(t) - p(a)$ and a is the endpoint or saddle point. Once the map **V** of the original domain **T** has been constructed, it is easy to ascertain whether the point $p(b) - p(a)$ can be joined to the origin by a path $\mathcal{Q}$ lying entirely in the intersection of **V** and the sector $|\text{ph}\,(e^{i\theta}v)| < \frac{1}{2}\pi$.[5] An admissible path $\mathcal{P}$ is the t-map of $\mathcal{Q}$, but its actual location need not be determined: an existence demonstration is sufficient for the purpose of applying Theorem I.

If a is a saddle point of order $\mu - 1$, $\mu \geqq 2$, that is, if

$$p'(a) = p''(a) = \cdots = p^{(\mu-1)}(a) = 0, \qquad p^{(\mu)}(a) \neq 0,$$

then the neighborhood of a is mapped on μ sheets in the v-plane. Fortunately, however, the full neighborhood of a need not be considered, because $\mathcal{Q}$ is confined to a single sheet.

5. Example. The Airy integral is given by

$$\text{(5.01)} \qquad \text{Ai}\,(z) = \frac{1}{2\pi i}\int_{\mathscr{C}} \exp\left(\frac{1}{3}t^3 - zt\right) dt,$$

where $\mathscr{C}$ is any contour which begins at infinity in the sector $-\frac{1}{2}\pi < \text{ph}\,t < -\frac{1}{6}\pi$,

---

[4] This situation has no analogue with real variables: when $p(t)$ is real and continuously differentiable it cannot attain a minimum at an interior point t_0 of (a, b), unless $p'(t_0) = 0$.

[5] If the method of steepest descents were being used, $\mathcal{Q}$ would have to be a ray.

and ends at infinity in the sector $\frac{1}{6}\pi < \text{ph}\, t < \frac{1}{2}\pi$. The integral converges for all z, real or complex, but we shall suppose at first that z is real and positive.

Replacement of t by $z^{1/2}t$ gives

$$\text{Ai}\,(z) = \frac{z^{1/2}}{2\pi i}\int_{\mathscr{C}} e^{-\xi p(t)}\,dt, \tag{5.02}$$

where $\xi = \frac{2}{3}z^{3/2}$ and $p(t) = \frac{3}{2}t - \frac{1}{2}t^3$. The saddle points are located at $t = \pm 1$, and the relevant Taylor series at $t = 1$ is simply

$$p(t) = 1 - \tfrac{3}{2}(t-1)^2 - \tfrac{1}{2}(t-1)^3.$$

Therefore in condition (iii) we have $\mu = 2$, $p_0 = -\frac{3}{2}$, and $\lambda = 1$. A possible path $\mathscr{P}$ is indicated in Fig. 1; thus

$$\text{Ai}\,(z) = \frac{z^{1/2}}{2\pi i}\{\mathscr{I}_1(\xi) - \mathscr{I}_{-1}(\xi)\}, \tag{5.03}$$

where

$$\mathscr{I}_{\pm 1}(\xi) = \int_1^{1+\infty e^{\pm\pi i/3}} e^{-\xi p(t)}\,dt.$$

It is easily perceived that all the conditions of § 2 are satisfied by $\mathscr{I}_1(\xi)$. Application of Theorem I yields

$$\mathscr{I}_1(\xi) \sim e^{-\xi}\sum_{s=0}^{\infty}\Gamma\left(\frac{s+1}{2}\right)\frac{a_s}{\xi^{(s+1)/2}}, \qquad \xi\to\infty. \tag{5.04}$$

The coefficients a_s are obtained by reverting the expression

$$v = -\frac{3}{2}(t-1)^2 - \frac{1}{2}(t-1)^3,$$

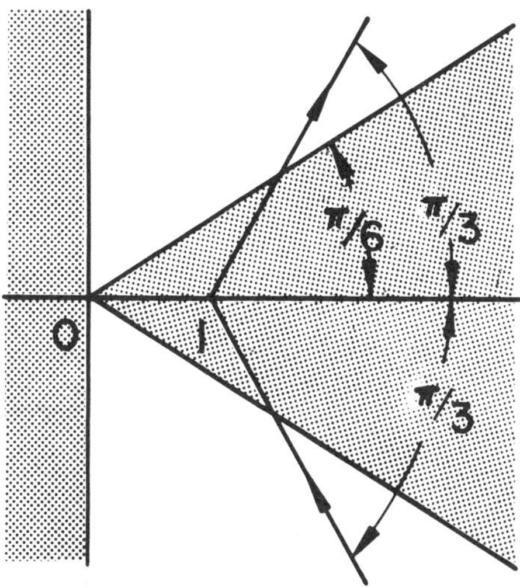

FIG. 1. *Integration path for* Ai(z), *z positive*

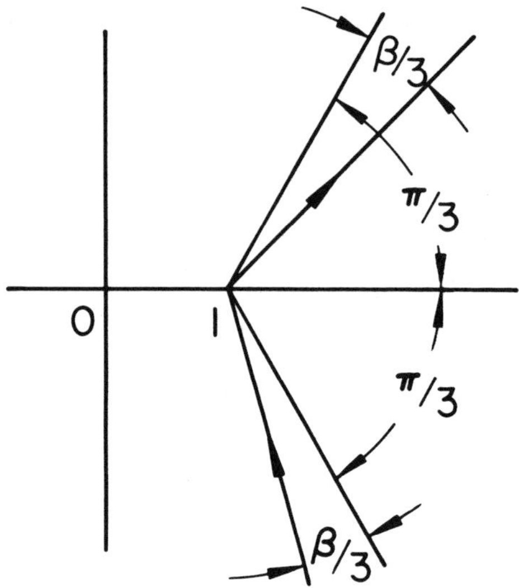

FIG. 2. *Integration path for* Ai (z), z *complex*

and then forming dt/dv. Since $\varpi = \frac{1}{3}\pi$ and $\theta = 0$, the correct choice of $\varpi_0 \equiv \mathrm{ph}\, p_0$ is $-\pi$; compare (3.01). With the aid of Lagrange's theorem on the reversion of series we find that

$$a_s = i^{s+1}\frac{2^{(s-1)/2}}{3^{(3s+1)/2}}\binom{-\frac{1}{2}s-\frac{1}{2}}{s};$$

compare [10, § 2.6]. The corresponding expansion for $\mathscr{I}_{-1}(\xi)$ is obtained on replacing a_s in (5.04) by $(-)^{s+1}a_s$. On substituting these results in (5.03) we obtain the well-known expansion

$$\mathrm{Ai}\,(z) \sim \frac{e^{-\xi}}{2\pi^{1/2}z^{1/4}}\sum_{s=0}^{\infty}(-)^s\frac{u_s}{\xi^s}, \qquad z \to \infty, \tag{5.05}$$

in which

$$u_s = \left(\frac{6}{\pi}\right)^{1/2}\frac{\Gamma(s+\frac{1}{2})a_{2s}}{i^{2s+1}} = \frac{(2s+1)(2s+3)\cdots(6s-1)}{(216)^s s!}.$$

Now suppose that z is a complex variable, and write $\vartheta \equiv \mathrm{ph}\,\xi = \frac{3}{2}\mathrm{ph}\,z$. The upper part of the path of Fig. 1 has the parametric equation

$$t = 1 + \tau e^{\pi i/3}, \qquad 0 \leqq \tau \leqq \infty.$$

Hence

$$p(t) - p(1) = \frac{3}{2}\tau^2 e^{-\pi i/3} + \frac{1}{2}\tau^3,$$

$$\mathrm{Re}\,\{e^{i\vartheta}p(t) - e^{i\vartheta}p(1)\} = \frac{3}{2}\tau^2\cos\left(\vartheta - \frac{1}{3}\pi\right) + \frac{1}{2}\tau^3\cos\vartheta.$$

When $-\frac{1}{6}\pi \leqq \vartheta \leqq \frac{1}{2}\pi$, one of the quantities $\cos(\vartheta - \frac{1}{3}\pi)$ and $\cos\vartheta$ is positive and the other nonnegative, ensuring that the minimum is attained at $\tau = 0$. The convergence condition (iv) of § 2 applied to $\mathscr{I}_1(\xi)$ demands that $|\vartheta| \leqq \frac{1}{2}\pi - \delta$, where δ is an arbitrarily small positive number. From these requirements, we deduce that the expansion (5.04) is uniformly valid in the sector $-\frac{1}{6}\pi \leqq \vartheta \leqq \frac{1}{2}\pi - \delta$. Similarly the expansion for $\mathscr{I}_{-1}(\xi)$ is valid in the conjugate sector. Accordingly, (5.05) holds when $|\vartheta| \leqq \frac{1}{6}\pi$, that is, when $|\mathrm{ph}\, z| \leqq \frac{1}{9}\pi$.

To extend the region of validity still further, we rotate the path of Fig. 1 around the point $t = 1$ through an arbitrary angle $-\frac{1}{3}\beta$, as indicated in Fig. 2. Use of Cauchy's theorem easily establishes that the integral

$$\int_1^{1+\infty e^{i(\pi-\beta)/3}} e^{-\xi p(t)}\, dt$$

converges uniformly when

$$-\frac{1}{2}\pi + \beta + \delta \leqq \vartheta \leqq \frac{1}{2}\pi + \beta - \delta, \tag{5.06}$$

and represents the analytic continuation of $\mathscr{I}_1(\xi)$ in this sector. Setting

$$t = 1 + \tau e^{i(\pi-\beta)/3}, \qquad 0 \leqq \tau \leqq \infty,$$

we find that

$$\mathrm{Re}\,\{e^{i\vartheta}p(t) - e^{i\vartheta}p(1)\} = \frac{3}{2}\tau^2 \cos\left(\vartheta - \frac{1}{3}\pi - \frac{2}{3}\beta\right) + \frac{1}{2}\tau^3 \cos(\vartheta - \beta).$$

If (5.06) holds, then $\cos(\vartheta - \beta) > 0$. The other cosine is nonnegative when

$$-\frac{1}{6}\pi + \frac{2}{3}\beta \leqq \vartheta \leqq \frac{5}{6}\pi + \frac{2}{3}\beta. \tag{5.07}$$

The range of values of ϑ for which β can be chosen to fulfil *both* (5.06) and (5.07) is obtained by eliminating β from these inequalities. This gives

$$-\frac{1}{6}\pi + \frac{2}{3}\left(\vartheta - \frac{1}{2}\pi + \delta\right) \leqq \vartheta \leqq \frac{5}{6}\pi + \frac{2}{3}\left(\vartheta + \frac{1}{2}\pi - \delta\right),$$

that is,

$$-\frac{3}{2}\pi + 2\delta \leqq \vartheta \leqq \frac{7}{2}\pi - 2\delta. \tag{5.08}$$

Again, application of Theorem I establishes that for large ξ the analytic continuation of $\mathscr{I}_1(\xi)$ has the expansion (5.04), uniformly valid in the sector (5.08). Similarly the expansion for $\mathscr{I}_{-1}(\xi)$ applies when $\vartheta \in [-\frac{7}{2}\pi + 2\delta, \frac{3}{2}\pi - 2\delta]$. Therefore *the final expansion* (5.05) *holds uniformly when* $|\vartheta| \leqq \frac{3}{2}\pi - 2\delta$, *that is, when* $|\mathrm{ph}\, z| \leqq \pi - \frac{4}{3}\delta$.

This example is purely illustrative, since the expansion (5.05) is derivable in simpler ways. For example, Copson [5, § 40] shows that the integral (5.01) may be transformed into

$$\mathrm{Ai}\,(z) = \frac{\exp\left(-\frac{2}{3}z^{3/2}\right)}{\pi}\int_0^\infty e^{-\tau^2 z^{1/2}} \cos\left(\frac{1}{3}\tau^3\right) d\tau, \qquad |\mathrm{ph}\, z| < \pi,$$

to which Watson's lemma is immediately applicable. The purpose of our illustration is to show how the asymptotic expansion of $\mathrm{Ai}\,(z)$ can be derived directly from its definition using simple integration paths. As Copson [5, § 39] observes, the steepest paths for (5.02) are quite complicated when the full phase range $|\mathrm{ph}\, z| < \pi$ has to be covered.

6. Error bounds. In the notation of § 3 let us assume, for simplicity, that λ is real and positive, and write

$$(6.01) \qquad Q_n(t) \equiv p'(t)v^{(n+\lambda-\mu)/\mu}f_n(v) = q(t) - p'(t)\sum_{s=0}^{n-1} a_s\{v(t)\}^{(s+\lambda-\mu)/\mu},$$

where, again,

$$(6.02) \qquad v(t) = p(t) - p(a).$$

Assume temporarily that $\mathscr{P}$ can be deformed in such a way that the only possible saddle point along its length is $t = a$, and its v-map lies along the ray $\mathrm{ph}\, v = -\beta$, where β is a constant. Thus if $\beta = 0$, then $\mathscr{P}$ would be part (or all) of a path of steepest descent. The original integral may be expressed as

$$(6.03) \qquad \int_a^b e^{-zp(t)}q(t)\, dt = e^{-zp(a)}\left\{\sum_{s=0}^{n-1}\Gamma\left(\frac{s+\lambda}{\mu}\right)\frac{a_s}{z^{(s+\lambda)/\mu}} - \varepsilon_{n,2}(z) + \varepsilon_{n,1}(z)\right\},$$

where $\varepsilon_{n,1}(z)$ is given by (3.12) with, now, $\kappa = p(b) - p(a)$, and

$$(6.04) \qquad \varepsilon_{n,2}(z) = \sum_{s=0}^{n-1}\Gamma\left\{\frac{s+\lambda}{\mu}, zp(b) - zp(a)\right\}\frac{a_s}{z^{(s+\lambda)/\mu}};$$

compare (3.05), (3.10) and (3.15).

In (6.04) we have $|\mathrm{ph}\,\{zp(b) - zp(a)\}| < \frac{1}{2}\pi$, and the incomplete gamma function takes its principal value. By transforming the integral (3.16) we obtain

$$(6.05) \qquad \Gamma(\alpha, \zeta) = e^{-\zeta}\zeta^\alpha \int_0^\infty e^{-\zeta t}(1+t)^{\alpha-1}\, dt, \qquad |\mathrm{ph}\, \zeta| < \tfrac{1}{2}\pi.$$

If $\vartheta \equiv \mathrm{ph}\, \zeta$, then rotation of the path through an angle $-\vartheta$ gives

$$(6.06) \qquad \Gamma(\alpha, \zeta) = e^{-\zeta}\zeta^\alpha e^{-i\vartheta}\int_0^\infty e^{-|\zeta|\tau}(1 + \tau e^{-i\vartheta})^{\alpha-1}\, d\tau,$$

this formula being valid for $|\vartheta| < \pi$, by analytic continuation. Since

$$1 \leqq |1 + \tau e^{-i\vartheta}| \leqq e^\tau, \qquad |\vartheta| \leqq \frac{1}{2}\pi, \tau \geqq 0,$$

it follows that

$$(6.07) \qquad |\Gamma(\alpha, \zeta)| \leqq |e^{-\zeta}\zeta^{\alpha-1}|, \qquad \alpha \leqq 1, |\mathrm{ph}\, \zeta| \leqq \frac{1}{2}\pi,$$

or

$$|\Gamma(\alpha, \zeta)| \leqq \frac{|e^{-\zeta}\zeta^{\alpha}|}{|\zeta| - \alpha + 1}, \qquad \alpha > 1, |\mathrm{ph}\, \zeta| \leqq \frac{1}{2}\pi, |\zeta| > \alpha - 1. \tag{6.08}$$

These inequalities immediately enable $|\varepsilon_{n,2}(z)|$ to be bounded in a realistic way.

For the other error term, we have for small $|p(t) - p(a)|$,

$$Q_n(t) = p'(t)[a_n\{v(t)\}^{(n+\lambda-\mu)/\mu} + a_{n+1}\{v(t)\}^{(n+1+\lambda-\mu)/\mu} + \cdots].$$

Define

$$\sigma_n(\beta) = \sup_{t \in \mathscr{P}} \left[\frac{1}{|v(t)|} \ln \left| \frac{Q_n(t)}{p'(t) a_n \{v(t)\}^{(n+\lambda-\mu)/\mu}} \right| \right], \tag{6.09}$$

so that

$$|f_n(v)| \leqq |a_n| e^{|v(t)|\sigma_n(\beta)}. \tag{6.10}$$

Bearing in mind that the integration path in (3.12) is $\mathrm{ph}\, v = -\beta$, we derive

$$|\varepsilon_{n,1}(z)| \leqq \Gamma\left(\frac{n+\lambda}{\mu}\right) \frac{|a_n|}{\{|z| \cos(\theta - \beta) - \sigma_n(\beta)\}^{(n+\lambda)/\mu}}, \tag{6.11}$$

$$|\theta - \beta| < \frac{1}{2}\pi, \quad |z| > \sigma_n(\beta) \sec(\theta - \beta).$$

It will be observed that if it is legitimate to take $\beta = 0$, *then this error bound is asymptotic to the absolute value of the actual error as* $|z| \to \infty$. In comparison, the bound for $|\varepsilon_{n,2}(z)|$ is exponentially small.

The bound (6.11) fails when $\sigma_n(\beta)$ is infinite. This obviously happens when $a_n = 0$, in which event we would simply proceed to a higher value of n. If $a_n \neq 0$, then in terms of v,

$$\sigma_n(\beta) = \sup_{t \in \mathscr{P}} \left\{ \frac{1}{|v|} \ln \left| \frac{f_n(v)}{a_n} \right| \right\}, \tag{6.12}$$

and failure occurs when the quantity in braces tends to $+\infty$ as $v \to 0$. For small v

$$f_n(v) = a_n + a_{n+1} v^{1/\mu} + a_{n+2} v^{2/\mu} + \cdots, \tag{6.13}$$

and

$$\frac{1}{|v|} \ln \left| \frac{f_n(v)}{a_n} \right| = \frac{1}{|v|} \mathrm{Re} \left\{ \frac{a_{n+1} v^{1/\mu}}{a_n} \right\} + O(v^{(2/\mu)-1}). \tag{6.14}$$

If $\mu > 1$ then $|v^{(1/\mu)-1}| \to \infty$. No problem arises when $\mathrm{Re}\,(a_{n+1} e^{-i\beta/\mu}/a_n)$ is negative, because the right-hand side of (6.14) tends to $-\infty$ as $v \to 0$. But if $\mu > 1$ and $\mathrm{Re}\,(a_{n+1} e^{-i\beta/\mu}/a_n) > 0$, then $\sigma_n(\beta) = \infty$.

A simple way of overcoming difficulties of this kind is to modify the majorant (6.10) by the inclusion of an arbitrary factor M which exceeds unity; $M = 2$ would be a realistic choice in many circumstances. Then in place of (6.11) we derive

$$|\varepsilon_{n,1}(z)| \leqq \Gamma\left(\frac{n+\lambda}{\mu}\right) \frac{M|a_n|}{\{|z| \cos(\theta - \beta) - \hat{\sigma}_n(\beta)\}^{(n+\lambda)/\mu}}, \tag{6.15}$$

where

$$\hat{\sigma}_n(\beta) = \sup_{t\in\mathscr{P}}\left[\frac{1}{|v(t)|}\ln\left|\frac{Q_n(t)}{p'(t)Ma_n\{v(t)\}^{(n+\lambda-\mu)/\mu}}\right|\right]. \tag{6.16}$$

This succeeds because as $t \to a$ the content of the square brackets in the last relation tends to $-\infty$.

Alternatively, let m be the largest integer such that $m < \mu$, and a_{n+j+1} be the first member of the set $a_{n+1}, a_{n+2}, \cdots, a_{n+m}$ for which $\operatorname{Re}\{(a_{n+j+1}/a_n)e^{-(j+1)i\beta/\mu}\}$ is negative, or, if no such member exists, set $j = m$. Define

$$\rho_n(\beta) = \sup_{t\in\mathscr{P}}\left\{\frac{1}{|v|}\ln\left|\frac{f_n(v)}{a_n + a_{n+1}v^{1/\mu} + \cdots + a_{n+j}v^{j/\mu}}\right|\right\}. \tag{6.17}$$

Then

$$|f_n(v)| \leqq |a_n + a_{n+1}v^{1/\mu} + \cdots + a_{n+j}v^{j/\mu}|e^{|v|\rho_n(\beta)}, \qquad t \in \mathscr{P}.$$

Substituting in (3.12) by means of this inequality, we derive

$$|\varepsilon_{n,1}(z)| \leqq \sum_{s=n}^{n+j}\Gamma\left(\frac{s+\lambda}{\mu}\right)\frac{|a_s|}{\{|z|\cos(\theta-\beta) - \rho_n(\beta)\}^{(s+\lambda)/\mu}}. \tag{6.18}$$

This bound is successful because

$$\frac{1}{|v|}\ln\left|\frac{f_n(v)}{a_n + a_{n+1}v^{1/\mu} + \cdots + a_{n+j}v^{j/\mu}}\right| = \frac{1}{|v|}\operatorname{Re}\left\{\frac{a_{n+j+1}v^{(j+1)/\mu}}{a_n}\right\} + O(v^{(j+2-\mu)/\mu})$$

and therefore tends to $-\infty$ ($j \leqq m-1$) or is bounded ($j = m$), as $v \to 0$. Moreover, $a_n + a_{n+1}v^{1/\mu} + \cdots + a_{n+j}v^{j/\mu}$ cannot vanish on $\mathscr{P}$.

The advantage of (6.18) is that its ratio to the absolute value of the actual error tends to unity as $|z| \to \infty$, unlike (6.15). The disadvantages of (6.18) lie in its more complicated form and the need to evaluate coefficients beyond a_n.

A third way of overcoming the difficulty is indicated by means of an example in the next section. It is applicable in the common case when there is a simple saddle point at an interior point of the integration path.

So far, we have assumed that a and b can be linked by a path having an equation of the form

$$\operatorname{ph}\{p(t) - p(a)\} = -\beta,$$

where β is a constant. When this is not the case we proceed from a along a path $\mathscr{P}$ of this type until a conveniently chosen point k is reached.[6] The journey from k to b is completed by any convenient path lying in $\mathbf{T}$ along which $\operatorname{Re}\{e^{i\theta}p(t) - e^{i\theta}p(a)\}$ is positive. The integral over $[a, k]_{\mathscr{P}}$ has the same asymptotic expansion as the integral over $[a, b]_{\mathscr{P}}$, and its error terms can be bounded by the methods just given. The contribution from $[k, b]_{\mathscr{P}}$ is bounded by an inequality of the form (3.24), in which η can be taken as the largest number fulfilling (3.23) when $t \in [k,b]_{\mathscr{P}}$.

[6] Section 3 shows that this is always possible. In the present context, however, k need not satisfy the criteria of § 3.

Since (3.24) is exponentially small compared with the bound for $|e^{-zp(a)}\varepsilon_{n,1}(z)|$, the choices of k and the path from k to b are not critical.

7. Gamma function of large argument. To illustrate the methods of the preceding section, we consider the well-known analysis used to derive the asymptotic expansion of the gamma function from its integral representation

$$\Gamma(z) = z^z e^{-z} \int_{-1}^{\infty} e^{-zp(t)}\, dt, \tag{7.01}$$

in which

$$p(t) = t - \ln(1 + t), \qquad p'(t) = t/(1 + t),$$

and the path of integration is the real axis. The integral converges when

$$\theta \equiv \operatorname{ph} z \in (-\tfrac{1}{2}\pi, \tfrac{1}{2}\pi).$$

The only saddle point is the origin, and in the interval $(-1, \infty)$ Re $\{e^{i\theta}p(t)\}$ attains its minimum at $t = 0$, as required by Theorem I.

Setting $v = p(t)$ and reverting, we find for positive t,

$$t = \sqrt{2}v^{1/2} + \frac{2}{3}v + \frac{\sqrt{2}}{18}v^{3/2} - \frac{2}{135}v^2 + \frac{\sqrt{2}}{1080}v^{5/2} + \cdots,$$

and thence

$$f(v) = \frac{dt}{dv} = a_0 v^{-1/2} + a_1 + a_2 v^{1/2} + a_3 v + \cdots; \tag{7.02}$$

compare (3.06), where

$$a_0 = \frac{\sqrt{2}}{2}, \quad a_1 = \frac{2}{3}, \quad a_2 = \frac{\sqrt{2}}{12}, \quad a_3 = -\frac{4}{135}, \quad a_4 = \frac{\sqrt{2}}{432}, \quad \cdots.$$

Accordingly, from Theorem I (or Watson's lemma) we derive

$$\int_0^{\infty} e^{-zp(t)}\, dt = \int_0^{\infty} e^{-zv} f(v)\, dv \sim \sum_{s=0}^{\infty} \Gamma\left(\frac{s+1}{2}\right) \frac{a_s}{z^{(s+1)/2}}, \tag{7.03}$$

as $|z| \to \infty$, uniformly in the sector $|\theta| \leqq \frac{1}{2}\pi - \delta < \frac{1}{2}\pi$.

For the other contribution to (7.01), let $t = \hat{t}$ be the point of $(-1, 0)$ which corresponds to a given positive value of v. Then

$$\hat{t} = -\sqrt{2}v^{1/2} + \frac{2}{3}v - \frac{\sqrt{2}}{18}v^{3/2} - \frac{2}{135}v^2 - \frac{\sqrt{2}}{1080}v^{5/2} + \cdots,$$

$$\hat{f}(v) \equiv \frac{d\hat{t}}{dv} = -a_0 v^{-1/2} + a_1 - a_2 v^{1/2} + a_3 v - \cdots \tag{7.04}$$

and

$$\int_0^{-1} e^{-zp(t)}\,dt = \int_0^{\infty} e^{-zv}\hat{f}(v)\,dv \sim \sum_{s=0}^{\infty} (-)^{s-1}\Gamma\left(\frac{s+1}{2}\right)\frac{a_s}{z^{(s+1)/2}}. \tag{7.05}$$

Substitution of (7.03) and (7.05) in (7.01) yields Stirling's series in the form

$$\Gamma(z) = z^z e^{-z}\left\{\sum_{s=0}^{n-1}\frac{2a_{2s}\Gamma(s+\frac{1}{2})}{z^{s+(1/2)}} + \varepsilon_n(z)\right\}, \tag{7.06}$$

where n is an arbitrary integer and $\varepsilon_n(z) = O(z^{-n-(1/2)})$ uniformly in the sector $|\theta| \leqq \frac{1}{2}\pi - \delta$.

To bound $\varepsilon_n(z)$, we express

$$\Gamma(z) = z^z e^{-z}\int_0^{\infty} e^{-zv}F(v)\,dv, \tag{7.07}$$

in which

$$F(v) = f(v) - \hat{f}(v) = \frac{1}{t} - \frac{1}{\hat{t}}. \tag{7.08}$$

For small v,

$$F(v) = 2a_0 v^{-1/2} + 2a_2 v^{1/2} + 2a_4 v^{3/2} + \cdots. \tag{7.09}$$

It is therefore possible to use a majorant of the form

$$\left|F(v) - \sum_{s=0}^{n-1} 2a_{2s}v^{s-(1/2)}\right| \leqq 2|a_{2n}|v^{n-(1/2)}e^{\sigma_n v}, \qquad 0 \leqq v < \infty, \tag{7.10}$$

in which σ_n is real and finite; compare (6.10). This leads to the bound

$$|\varepsilon_n(z)| \leqq \frac{2|a_{2n}|\Gamma(n+\frac{1}{2})}{\{|z|\cos\theta - \sigma_n\}^{n+(1/2)}}, \qquad |\theta| < \tfrac{1}{2}\pi, |z| > \sigma_n \sec\theta. \tag{7.11}$$

Numerical values of the minimum σ_n satisfying (7.10) were given in [7] for $\theta = 0$ and $n = 0, 1, 2, 3$, the notation in this reference for σ_n being ψ_{2n}.

The bound (7.11) fails as $\theta \to \pm\frac{1}{2}\pi$, although the asymptotic expansion (7.06) is known to be uniformly valid when $|\theta| \leqq \pi - \delta$. To extend the error analysis, we consider the paths on which

$$\operatorname{ph} v = -\beta,$$

where β is a disposable constant whose value will be assigned later.

The mapping between the planes of t and v is indicated in Figs. 3 through 6, passage from one sheet to the other taking place across the dashed lines OA on the real axis. The dotted curves indicate the location of the steepest paths and their maps in the case $\beta = \frac{1}{6}\pi$. The continuation of Fig. 6 is readily perceived from the fact that a single positive circuit of $t = -1$ decreases v by $2\pi i$. In consequence, when $-\frac{1}{2}\pi < \beta < 0$ the continuation of the path OXY of Fig. 5 spirals in on $t = -1$, crossing the negative real axis at angle $-\beta$ twice on each circuit.

Provided that $|\beta| < \frac{1}{2}\pi$ and $|\theta - \beta| < \frac{1}{2}\pi$, the analytic continuation of the integrals on the left-hand sides of the relations (7.03) and (7.05) may be obtained

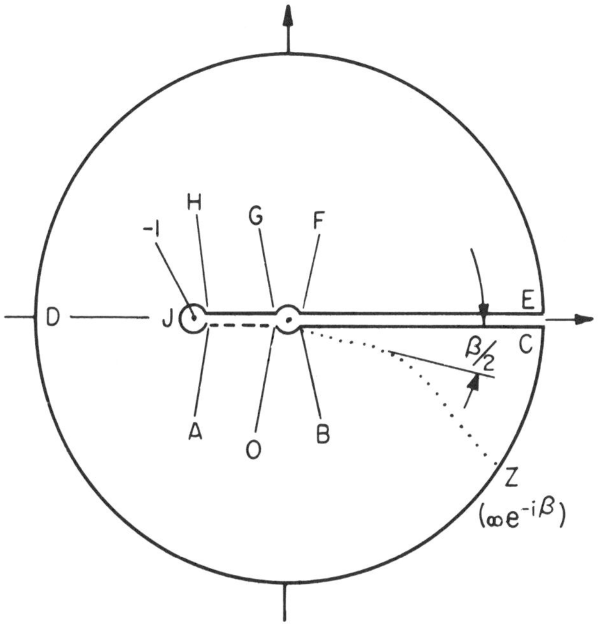

FIG. 3. *t-plane* (i)

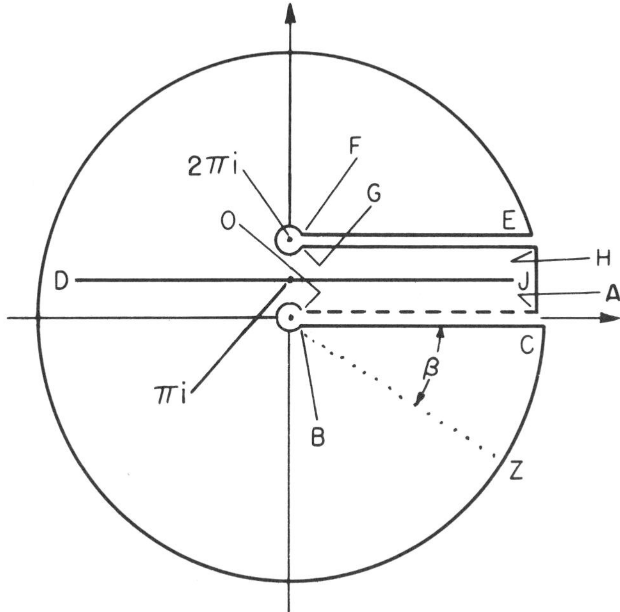

FIG. 4. *v-plane* (i)

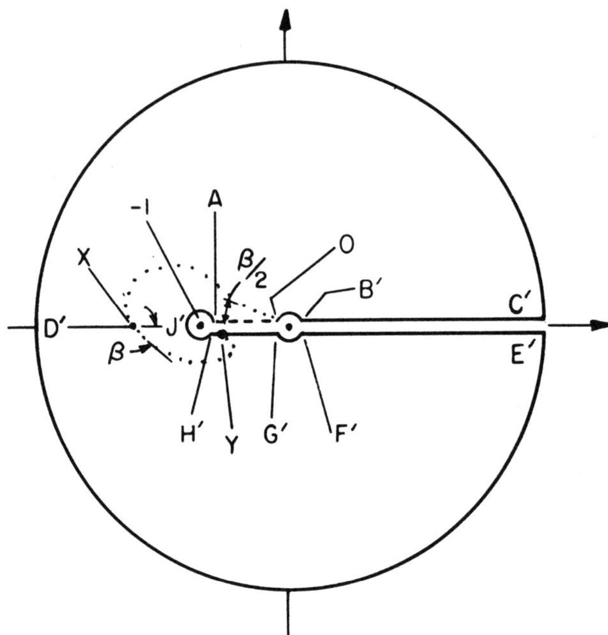

FIG. 5. *t-plane* (ii)

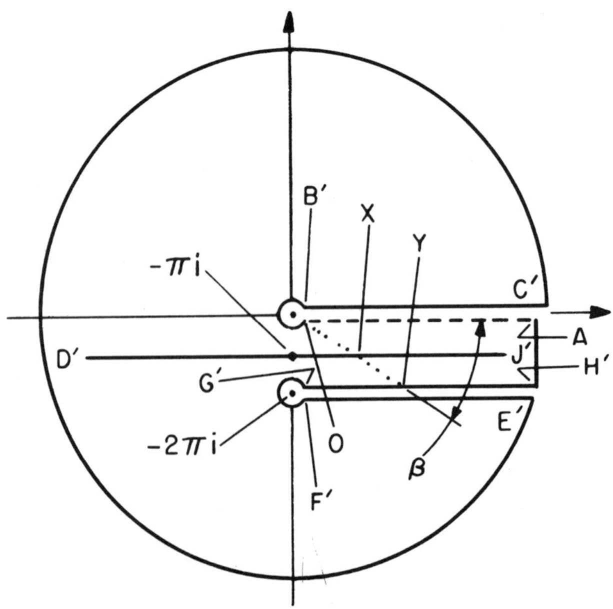

FIG. 6. *v-plane* (ii)

by deforming the integration paths into steepest paths. Setting $v = \tau e^{-i\beta}$, where the parameter τ runs from 0 to ∞, we arrive at

$$(7.12)\qquad \Gamma(z) = z^z e^{-z} e^{-i\beta} \int_0^\infty \exp(-z\tau e^{-i\beta}) F(\tau e^{-i\beta})\, d\tau,$$

in which $F(v)$ is again defined by (7.08), t and $\hat{t}$ being the maps of the point $v = \tau e^{-i\beta}$ on the curves BZ (Fig. 3) and OXY $\cdots$ (Fig. 5), respectively. The extension of (7.11) is given by

$$(7.13)\qquad |\varepsilon_n(z)| \leqq \frac{2|a_{2n}|\Gamma(n+\frac{1}{2})}{\{|z|\cos(\theta-\beta) - \sigma_n(\beta)\}^{n+(1/2)}},$$

valid when

$$|\beta| < \frac{1}{2}\pi, \quad |\theta - \beta| < \frac{1}{2}\pi, \quad |z| > \sigma_n(\beta)\sec(\theta - \beta),$$

where

$$(7.14)\ \sigma_n(\beta) = \sup_{\mathrm{ph}\,v = -\beta} \left\{ \frac{1}{|v|} \ln \left| \frac{v^{1/2}F(v) - 2a_0 - 2a_2 v - \cdots - 2a_{2n-2}v^{n-1}}{2a_{2n}v^n} \right| \right\}.$$

For $n = 1$, numerical values to two decimal places are as follows:

$$\sigma_1(0) = 0.03, \quad \sigma_1\left(\frac{1}{4}\pi\right) = 0.02, \quad \sigma_1\left(\frac{3}{8}\pi\right) = 0.03, \quad \sigma_1\left(\frac{7}{16}\pi\right) = 0.06,$$

$$\sigma_1\left(\frac{15}{32}\pi\right) = 0.09, \quad \sigma_1\left(\frac{31}{64}\pi\right) = 0.13, \quad \sigma_1\left(\frac{63}{128}\pi\right) = 0.18.$$

They were obtained by evaluating the quantity in braces in (7.14) for a suitably chosen discrete range of values on the ray $\mathrm{ph}\, v = -\beta$. The values of $F(v)$ were obtained from (7.08), the necessary values of t being found by numerical solution of the differential equation

$$\frac{dt}{d\omega} = \frac{1+t}{t} 2\omega e^{-i\beta}, \qquad v = \omega^2 e^{-i\beta}, \qquad 0 \leqq \omega < \infty,$$

by the Runge-Kutta process, and checked by substitution in the equation

$$v = t - \ln(1 + t)$$

for various spot values of ω. Similarly for $\hat{t}$.

As $\beta \to \frac{1}{2}\pi$ the value of $\sigma_n(\beta)$ tends to infinity. This is because the point Y of Fig. 5 approaches the origin, causing the corresponding value of $|F(v)|$ to tend to infinity. The growth in $\sigma_n(\beta)$ is only logarithmic in character, however.

By appropriate choice of β it is possible to arrange for the bound (7.13) to apply to any θ in the interval $(-\pi, \pi)$. With $\beta = \theta$, for example, the ratio of the bound to the actual value of $|\varepsilon_n(z)|$ tends to unity as $|z| \to \infty$. However, it is possible to choose $\beta = \theta$ only when $|\theta| < \frac{1}{2}\pi$, and even in this range it is undesirable to take θ too close to $\pm\frac{1}{2}\pi$ because of the growth of $\sigma_n(\theta)$.

A useful form of (7.13) is obtained on setting $\beta = \theta\,(>0)$ and replacing $\sigma_n(\theta)$ by

$$\chi_n(\theta) \equiv \sup_{0 \leqq \theta' \leqq \theta} \{\sigma_n(\theta')\}.$$

The numerical values given above indicate, for example, that $\chi_1(\frac{31}{64}\pi) = 0.13$. Thus we derive

$$|\varepsilon_1(z)| \leqq \frac{2|a_2|\Gamma(\frac{3}{2})}{(|z|-0.13)^{3/2}}, \qquad |\theta| \leqq \frac{31}{64}\pi,$$

and

$$|\varepsilon_1(z)| \leqq \frac{2|a_2|\Gamma(\frac{3}{2})}{\{|z| \cos(\theta - \frac{31}{64}\pi) - 0.13\}^{3/2}}, \qquad \frac{31}{64}\pi \leqq |\theta| < \frac{63}{64}\pi,$$

provided that the denominators are positive in each case. Another interesting form of bound is given by

$$|\varepsilon_1(z)| \leqq \frac{1.03 \times 2|a_2|\Gamma(\frac{3}{2})}{(|z| - 0.06)^{3/2}}, \qquad |\theta| \leqq \frac{1}{2}\pi,$$

which is obtained by taking $\beta = \frac{7}{16}\pi$ and using the fact that $\sec^{3/2}(\frac{1}{16}\pi) = 1.03$.

Error bounds for higher values of n are calculable in a similar way. The results already given are sufficient illustration of the method, however, because satisfactory bounds for the error terms in the corresponding expansion for $\ln \Gamma(z)$ have been derived analytically by Nielsen [11, § 79].

8. Summary. In §§ 2 and 3 we state and prove a theorem concerning the asymptotic expansion of contour integrals of the form

$$\int_a^b e^{-zp(t)} q(t)\, dt,$$

in which z is a large complex parameter. In comparison with the method of steepest descents, the essential condition imposed on the integration path by the present theorem is simply that $\operatorname{Re}\{zp(t)\}$ attains its minimum at one of the endpoints. The application of this result is discussed in general terms in § 4 and illustrated by an example in § 5.

In § 6 it is shown how to calculate realistic bounds for the error terms in the asymptotic expansion. In this analysis the use of paths of steepest descent is necessary, at least in the neighborhood of the minimum of $\operatorname{Re}\{zp(t)\}$. In the concluding section, § 7, the error analysis is illustrated by some new results concerning Stirling's series.

Acknowledgments. The writer is pleased to acknowledge the help of G. F. Miller, who read and criticized the manuscript, and of J. M. Smith, who carried out the numerical computations of § 7.

REFERENCES

[1] G. N. WATSON, *A Treatise on the Theory of Bessel Functions*, 2nd ed., Cambridge University Press, Cambridge, 1944.

[2] B. Riemann, *Sullo svolgimento del quotiente di due serie ipergeometriche in frazione continua infinita*, Collected Works; 2nd ed., Dover, New York, 1953, pp. 424–430.
[3] P. Debye, *Näherungsformeln für die Zylinderfunktionen für grosse Werte des Arguments und unbeschränkt veränderliche Werte des Index*, Math. Ann., 67 (1909), pp. 535–558.
[4] N. G. De Bruijn, *Asymptotic Methods in Analysis*, North-Holland, Amsterdam, 1961.
[5] E. T. Copson, *Asymptotic Expansions*, Cambridge University Press, Cambridge, 1965.
[6] M. Wyman, *The method of Laplace*, Trans. Roy. Soc. Canad., 2 (1964), pp. 227–256.
[7] F. W. J. Olver, *Error bounds for the Laplace approximation for definite integrals*, J. Approximation Theor., 1 (1968), pp. 293–313.
[8] E. T. Copson, *Theory of Functions of a Complex Variable*, Oxford University Press, Oxford, 1935.
[9] A. Erdélyi, W. Magnus, F. Oberhettinger and F. G. Tricomi, *Higher Transcendental Functions*, vol. 1, McGraw-Hill, New York, 1953.
[10] A. Erdélyi, *Asymptotic Expansions*, Dover, New York, 1956.
[11] N. Nielsen, *Handbuch der Theorie der Gammafunktion*, Teubner, Leipzig, 1906.

SIAM J. MATH. ANAL.
Vol. 1, No. 4, November 1970

A PARADOX IN ASYMPTOTICS*

F. W. J. OLVER†

Abstract. An example is given of a convergent series expansion which has twice itself as its own asymptotic expansion.

Can a convergent series possess a sum which differs from its asymptotic sum? For ordinary Poincaré expansions in powers of the asymptotic variable the answer is always no. In the case of generalized asymptotic expansions [1], however, the answer can be affirmative. This phenomenon was pointed out by van der Corput in 1962 [2]. The principal examples in this reference are somewhat artificial, however, in the sense that the terms of the series are discontinuous. The purpose of the present note is to draw attention to an example which occurs naturally in special function theory and has been discussed erroneously in the literature.

Consider the Legendre polynomial $P_n(\cos\theta)$ for large n and fixed θ, with $0 < \theta < \pi$. We have

$$P_n(\cos\theta) \sim \left(\frac{2}{\sin\theta}\right)^{1/2} \sum_{\nu=0}^{\infty} \binom{-\frac{1}{2}}{\nu}\binom{\nu-\frac{1}{2}}{n} \frac{\cos\theta_{n,\nu}}{(2\sin\theta)^{\nu}}, \qquad n \to \infty, \tag{1}$$

where

$$\theta_{n,\nu} = (n - \nu + \tfrac{1}{2})\theta + (n - \tfrac{1}{2}\nu - \tfrac{1}{4})\pi,$$

in the sense that the difference between $P_n(\cos\theta)$ and the pth partial sum of the series is $O\left\{\binom{p-\frac{1}{2}}{n}\right\}$ which is equivalent to $O(n^{-p-(1/2)})$. This result is derivable by the method of Darboux;[1] the leading term is, in effect, Laplace's well-known approximation.

When $2\sin\theta > 1$, that is, when $\frac{1}{6}\pi < \theta < \frac{5}{6}\pi$, the series on the right of (1) converges. But contrary to our natural expectations, and also to a statement in Szegö's comprehensive treatise,[2] the sum is not $P_n(\cos\theta)$ but $2P_n(\cos\theta)$. This may be verified as follows. By expansion in Taylor series at the point $t = e^{-i\theta}$, we have

$$\frac{1}{(1 - 2t\cos\theta + t^2)^{1/2}} = \frac{e^{-\pi i/4}}{(2\sin\theta)^{1/2}} \sum_{\nu=0}^{\infty} \binom{-\frac{1}{2}}{\nu} \frac{(e^{-i\theta} - t)^{\nu-(1/2)}}{(2i\sin\theta)^{\nu}}.$$

This series converges uniformly in the disc $|e^{-i\theta} - t| \leqq 2\sin\theta - \delta$, where δ is an arbitrary small positive number. If $2\sin\theta > 1$, then $t = 0$ is includable in the region of uniform convergence. Differentiating n times, setting $t = 0$, and equating

* Received by the editors April 3, 1970.

† Institute for Fluid Dynamics and Applied Mathematics, University of Maryland, College Park, Maryland 20742. This research was supported by the U.S. Army Research Office, Durham, under Contract 67-C-0062.

[1] See [3, § 8.4]. The statement of the result in [4, p. 534] contains a misprint in the cosine term.

[2] [3, § 8.4, (3)].

real parts, we obtain

$$P_n(\cos\theta) = \frac{1}{(2\sin\theta)^{1/2}} \sum_{\nu=0}^{\infty} \binom{-\frac{1}{2}}{\nu}\binom{\nu-\frac{1}{2}}{n} \frac{\cos\theta_{n,\nu}}{(2\sin\theta)^{\nu}}, \qquad \tfrac{1}{6}\pi < \theta < \tfrac{5}{6}\pi, \tag{2}$$

as asserted. The result (2) can be easily checked in the case $n = 0$, for example.

The explanation of the paradox is that the tail of the series (2), that is, the sum from $\nu = p$ (fixed) to $\nu = \infty$, is not of the same O-order as the first neglected term (as it always is with Poincaré expansions). For example, apart from the oscillatory factor $\cos\theta_{n,\nu}$ the term for which $\nu = 2n$ contributes

$$\binom{-\frac{1}{2}}{\nu}\binom{\nu-\frac{1}{2}}{n} \frac{1}{(2\sin\theta)^{\nu+(1/2)}} = \frac{\Gamma(2n+\frac{1}{2})\Gamma(n+\frac{1}{4})\Gamma(n+\frac{3}{4})}{2\pi\Gamma(2n+1)\Gamma(n+\frac{1}{2})\Gamma(n+1)(\sin\theta)^{2n+(1/2)}}$$

$$\sim \frac{1}{2^{3/2}\pi n(\sin\theta)^{2n+(1/2)}},$$

as $n \to \infty$. This is infinitely large compared with the $(p + 1)$th term, whatever the value of p.

REFERENCES

[1] A. Erdélyi and M. Wyman, *The asymptotic evaluation of certain integrals*, Arch. Rational Mech. Anal., 14 (1963), pp. 217–260.

[2] J. G. van der Corput, *Can a series possess a sum 1 and an asymptotic sum 0*? Studies in Mathematical Analysis and Related Topics, Stanford University Press, Stanford, 1962, pp. 410–418.

[3] G. Szegö, *Orthogonal Polynomials*, Colloquium Publications, vol. 23, 3rd ed., American Mathematical Society, New York, 1967.

[4] R. Courant and D. Hilbert, *Methods of Mathematical Physics*, vol. 1, Interscience, New York, 1953.

MATHEMATICS OF COMPUTATION, VOLUME 26, NUMBER 120, OCTOBER 1972

Note on Backward Recurrence Algorithms

By F. W. J. Olver and D. J. Sookne

Abstract. An algorithm is given for the computation of the recessive solution of a second-order linear difference equation, based upon a combination of algorithms due to J.C.P. Miller and F.W.J. Olver. A special feature is automatic and rigorous control of truncation error.

The method is illustrated by application to the well-used example of the Bessel functions $J_r(x)$.

1. Introduction and Summary. Let

$$a_r y_{r-1} - b_r y_r + c_r y_{r+1} = 0 \qquad (r = 1, 2, \cdots) \tag{1}$$

be a given difference equation in which the coefficients a_r and c_r do not vanish. Suppose that the equation has a pair of solutions f_r and g_r such that $f_r/g_r \to 0$ as $r \to \infty$. Then f_r is said to be a *recessive* (or *subdominant* or *distinguished*) solution of the difference equation at $r = \infty$, and g_r is said to be *dominant*. The recessive solution is unique, apart from a constant factor. The dominant solution is not unique, however, since any constant multiple of f_r may be added to g_r without affecting the asymptotic form of g_r.

Computation of f_r from (1) by forward recurrence is usually impractical owing to strong instability. On the other hand, backward application of (1) provides a stable way of computing f_r (but not g_r), since rounding errors grow no faster than the wanted solution, as a rule.* In the next section, we describe briefly two published algorithms which enable f_r to be computed without the need for accurate starting values at high values of r.

In Section 3, certain difficulties in the implementation of the algorithms are described, and in the next section, it is shown how these difficulties can be overcome by combining the algorithms.

In Section 5, the well-used Bessel function example is considered. A computing routine is described in which the truncation error is bounded rigorously, without loss of efficiency. The method is compared with methods of earlier writers.

The concluding section, Section 6, gives proofs of certain results used in earlier sections.

2. The Two Algorithms. The first algorithm, which we shall refer to as *Algorithm* I, is due to J. C. P. Miller [2] and has been used extensively in the computation

Received August 23, 1971.

AMS 1970 *subject classifications.* Primary 65Q05; Secondary 15A06, 33A40, 39A10, 40A25, 65D20.

Key words and phrases. Bessel functions, difference equations, error bounds, FORTRAN, Miller algorithm, recursion.

* Full discussions of this aspect and related matters are given by Gautschi in [1].

of special functions and in other contexts. It proceeds as follows. For a suitably chosen large integer N, a "trial" solution $y_r^{(N)}$ of (1) is generated recursively for $r = N, N-1, \cdots, 0$, beginning with $y_N^{(N)} = 0$ and $y_{N-1}^{(N)} = 1$. Then f_r is found by multiplying the $y_r^{(N)}$ by a normalizing factor λ_N. For example, if the value of f_0 is given then $\lambda_N = f_0/y_0^{(N)}$. More generally, if f_r satisfies a condition of the form

$$m_0 f_0 + m_1 f_1 + m_2 f_2 + \cdots = 1, \tag{2}$$

with given coefficients m_r, then

$$\lambda_N = 1/(m_0 y_0^{(N)} + m_1 y_1^{(N)} + \cdots + m_{N-1} y_{N-1}^{(N)}).$$

The value of N can be estimated from the asymptotic form of f_r for large r. But often such information is unavailable, in which event N is assigned arbitrarily. The adequacy of the guess is tested by repetition of the algorithm with a higher value and comparing results. If agreement is inadequate, then additional higher values must be tried.

Algorithm II was proposed by one of the present writers [3] primarily for the solution of inhomogeneous difference equations of the second order.** It is, in part, a forward recurrence procedure. Beginning with $p_0 = 0$ and $p_1 = 1$, we compute a solution p_r of (1) for $r = 1, 2, \cdots$. Also computed is a sequence $\{e_r\}$ defined by

$$e_0 = f_0, \qquad e_r = a_r e_{r-1}/c_r \qquad (r \geqq 1).$$

Here, f_0 is either the given value of the wanted solution or an arbitrary value; in the latter event a final normalization has to be effected by use of a relation of the form (2), as in Algorithm I. Computation of p_r and e_r is terminated automatically at a certain value of r, which we denote by $N + 1$. The determination of N is described below. The required approximation $f_r^{(N)}$ to the wanted solution f_r is then generated according to the equations $f_N^{(N)} = 0$ and

$$p_{r+1} f_r^{(N)} - p_r f_{r+1}^{(N)} = e_r \qquad (r = N-1, N-2, \cdots, 0). \tag{3}$$

This algorithm is based on the solution of the simultaneous set of equations (1) for $r = 1, 2, \cdots, N-1$, with the conditions $y_0 = f_0$ and $y_N = 0$. Computation of p_r and e_r represents the elimination stage, and (3) is the process of back-substitution.

The value of N is found in the following way. We have from [3, Section 5]

$$f_r - f_r^{(N)} = E_N p_r, \tag{4}$$

where

$$E_N = \frac{e_N}{p_N p_{N+1}} + \frac{e_{N+1}}{p_{N+1} p_{N+2}} + \frac{e_{N+2}}{p_{N+2} p_{N+3}} + \cdots . \tag{5}$$

Generally, this series converges fairly rapidly and the sum is of the same order of magnitude as the first term. Suppose, for example, that the final solution is required to D decimal places. Then the size of the function $e_N p_r/(p_N p_{N+1})$ is examined as the computations proceed. As soon as this test function falls below $\frac{1}{2} \times 10^{-D}$ for all values of r in the range of interest, the corresponding value of N is accepted.

In cases where the series (5) does not converge rapidly, the actual value of E_N

** A similar method has been described in [4], written apparently without knowledge of [3].

can be found *a posteriori* by computing values of p_r beyond $r = N + 1$ and summing (5) numerically. Multiplication of E_N by the various p_r then enables the truncation errors (4) to be estimated reliably.

3. Implementation. The principal difficulty associated with Algorithm I is the estimation of N. Computing time is wasted if either the asymptotic estimate or the initial guess is much too low or much too high. Another difficulty is a slight uncertainty associated with the acceptance criterion: mere numerical agreement of solutions computed with two different values of N does not guarantee their accuracy.

Neither of these problems attends Algorithm II. The optimum N is determined automatically, and the expansion (5) is available to bound truncation errors in the final solution. In consequence, although Algorithm II entails the more complicated computing procedure, the fact that only one application is needed may make it faster in practice.

There is a difficulty, however, in constructing fully satisfactory computing programs based on Algorithm II stemming from possible loss in accuracy in the formation of the sequence p_r. Since p_r is a linear combination of f_r and g_r, it increases ultimately in proportion to g_r.*** Initially, however, p_r may behave more like a multiple of f_r, in which event precision is lost by cancellation. Whether this affects the accuracy of the final solution depends on the normalization condition being used. There *is* a final loss if the condition prescribes the value of f_0, but not, as a rule, with a more general condition of the form (2); compare [3, Section 7].

4. Combined Algorithm. Although none of the drawbacks mentioned in Section 3 is catastrophic in practice, they can be overcome altogether—without sacrifice of speed—by judicious combination of Algorithms I and II. The modification applies when the dominant solution g_r tends to infinity with r in such a way that ultimately the absolute value of g_r is monotonic.

First, an integer M is chosen large enough to insure that if the sequence p_r is computed from (1), beginning with $p_M = 0$ and $p_{M+1} = 1$, then its members are nondecreasing in absolute value. Thus, cancellation is completely precluded. The selection of M is discussed below.

Second, Algorithm II is applied to compute, for $r \geqq M + 1$, the recessive solution which satisfies the condition $y_M = 1$.

Third, the values of y_r for $r = M - 1, M - 2, \cdots, 0$ are computed by backward application of (1). As in the case of Algorithm I, this is a stable procedure.

Finally, the wanted solution f_r is found by multiplying the y_r by a normalizing constant determined from (2).

The exact choice of M is not critical, and an acceptable value can often be determined by application of the following result:

LEMMA 1. *If* $|b_r| \geqq |a_r| + |c_r|$ *when* $r \geqq M + 1$, *and* p_r *is the solution of* (1) *satisfying* $p_M = 0$ *and* $p_{M+1} = 1$, *then*

$$|p_r| \geqq |p_{r-1}| \qquad (r \geqq M + 1). \tag{6}$$

This is proved in Section 6.

*** Even in a case in which, initially, p_r is *exactly* a multiple of f_r, it behaves eventually as a multiple of g_r owing to the introduction of rounding errors.

In practice, the procedure lends itself to two improvements, as follows.

(i) Suppose that the solution y_r is required in floating-point form. That is, y_r is assumed to be needed to S significant figures for the range $(0, L)$, both S and L being given. From [3, (5.03)], we have

$$y_r = p_r \sum_{s=r}^{\infty} \frac{e_s}{p_s p_{s+1}}. \tag{7}$$

As in the case of (5), it is reasonable, for large r, to approximate this expansion by its first term. Then referring to (4), we see that the relative error of $y_r^{(N)}$ is approximately

$$(e_N/p_N p_{N+1})(p_r p_{r+1}/e_r). \tag{8}$$

The value of N $(> L)$ is found by insuring that this quantity is bounded by $\frac{1}{2} \times 10^{-S}$ for all values of r in (M, L). The back-substitution begins with $y_N = 0$ and $y_{N-1} = e_{N-1}/p_N$, but, instead of continuing by application of (3), we revert to the original difference equation (1), generating y_r from $r = N - 2$ through $r = M$ down to $r = 0$. This is allowable, since errors in the backward recursion grow no faster than the wanted solution.

This refinement simplifies the programming because the formula used for y_r is the same in the ranges $(0, M)$ and (M, L). Moreover, in generating the sequence p_r, only current values need be stored. In this form, *the role of Algorithm* II *may be regarded simply as a procedure for finding the optimum N for use with Algorithm* I.

Perhaps, it should be noted that this refinement applies only to homogeneous difference equations as a rule; in the case of an inhomogeneous equation, both forward *and* backward recursion may be unstable.

(ii) In [5], it is shown how to construct lower bounds for the $|p_r|$, and thence an upper bound for E_N. By using these bounds, it is possible to choose N in such a way that the truncation error $E_N p_r$ falls below the specified tolerance in f_r throughout the given range of values of r. In other words, *all* terms in the expansion (5) are automatically taken into account in determining N, and not merely the first term.

Both improvements (i) and (ii) are incorporated in the example given in the next section.

5. Example. Consider the recurrence relation

$$y_{r-1} - (2r/x)y_r + y_{r+1} = 0 \tag{9}$$

satisfied by the Bessel functions $J_r(x)$ and $Y_r(x)$, and suppose that x is real and positive. It is well known that for fixed x and varying r the behavior of the Bessel functions is quite different in the ranges $0 \leqq r < x$ and $x < r < \infty$. In the former range, the functions oscillate with slowly changing amplitude, whereas in the latter range, $Y_r(x)$ tends rapidly to $-\infty$ and $J_r(x)$ tends rapidly to zero. Let us suppose that $J_r(x)$ is required to fixed-point accuracy for $r \leqq x$ and floating-point accuracy for $r > x$. More precisely, if $[x]$ denotes the integer part of x then $J_r(x)$ is required to D decimal places for $0 \leqq r \leqq [x]$, and S significant figures for $[x] \leqq r \leqq L$. Here, S and L are prescribed, and D is the number of decimal places in $J_{[x]}(x)$ corresponding to S significant figures in the same quantity. Since $|J_r(x)|$ is bounded by unity,† it follows that $D \geqq S$.

A sequence $p_M, p_{M+1}, \cdots$ is computed by forward application of (9), beginning with $p_M = 0$ and $p_{M+1} = 1$. In the notation of Lemma 1, we have $|b_r| = 2r/x$ and $|a_r| + |c_r| = 2$. Hence, with $M = [x]$, we can be sure that $p_M, p_{M+1}, \cdots$ is a non-decreasing sequence. Since all the quantities e_r in this example are unity, computation of p_r is terminated at $r = N$, where $N\,(>L)$ is the least odd integer for which

$$p_N p_{N+1} > (2 \times 10^S) p_L p_{L+1}; \tag{10}$$

compare (8). Values of y_r for $r = N, N-1, \cdots, 0$ are found by backward application of (9), beginning with $y_N = 0$ and $y_{N-1} = 1/p_N$. Finally, the desired $J_r(x)$ are obtained by multiplying the y_r by the factor $1/(y_0 + 2y_2 + 2y_4 + \cdots + 2y_{N-1})$, since††

$$J_0(x) + 2J_2(x) + 2J_4(x) + \cdots = 1.$$

Because $y_M = 1$, this normalizing factor equals $J_M(x)$, approximately, and therefore cannot exceed unity in absolute value.

The actual truncation error in each y_r for the range (M, L) could be found from (4) and (5). Instead, however, we adopt the suggestion made in Section 4(ii) and increase N to $\bar{N}$, say, to *guarantee* that $\frac{1}{2} \times 10^{-S}$ is an upper bound for the relative truncation errors in $y_M, y_{M+1}, \cdots, y_L$. This depends on the following result, which is a refinement of Theorem 2 of [5] in the present case.

LEMMA 2. *Let $x > 0$, $M = [x]$, and the sequence $p_M, p_{M+1}, \cdots$ be defined by $p_M = 0$, $p_{M+1} = 1$, and*

$$p_{r+1} = (2r/x)p_r - p_{r-1} \qquad (r \geqq M+1).$$

Then

$$p_{s+1}/p_s \geqq \min(p_{r+1}/p_r, \lambda_r) \qquad (s \geqq r \geqq M),$$

where λ_r is the largest zero of the quadratic $\lambda^2 - 2(r+1)x^{-1}\lambda + 1$.

The proof of this result is given in Section 6. To apply the lemma, write

$$\kappa_r = p_{r+1}/p_r, \qquad \rho_r = \min(\kappa_r, \lambda_r), \tag{11}$$

and let N be the least integer for which (10) is satisfied. The value of ρ_N is found by testing the sign of

$$\Delta_N \equiv p_{N+1}^2 - 2(N+1)x^{-1}p_{N+1}p_N + p_N^2.$$

If $\Delta_N \leqq 0$, then $\rho_N = p_{N+1}/p_N$, otherwise

$$\rho_N = \lambda_N = (N+1)/x + \{((N+1)/x)^2 - 1\}^{1/2}.$$

Having obtained ρ_N, we compute $\bar{N}$ as the least integer exceeding N for which

$$\rho_N^{\bar{N}} > (2 \times 10^S) p_L p_{L+1} \rho_N / (\rho_N^2 - 1). \tag{12}$$

† [6, Section 2.5].
†† [6, Section 2.22].

To verify that $\bar{N}$ has the desired property when defined in this way, we have, from (5) and (11),

$$E_{\bar{N}} = \frac{1}{p_{\bar{N}}^2}\left(\frac{1}{\kappa_{\bar{N}}} + \frac{1}{\kappa_{\bar{N}}^2\kappa_{\bar{N}+1}} + \frac{1}{\kappa_{\bar{N}}^2\kappa_{\bar{N}+1}^2\kappa_{\bar{N}+2}} + \cdots\right).$$

Use of Lemma 2 gives

$$E_{\bar{N}} \leqq \frac{1}{p_{\bar{N}}^2}\left(\frac{1}{\rho_N} + \frac{1}{\rho_N^3} + \frac{1}{\rho_N^5} + \cdots\right) = \frac{\rho_N}{p_{\bar{N}}^2(\rho_N^2 - 1)}. \tag{13}$$

Since y_r is at least $1/p_{r+1}$ (compare (7)), the truncation error of $y_r^{(N)}$ relative to the magnitude of y_r is bounded by $E_{\bar{N}}p_rp_{r+1}$. If $M \leqq r \leqq L$, then from (12), (13), and Lemma 1, we see that

$$E_{\bar{N}}p_rp_{r+1} \leqq E_{\bar{N}}p_Lp_{L+1} \leqq \rho_Np_Lp_{L+1}/p_{\bar{N}}^2(\rho_N^2 - 1) \leqq \tfrac{1}{2} \times 10^{-S},$$

as required.

Remarks (i). In an unpublished paper [7], Kahan proposed the following method for estimating N. Starting with $y_{[x]} = 0$ and $y_{[x]+1} = \beta$, where β is an arbitrary positive number (though small in practice) the sequence y_r is computed by forward application of (9). Then, for fixed-point computation to D decimals, N is the least integer for which $y_{N+1} \geqq (2 \times 10^D)\beta$. FORTRAN programs based on this criterion have been constructed [8]. In spirit, this procedure is the same as in the present paper, but lacks the precise control of error.

(ii) Asymptotic estimates of an acceptable N are given in [1, Section 5]. These are somewhat complicated to reproduce in full, but it is to be noted that they give a value exceeding $\frac{1}{2}ex$, where e is the base of natural logarithms. For large x, this is a considerable overestimate. For example, with $L = x = 1024$ and $S = 19$, the criterion of the present section gives $\bar{N} - x = 130$, compared with $\frac{1}{2}ex - x = 368$.

(iii) Numerical tables of the optimum N for floating-point accuracy (S significant figures) are given in [9] for the ranges $x \leqq 100$ and $S \leqq 30$. Apart from the fact that the range of x is somewhat restrictive, tables of this kind are cumbersome to incorporate in a flexible computing program.

(iv) FORTRAN programs for this example, together with extensions to complex x, are being prepared.

6. Proofs of Lemmas.

LEMMA 1. If (6) holds for a certain value of r—as is the case when $r = M + 1$—then

$$|p_{r+1}| = \left|\frac{b_rp_r - a_rp_{r-1}}{c_r}\right| \geqq \frac{(|b_r| - |a_r|)\,|p_r|}{|c_r|} \geqq |p_r|.$$

LEMMA 2. By use of Lemma 1, it is seen that p_r is positive for $r \geqq M + 1$. From the definitions (11) and the given conditions, we have

$$\kappa_{r+1} = 2(r + 1)/x - 1/\kappa_r, \tag{14}$$

and

$$\lambda_r = 2(r + 1)/x - 1/\lambda_r. \tag{15}$$

(i) Suppose first that $\kappa_r \geqq \lambda_r$. Then (14) and (15) imply that $\kappa_{r+1} \geqq \lambda_r$. Next,

$$\kappa_{r+2} = \frac{2(r+2)}{x} - \frac{1}{\kappa_{r+1}} \geqq \frac{2(r+1)}{x} - \frac{1}{\kappa_{r+1}} \geqq \lambda_r.$$

Continuation of this argument shows that $\kappa_{r+s} \geqq \lambda_r$ for $s = 0, 1, \cdots$, as required.

(ii) Alternatively, suppose that $\kappa_r < \lambda_r$. Because $\kappa_r \geqq 1$ (Lemma 1), it follows that κ_r separates the two λ-zeros of the function

$$\lambda - 2(r+1)/x + 1/\lambda.$$

Hence

$$\kappa_r - 2(r+1)/x + 1/\kappa_r < 0,$$

and therefore, from (14), $\kappa_r < \kappa_{r+1}$.

Next, either $\kappa_{r+1} \geqq \lambda_{r+1}$ or $\kappa_{r+1} < \lambda_{r+1}$. As in (i) the first alternative implies that $\kappa_{r+s} \geqq \lambda_{r+1}$ ($s \geqq 1$). Since λ_r is an increasing function of r this gives $\kappa_{r+s} > \lambda_r$ and thence $\kappa_{r+s} \geqq \kappa_r$, $s = 0, 1, \cdots$, as required. For the second alternative, we reason as in the preceding paragraph that $\kappa_{r+1} < \kappa_{r+2}$ and then examine the alternatives $\kappa_{r+2} \geqq \lambda_{r+2}$ and $\kappa_{r+2} < \lambda_{r+2}$. And so on.

Acknowledgments. The writers are pleased to acknowledge helpful discussions with their colleague Dr. H. J. Oser. The work was funded in part by the Center for Computer Sciences and Technology at N.B.S.

Institute for Fluid Dynamics and Applied Mathematics
University of Maryland
College Park, Maryland 20742

Mathematical Analysis Section, IBS
National Bureau of Standards
Washington, D. C. 20234

1. W. GAUTSCHI, "Computational aspects of three-term recurrence relations," *SIAM Rev.*, v. 9, 1967, pp. 24–82. MR **34** #3927.

2. BRITISH ASSOCIATION FOR THE ADVANCEMENT OF SCIENCE, "Bessel functions. Part II," *Mathematical Tables*, v. 10, Cambridge University Press, Cambridge, 1952.

3. F. W. J. OLVER, "Numerical solution of second-order linear difference equations," *J. Res. Nat. Bur. Standards Sect. B*, v. 71, 1967, pp. 111–129. MR **36** #4841.

4. J. G. WILLS, "On the use of recursion relations in the numerical evaluation of spherical Bessel functions and Coulomb functions," *J. Computational Phys.*, v. 8, 1971, pp. 162–166.

5. F. W. J. OLVER, "Bounds for the solutions of second-order linear difference equations," *J. Res. Nat. Bur. Standards Sect. B*, v. 71, 1967, pp. 161–166. MR **37** #4981

6. G. N. WATSON, *A Treatise on the Theory of Bessel Functions*, 2nd ed., Cambridge Univ. Press, Cambridge, 1944.

7. W. KAHAN, "Note on bounds for generating Bessel functions by recurrence." (Unpublished.)

8. D. JORDAN, *Argonne National Laboratory Library Routine*, ANL C370S—BESJY, October 1967.

9. S. MAKINOUCHI, "Note on the recurrence techniques for the calculation of Bessel functions $J_\nu(x)$," *Tech. Rep. Osaka Univ.*, v. 15, 1965, pp. 185–201. MR **33** #6813.

SIAM J. MATH. ANAL.
Vol. 5, No. 1, February 1974

ERROR BOUNDS FOR STATIONARY PHASE APPROXIMATIONS*

F. W. J. OLVER†

Abstract. An error theory is constructed for the method of stationary phase for integrals of the form

$$I(x) = \int_a^b e^{ixp(t)} q(t)\, dt.$$

Here x is a large real parameter, the function $p(t)$ is real, and neither $p(t)$ nor $q(t)$ need be analytic in t. For both finite and infinite ranges of integration, explicit expressions are derived for the truncation errors associated with the asymptotic expansion of $I(x)$. The use of these explicit expressions for the computation of realistic error bounds is illustrated by means of an example.

1. Introduction and summary. The most comprehensive practicable theory of the method of stationary phase for single integrals of the form

$$I(x) = \int_a^b e^{ixp(t)} q(t)\, dt \tag{1.1}$$

appears to be that of Erdélyi [4], [5]. In this integral a, b, and the function $p(t)$ are real, and x is a large real parameter. Erdélyi's first paper concerns the case in which $p(t)$ and $q(t)$ are expansible at a and b in series of fractional powers of $t - a$ and $b - t$. The second paper extends the analysis to singularities of logarithmic type.[1]

The main purpose of the present paper is to supply explicit expressions for the error terms associated with the expansions of [4] from which realistic bounds are readily computable. The derivations of Erdélyi do not lend themselves readily to the construction of error bounds owing to the somewhat artificial nature of the neutralizer functions employed in the analysis. Our approach is based instead on Hardy's theory of generalized integrals [7], [8].

A secondary purpose of the present paper is to facilitate the application of the method of stationary phase to integrals having an infinite range of integration. These integrals can be treated by combination of results given in [4]. In the present account, however, they are analyzed directly, and in certain ways results for infinite integrals are simpler than those for finite integrals. In applications, writers often shun the method of stationary phase for infinite integrals. Instead, the path is deformed into the complex plane and the method of steepest descents invoked; see, for example, [2, Chap. 4], and [3, Chap. 5]. Such deformations presuppose that $p(t)$ and $q(t)$ are analytic functions of the complex variable t, which is not an

* Received by the editors August 4, 1972, and in revised form November 10, 1972.

† Institute for Fluid Dynamics and Applied Mathematics, University of Maryland, and National Bureau of Standards, Washington, D.C. This research was supported by the National Science Foundation under Grant GP 20529, and the U.S. Army Research Office, Durham, under Contract DA ARO D 31 124 71 G133.

[1] For a brief history of the method of stationary phase, see Jones [10] and another paper by Erdélyi [6], and for an important correction to [5], see [11].

inherent feature of the method nor one which is needed in the present theory or that of Erdélyi.

The paper is arranged as follows. Assumptions are listed and discussed in § 2, and the main theorem stated in § 3. This theorem gives the expansion of the integral (1.1) for large x, complete with explicit error terms. The proof follows in §§ 4 and 5. In § 6 the asymptotic nature of the expansion is discussed, including the derivation of a stationary phase analogue of Watson's lemma. The same section also indicates how bounds for the error terms can be obtained. The concluding section, § 7, contains an illustrative example concerning the Anger–Weber functions.

2. Assumptions. The following conditions and notations are similar to those adopted for Laplace's method in [12] and [13]. In (1.1) the limits a and b are independent of the positive parameter x, a being finite and $b\,(>a)$ finite or infinite. The functions $p(t)$ and $q(t)$ are independent of x, $p(t)$ being real and $q(t)$ real or complex. They have the properties:

(i) In (a, b), $p^{(m+1)}(t)$ and $q^{(m)}(t)$ are continuous, m being a nonnegative integer, and $p'(t) > 0$.

(ii) As $t \to a$ from the right,

$$p(t) \sim p(a) + \sum_{s=0}^{\infty} p_s(t-a)^{s+\mu}, \qquad q(t) \sim \sum_{s=0}^{\infty} q_s(t-a)^{s+\lambda-1}, \tag{2.1}$$

where the coefficients p_0 and q_0 are nonzero, and μ and λ are constants such that

$$\mu > 0, \quad (m+1)\mu + 1 > \operatorname{Re}\lambda > 0. \tag{2.2}$$

Moreover, the first of these expansions is differentiable $m + 1$ times and the second m times.

(iii) When $p(b) \equiv \lim_{t\to b-}\{p(t)\}$ is finite, each of the functions

$$P_s(t) \equiv \left\{\frac{1}{p'(t)}\frac{d}{dt}\right\}^s \frac{q(t)}{p'(t)}, \qquad s = 0, 1, \cdots, m, \tag{2.3}$$

tends to a finite limit as $t \to b-$.

(iv) When $p(b) = \infty$, $\lim_{t\to b-}\{q(t)/p'(t)\} = 0$ and each of the integrals

$$\int e^{ixp(t)} P_s(t) p'(t)\, dt, \qquad s = 0, 1, \cdots, m,$$

converges at $t = b$ uniformly for all sufficiently large x.

Remarks. (a) Cases in which x is a negative parameter, or $p'(t)$ is negative, can be included by changing the sign of i throughout. Cases in which $p'(t)$ has zeros in (a, b), that is, cases in which the integral (1.1) has interior stationary points, are treatable by subdividing the range at the stationary points and intermediate points. Similar subdivisions may also be made when b is finite and $p(t)$ and $q(t)$ have expansions at b in fractional powers of $b - t$ of the type (2.1).

(b) Condition (iii) is fulfilled in the common case in which $p^{(m+1)}(t)$ and $q^{(m)}(t)$ are continuous at b and $p'(b) \neq 0$.

(c) Condition (iv), with $s = 0$, implies that the original integral converges uniformly at its upper limit.

3. Main result. In consequence of condition (i) there is a one-to-one relationship between t and the variable v, defined by

$$v = p(t) - p(a). \tag{3.1}$$

In terms of this variable the integral (1.1) transforms into

$$\int_a^b e^{ixp(t)}q(t)\,dt = e^{ixp(a)}\int_0^{p(b)-p(a)} e^{ixv}f(v)\,dv, \tag{3.2}$$

in which

$$f(v) = q(t)/p'(t) = P_0(t). \tag{3.3}$$

Again, condition (i) shows that $f(v)$ and its first m derivatives are continuous when $0 < v < p(b) - p(a)$. For small v, $t - a$ and $f(v)$ can be expanded in asymptotic series of the form

$$t - a \sim \sum_{s=1}^{\infty} c_s v^{s/\mu}, \qquad f(v) \sim \sum_{s=0}^{\infty} a_s v^{(s+\lambda-\mu)/\mu}. \tag{3.4}$$

The coefficients c_s and a_s depend on p_s and q_s, and may be found by standard procedures for reverting series. In particular,[2]

$$c_1 = \frac{1}{p_0^{1/\mu}}, \qquad c_2 = -\frac{p_1}{\mu p_0^{1+(2/\mu)}},$$

$$a_0 = \frac{q_0}{\mu p_0^{\lambda/\mu}}, \qquad a_1 = \left\{\frac{q_1}{\mu} - \frac{(\lambda+1)p_1 q_0}{\mu^2 p_0}\right\}\frac{1}{p_0^{(\lambda+1)/\mu}}.$$

THEOREM 1. *Assume the conditions and notation of § 2 and the present section, and let n be a nonnegative integer satisfying*[3]

$$m\mu - \lambda \leqq n < (m+1)\mu - \lambda + 1 \qquad (\lambda \text{ real}), \tag{3.5}$$

or

$$m\mu - \operatorname{Re}\lambda < n < (m+1)\mu - \operatorname{Re}\lambda + 1 \qquad (\lambda \text{ complex}). \tag{3.6}$$

Then

$$\begin{aligned}\int_a^b e^{ixp(t)}q(t)\,dt = {} & e^{ixp(a)}\sum_{s=0}^{n-\nu} \exp\left\{\frac{(s+\lambda)\pi i}{2\mu}\right\}\Gamma\left(\frac{s+\lambda}{\mu}\right)\frac{a_s}{x^{(s+\lambda)/\mu}} \\ & - e^{ixp(b)}\sum_{s=0}^{m-1} P_s(b)\left(\frac{i}{x}\right)^{s+1} + \delta_{m,n}(x) - \varepsilon_{m,n}(x)\end{aligned} \tag{3.7}$$

---

[2] With the assumed conditions, p_0 is necessarily positive. Corresponding expressions for c_3 and a_2 are given in [13].

[3] The inequalities (2.2) guarantee that at least one value of n can be found.

if $p(b) < \infty$, *or*

$$\int_a^b e^{ixp(t)}q(t)\,dt$$

(3.8)

$$= e^{ixp(a)} \sum_{s=0}^{n-\nu} \exp\left\{\frac{(s+\lambda)\pi i}{2\mu}\right\} \Gamma\left(\frac{s+\lambda}{\mu}\right) \frac{a_s}{x^{(s+\lambda)/\mu}} + \delta_{m,n}(x)$$

if $p(b) = \infty$. *Here* $\nu = 0$ *when* $n = m\mu - \lambda$, *and* $\nu = 1$ *in all other cases.*[4] *The error terms are given by*

(3.9)

$$\varepsilon_{m,n}(x) = e^{ixp(a)} \sum_{s=0}^{n-1} \exp\left\{\frac{(s+\lambda)\pi i}{2\mu}\right\} \frac{\Gamma\{(s+\lambda)/\mu\}}{\Gamma\{(s+\lambda-m\mu)/\mu\}}$$
$$\times \Gamma\left\{\frac{s+\lambda-m\mu}{\mu}, ixp(a) - ixp(b)\right\} \frac{a_s}{x^{(s+\lambda)/\mu}},$$

and

(3.10)
$$\delta_{m,n}(x) = \left(\frac{i}{x}\right)^m \int_a^b e^{ixp(t)} Q'_{m,n}(t)\,dt,$$

where[5]

(3.11)
$$Q_{m,n}(t) = P_{m-1}(t) - \sum_{s=0}^{n-1} \frac{\Gamma\{(s+\lambda)/\mu\}}{\Gamma\{(s+\lambda+\mu-m\mu)/\mu\}} \frac{a_s}{\{p(t)-p(a)\}^{(m\mu-s-\lambda)/\mu}}.$$

In (3.9) the incomplete gamma function takes its principal value, that is,

(3.12)
$$\Gamma(\alpha, z) = \int_z^\infty e^{-t} t^{\alpha-1}\,dt,$$

where the path does not intersect the negative real axis, and $t^{\alpha-1}$ has its principal value.

4. Preliminary lemmas.

LEMMA 1. *When* $x > 0$ *and* $\operatorname{Re}\alpha > 0$,

(4.1)
$$\lim_{\eta\to 0+} \int_0^\infty e^{-\eta v} e^{ixv} v^{\alpha-1}\,dv = \frac{e^{\alpha\pi i/2}\Gamma(\alpha)}{x^\alpha}.$$

This may be proved by rotation of the path of integration until it coincides with the ray $\operatorname{ph} v = \tan^{-1}(x/\eta)$, the deformation being easily justified by means of Cauchy's theorem. On the new path set

$$v = \tau/(\eta - ix),$$

so that τ is real and ranges from 0 to ∞. Then

$$\int_0^\infty e^{-\eta v} e^{ixv} v^{\alpha-1}\,dv = \frac{1}{(\eta - ix)^\alpha} \int_0^\infty e^{-\tau} \tau^{\alpha-1}\,d\tau = \frac{\Gamma(\alpha)}{(\eta - ix)^\alpha}.$$

Passage to the limit yields (4.1).

[4] As usual, empty sums are understood to be zero.

[5] In the case $m = 0$, $P_{-1}(t)$ is defined to be $\int q(t)\,dt$, consistent with (2.3).

The left-hand side of (4.1) is a *generalized integral* in the sense of Hardy [7], [8]. Its importance lies in assigning a meaning to Euler's integral

$$\int_0^\infty e^{ixv} v^{\alpha-1}\, dv \tag{4.2}$$

when Riemann's definition is inapplicable owing to divergence at the upper limit.

In the case $0 < \operatorname{Re} \alpha < 1$, (4.2) converges and equals $e^{\alpha\pi i/2}\Gamma(\alpha)/x^\alpha$. In other words, Hardy's generalization is consistent with the ordinary definition. This is a special case of the following result, included in Hardy's theory.[6]

LEMMA 2. *If $\phi(v)$ is sectionally (piecewise) continuous in $(0, \infty)$ and $\int_0^\infty \phi(v)\, dv$ converges, then $\int_0^\infty e^{-\eta v}\phi(v)\, dv$ converges for every positive number η and tends to $\int_0^\infty \phi(v)\, dv$ as $\eta \to 0+$.*

Lastly we shall need the following result, which is provable in a manner similar to Lemma 1.

LEMMA 3. *When $x > 0$ and $\beta > 0$,*

$$\lim_{\eta \to 0+} \int_\beta^\infty e^{-\eta v} e^{ixv} v^{\alpha-1}\, dv = \frac{e^{\alpha\pi i/2}}{x^\alpha} \Gamma(\alpha, -ix\beta). \tag{4.3}$$

5. Proof of Theorem 1. Write

$$\beta \equiv p(b) - p(a),$$

and for each nonnegative integer n, define $\phi_n(v)$ by

$$f(v) = \sum_{s=0}^{n-1} a_s v^{(s+\lambda-\mu)/\mu} + \phi_n(v), \qquad 0 < v < \beta, \tag{5.1}$$

so that

$$\phi_n(v) \sim a_n v^{(n+\lambda-\mu)/\mu} + a_{n+1} v^{(n+1+\lambda-\mu)/\mu} + \cdots, \qquad v \to 0+, \tag{5.2}$$

and

$$\phi_n^{(j)}(v) = f^{(j)}(v) - \sum_{s=0}^{n-1} \frac{\Gamma\{(s+\lambda)/\mu\}}{\Gamma\{(s+\lambda-j\mu)/\mu\}} a_s v^{(s+\lambda-\mu-j\mu)/\mu}, \qquad j = 0, 1, \cdots, m. \tag{5.3}$$

Assume first that β is finite. Since

$$f^{(j)}(v) = P_j(t), \tag{5.4}$$

conditions (i) and (iii) of § 2 show that $f(v)$, $\phi_n(v)$, and their first m derivatives are continuous when $v \in (0, \beta]$. With η denoting an arbitrary positive number, we have

$$\begin{aligned} &\int_0^\beta e^{-\eta v + ixv} f(v)\, dv \\ &\quad = \sum_{s=0}^{n-1} a_s \int_0^\infty e^{-\eta v + ixv} v^{(s+\lambda-\mu)/\mu}\, dv + E_n(\eta, x) - F_n(\eta, x), \end{aligned} \tag{5.5}$$

[6] [7, p. 51]. It is important to notice that $\int_0^\infty \phi(v)\, dv$ need not converge absolutely at either limit.

where

$$E_n(\eta, x) = \int_0^\beta e^{-\eta v + ixv}\phi_n(v)\,dv, \tag{5.6}$$

and

$$F_n(\eta, x) = \sum_{s=0}^{n-1} a_s \int_\beta^\infty e^{-\eta v + ixv} v^{(s+\lambda-\mu)/\mu}\,dv. \tag{5.7}$$

Letting $\eta \to 0$ and applying Lemmas 1 and 2, we obtain[7]

$$\int_0^\beta e^{ixv} f(v)\,dv = \sum_{s=0}^{n-1} \exp\left\{\frac{(s+\lambda)\pi i}{2\mu}\right\}\Gamma\left(\frac{s+\lambda}{\mu}\right)\frac{a_s}{x^{(s+\lambda)/\mu}} + E_n(x) - F_n(x), \tag{5.8}$$

where

$$E_n(x) = \lim_{\eta \to 0+} E_n(\eta, x), \qquad F_n(x) = \lim_{\eta \to 0+} F_n(\eta, x).$$

By integrating by parts, we have

$$E_n(\eta, x) = \frac{\phi_n(0)}{\eta - ix} - \frac{e^{-\eta\beta + ix\beta}}{\eta - ix}\phi_n(\beta) + \frac{1}{\eta - ix}\int_0^\beta e^{-\eta v + ixv}\phi_n'(v)\,dv. \tag{5.9}$$

Since $\phi_n^{(m)}(v)$ is continuous in $(0, \beta]$ the process of partial integration may be repeated $m - 1$ times. The conditions of § 2 show that the expansion (3.4) for $f(v)$ may be differentiated m times, hence the same is true of (5.2). In consequence of the conditions (3.5) and (3.6), all needed derivatives of $\phi_n(v)$ at $v = 0$ vanish, except possibly $\phi_n^{(m-1)}(v)$; accordingly

$$\begin{aligned} E_n(\eta, x) = {} & \frac{\phi_n^{(m-1)}(0)}{(\eta - ix)^m} - e^{-\eta\beta + ix\beta}\sum_{j=0}^{m-1}\frac{\phi_n^{(j)}(\beta)}{(\eta - ix)^{j+1}} \\ & + \frac{1}{(\eta - ix)^m}\int_0^\beta e^{-\eta v + ixv}\phi_n^{(m)}(v)\,dv. \end{aligned} \tag{5.10}$$

We propose to let $\eta \to 0$ in this result to obtain

$$E_n(x) = \left(\frac{i}{x}\right)^m \phi_n^{(m-1)}(0) - e^{ix\beta}\sum_{j=0}^{m-1}\phi_n^{(j)}(\beta)\left(\frac{i}{x}\right)^{j+1} + \left(\frac{i}{x}\right)^m \int_0^\beta e^{ixv}\phi_n^{(m)}(v)\,dv. \tag{5.11}$$

This step is justifiable by Lemma 2, provided that the integral in (5.11) converges. Now from the differentiated forms of (5.2) we see that as $v \to 0$, $\phi_n^{(m)}(v)$ is $O\{v^{(n + \mathrm{Re}\,\lambda - \mu - m\mu)/\mu}\}$ or $O\{v^{(1/\mu) - 1}\}$ according as $n > m\mu - \mathrm{Re}\,\lambda$ or $n = m\mu - \lambda$. In either event, the integral in question converges absolutely. Thus (5.11) is established.

Next, consider $F_n(x)$. Application of Lemma 3 to (5.7) gives

$$F_n(x) = \sum_{s=0}^{n-1} a_s \frac{e^{(s+\lambda)\pi i/(2\mu)}}{x^{(s+\lambda)/\mu}}\Gamma\left(\frac{s+\lambda}{\mu}, -ix\beta\right).$$

[7] In applying Lemma 2, $\phi(v)$ is taken to be $e^{ixv}f(v)$ when $0 < v \leqq \beta$ and 0 when $v > \beta$.

From (3.12) we derive by repeated integrations by parts,

$$\Gamma(\alpha, z) = e^{-z}z^{\alpha-1}\sum_{j=0}^{m-1}\frac{\Gamma(\alpha)}{\Gamma(\alpha-j)}\frac{1}{z^j} + \frac{\Gamma(\alpha)}{\Gamma(\alpha-m)}\Gamma(\alpha-m, z).$$

Hence

$$F_n(x) = e^{ix\beta}\sum_{s=0}^{n-1}\sum_{j=0}^{m-1} a_s \frac{\Gamma\{(s+\lambda)/\mu\}}{\Gamma\{(s+\lambda-j\mu)/\mu\}}\left(\frac{i}{x}\right)^{j+1}\beta^{(s+\lambda-\mu-j\mu)/\mu} + e^{-ixp(a)}\varepsilon_{m,n}(x), \tag{5.12}$$

where $\varepsilon_{m,n}(x)$ is defined by (3.9).

We now substitute in (5.11) by means of (5.3), with $v = \beta$, and subtract (5.12). The double sum cancels, and we are left with

$$E_n(x) - F_n(x) = -e^{ix\beta}\sum_{j=0}^{m-1} f^{(j)}(\beta)\left(\frac{i}{x}\right)^{j+1} + \left(\frac{i}{x}\right)^m \phi_n^{(m-1)}(0) + e^{-ixp(a)}\{\delta_{m,n}(x) - \varepsilon_{m,n}(x)\}, \tag{5.13}$$

where

$$\delta_{m,n}(x) = e^{ixp(a)}\left(\frac{i}{x}\right)^m \int_0^\beta e^{ixv}\phi_n^{(m)}(v)\,dv. \tag{5.14}$$

From (2.3), (3.11), and (5.3), it is verifiable that

$$\phi_n^{(m)}(v) = Q'_{m,n}(t)/p'(t), \tag{5.15}$$

and thence that (5.14) agrees with (3.10). From (3.5), (3.6) and (5.2) it is seen that $\phi_n^{(m-1)}(0)$ vanishes unless λ is real and $n = m\mu - \lambda$, in which event it equals $(m-1)!a_n$. Combination of (3.2), (5.4), (5.8) and (5.13) yields the first of the desired results (3.7).

The proof of (3.8) is similar. First, we observe that when $0 \leqq s \leqq m-1$,

$$\int e^{ixp(t)}P_s(t)p'(t)\,dt = \frac{e^{ixp(t)}}{ix}P_s(t) - \frac{1}{ix}\int e^{ixp(t)}P_{s+1}(t)p'(t)\,dt.$$

By hypothesis, as $t \to b-$, both integrals converge and $p(t) \to \infty$. Therefore $P_s(t) \to 0$. In terms of v this implies that $\int^\infty e^{ixv}f^{(s)}(v)\,dv$ converges when $s \leqq m$, and $f^{(s)}(v) \to 0$ when $s \leqq m-1$.

In (5.5) and (5.6), β is replaced by ∞ and the term $F_n(\eta, x)$ is absent. The convergence of the integrals is assured by Lemma 2 and the convergence of $\int_0^\infty e^{ixv}f(v)\,dv$. From (5.1) and the fact that $f(v) \to 0$ as $v \to \infty$ (condition (iv)), it follows that $e^{-\eta v+ixv}\phi_n(v) \to 0$ as $v \to \infty$. Hence (5.9) becomes

$$E_n(\eta, x) = \frac{\phi_n(0)}{\eta - ix} + \frac{1}{\eta - ix}\int_0^\infty e^{-\eta v+ixv}\phi'_n(v)\,dv.$$

Similarly, in place of (5.10) we have

$$E_n(\eta, x) = \frac{\phi_n^{(m-1)}(0)}{(\eta - ix)^m} + \frac{1}{(\eta - ix)^m}\int_0^\infty e^{-\eta v+ixv}\phi_n^{(m)}(v)\,dv.$$

In (5.3) with $j = m$, all powers of v in the sum have negative real part: this is a consequence of (3.5) and (3.6). Hence $\int_0^\infty e^{ixv}\phi_n^{(m)}(v)\,dv$ converges, and by application of Lemma 2 we obtain

$$E_n(x) = \left(\frac{i}{x}\right)^m \phi_n^{(m-1)}(0) + \left(\frac{i}{x}\right)^m \int_0^\infty e^{ixv}\phi_n^{(m)}(v)\,dv;$$

compare (5.11). Combination of this result with (5.8) (again with $\beta = \infty$) leads to (3.8).

The proof of Theorem 1 is complete.

6. Asymptotic properties and error bounds. From (3.10) we immediately have

$$|\delta_{m,n}(x)| \leqq \frac{1}{x^m}\int_a^b |Q'_{m,n}(t)|\,dt = \frac{\mathscr{V}_{a,b}\{Q_{m,n}(t)\}}{x^m}, \tag{6.1}$$

where $\mathscr{V}_{a,b}$ denotes the total variation of the function within the braces over the closure of the given interval. Another bound for $\delta_{m,n}(x)$, which has the advantage of being $O(x^{-m-1})$ for large x, may be found by partial integration of (3.10), using the identity

$$Q'_{m,n}(t)/p'(t) = Q_{m+1,n}(t). \tag{6.2}$$

Thus

$$|\delta_{m,n}(x)| \leqq [|Q_{m+1,n}(a)| + |Q_{m+1,n}(b)| + \mathscr{V}_{a,b}\{Q_{m+1,n}(t)\}]x^{-m-1}, \tag{6.3}$$

provided that the right-hand side is finite: in contrast to (6.1), finiteness is not guaranteed by the conditions adopted in §§ 2 and 3.

Now consider the other error term. For fixed α we have

$$\Gamma(\alpha, z) \sim e^{-z}z^{\alpha-1}$$

as $z \to \infty$ in a sector which includes $|\arg z| \leqq \pi$; see [1, § 6.5]. Application to (3.9) shows that

$$\varepsilon_{m,n}(x) = O(x^{-m-1}), \qquad x \to \infty. \tag{6.4}$$

Relations (6.1), (6.3), and (6.4) confirm the asymptotic nature of the expansions (3.7) and (3.8).

An interesting special case of (3.8) is obtained on taking $a = 0$, $b = \infty$, $m = \infty$, $p(t) = t^\mu$, $q(t) = \mu t^{\mu-1}g(t^\mu)$, and subsequently replacing t^μ by t.

THEOREM 2. *Assume that*

(i) *$g(t)$ is infinitely differentiable in* $(0, \infty)$;

(ii)

$$g(t) \sim \sum_{s=0}^\infty g_s t^{(s+\lambda-\mu)/\mu}, \qquad t \to 0+, \tag{6.5}$$

where Re λ *and* μ *are positive, this asymptotic expansion being differentiable any number of times*;

(iii) *Each of the integrals*

$$\int e^{ixt} g^{(s)}(t)\, dt, \qquad s = 0, 1, \cdots,$$

converges at $t = \infty$ *uniformly for sufficiently large* x.[8]

Then for large positive values of x, *the asymptotic expansion of the integral*

$$\int_0^\infty e^{ixt} g(t)\, dt$$

is obtained by substituting (6.5) *and integrating formally term by term in Hardy's generalized sense.*

Although this theorem can be derived from existing results, for example [9], it does not appear to have been emphasized in the literature. This is somewhat surprising since it is the natural analogue for Fourier integrals of Watson's lemma for Laplace integrals.

Let us return to Theorem 1 and consider the actual calculation of bounds for the error terms. For $\delta_{m,n}(x)$ the inequality (6.1) (or (6.3)) may be used as it stands. For $\varepsilon_{m,n}(x)$ we make the simplifying assumption *that* λ *is real.* We may then apply the inequality

$$|\Gamma(\alpha, \pm iy)| \leqq 2y^{\alpha - 1}, \qquad \alpha < 1, \quad y > 0, \tag{6.6}$$

which is itself establishable by deforming the path in (3.12) until it lies along the imaginary axis and then integrating by parts. Thus we have

$$|\varepsilon_{m,n}(x)| \leqq \frac{2}{x^{m+1}} \sum_{s=0}^{n-1} \frac{\Gamma\{(s+\lambda)/\mu\}}{|\Gamma\{(s+\lambda-m\mu)/\mu\}|} \frac{|a_s|}{\{p(b)-p(a)\}^{(m\mu+\mu-s-\lambda)/\mu}}. \tag{6.7}$$

In the case $p(b) = \infty$, this bound vanishes and only the error term $\delta_{m,n}(x)$ survives.

7. Example. As a simple example of the calculation of error bounds, consider the functions of Anger and Weber with equal argument and order:

$$\mathbf{J}_x(x) = \frac{1}{\pi}\int_0^\pi \cos(xt - x\sin t)\, dt, \qquad \mathbf{E}_x(x) = \frac{1}{\pi}\int_0^\pi \sin(xt - x\sin t)\, dt;$$

see [14, § 10.1]. In the notation of §§ 2 and 3 we have

$$\pi\{\mathbf{J}_x(x) + i\mathbf{E}_x(x)\} = \int_0^\pi e^{ixp(t)}\, dt,$$

where

$$p(t) = t - \sin t = \frac{t^3}{3!} - \frac{t^5}{5!} + \frac{t^7}{7!} - \cdots. \tag{7.1}$$

Thus $a = 0$, $b = \pi$, $q(t) = 1$, $\mu = 3$, and $\lambda = 1$.

[8] Convergence at $t = 0$ is not required, however, except for $s = 0$.

The definition (2.3) gives

$$P_s(t) = \frac{1}{2^{s+1}}\left(\frac{1}{\sin^2 \frac{1}{2}t}\frac{d}{dt}\right)^s \frac{1}{\sin^2 \frac{1}{2}t}. \tag{7.2}$$

In particular,

$$P_0(t) = \frac{1}{2\sin^2 \frac{1}{2}t}, \qquad P_1(t) = -\frac{\cos \frac{1}{2}t}{4\sin^5 \frac{1}{2}t}, \qquad P_2(t) = \frac{5 - 4\sin^2 \frac{1}{2}t}{16 \sin^8 \frac{1}{2}t}. \tag{7.3}$$

By reversion of (7.1) we have for small $v \equiv p(t)$,

$$t = (6v)^{1/3} + \tfrac{1}{60}(6v) + \tfrac{1}{1400}(6v)^{5/3} + \tfrac{1}{25200}(6v)^{7/3} + \tfrac{43}{17248000}(6v)^3 + \cdots;$$

compare [14, § 8.21]. Differentiation yields

$$f(v) = \frac{dt}{dv} = \sum_{s=0}^{\infty} a_s v^{(s-2)/3};$$

compare the second of (3.4). In this expansion the coefficients a_s of odd suffix vanish, and

$$a_0 = \tfrac{1}{3}6^{1/3}, \quad a_2 = \tfrac{1}{60}6, \quad a_4 = \tfrac{1}{840}6^{5/3}, \quad a_6 = \tfrac{1}{10800}6^{7/3}, \quad a_8 = \tfrac{129}{17248000}6^3, \cdots.$$

Using the asymptotic results of § 6, we derive from (3.7)

$$\pi\{\mathbf{J}_x(x) + i\mathbf{E}_x(x)\} \sim \sum_{s=0}^{\infty} e^{(2s+1)\pi i/6}\Gamma(\tfrac{2}{3}s + \tfrac{1}{3})\frac{a_{2s}}{x^{(2s+1)/3}} - e^{\pi i x}\sum_{s=0}^{\infty} P_s(\pi)\left(\frac{i}{x}\right)^{s+1}. \tag{7.4}$$

Explicitly, the first few terms on the right-hand side are given by

$$e^{\pi i/6}\Gamma\left(\frac{1}{3}\right)\frac{1}{3}\left(\frac{6}{x}\right)^{1/3} + i\frac{1}{60}\left(\frac{6}{x}\right) + e^{5\pi i/6}\Gamma\left(\frac{2}{3}\right)\frac{1}{1260}\left(\frac{6}{x}\right)^{5/3} - e^{\pi i/6}\Gamma\left(\frac{1}{3}\right)\frac{1}{24300}\left(\frac{6}{x}\right)^{7/3}$$
$$-i\frac{129}{8624000}\left(\frac{6}{x}\right)^3 + \cdots + i\,e^{\pi i x}\left(-\frac{1}{2x} + \frac{1}{16x^3} + \cdots\right).$$

To evaluate error bounds, we note that the condition (3.5) requires

$$3m - 1 \leqq n < 3m + 3.$$

Therefore from (3.7),

$$\begin{aligned}\pi\{\mathbf{J}_x(x) + i\mathbf{E}_x(x)\} &= \sum_{s=0}^{n-1} e^{(s+1)\pi i/6}\Gamma(\tfrac{1}{3}s + \tfrac{1}{3})\frac{a_s}{x^{(s+1)/3}} \\ &\quad - e^{\pi i x}\sum_{s=0}^{m-1} P_s(\pi)\left(\frac{i}{x}\right)^{s+1} + \delta_{m,n}(x) - \varepsilon_{m,n}(x),\end{aligned} \tag{7.5}$$

where m is an arbitrary nonnegative integer, and $n = 3m$, $3m + 1$, or $3m + 2$.

Suppose, for example, that $m = 0$ and $n = 1$, that is, the expansion comprises the single term $e^{\pi i/6}\Gamma(\frac{1}{3})\frac{1}{3}(6/x)^{1/3}$. Equations (3.11) and (7.3) give

$$Q_{1,1}(t) = \frac{1}{2\sin^2 \frac{1}{2}t} - \frac{2}{\{6(t - \sin t)\}^{2/3}}. \tag{7.6}$$

The bound (6.3) yields

$$|\delta_{0,1}(x)| \leqq \left[\frac{1}{10} + \left\{\frac{1}{2} - \frac{2}{(6\pi)^{2/3}}\right\} + \mathscr{V}_{0,\pi}(Q_{1,1})\right]\frac{1}{x}.$$

Also, from (6.7) we derive

$$|\varepsilon_{0,1}(x)| \leqq 4(6\pi)^{-2/3}x^{-1}.$$

Numerical calculation shows that

$$2(6\pi)^{-2/3} = 0.28, \qquad \mathscr{V}_{0,\pi}(Q_{1,1}) = 0.12,$$

to two decimal places, and thence

$$|\delta_{0,1}(x)| \leqq 0.44x^{-1}, \qquad |\varepsilon_{0,1}(x)| \leqq 0.56x^{-1}.$$

That the combined bound $1.00x^{-1}$ for $|\delta_{0,1}(x)| + |\varepsilon_{0,1}(x)|$ is quite realistic can be seen by comparing it with the first neglected terms of the asymptotic expansion, given by

$$\frac{i}{10x} - \frac{i\,e^{\pi i x}}{2x}.$$

Similar results can be obtained for higher error terms, provided that in calculating the functions $Q_{m,n}(t)$ from expressions of the type (7.6), sufficient precision is maintained to overcome numerical cancellation for small t.

REFERENCES

[1] M. ABRAMOWITZ AND I. A. STEGUN, eds., *Handbook of Mathematical Functions*, N.B.S. Applied Mathematics Series 55, U.S. Government Printing Office, Washington, D.C., 1964.
[2] E. T. COPSON, *Asymptotic Expansions*, Cambridge University Press, London, 1965.
[3] N. G. DE BRUIJN, *Asymptotic Methods in Analysis*, 2nd ed., Interscience, New York, 1961.
[4] A. ERDÉLYI, *Asymptotic representations of Fourier integrals and the method of stationary phase*, SIAM J. Appl. Math., 3 (1955), pp. 17–27.
[5] ———, *Asymptotic expansions of Fourier integrals involving logarithmic singularities*, SIAM J. Appl. Math., 4 (1956), pp. 38–47.
[6] ———, *On the principle of stationary phase*, Proc. 4th Canadian Math. Congress, Toronto, 1959, pp. 137–146.
[7] G. H. HARDY, *Researches in the theory of divergent series and divergent integrals*, Quart. J. Pure Appl. Math., 35 (1904), pp. 22–66.
[8] ———, *Further researches in the theory of divergent series and integrals*, Trans. Camb. Philos. Soc., 21 (1908), pp. 1–48.
[9] D. S. JONES, *Fourier transforms and the method of stationary phase*, J. Inst. Math. Appl., 2 (1966), pp. 197–222.
[10] ———, *Asymptotic behavior of integrals*, SIAM Rev., 14 (1972), pp. 286–317.
[11] J. MCKENNA, *Note on asymptotic expansions of Fourier integrals involving logarithmic singularities*, SIAM J. Appl. Math., 15 (1967), pp. 810–812.
[12] F. W. J. OLVER, *Error bounds for the Laplace approximation for definite integrals*, J. Approximation Theory, 1 (1968), pp. 293–313.
[13] ———, *Why steepest descents?* SIAM Rev., 12 (1970), pp. 228–247.
[14] G. N. WATSON, *Theory of Bessel Functions*, 2nd ed., Cambridge University Press, London, 1944.

Reprinted from:
THEORY AND APPLICATION OF SPECIAL FUNCTIONS

ACADEMIC PRESS, INC.

Unsolved Problems in the Asymptotic Estimation of Special Functions

F. W. J. Olver

Abstract

The first part of this paper surveys the tools of asymptotic analysis that are presently available and those that are needed for the next stages of development of the asymptotic theory of special functions. Methods are grouped according to the number of free variables to which they achieve a uniform reduction. Topics include the approximation of functions defined parametrically by a definite integral or infinite sum, and approximate solutions of linear ordinary differential equations. The main areas in which work appears to be needed are (i) problems of confluence, that is, coalescing saddle points and singularities of integrals, or coalescing turning points and singularities of differential equations; (ii) rigorous error analysis.

The second part of the paper discusses the more important special functions in increasing order of the total number of variables and parameters involved. Almost all asymptotic problems concerning special functions of one variable have been solved. For functions of two variables the problems are solved, or can be solved by use of existing asymptotic tools. It is in the case of functions depending on three variables (including parameters) that the most significant work can be expected in the immediate future. Two new asymptotic tools have been developed very recently, one for integrals with three coalescing saddle points, the other for second-order differential equations having two coalescing turning points. Applications of these tools are needed, as well as the development of similar tools for other three-variable problems involving confluence. As the total number of variables increases beyond three, our knowledge of the asymptotic behavior of the special functions becomes more fragmentary. It will undoubtedly be many years before we have complete knowledge of the asymptotic behavior of any of the functions of four variables, including for example, the hypergeometric function $F(a, b;c;z)$.

CONTENTS

1. INTRODUCTION

1.1. Classification of problems. The problems referred to in the title of this paper are of two types: problems for which the tools of asymptotic analysis are already available, and problems for which the asymptotic tools need to be developed. For the most part, problems of the first type are the easier to solve. I hasten to add, however, that in most instances it would be wrong to replace the words "the easier" by "easy". Quite apart from the heavy algebraic computations that plague applications of asymptotic analysis, incorrect choice of tool or poor application of a good tool may lead to results that are meaningless, crude, or at best unnecessarily restricted. Judgment and experience are of great importance.

Problems in the asymptotic estimation of special functions for which the asymptotic tools are not on the shelf vary enormously in difficulty. For some, the method is known in principle and merely needs development and refinement. For others, the difficulties are profound and await new inspirations. An index of the difficulty of a given problem is the number of free variables (including parameters) that are involved. The simplest type is the determination of an asymptotic approximation, in terms of elementary functions, for a function $f(x)$ of a single real or complex variable x, as x tends to infinity or a finite singularity of $f(x)$.

For a function $f(x, y)$ of two variables, we consider its behavior as one of the variables x, say, tends to infinity or a singularity, and seek asymptotic approximations that are uniform with respect to the second variable y when $y \in Y$. Here Y is a given interval, complex domain, or other infinite point set which may depend on x, and may contain singularities of $f(x, y)$ regarded as a function of y. The concept of uniformity is of utmost importance; without it we would be considering a finite set of fixed values of y in effect, and the problem reverts to the type mentioned in the preceding paragraph. With the formulation just given, x is called the asymptotic variable and y the nonasymptotic variable. The roles are interchangeable, that is, we may require asymptotic approximations as y tends to infinity or a singularity that are uniform with respect to x. As approximants we naturally try to use elementary functions wherever possible. Sometimes, however, the neighborhoods of certain points in the plane of the nonasymptotic variable can be included only by employing transcendental functions as approximants. Furthermore, these exceptional points need not be singularities of $f(x, y)$.†

† An example is furnished by the Bessel functions $J_n(x)$ and $Y_n(x)$. Asymptotic expansions for large n, which are uniform with respect to unrestricted positive real values of x, can be found in terms of Airy

For a function $f(x, y, z)$ of three variables, the problem is to construct asymptotic approximations as one of the variables x, say, tends to infinity or a singularity, which are uniform with respect to both y and z. Thus there is one asymptotic variable and two nonasymptotic variables; again, the roles are interchangeable. In some regions of the $Y \times Z$ space it will be possible to employ elementary functions as approximants. In others, transcendental functions that depend on a single variable will suffice. To cover the whole of the $Y \times Z$ space, however, it will be necessary in some regions to resort to transcendental functions of two variables as approximants.

For a function of r free variables, approximants that depend on as many as $r - 1$ variables will be needed, as a rule, in order to have a complete description of the function when any one of its variables tends to infinity or a singularity. To stress the importance of the concept of uniformity with respect to the nonasymptotic variables, we introduce the following terminology. An asymptotic formula that approximates a function of r variables is said to achieve a uniform reduction of free variables if it satisfies both of the following conditions: (i) the approximation is uniform with respect to all $r - 1$ nonasymptotic variables as each ranges independently over an infinite set of values; (ii) the approximation involves only elementary functions or transcendental functions that depend on fewer than r free variables, and the total number of such functions is finite.

As an illustration† if we approximate Legendre functions $P_n^m(z)$ and $Q_n^m(z)$ for large n by an asymptotic formula which involves the parabolic cylinder function $U(a, x)$ and is uniform with respect to both m and z, then we achieve a uniform reduction from three free variables to two. But if this problem were modified by fixing m, or fixing m/n,

functions, but not in terms of elementary functions. The troublesome point here is $x = n$, which is not a singularity of $J_n(x)$ or $Y_n(x)$.

† This example is discussed more fully in §6.1 below.

then it would not be permissible to use a parabolic cylinder function to achieve a uniform reduction of variables, unless for example, the variable a in the approximant were fixed.

For the purpose of providing analytical insight into the nature of a given function, or the construction of algorithms for computing numerical values, asymptotic approximations that achieve a uniform reduction of free variables are much superior to other kinds. For if the number of free variables in an approximant equals or exceeds the number in the original function, then it is just as easy, as a rule, to evaluate the original function by direct numerical methods (such as numerical integration of an ordinary differential equation) as it would be to compute the approximant.

1.2. Arrangement of the paper. The first part of the present paper describes the tools of asymptotic analysis that are presently available, and indicates extensions and new tools that are needed. Although any method of asymptotic analysis may eventually prove to have applications to the special functions, the account is restricted to methods that strike the author as having significant applications in the next stages of development. The topics treated are the approximation of functions defined parametrically by a definite integral or infinite sum, and approximate solutions of ordinary linear differential equations. For each topic, methods are grouped according to the number of free variables to which they achieve uniform reduction.

In the second part of the paper, the commoner special functions are discussed in increasing order of the number of variables on which they depend. Perhaps it should be clarified that by "special functions" we mean throughout the functions defined and discussed in the Bateman Manuscript Project [1] and the N. B. S. Handbook [35]. Among the more important omissions in the present paper are the Mathieu and Lamé functions. Other omissions that have been made in order to keep the paper to a reasonable length are associated problems, such as asymptotic approximations of integrals of the special functions, asymptotic approximations of zeros, and the construction of converging factors.†

Some special functions, for example the generalized hypergeometric function, have an arbitrarily large number of variables and it is possible that a complete description of their asymptotic behavior will have to proceed through a chain of lower functions, successive functions in the chain having fewer and fewer variables. As a corollary, from the standpoint of asymptotic analysis attempts to unify the theory of special functions within the framework of a single function of many variables, for example Meijer's G-function, are usually unfruitful because we know so little about the general asymptotic behavior of such functions.‡ On the whole it is the particular specializations of the parameters in the more general functions – in other words the special properties of the special functions – that lead to solvable problems concerning their asymptotic behavior.

1.3. References. In the rest of this paper references are given in the usual manner, except that to avoid frequent repetition almost all references to the Bateman Manuscript Project [1] and the N. B. S. Handbook [35], and some references to the author's recent book [42] are omitted. The first two supply all standard properties of the special functions that are needed, and the third may be consulted for further details concerning standard asymptotic methods.

PART I. THE TOOLS OF ASYMPTOTIC ANALYSIS

2. INTEGRALS

2.1. Standard methods. The approach that has been used most frequently for deriving asymptotic approximations for the special functions is via

† An extensive (albeit formal) treatment of converging factors for asymptotic expansions of the special functions is included in Dingle's recent book [10]. Dingle also introduces the name "terminant" in preference to "converging factor".

‡ Compare §7.3 below.

integral representations of the form

$$I(z) = \int_P K(z,t)dt \quad . \tag{2.1}$$

Here P is an interval of the real axis or a contour in the complex plane, often infinite in length, and z is the asymptotic variable under consideration. Sometimes repeated integration by parts is an effective procedure: each integration yields a new term in the asymptotic expansion of $I(z)$, and the error term is given explicitly as an integral. More often than not, however, other methods are needed.

For real variables, standard procedures are the method of Laplace, which is applicable to integrands of the form

$$K(z,t) = e^{-zp(t)} q(t) \quad , \tag{2.2}$$

where $p(t)$ and $q(t)$ are independent of z, and the method of stationary phase, applicable when

$$K(z,t) = e^{izp(t)} q(t) \quad . \tag{2.3}$$

Both methods apply to the case $z \to \infty$ and yield asymptotic approximations for $I(z)$ in terms of elementary functions. Often the approximations are extendable to asymptotic expansions in descending fractional powers of z .

For contour integrals, there is the complex-variable analogue of Laplace's method, which is often called the method of steepest descents, though this name is not entirely appropriate [41]. Probably the most frequently used method of all is Watson's lemma, which may be regarded as the special case of Laplace's method in which $p(t) = t$ and P is the positive real axis.

All of the above methods date back at least half a century, and although several papers have been published showing how to estimate or otherwise control the error terms associated with the resulting approximations, for example [21], [37], and [28], it is only quite recently that theories have been devised to enable the error terms to be bounded in a systematic and rigorous manner; see [40], [41], [43], and [22]. Mention may also be made here of the work of Stenger [52] in adapting quadrature formulas to the derivation of asymptotic approximations, complete with strict error bounds. All of these error analyses need extensive applications in order to test their effectiveness.

2.2. Movable saddle points and singularities. Of decisive importance in the construction of asymptotic approximations for functions $I(z)$ represented by integrals of the form (2.1) are the points at which $\partial K(z, t)/\partial t$ vanishes. These are the so-called stationary points (real variables) or saddle points (complex variables). Their importance stems from the fact that for large $|z|$ the major contribution to the integral comes from the neighborhoods of these points or the endpoints. In the case of a contour integral it may be necessary to deform the contour to pass through an appropriate saddle point before Laplace's method can be applied successfully. In the next class of problems the integral depends on a parameter α; thus

$$I(\alpha, z) = \int_P K(\alpha, z, t)dt \quad . \tag{2.4}$$

Asymptotic approximations are needed for large $|z|$ that are uniform with respect to α when α ranges over a given infinite point set. In general, the saddle points (or stationary points) depend on α. If, as α varies continuously, the saddle points neither coalesce with each other nor coincide with an endpoint, then the modifications to the methods of §2.1 are straightforward and the same approximations are obtained. When these conditions are not fulfilled the problem is more profound, and it is no longer possible to construct asymptotic approximations that are uniform with respect to α in terms of elementary functions. Surveys of methods for treating these problems have been made by Erdélyi [12] and Jones [22]. We describe these methods briefly and mention some more recent developments.

Suppose first that the variables are real, P is a fixed finite or infinite interval $(0, T)$, and the integrand in (2.4) has the form

$$K(\alpha, z, t) = e^{-zp(\alpha, t)} q(\alpha, t) , \tag{2.5}$$

where $p(\alpha, t)$ and $q(\alpha, t)$ are independent of z. Furthermore, assume that $\partial p(\alpha, t)/\partial t$ has exactly one zero in $[0, T)$ which varies continuously with α, and coincides with the endpoint $t = 0$ for a certain critical value $\alpha = \alpha_0$, say. Following earlier writers, especially van der Waerden [56], Bleistein [3] has shown that by making a quadratic change of integration variable it is possible to obtain an asymptotic approximation for large $|z|$ that is uniform with respect to α in a closed interval containing α_0. The approximant employed is the simplest integral possessing the essential features of the problem, given by

$$\int_0^\infty \exp\{-z(\tfrac{1}{2}\tau^2 - a\tau)\} d\tau . \tag{2.6}$$

Here a is a function of α that vanishes when $\alpha = \alpha_0$. The integral (2.6) is transformable into the error function, accordingly the method achieves a uniform reduction from two free variables to one.

The foregoing discussion is predicated on the assumption that $p(\alpha, t)$ and $q(\alpha, t)$ are free from singularity when $t \in [0, T)$. If, for example, $q(\alpha, t)$ behaves as a multiple of $t^{\lambda-1}$ when $t \to 0+$, where λ denotes a positive constant, then the uniform approximant (2.6) is replaced by

$$\int_0^\infty \exp\{-z(\tfrac{1}{2}\tau^2 - a\tau)\}\tau^{\lambda-1} d\tau ,$$

which is expressible in terms of the parabolic cylinder function

$U(\lambda - \frac{1}{2}, -az^{\frac{1}{2}})$. Since λ is fixed, this is a function of the single variable $az^{\frac{1}{2}}$.

Next, consider the case in which P is a real interval or contour in the complex plane, $K(\alpha, z, t)$ again has the form (2.5), and $\partial p(\alpha, t)/\partial t$ has exactly two zeros which coalesce into a double zero lying on P for a certain critical value of α . To handle this case a cubic change of integration variable is needed, and the uniform approximant is either

$$\int_{\infty e^{-\pi i/3}}^{\infty e^{\pi i/3}} \exp\{z(\frac{1}{3}\tau^3 - a\tau)\} d\tau \ , \tag{2.7}$$

or

$$\int_0^{\infty} \exp\{z(-\frac{1}{3}\tau^3 + a\tau)\} d\tau \ , \tag{2.8}$$

depending whether the limiting double zero is located at an interior point of P, or happens to be an endpoint. The integral (2.7) is transformable into the Airy function, and (2.8) is transformable into Scorer's function; furthermore, each of these functions depends on a single variable. This is the method of Chester, Friedman, and Ursell [7].

Other methods of similar type that achieve a uniform reduction from two free variables to one have been described by Wong [64] and Erdélyi [13]. In Wong's problem there are no stationary points on the path of integration, except for a critical value of the nonasymptotic variable α when an exterior saddle point coincides with an endpoint. Furthermore, this endpoint is permitted to be an algebraic singularity of the function $q(\alpha, t)$. In Erdélyi's problem there are no stationary points or singularities on the path, except for a critical value of α when an exterior algebraic singularity coincides with an endpoint.

To summarize the present subsection, the main problems associated with Laplace's method concerning confluence of saddle points, algebraic singularities, and endpoints have been solved, at least in principle, for integrals representing functions of two free variables. Subsidiary problems of varying difficulty remain, however, including extensions to logarithmic singularities, determination of maximal regions of validity when the variables are complex, and the construction of strict and realistic error bounds.

2.3. Several movable saddle points and singularities. The next level of difficulty is presented by functions of three variables. It is now possible for any three saddle points or singularities to move independently and coincide with each other, or for any two of these points to move independently and coincide with each other and also with an endpoint of the path of integration.

The only problem of this type that has been solved successfully to date is Ursell's treatment [55] of the case in which three saddle points are allowed to coalesce into a triple saddle point at an interior point of the path of integration. By introducing a quartic change of integration variable Ursell is able to construct uniform asymptotic approximations in terms of functions of the form

$$\int_{\infty e^{-\pi i/4}}^{\infty e^{\pi i/4}} \tau^j \exp\{z(\tfrac{1}{4}\tau^4 - a\tau^2 - b\tau)\}d\tau, \qquad j = 0, 1, 2 \quad , \tag{2.9}$$

where a and b are parameters that vanish in the critical situation. Since each of these integrals is effectively a function of two variables, the method achieves a uniform reduction from three free variables to two. Moreover, in the same paper Ursell notes that his method can be extended in a straightforward manner to the case of a function of r variables expressed as an integral with s coalescing saddle points. Provided that $s \leq r$ this extension, also, achieves a uniform reduction of free variables.

Some of the unsolved problems involving three or more free variables would have immediate applications to the special functions. An example is provided by the case of two saddle points coalescing into a double saddle point which, in turn, may coincide with a singularity at an endpoint for an appropriate combination of variables. This problem arises with Whittaker functions with large parameters; see §6.2 below.

The general problem of an arbitrary number of saddle points coinciding with an arbitrary number of algebraic singularities has been studied by Bleistein [4]. His analysis is incomplete and Erdélyi and Ursell have suggested that there are severe difficulties in rendering it complete [12], [55]. It may therefore be worthwhile to pursue special cases, such as that mentioned in the preceding paragraph, both for their own applications and for any light that may be shed on the general problem. Subsidiary problems of the kind mentioned in the closing paragraph of §2.2 also arise in connection with the methods of the present subsection.

2.4. Other kernels. Another direction for the generalization of Laplace's method and the method of stationary phase is to replace the exponential function in (2.2), (2.3), or (2.5), by a transcendental function, for example, an Airy or Bessel function. Some of these problems are amenable to fairly straightforward extensions of the standard methods. Other cases, particularly real integrals with oscillatory integrands, require more sophisticated techniques such as the use of the Parseval formula for the Mellin transform. For introductions to these generalizations the reader should consult the recent books by Bleistein and Handelsman [5] and Riekstinš [47]. All the extensions mentioned in §§2.2 and 2.3 also present themselves for integrals of the more general type.

Strict error bounds are lacking for most approximations obtained by these methods. An exception is the Hankel transform, which has recently been studied by Wong [65].

3. SUMS AND SEQUENCES

3.1. Sums. Most methods for the asymptotic estimation of a function defined as the sum of an infinite power series depend upon the replacement, or approximate replacement, of the sum by an integral, which is then estimated by the residue calculus or the methods mentioned in Section 2.

The Euler-Maclaurin formula estimates the sum

$$f(0) + f(1) + \dots + f(n)$$

by the integral

$$\int_0^n f(t)dt \ ,$$

with a correction expressed as a finite series involving the values of $f(t)$ and its odd derivatives at $t = 0$ and n, together with an explicit integral representation for the error term.

Methods of Barnes' type depend on the use of the residue theorem to express an infinite sum depending on a variable z

$$S(z) = \sum_{n=0}^{\infty} f(z, n)$$

as an integral of the type

$$S(z) = \frac{1}{2i} \int_P \cot(\pi t)\ f(z, t)dt \ ,$$

in which P is a loop contour enclosing the points $t = 0, 1, 2, \cdots,$ but not enclosing $t = -1, -2, -3, \cdots,$ or the singularities of $f(z, t)$.

3.2. Sequences. The commonest way in which special functions, especially the orthogonal polynomials, appear as members of an infinite sequence is in the expansion of a generating function, thus

$$G(z,t) = \sum_{n=0}^{\infty} \phi_n(z)\, t^n \ .$$

The problem of determining the asymptotic behavior of $\phi_n(z)$ as $n \to \infty$ can be attacked by use of Cauchy's integral formula

$$\phi_n(z) = \frac{1}{2\pi i} \int_P \frac{G(z,t)}{t^{n+1}} dt , \tag{3.1}$$

where P is a simple closed contour encircling $t = 0$, but not any singularity of $G(z,t)$. Thus the methods mentioned in Section 2 again become relevant.

A different procedure for approximating integrals of the form (3.1) for large n, which is applicable when $G(z,t)$ is not an entire function of t, is Darboux's method, as follows. The generating function $G(z,t)$ is suitably matched by another function $H(z,t)$ for which the coefficients $\psi_n(z)$ in its Maclaurin expansion

$$H(z,t) = \sum_{n=0}^{\infty} \psi_n(z) t^n$$

are known explicitly. Then

$$\phi_n(z) \sim \psi_n(z), \quad n \to \infty .$$

The matching process consists of identifying the singularities of $G(z,t)$ on its circle of convergence in the t-plane, and arranging for $H(z,t)$ to have the same kind of singularity at exactly the same points.

Darboux's method dates back almost a century, and although often used it does not appear to have been extended until quite recently. Wong and Wyman [66] have shown how to treat the case in which the singularities of $G(z,t)$ on its circle of convergence are of logarithmic type. Fields [16] has examined the confluent case in which two algebraic singularities on the circle of convergence coalesce. Fields' asymptotic expansions are in terms of certain transcendental functions of one variable, and are uniform with respect to the distance separating the singularities, including zero distance. Accordingly, the expansions achieve a uniform reduction from two free variables to one.

Fields' work has pointed the way to further problems of confluence, for example, three coalescing singularities on the circle of convergence. First, however, it may be desirable to investigate whether any simplifications are feasible, since the results in [16] are rather complicated to apply in their present form.

Other unsolved problems in the asymptotic approximation of sequences are discussed in a recent survey by Bender [2], written from a broader standpoint than the estimation of the special functions.

4. LINEAR ORDINARY DIFFERENTIAL EQUATIONS

4.1. Introduction. Many of the special functions originated as solutions of ordinary differential equations after separating variables in Laplace's equation. These equations are therefore linear and of the second order, and as we know now general methods for constructing asymptotic solutions can be devised in a variety of circumstances. The development of these methods lagged well behind that of methods for the asymptotic approximation of integrals, however. Consequently, in the early history of special functions asymptotic approximations were almost invariably derived from integral representations rather than directly from the defining differential equation. An impression of the meager state of the art in the 1920's can be formed by perusal of Ince's classic text [20]. Well before then, the powerful methods for integrals mentioned in §2.1 were established and applied frequently.

The situation has changed in recent years, and certainly as far as differential equations of the second order are concerned, a considerable range of powerful tools is available. Furthermore, compared with methods for integrals the differential-equation approach often has the advantages of easing the determination of maximal regions of validity, easing the calculation of higher terms in asymptotic expansions, and yielding simpler and sharper error bounds.

4.2. Irregular singularities. The first asymptotic problem associated with an ordinary differential equation is the nature of its solutions in the neighborhood of a singularity, particularly an irregular singularity. Here the theory of error bounds has paid a remarkable dividend. Bounds that were constructed for the errors associated with the Liouville-Green approximation (or JWKB approximation in the terminology of theoretical physicists) have established that this approximation furnishes an asymptotic representation of solutions of second-order differential equations in the neighborhood of virtually all kinds of irregular singularity, not merely those of finite rank. Accordingly, when a new type of irregular singularity is encountered we no longer need follow the approach of most standard texts and begin by trying to guess the type of asymptotic approximation that is appropriate. A further consequence is that the concept of rank is of diminished importance, because there are now numerous treatable types of singularity that do not have classifiable rank.

4.3. Two-variable problems. In the next class of problems the differential equation contains a real or complex parameter u, say; thus

$$d^2w/dz^2 = f(u, z)w \quad . \tag{4.1}$$

We seek asymptotic approximations for the solutions when $u \to \infty$, or some other distinguished point, that are uniform with respect to $z \in Z$, where Z is a given real interval or complex domain. The form of the asymptotic solutions is governed by the number and nature of the singularities and zeros of $f(u, z)$ in Z.

In the case of equations satisfied by the special functions, the coefficient $f(u, z)$ often can be decomposed in the form

$$f(u, z) = u^2 f(z) + g(u, z) \quad , \tag{4.2}$$

where $f(z)$ is independent of u, and $|g(u, z)|$ is uniformly small compared with $|u^2 f(z)|$ when $|u|$ is large, except possibly in fixed neighborhoods of the singularities and zeros of $f(z)$. If the only singularities of the differential equation in Z are irregular and Z contains no zeros of $f(z)$, then the Liouville-Green functions

$$f^{-\frac{1}{4}}(z) \exp\{\pm u \int f^{\frac{1}{2}}(z)dz\} \tag{4.3}$$

furnish uniform asymptotic approximations to the solutions as $|u| \to \infty$. Furthermore, when $g(u, z)$ can be expanded uniformly in descending powers of u, the asymptotic approximations to the solutions can be extended to asymptotic expansions; in particular, this includes the case in which $g(u, z)$ is independent of u.

Zeros of $f(z)$ are called turning points or transition points of the differential equation. At these points the Liouville-Green functions (4.3) become infinite and no longer represent the solutions. In order to approximate the solutions uniformly in a region containing a simple turning point we employ, as approximants, solutions of the simplest kind of differential equation possessing the essential features, given by

$$d^2w/dz^2 = u^2 zw \ . \tag{4.4}$$

Solutions of (4.4) are $Ai(u^{2/3}z)$ and $Bi(u^{2/3}z)$, where Ai and Bi denote the Airy functions. Since these approximants depend on the single variable $u^{2/3}z$, a uniform reduction from two free variables to one is achieved by this procedure. Again, the asymptotic approximations to the solutions can be extended to asymptotic expansions in descending powers of u when $g(u, z)$ depends on u in an appropriate manner.

In a similar way, in a region containing a turning point of multiplicity ℓ, uniform asymptotic approximations to the solutions can be constructed in terms of solutions of the equation

$$d^2w/dz^2 = u^2 z^\ell w \ .$$

Again, these approximants are functions of a single variable, namely $u^{2/(\ell+2)}z$; they can be expressed as Bessel functions of order $1/(\ell+2)$. When $\ell > 1$, however, there is no straightforward extension from asymptotic approximations to asymptotic expansions in descending powers of u [46].

Returning from turning points to singularities, we have already noted that the Liouville-Green functions furnish uniform asymptotic approximations for large $|u|$ in regions containing irregular singularities. This is also the case when the region contains a regular singularity which is a double pole of $f(z)$, provided that the parameter u is suitably modified. In the case of a simple pole of $f(z)$, however, it is no longer possible to use elementary functions as approximants. Instead, we employ Bessel functions, the order of which depends on the behavior of the term $g(u, z)$ at the pole of $f(z)$. Provided that the order is independent of u, and this is the usual case, we do not violate the principle of reduction of free variables.

For the most part, the methods mentioned in the present subsection are well developed, both for real and complex variables, including the construction of error bounds. Among the problems that need more detailed treatment is the case in which $f(z)$ has a simple pole and the order of the approximating Bessel functions is purely imaginary.

4.4. Three or more variables. With the inclusion of an extra parameter α, say, in (4.2), the differential equation (4.1) takes the form

$$d^2w/dz^2 = \{u^2 f(\alpha, z) + g(u, \alpha, z)\} w \ .$$

The turning points and singularities now depend on α, and may coalesce as α assumes certain critical values. The situation is analogous to that for integrals with movable saddle points and singularities discussed in §2.2, and the problem is to obtain asymptotic solutions for large $|u|$ that are uniform with respect to α, including the critical values, as well as with respect to z.

The only problem of this kind that has been solved successfully so far is the case of two coalescing simple turning points, treated by the present writer in [44]. This reference gives asymptotic approximations, complete with error bounds, these approximations being uniform with respect to α in closed intervals containing the critical value, and also uniform with respect to z in intervals that extend to infinity or to the nearest singularities of the differential equation. The approximants that are used are the solutions of the simplest form of differential equation having the essential features, given by

$$d^2w/dz^2 = u^2(z^2-a^2)w \ . \tag{4.5}$$

Here a is a parameter that depends on α, and vanishes at the critical value. Solutions of (4.5) are the parabolic cylinder functions $U\{-\frac{1}{2}ua^2, \pm(2u)^{\frac{1}{2}}z\}$, consequently the analysis achieves a uniform reduction from three free variables to two. The investigation in [44] is restricted to real values of the variables; a significant open problem is to extend the results to complex variables.

Similar problems arise when a turning point and pole coalesce. This occurs, for example, in the case of the associated Legendre equation, confluent hypergeometric equation, and hypergeometric equation; see §§6.1, 6.2, and 7.2 below. If the pole is double, then the natural choice of approximating differential equation is given by

$$\frac{d^2w}{dz^2} = \left(u^2\frac{a-z}{z^2} + \frac{\rho}{z^2}\right)w \ , \tag{4.6}$$

where a is a parameter that depends on α and vanishes at the critical value, and ρ is a constant. Solutions of (4.6) are the two-variable functions $z^{\frac{1}{2}}C_\nu(2uz^{\frac{1}{2}})$, where

$$\nu^2 = 1 + 4\rho + 4au^2 \ ,$$

and C_ν denotes any cylinder function of order ν. Alternatively, if the pole that coalesces with the turning point is simple, then the appropriate approximating differential equation would appear to be

$$\frac{d^2w}{dz^2} = \left(u^2\frac{z-a}{z} + \frac{\rho}{z^2}\right)w \ , \tag{4.7}$$

where a and ρ have the same significance as in (4.6). Solutions of (4.7) are the Whittaker functions $M_{au/2,\mu}(2uz)$ and $W_{au/2,\mu}(2uz)$, where

$$\mu^2 = \rho + \frac{1}{4} .$$

Again, with the given conditions they are functions of two variables. As far as the present writer is aware, neither of these problems has been attempted. For real variables they may be solvable by methods similar to those of [44].

As the number of free variables in the original differential equation advances from three to four and higher, the number of possible problems involving coalescing turning points and singularities proliferates. None has been solved, or is likely to be solved in the near future.

4.5. Inhomogeneous equations. For each problem mentioned in the preceding subsections there are corresponding problems for inhomogeneous differential equations, the solution of which often requires considerably more effort than straightforward application of the method of variation of parameters. Naturally the nature of the inhomogeneous term affects the problem. In the case of irregular singularities, mentioned in §4.2, and also in the simplest case involving a large parameter, mentioned in the second paragraph of §4.3, some fairly general results are included in [42] for the types of problem that arise with the special functions. The case of a simple turning point is also treated briefly in [42], and more completely in [9]. Inhomogeneous equations with double turning points are the subject of [29], but the asymptotic expansions obtained in this reference do not achieve a uniform reduction of free variables.

4.6. Differential equations of higher order. Another direction for generalization is to differential equations of order higher than two, or matrix systems of first-order differential equations. There is an extensive body of literature on this topic; see Wasow's book [57], and also [26], [19], [30], and [36]. The present writer is not aware of any significant applications to the special functions so far, however.

Other open problems associated with higher-order equations, or systems, include the development of error analyses. Stenger [50], [51] has given bounds for the error terms in the asymptotic expansions of solutions in the neighborhood of an irregular singularity of finite rank. The next problem ripe for error analysis would appear to be the generalization of the problem mentioned in the second paragraph of §4.3, that is, the case of a large parameter with elementary functions used as approximants.

PART II. ASYMPTOTIC ESTIMATES OF THE SPECIAL FUNCTIONS

5. FUNCTIONS OF ONE OR TWO VARIABLES

5.1. Functions of one variable. Included in this category are the Gamma function, exponential integral, sine and cosine integrals, error function, Fresnel integrals, Airy functions, and related functions. Each has an essential singularity, which is located at infinity. In all cases asymptotic expansions are available, these expansions being in series of descending powers of the argument and involving only elementary functions. Furthermore, in almost all cases satisfactory error bounds have been found, both for real and complex arguments.

5.2. Incomplete Gamma function. The asymptotic behavior of the incomplete Gamma function $\Gamma(\alpha, x)$ for large values of the variable x and fixed (or bounded) values of the parameter α is well known, as is the behavior for large α and fixed (or bounded) x. Asymptotic expansions for large values of the parameter that are uniform with respect to unbounded values of the variable (or vice versa) can be found as follows.

Suppose first we are considering $\Gamma(1-\alpha, x)$ for positive values of α and x. Then we have

$$x^{\alpha-1} e^{x} \Gamma(1-\alpha, x) = \int_0^\infty e^{-\alpha\{yt+\ell n(1+t)\}} dt \quad ,$$

where $y = x/\alpha$. Since $y > 0$ the function $yt + \ell n(1+t)$ attains its minimum in the interval of integration at the left endpoint $t = 0$. Accordingly, Laplace's method is applicable, and yields an expansion in descending powers of α. This expansion begins

$$x^{\alpha-1} e^{x} \Gamma(1-\alpha, x) \sim \frac{1}{1+y} \frac{1}{\alpha} + \frac{1}{(1+y)^3} \frac{1}{\alpha^2} + \ldots, \quad \alpha \to \infty ,$$

and is uniformly valid for $y \in [0, \infty)$, or, correspondingly, $x \in [0, \infty)$. The result is not new; it is in fact Blanch's uniform expansion for the generalized exponential integral $E_n(x)$ for large n; see [34], p. 61. This integral is related to the incomplete Gamma function by the formula

$$E_n(x) = x^{n-1} \Gamma(1-n, x) .$$

Next, consider $\Gamma(1+\alpha, x)$, where α and x again are positive. In this case

$$x^{-1-\alpha} e^{x} \Gamma(1+\alpha, x) = \int_0^\infty e^{-\alpha \{yt - \ell n(1+t)\}} dt ,$$

where, as before, $y = x/\alpha$. The function $yt - \ell n(1+t)$ has a minimum at $t = (1/y) - 1$. This lies inside the interval of integration when $0 < y < 1$, coincides with the left endpoint when $y = 1$, and is outside the interval of integration when $y > 1$. Since y is a free variable we have exactly the situation described in §2.2 in which a stationary point coalesces with an endpoint. In consequence it is possible to construct uniform asymptotic expansions in terms of the error function. These expansions have been supplied recently by Wong [64], §4 for the interval $y \in [\delta, 1]$, or equivalently $x \in [\delta\alpha, \alpha]$, where δ is a positive constant, and by Temme [53] for the more extensive range $y \in [0, \infty)$, or equivalently $x \in [0, \infty)$. Extensions to complex variables and the construction of error bounds are still needed.

5.3. Bessel functions and associated functions. Among all functions of two variables those having the most thoroughly studied asymptotic behavior are undoubtedly the Bessel functions. All of the methods sketched in §§2 and 4 that can be applied to these functions have been applied; see, for example, [42], Chapters 4 and 7 to 11.

For large argument, asymptotic expansions in terms of elementary functions are available, complete with error bounds. These expansions are uniformly valid with respect to bounded values of the order. For large order, asymptotic expansions are available in terms of elementary functions or in terms of Airy functions. These expansions are uniformly valid in restricted, though unbounded, regions of the plane of the argument; the regions overlap and their union covers the whole complex plane. The only essential problem that remains is the evaluation of bounds for the error terms in the uniform expansions for large order in the case when the order is complex.

A similar, though less complete, situation exists for the Struve functions, Anger functions, and functions of H. F. Weber, each of which satisfies an inhomogeneous form of Bessel's equation. Asymptotic expansions for large argument are available in terms of elementary functions, and uniform expansions for large order are available in terms of elementary functions or transcendental functions of a single variable; see [60], Chapter 10 and [42], Chapters 7 and 9. In most cases, however, error bounds have been calculated only for real variables.

5.4. Parabolic cylinder functions. These functions are entire in the argument and entire in the parameter. The situation is much the same as with the Bessel functions. Asymptotic expansions for large argument and fixed (or bounded) values of the parameter are available in terms of elementary functions. Asymptotic expansions for large values of the parameter are available in terms of elementary functions or Airy functions, these expansions being uniformly valid in overlapping regions whose union covers the entire complex plane of the argument. These results are given in [14], [33], [38], and [25]. Some error bounds have been calculated for the approximations in terms of elementary functions when the variables are real; see [42], pp. 206-208. These need extending to complex variables, and similar bounds also need to be calculated for the approximations in terms of Airy functions.

An important specialization is to the Hermite polynomials, via the formula

$$H_n(x) = 2^{n/2} \exp(\tfrac{1}{2}x^2)\, U(-n-\tfrac{1}{2},\ x\sqrt{2}\,) .$$

In consequence, uniform asymptotic expansions of $H_n(x)$ for large n are available in terms of Airy functions, complete with error bounds when x is real; see [42], pp. 403 and 412.

6. FUNCTIONS OF THREE VARIABLES

6.1. Legendre functions. The singularities of the associated Legendre equation

$$(1-z^2)\frac{d^2L}{dz^2} - 2z\frac{dL}{dz} + \left\{n(n+1) - \frac{m^2}{1-z^2}\right\}L = 0$$

are located at $z = \pm 1$ and ∞. Since each is regular, no problems of an asymptotic nature arise when z lies in their neighborhoods.

To discuss cases in which the parameters m and n are large we begin by removing the term in the first derivative by the substitution $w = (z^2-1)^{\frac{1}{2}} L$; thus

$$\frac{d^2w}{dz^2} = \left\{\frac{n(n+1)}{z^2-1} + \frac{m^2-1}{(z^2-1)^2}\right\} w . \tag{6.1}$$

For simplicity in the main discussion we shall assume that the symbols m and n denote real numbers such that $m \geq 0$ and $n \geq -\frac{1}{2}$; neither need be an integer, however. The variable z may be real or complex.

When n is large and m is fixed (or bounded) the coefficient of the large parameter in (6.1) has simple poles at $z = \pm 1$ and no zeros. Hence from §4.3 uniform asymptotic expansions can be constructed in terms of Bessel functions of fixed order. These expansions are given in

[42], Chapter 12, complete with error bounds; the order of the Bessel functions is, in fact, m, and the expansions descend in powers of $n+\frac{1}{2}$. Each expansion has a domain of validity that includes one or other of the singularities $z = \pm 1$, and extends to $z = \infty$.

In the complementary situation in which m is large and n is fixed (or bounded) the coefficient of the large parameter in (6.1) has double poles at $z = \pm 1$, and again no zeros. Hence from §4.3 asymptotic expansions can be constructed in terms of elementary functions that are uniform in a domain containing one or other of the singularities $z = \pm 1$. These expansions are none other than the well-known representations in hypergeometric series, typified by

$$P_n^{-m}(z) = \frac{(z-1)^{m/2}}{\Gamma(m+1)(z+1)^{m/2}} \sum_{s=0}^{\infty} \frac{(n+1)_s(-n)_s}{(m+1)_s \; s!}\left(\frac{1-z}{2}\right)^s ,$$

where, for example, $(n+1)_s$ is Pochhammer's symbol for $(n+1)(n+2)\ldots(n+s)$. This series converges when $|z-1| < 2$, and as $m \to \infty$ it furnishes an asymptotic expansion that is uniformly valid in any compact z-domain lying within the sector $|\mathrm{ph}(z+1)| < \pi$.

Asymptotic expansions for large m that are uniform in <u>unbounded</u> z-domains can be found as follows. We introduce a new variable ζ related to z by

$$\frac{\dot{z}^2}{(z^2-1)^2} = \frac{1}{\zeta} ,$$

where the dot signifies differentiation with respect to ζ. Integrating and making $\zeta = 0$ correspond to $z = \infty$, we obtain

$$z = \coth(2\sqrt{\zeta}) .$$

The Liouville transformation is completed by introducing the new dependent variable

$$W = \dot{z}^{-\frac{1}{2}} w \quad .$$

With these substitutions equation (6.1) becomes

$$\frac{d^2W}{d\zeta^2} = \left[\frac{m^2}{\zeta} + \frac{(n+\frac{1}{2})^2 - 1}{4\zeta^2} + \frac{n(n+1)}{\zeta}\left\{\operatorname{cosech}^2(2\sqrt{\zeta}) - \frac{1}{4\zeta}\right\}\right] W \; .$$

In this form the differential equation is ready for direct application of the approximation theorems given in [42], Chapter 12. Identifying the solutions in the usual manner, we arrive at

$$(6.2)\quad \begin{cases} P_n^{-m}(z) = \left(\dfrac{m}{\pi}\right)^{\frac{1}{2}} \dfrac{2\zeta^{\frac{1}{4}}}{\Gamma(m+1)} K_{n+\frac{1}{2}}(2m\zeta^{\frac{1}{2}})\{1+\varepsilon_1(\zeta)\} \; , \\[2ex] Q_n^m(z) = \pi^{\frac{1}{2}} e^{m\pi i} \dfrac{\Gamma(m+n+1)}{m^{n+\frac{1}{2}}} \zeta^{\frac{1}{4}} I_{n+\frac{1}{2}}(2m\zeta^{\frac{1}{2}})\{1+\varepsilon_2(\zeta)\} \; , \end{cases}$$

where the error terms have the properties

$$\varepsilon_1(\zeta) \to 0, \quad \zeta \to \infty; \; \varepsilon_2(\zeta) \to 0, \quad \zeta \to 0 \; .$$

Furthermore, as $m \to \infty$

$$(6.3)\qquad \varepsilon_1(\zeta) = O(m^{-1}), \quad \varepsilon_2(\zeta) = O(m^{-1}) \; ,$$

uniformly in unbounded domains containing $\zeta = 0$, or, correspondingly, in unbounded z-domains containing $z = 1$. Relations (6.3) are the desired asymptotic properties, and the results (6.2) appear to be new. Further details need working out, such as the maximal domains of validity, explicit expressions for higher terms, and error bounds.

The next problem to be considered is the construction of asymptotic approximations for large n that are uniform with respect to unbounded values of m, as well as being uniform with respect to z. To this end we set

$$u = n + \frac{1}{2}, \quad \alpha = \frac{m}{u} \quad .$$

Then equation (6.1) becomes

$$\frac{d^2w}{dz^2} = \left\{ u^2 \frac{z^2-1+\alpha^2}{(z^2-1)^2} - \frac{z^2+3}{4(z^2-1)^2} \right\} w ,$$

in which u is a large positive parameter and α is a nonnegative real parameter. In this form the characterizing features are the double poles at $z = \pm 1$ and turning points at $z = \pm(1-\alpha^2)^{\frac{1}{2}}$. Four cases arise, which we discuss in turn.

(i) If the turning points are bounded away from each other, and also bounded away from the singularities at $z = \pm 1$ and ∞, that is, if

$$\delta \leq \alpha \leq 1-\delta \quad \text{or} \quad 1+\delta \leq \alpha \leq 1+\Delta ,$$

where δ and Δ are positive constants such that $\delta < \frac{1}{2}$ and $\delta < \Delta$, then asymptotic expansions that are uniform with respect to α, as well as z, can be constructed in terms of Airy functions (§4. 3). Moreover, the regions of validity of these expansions will extend to the singularities at ± 1 and ∞. For the case $\delta \leq \alpha \leq 1-\delta$ these expansions have been investigated by Thorne [54].

(ii) When $1-\delta \leq \alpha \leq 1+\delta$, we have the problem of coalescing turning points. Accordingly, asymptotic approximations that are uniform with respect to α and z can be found in terms of parabolic cylinder functions (§4.4). For real z these approximations have been given recently by the present writer, complete with error bounds [45].

(iii) When $0 \leq \alpha \leq \delta$, each turning point may coalesce with one of the double poles at $z = \pm 1$. As we noted in §4.4, a general theory for this case is not yet available. When the theory has been supplied, we should be able to construct uniform asymptotic approximations in terms of Bessel functions of variable order. Some of the results of Thorne given in [54], §5 may prove to be pertinent to this problem, even though Thorne considered only fixed values of α in his analysis.

(iv) The remaining interval to be covered is $\alpha \in [1+\Delta, \infty)$. In this case both turning points may coalesce with the singularity at infinity.

To treat this problem, we interchange the roles of m and $n + \frac{1}{2}$, regarding m now as the large parameter, and $n + \frac{1}{2}$ as a free variable such that

$$0 \leq n + \frac{1}{2} \leq (1+\Delta)^{-1} m \ .$$

Writing

$$\beta = \frac{1}{\alpha} = \frac{n+\frac{1}{2}}{m} \ , \tag{6.4}$$

and applying the Liouville transformation

$$\xi = \frac{1}{z^2} \ , \quad \hat{W} = \left(\frac{dz}{d\xi}\right)^{-\frac{1}{2}} w \ ,$$

to (6.1), we obtain

$$\frac{d^2\hat{W}}{d\xi^2} = \left\{ m^2 \frac{(1-\beta^2)\xi+\beta^2}{4\xi^2(1-\xi)^2} - \frac{3\xi^2-3\xi+4}{16\xi^2(1-\xi)^2} \right\} \hat{W} \ .$$

This equation has a single turning point, located at $\xi = -\beta^2/(1-\beta^2)$. As $\beta \to 0$ this turning point coalesces with the double pole at $\xi = 0$. Hence from §4.4, again, we expect to be able to construct approximations in terms of Bessel functions of variable order† that are uniform with respect to $\beta \in [0, (1+\Delta)^{-1}]$ and ξ lying in a domain that includes $\xi = 0$. This domain may possibly extend to $\xi = 1$ and $\xi = \infty$.

Once the problems (i) to (iv) have all been solved we shall have a complete picture of the behavior of the Legendre functions not only for large n, but also for large m. This can be seen as follows. Let us again define β by (6.4), so that $\beta \in [0,\infty)$. The case $\beta \in [0, (1+\Delta)^{-1}]$ has been treated directly in (iv). Alternatively, when $\beta \geq (1+\Delta)^{-1}$ it

† The order is in fact $n + \frac{1}{2}$.

follows that $n+\frac{1}{2}\to\infty$ as $m\to\infty$, consequently the results of (i), (ii), and (iii) supply the required information.

Each of the problems discussed in this subsection is capable of generalization to complex values of m and n. The effort required to solve such problems completely would be very substantial, however, and would seem to be justifiable at present only in cases for which physical applications are known. This includes purely imaginary values of m and $n+\frac{1}{2}$, for which some results have already been obtained; see [8], [42], pp. 473-474, and [45].

6.2. Confluent hypergeometric functions. Whittaker's form of the confluent hypergeometric equation is given by

$$\frac{d^2w}{dz^2}=\left(\frac{1}{4}-\frac{k}{z}+\frac{m^2-\frac{1}{4}}{z^2}\right)w\,. \tag{6.5}$$

This differential equation has a regular singularity at $z=0$, and an irregular singularity at infinity. Asymptotic expansions of the solutions for large $|z|$ and fixed (or bounded) values of the parameters k and m are well known, and error bounds for real or complex values of k, m, and z are supplied in [39].

When $|k|$ is large and m is fixed (or bounded) we replace z by $4k\xi$; thus

$$\frac{d^2w}{d\xi^2}=\left(4k^2\,\frac{\xi-1}{\xi}+\frac{m^2-\frac{1}{4}}{\xi^2}\right)w\,.$$

The coefficient of k^2 has a zero at $\xi=1$ and a simple pole at $\xi=0$. In consequence (§4.3) asymptotic expansions can be found in terms of Airy functions that are uniform in an unbounded domain containing $\xi=1$ but not $\xi=0$, or in terms of Bessel functions of order $2m$ that are uniform in an unbounded domain containing $\xi=0$ but not $\xi=1$. For real or complex values of k, m, and z these expansions have been supplied by Erdélyi and Swanson [15] and, more generally, by Skovgaard [48]. Error bounds, mainly for real variables, are given in [42], pp. 412 and 446-447.

We turn now to asymptotic approximations for large m that are uniform with respect to k, as well as z. For simplicity in the main discussion we assume that k is real, and m and z are real and positive. We write

$$\alpha = k/m, \quad \zeta = z/m \tag{6.6}$$

so that $\alpha \in (-\infty, \infty)$ and $\zeta \in (0, \infty)$. Equation (6.5) becomes

$$\frac{d^2 w}{d\zeta^2} = \left(m^2 \frac{\zeta^2 - 4\alpha\zeta + 4}{4\zeta^2} - \frac{1}{4\zeta^2} \right) w .$$

In this form the characterizing features are the singularities at $\zeta = 0$ and ∞ and the turning points at

$$\zeta = 2\alpha \pm 2(\alpha^2 - 1)^{\frac{1}{2}} .$$

Four cases arise, which we discuss in turn.

(i) Assume that $-1-\Delta \leq \alpha \leq 1-\delta$, where δ and Δ are positive constants such that $\delta < 1$ and $\delta < \Delta$. Both turning points are then bounded away from the positive real axis. Consequently asymptotic approximations can be found in terms of elementary functions that are uniform with respect to $\alpha \in [-1-\Delta, 1-\delta]$ as well as $\zeta \in (0, \infty)$. These approximations are given in [42], pp. 260-261.† It should be observed that the problem in which m is large and k is fixed (or bounded) is included as a special case of these results.

† This reference discusses explicitly only values of α in the interval $[0, 1-\delta]$, but the results do in fact hold in the larger interval $[-1-\Delta, 1-\delta]$.

(ii) When $1+\delta \leq \alpha \leq 1+\Delta$ both turning points lie in the interval $(0,\infty)$ and are bounded away from each other and also away from the singularities at $\zeta = 0$ and ∞. Consequently two sets of asymptotic expansions can be constructed in terms of Airy functions. The first set is valid in an interval that includes $\zeta = 2\alpha-2(\alpha^2-1)^{\frac{1}{2}}$ and extends to $\zeta = 0$; the second set is uniformly valid in an interval that includes $\zeta = 2\alpha+2(\alpha^2-1)^{\frac{1}{2}}$ and extends to infinity. The intervals overlap and each expansion is uniformly valid with respect to α as well as ζ. At present, these expansions do not appear to be in the literature, but some related results are given in [42], pp. 401-403.

(iii) When $1-\delta \leq \alpha \leq 1+\delta$ we have the problem of coalescing turning points. The theory of [44] is relevant, and has been applied by the present writer to derive uniform approximations in terms of parabolic cylinder functions similar to those mentioned in §6.1(ii) for Legendre functions. This work is unpublished.†

(iv) Lastly, when $\alpha \in (-\infty, -1-\Delta]$ or $\alpha \in [1+\Delta, \infty)$, one turning point may coincide with the irregular singularity at infinity whilst the other simultaneously coincides with the regular singularity at the origin. To treat these cases we regard $|k|$ as the large parameter, set

$$z = 4k\xi, \quad m = k\beta \quad ,$$

and restrict

$$-(1+\Delta)^{-1} \leq \beta \leq (1+\Delta)^{-1} \quad .$$

Equation (6.5) becomes

† Kazarinoff [24] has given asymptotic approximations in terms of parabolic cylinder functions for large m that are uniform with respect to bounded values of $(m^2-k^2)/k$. These approximations are less powerful than those derived by the present writer for the following reason. If $(m^2-k^2)/k$ is bounded, then $(1-\alpha^2)/\alpha = O(1/m)$, implying that the α-interval of uniform validity is not fixed; indeed it shrinks on $\alpha = 1$ as $m \to \infty$. We may also note that some formal results pertinent to the problem have been given by Jorna [23].

$$\frac{d^2w}{d\xi^2} = \left(k^2 \frac{4\xi^2-4\xi+\beta^2}{\xi^2} - \frac{1}{4\xi^2}\right) w \; .$$

This has turning points at

$$\xi = \frac{1}{2} \pm \frac{1}{2}(1-\beta^2)^{\frac{1}{2}} \; .$$

Since

$$\beta^2 \leq (1+\Delta)^{-2} < 1 \; ,$$

the turning points are bounded away from each other, but the smaller one may coalesce with the double pole at $\xi = 0$. Therefore from §4.4 we expect to be able to construct asymptotic approximations in terms of Bessel functions of variable order† that are uniform with respect to β, and also with respect to ξ lying in a compact interval that includes the origin, but not the point $\frac{1}{2} + \frac{1}{2}(1-\beta^2)^{\frac{1}{2}}$. ‡

When the problems (i), (ii), (iii), and (iv) have been solved, we shall also have a complete picture of the asymptotic behavior of the Whittaker functions for large positive or negative values of k; compare the penultimate paragraph of §6.1. Again, as in the case of the Legendre functions, each of the problems generalizes to complex values of k, m, and z, but first efforts in this direction should be aimed at the physically important combinations of purely imaginary values of k, m, and z, for which Whittaker's equation assumes a real form; compare [42], pp. 401-403 and 433.

It is instructive to look briefly at the approach to Cases (i), (ii), and (iii) via integral representations. We have

† The order is in fact $2m$.

‡ If we were considering complex values of ξ, then the domain of validity may well extend to $\xi = \infty$, but the second turning point would still be excluded.

$$\Gamma(m-k+\tfrac{1}{2})\, e^{z/2}\, z^{-m-\frac{1}{2}}\, W_{k,m}(z) = \int_0^\infty e^{-zt}\, t^{m-k-\frac{1}{2}}(1+t)^{m+k-\frac{1}{2}} dt \ .$$

With the substitutions (6.6), the integral on the right-hand side becomes

$$\int_0^\infty e^{-m\zeta t}\, t^{m(1-\alpha)-\frac{1}{2}}(1+t)^{m(1+\alpha)-\frac{1}{2}}\, dt \ .$$

Since m is the large parameter, the stationary points are given by the equation

$$-\zeta + \frac{1-\alpha}{t} + \frac{1+\alpha}{1+t} = 0 \ ,$$

the roots of which are

$$t = \frac{2-\zeta \pm (\zeta^2 - 4\alpha\zeta + 4)^{\frac{1}{2}}}{2\zeta} \ .$$

It is easily verified that in Case (i) one stationary point lies within the interval of integration and the other stationary point lies outside; compare §2.1. In Case (ii) either both stationary points lie within the interval $(0, \infty)$ or both lie outside this interval, furthermore they coalesce at an interior point of the interval when

$$\zeta = 2\alpha - 2(\alpha^2 - 1)^{\frac{1}{2}} \ ;$$

compare §2.2. In Case (iii) the stationary points coalesce with each other and also with the endpoint $t = 0$ when $\alpha = 1$ and $\zeta = 2$; as we saw in §2.3 this type of problem is unsolved.

The results for Whittaker's equation (6.5) can be reformulated for the solutions of the standard confluent hypergeometric equation

$$z \frac{d^2 W}{dz^2} + (b-z)\frac{dW}{dz} - aW = 0 \ , \tag{6.7}$$

by means of the substitutions

$$k = \frac{1}{2}b - a, \quad m = \frac{1}{2}b - \frac{1}{2}, \quad w = e^{-z/2} z^{b/2} W .$$

Many kinds of asymptotic expansions for the solutions of (6.7) have been given explicitly by Slater [49], pp. 65-88 and Luke [27], pp. 129-134. On the whole, however, because of their more restricted regions of validity these expansions are less powerful than the transformed versions of the expansions and approximations described above.

Another important renormalization is to the Laguerre polynomials $L_n^{(\sigma)}(z)$, by means of the substitutions $k = n + \frac{1}{2}\sigma + \frac{1}{2}$, $m = \frac{1}{2}\sigma$.

6.3. Unnamed special functions. A somewhat different type of problem is posed by the differential equation

$$\frac{d^2 w}{dz^2} = z(z^2+az+b)w \quad , \tag{6.8}$$

in which a and b are parameters. This is the simplest type of second-order differential equation having three free turning points, just as Airy's equation is the simplest equation with one turning point, and Weber's equation is the simplest equation having two turning points. The solutions of (6.8) do not appear to be expressible in closed form in terms of named special functions. In order to attack the problem of constructing asymptotic solutions of a general second-order differential equation having three turning points that may coalesce in any manner, a thorough study of equation (6.8) would be needed. This includes the definition of standard solutions that are numerically satisfactory in the sense of Miller [32], the determination of connection formulas, and the determination of the uniform asymptotic behavior of the solutions for large a, b, or z .

In a similar way, a study of the solutions of the equation

$$\frac{d^2 w}{dz^2} = \frac{z^2+az+b}{z} w$$

appears to be a necessary preliminary to an attack on the general problem of second-order differential equations having two turning points which may coalesce with a simple pole. On the other hand, the prototype equation for the case of two simple turning points coalescing with a double pole, given by

$$\frac{d^2w}{dz^2} = \frac{z^2+az+b}{z^2} w ,$$

is solvable in terms of Whittaker functions.

Analogous problems also arise in connection with methods for approximating integrals. For example, we know very little about the functions (2.9) that are used as uniform approximants in Ursell's method for integrals with three coalescing saddle points.

7. FUNCTIONS OF FOUR OR MORE VARIABLES

7.1. General remarks. The dividing line between functions of three variables and functions of four variables can be regarded as a watershed in our present stage of knowledge of asymptotic analysis. Although a great deal of work remains to be completed, most of the necessary tools are available to describe completely the asymptotic behavior, as the variables change in any manner, of several special functions of three variables, including the Legendre functions (§6.1) and the Whittaker functions (§6.2). The remaining tools that are needed probably can be constructed by methods similar to those used to develop existing tools; compare §2.3 and §4.4. The prospect of achieving a complete description of the asymptotic behavior of any special function of four variables, however, appears quite remote at present. Available methods will not suffice and fundamentally new work is needed.

7.2. Hypergeometric function. Consider, for example, one of the simplest special functions of four variables, the hypergeometric function. Using the differential-equation approach, we start with the hypergeometric equation

$$z(1-z)\frac{d^2w}{dz^2}+\{c-(a+b+1)z\}\frac{dw}{dz}-abw=0 \ .$$

Removing the term in the first derivative by setting

$$W=z^{c/2}(1-z)^{(a+b-c+1)/2}\,w \ ,$$

we arrive at

$$\frac{d^2W}{dz^2}=\frac{\{(a-b)^2-1\}z^2+\{4ab+2c(1-a-b)\}z+c(c-2)}{4z^2(1-z)^2}\,W \ . \tag{7.1}$$

This equation has regular singularities at $z = 0, 1$, and ∞, and turning points at the zeros of the numerator on the right-hand side.

If, as the parameters a, b, and c vary, the turning points are bounded away from the singularities and also away from each other, then uniform asymptotic expansions can be found in terms of elementary functions or Airy functions. Moreover, the regions of validity of these expansions will overlap and cover the whole z-plane, including the singularities. Secondly, if the turning points coalesce, then the methods described in §4.4 will yield asymptotic approximations in terms of parabolic cylinder functions that are uniform in intervals or domains containing the singularities. Thirdly, if one of the turning points coalesces with a singularity, then, as we also noted in §4.4, uniform asymptotic approximations probably can be constructed in terms of Bessel functions of variable order. Lastly, however, it may happen that both turning points coalesce with one of the singularities. In this event, as we suggested in §6.3, it is probably necessary to employ functions of three variables (in fact, the Whittaker functions) in order to obtain uniform approximations, but we are unable to cope with this complicated problem with our present knowledge.

Results that have been found so far for the hypergeometric function permit at most three degrees of freedom in the four variables. For example, Watson [59], followed by Luke [27], Chapter 7, used the integral representation

$$F(a,b;c;z) = \frac{\Gamma(c)}{\Gamma(b)\Gamma(c-b)} \int_0^1 t^{b-1}(1-t)^{c-b-1}(1-zt)^{-a}\,dt$$

and Laplace's method to obtain asymptotic expansions of $F(a+\theta_1 u,\ b+\theta_2 u;\ c+\theta_3 u;\ z)$ as $u \to +\infty$, where $\theta_1, \theta_2, \theta_3 = 0$ or ± 1. These expansions are in terms of elementary functions and are uniform with respect to z in specified complex domains.

Another example permitting two degrees of freedom is provided by Elliott's expansions for Jacobi polynomials of large degree [11]. These polynomials are related to the hypergeometric function by

$$P_n^{(\alpha,\beta)}(x) = \binom{n+\alpha}{n} F(a,b;c;z) \quad ,$$

where

$$a = -n, \quad b = \alpha+\beta+n+1, \quad c = \alpha+1, \quad z = \frac{1}{2} - \frac{1}{2}x \ .$$

In Elliott's analysis n is large, and α and β are fixed (or bounded). Thus on the right-hand side of (7.1) the only terms in the numerator in which the large parameter appears are those in z^2 and z. In consequence, the singularity at $z = 0$ of the term in the differential equation involving the large parameter reduces to a simple pole. Similarly, this term has a simple pole at $z = 1$. Hence from §4.3, we can construct expansions that are uniform with respect to z in unbounded domains containing one or other of the singularities by using Bessel functions of fixed order (in fact, α or β). These expansions have been found by Elliott. Similar results for the same problem are sketched by Fields in [16], the approach being via Fields' extension of Darboux's method (§3.2).

Examples of uniform asymptotic approximations for the hypergeometric function that permit three degrees of freedom among the four variables are supplied by the results mentioned in §6.1 for Legendre functions. For example, we have

$$P_n^{-m}(z) = \frac{(z-1)^{m/2}}{\Gamma(m+1)(z+1)^{m/2}} F(n+1, -n; m+1; \tfrac{1}{2} - \tfrac{1}{2}z) .$$

7.3. Generalizations of the hypergeometric function. The situation is similar for generalized hypergeometric functions, basic hypergeometric functions, and the G-function. Apart from the specializations to the confluent hypergeometric function and the ordinary hypergeometric function which we discussed in §6.2 and the present section, almost all existing asymptotic approximations and expansions either pertain to the case in which the variable z tends to infinity and the parameters are kept fixed or bounded (for example, [58], [63], [31], [6], [27], Chapter 5, and [17]), or permit some or all of the parameters to depend linearly on a single variable which tends to infinity ([62], [27], Chapter 7, [61], and [18]). The asymptotic expansions are in terms of elementary functions, and in most cases the derivations proceed via integral representations using Laplace's method or the residue calculus.

ACKNOWLEDGMENTS

This research was supported in part by the U. S. Army Research Office, Durham under Contract DA ARO D 31 124 73 G204, and the National Science Foundation under Grant GP 32841X. The writer is indebted to Dr. L. Maximon for suggestions for improving the presentation of the results.

REFERENCES

[1] Bateman Manuscript Project, Higher transcendental functions (A. Erdélyi, ed.), Vols. 1, 2, 3, McGraw-Hill, New York (1953, 1953, 1955).

[2] E. A. Bender, Asymptotic methods in enumeration, SIAM Rev., 16 (1974), 485-515.

[3] N. Bleistein, Uniform asymptotic expansions of integrals with stationary point near algebraic singularity, Comm. Pure Appl. Math., 19 (1966), 353-370.

[4] N. Bleistein, Uniform asymptotic expansions of integrals with many nearby stationary points and algebraic singularities, J. Math. Mech., 17 (1967), 533-559.

[5] N. Bleistein and R. A. Handelsman, Asymptotic expansions of integrals, Holt, Rinehart, and Winston, New York (1975).

[6] B. L. J. Braaksma, Asymptotic expansions and analytic continuations for a class of Barnes-integrals, Noordhoff, Groningen (1963).

[7] C. Chester, B. Friedman, and F. Ursell, An extension of the method of steepest descents, Proc. Cambridge Philos. Soc., 53 (1957), 599-611.

[8] N. K. Chuhrukidze, Asymptotic expansions for spherical Legendre functions of imaginary argument (Russian), Ž. Vyčisl. Mat. i Mat. Fiz., 8 (1968), 3-12. [English translation: U. S. S. R. Computational Math. and Math. Phys., 8 (1968), No. 1, 1-13.]

[9] R. A. Clark, Asymptotic solutions of a nonhomogeneous differential equation with a turning point, Arch. Rational Mech. Anal., 12 (1963), 34-51.

[10] R. B. Dingle, Asymptotic expansions: Their derivation and interpretation, Academic Press, New York (1973).

[11] D. Elliott, Uniform asymptotic expansions of the Jacobi polynomials and an associated function, Math. Comp., 25 (1971), 309-315.

[12] A. Erdélyi, Uniform asymptotic expansion of integrals, Analytic methods in mathematical physics (R. P. Gilbert and R. G. Newton, eds.), 149-168, Gordon and Breach, New York (1970).

[13] A. Erdélyi, Asymptotic evaluation of integrals involving a fractional derivative, SIAM J. Math. Anal., 5 (1974), 159-171.

[14] A. Erdélyi, M. Kennedy, and J. L. McGregor, Parabolic cylinder functions of large order, J. Rational Mech. Anal., 3 (1954), 459-485.

[15] A. Erdélyi and C. A. Swanson, Asymptotic forms of Whittaker's confluent hypergeometric functions, Mem. Amer. Math. Soc., No. 25 (1957).

[16] J. L. Fields, A uniform treatment of Darboux's method, Arch. Rational Mech. Anal., 27 (1968), 289-305.

[17] J. L. Fields, The asymptotic expansion of the Meijer G-function, Math. Comp., 26 (1972), 757-765.

[18] J. L. Fields, Uniform asymptotic expansions of certain classes of Meijer G-functions for a large parameter, SIAM J. Math. Anal., 4 (1973), 482-507.

[19] R. J. Hanson, Reduction theorems for systems of ordinary differential equations with a turning point, J. Math. Anal. Appl., 16 (1966), 280-301.

[20] E. L. Ince, Ordinary differential equations, Longmans, Green, London (1927). Reprinted by Dover, New York (1956).

[21] H. Jeffreys, The remainder in Watson's lemma, Proc. Roy. Soc. London Ser. A, 248 (1958), 88-92.

[22] D. S. Jones, Asymptotic behavior of integrals, SIAM Rev., 14 (1972), 286-317.

[23] S. Jorna, Derivation of Green-type, transitional, and uniform asymptotic expansions from differential equations III. The confluent hypergeometric function $U(a, c, z)$ for large $|c|$. Proc. Roy. Soc. London Ser. A, 284 (1965), 531-539.

[24] N. D. Kazarinoff, Asymptotic forms for the Whittaker functions with both parameters large, J. Math. Mech., 6 (1957), 341-360.

[25] M. K. Kerimov, Certain new results of the theory of Weber functions (Russian), Biblioteka Mat. Tablits, No. 45, Vyčisl. Tsentr SSSR, Moscow (1968), 88-116. [Translated into English by A. Werner and issued as Report No. TR 71-86 of the Bell Laboratories.]

[26] K. H. Kiyek, Zur Theorie der linearen Differentialgleichungssysteme mit einem grossen Parameter, Doctoral thesis, Univ. of Würzburg (1963).

[27] Y. L. Luke, The special functions and their approximations, Vol. 1, Academic Press, New York (1969).

[28] J. J. Mahony, Error estimates for asymptotic representations of integrals, J. Austral. Math. Soc., 13 (1972), 395-410.

[29] D. J. McGuinness, Nonhomogeneous differential equation with a second-order turning point, J. Mathematical Phys., 7 (1966), 1030-1037.

[30] J. A. M. McHugh, An historical survey of ordinary linear differential equations with a large parameter and turning points, Arch. History Exact Sci., 7 (1971), 277-324.

[31] C. S. Meijer, On the G-function. Parts I, VI, VII, and VIII, Nederl. Akad. Wetensch. Proc. Ser. A, 49 (1946), 227-237, 936-943, 1063-1072, and 1165-1175.

[32] J. C. P. Miller, On the choice of standard solutions for a homogenous linear differential equation of the second order, Quart. J. Mech. Appl. Math., 3 (1950), 225-235.

[33] J. C. P. Miller, Tables of Weber parabolic cylinder functions, H. M. Stationery Office, London (1955).

[34] National Bureau of Standards, Tables of functions and of zeros of functions, Appl. Math. Ser. No. 37, U. S. Govt. Printing Office, Washington, D. C. (1954).

[35] National Bureau of Standards, Handbook of mathematical functions, Appl. Math. Ser. No. 55 (M. Abramowitz and I. A. Stegun, eds.), U. S. Govt. Printing Office, Washington, D. C. (1964).

[36] A. H. Nayfeh, Perturbation methods, Wiley (Interscience), New York (1973).

[37] F. Oberhettinger, On a modification of Watson's lemma, J. Res. Nat. Bur. Standards Sect. B, 63 (1959), 15-17.

[38] F. W. J. Olver, Uniform asymptotic expansions for Weber parabolic cylinder functions of large orders, J. Res. Nat. Bur. Standards Sect. B, 63 (1959), 131-169.

[39] F. W. J. Olver, On the asymptotic solutions of second-order differential equations having an irregular singularity of rank one, with an application to Whittaker functions, SIAM J. Numer. Anal. Ser. B, 2 (1965), 225-243.

[40] F. W. J. Olver, Error bounds for the Laplace approximation for definite integrals, J. Approximation Theory, 1 (1968), 293-313.

[41] F. W. J. Olver, Why steepest descents?, Studies in applied mathematics, No. 6 (D. Ludwig and F. W. J. Olver, eds.), 44-63, Soc. Indust. and Appl. Math., Philadelphia (1970). Reprinted in SIAM Rev., 12 (1970), 228-247.

[42] F. W. J. Olver, Asymptotics and special functions, Academic Press, New York (1974).

[43] F. W. J. Olver, Error bounds for stationary phase approximations, SIAM J. Math. Anal., 5 (1974), 19-29.

[44] F. W. J. Olver, Second-order linear differential equations with two turning points, Philos. Trans. Roy. Soc. London Ser. A, 278 (1975), 137-174.

[45] F. W. J. Olver, Legendre functions with both parameters large, Philos. Trans. Roy. Soc. London Ser. A, 278 (1975), 175-185.

[46] F. W. J. Olver, Second-order differential equations with fractional transition points. [To be published.]

[47] E. Ya. Riekstinš, Asymptotic expansions of integrals, Vol. 1 (Russian), Zinātne, Riga (1974).

[48] H. Skovgaard, Uniform asymptotic expansions of confluent hypergeometric functions and Whittaker functions, Gjellerups Publ., Copenhagen (1966).

[49] L. J. Slater, Confluent hypergeometric functions, Cambridge Univ. Press, London and New York (1960).

[50] F. Stenger, Error bounds for asymptotic solutions of differential equations I. The distinct eigenvalue case, J. Res. Nat. Bur. Standards Sect. B, 70 (1966), 167-186.

[51] F. Stenger, Error bounds for asymptotic solutions of differential equations II. The general case, J. Res. Nat. Bur. Standards Sect. B, 70 (1966), 187-210.

[52] F. Stenger, The asymptotic approximation of certain integrals, SIAM J. Math. Anal., 1 (1970), 392-404.

[53] N. M. Temme, Uniform asymptotic expansions of the incomplete Gamma functions and the incomplete Beta function, Math. Comp. [In press.]

[54] R. C. Thorne, The asymptotic expansion of Legendre functions of large degree and order, Philos. Trans. Roy. Soc. London Ser. A, 249 (1957), 597-620.

[55] F. Ursell, Integrals with a large parameter. Several nearly coincident saddle-points, Proc. Cambridge Philos. Soc., 72 (1972), 49-65.

[56] B. L. van der Waerden, On the method of saddle points, Appl. Sci. Res. Ser. B, 2 (1951), 33-45.

[57] W. Wasow, Asymptotic expansions for ordinary differential equations, Wiley (Interscience), New York (1965).

[58] G. N. Watson, The continuations of functions defined by generalised hypergeometric series, Trans. Cambridge Philos. Soc., 21 (1910), 281-299.

[59] G. N. Watson, Asymptotic expansions of hypergeometric functions, Trans. Cambridge Philos. Soc., 22 (1918), 277-308.

[60] G. N. Watson, A treatise on the theory of Bessel functions, 2nd ed., Cambridge Univ. Press, London and New York (1944).

[61] M. W. Wilson, On the Hahn polynomials, SIAM J. Math. Anal., 1 (1970), 131-139.

[62] J. Wimp, The asymptotic representation of a class of G-functions for large parameter, Math. Comp., 21 (1967), 639-646.

[63] E. M. Wright, The asymptotic expansion of the generalized hypergeometric function, Proc. London Math. Soc., 46 (1940), 389-408.

[64] R. Wong, On uniform asymptotic expansion of definite integrals, J. Approximation Theory, 7 (1973), 76-86.

[65] R. Wong, Error bounds for asymptotic expansions of Hankel transforms. [To be published.]

[66] R. Wong and M. Wyman, The method of Darboux, J. Approximation Theory, 10 (1974), 159-171.

University of Maryland and
National Bureau of Standards

[137]

SECOND-ORDER LINEAR DIFFERENTIAL EQUATIONS WITH TWO TURNING POINTS

By F. W. J. OLVER
The University of Maryland and *The National Bureau of Standards*

(*Communicated by Sir Harold Jeffreys, F.R.S. – Received* 8 *May* 1974)

CONTENTS

Differential equations of the form

$$\mathrm{d}^2w/\mathrm{d}x^2 = \{u^2f(u, a, x) + g(u, a, x)\}w$$

are considered for large values of the real parameter u. Here x is a real variable ranging over an open, possibly infinite, interval (x_1, x_2), and a is a bounded real parameter. It is assumed that $f(u, a, x)$ and $g(u, a, x)$ are free from singularity within (x_1, x_2), and $f(u, a, x)$ has exactly two zeros, which depend continuously on a and coincide for a certain value of a. Except in the neighbourhoods of the zeros, $g(u, a, x)$ is small in absolute value compared with $u^2f(u, a, x)$.

By application of the Liouville transformation, the differential equation is converted into one of four standard forms, with continuous coefficients. Asymptotic approximations for the solutions are then constructed in terms of parabolic cylinder functions.

These approximations are valid for large u, uniformly with respect to $x \in (x_1, x_2)$ and also uniformly with respect to a. Each approximation is accompanied by a strict and realistic error bound.

The paper also includes some new properties of parabolic cylinder functions.

PART A. INTRODUCTION AND PRELIMINARY TRANSFORMATIONS

1. Introduction

This paper is concerned with differential equations of the form

$$\mathrm{d}^2w/\mathrm{d}x^2 = \{u^2 f(u, a, x) + g(u, a, x)\}w. \tag{1.1}$$

Here u is a large positive parameter, and a is a bounded real parameter (which may be a function of u). The independent variable x ranges over a given finite or infinite open interval (x_1, x_2), the endpoints x_1 and x_2 possibly being functions of u and a. For each value of u, the functions $f(u, a, x)$, $g(u, a, x)$, and certain derivatives of $f(u, a, x)$ are continuous in a and x.† The function $f(u, a, x)$ is non-vanishing in (x_1, x_2) save for distinct simple zeros at $x = z_1$ and $x = z_2$ with $z_1 < z_2$ or, alternatively, a single double zero at $x = z_0$. Zeros of $f(u, a, x)$ are the so-called turning points or transition points of the differential equation. Both z_1 and z_2 are continuous functions of the parameter a, and tend to z_0 as a tends to a certain critical value a_0, say. The function $g(u, a, x)$ is small in absolute value compared with $u^2 f(u, a, x)$, except in the neighbourhoods of the turning points where it is small compared with $u^2(x-z_1)^{-1}(x-z_2)^{-1}f(u, a, x)$. In particular, $g(u, a, x)$ may be identically zero.

We shall establish asymptotic approximations for solutions of (1.1) in terms of parabolic cylinder functions which are uniform for $x \in (x_1, x_2)$. Rigorous investigations of this problem have been undertaken previously by Langer (1934, 1959), McKelvey (1955) and Kazarinoff (1958). Several formal investigations are also pertinent, including those of Moriguchi (1959), Heading (1962) and Lynn & Keller (1970).

The present investigation differs from earlier ones in three major respects. First, the asymptotic approximations are uniform with respect to the parameter a in a closed interval which includes the critical value a_0.‡ Secondly, the endpoints x_1 and x_2 may be located at finite singularities of the differential equation, or at $\pm\infty$.§ Thirdly, explicit bounds are constructed for the error terms associated with the approximations.

The importance of the first of these extensions can be judged in the following way. The papers by Kazarinoff (1958) and Langer (1959) treat the case of distinct turning points z_1 and z_2, whereas those by Langer (1934) and McKelvey (1955) treat a double turning point z_0. If z_1 is allowed to approach z_2, then the analyses of Kazarinoff and Langer are inapplicable because their asymptotic approximations are non-uniform for arbitrarily small values of $|z_2 - z_1|$. Thus the theories in these four references essentially apply only to differential equations having *fixed* turning points. This severely limits the value of the results since for fixed a the solutions of (1.1) are functions of two variables (u and x) *as are the approximating parabolic cylinder functions.* This contrasts with the well-known case of a simple turning point, for which functions of two variables are

† This means, as usual, in a and x simultaneously and not separately.

‡ The feasibility of this extension is noted in the formal treatment of Lynn & Keller (1970, p. 393).

§ The conditions in the paper of Kazarinoff also admit unbounded domains for the independent variable. These results are discussed in §11.

approximated uniformly by functions of a single variable (Airy functions); see, for example, Olver (1974, chapter 11). In the present paper we approximate, in a uniform manner, functions of *three* variables (u, a and x) by functions of two variables (again, parabolic cylinder functions).†

The second extension in effect maximizes the x-intervals of validity. It also leads indirectly to increased accuracy *throughout* the interval (x_1, x_2), and not merely in the neighbourhoods of the singularities at x_1 and x_2. Another important consequence in applications is a simplification of the task of identifying standard solutions of the given differential equation in terms of the asymptotic solutions.

The third extension—availability of strict error bounds—is of obvious value from a computational standpoint. An important analytical advantage is a considerable flexibility in the way the parameter u is permitted to enter the differential equation (1.1). Thus we do not need to decide *a priori* that we are seeking asymptotic solutions of equations having the form, for example,

$$\mathrm{d}^2w/\mathrm{d}x^2 = \{u^2f(a,x) + g(a,x)\}\,w, \tag{1.2}$$

in which $f(a,x)$ and $g(a,x)$ are independent of u, or, as another example,

$$\mathrm{d}^2w/\mathrm{d}x^2 = \{u^2f(a,x) + uh(a,x) + g(a,x)\}\,w, \tag{1.3}$$

in which $h(a,x)$ also is independent of u. In all cases asymptotic properties of the approximations can be investigated *a posteriori* by examining the asymptotic form of the error bound.‡

The methods we shall use apply in principle to real or complex variables. We concentrate on the former in the present paper for the following reasons. First, many (if not most) physical applications are restricted to real variables. Secondly, continuity and differentiability requirements imposed on $f(u,a,x)$ and $g(u,a,x)$ are less restrictive than holomorphicity conditions needed with complex variables. Thirdly, error bounds for complex variables will not specialize into the sharper bounds available from real-variable theory. There is, however, one special extension that is of considerable importance in applications. In this case the variables are real, but the turning points are conjugate purely imaginary numbers that coalesce at the origin as $a \to a_0$. Provided that $f(u,a,x)$ is even in x and takes real values on the real and imaginary axes, this case is treatable by exactly similar methods, and full details are included.

In employing parabolic cylinder functions as approximants we shall adopt the standardizations of Miller (1955) in preference to the older, and commoner, standardization of Whittaker (1903) Miller's functions are more convenient when the variables are real, and the present investigation owes much to Miller's thorough groundwork in the choice of standard solutions and investigation of their main properties.

An application of the theory of the present paper to the associated Legendre equation, with both degree and order large, is given in the paper immediately following (Olver 1975). A similar application to Whittaker's equation, with both parameters large, will be published in due course. Potential physical applications include unsolved wave-scattering problems involving elliptic cylinders or prolate spheroids in regions of resonance (see, for example, Kazarinoff & Ritt (1960), pp. 183–185), and new uniform asymptotic approximations for the eigenfunctions of a symmetric rotator (see, for example, Brussaard & Tolhoek (1957), p. 961).

† Although the *results* of Kazarinoff, Langer and McKelvey are of limited usefulness in the view of the present writer, the *methods* they employ are valuable and are followed in parts of the present analysis.

‡ Similar considerations hold for differential equations having no turning point or one turning point; see Olver (1974, pp. 205 and 401).

2. Transformation to standard form

2.1. *The Liouville transformation*

The principal theorems in this paper apply to equation (1.1) with $f(u,a,x)$ taken to be a quadratic polynomial in a and x. Other equations can be converted into this form by means of the Liouville transformation (see, for example, Olver (1974), p. 363), as follows. New variables W and ζ are introduced, related by

$$W = \dot{x}^{-\frac{1}{2}} w,$$

the dot signifying differentiation with respect to ζ. Then

$$\frac{d^2 W}{d\zeta^2} = \left\{u^2 \dot{x}^2 f(u,a,x) + \dot{x}^2 g(u,a,x) + \dot{x}^{\frac{1}{2}} \frac{d^2}{d\zeta^2}(\dot{x}^{-\frac{1}{2}})\right\} W. \tag{2.1}$$

There are four cases to consider, which we designate I, II, III and IV. *In cases I and III the zeros* z_1 *and* z_2 *of* $f(u,a,x)$ *are real. In case I* $f(u,a,x)$ *is negative between the zeros* (and therefore positive in the exterior intervals $x_1 < x < z_1$ and $z_2 < x < x_2$). *In case III* $f(u,a,x)$ *is positive between the zeros. In cases II and IV the zeros* z_1 *and* z_2 *are purely imaginary and* $f(u,a,x)$ *is an even function of* x *that is real on the real and imaginary axes, its sign on the real axis and the join of* z_1 *and* z_2 *being positive in case II and negative in case IV.*

2.2. *Case I*

We prescribe

$$\dot{x}^2 f(u,a,x) = \zeta^2 - \alpha^2, \tag{2.2}$$

where α is a non-negative number, chosen to make z_1 and z_2 correspond to $\zeta = -\alpha$ and $\zeta = \alpha$, respectively. Equation (2.2) is a first-order differential equation for ζ, which yields on integration

$$\int_{z_1}^{x} \{-f(u,a,t)\}^{\frac{1}{2}} dt = \int_{-\alpha}^{\zeta} (\alpha^2 - \tau^2)^{\frac{1}{2}} d\tau, \tag{2.3}$$

provided that $z_1 \leqslant x \leqslant z_2$. With these integration limits z_1 obviously corresponds to $-\alpha$. The condition that z_2 and α correspond requires

$$\int_{z_1}^{z_2} \{-f(u,a,t)\}^{\frac{1}{2}} dt = \int_{-\alpha}^{\alpha} (\alpha^2 - \tau^2)^{\frac{1}{2}} d\tau. \tag{2.4}$$

Thus

$$\alpha^2 = \frac{2}{\pi} \int_{z_1}^{z_2} \{-f(u,a,t)\}^{\frac{1}{2}} dt. \tag{2.5}$$

For each value of u, this equation defines α as a continuous function of a which vanishes as $a \rightarrow a_0$.

Next, from (2.3) we derive

$$\int_{z_1}^{x} \{-f(u,a,t)\}^{\frac{1}{2}} dt = \tfrac{1}{2}\alpha^2 \arccos\left(-\frac{\zeta}{\alpha}\right) + \tfrac{1}{2}\zeta(\alpha^2 - \zeta^2)^{\frac{1}{2}} \quad (z_1 \leqslant x \leqslant z_2), \tag{2.6}$$

the inverse cosine having its principal value, that is, the value in $[0, \pi]$.† For the other x-intervals, integration of (2.2) yields

$$\int_{x}^{z_1} f^{\frac{1}{2}}(u,a,t)\, dt = \int_{\zeta}^{-\alpha} (\tau^2 - \alpha^2)^{\frac{1}{2}} d\tau = -\tfrac{1}{2}\alpha^2 \operatorname{arcosh}\left(-\frac{\zeta}{\alpha}\right) - \tfrac{1}{2}\zeta(\zeta^2 - \alpha^2)^{\frac{1}{2}} \quad (x_1 < x \leqslant z_1), \tag{2.7}$$

and

$$\int_{z_2}^{x} f^{\frac{1}{2}}(u,a,t)\, dt = \int_{\alpha}^{\zeta} (\tau^2 - \alpha^2)^{\frac{1}{2}} d\tau = -\tfrac{1}{2}\alpha^2 \operatorname{arcosh}\left(\frac{\zeta}{\alpha}\right) + \tfrac{1}{2}\zeta(\zeta^2 - \alpha^2)^{\frac{1}{2}} \quad (z_2 \leqslant x < x_2). \tag{2.8}$$

† As is customary, square brackets denote a closed interval.

Equations (2.6), (2.7) and (2.8) define ζ as a continuous increasing function of x in (x_1, x_2); accordingly, there is a one-to-one correspondence between these variables. On substituting in (2.1) by means of (2.2), we obtain the desired form of differential equation, given by

$$d^2W/d\zeta^2 = \{u^2(\zeta^2-\alpha^2)+\psi(u,\alpha,\zeta)\}\,W, \tag{2.9}$$

where

$$\psi(u,\alpha,\zeta) = \dot{x}^2 g(u,a,x)+\dot{x}^{\frac{1}{2}}\,d^2(\dot{x}^{-\frac{1}{2}})/d\zeta^2. \tag{2.10}$$

In the critical situation in which the zeros z_1 and z_2 coalesce into a double zero, we employ the limiting form of the foregoing transformations obtained by setting $z_1 = z_2 = z_0$. Thus

$$\int_x^{z_0} f^{\frac{1}{2}}(u,a_0,t)\,dt = -\int_\zeta^0 \tau\,d\tau = \tfrac{1}{2}\zeta^2 \quad (x_1 < x \leqslant z_0),$$

$$\int_{z_0}^{x} f^{\frac{1}{2}}(u,a_0,t)\,dt = \int_0^\zeta \tau\,d\tau = \tfrac{1}{2}\zeta^2 \quad (z_0 \leqslant x < x_2),$$

and equations (2.9) and (2.10) apply with $\alpha = 0$.

2.3. *Case II*

We prescribe

$$\dot{x}^2 f(u,a,x) = \alpha^2+\zeta^2,$$

where α is chosen to make z_1 and z_2 correspond to $\zeta = -i\alpha$ and $\zeta = i\alpha$, respectively. Corresponding to (2.4), we prescribe that

$$\int_{z_1}^{z_2} f^{\frac{1}{2}}(u,a,t)\,dt = \int_{-i\alpha}^{i\alpha}(\alpha^2+\tau^2)^{\frac{1}{2}}\,d\tau = \tfrac{1}{2}i\pi\alpha^2. \tag{2.11}$$

In the left-hand integral the path lies along the imaginary axis, and the branch of $f^{\frac{1}{2}}(u,a,t)$ is real and positive. Thus α is a continuous real function of a. To remove all ambiguity we again require α to be non-negative.

Since, by hypothesis, $f(u,a,-t) = f(u,a,t)$ both for real and purely imaginary values of t we have

$$\int_{z_1}^{0} f^{\frac{1}{2}}(u,a,t)\,dt = \int_{-i\alpha}^{0}(\alpha^2+\tau^2)^{\frac{1}{2}}\,d\tau = \tfrac{1}{4}i\pi\alpha^2.$$

Hence when x is real

$$\int_0^x f^{\frac{1}{2}}(u,a,t)\,dt = \int_0^\zeta(\alpha^2+\tau^2)^{\frac{1}{2}}\,d\tau$$

$$= \tfrac{1}{2}\alpha^2\ln\{\zeta+(\alpha^2+\zeta^2)^{\frac{1}{2}}\}-\tfrac{1}{2}\alpha^2\ln\alpha+\tfrac{1}{2}\zeta(\alpha^2+\zeta^2)^{\frac{1}{2}}. \tag{2.12}$$

This equation takes the place of (2.6), (2.7) and (2.8). Clearly ζ is real when $x \in (x_1, x_2)$; moreover, ζ increases continuously as x increases continuously. The transformed differential equation becomes

$$d^2W/d\zeta^2 = \{u^2(\alpha^2+\zeta^2)+\psi(u,\alpha,\zeta)\}\,W,$$

with $\psi(u,\alpha,\zeta)$ given by (2.10). In the critical case $a = a_0$, the foregoing equations apply with $\alpha = 0$; in particular,

$$\int_0^x f^{\frac{1}{2}}(u,a_0,t)\,dt = \pm\tfrac{1}{2}\zeta^2 \quad (x \gtrless 0).$$

2.4. *Case III*

This is similar to case I. In equations (2.2) to (2.8) f is replaced everywhere by $-f$. The new differential equation is

$$d^2W/d\zeta^2 = \{u^2(\alpha^2-\zeta^2)+\psi(u,\alpha,\zeta)\}\,W,$$

with $\psi(u,\alpha,\zeta)$ given by (2.10).

2.5. *Case IV*

This parallels case II. In equations (2.11) and (2.12) f is replaced by $-f$. The new differential equation is

$$d^2W/d\zeta^2 = \{-u^2(\alpha^2+\zeta^2)+\psi(u,\alpha,\zeta)\}\,W,$$

with $\psi(u,\alpha,\zeta)$ again given by (2.10).

3. Continuity of $\psi(u,\alpha,\zeta)$: cases I and III

3.1. *Statement of lemma*

In later sections it is assumed that for each value of u, the function $\psi(u,\alpha,\zeta)$ resulting from the Liouville transformations of § 2 is continuous in α and ζ. In this section and the next one we establish that this is the case when $f(u,a,x)$, $g(u,a,x)$, and certain derivatives of the former function satisfy suitable conditions concerning continuity and boundedness. The present section treats case I, and § 4 treats case II. The proofs for cases III and IV are almost identical to those for I and II, respectively.

By redefining the parameter a and making a linear change in x we may suppose, without loss of generality, that the zeros z_1 and z_2 of $f(u,a,x)$ are located at $x = \pm a$. The critical value of a is then zero. We shall suppose that a ranges over a compact interval $[0,A]$, where A may depend on u, and that the endpoints x_1 and x_2 of the x-interval satisfy $x_1 < -a$ and $x_2 > a$. As stated in § 1, x_1 and x_2 may depend on u and a, provided that the dependence on a is continuous, or they can be infinite. For convenience we introduce the notation

$$f(u,a,x) = (x^2-a^2)\,p(u,a,x). \tag{3.1}$$

Lemma I. *For each value of u assume that:*

(i) $p(u,a,x)$, $\partial p(u,a,x)/\partial x$, $\partial^2 p(u,a,x)/\partial x^2$ *and* $g(u,a,x)$ *are continuous functions of a and x in the region* $a\in[0,A]$, $x\in(x_1,x_2)$.

(ii) $p(u,a,x)$ *is positive throughout the same region.*

(iii) $|\partial^3 p(u,a,x)/\partial x^3|$ *is bounded in a neighbourhood of the point* $a = x = 0$.

(iv) $f(u,a,x)$ *is a non-increasing function of a when* $a\in[0,A]$ *and* $x\in[-a,a]$.

Then the function $\psi(u,\alpha,\zeta)$ defined by (2.10) *is continuous in the corresponding region of the* (α,ζ)*-plane.*

This result is fairly easy to prove when x ranges over a complex domain instead of a real interval and the functions involved are holomorphic in x as well as being continuous in a and x. This is because in these circumstances x and its partial derivatives with respect to ζ can be represented as contour integrals by means of Cauchy's formulae. The proof that occupies the remainder of the present section employs only real-variable analysis and is much more difficult.†

† To avoid being distracted from the main theme of the present paper the reader may by-pass the proof and proceed directly to § 5. It is in applications (for example Olver 1975) that lemma I and the corresponding result in § 4 are needed.

3.2. *The Schwarzian derivative*

We need the following explicit expression for the Schwarzian derivative of x with respect to ζ:

$$\dot{x}^{\frac{1}{2}}\frac{\mathrm{d}^2(\dot{x}^{-\frac{1}{2}})}{\mathrm{d}\zeta^2} = \left\{\left(\frac{p''(x)}{4p(x)} - \frac{5}{16}\frac{p'^2(x)}{p^2(x)}\right)(a^2-x^2)^2 + \frac{p'(x)}{4p(x)}x(a^2-x^2) - \tfrac{1}{4}(2a^2+3x^2)\right\}$$
$$\times\frac{\alpha^2-\zeta^2}{p(x)\,(a^2-x^2)^3} + \frac{2\alpha^2+3\zeta^2}{4(\alpha^2-\zeta^2)^2}. \tag{3.2}$$

Here, and in much subsequent analysis, the notation has been simplified by suppression of the first two arguments of the function $p(u,a,x)$ and its partial derivatives; thus $p(x) \equiv p(u,a,x)$, $p'(x) \equiv \partial p(u,a,x)/\partial x$, and so on. The expression (3.2) is verifiable in a straightforward manner by use of the relation

$$\dot{x} = \left(\frac{\zeta^2-\alpha^2}{x^2-a^2}\right)^{\frac{1}{2}}\frac{1}{p^{\frac{1}{2}}(x)} \tag{3.3}$$

obtained from (2.2) and (3.1).

3.3. *Relation between α and a*

From (2.5) and (3.1) we have

$$\alpha^2 = \frac{2}{\pi}\int_{-a}^{a}(a^2-t^2)^{\frac{1}{2}}p^{\frac{1}{2}}(t)\,\mathrm{d}t = \frac{2a^2}{\pi}\int_{-1}^{1}(1-v^2)^{\frac{1}{2}}p^{\frac{1}{2}}(av)\,\mathrm{d}v. \tag{3.4}$$

From condition (iv) we see that α is an increasing function of a, and if $\alpha = \mathrm{A}$ corresponds to $a = A$, then there is a one-to-one continuous correspondence between these variables in the intervals $[0,\mathrm{A}]$ and $[0,A]$.

We shall need to know the behaviour of α as $a\to 0$. Since $|p'''(x)|$ is bounded in a neighbourhood of $a = x = 0$ and $p(x)$ is non-vanishing, application of Taylor's theorem yields

$$p^{\frac{1}{2}}(x) = p^{\frac{1}{2}}(0) + \frac{p'(0)}{2p^{\frac{1}{2}}(0)}x + \left(\frac{p''(0)}{4p^{\frac{1}{2}}(0)} - \frac{p'^2(0)}{8p^{\frac{3}{2}}(0)}\right)x^2 + x^3O(1). \tag{3.5}$$

Substitution in (3.4) yields

$$\alpha^2 = p^{\frac{1}{2}}(0)\,a^2 + \left(\frac{p''(0)}{4p^{\frac{1}{2}}(0)} - \frac{p'^2(0)}{8p^{\frac{3}{2}}(0)}\right)\frac{a^4}{4} + O(a^5) \quad (a\to 0). \tag{3.6}$$

The required formula is therefore given by

$$\alpha = p^{\frac{1}{4}}(0)\,a\left\{1 + \left(\frac{p''(0)}{32p(0)} - \frac{p'^2(0)}{64p^2(0)}\right)a^2 + O(a^3)\right\} \quad (a\to 0). \tag{3.7}$$

Since $p(0)$, $p'(0)$ and $p''(0)$ may depend on a, the last expansion is not the beginning of a power-series expansion, in general, but this is immaterial in subsequent analysis.

3.4. *Proof of lemma when α and a are positive*

When $\alpha > 0$ the zeros $\pm a$ of $p(x)$ are distinct. Let ξ be the real number defined by

$$\tfrac{2}{3}\xi^{\frac{3}{2}} = \int_a^x (t^2-a^2)^{\frac{1}{2}}p^{\frac{1}{2}}(t)\,\mathrm{d}t \quad (x \geqslant a), \tag{3.8}$$

$$\tfrac{2}{3}(-\xi)^{\frac{3}{2}} = \int_x^a (a^2-t^2)^{\frac{1}{2}}p^{\frac{1}{2}}(t)\,\mathrm{d}t \quad (-a \leqslant x \leqslant a). \tag{3.9}$$

Clearly ξ is an increasing function of x, with $\xi = 0$ corresponding to $x = a$, and $\xi = -(\tfrac{3}{4}\pi\alpha^2)^{\frac{2}{3}}$ corresponding to $x = -a$; compare (3.4).

By following the proof of the corresponding lemma for the case of a simple turning point, given on page 399 of Olver (1974), and verifying that the functions that appear at each step are continuous in a and x, it is readily shown that $\xi/(x-a)$ is positive, and $\xi/(x-a)$ and its first two partial x-derivatives are continuous for $a > 0$ and $x > -a$. We now need to invert this result by considering x and $\xi/(x-a)$ as functions of α and ξ. This step is facilitated by the following form of the elementary implicit function theorem:

Let $\Xi(a, x)$ be continuous and $\partial\Xi(a,x)/\partial x$ be positive and continuous in the compact region $A_1 \leqslant a \leqslant A_2$, $\chi_1(a) \leqslant x \leqslant \chi_2(a)$, where A_1 and A_2 are unequal constants, and the functions $\chi_1(a)$ and $\chi_2(a)$ are continuous and unequal when $a \in [A_1, A_2]$. Then the equation $\xi = \Xi(a, x)$ has a unique solution $x = X(a, \xi)$ which is a strictly increasing function of ξ, and continuous in the region $A_1 \leqslant a \leqslant A_2$, $\Xi\{a, \chi_1(a)\} \leqslant \xi \leqslant \Xi\{a, \chi_2(a)\}$.

In the present instance we have from (3.8) and (3.9)

$$\frac{\partial\xi}{\partial x} = \left(\frac{x-a}{\xi}\right)^{\frac{1}{2}} (x+a)^{\frac{1}{2}} p^{\frac{1}{2}}(x).$$

Clearly $\partial\xi/\partial x$ is positive, hence x is an increasing function of ξ and a continuous function of a and ξ. We saw in § 3.3 that a is a continuous function of α, hence x may be regarded as a continuous function of α and ξ when $\alpha > 0$ and $\xi > -(\frac{3}{4}\pi\alpha^2)^{\frac{2}{3}}$. Combination with the results of the preceding paragraph shows that $\xi/(x-a)$ and its first two partial x-derivatives are continuous in the same circumstances.

Next, we have from (2.3), (2.4), (2.8), (3.1), (3.8) and (3.9)

$$\tfrac{2}{3}\xi^{\frac{3}{2}} = \int_\alpha^\zeta (\tau^2-\alpha^2)^{\frac{1}{2}}\,d\tau \quad (\zeta \geqslant \alpha),$$

$$\tfrac{2}{3}(-\xi)^{\frac{3}{2}} = \int_\zeta^\alpha (\alpha^2-\tau^2)^{\frac{1}{2}}\,d\tau \quad (-\alpha \leqslant \zeta \leqslant \alpha).$$

By reasoning similar to that used for equations (3.8) and (3.9), we see that $\xi/(\zeta-\alpha)$ is positive and $\xi/(\zeta-\alpha)$ and all its partial ζ-derivatives are continuous when $\alpha > 0$ and $\zeta > -\alpha$. As a corollary, ξ is continuous in these circumstances. Combination with the results of the preceding paragraph shows that x, together with the function $\xi/(x-a)$ and its first two partial x-derivatives, is continuous in the region $\alpha > 0$, $\zeta > -\alpha$.

Lastly, equation (3.3) may be rearranged in the form

$$\dot{x} = \left(\frac{\xi}{x-a}\right)^{\frac{1}{2}} \left(\frac{\zeta-\alpha}{\xi}\right)^{\frac{1}{2}} \left(\frac{\zeta+\alpha}{x+a}\right)^{\frac{1}{2}} \frac{1}{p^{\frac{1}{2}}(x)},$$

from which it is seen that $\dot{x}$ is a positive, continuous function of α and ζ. Furthermore, since $d/d\zeta = \dot{x}\,d/dx$ it is also seen that $\dot{x}^{\frac{1}{2}}\,d^2(\dot{x}^{-\frac{1}{2}})/d\zeta^2$ is continuous in the same circumstances. This yields the desired result concerning $\psi(u, \alpha, \zeta)$ when $\alpha > 0$ and $\zeta > -\alpha$. A similar proof applies to the overlapping region $\alpha > 0$ and $\zeta < \alpha$.

3.5. *Proof of lemma when $\alpha = a = 0$ and $\zeta \neq 0$*

By elementary analysis it is verifiable that the quantity ξ defined by (3.8) is a continuous function of a and x on the line segment $a = 0$ and $x > 0$, the continuity with respect to a being one-sided since negative values are not considered. Then by analysis similar to that used in § 3.4 we may show that x is a continuous function of α and ζ on the line segment $\alpha = 0$ and $\zeta > 0$.

Continuity of $\psi(u, \alpha, \zeta)$ then follows directly from the expressions (3.2) and (3.3). The proof when $\alpha = 0$ and $\zeta < 0$ is similar.

The remaining point of the (α, ζ)-plane to be considered is the critical point $\alpha = \zeta = 0$. The purpose of §§ 3.6–3.8 is to establish the continuity of $\psi(u, \alpha, \zeta)$ at this point.

3.6. *Expansion of x for small $\alpha - \zeta$*

Let $x = x_0$ correspond to $\zeta = 0$, so that $x_0 = 0$ when $\alpha = 0$. As in § 3.4 we have

$$\int_x^a (a^2 - t^2)^{\frac{1}{2}} p^{\frac{1}{2}}(t)\,\mathrm{d}t = \int_\zeta^\alpha (\alpha^2 - \tau^2)^{\frac{1}{2}}\,\mathrm{d}\tau \quad (-\alpha \leqslant \zeta \leqslant \alpha). \tag{3.10}$$

Setting $\zeta = 0$ and rearranging, we obtain

$$\int_0^{x_0} (a^2 - t^2)^{\frac{1}{2}} p^{\frac{1}{2}}(t)\,\mathrm{d}t = \int_0^a (a^2 - t^2)^{\frac{1}{2}} p^{\frac{1}{2}}(t)\,\mathrm{d}t - \int_0^\alpha (\alpha^2 - \tau^2)^{\frac{1}{2}}\,\mathrm{d}\tau = O(a^3),$$

as $a \to 0$; compare (3.5) and (3.6). Applying the mean-value theorem to the integral on the left-hand side and bearing in mind that $p^{\frac{1}{2}}(t)$ is bounded away from zero, we deduce that

$$x_0 = O(a^2) \quad (a \to 0). \tag{3.11}$$

From here until § 3.9 we shall suppose that $\alpha > 0$ and $0 \leqslant \zeta \leqslant \alpha$, or, equivalently, $a > 0$ and $x_0 \leqslant x \leqslant a$.

When α is fixed and $\zeta \to \alpha-$, the right-hand side of (3.10) is expansible in odd powers of $(\alpha - \zeta)^{\frac{1}{2}}$. Since $x \to a-$ in these circumstances and $|p'''(x)|$ is bounded, we can apply Taylor's theorem to construct the first three terms in the corresponding expansion of the left-hand side in powers of $(a - x)^{\frac{1}{2}}$. Then by reversion we arrive at an expansion of the form

$$a - x = X_1(\alpha - \zeta) + X_2(\alpha - \zeta)^2 + X_3(\alpha - \zeta)^3 + O\{(\alpha - \zeta)^4\} \quad (\zeta \to \alpha -), \tag{3.12}$$

in which the coefficients X_1, X_2 and X_3 depend on u and α, or equivalently, u and a. Straightforward calculation† gives

$$X_1 = \left(\frac{\alpha}{ap(a)}\right)^{\frac{1}{2}}, \quad X_2 = \left(\frac{1}{10a} + \frac{p'(a)}{5p(a)}\right) X_1^2 - \frac{X_1}{10\alpha}, \tag{3.13}$$

and

$$X_3 = \left(\frac{11}{350}\frac{1}{a^2} + \frac{19}{350}\frac{p'(a)}{ap(a)} + \frac{22}{175}\frac{p'^2(a)}{p^2(a)} - \frac{p''(a)}{14p(a)}\right) X_1^3 - \left(\frac{1}{10a} + \frac{p'(a)}{5p(a)}\right)\frac{X_1^2}{5\alpha} - \frac{2}{175}\frac{X_1}{\alpha^2}. \tag{3.14}$$

3.7. *Uniformity of the error term in the expansion of* § 3.6

Of crucial importance is the fact that error term $O\{(\alpha - \zeta)^4\}$ in (3.12) is uniform in a triangular region $\boldsymbol{T}$, defined by $0 < \alpha \leqslant \delta$, $0 \leqslant \zeta \leqslant \alpha$, where δ is an assignable positive constant. This result is not easily deducible from the analysis already given. We begin by determining the behaviour of the coefficients X_s as $a \to 0$.

From Taylor's theorem

$$p(a) = p(0) + ap'(0) + \tfrac{1}{2}a^2 p''(0) + O(a^3),$$

$$p'(a) = p'(0) + ap''(0) + O(a^2), \quad p''(a) = p''(0) + O(a).$$

† Albeit rather lengthy; this observation also applies to several subsequent derivations in this section.

Substituting these expansions and (3.7) in the expressions (3.13) and (3.14), we find that terms in a^{-1} and a^{-2} cancel, leaving

$$X_1 = \frac{1}{p^{\frac{1}{4}}(0)} - \frac{p'(0)}{3p^{\frac{5}{4}}(0)}a + \left(-\frac{5}{32}\frac{p''(0)}{p^{\frac{5}{4}}(0)} + \frac{125}{576}\frac{p'^2(0)}{p^{\frac{9}{4}}(0)}\right)a^2 + O(a^3), \tag{3.15}$$

$$X_2 = \frac{p'(0)}{6p^{\frac{3}{2}}(0)} + \left(\frac{3}{16}\frac{p''(0)}{p^{\frac{3}{2}}(0)} - \frac{29}{96}\frac{p'^2(0)}{p^{\frac{5}{2}}(0)}\right)a + O(a^2), \tag{3.16}$$

$$X_3 = -\frac{p''(0)}{16p^{\frac{7}{4}}(0)} + \frac{29}{288}\frac{p'^2(0)}{p^{\frac{11}{4}}(0)} + O(a). \tag{3.17}$$

These results immediately show that each coefficient is continuous at $a = 0$.

The next step is to rearrange the integrals (3.10). Denote the error term in (3.12) by $-\eta$, so that

$$a - x = X_1(\alpha-\zeta) + X_2(\alpha-\zeta)^2 + X_3(\alpha-\zeta)^3 - \eta.$$

Then

$$I = \int_\zeta^\alpha (\alpha^2-\tau^2)^{\frac{1}{2}}\,\mathrm{d}\tau - J, \tag{3.18}$$

where

$$I = \int_{a-X_1(\alpha-\zeta)-X_2(\alpha-\zeta)^2-X_3(\alpha-\zeta)^3+\eta}^{a-X_1(\alpha-\zeta)-X_2(\alpha-\zeta)^2-X_3(\alpha-\zeta)^3} (a^2-t^2)^{\frac{1}{2}}p^{\frac{1}{2}}(t)\,\mathrm{d}t, \tag{3.19}$$

$$J = \int_{a-X_1(\alpha-\zeta)-X_2(\alpha-\zeta)^2-X_3(\alpha-\zeta)^3}^{a} (a^2-t^2)^{\frac{1}{2}}p^{\frac{1}{2}}(t)\,\mathrm{d}t. \tag{3.20}$$

In the last integral we take a new integration variable y, related to t by

$$a - t = X_1(\alpha-y) + X_2(\alpha-y)^2 + X_3(\alpha-y)^3. \tag{3.21}$$

Thus $y = \zeta$ and $y = \alpha$ correspond respectively to the lower and upper limits in (3.20). We also have

$$\mathrm{d}t/\mathrm{d}y = X_1 + 2X_2(\alpha-y) + 3X_3(\alpha-y)^2, \tag{3.22}$$

and from (3.15), (3.16) and (3.17) it is seen that if $\zeta \leqslant y \leqslant \alpha$ and α and ζ lie in the triangular region $\boldsymbol{T}$, then $\mathrm{d}t/\mathrm{d}y$ is positive, provided that δ is sufficiently small. In these circumstances the $t \leftrightarrow y$ transformation is one-to-one. We now proceed to expand the factors in the integrand uniformly in powers of $\alpha - y$.

From (3.21) we have

$$a + t = 2a - X_1(\alpha-y) - X_2(\alpha-y)^2 - X_3(\alpha-y)^3. \tag{3.23}$$

With a view to expressing the right-hand side as a *uniform* multiple of $\alpha + y$, we substitute for the first term by means of the identity

$$2a = Y_0(\alpha+y) + Y_0(\alpha-y),$$

in which $Y_0 = a/\alpha$. Similarly

$$(Y_0 - X_1)(\alpha-y) = Y_1(\alpha+y)(\alpha-y) + Y_1(\alpha-y)^2,$$

where $2\alpha Y_1 = Y_0 - X_1$. Repeating the process and combining the results, we arrive at

$$a + t = (\alpha+y)\{Y_0 + Y_1(\alpha-y) + Y_2(\alpha-y)^2 + Y_3(\alpha-y)^3\} + Y_3(\alpha-y)^4, \tag{3.24}$$

where

$$2\alpha Y_s = Y_{s-1} - X_s \quad (s = 1, 2, 3).$$

What is the behaviour of Y_0, Y_1, Y_2 and Y_3 for small a? This can be determined by substitution of (3.7), (3.15), (3.16) and (3.17). Again, we find that terms involving negative powers of a cancel; thus

$$Y_0 = \frac{1}{p^{\frac{1}{4}}(0)} + \left(-\frac{p''(0)}{32p^{\frac{5}{4}}(0)} + \frac{p'^2(0)}{64p^{\frac{9}{4}}(0)}\right) a^2 + O(a^3),$$

$$Y_1 = \frac{p'(0)}{6p^{\frac{3}{2}}(0)} + \left(\frac{p''(0)}{16p^{\frac{3}{2}}(0)} - \frac{29}{288}\frac{p'^2(0)}{p^{\frac{5}{2}}(0)}\right) a + O(a^2),$$

$$Y_2 = -\frac{p''(0)}{16p^{\frac{7}{4}}(0)} + \frac{29}{288}\frac{p'^2(0)}{p^{\frac{11}{4}}(0)} + O(a), \quad Y_3 = O(1).$$

Substituting this estimate for Y_3 in (3.24) and bearing in mind that $(\alpha-y)/(\alpha+y)$ is non-negative and bounded by unity, we see that

$$a+t = (\alpha+y)\,[Y_0+Y_1(\alpha-y)+Y_2(\alpha-y)^2+O\{(\alpha-y)^3\}], \tag{3.25}$$

uniformly when α and y lie in $\boldsymbol{T}$.

The remaining factor to be considered is $p^{\frac{1}{2}}(t)$. From Taylor's theorem we have uniformly

$$p(t) = p(a) - p'(a)\,(a-t) + \tfrac{1}{2}p''(a)\,(a-t)^2 + O\{(a-t)^3\},$$

and substituting for $a-t$ by means of (3.21) we derive

$$p(t) = p(a)\left[1 - \frac{p'(a)}{p(a)}X_1(\alpha-y) + \left(\frac{p''(a)}{2p(a)}X_1^2 - \frac{p'(a)}{p(a)}X_2\right)(\alpha-y)^2 + O\{(\alpha-y)^3\}\right]. \tag{3.26}$$

Let us assemble these results. We multiply together the series (3.21), (3.25) and (3.26), take the square root of the product and then multiply by (3.22).† In the final result the coefficients of the second and third terms vanish, leaving

$$(a^2-t^2)^{\frac{1}{2}}p^{\frac{1}{2}}(t)\,(\mathrm{d}t/\mathrm{d}y) = (\alpha^2-y^2)^{\frac{1}{2}}[1+O\{(\alpha-y)^3\}],$$

uniformly in $\boldsymbol{T}$. Substituting in (3.20) and changing the symbol for the dummy integration variable from y to τ, we obtain

$$J = \int_\zeta^\alpha (\alpha^2-\tau^2)^{\frac{1}{2}}[1+O\{(\alpha-\tau)^3\}]\,\mathrm{d}\tau.$$

Now consider the integral I. Combination of (3.18) with the result just obtained gives

$$I = \int_\zeta^\alpha (\alpha^2-\tau^2)^{\frac{1}{2}}\,O\{(\alpha-\tau)^3\}\,\mathrm{d}\tau = \alpha^{\frac{1}{2}}O\{(\alpha-\zeta)^{\frac{9}{2}}\},$$

uniformly when α and ζ lie in $\boldsymbol{T}$. The desired estimate for η can be found from this result by applying the mean-value theorem to the integral (3.19), as follows. Provided that δ is sufficiently small, the upper limit changes monotonically when ζ ranges over the interval $[0,\alpha]$. When $\zeta=\alpha$ this limit is a, and when $\zeta=0$ it is $a-X_1\alpha-X_2\alpha^2-X_3\alpha^3$. The last quantity is $O(a^2)$: this is easily verifiable from (3.7) and (3.15). Therefore

$$O(a^2) \leqslant a - X_1(\alpha-\zeta) - X_2(\alpha-\zeta)^2 - X_3(\alpha-\zeta)^3 \leqslant a.$$

The lower limit in (3.19) equals x. Since $x \in [x_0, a]$, where x_0 is defined in § 3.6, we have

$$O(a^2) \leqslant a - X_1(\alpha-\zeta) - X_2(\alpha-\zeta)^2 - X_3(\alpha-\zeta)^3 + \eta \leqslant a; \tag{3.27}$$

† This calculation is simplified on working with the logarithmic forms of the various expansions.

compare (3.11). The mean-value theorem accordingly yields

$$I = (a+t_1)^{\frac{1}{2}} p^{\frac{1}{2}}(t_1) \int_{a - X_1(\alpha-\zeta) - X_2(\alpha-\zeta)^2 - X_3(\alpha-\zeta)^3 + \eta}^{a - X_1(\alpha-\zeta) - X_2(\alpha-\zeta)^2 - X_3(\alpha-\zeta)^3} (a-t)^{\frac{1}{2}}\, dt,$$

where $O(a^2) < t_1 < a$. Therefore

$$I = \tfrac{2}{3}(a+t_1)^{\frac{1}{2}} p^{\frac{1}{2}}(t_1)\, \{X_1(\alpha-\zeta) + X_2(\alpha-\zeta)^2 + X_3(\alpha-\zeta)^3\}^{\frac{3}{2}} \{(1-\chi)^{\frac{3}{2}} - 1\},$$

where
$$\chi = \eta / \{X_1(\alpha-\zeta) + X_2(\alpha-\zeta)^2 + X_3(\alpha-\zeta)^3\}.$$

From (3.27) it is seen that $\chi \in (-\infty, 1]$. In this range

$$|\chi/\{(1-\chi)^{\frac{3}{2}} - 1\}| \leqslant C,$$

where C is an assignable positive constant. Hence

$$|I| \geqslant \tfrac{2}{3}|\eta|\, C^{-1}(a+t_1)^{\frac{1}{2}} p^{\frac{1}{2}}(t_1)\, \{X_1(\alpha-\zeta) + X_2(\alpha-\zeta)^2 + X_3(\alpha-\zeta)^3\}^{\frac{1}{2}}.$$

We have already proved that $I = \alpha^{\frac{1}{2}} O\{(\alpha-\zeta)^{\frac{9}{2}}\}$. Since $(a+t_1)^{-\frac{1}{2}}$ is $O(a^{-\frac{1}{2}})$, that is, $O(\alpha^{-\frac{1}{2}})$, and $p^{-\frac{1}{2}}(t_1)$ is bounded, it is clear that $\eta = O\{(\alpha-\zeta)^4\}$, uniformly in T. This is the desired estimate.

3.8. *Expansions of $\dot{x}$, $\ddot{x}$ and $\dddot{x}$ in powers of $\alpha-\zeta$*

In § 3.7 we proved that

$$a - x = X_1(\alpha-\zeta) + X_2(\alpha-\zeta)^2 + X_3(\alpha-\zeta)^3 + O\{(\alpha-\zeta)^4\}, \tag{3.28}$$

uniformly in a region T given by $0 < \alpha \leqslant \delta$, $0 \leqslant \zeta \leqslant \alpha$. We also require the corresponding expansions

$$\dot{x} = X_1 + 2X_2(\alpha-\zeta) + 3X_3(\alpha-\zeta)^2 + O\{(\alpha-\zeta)^3\}, \tag{3.29}$$

$$\ddot{x} = -2X_2 - 6X_3(\alpha-\zeta) + O\{(\alpha-\zeta)^2\}, \tag{3.30}$$

$$\dddot{x} = 6X_3 + O\{(\alpha-\zeta)\}. \tag{3.31}$$

These results cannot, of course, be established simply by differentiating (3.28) with respect to ζ.

To prove (3.29) we use (3.3). Expanding $p(x)$ at $x = a$ by Taylor's theorem and substituting in the result by means of (3.28), we obtain

$$p(x) = p(a)\left[1 - \frac{p'(a)}{p(a)} X_1(\alpha-\zeta) + \left(\frac{p''(a)}{2p(a)} X_1^2 - \frac{p'(a)}{p(a)} X_2\right)(\alpha-\zeta)^2 + O\{(\alpha-\zeta)^3\}\right]; \tag{3.32}$$

compare (3.26). Next, from (3.28) we have

$$a + x = 2a - X_1(\alpha-\zeta) - X_2(\alpha-\zeta)^2 - X_3(\alpha-\zeta)^3 + O\{(\alpha-\zeta)^4\}.$$

In order to divide the right-hand side by $\alpha+\zeta$ and arrive at a uniform error term we follow the procedure used to derive (3.25) from (3.23). In this way we find that

$$a + x = (\alpha+\zeta)\,[Y_0 + Y_1(\alpha-\zeta) + Y_2(\alpha-\zeta)^2 + O\{(\alpha-\zeta)^3\}], \tag{3.33}$$

where the coefficients Y_s are defined as before. We now multiply together (3.28), (3.32), and (3.33), and divide the product by $\alpha^2 - \zeta^2$. Then taking the square root of the reciprocal of the answer, we arrive at the desired formula (3.29).†

† Again, use of the logarithmic forms of the various expansions eases the calculations.

To verify (3.30) we employ the identity

$$\frac{2\ddot{x}}{\dot{x}} = \frac{1}{\alpha+\zeta} - \frac{1}{\alpha-\zeta} + \left\{\frac{1}{a-x} - \frac{1}{a+x} - \frac{p'(x)}{p(x)}\right\}\dot{x}, \tag{3.34}$$

obtained by logarithmic differentiation of (3.3). The only terms presenting difficulty are

$$\frac{1}{\alpha+\zeta} - \frac{\dot{x}}{a+x}.$$

They can be treated as follows. From (3.29) and (3.33) we derive

$$\frac{\alpha+\zeta}{a+x}\dot{x} - 1 = Z_0 + Z_1(\alpha-\zeta) + Z_2(\alpha-\zeta)^2 + O\{(\alpha-\zeta)^3\}, \tag{3.35}$$

where $$Z_0 = \frac{X_1}{Y_0} - 1, \quad Z_1 = -\frac{X_1 Y_1}{Y_0^2} + \frac{2X_2}{Y_0}, \quad Z_2 = \frac{X_1 Y_1^2}{Y_0^3} - \frac{X_1 Y_2 + 2X_2 Y_1}{Y_0^2} + \frac{3X_3}{Y_0}.$$

As in the derivation of (3.25) and (3.33) we rearrange the right-hand side of (3.35) in the form

$$\frac{\alpha+\zeta}{a+x}\dot{x} - 1 = (\alpha+\zeta)\{\hat{Z}_0 + \hat{Z}_1(\alpha-\zeta) + \hat{Z}_2(\alpha-\zeta)^2\} + \hat{Z}_2(\alpha-\zeta)^3 + O\{(\alpha-\zeta)^3\}, \tag{3.36}$$

where $$2\alpha\hat{Z}_0 = Z_0, \quad 2\alpha\hat{Z}_1 = \hat{Z}_0 + Z_1, \quad 2\alpha\hat{Z}_2 = \hat{Z}_1 + Z_2.$$

From the expansions of the X_s and Y_s given in § 3.7 for small a we calculate

$$\hat{Z}_0 = -\frac{p'(0)}{6p^{\frac{3}{2}}(0)} + \left(-\frac{p''(0)}{16p^{\frac{5}{4}}(0)} + \frac{29}{288}\frac{p'^2(0)}{p^{\frac{9}{4}}(0)}\right)a + O(a^2),$$

$$\hat{Z}_1 = \frac{p''(0)}{8p^{\frac{3}{2}}(0)} - \frac{25}{144}\frac{p'^2(0)}{p^{\frac{5}{2}}(0)} + O(a), \quad \hat{Z}_2 = O(1),$$

terms in negative powers of a again disappearing. Since $(\alpha-\zeta)/(\alpha+\zeta)$ is bounded, division of (3.36) by $-(\alpha+\zeta)$ yields

$$\frac{1}{\alpha+\zeta} - \frac{\dot{x}}{a+x} = -\hat{Z}_0 - \hat{Z}_1(\alpha-\zeta) + O\{(\alpha-\zeta)^2\}.$$

The expansion of the remaining terms in (3.34) is straightforward, and on combining the results we eventually arrive at the desired expansion (3.29).

The proof of (3.31) proceeds from the differentiated form of (3.34) and is similar.

3.9. *Completion of proof of lemma I*

The expansions (3.28) to (3.31) have been shown to hold uniformly in the triangular region of the (α, ζ)-plane given by $0 < \alpha \leqslant \delta$, $0 \leqslant \zeta \leqslant \alpha$. In a similar way, starting from the relation

$$\int_a^x (t^2-a^2)^{\frac{1}{2}} p^{\frac{1}{2}}(t)\,dt = \int_\alpha^\zeta (\tau^2-\alpha^2)^{\frac{1}{2}}\,d\tau \quad (\zeta \geqslant \alpha)$$

in place of (3.10), we can show that these expansions also hold in the triangle $0 \leqslant \alpha \leqslant \zeta$, $0 < \zeta \leqslant \delta$, again provided that the positive constant δ is sufficiently small. The union of the two triangles is the square $0 \leqslant \alpha \leqslant \delta$, $0 \leqslant \zeta \leqslant \delta$, with the point $\alpha = \zeta = 0$ excluded. From these results it is evident that if α and ζ are restricted to non-negative values, then x, $\dot{x}$, $\ddot{x}$ and $\dddot{x}$ are continuous at $\alpha = \zeta = 0$. And by constructing expansions in powers of $\alpha+\zeta$ in place of $\alpha-\zeta$ we can show in an exactly similar manner that x, $\dot{x}$, $\ddot{x}$ and $\dddot{x}$ are continuous at $\alpha = \zeta = 0$ when α is restricted to non-negative values and ζ to non-positive values. Therefore each function is continuous at $\alpha = \zeta = 0$, provided only that α is non-negative. Since

$$\dot{x}^{\frac{1}{2}} \frac{\mathrm{d}^2}{\mathrm{d}\zeta^2}(\dot{x}^{-\frac{1}{2}}) = \frac{3\ddot{x}^2}{4\dot{x}^2} - \frac{\dddot{x}}{2\dot{x}}$$

and $\dot{x}$ is non-vanishing, the Schwarzian is continuous in these circumstances. Therefore from (2.10) $\psi(u, \alpha, \zeta)$ is continuous at $\alpha = \zeta = 0$. This completes the proof of lemma I.

In applications the values of $\dot{x}$ and the Schwarzian at the critical point are sometimes needed. Using (3.15) to (3.17) and (3.29) to (3.31), we find that

$$[\dot{x}]_{\alpha=\zeta=0} = \frac{1}{p^{\frac{1}{3}}(u,0,0)}, \quad \left[\dot{x}^{\frac{1}{2}} \frac{\mathrm{d}^2}{\mathrm{d}\zeta^2}(\dot{x}^{-\frac{1}{2}})\right]_{\alpha=\zeta=0} = \frac{3}{16} \frac{p''(u,0,0)}{p^{\frac{4}{3}}(u,0,0)} - \frac{7}{32} \frac{p'^2(u,0,0)}{p^{\frac{5}{3}}(u,0,0)}. \tag{3.37}$$

4. Continuity of $\psi(u, \alpha, \zeta)$: cases II and IV

4.1. *Statement of lemma*

In the treatment of case II (case IV is exactly similar), we suppose that the zeros of $f(u, a, x)$ are located at $x = \pm ia$, where a is again a real parameter ranging over a compact interval $[0, A]$. The range of x is the point set $\boldsymbol{X}$ comprising the union of a real interval $(-x_1, x_1)$ and an open segment of the imaginary axis joining two points $-i\hat{x}_1$ and $i\hat{x}_1$. Both x_1 and $\hat{x}_1$ may depend on a and u, provided that the dependence on a is continuous, or they can be infinite; in any case it is assumed that $a < \hat{x}_1$. Corresponding to (3.1) we define $p(u, a, x)$ by the equation

$$f(u, a, x) = (a^2 + x^2)\, p(u, a, x).$$

Lemma II. *For each value of u assume that:*

(i) *$p(u, a, x)$ is continuous, positive, and even in x when $a \in [0, A]$, $x \in \boldsymbol{X}$.*

(ii) *$\partial p(u, a, x)/\partial x$, $\partial^2 p(u, a, x)/\partial x^2$ and $g(u, a, x)$ are continuous when $a \in [0, A]$, $x \in (-x_1, x_1)$.*

(iii) *$\partial^3 p(u, a, x)/\partial x^3$ is continuous and $|\partial^4 p(u, a, x)/\partial x^4|$ is bounded in a neighbourhood of the point $a = x = 0$ for real and purely imaginary values of x.*†

(iv) *$f(u, a, x)$ is a non-decreasing function of a when $a \in [0, A]$ and $ix \in [-a, a]$.*

Then the function $\psi(u, \alpha, \zeta)$ defined by (2.10) is continuous in the region of the (α, ζ)-plane corresponding to $a \in [0, A]$, $x \in (-x_1, x_1)$.

4.2. *Proof*

The proof of lemma II is somewhat simpler than that of lemma I, and we merely draw attention to the main differences. Since $p(x)$ is even in x, we have $p'(0) = p'''(0) = 0$. From this result and the assumption that $|p^{\mathrm{iv}}(x)|$ is bounded in a neighbourhood of $a = x = 0$, it follows that the expansion corresponding to (3.7) of § 3.3 is given by

$$\alpha = p^{\frac{1}{2}}(0)\, a \left\{1 - \frac{p''(0)}{32 p(0)} a^2 + O(a^4)\right\} \quad (a \to 0).$$

The proof when α and a are positive (§ 3.4) is simplified because in these circumstances $f(u, a, x)$ has no zeros on the real axis. § 3.5 is almost unchanged.

In the analysis corresponding to § 3.6, the expansion corresponding to (3.12) ascends in powers of ζ:

$$x = X_1 \zeta + X_3 \zeta^3 + O(\zeta^5) \quad (\zeta \to 0),$$

† When x is imaginary, partial differentiations with respect to x are to be performed along the imaginary axis.

where
$$X_1 = \frac{\alpha}{ap^{\frac{1}{2}}(0)}, \quad X_3 = \frac{1}{6a\alpha p^{\frac{1}{2}}(0)} - \frac{\alpha^3}{6a^5 p^{\frac{3}{2}}(0)} - \frac{\alpha^3 p''(0)}{12a^3 p^{\frac{5}{2}}(0)}.$$

To establish that the error term in the last expansion is uniform with respect to α in an assignable interval $0 < \alpha \leqslant \delta$, we denote it by η and rearrange (2.12) in the form

$$I = \int_0^\zeta (\alpha^2 + \tau^2)^{\frac{1}{2}}\,\mathrm{d}\tau - J,$$

where
$$I = \int_{X_1\zeta + X_3\zeta^3}^{X_1\zeta + X_3\zeta^3 + \eta} (a^2 + t^2)^{\frac{1}{2}} p^{\frac{1}{2}}(t)\,\mathrm{d}t, \quad J = \int_0^{X_1\zeta + X_3\zeta^3} (a^2 + t^2)^{\frac{1}{2}} p^{\frac{1}{2}}(t)\,\mathrm{d}t;$$

compare (3.18) to (3.20). Corresponding to (3.21) we define a new integration variable y by

$$t = X_1 y + X_3 y^3,$$

and express $a^2 + t^2$ in the form

$$a^2 + t^2 = (\alpha^2 + y^2)(Y_0 + Y_2 y^2 + Y_4 y^4) + Y_6 y^6;$$

compare (3.24). Again, the coefficients Y_0, Y_2, Y_4 and Y_6 can be shown to be bounded as $a \to 0$. Using analysis similar to that of § 3.7 we find that

$$J = \int_0^\zeta (\alpha^2 + \tau^2)^{\frac{1}{2}} \{1 + O(\tau^4)\}\,\mathrm{d}\tau,$$

and thence that

$$I = (\alpha^2 + \zeta^2)^{\frac{1}{2}}\, O(\zeta^5),$$

the O-term in each relation being uniform with respect to α. From this result and the definition of I it is an elementary deduction that η is uniformly $O(\zeta^5)$.

The remaining analysis corresponding to §§ 3.8–3.9 proceeds in a similar manner and it is unnecessary to record further details.

PART B. APPROXIMATIONS IN TERMS OF $U(b, x)$ AND $\bar{U}(b, x)$

5. Relevant properties of parabolic cylinder functions

5.1. *Introduction*

We shall follow the notation of Miller (1955, 1964) and adopt

$$\mathrm{d}^2 w/\mathrm{d}x^2 = (\tfrac{1}{4}x^2 + b)\, w \tag{5.1}$$

as the standard form of differential equation for the parabolic cylinder functions.† The character of the solutions depends on the sign of b. When $b < 0$ there are turning points at $x = \pm 2(-b)^{\frac{1}{2}}$. The solutions oscillate in the interval between these points, but not in the exterior intervals. When $b > 0$ there are no real turning points and no oscillations.

Again, following Miller (1955), standard solutions of (5.1) are $U(b, \pm x)$ and $\bar{U}(b, \pm x)$, defined by

$$U(b, \pm x) = \frac{\pi^{\frac{1}{2}} 2^{-\frac{1}{4}(2b+1)}}{\Gamma(\frac{3}{4} + \frac{1}{2}b)} \mathrm{e}^{-\frac{1}{4}x^2}\, {}_1F_1(\tfrac{1}{2}b + \tfrac{1}{4}; \tfrac{1}{2}; \tfrac{1}{2}x^2) \mp \frac{\pi^{\frac{1}{2}} 2^{-\frac{1}{4}(2b-1)}}{\Gamma(\frac{1}{4} + \frac{1}{2}b)} \mathrm{e}^{-\frac{1}{4}x^2} x\, {}_1F_1(\tfrac{1}{2}b + \tfrac{3}{4}; \tfrac{3}{2}; \tfrac{1}{2}x^2),$$

$$\begin{aligned}\bar{U}(b, \pm x) = {} & \pi^{-\frac{1}{2}} 2^{-\frac{1}{4}(2b+1)} \Gamma(\tfrac{1}{4} - \tfrac{1}{2}b) \sin(\tfrac{3}{4}\pi - \tfrac{1}{2}b\pi)\, \mathrm{e}^{-\frac{1}{4}x^2}\, {}_1F_1(\tfrac{1}{2}b + \tfrac{1}{4}; \tfrac{1}{2}; \tfrac{1}{2}x^2) \\ & \mp \pi^{-\frac{1}{2}} 2^{-\frac{1}{4}(2b-1)} \Gamma(\tfrac{3}{4} - \tfrac{1}{2}b) \sin(\tfrac{5}{4}\pi - \tfrac{1}{2}b\pi)\, \mathrm{e}^{-\frac{1}{4}x^2} x\, {}_1F_1(\tfrac{1}{2}b + \tfrac{3}{4}; \tfrac{3}{2}; \tfrac{1}{2}x^2),\end{aligned}$$

† Miller's symbol a is replaced here by b to avoid a clash with the notation of earlier sections.

where ${}_1F_1$ denotes the confluent hypergeometric function in the usual notation. The solutions $U(b,x)$ and $\bar{U}(b,x)$ comprise a numerically satisfactory pair, in the sense of Miller (1950), when $b \leqslant 0$ and $x \in [0,\infty)$; moreover, each is continuous in b and x in this region. For $b \geqslant 0$ and, in this case, $x \in (-\infty,\infty)$ the solutions $U(b,x)$ and $U(b,-x)$ are continuous and comprise a numerically satisfactory pair. In the present section we collect properties of these solutions that will be needed later. We continue to use primes to denote partial differentiations with respect to the argument x.

5.2. *Basic properties*

Values at $x = 0$

$$\left.\begin{aligned} U(b,0) &= \pi^{-\frac{1}{2}}2^{-\frac{1}{4}(2b+1)}\Gamma(\tfrac{1}{4}-\tfrac{1}{2}b)\sin(\tfrac{1}{4}\pi-\tfrac{1}{2}b\pi), \\ U'(b,0) &= -\pi^{-\frac{1}{2}}2^{-\frac{1}{4}(2b-1)}\Gamma(\tfrac{3}{4}-\tfrac{1}{2}b)\sin(\tfrac{3}{4}\pi-\tfrac{1}{2}b\pi), \end{aligned}\right\} \tag{5.2}$$

$$\left.\begin{aligned} \bar{U}(b,0) &= \pi^{-\frac{1}{2}}2^{-\frac{1}{4}(2b+1)}\Gamma(\tfrac{1}{4}-\tfrac{1}{2}b)\sin(\tfrac{3}{4}\pi-\tfrac{1}{2}b\pi), \\ \bar{U}'(b,0) &= -\pi^{-\frac{1}{2}}2^{-\frac{1}{4}(2b-1)}\Gamma(\tfrac{3}{4}-\tfrac{1}{2}b)\sin(\tfrac{5}{4}\pi-\tfrac{1}{2}b\pi). \end{aligned}\right\} \tag{5.3}$$

For positive b it is sometimes convenient to use the alternative forms

$$U(b,0) = \pi^{\frac{1}{2}}/\{2^{\frac{1}{4}(2b+1)}\Gamma(\tfrac{3}{4}+\tfrac{1}{2}b)\}, \quad U'(b,0) = -\pi^{\frac{1}{2}}/\{2^{\frac{1}{4}(2b-1)}\Gamma(\tfrac{1}{4}+\tfrac{1}{2}b)\}. \tag{5.4}$$

Behaviour as $x \to \infty$

$$U(b,x) \sim x^{-b-\frac{1}{2}}\,e^{-\frac{1}{4}x^2}, \quad U'(b,x) \sim -\tfrac{1}{2}x^{-b+\frac{1}{2}}\,e^{-\frac{1}{4}x^2}, \tag{5.5}$$

$$\bar{U}(b,x) \sim 2^{\frac{1}{2}}\pi^{-\frac{1}{2}}\Gamma(\tfrac{1}{2}-b)\,x^{b-\frac{1}{2}}\,e^{\frac{1}{4}x^2}, \quad \bar{U}'(b,x) \sim (2\pi)^{-\frac{1}{2}}\Gamma(\tfrac{1}{2}-b)\,x^{b+\frac{1}{2}}\,e^{\frac{1}{4}x^2}, \tag{5.6}$$

$$U(b,-x) \sim (2\pi)^{\frac{1}{2}}x^{b-\frac{1}{2}}\,e^{\frac{1}{4}x^2}/\Gamma(\tfrac{1}{2}+b), \quad U'(b,-x) \sim -2^{-\frac{1}{2}}\pi^{\frac{1}{2}}x^{b+\frac{1}{2}}\,e^{\frac{1}{4}x^2}/\Gamma(\tfrac{1}{2}+b). \tag{5.7}$$

These relations hold for fixed b, other than those values for which the Gamma functions become infinite. They are also uniform with respect to b when b ranges over a fixed compact interval that does not include any of the exceptional values.

Wronskians

$$\mathscr{W}\{U(b,x),\bar{U}(b,x)\} = 2^{\frac{1}{2}}\pi^{-\frac{1}{2}}\Gamma(\tfrac{1}{2}-b), \quad \mathscr{W}\{U(b,x),U(b,-x)\} = (2\pi)^{\frac{1}{2}}/\Gamma(\tfrac{1}{2}+b). \tag{5.8}$$

Connection formulae

$$\left.\begin{aligned} U(b,-x) &= \cos(b\pi)\,\bar{U}(b,x) - \sin(b\pi)\,U(b,x), \\ \bar{U}(b,-x) &= \sin(b\pi)\,\bar{U}(b,x) + \cos(b\pi)\,U(b,x). \end{aligned}\right\} \tag{5.9}$$

Case $b = 0$: *relation to modified Bessel functions.* When $x \geqslant 0$

$$U(0,x) = (2\pi)^{-\frac{1}{2}}x^{\frac{1}{2}}K_{\frac{1}{4}}(\tfrac{1}{4}x^2), \tag{5.10}$$

$$\begin{aligned} \bar{U}(0,x) = U(0,-x) &= \tfrac{1}{2}(\pi x)^{\frac{1}{2}}\{I_{\frac{1}{4}}(\tfrac{1}{4}x^2) + I_{-\frac{1}{4}}(\tfrac{1}{4}x^2)\} \\ &= (\pi x)^{\frac{1}{2}}I_{\frac{1}{4}}(\tfrac{1}{4}x^2) + (2\pi)^{-\frac{1}{2}}x^{\frac{1}{2}}K_{\frac{1}{4}}(\tfrac{1}{4}x^2). \end{aligned} \tag{5.11}$$

5.3. *Behaviour for large negative b*

Write $b = -\frac{1}{2}\mu^2$ and $x = \mu y\sqrt{2}$. Then

$$U(-\tfrac{1}{2}\mu^2,\mu y\sqrt{2}) = \frac{2^{\frac{1}{2}}\pi^{\frac{1}{4}}\{\Gamma(\frac{1}{2}+\frac{1}{2}\mu^2)\}^{\frac{1}{2}}}{\mu^{\frac{1}{6}}}\left(\frac{\eta}{y^2-1}\right)^{\frac{1}{4}}\left\{\mathrm{Ai}(\mu^{\frac{4}{3}}\eta) + E^{-1}(\mu^{\frac{4}{3}}\eta)\,M(\mu^{\frac{4}{3}}\eta)\,O\left(\frac{1}{\mu^2}\right)\right\}, \tag{5.12}$$

$$\bar{U}(-\tfrac{1}{2}\mu^2,\mu y\sqrt{2}) = \frac{2^{\frac{1}{2}}\pi^{\frac{1}{4}}\{\Gamma(\frac{1}{2}+\frac{1}{2}\mu^2)\}^{\frac{1}{2}}}{\mu^{\frac{1}{6}}}\left(\frac{\eta}{y^2-1}\right)^{\frac{1}{4}}\left\{\mathrm{Bi}(\mu^{\frac{4}{3}}\eta) + E(\mu^{\frac{4}{3}}\eta)\,M(\mu^{\frac{4}{3}}\eta)\,O\left(\frac{1}{\mu^2}\right)\right\}, \tag{5.13}$$

as $\mu \to \infty$, uniformly in a y-interval that includes $[0, \infty)$. Here

$$\eta = \left\{\frac{3}{2}\int_1^y (\tau^2-1)^{\frac{1}{2}}\,d\tau\right\}^{\frac{2}{3}} \quad (y \geqslant 1), \qquad \eta = -\left\{\frac{3}{2}\int_y^1 (1-\tau^2)^{\frac{1}{2}}\,d\tau\right\}^{\frac{2}{3}} \quad (0 \leqslant y \leqslant 1), \tag{5.14}$$

Ai and Bi denote the Airy functions, and E and M are auxiliary functions introduced in Olver (1974, pp. 394–395); thus

$$E(x) = \{\mathrm{Bi}\,(x)/\mathrm{Ai}\,(x)\}^{\frac{1}{2}}, \quad M(x) = \{2\,\mathrm{Ai}\,(x)\,\mathrm{Bi}\,(x)\}^{\frac{1}{2}} \quad (x \geqslant c), \tag{5.15}$$

$$E(x) = 1, \quad M(x) = \{\mathrm{Ai}^2(x) + \mathrm{Bi}^2(x)\}^{\frac{1}{2}} \quad (x \leqslant c), \tag{5.16}$$

where $c(= -0.36605\ldots)$ is the smallest root, in absolute value, of the equation $\mathrm{Ai}\,(x) = \mathrm{Bi}\,(x)$.

5.4. *Behaviour for large positive b*

Here we write $b = \frac{1}{2}\mu^2$ and $x = \mu y\sqrt{2}$. Then

$$U(\tfrac{1}{2}\mu^2, \mu y\sqrt{2}) = \frac{2^{\frac{1}{4}(\mu^2-1)}\,e^{\frac{1}{4}\mu^2}}{\mu^{\frac{1}{2}(\mu^2+1)}}\,\frac{e^{-\mu^2\bar{\eta}}}{(y^2+1)^{\frac{1}{4}}}\left\{1 + O\left(\frac{1}{\mu^2}\right)\right\}, \tag{5.17}$$

$$U'(\tfrac{1}{2}\mu^2, \mu y\sqrt{2}) = -\frac{2^{\frac{1}{4}(\mu^2-3)}\,e^{\frac{1}{4}\mu^2}}{\mu^{\frac{1}{2}(\mu^2-1)}}\,(y^2+1)^{\frac{1}{4}}\,e^{-\mu^2\bar{\eta}}\left\{1 + O\left(\frac{1}{\mu^2}\right)\right\}, \tag{5.18}$$

as $\mu \to \infty$, uniformly for $y \in (-\infty, \infty)$, where

$$\bar{\eta} = \int_0^y (\tau^2+1)^{\frac{1}{2}}\,d\tau. \tag{5.19}$$

Proofs of the foregoing results will be found in Miller (1955) in the case of §§ 5.1 and 5.2 and Olver (1959) in the case of §§ 5.3 and 5.4. We shall need some additional properties not found in these references, as follows.

5.5. *Zeros*

Throughout this subsection we suppose that $b \leqslant 0$. For fixed b, the zeros of any nontrivial solution of (5.1) are all simple. This is easily seen by considering the Taylor-series expansion of the solution at the zero.

From (5.10) it is known that when $b = 0$ the function $U(b, x)$ has no zeros in the interval $x \in [0, \infty)$. As b decreases in a continuous manner, zeros can enter or leave this interval only at the left endpoint. This is because the graph of $U(b, x)$, as a function of x, shows that within $(0, \infty)$ an emerging (or disappearing) zero would have to begin (or end) its existence as a multiple zero, which is impossible. From the first of (5.2) it is seen that the values of b for which $U(b, 0)$ vanishes are $-\frac{3}{2}, -\frac{7}{2}, -\frac{11}{2}, \ldots$; moreover, in a neighbourhood of $x = 0$ the graph of $U(b, x)$, again as a function of x, is moving downwards as b decreases through the value $-\frac{3}{2}$. The second of (5.2) shows that $U'(-\frac{3}{2}, 0) > 0$, hence one zero of $U(b, x)$ enters $[0, \infty)$ as b passes through $-\frac{3}{2}$. Similarly a zero also enters as b decreases through each of the values $-\frac{7}{2}, -\frac{11}{2}, \ldots$. Therefore *when* $b \leqslant 0$ *the number of zeros of* $U(b, x)$ *in the interval* $[0, \infty)$ *is* $[\frac{1}{4} - \frac{1}{2}b]$, *that is, the integer part of* $\frac{1}{4} - \frac{1}{2}b$.

Similar reasoning shows that *when* $b \leqslant 0$ *the function* $\bar{U}(b, x)$ *has* $[\frac{3}{4} - \frac{1}{2}b]$ *zeros in the interval* $[0, \infty)$.

Next, as b changes in a continous manner, the zeros of $U(b, x)$ and $\bar{U}(b, x)$ cannot pass each other, otherwise these functions would have a common zero for certain values of b, a situation precluded by the first of the Wronksian relations (5.8). Therefore *in* $[0, \infty)$ *the zeros of* $U(b, x)$ *and* $\bar{U}(b, x)$ *interlace, the largest zero belonging to* $\bar{U}(b, x)$.

The real zeros of $U(b,x)$ *all lie to the left of the turning point* $x = 2(-b)^{\frac{1}{2}}$. For suppose that this statement were false for a certain value of b, of necessity satisfying $b < -\frac{3}{2}$. The preceding paragraphs show that there are only a finite number of positive zeros; let x_b denote the largest. In the interval $2(-b)^{\frac{1}{2}} \leqslant x < \infty$ the coefficient $\frac{1}{4}x^2+b$ in (5.1) is non-negative, as are all its x-derivatives. The sign of $U'(b,x_b)$ cannot be positive because the Taylor-series expansion of $U(b,x)$ at $x = x_b$ (which converges for all finite x) would then require all derivatives of $U(b,x)$ to be positive when $x > x_b$, contradicting the second of (5.5). Similarly the sign of $U'(b,x_b)$ cannot be negative. The result stated at the beginning of this paragraph now follows.†

We shall also need properties of the solution $\bar{U}(b,x)-U(b,x)$. From (5.2) and (5.3) we derive

$$\bar{U}(b,0)-U(b,0) = \pi^{-\frac{1}{2}}2^{-\frac{1}{4}(2b-1)}\Gamma(\tfrac{1}{4}-\tfrac{1}{2}b)\sin(\tfrac{1}{2}b\pi),$$

$$\bar{U}'(b,0)-U'(b,0) = \pi^{-\frac{1}{2}}2^{-\frac{1}{4}(2b-3)}\Gamma(\tfrac{3}{4}-\tfrac{1}{2}b)\cos(\tfrac{1}{2}b\pi).$$

Similar reasoning shows that *when* $b \leqslant 0$ *the number of non-negative zeros of* $\bar{U}(b,x)-U(b,x)$ *in* $[0,\infty)$ *is* $[1-\frac{1}{2}b]$. These zeros interlace with those of $U(b,x)$ and also with those of $\bar{U}(b,x)$, the largest zero belonging to $\bar{U}(b,x)-U(b,x)$.

5.6. *Monotonicity properties*

In this subsection we suppose that $b \geqslant 0$. From (5.4) it is seen that both $U(b,-x)$ and $\mathrm{d}U(b,-x)/\mathrm{d}x$ are positive at $x = 0$. Beginning with this result and repeatedly differentiating (5.1), we find that all coefficients in the Maclaurin series of $U(b,-x)$ are positive. Since this series has an infinite radius of convergence the function $U(b,-x)$ is absolutely monotonic in the interval $[0,\infty)$, that is, $U(b,-x)$ and all its x-derivatives are positive.

Now consider $U(b,x)$ in the same interval. At $x = 0$ this function is positive and its derivative negative. Let x_b be the smallest zero, if any, of $U(b,x)$ in $(0,\infty)$. Then from graphical considerations we see that $U'(b,x_b) \leqslant 0$. Again, repeated differentiations of (5.1) then show that all x-derivatives of $U(b,x)$ at x_b are non-positive, implying that $U(b,x)$ is negative in (x_b,∞) and therefore contradicting the first of (5.5). Therefore for non-negative b, $U(b,x)$ has no zeros in $[0,\infty)$. As a corollary, both $U(b,x)$ and $U''(b,x)$ are positive. Similar arguments show that $U'(b,x)$ is negative in the same interval.

On combining the results of this subsection, we have: *if* b $(\geqslant 0)$ *is fixed and* x *increases from* $-\infty$ *to* ∞, *then the solution* $U(b,x)$ *decreases strictly from* ∞ *to* 0, *and* $U'(b,x)$ *increases strictly from* $-\infty$ *to* 0.

5.7. *Graphs*

The graphs are plotted in figures 5.1–5.3. The vertical scale in figure 5.1 exceeds that in figures 5.2 and 5.3 by a factor of 500. Horizontal scales are the same in each figure.

5.8. *Auxiliary functions*

When b *is negative* the character of both $U(b,x)$ and $\bar{U}(b,x)$ is partly oscillatory and partly exponential. As in the case of the Airy functions (§ 5.3) we introduce a weight function $\mathsf{E}(b,x)$, modulus functions $\mathsf{M}(b,x)$ and $\mathsf{N}(b,x)$, and phase functions $\theta(b,x)$ and $\omega(b,x)$, defined as follows.

† Numerical computations indicate that the zeros of $\bar{U}(b,x)$, also, lie to the left of $x = 2(-b)^{\frac{1}{2}}$. The method of proof used for $U(b,x)$ fails, but for sufficiently large $-b$ the result follows from (5.13) and known properties of the zeros of the Bi function.

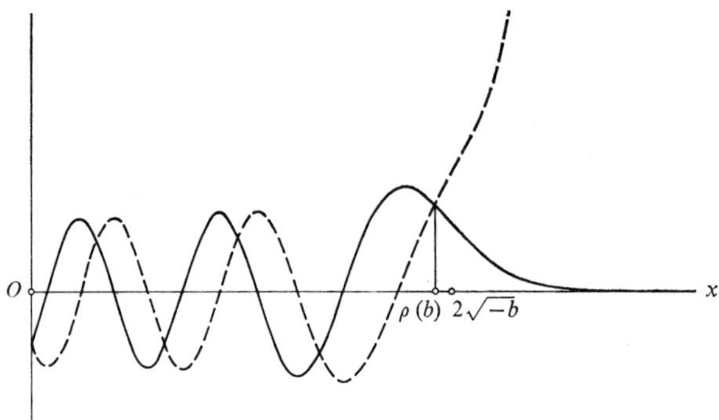

FIGURE 5.1. —— $U(-10, x)$; - - - $\bar{U}(-10, x)$.

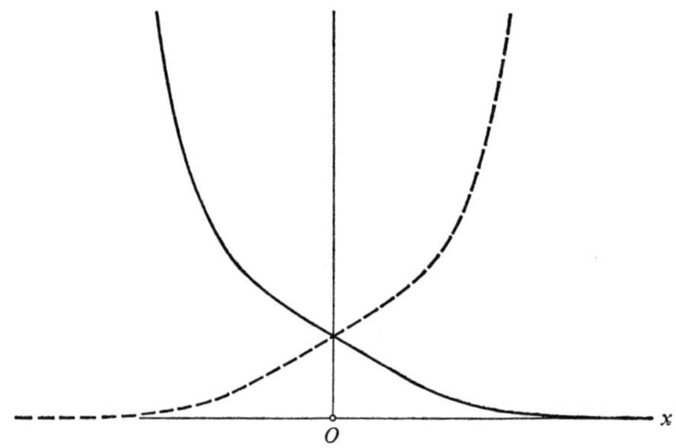

FIGURE 5.2. —— $U(0, x)$; - - - $\bar{U}(0, x)$ ($\equiv U(0, -x)$).

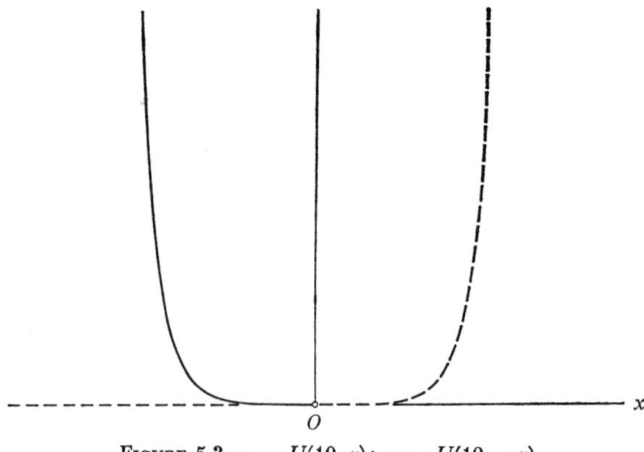

FIGURE 5.3. —— $U(10, x)$; - - - $U(10, -x)$.

First, we denote by $x = \rho(b)$ the (algebraically) largest real root of the equation

$$\bar{U}(b,x) = U(b,x).$$

From § 5.5 we already know that $\rho(0) = 0$ and $\rho(b) > 0$ when $b < 0$. That $\rho(b)$ is continuous when $b \in (-\infty, 0]$ follows by similar analysis to that on page 246 of Olver (1974). An asymptotic estimate of $\rho(b)$ for large negative b can be found from (5.12) and (5.13), using the approximation

$$\eta = 2^{\frac{1}{3}}(y-1) + O\{(y-1)^2\} \quad (y \to 1), \tag{5.20}$$

obtained from (5.14). Thus

$$\rho(b) = 2(-b)^{\frac{1}{2}} + c(-b)^{-\frac{1}{6}} + O(b^{-\frac{5}{6}}) \quad (b \to -\infty), \tag{5.21}$$

where c is defined in § 5.3.

We now define, for $b \leqslant 0$,

$$\mathsf{E}(b,x) = 1 \quad (0 \leqslant x \leqslant \rho(b)), \qquad \mathsf{E}(b,x) = \{\bar{U}(b,x)/U(b,x)\}^{\frac{1}{2}} \quad (x \geqslant \rho(b)). \tag{5.22}$$

It is easily seen that $\mathsf{E}(b,x)$ is continuous in the region $-\infty < b \leqslant 0$, $0 \leqslant x < \infty$. Also, from (5.8) we have

$$\frac{\mathrm{d}}{\mathrm{d}x}\frac{\bar{U}(b,x)}{U(b,x)} = \left(\frac{2}{\pi}\right)^{\frac{1}{2}} \frac{\Gamma(\frac{1}{2}-b)}{U^2(b,x)}.$$

Therefore *when b is fixed, $\mathsf{E}(b,x)$ is non-decreasing in the interval* $0 \leqslant x < \infty$.

Again, for $b \leqslant 0$ and $x \geqslant 0$, we define

$$U(b,x) = \mathsf{E}^{-1}(b,x)\,\mathsf{M}(b,x)\sin\theta(b,x), \quad \bar{U}(b,x) = \mathsf{E}(b,x)\,\mathsf{M}(b,x)\cos\theta(b,x),$$

$$U'(b,x) = \mathsf{E}^{-1}(b,x)\,\mathsf{N}(b,x)\sin\omega(b,x), \quad \bar{U}'(b,x) = \mathsf{E}(b,x)\,\mathsf{N}(b,x)\cos\omega(b,x),$$

where $\mathsf{E}^{-1}(b,x) \equiv 1/\mathsf{E}(b,x)$. Thus for $0 \leqslant x \leqslant \rho(b)$

$$\mathsf{M}(b,x) = \{U^2(b,x) + \bar{U}^2(b,x)\}^{\frac{1}{2}}, \quad \theta(b,x) = \tan^{-1}\{U(b,x)/\bar{U}(b,x)\},$$

$$\mathsf{N}(b,x) = \{U'^2(b,x) + \bar{U}'^2(b,x)\}^{\frac{1}{2}}, \quad \omega(b,x) = \tan^{-1}\{U'(b,x)/\bar{U}'(b,x)\};$$

whereas for $x \geqslant \rho(b)$

$$\mathsf{M}(b,x) = \{2U(b,x)\,\bar{U}(b,x)\}^{\frac{1}{2}}, \quad \theta(b,x) = \tfrac{1}{4}\pi,$$

$$\mathsf{N}(b,x) = \left\{\frac{U'^2(b,x)\,\bar{U}^2(b,x) + \bar{U}'^2(b,x)\,U^2(b,x)}{U(b,x)\,\bar{U}(b,x)}\right\}^{\frac{1}{2}}, \quad \omega(b,x) = \arctan\left\{\frac{U'(b,x)\,\bar{U}(b,x)}{\bar{U}'(b,x)\,U(b,x)}\right\}.$$

As in the case of $\mathsf{E}(b,x)$, both $\mathsf{M}(b,x)$ and $\mathsf{N}(b,x)$ are continuous in the region $-\infty < b \leqslant 0$, $0 \leqslant x < \infty$.

For large x, we obtain from (5.5) and (5.6)

$$\mathsf{M}(b,x) \sim \left(\frac{8}{\pi}\right)^{\frac{1}{4}} \frac{\{\Gamma(\frac{1}{2}-b)\}^{\frac{1}{2}}}{x^{\frac{1}{2}}}, \quad \mathsf{N}(b,x) \sim \frac{\{\Gamma(\frac{1}{2}-b)\}^{\frac{1}{2}}}{(2\pi)^{\frac{1}{4}}} x^{\frac{1}{2}}. \tag{5.23}$$

These relations hold for fixed b, and are also uniform for b ranging over any compact interval in $(-\infty, 0]$.

For positive b, the monotonicity properties established in § 5.6 obviate any need for auxiliary functions.

6. Case I

6.1. *Main theorem*

We return to the central problem of the present investigation, that is, the approximate solution of the various forms of differential equation resulting from the transformations of § 2. In the present section we treat

$$d^2w/d\zeta^2 = \{u^2(\zeta^2-\alpha^2)+\psi(u,\alpha,\zeta)\}\,w. \tag{6.1}$$

This is (2.9) with the symbol W changed to w (to avoid later conflict). *In subsequent analysis* the symbol A denotes an endpoint of an interval containing the parameter α; A is a finite constant and is always positive. Similarly ζ_1 and ζ_2 denote endpoints of an interval containing ζ; ζ_1 is negative and ζ_2 is positive, and both may depend continuously on α or be infinite. In addition, $\Omega(x)$ denotes a conveniently chosen continuous function of the real variable x that is positive, except possibly when $x = 0$, and satisfies

$$\Omega(x) = O(x) \quad (x \to \pm\infty). \tag{6.2}$$

With this understanding, we introduce the error-control function

$$F(u,\alpha,\zeta) = \int \frac{\psi(u,\alpha,\zeta)}{\Omega(\zeta\sqrt{2u})}\,d\zeta, \tag{6.3}$$

the choice of integration constant being immaterial.

Theorem I. *Assume that for each value of u, the function* $\psi(u,\alpha,\zeta)$ *is continuous in the region* $\alpha\in[0,\mathrm{A}]$, $\zeta\in[0,\zeta_2)$, *and*†

$$\mathscr{V}_{0,\zeta_2}(F) \equiv \int_0^{\zeta_2} \frac{|\psi(u,\alpha,t)|}{\Omega(t\sqrt{2u})}\,dt$$

converges uniformly with respect to α. *Then in this region equation* (6.1) *has solutions* $w_1(u,\alpha,\zeta)$ *and* $w_2(u,\alpha,\zeta)$ *which are continuous, have continuous first and second partial* ζ*-derivatives, and are given by*

$$w_1(u,\alpha,\zeta) = U(-\tfrac{1}{2}u\alpha^2, \zeta\sqrt{2u}) + \epsilon_1(u,\alpha,\zeta), \quad w_2(u,\alpha,\zeta) = \bar{U}(-\tfrac{1}{2}u\alpha^2, \zeta\sqrt{2u}) + \epsilon_2(u,\alpha,\zeta), \tag{6.4}$$

where

$$\frac{|\epsilon_1(u,\alpha,\zeta)|}{\mathsf{M}(-\frac{1}{2}u\alpha^2,\zeta\sqrt{2u})},\ \frac{|\partial\epsilon_1(u,\alpha,\zeta)/\partial\zeta|}{(2u)^{\frac{1}{2}}\mathsf{N}(-\frac{1}{2}u\alpha^2,\zeta\sqrt{2u})} \leqslant \mathsf{E}^{-1}(-\tfrac{1}{2}u\alpha^2,\zeta\sqrt{2u})\left[\exp\left\{\frac{1}{2}\left(\frac{\pi}{u}\right)^{\frac{1}{2}} l_1(-\tfrac{1}{2}u\alpha^2)\,\mathscr{V}_{\zeta,\zeta_2}(F)\right\}-1\right], \tag{6.5}$$

$$\frac{|\epsilon_2(u,\alpha,\zeta)|}{\mathsf{M}(-\frac{1}{2}u\alpha^2,\zeta\sqrt{2u})},\ \frac{|\partial\epsilon_2(u,\alpha,\zeta)/\partial\zeta|}{(2u)^{\frac{1}{2}}\mathsf{N}(-\frac{1}{2}u\alpha^2,\zeta\sqrt{2u})} \leqslant \mathsf{E}(-\tfrac{1}{2}u\alpha^2,\zeta\sqrt{2u})\left[\exp\left\{\frac{1}{2}\left(\frac{\pi}{u}\right)^{\frac{1}{2}} l_1(-\tfrac{1}{2}u\alpha^2)\,\mathscr{V}_{0,\zeta}(F)\right\}-1\right], \tag{6.6}$$

and

$$l_1(b) = \sup_{x\in(0,\infty)} \{\Omega(x)\,\mathsf{M}^2(b,x)/\Gamma(\tfrac{1}{2}-b)\} \quad (b \leqslant 0). \tag{6.7}$$

In this theorem the auxiliary functions E, M, and N are defined as in § 5.8.

Before starting the proof, we note that as a consequence of (5.23) and (6.2), the supremum in (6.7) is finite for each value of b. Moreover, because the relations (5.23) hold uniformly in compact b-intervals, $l_1(b)$ is continuous. The behaviour of $l_1(b)$ as $b\to-\infty$ is considered in § 6.2.

† $\mathscr{V}$ denotes the variational operator, as in Olver (1974, pp. 27–28).

To begin the proof, we observe that the approximant $U(-\frac{1}{2}u\alpha^2, \zeta\sqrt{2u})$ satisfies the differential equation

$$d^2U/d\zeta^2 = u^2(\zeta^2-\alpha^2)\,U;$$

compare (5.1). Subtraction from (6.1) gives the following inhomogeneous differential equation for the error term:

$$(d^2\epsilon_1/d\zeta^2) - u^2(\zeta^2-\alpha^2)\,\epsilon_1 = \psi(u,\alpha,\zeta)\,\{\epsilon_1 + U(-\tfrac{1}{2}u\alpha^2, \zeta\sqrt{2u})\}.$$

An equivalent integral equation is found by applying the method of variation of parameters and using the first of (5.8). Thus

$$\epsilon_1(u,\alpha,\zeta) = \frac{1}{2}\left(\frac{\pi}{u}\right)^{\frac{1}{2}} \frac{1}{\Gamma(\frac{1}{2}+\frac{1}{2}u\alpha^2)} \int_\zeta^{\zeta_2} \mathsf{K}(\zeta,t)\,\psi(u,\alpha,t)\,\{\epsilon_1(u,\alpha,t) + U(-\tfrac{1}{2}u\alpha^2, t\sqrt{2u})\}\,dt, \qquad (6.8)$$

where

$$\mathsf{K}(\zeta,t) = U(\zeta\sqrt{2u})\,\bar{U}(t\sqrt{2u}) - \bar{U}(\zeta\sqrt{2u})\,U(t\sqrt{2u}). \qquad (6.9)$$

In the last equation the parameter $-\frac{1}{2}u\alpha^2$ has been suppressed for conciseness; for example, $U(\zeta\sqrt{2u})$ means $U(-\frac{1}{2}u\alpha^2, \zeta\sqrt{2u})$. On expressing the kernel in terms of the auxiliary functions of § 5.8 and again suppressing the parameter $-\frac{1}{2}u\alpha^2$, we have

$$\mathsf{K}(\zeta,t) = \mathsf{E}^{-1}(\zeta\sqrt{2u})\,\mathsf{E}(t\sqrt{2u})\,\mathsf{M}(\zeta\sqrt{2u})\,\mathsf{M}(t\sqrt{2u})$$
$$\times\{\sin\theta(\zeta\sqrt{2u})\cos\theta(t\sqrt{2u}) - \mathsf{E}^2(\zeta\sqrt{2u})\,\mathsf{E}^{-2}(t\sqrt{2u})\cos\theta(\zeta\sqrt{2u})\sin\theta(t\sqrt{2u})\}.$$

Since $0 \leqslant \zeta \leqslant t$ and $\mathsf{E}(t\sqrt{2u})$ is non-decreasing, it follows that

$$\mathsf{E}^2(\zeta\sqrt{2u})\,\mathsf{E}^{-2}(t\sqrt{2u}) \leqslant 1,$$

and thence

$$|\mathsf{K}(\zeta,t)| \leqslant \mathsf{E}^{-1}(\zeta\sqrt{2u})\,\mathsf{E}(t\sqrt{2u})\,\mathsf{M}(\zeta\sqrt{2u})\,\mathsf{M}(t\sqrt{2u}).$$

Similarly,

$$|\partial\mathsf{K}(\zeta,t)/\partial\zeta| \leqslant (2u)^{\frac{1}{2}}\,\mathsf{E}^{-1}(\zeta\sqrt{2u})\,\mathsf{E}(t\sqrt{2u})\,\mathsf{N}(\zeta\sqrt{2u})\,\mathsf{M}(t\sqrt{2u}).$$

The bounds (6.5) and (6.6) are obtained on solving (6.8) by the standard procedure of successive approximations, for example, by application of theorem 10.2 of Olver (1974, chapter 6). In the notation of this theorem we set

$$\phi(t) = \psi_0(t) = \psi(u,\alpha,t)/\Omega(t\sqrt{2u}), \quad \psi_1(t) = 0, \quad J(t) = U(t\sqrt{2u}).$$

The kernel $\mathsf{K}(\zeta,t)$ then becomes $-\frac{1}{2}(\pi/u)^{\frac{1}{2}}\,\Omega(t\sqrt{2u})\,\mathsf{K}(\zeta,t)/\Gamma(\frac{1}{2}+\frac{1}{2}u\alpha^2)$. Also,

$$Q(t) = \tfrac{1}{2}(\pi/u)^{\frac{1}{2}}\,\Omega(t\sqrt{2u})\,\mathsf{E}(t\sqrt{2u})\,\mathsf{M}(t\sqrt{2u})/\Gamma(\tfrac{1}{2}+\tfrac{1}{2}u\alpha^2),$$

$$P_0(\zeta) = \mathsf{E}^{-1}(\zeta\sqrt{2u})\,\mathsf{M}(\zeta\sqrt{2u}), \quad P_1(\zeta) = (2u)^{\frac{1}{2}}\,\mathsf{E}^{-1}(\zeta\sqrt{2u})\,\mathsf{N}(\zeta\sqrt{2u}).$$

The role of the integration limit α is played here by ζ_2; this gives

$$\Phi(\zeta) = \Psi_0(\zeta) = \mathcal{V}_{\zeta,\zeta_2}(F), \quad \Psi_1(\zeta) = 0, \quad \kappa_0 \leqslant \tfrac{1}{2}(\pi/u)^{\frac{1}{2}}\,l_1(-\tfrac{1}{2}u\alpha^2).$$

The remaining quantity needed is κ; for simplicity we replace it by the upper bound κ_0. The bounds (6.5) and (6.6) then follow from the theorem.

To prove that $w_1(u,\alpha,\zeta)$ and its first two partial ζ-derivatives are continuous in α and ζ, it suffices to show that the same is true of $\epsilon_1(u,\alpha,\zeta)$. This is achievable in a straightforward manner by verifying that the integrals which occur in the analysis converge uniformly when $\alpha\in[0,\mathrm{A}]$ and ζ lies in any compact interval in $[0,\zeta_2)$.

The proof for the second solution is similar. This completes the proof of theorem I.

It may be noted that theorem I remains valid if the function $\Omega(x)$ is permitted to depend on the parameters u and α, provided that the relation (6.2) still holds. For example, we might take

$$\Omega(x) = \Gamma(\tfrac{1}{2}+\tfrac{1}{2}u\alpha^2)/\mathsf{M}^2(-\tfrac{1}{2}u\alpha^2, |x|);$$

compare the first of (5.23). In this event $l_1(-\tfrac{1}{2}u\alpha^2) = 1$. We shall not need this increased flexibility in the subsequent analysis, however.

6.2. *Behaviour of $l_1(b)$ as $b \to -\infty$*

As a preliminary to the determination of the asymptotic form of the error bounds (6.5) and (6.6) for large u, we investigate the behaviour of the quantity defined by (6.7) for large negative values of b.

For large positive μ we have from §§ 5.3 and 5.8

$$\frac{\mathsf{M}^2(-\tfrac{1}{2}\mu^2, \mu y\sqrt{2})}{\Gamma(\tfrac{1}{2}+\tfrac{1}{2}\mu^2)} = \frac{4\pi^{\frac{1}{2}}}{\mu^{\frac{1}{3}}}\left(\frac{\eta}{y^2-1}\right)^{\frac{1}{2}}\left\{\mathrm{Ai}\,(\mu^{\frac{4}{3}}\eta)\,\mathrm{Bi}\,(\mu^{\frac{4}{3}}\eta) + M^2(\mu^{\frac{4}{3}}\eta)\,O\left(\frac{1}{\mu^2}\right)\right\}, \tag{6.10}$$

or

$$\frac{\mathsf{M}^2(-\tfrac{1}{2}\mu^2, \mu y\sqrt{2})}{\Gamma(\tfrac{1}{2}+\tfrac{1}{2}\mu^2)} = \frac{2\pi^{\frac{1}{2}}}{\mu^{\frac{1}{3}}}\left(\frac{\eta}{y^2-1}\right)^{\frac{1}{2}}\left\{\mathrm{Ai}^2\,(\mu^{\frac{4}{3}}\eta) + \mathrm{Bi}^2\,(\mu^{\frac{4}{3}}\eta) + E^2(\mu^{\frac{4}{3}}\eta)\,M^2(\mu^{\frac{4}{3}}\eta)\,O\left(\frac{1}{\mu^2}\right)\right\}, \tag{6.11}$$

according as $\mu y\sqrt{2} > \rho(-\tfrac{1}{2}\mu^2)$ or $0 \leqslant \mu y\sqrt{2} \leqslant \rho(-\tfrac{1}{2}\mu^2)$. In each case the estimate $O(1/\mu^2)$ is uniform with respect to y. From (5.21) it is seen that the changeover point from (6.10) to (6.11) is given by

$$y = \rho(-\tfrac{1}{2}\mu^2)/(\mu\sqrt{2}) = 1 + 2^{-\frac{1}{3}}c\mu^{-\frac{4}{3}} + O(\mu^{-\frac{8}{3}}) \quad (\mu \to \infty).$$

Correspondingly, $\mu^{\frac{4}{3}}\eta = c + O(\mu^{-\frac{4}{3}})$; compare (5.20). Since $E(x)$ is bounded for $0 \leqslant x \leqslant c + O(\mu^{-\frac{4}{3}})$ equation (6.11) may be rewritten in the form

$$\frac{\mathsf{M}^2(-\tfrac{1}{2}\mu^2, \mu y\sqrt{2})}{\Gamma(\tfrac{1}{2}+\tfrac{1}{2}\mu^2)} = \frac{2\pi^{\frac{1}{2}}}{\mu^{\frac{1}{3}}}\left(\frac{\eta}{y^2-1}\right)^{\frac{1}{2}}\left\{\mathrm{Ai}^2\,(\mu^{\frac{4}{3}}\eta) + \mathrm{Bi}^2\,(\mu^{\frac{4}{3}}\eta) + M^2(\mu^{\frac{4}{3}}\eta)\,O\left(\frac{1}{\mu^2}\right)\right\}, \tag{6.12}$$

the O-term again being uniform for $0 \leqslant \mu y\sqrt{2} \leqslant \rho(-\tfrac{1}{2}\mu^2)$.

By referring to the asymptotic approximations of the functions Ai, Bi and M,† we derive from (6.10) and (6.12)

$$\frac{\mathsf{M}^2(-\tfrac{1}{2}\mu^2, \mu y\sqrt{2})}{\Gamma(\tfrac{1}{2}+\tfrac{1}{2}\mu^2)} \leqslant \frac{K}{\mu^{\frac{1}{3}}}\left(\frac{\eta}{y^2-1}\right)^{\frac{1}{2}}\frac{1}{1+\mu^{\frac{2}{3}}|\eta|^{\frac{1}{2}}} \quad (\mu \geqslant 1,\ y \geqslant 0),$$

where K is an assignable constant. When $y \to \infty$ the first of (5.14) yields $\eta \sim (\tfrac{3}{4})^{\frac{2}{3}}y^{\frac{4}{3}}$. Hence

$$\left(\frac{\eta}{y^2-1}\right)^{\frac{1}{2}} \leqslant \frac{K}{1+|\eta|^{\frac{1}{4}}} \quad (y \geqslant 0),$$

where the symbol K is now being used generically.‡ Also, as a consequence of the assumption that Ω is continuous and satisfies (6.2), we have

$$\Omega(\mu y\sqrt{2}) \leqslant K(1+\mu y) \leqslant K\mu(1+|\eta|^{\frac{3}{4}}) \quad (\mu \geqslant 1,\ y \geqslant 0). \tag{6.13}$$

Combination of the last three inequalities yields

$$\Omega(\mu y\sqrt{2})\,\frac{\mathsf{M}^2(-\tfrac{1}{2}\mu^2, \mu y\sqrt{2})}{\Gamma(\tfrac{1}{2}+\tfrac{1}{2}\mu^2)} \leqslant K\mu^{\frac{2}{3}}\frac{1+|\eta|^{\frac{3}{4}}}{1+|\eta|^{\frac{1}{4}}}\,\frac{1}{1+\mu^{\frac{2}{3}}|\eta|^{\frac{1}{2}}} \leqslant K\mu^{\frac{2}{3}} \quad (\mu \geqslant 1,\ y \geqslant 0).$$

† Olver (1974, pp. 392–395).

‡ That is, K does not necessarily denote the same constant at each appearance.

Then referring to (6.7) and replacing $\frac{1}{2}\mu^2$ by $-b$, we arrive at the desired result

$$l_1(b) = O(b^{\frac{1}{3}}) \quad (b \to -\infty). \tag{6.14}$$

In the absence of any further restrictions on $\Omega(x)$ the estimate (6.14) is the best possible: this is easily verifiable by setting $y = 1$ and $\eta = 0$. If, however, we prescribe $\Omega(x) = |x|^{\frac{1}{3}}$, then (6.13) is replaceable by

$$\Omega(\mu y \sqrt{2}) = 2^{\frac{1}{6}}\mu^{\frac{1}{3}}y^{\frac{1}{3}} \leqslant K\mu^{\frac{1}{3}}(1+|\eta|^{\frac{1}{4}}),$$

which immediately yields the improvement

$$l_1(b) = O(1) \quad (b \to -\infty). \tag{6.15}$$

6.3. *Uniform asymptotic properties of the error term*

The asymptotic behaviour of the solutions $w_1(u, \alpha, \zeta)$ and $w_2(u, \alpha, \zeta)$ for large u is deducible from (6.5), (6.6) and the results just obtained concerning $l_1(b)$. For simplicity, we confine attention to cases in which $\psi(u, \alpha, \zeta)$ is independent of u, but this restriction is not essential: other cases, including, for example, the transformed form of (1.3), can be attacked in a similar manner. We shall also suppose that $\zeta_2 = \infty$, which is the most interesting situation in applications.

Assume first that $|\psi(\alpha, \zeta)|/\zeta$ is integrable at $\zeta = \infty$, uniformly with respect to α. Then we may prescribe $\Omega(x) = 1 + |x|$, yielding the finite variation

$$\mathscr{V}_{0,\infty}(F) = \int_0^\infty \frac{|\psi(\alpha, t)|}{1 + t\sqrt{2u}}\,\mathrm{d}t.$$

Subdividing the integration range at $t = 1$ and again using the symbol K generically to denote a constant, we obtain

$$\mathscr{V}_{0,\infty}(F) \leqslant K\int_0^1 \frac{\mathrm{d}t}{1+t\sqrt{2u}} + \frac{1}{\sqrt{2u}}\int_1^\infty \frac{|\psi(\alpha, t)|}{t}\,\mathrm{d}t = O\left(\frac{\ln u}{u^{\frac{1}{2}}}\right). \tag{6.16}$$

Next, equation (6.14) gives $l_1(-\frac{1}{2}u\alpha^2) = O(u^{\frac{1}{3}})$. Substitution of these results in (6.5) and (6.6) leads to the required results

$$\epsilon_1(u, \alpha, \zeta) = \mathsf{E}^{-1}(-\tfrac{1}{2}u\alpha^2, \zeta\sqrt{2u})\,\mathsf{M}(-\tfrac{1}{2}u\alpha^2, \zeta\sqrt{2u})\,O(u^{-\frac{2}{3}}\ln u), \tag{6.17}$$

$$\epsilon_2(u, \alpha, \zeta) = \mathsf{E}(-\tfrac{1}{2}u\alpha^2, \zeta\sqrt{2u})\,\mathsf{M}(-\tfrac{1}{2}u\alpha^2, \zeta\sqrt{2u})\,O(u^{-\frac{2}{3}}\ln u), \tag{6.18}$$

as $u \to \infty$, uniformly for $\alpha \in [0, \mathrm{A}]$ and $\zeta \in [0, \infty)$. Similarly,

$$\partial\epsilon_1(u, \alpha, \zeta)/\partial\zeta = \mathsf{E}^{-1}(-\tfrac{1}{2}u\alpha^2, \zeta\sqrt{2u})\,\mathsf{N}(-\tfrac{1}{2}u\alpha^2, \zeta\sqrt{2u})\,O(u^{-\frac{1}{6}}\ln u), \tag{6.19}$$

$$\partial\epsilon_2(u, \alpha, \zeta)/\partial\zeta = \mathsf{E}(-\tfrac{1}{2}u\alpha^2, \zeta\sqrt{2u})\,\mathsf{N}(-\tfrac{1}{2}u\alpha^2, \zeta\sqrt{2u})\,O(u^{-\frac{1}{6}}\ln u). \tag{6.20}$$

Without further restrictions on $\psi(\alpha, \zeta)$ these O-estimates are the best attainable. We mention, however, two interesting special cases in which the factor $\ln u$ can be discarded from (6.17) to (6.20). The first occurs when $|\psi(\alpha, \zeta)|/\zeta^{\frac{2}{3}}$ is uniformly integrable at $\zeta = \infty$. In this event we merely prescribe $\Omega(x) = |x|^{\frac{1}{3}}$ and use (6.15) in place of (6.14). The second case occurs when $|\psi(\alpha, \zeta)|/\zeta$ is uniformly integrable at $\zeta = 0$ (as well as at $\zeta = \infty$). With this extra condition it is permissible to prescribe $\Omega(x) = |x|$, which leads to the stated improvement.

Another interesting situation arises when $\alpha = 0$, that is, when equation (6.1) has a double turning point at $\zeta = 0$. We assume again that $|\psi(0, \zeta)|/\zeta$ is integrable at infinity. Then $l_1(-\frac{1}{2}u\alpha^2)$

is independent of u, and by taking $\Omega(x)$ to be $|x|$ and $1+|x|$ in turn, we see that the term $O(u^{-\frac{2}{3}}\ln u)$ in (6.17) and (6.18) may be strengthened to $O(u^{-1})$ or $O(u^{-1}\ln u)$, depending whether or not $|\psi(0,\zeta)|/\zeta$ is also integrable at $\zeta = 0$. Similarly for (6.19) and (6.20).

6.4. *Solutions when* $\zeta \leqslant 0$

Theorem I applies directly only to non-negative values of ζ: it is not possible to construct error bounds similar to (6.5) and (6.6) which hold for negative ζ. This is because as u increases in a continuous manner the asymptotic character of the approximant $U(-\frac{1}{2}u\alpha^2, \zeta\sqrt{2u})$ at $\zeta = -\infty$ changes abruptly as $u\alpha^2$ passes through odd integer values; compare (5.5), (5.6) and (5.9). We cannot expect $w_1(u,\alpha,\zeta)$ to exhibit this change in behaviour at *exactly* the same values of $u\alpha^2$. Similar considerations also apply to $w_2(u,\alpha,\zeta)$.

To a certain extent, however, the asymptotic behaviour of $w_1(u,\alpha,\zeta)$ and $w_2(u,\alpha,\zeta)$ for large u and negative values of ζ can be determined by use of connection formulae. For simplicity, suppose that $\zeta_1 = -\infty$, $\zeta_2 = \infty$, and the function $|\psi(u,\alpha,\zeta)/\zeta|$ is independent of u and integrable at $\zeta = -\infty$ and $+\infty$, uniformly with respect to α in both cases. Then, as we have already seen, (6.17) and (6.18) apply as $u\to\infty$, uniformly with respect to $\alpha\in[0,\mathrm{A}]$ and $\zeta\in[0,\infty)$. Moreover, on replacing ζ by $-\zeta$ and again applying theorem I, we immediately see that there exist two further solutions of (6.1), given by

$$w_3(u,\alpha,\zeta) = U(-\zeta\sqrt{2u}) + \mathsf{E}^{-1}(-\zeta\sqrt{2u})\,\mathsf{M}(-\zeta\sqrt{2u})\,O(u^{-\frac{2}{3}}\ln u), \tag{6.21}$$

$$w_4(u,\alpha,\zeta) = \bar{U}(-\zeta\sqrt{2u}) + \mathsf{E}(-\zeta\sqrt{2u})\,\mathsf{M}(-\zeta\sqrt{2u})\,O(u^{-\frac{2}{3}}\ln u), \tag{6.22}$$

as $u\to\infty$, uniformly with respect to $\alpha\in[0,\mathrm{A}]$ and $\zeta\in(-\infty,0]$. Again, for brevity the parameter $-\frac{1}{2}u\alpha^2$ has been suppressed in the functions on the right-hand sides.

It suffices to determine approximations for the coefficients A_1, B_1, A_2 and B_2 in the linear relations

$$w_1(u,\alpha,\zeta) = A_1 w_3(u,\alpha,\zeta) + B_1 w_4(u,\alpha,\zeta), \tag{6.23}$$

$$w_2(u,\alpha,\zeta) = A_2 w_3(u,\alpha,\zeta) + B_2 w_4(u,\alpha,\zeta). \tag{6.24}$$

Clearly,

$$A_1 = \frac{w_4'(u,\alpha,0)\,w_1(u,\alpha,0) - w_1'(u,\alpha,0)\,w_4(u,\alpha,0)}{w_4'(u,\alpha,0)\,w_3(u,\alpha,0) - w_3'(u,\alpha,0)\,w_4(u,\alpha,0)},$$

$$B_1 = -\frac{w_3'(u,\alpha,0)\,w_1(u,\alpha,0) - w_1'(u,\alpha,0)\,w_3(u,\alpha,0)}{w_4'(u,\alpha,0)\,w_3(u,\alpha,0) - w_3'(u,\alpha,0)\,w_4(u,\alpha,0)},$$

with similar expressions for A_2 and B_2. Using the results of §§ 5.2 and 5.8, we have $\mathsf{E}(0) = 1$ and

$$w_1(u,\alpha,0) = \mathsf{M}(0)\,\{\sin\beta + O(u^{-\frac{2}{3}}\ln u)\}, \quad w_1'(u,\alpha,0) = -(2u)^{\frac{1}{2}}\mathsf{N}(0)\,\{\cos\beta + O(u^{-\frac{2}{3}}\ln u)\},$$

$$w_2(u,\alpha,0) = \mathsf{M}(0)\,\{\cos\beta + O(u^{-\frac{2}{3}}\ln u)\}, \quad w_2'(u,\alpha,0) = (2u)^{\frac{1}{2}}\mathsf{N}(0)\,\{\sin\beta + O(u^{-\frac{2}{3}}\ln u)\},$$

$$w_3(u,\alpha,0) = \mathsf{M}(0)\,\{\sin\beta + O(u^{-\frac{2}{3}}\ln u)\}, \quad w_3'(u,\alpha,0) = (2u)^{\frac{1}{2}}\mathsf{N}(0)\,\{\cos\beta + O(u^{-\frac{2}{3}}\ln u)\},$$

$$w_4(u,\alpha,0) = \mathsf{M}(0)\,\{\cos\beta + O(u^{-\frac{2}{3}}\ln u)\}, \quad w_4'(u,\alpha,0) = -(2u)^{\frac{1}{2}}\mathsf{N}(0)\,\{\sin\beta + O(u^{-\frac{2}{3}}\ln u)\},$$

where $\beta \equiv (\frac{1}{4} + \frac{1}{4}u\alpha^2)\,\pi$. Substituting in the above expressions for A_1 and B_1 and the corresponding expressions for A_2 and B_2 we obtain the required approximations

$$A_1 = \sin(\tfrac{1}{2}\pi u\alpha^2) + O(u^{-\frac{2}{3}}\ln u), \quad B_1 = \cos(\tfrac{1}{2}\pi u\alpha^2) + O(u^{-\frac{2}{3}}\ln u),$$

$$A_2 = \cos(\tfrac{1}{2}\pi u\alpha^2) + O(u^{-\frac{2}{3}}\ln u), \quad B_2 = -\sin(\tfrac{1}{2}\pi u\alpha^2) + O(u^{-\frac{2}{3}}\ln u),$$

as $u\to\infty$, uniformly for $\alpha\in[0,\mathrm{A}]$.

An important deduction from these results concerns the eigenvalues of equation (6.1), that is, those values of α for which there exists a solution that is recessive at both $\zeta=-\infty$ and $\zeta=\infty$. By referring to the asymptotic form of $w_1(u,\alpha,\zeta)$ as $\zeta\to+\infty$, and the asymptotic forms of $w_3(u,\alpha,\zeta)$ and $w_4(u,\alpha,\zeta)$ as $\zeta\to-\infty$ (compare (6.21) and (6.22)), we see from (6.23) that the condition for the existence of an eigenvalue is given by $B_1=0$. Accordingly,

$$\alpha=\left(\frac{2s+1}{u}\right)^{\frac{1}{2}}\left\{1+\frac{1}{2s+1}O\left(\frac{\ln u}{u^{\frac{2}{3}}}\right)\right\}\quad(u\to\infty),$$

where s is a non-negative integer. This result is valid for fixed s. It is also uniformly valid for values of s that satisfy $2s+1\leqslant\kappa A^2u$, where $[0,A]$ is the range of α, and κ is any constant in $(0,1)$.

7. Case II

7.1. *Main theorem*

In this section we consider the differential equation

$$d^2w/d\zeta^2=\{u^2(\alpha^2+\zeta^2)+\psi(u,\alpha,\zeta)\}w. \tag{7.1}$$

The analysis is easier than for case I, because the approximants $U(\frac{1}{2}u\alpha^2,\pm\zeta\sqrt{2u})$ are monotonic and non-vanishing throughout the doubly infinite interval $-\infty<\zeta<\infty$. We specify the functions $\Omega(x)$ and $F(u,\alpha,\zeta)$ exactly as in § 6.1.

Theorem II. *Assume that for each value of u, the function $\psi(u,\alpha,\zeta)$ is continuous in the region $\alpha\in[0,A]$, $\zeta\in(\zeta_1,\zeta_2)$, and*

$$\mathscr{V}_{\zeta_1,\zeta_2}(F)\equiv\int_{\zeta_1}^{\zeta_2}\frac{|\psi(u,\alpha,t)|}{\Omega(t\sqrt{2u})}dt$$

converges uniformly with respect to α. Then in this region equation (7.1) has a continuous solution $w(u,\alpha,\zeta)$, with continuous first and second partial ζ-derivatives, given by

$$w(u,\alpha,\zeta)=U(\tfrac{1}{2}u\alpha^2,\zeta\sqrt{2u})+\epsilon(u,\alpha,\zeta), \tag{7.2}$$

where

$$|\epsilon(u,\alpha,\zeta)|\leqslant U(\tfrac{1}{2}u\alpha^2,\zeta\sqrt{2u})\left[\exp\left\{\frac{l_2(\frac{1}{2}u\alpha^2)}{2(\pi u)^{\frac{1}{2}}}\mathscr{V}_{\zeta,\zeta_2}(F)\right\}-1\right], \tag{7.3}$$

$$\left|\frac{\partial\epsilon(u,\alpha,\zeta)}{\partial\zeta}\right|\leqslant(2u)^{\frac{1}{2}}\hat{l}_2(\tfrac{1}{2}u\alpha^2)\,|U'(\tfrac{1}{2}u\alpha^2,\zeta\sqrt{2u})|\left[\exp\left\{\frac{l_2(\frac{1}{2}u\alpha^2)}{2(\pi u)^{\frac{1}{2}}}\mathscr{V}_{\zeta,\zeta_2}(F)\right\}-1\right], \tag{7.4}$$

and

$$l_2(b)=\sup_{x\in(-\infty,\infty)}\{\Gamma(\tfrac{1}{2}+b)\,\Omega(x)\,U(b,x)\,U(b,-x)\}\quad(b\geqslant0), \tag{7.5}$$

$$\hat{l}_2(b)=1+\sup_{x\in(-\infty,\infty)}\left\{\frac{U'(b,-x)\,U(b,x)}{U'(b,x)\,U(b,-x)}\right\}\quad(b\geqslant0). \tag{7.6}$$

Considerations similar to those of § 6.1 show that $l_2(b)$ and $\hat{l}_2(b)$ are positive and continuous when $b\in[0,\infty)$.

The proof of theorem II parallels that of theorem I. The inhomogeneous differential equation for the error term is

$$(d^2\epsilon/d\zeta^2)-u^2(\alpha^2+\zeta^2)\epsilon=\psi(u,\alpha,\zeta)\{\epsilon+U(\tfrac{1}{2}u\alpha^2,\zeta\sqrt{2u})\},$$

with corresponding integral equation

$$\epsilon(u,\alpha,\zeta)=\frac{\Gamma(\frac{1}{2}+\frac{1}{2}u\alpha^2)}{2(\pi u)^{\frac{1}{2}}}\int_{\zeta}^{\zeta_2}\mathsf{K}(\zeta,t)\,\psi(u,\alpha,t)\{\epsilon(u,\alpha,t)+U(\tfrac{1}{2}u\alpha^2,t\sqrt{2u})\}\,dt, \tag{7.7}$$

in which $$\mathsf{K}(\zeta,t) = U(\zeta\sqrt{2u})\,U(-t\sqrt{2u}) - U(-\zeta\sqrt{2u})\,U(t\sqrt{2u});$$

compare (6.8) and (6.9). The suppressed parameter is now $\frac{1}{2}u\alpha^2$. From the monotonicity properties of U and U' established in § 5.6 it is seen that

$$0 \leqslant \mathsf{K}(\zeta,t) < U(\zeta\sqrt{2u})\,U(-t\sqrt{2u}) \quad (\zeta \leqslant t).$$

Similarly,

$$0 \leqslant -\partial\mathsf{K}(\zeta,t)/\partial\zeta \leqslant -(2u)^{\frac{1}{2}}\hat{l}_2(\tfrac{1}{2}u\alpha^2)\,U'(\zeta\sqrt{2u})\,U(-t\sqrt{2u}) \quad (\zeta \leqslant t);$$

compare (7.6). The bounds (7.3) and (7.4) follow on solving (7.7) in the manner of § 6.1, and the rest of the proof is completed in a similar way.

Theorem II gives explicitly only one solution of (7.1). To obtain a second solution we apply the theorem to the same differential equation with ζ replaced by $-\zeta$ throughout.

7.2. *Behaviour of* $l_2(b)$ *and* $\hat{l}_2(b)$ *as* $b \to \infty$

Consider the asymptotic approximation (5.17). From the definition (5.19) it is evident that $\bar{\eta}$ is an odd function of y. Hence

$$U(\tfrac{1}{2}\mu^2, \mu y\sqrt{2})\,U(\tfrac{1}{2}\mu^2, -\mu y\sqrt{2}) = \frac{2^{\frac{1}{2}(\mu^2-1)}\,e^{\frac{1}{2}\mu^2}}{\mu^{\mu^2+1}(y^2+1)^{\frac{1}{2}}}\left\{1 + O\left(\frac{1}{\mu^2}\right)\right\}$$

as $\mu \to \infty$, uniformly for $y \in (-\infty, \infty)$.

Next, if μ is bounded away from zero, then

$$\Omega(\mu y\sqrt{2}) \leqslant K(1+\mu|y|) \leqslant K\mu(1+|y|),$$

where K again denotes a generic constant. Lastly, Stirling's formula yields

$$\Gamma(\tfrac{1}{2}+\tfrac{1}{2}\mu^2)\,\frac{2^{\frac{1}{2}(\mu^2-1)}\,e^{\frac{1}{2}\mu^2}}{\mu^{\mu^2+1}} \sim \frac{\pi^{\frac{1}{2}}}{\mu} \quad (\mu \to \infty).$$

Combining the foregoing results and replacing $\frac{1}{2}\mu^2$ by b, we see that

$$l_2(b) = O(1) \quad (b \to \infty). \tag{7.8}$$

Similarly,

$$\hat{l}_2(b) = O(1) \quad (b \to \infty). \tag{7.9}$$

7.3. *Uniform asymptotic properties of the error term*

Following § 6.3 we assume that $|\psi(u,\alpha,\zeta)/\zeta|$ is independent of u and integrable at $\zeta = -\infty$ and $\zeta = +\infty$, uniformly with respect to α. Then by taking $\Omega(x) = 1+|x|$, we derive

$$\mathscr{V}_{-\infty,\infty}(F) = O(u^{-\frac{1}{2}}\ln u);$$

compare (6.16). Substituting in (7.3) and (7.4) by means of this relation and (7.8) and (7.9), we find that

$$\epsilon(u,\alpha,\zeta) = U(\tfrac{1}{2}u\alpha^2, \zeta\sqrt{2u})\,O(u^{-1}\ln u), \quad \partial\epsilon(u,\alpha,\zeta)/\partial\zeta = U'(\tfrac{1}{2}u\alpha^2, \zeta\sqrt{2u})\,O(u^{-\frac{1}{2}}\ln u), \tag{7.10}$$

as $u \to \infty$, uniformly for $\alpha \in [0, \mathrm{A}]$ and $\zeta \in (-\infty, \infty)$.

Again, when $|\psi(u,\alpha,\zeta)/\zeta|$ is uniformly integrable at $\zeta = 0$ (as well as at $\zeta = \pm\infty$), we may set $\Omega(x) = |x|$ and sharpen both relations (7.10) by removal of the factor $\ln u$.

It is noteworthy that the O-terms in (7.10) are of smaller order than the corresponding O-terms in the results (6.17) and (6.19) for case I.

PART C. APPROXIMATIONS IN TERMS OF $W(b, \pm x)$

8. Relevant properties of modified parabolic cylinder functions

8.1. *Introduction*

Again following the notation of Miller (1955, 1964), we take the standard form of approximating differential equation in cases III and IV to be

$$d^2w/dx^2 = (b - \tfrac{1}{4}x^2)\,w. \tag{8.1}$$

When $b > 0$ there are turning points at $x = \pm 2b^{\frac{1}{2}}$; they separate a finite interval in which the solutions are monotonic, from two infinite intervals in which the solutions oscillate. When $b \leqslant 0$ the turning points are $\pm 2i(-b)^{\frac{1}{2}}$, and the entire real axis is an interval of oscillation.

Standard solutions of (8.1) are $W(b, x)$ and $W(b, -x)$, defined in terms of the confluent hypergeometric function by

$$\begin{aligned} W(b, \pm x) = {} & 2^{-\frac{3}{4}} |\Gamma(\tfrac{1}{4}+\tfrac{1}{2}ib)/\Gamma(\tfrac{3}{4}+\tfrac{1}{2}ib)|^{\frac{1}{2}} e^{\frac{1}{4}ix^2} {}_1F_1(\tfrac{1}{2}ib+\tfrac{1}{4}; \tfrac{1}{2}; -\tfrac{1}{2}ix^2) \\ & \mp 2^{-\frac{1}{4}} |\Gamma(\tfrac{3}{4}+\tfrac{1}{2}ib)/\Gamma(\tfrac{1}{4}+\tfrac{1}{2}ib)|^{\frac{1}{2}} e^{\frac{1}{4}ix^2} x\, {}_1F_1(\tfrac{1}{2}ib+\tfrac{3}{4}; \tfrac{3}{2}; -\tfrac{1}{2}ix^2). \end{aligned}$$

They are real and continuous for all real values of b and x. A numerically satisfactory pair of solutions is obtained by taking appropriate multiples of $W(b, \pm x)$, although the actual multiples differ for the x-intervals $[0, \infty)$ and $(-\infty, 0]$. As in § 5, we collect here relevant properties of the standard solutions. It is convenient to employ the notations

$$k(b) = (1+e^{2\pi b})^{\frac{1}{2}} - e^{\pi b}, \quad \phi(b) = \tfrac{1}{4}\pi + \tfrac{1}{2}\,\mathrm{ph}\,\Gamma(\tfrac{1}{2}+ib), \tag{8.2}$$

it being understood that the phase of $\Gamma(\frac{1}{2}+ib)$ in the second relation is continuous and vanishes at $b = 0$.† As b increases from $-\infty$ to ∞, $k(b)$ decreases monotonically from 1 to 0.

8.2. *Basic properties*

Values at $x = 0$

$$W(b, 0) = \frac{1}{2^{\frac{3}{4}}} \left| \frac{\Gamma(\frac{1}{4}+\frac{1}{2}ib)}{\Gamma(\frac{3}{4}+\frac{1}{2}ib)} \right|^{\frac{1}{2}}, \quad W'(b, 0) = -\frac{1}{2^{\frac{1}{4}}} \left| \frac{\Gamma(\frac{3}{4}+\frac{1}{2}ib)}{\Gamma(\frac{1}{4}+\frac{1}{2}ib)} \right|^{\frac{1}{2}}. \tag{8.3}$$

Behaviour as $x \to \infty$

$$\left.\begin{aligned} W(b, x) &= \left\{\frac{2k(b)}{x}\right\}^{\frac{1}{2}} \cos\{\tfrac{1}{4}x^2 - b\ln x + \phi(b)\} + O\left(\frac{1}{x^{\frac{5}{2}}}\right), \\ W'(b, x) &= -\left\{\frac{k(b)\,x}{2}\right\}^{\frac{1}{2}} \sin\{\tfrac{1}{4}x^2 - b\ln x + \phi(b)\} + O\left(\frac{1}{x^{\frac{3}{2}}}\right), \end{aligned}\right\} \tag{8.4}$$

$$\left.\begin{aligned} W(b, -x) &= \left\{\frac{2}{k(b)\,x}\right\}^{\frac{1}{2}} \sin\{\tfrac{1}{4}x^2 - b\ln x + \phi(b)\} + O\left(\frac{1}{x^{\frac{5}{2}}}\right), \\ W'(b, -x) &= -\left\{\frac{x}{2k(b)}\right\}^{\frac{1}{2}} \cos\{\tfrac{1}{4}x^2 - b\ln x + \phi(b)\} + O\left(\frac{1}{x^{\frac{3}{2}}}\right). \end{aligned}\right\} \tag{8.5}$$

These relations are uniform for b lying in any fixed compact interval.

Wronskian

$$\mathscr{W}\{W(b, x), W(b, -x)\} = 1. \tag{8.6}$$

† Our symbol $\phi(b)$ differs in meaning from the $\phi(b)$ of Miller; he writes

$$\phi(b) = \mathrm{ph}\,\Gamma(ib) \quad \text{and} \quad \phi_2(b) = \mathrm{ph}\,\Gamma(\tfrac{1}{2}+ib).$$

Case $b = 0$: *relation to Bessel functions.* When $b = 0$ we have $k(0) = 2^{\frac{1}{2}} - 1$, $\phi(0) = \frac{1}{4}\pi$, and

$$W(0, \pm x) = 2^{-\frac{5}{4}}(\pi x)^{\frac{1}{2}}\{J_{-\frac{1}{4}}(\tfrac{1}{4}x^2) \mp J_{\frac{1}{4}}(\tfrac{1}{4}x^2)\} \quad (x \geqslant 0), \tag{8.7}$$

$$W'(0, \pm x) = 2^{-\frac{9}{4}}\pi^{\frac{1}{2}}x^{\frac{3}{2}}\{\mp J_{\frac{3}{4}}(\tfrac{1}{4}x^2) - J_{-\frac{3}{4}}(\tfrac{1}{4}x^2)\} \quad (x \geqslant 0). \tag{8.8}$$

8.3. *Behaviour for large positive b*

Write $b = \frac{1}{2}\mu^2$ and $x = \mu y\sqrt{2}$. Then for large positive μ

$$k(\tfrac{1}{2}\mu^2) = \tfrac{1}{2}\,\mathrm{e}^{-\frac{1}{2}\pi\mu^2} + O(\mathrm{e}^{-\frac{3}{2}\pi\mu^2}), \quad \phi(\tfrac{1}{2}\mu^2) = \tfrac{1}{4}\mu^2\ln(\tfrac{1}{2}\mu^2) - \tfrac{1}{4}\mu^2 + \tfrac{1}{4}\pi + O(\mu^{-2}), \tag{8.9}$$

$$k^{-\frac{1}{2}}(\tfrac{1}{2}\mu^2)\, W(\tfrac{1}{2}\mu^2, \mu y\sqrt{2}) = \frac{2^{\frac{1}{4}}\pi^{\frac{1}{2}}}{\mu^{\frac{1}{6}}}\left(\frac{\eta}{y^2-1}\right)^{\frac{1}{4}}\left\{\mathrm{Bi}\,(-\mu^{\frac{4}{3}}\eta) + E(-\mu^{\frac{4}{3}}\eta)\, M(-\mu^{\frac{4}{3}}\eta)\, O\!\left(\frac{1}{\mu^2}\right)\right\}, \tag{8.10}$$

$$k^{\frac{1}{2}}(\tfrac{1}{2}\mu^2)\, W(\tfrac{1}{2}\mu^2, -\mu y\sqrt{2}) = \frac{2^{\frac{1}{4}}\pi^{\frac{1}{2}}}{\mu^{\frac{1}{6}}}\left(\frac{\eta}{y^2-1}\right)^{\frac{1}{4}}\left\{\mathrm{Ai}\,(-\mu^{\frac{4}{3}}\eta) + E^{-1}(-\mu^{\frac{4}{3}}\eta)\, M(-\mu^{\frac{4}{3}}\eta)\, O\!\left(\frac{1}{\mu^2}\right)\right\}. \tag{8.11}$$

In the last two relations the O-terms are uniformly valid in a y-interval that includes $[0, \infty)$, and η and the auxiliary Airy functions are the same as in § 5.3.

8.4. *Behaviour for large negative b*

Here we write $b = -\frac{1}{2}\mu^2$ and $x = \mu y\sqrt{2}$. Then

$$W(-\tfrac{1}{2}\mu^2, \mu y\sqrt{2}) = \frac{2^{\frac{1}{4}}}{\mu^{\frac{1}{2}}(y^2+1)^{\frac{1}{4}}}\left\{\cos(\mu^2\bar{\eta} + \tfrac{1}{4}\pi) + O\!\left(\frac{1}{\mu^2}\right)\right\}, \tag{8.12}$$

as $\mu \to \infty$, uniformly for $y \in (-\infty, \infty)$, where $\bar{\eta}$ is defined by (5.19).

Derivations of the results of §§ 8.1–8.2 will be found in Miller (1955), and of §§ 8.3–8.4 in Olver (1959).

8.5. *Graphs*

Figures 8.1–8.3 are based on the tables in Miller (1955), and are all drawn to the same scale.

8.6. *Auxiliary functions for* $b \geqslant 0$

For non-negative b the x-intervals $[0, \infty)$ and $(-\infty, 0]$ are treated separately. This is because solutions having equal amplitudes of oscillation as $x \to \infty$ have unequal amplitudes as $x \to -\infty$; compare (8.4) and (8.5). The solutions that will be used as approximants are $k^{-\frac{1}{2}}(b)\, W(b, x)$ and $k^{\frac{1}{2}}(b)\, W(b, -x)$ when $x \in [0, \infty)$, and $k^{\frac{1}{2}}(b)\, W(b, x)$ and $k^{-\frac{1}{2}}(b)\, W(b, -x)$ when $x \in (-\infty, 0]$.

From (8.6) we derive

$$\frac{\mathrm{d}}{\mathrm{d}x}\frac{k^{\frac{1}{2}}(b)\, W(b, -x)}{k^{-\frac{1}{2}}(b)\, W(b, x)} = \frac{k(b)}{W^2(b, x)} > 0. \tag{8.13}$$

At $x = 0$, $k^{-\frac{1}{2}}(b)\, W(b, x)$ and $k^{\frac{1}{2}}(b)\, W(b, -x)$ are both positive, and because $0 < k(b) < 1$ the former exceeds the latter. The relation (8.13) shows that as x increases, the first zero of $k^{-\frac{1}{2}}(b)\, W(b, x)$ is reached before $k^{\frac{1}{2}}(b)\, W(b, -x)$ vanishes. Denote by $\sigma(b)$ the smallest positive root of the equation

$$k^{-\frac{1}{2}}(b)\, W(b, x) = k^{\frac{1}{2}}(b)\, W(b, -x).$$

Then the result just given establishes that

$$k^{-\frac{1}{2}}(b)\, W(b, x) > k^{\frac{1}{2}}(b)\, W(b, -x) > 0 \quad (0 \leqslant x < \sigma(b)).$$

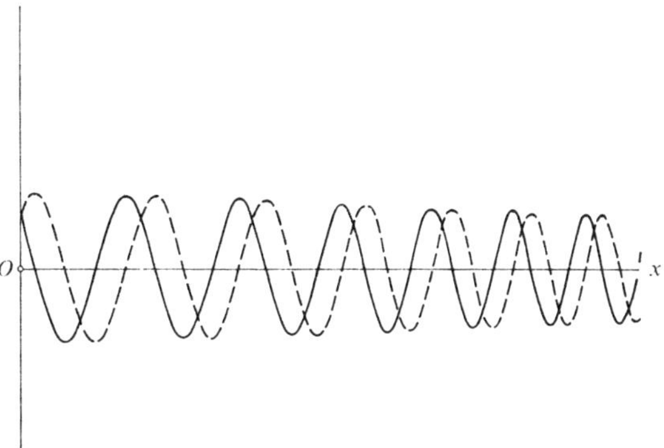

FIGURE 8.1. —— $k^{-\frac{1}{2}}W(-10, x)$; - - - $k^{\frac{1}{2}}W(-10, -x)$.

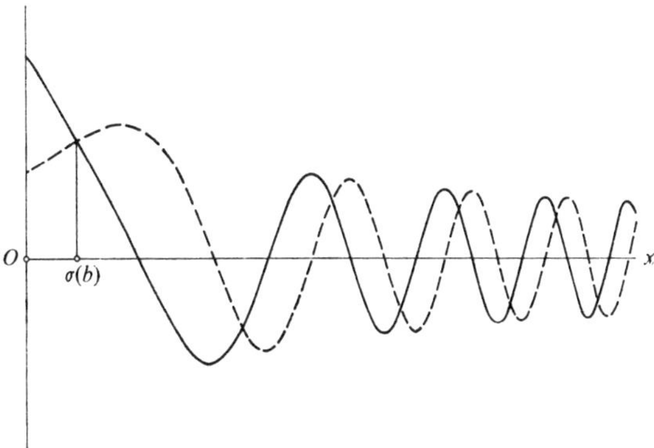

FIGURE 8.2. —— $k^{-\frac{1}{2}}W(0, x)$; - - - $k^{\frac{1}{2}}W(0, -x)$.

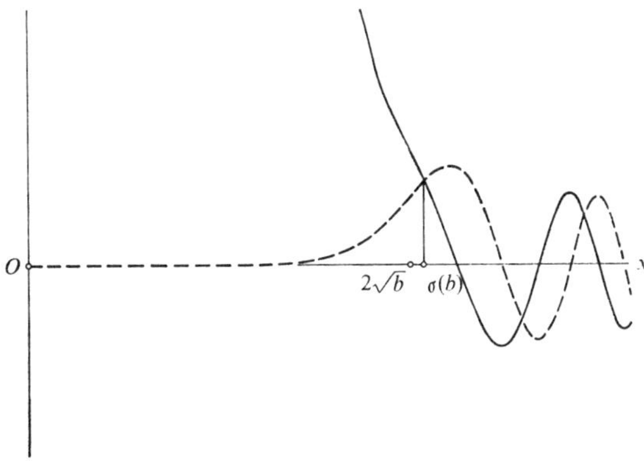

FIGURE 8.3. —— $k^{-\frac{1}{2}}W(10, x)$; - - - $k^{\frac{1}{2}}W(10, -x)$.

As in the case of the function $\rho(b)$ defined in § 5.8, $\sigma(b)$ is a continuous function of b. An asymptotic approximation corresponding to (5.21) can be derived from the results of § 8.3; thus

$$\sigma(b) = 2b^{\frac{1}{2}} - cb^{-\frac{1}{6}} + O(b^{-\frac{5}{6}}) \quad (b \to \infty), \tag{8.14}$$

where c is defined in § 5.3.†

The weight function we employ is defined by

$$\mathsf{E}(b,x) = \left\{\frac{k(b)\,W(b,-x)}{W(b,x)}\right\}^{\frac{1}{2}} \quad (0 \leqslant x \leqslant \sigma(b)), \qquad \mathsf{E}(b,x) = 1 \quad (x \geqslant \sigma(b)), \tag{8.15}$$

and $\mathsf{E}(b,x) = \mathsf{E}(b,-x)$ when $x < 0$. It is easily seen that $\mathsf{E}(b,x)$ is continuous in the region $0 \leqslant b < \infty$, $-\infty < x < \infty$. Also, from (8.13) it is clear that *when b is fixed, $\mathsf{E}(b,x)$ is non-decreasing for $x \in [0,\infty)$, and*

$$k^{\frac{1}{2}}(b) \leqslant \mathsf{E}(b,x) \leqslant 1. \tag{8.16}$$

Modulus and phase functions are defined by

$$\left.\begin{aligned} k^{-\frac{1}{2}}(b)\,W(b,x) &= \mathsf{E}^{-1}(b,x)\,\mathsf{M}(b,x)\sin\theta(b,x), \\ k^{\frac{1}{2}}(b)\,W(b,-x) &= \mathsf{E}(b,x)\,\mathsf{M}(b,x)\cos\theta(b,x), \\ k^{-\frac{1}{2}}(b)\,W'(b,x) &= \mathsf{E}^{-1}(b,x)\,\mathsf{N}(b,x)\sin\omega(b,x), \\ k^{\frac{1}{2}}(b)\,W'(b,-x) &= -\mathsf{E}(b,x)\,\mathsf{N}(b,x)\cos\omega(b,x), \end{aligned}\right\} \tag{8.17}$$

when $x \geqslant 0$, and by

$$\left.\begin{aligned} k^{\frac{1}{2}}(b)\,W(b,x) &= \mathsf{E}(b,x)\,\mathsf{M}(b,x)\cos\theta(b,x), \\ k^{-\frac{1}{2}}(b)\,W(b,-x) &= \mathsf{E}^{-1}(b,x)\,\mathsf{M}(b,x)\sin\theta(b,x), \\ k^{\frac{1}{2}}(b)\,W'(b,x) &= -\mathsf{E}(b,x)\,\mathsf{N}(b,x)\cos\omega(b,x), \\ k^{-\frac{1}{2}}(b)\,W'(b,-x) &= \mathsf{E}^{-1}(b,x)\,\mathsf{N}(b,x)\sin\omega(b,x), \end{aligned}\right\} \tag{8.18}$$

when $x \leqslant 0$. As usual, $\mathsf{E}^{-1}(b,x)$ denotes $1/\mathsf{E}(b,x)$. It is readily verified that each function is even in x, and $\mathsf{M}(b,x)$ and $\mathsf{N}(b,x)$ are continuous in the region $0 \leqslant b < \infty$, $-\infty < x < \infty$.

When $0 \leqslant x \leqslant \sigma(b)$, we find that

$$\mathsf{M}(b,x) = \{2W(b,x)\,W(b,-x)\}^{\frac{1}{2}}, \quad \theta(b,x) = \tfrac{1}{4}\pi, \tag{8.19}$$

$$\left.\begin{aligned} \mathsf{N}(b,x) &= \left\{\frac{W'^2(b,x)\,W^2(b,-x) + W'^2(b,-x)\,W^2(b,x)}{W(b,x)\,W(b,-x)}\right\}^{\frac{1}{2}}, \\ \omega(b,x) &= -\arctan\left\{\frac{W'(b,x)\,W(b,-x)}{W'(b,-x)\,W(b,x)}\right\}. \end{aligned}\right\} \tag{8.20}$$

And when $x \geqslant \sigma(b)$

$$\left.\begin{aligned} \mathsf{M}(b,x) &= \{k^{-1}(b)\,W^2(b,x) + k(b)\,W^2(b,-x)\}^{\frac{1}{2}}, \\ \theta(b,x) &= \arctan\{k^{-1}(b)\,W(b,x)/W(b,-x)\}, \end{aligned}\right\} \tag{8.21}$$

$$\left.\begin{aligned} \mathsf{N}(b,x) &= \{k^{-1}(b)\,W'^2(b,x) + k(b)\,W'^2(b,-x)\}^{\frac{1}{2}}, \\ \omega(b,x) &= -\arctan\{k^{-1}(b)\,W'(b,x)/W'(b,-x)\}. \end{aligned}\right\} \tag{8.22}$$

For $x \to \pm\infty$,

$$\mathsf{M}(b,x) \sim |2/x|^{\frac{1}{4}}, \quad \mathsf{N}(b,x) \sim |x/2|^{\frac{1}{4}}, \tag{8.23}$$

uniformly with respect to b in any compact interval in $[0,\infty)$.

† Numerical computations suggest that $\sigma(b) > 2b^{\frac{1}{2}}$ for all $b \in [0, \infty)$, but this inequality has not been proved.

8.7. *Auxiliary functions for $b \leqslant 0$*

For non-positive b there is no x-interval in which the solutions of (8.1) exhibit exponential-type behaviour. In consequence, a weight function is not needed. Modulus and phase functions are defined by (8.17) when $x \geqslant 0$ and by (8.18) when $x \leqslant 0$, with $\mathsf{E}(b,x)$ replaced by unity throughout.† Again, each function is even in x. Furthermore, equations (8.21) and (8.22) now apply when $x \geqslant 0$, and the asymptotic relations (8.23) are valid as $x \to \pm\infty$, uniformly with respect to b in any compact interval in $(-\infty, 0]$.

9. Case III

9.1. *Main theorem*

In this section we treat the equation

$$\mathrm{d}^2w/\mathrm{d}\zeta^2 = \{u^2(\alpha^2-\zeta^2)+\psi(u,\alpha,\zeta)\}w. \tag{9.1}$$

We specify $\Omega(x)$ and $F(u,\alpha,\zeta)$ exactly as in § 6.1.

Theorem III. *Assume that for $\alpha \in [0, \mathrm{A}]$, $\zeta \in [0, \zeta_2)$, and each value of u, the function $\psi(u,\alpha,\zeta)$ is continuous and $\mathscr{V}_{0,\zeta_2}(F)$ is bounded. Then in this region equation (9.1) has solutions $w_1(u,\alpha,\zeta)$ and $w_2(u,\alpha,\zeta)$, which are continuous, have continuous first and second partial ζ-derivatives, and are given by*

$$w_1(u,\alpha,\zeta) = k^{-\frac{1}{2}}(\tfrac{1}{2}u\alpha^2)\, W(\tfrac{1}{2}u\alpha^2, \zeta\sqrt{2u}) + \epsilon_1(u,\alpha,\zeta), \tag{9.2}$$

$$w_2(u,\alpha,\zeta) = k^{\frac{1}{2}}(\tfrac{1}{2}u\alpha^2)\, W(\tfrac{1}{2}u\alpha^2, -\zeta\sqrt{2u}) + \epsilon_2(u,\alpha,\zeta), \tag{9.3}$$

where

$$\frac{|\epsilon_1(u,\alpha,\zeta)|}{\mathsf{M}(\frac{1}{2}u\alpha^2,\zeta\sqrt{2u})},\ \frac{|\partial\epsilon_1(u,\alpha,\zeta)/\partial\zeta|}{(2u)^{\frac{1}{2}}\,\mathsf{N}(\frac{1}{2}u\alpha^2,\zeta\sqrt{2u})} \leqslant \mathsf{E}^{-1}(\tfrac{1}{2}u\alpha^2,\zeta\sqrt{2u})\left[\exp\left\{\frac{l_3(\frac{1}{2}u\alpha^2)}{(2u)^{\frac{1}{2}}}\mathscr{V}_{\zeta,\zeta_2}(F)\right\}-1\right], \tag{9.4}$$

$$\frac{|\epsilon_2(u,\alpha,\zeta)|}{\mathsf{M}(\frac{1}{2}u\alpha^2,\zeta\sqrt{2u})},\ \frac{|\partial\epsilon_2(u,\alpha,\zeta)/\partial\zeta|}{(2u)^{\frac{1}{2}}\,\mathsf{N}(\frac{1}{2}u\alpha^2,\zeta\sqrt{2u})} \leqslant \mathsf{E}(\tfrac{1}{2}u\alpha^2,\zeta\sqrt{2u})\left[\exp\left\{\frac{l_3(\frac{1}{2}u\alpha^2)}{(2u)^{\frac{1}{2}}}\mathscr{V}_{0,\zeta}(F)\right\}-1\right], \tag{9.5}$$

and

$$l_3(b) = \sup_{x\in(0,\infty)}\{\Omega(x)\,\mathsf{M}^2(b,x)\}\quad (b \geqslant 0). \tag{9.6}$$

In this theorem the auxiliary functions E, M and N are defined in § 8.6. The proof is similar to that of theorem I and details need not be recorded.

9.2. *Behaviour of $l_3(b)$ as $b \to \infty$*

As in the case of the functions $l_1(b)$ and $l_2(b)$ of §§ 6 and 7, $l_3(b)$ is continuous when $b \in [0,\infty)$. Analysis similar to § 6.2 yields

$$l_3(b) = O(b^{\frac{1}{3}})\quad (b\to\infty). \tag{9.7}$$

Furthermore, without specialization of $\Omega(x)$ this estimate cannot be sharpened. With $\Omega(x) = |x|^{\frac{1}{3}}$, however, we find that, as in the case of (6.15),

$$l_3(b) = O(1)\quad (b\to\infty). \tag{9.8}$$

9.3. *Uniform asymptotic properties of the error terms*

Following § 6.3 we assume that $|\psi(u,\alpha,\zeta)|/\zeta$ is independent of u and integrable at $\zeta=\infty$, uniformly with respect to α. Then by taking $\Omega(x) = 1+|x|$ we find that

$$\epsilon_1(u,\alpha,\zeta) = \mathsf{E}^{-1}(\tfrac{1}{2}u\alpha^2,\zeta\sqrt{2u})\,\mathsf{M}(\tfrac{1}{2}u\alpha^2,\zeta\sqrt{2u})\,O(u^{-\frac{2}{3}}\ln u),$$

$$\epsilon_2(u,\alpha,\zeta) = \mathsf{E}(\tfrac{1}{2}u\alpha^2,\zeta\sqrt{2u})\,\mathsf{M}(\tfrac{1}{2}u\alpha^2,\zeta\sqrt{2u})\,O(u^{-\frac{2}{3}}\ln u),$$

† In Miller's notation $\mathsf{M}(b,x)$ is denoted by $|E(b,x)|$ when $x \geqslant 0$.

as $u \to \infty$, uniformly for $\alpha \in [0, A]$ and $\zeta \in [0, \infty)$. For the ζ-derivatives of these error terms the symbol M is changed to N and the estimate $O(u^{-\frac{2}{3}} \ln u)$ replaced by $O(u^{-\frac{1}{6}} \ln u)$. Furthermore, in consequence of (9.8) refinements of the kind described in the last two paragraphs of § 6.3 are again available.

9.4. *Solutions when* $\zeta \leqslant 0$†

As in § 6.4, assume that $|\psi(u, \alpha, \zeta)/\zeta|$ is independent of u and integrable at $\zeta = -\infty$ and $+\infty$, uniformly with respect to α. By applying theorem III to the differential equation obtained from (9.1) on replacing ζ by $-\zeta$, and recollecting that $\mathsf{E}(\frac{1}{2}u\alpha^2, \zeta\sqrt{2u})$ and $\mathsf{M}(\frac{1}{2}u\alpha^2, \zeta\sqrt{2u})$ are even in ζ, we obtain the following solutions of (9.1):

$$w_3(u, \alpha, \zeta) = k^{-\frac{1}{2}}(\tfrac{1}{2}u\alpha^2) W(\tfrac{1}{2}u\alpha^2, -\zeta\sqrt{2u}) + \mathsf{E}^{-1}(\tfrac{1}{2}u\alpha^2, \zeta\sqrt{2u})\, \mathsf{M}(\tfrac{1}{2}u\alpha^2, \zeta\sqrt{2u})\, O(u^{-\frac{2}{3}} \ln u), \quad (9.9)$$

$$w_4(u, \alpha, \zeta) = k^{\frac{1}{2}}(\tfrac{1}{2}u\alpha^2) W(\tfrac{1}{2}u\alpha^2, \zeta\sqrt{2u}) + \mathsf{E}(\tfrac{1}{2}u\alpha^2, \zeta\sqrt{2u})\, \mathsf{M}(\tfrac{1}{2}u\alpha^2, \zeta\sqrt{2u})\, O(u^{-\frac{2}{3}} \ln u), \quad (9.10)$$

valid when $\zeta \leqslant 0$.

Estimates for the coefficients in the connection formulae

$$w_1(u, \alpha, \zeta) = A_1 w_3(u, \alpha, \zeta) + B_1 w_4(u, \alpha, \zeta), \quad w_2(u, \alpha, \zeta) = A_2 w_3(u, \alpha, \zeta) + B_2 w_4(u, \alpha, \zeta), \quad (9.11)$$

can be found by comparing the solutions and their derivatives at $\zeta = 0$ in the manner of § 6.4; thus

$$A_1 = O(u^{-\frac{2}{3}} \ln u), \quad B_1 = k^{-1}(\tfrac{1}{2}u\alpha^2)\{1 + O(u^{-\frac{2}{3}} \ln u)\}, \quad (9.12)$$

$$A_2 = k(\tfrac{1}{2}u\alpha^2)\{1 + O(u^{-\frac{2}{3}} \ln u)\}, \quad B_2 = O(u^{-\frac{2}{3}} \ln u). \quad (9.13)$$

Substituting in the first of (9.11) by means of (9.9), (9.10) and (9.12), and using (8.16), we find that

$$w_1(u, \alpha, \zeta) = k^{-1}(\tfrac{1}{2}u\alpha^2)\{k^{\frac{1}{2}}(\tfrac{1}{2}u\alpha^2)\, W(\tfrac{1}{2}u\alpha^2, \zeta\sqrt{2u}) + \mathsf{E}(\tfrac{1}{2}u\alpha^2, \zeta\sqrt{2u})\, \mathsf{M}(\tfrac{1}{2}u\alpha^2, \zeta\sqrt{2u})\, O(u^{-\frac{2}{3}} \ln u)\} \quad (\zeta \leqslant 0), \quad (9.14)$$

uniformly with respect to α and ζ. A similar result is unavailable for $w_2(u, \alpha, \zeta)$, however. Indeed, because of the smallness of A_2 compared with B_2, the second of (9.11) yields little information concerning $w_2(u, \alpha, \zeta)$ for negative values of ζ. This is to be expected, since $w_2(u, \alpha, \zeta)$ exhibits recessive behaviour as $\zeta \to 0$ through positive values. The 'correct' companion formula to (9.14) is supplied by

$$w_3(u, \alpha, \zeta) = k^{-1}(\tfrac{1}{2}u\alpha^2)\{k^{\frac{1}{2}}(\tfrac{1}{2}u\alpha^2)\, W(\tfrac{1}{2}u\alpha^2, -\zeta\sqrt{2u}) + \mathsf{E}(\tfrac{1}{2}u\alpha^2, \zeta\sqrt{2u})\, \mathsf{M}(\tfrac{1}{2}u\alpha^2, \zeta\sqrt{2u})\, O(u^{-\frac{2}{3}} \ln u)\} \quad (\zeta \geqslant 0). \quad (9.15)$$

The coefficient $\mathsf{E}(\frac{1}{2}u\alpha^2, \zeta\sqrt{2u})\, \mathsf{M}(\frac{1}{2}u\alpha^2, \zeta\sqrt{2u})$ of the error terms in (9.14) and (9.15) is of the same order of magnitude as $k^{\frac{1}{2}}(\frac{1}{2}u\alpha^2) W(\frac{1}{2}u\alpha^2, \zeta\sqrt{2u})$ in (9.14) and $k^{\frac{1}{2}}(\frac{1}{2}u\alpha^2) W(\frac{1}{2}u\alpha^2, -\zeta\sqrt{2u})$ in (9.15), except of course in the neighbourhoods of the zeros of the W-functions.

10. Case IV

10.1. *Main theorem*

The last form of differential equation to be considered is given by

$$\mathrm{d}^2 w/\mathrm{d}\zeta^2 = \{-u^2(\alpha^2 + \zeta^2) + \psi(u, \alpha, \zeta)\} w. \quad (10.1)$$

Again $\Omega(x)$ and $F(u, \alpha, \zeta)$ are specified as in § 6.1.

† Another way of extending the solutions to negative values of ζ will be given in a subsequent paper.

THEOREM IV. *Assume that for $\alpha \in [0, A]$, $\zeta \in (\zeta_1, \zeta_2)$, and each value of u, the function $\psi(u, \alpha, \zeta)$ is continuous and $\mathscr{V}_{\zeta_1, \zeta_2}(F)$ is bounded. Then in this region equation (10.1) has a continuous solution $w(u, \alpha, \zeta)$, with continuous first and second partial ζ-derivatives, given by*

$$w(u, \alpha, \zeta) = W(-\tfrac{1}{2}u\alpha^2, \zeta\sqrt{2u}) + \epsilon(u, \alpha, \zeta), \tag{10.2}$$

where

$$\frac{|\epsilon(u, \alpha, \zeta)|}{\mathsf{M}(-\frac{1}{2}u\alpha^2, \zeta\sqrt{2u})}, \quad \frac{|\partial\epsilon(u, \alpha, \zeta)/\partial\zeta|}{(2u)^{\frac{1}{2}}\mathsf{N}(-\frac{1}{2}u\alpha^2, \zeta\sqrt{2u})} \leqslant \frac{1}{k^{\frac{1}{2}}(-\frac{1}{2}u\alpha^2)}\left[\exp\left\{\frac{l_4(-\frac{1}{2}u\alpha^2)}{(2u)^{\frac{1}{2}}k(-\frac{1}{2}u\alpha^2)}\mathscr{V}_{\zeta_0, \zeta}(F)\right\} - 1\right]. \tag{10.3}$$

Here ζ_0 is an arbitrary number in the closure of (ζ_1, ζ_2), and

$$l_4(b) = \sup_{x \in (-\infty, \infty)} \{\Omega(x)\, \mathsf{M}^2(b, x)\} \quad (b \leqslant 0). \tag{10.4}$$

In this result the modulus functions M and N are defined as in § 8.7. The proof is similar to the proofs of theorems I, II and III, and is omitted.

The solution $w(u, \alpha, \zeta)$ depends on ζ_0: this is evident on setting $\zeta = \zeta_0$ in (10.3). Also, when ζ_0 and ζ have the same sign the right-hand side of (10.3) can be replaced by the slightly sharper bound

$$k^{\pm\frac{1}{2}}(-\tfrac{1}{2}u\alpha^2)\left[\exp\left\{\frac{l_4(-\frac{1}{2}u\alpha^2)}{(2u)^{\frac{1}{2}}}\mathscr{V}_{\zeta_0, \zeta}(F)\right\} - 1\right]$$

in which the upper or lower sign is taken according as ζ_0 and ζ are both non-negative or both non-positive. To obtain a numerically satisfactory companion to the solution $w(u, \alpha, \zeta)$ we apply theorem IV to the equation obtained from (10.1) by replacing ζ with $-\zeta$.

10.2. *Behaviour of $l_4(b)$ as $b \to -\infty$*

By using (8.12) and similar analysis to that of § 7.2 it is verifiable that $l_4(b)$ is continuous and bounded when $b \in (-\infty, 0]$.

10.3. *Uniform asymptotic properties of the error terms*

If $|\psi(u, \alpha, \zeta)/\zeta|$ is independent of u and integrable at $\zeta = -\infty$ and $\zeta = +\infty$, uniformly with respect to α, then

$$\epsilon(u, \alpha, \zeta) = \mathsf{M}(-\tfrac{1}{2}u\alpha^2, \zeta\sqrt{2u})\, O(u^{-1}\ln u), \quad \partial\epsilon(u, \alpha, \zeta)/\partial\zeta = \mathsf{N}(-\tfrac{1}{2}u\alpha^2, \zeta\sqrt{2u})\, O(u^{-\frac{1}{2}}\ln u), \tag{10.5}$$

as $u \to \infty$, uniformly for $\alpha \in [0, A]$ and $\zeta \in (-\infty, \infty)$; compare (7.10). If, also, $|\psi(u, \alpha, \zeta)/\zeta|$ is uniformly integrable at $\zeta = 0$, then the logarithmic factors in (10.5) can be omitted.

PART D. EXTENSIONS AND CONCLUSIONS

11. EXPANSIONS IN DESCENDING POWERS OF THE LARGE PARAMETER

In §§ 1 and 3 it was indicated that an extension of the theory in this paper to complex values of the variables is feasible. We shall not, however, attempt to supply details of the analysis. The purpose of the present section is to consider the following question. Is it possible to construct asymptotic solutions of the given differential equations in series of descending powers of the large parameter u?

In discussing this question, we assume for simplicity that the function $\psi(u, \alpha, \zeta)$ is independent of u and α. It will also suffice to restrict the discussion to case I, for which the standard differential equation is given by

$$d^2w/d\zeta^2 = \{u^2(\zeta^2-\alpha^2)+\psi(\zeta)\}w. \tag{11.1}$$

Formal series solutions can be constructed by the methods of McKelvey (1955) and Kazarinoff (1958), as follows. We first introduce an auxiliary parameter β related to α by the expansion

$$\alpha^2 = \beta^2+\frac{\beta_0(\beta)}{u^2}+\frac{\beta_1(\beta)}{u^4}+\ldots, \tag{11.2}$$

in which the functions $\beta_0(\beta), \beta_1(\beta), \ldots$ are at our disposal. Then we set

$$w = V(u,\beta,\zeta)\sum_{s=0}^{\infty}\frac{A_s(\beta,\zeta)}{u^{2s}}+\frac{1}{u^2}\frac{\partial V(u,\beta,\zeta)}{\partial\zeta}\sum_{s=0}^{\infty}\frac{B_s(\beta,\zeta)}{u^{2s}}, \tag{11.3}$$

where $V(u, \beta, \zeta)$ is a solution of the equation

$$d^2V/d\zeta^2 = u^2(\zeta^2-\beta^2)\,V;$$

for example, $V = U(-\tfrac{1}{2}u\beta^2, \zeta\sqrt{2u})$ or $\bar{U}(-\tfrac{1}{2}u\beta^2, \zeta\sqrt{2u})$.

The functions $A_s(\beta, \zeta)$, $B_s(\beta, \zeta)$ and $\beta_s(\beta)$ are determinable by formally differentiating (11.3) to obtain a similar expansion for $d^2w/d\zeta^2$, substituting the result and (11.2) and (11.3) in (11.1), and equating coefficients of like powers of u. We find first that $A_0(\beta, \zeta)$ is independent of ζ, and without loss of generality set $A_0(\beta, \zeta) = 1$. The next equation to be satisfied is given by

$$\zeta B_0(\beta,\zeta)+(\zeta^2-\beta^2)\{\partial B_0(\beta,\zeta)/\partial\zeta\} = \tfrac{1}{2}\psi(\zeta)-\tfrac{1}{2}\beta_0(\beta).$$

Integration yields

$$B_0(\beta,\zeta) = \tfrac{1}{2}(\beta^2-\zeta^2)^{-\frac{1}{2}}\int(\beta^2-\zeta^2)^{-\frac{1}{2}}\{\beta_0(\beta)-\psi(\zeta)\}\,d\zeta. \tag{11.4}$$

If the lower limit in this integral is taken as $-\beta$, then $B_0(\beta, \zeta)$ is analytic at $\zeta = -\beta$, provided, of course, that the same is true of $\psi(\zeta)$. Similarly, if the lower limit were taken to be β, then $B_0(\beta, \zeta)$ would be analytic at $\zeta = \beta$. To render $B_0(\beta, \zeta)$ analytic at *both* turning points $\zeta = \pm\beta$, we clearly require

$$\int_{-\beta}^{\beta}(\beta^2-\zeta^2)^{-\frac{1}{2}}\{\beta_0(\beta)-\psi(\zeta)\}\,d\zeta = 0.$$

This yields an explicit formula for the function $\beta_0(\beta)$:

$$\beta_0(\beta) = \frac{1}{\pi}\int_{-\beta}^{\beta}(\beta^2-\zeta^2)^{-\frac{1}{2}}\psi(\zeta)\,d\zeta. \tag{11.5}$$

Continuation of the process leads to the relations

$$A_s = -\tfrac{1}{2}B'_{s-1}+\tfrac{1}{2}\int(\psi B_{s-1}-\beta_0 B_{s-1}-\beta_1 B_{s-2}-\ldots-\beta_{s-1}B_0)\,d\zeta, \tag{11.6}$$

$$\beta_s = -\frac{1}{\pi}\int_{-\beta}^{\beta}(\beta^2-\zeta^2)^{-\frac{1}{2}}(A''_s-\psi A_s+\beta_0 A_s+\beta_1 A_{s-1}+\ldots+\beta_{s-1}A_1)\,d\zeta, \tag{11.7}$$

and

$$B_s = \tfrac{1}{2}(\beta^2-\zeta^2)^{-\frac{1}{2}}\int_{\pm\beta}^{\zeta}(\beta^2-\zeta^2)^{-\frac{1}{2}}(A''_s-\psi A_s+\beta_0 A_s+\beta_1 A_{s-1}+\ldots+\beta_s A_0)\,d\zeta, \tag{11.8}$$

with $s = 1, 2, 3, \ldots$. Here the arguments β and ζ have been suppressed and primes signify differentiations with respect to ζ. The choice of integration constant in (11.6) is at our disposal, and either sign may be used for the lower limit in (11.8). These three formulas determine the terms in (11.2) and (11.3) in a recursive manner.

The asymptotic nature of the formal expansion (11.3), for large u, can be investigated by the standard procedure of constructing a differential equation for the error term and solving an equivalent integral equation by successive approximations. An essential limitation, however, is that the asymptotic property of (11.3) cannot be uniformly valid in *unbounded* ζ-intervals. To understand this, let us make the assumption, as in earlier sections, that $\psi(\zeta)/\zeta$ is absolutely integrable at $\zeta = \infty$. When $\zeta > \beta$, the real form of equation (11.4) is given by

$$B_0(\beta, \zeta) = \tfrac{1}{2}(\zeta^2 - \beta^2)^{-\frac{1}{2}} \int_\beta^\zeta (\tau^2 - \beta^2)^{-\frac{1}{2}} \{\psi(\tau) - \beta_0(\beta)\} \, d\tau.$$

Letting $\zeta \to \infty$ we find that

$$B_0(\beta, \zeta) \sim -\tfrac{1}{2}\beta_0(\beta)\, \zeta^{-1} \ln \zeta,$$

provided that $\beta_0(\beta) \neq 0$. If we further assume that $\zeta^{-1} \ln \zeta\, \psi(\zeta)$ is integrable at $\zeta = \infty$, then from (11.6) with $s = 1$ we obtain

$$A_1(\beta, \zeta) \sim \tfrac{1}{8}\{\beta_0(\beta) \ln \zeta\}^2.$$

By using these results and (5.5) and (5.6), it is clear that as $\zeta \to \infty$ the terms

$$\frac{1}{u^2} \frac{\partial V(u, \beta, \zeta)}{\partial \zeta} B_0(\beta, \zeta), \quad V(u, \beta, \zeta) \frac{A_1(\beta, \zeta)}{u^2},$$

in (11.3) become infinitely large compared with the leading term $V(u, \beta, \zeta)$. And even in fortuitous cases in which $\beta_0(\beta) = 0$, higher terms in (11.3) will give rise to the same difficulty.

Another way of perceiving the inherent lack of uniformity of (11.3) at $\zeta = \infty$ is to consider the relation

$$U(-\tfrac{1}{2}u\alpha^2, \zeta\sqrt{2u}) \sim e^{-\frac{1}{2}u\zeta^2} (\zeta\sqrt{2u})^{\frac{1}{2}(u\alpha^2 - 1)} \quad (\zeta \to \infty), \tag{11.9}$$

obtained from (5.5). Theorem I shows that when $|\psi(\zeta)|/\zeta$ is integrable at $\zeta = \infty$ the solution $w_1(u, \alpha, \zeta)$ shares the asymptotic form (11.9) as $\zeta \to \infty$. Changing from the parameter α to a new parameter β, as we did in constructing (11.3), obviously destroys the agreement of the asymptotic form of the approximant with that of the solution.†

A consequence of lack of uniformity at $\zeta = \infty$ is that although inclusion of higher terms in (11.3) leads to a more accurate representation of the solution w in the neighbourhood of the turning points, the *single* term supplied by Theorem I furnishes a better approximation for all sufficiently large $|\zeta|$. Furthermore, identification of standard solutions of the differential equation in terms of the asymptotic solutions is greatly facilitated when comparisons can be made at $\zeta = \infty$.‡

† In the paper by Kazarinoff (1958), which was discussed in §1 of the present paper, the author states that the asymptotic expansions he obtains are uniformly valid in unbounded regions of the complex plane. However, careful examination of condition (iv) on page 137 of this reference reveals that it is satisfiable only when the asymptotic expansions are truncated at their leading term. This is because Kazarinoff's coefficients $a_j(z)$ and $b_j(z)$ are unbounded as $z \to \infty$, in the same way that our $A_s(\beta, \zeta)$ and $B_s(\beta, \zeta)$ are unbounded for large ζ. Therefore neither Kazarinoff's theory nor that of the present paper supplies asymptotic approximations which consist of more than a single term and are uniformly valid in unbounded ζ-regions.

‡ For illustrations of the process of identification see the paper that follows (Olver 1975).

12. Summary

In part A it was shown how to transform linear second-order differential equations having two simple turning points, or one double turning point, into forms closely resembling differential equations satisfied by the parabolic cylinder functions. It was also proved that the coefficients in the transformed equations are continuous functions of both the independent variable and a parameter governing the distance separating the turning points.

In parts B and C relevant properties of the parabolic cylinder functions were assembled, and certain auxiliary functions introduced to facilitate estimation of the parabolic cylinder functions throughout oscillatory and exponential regions. Then approximate solutions of the transformed differential equations mentioned in the preceding paragraph were constructed in terms of the parabolic cylinder functions. These approximations are asymptotic in nature for large values of a certain parameter, uniformly with respect to the distance separating the turning points, including zero distance, and also with respect to the independent variable in unbounded intervals containing the turning points. Rigorous and realistic bounds for the errors in the asymptotic approximations were supplied in terms of the auxiliary functions.

Also considered in parts B and C was the problem of extending the asymptotic solutions to negative values of the independent variable in cases where the direct results apply only to positive values.

Lastly, part D revealed a basic difficulty in the extension of the asymptotic approximations obtained in earlier sections into asymptotic expansions in series of descending powers of the large parameter.

This research was supported by the U.S. Army Research Office, Durham under Contract DA ARO D 31 124 72 G49, and the National Science Foundation under Grant GP 32841X. The writer is indebted to Dr L. Maximon for many suggestions for improving the presentation of the results, and is also glad to acknowledge the assistance of Mr R. E. Kaylor and Mrs G. E. Olver with various numerical computations, including those needed for the graphs in § 5.

References

Brussaard, P. J. & Tolhoek, H. A. 1957 Classical limits of Clebsch–Gordan coefficients, Racah coefficients and $D^l_{mn}(\phi, \vartheta, \psi)$-functions. *Physica* **23**, 955–971.

Heading, J. 1962 Phase-integral methods I. *Q. Jl mech. appl. Math.* **15**, 215–244.

Kazarinoff, N. D. 1958 Asymptotic theory of second order differential equations with two simple turning points. *Archs ration. Mech. Analysis* **2**, 129–150.

Kazarinoff, N. D. & Ritt, R. K. 1960 Scalar diffraction theory and turning-point problems. *Archs ration. Mech. Analysis* **5**, 177–186.

Langer, R. E. 1934 The asymptotic solutions of certain linear ordinary differential equations of the second order. *Trans. Am. Math. Soc.* **36**, 90–106.

Langer, R. E. 1959 The asymptotic solutions of a linear differential equation of the second order with two turning points. *Trans. Am. Math. Scc.* **90**, 113–142.

Lynn, R. Y. S. & Keller, J. B. 1970 Uniform asymptotic solutions of second order linear ordinary differential equations with turning points. *Communs pure appl. Math.* **23**, 379–408.

McKelvey, R. W. 1955 The solutions of second order linear ordinary differential equations about a turning point of order two. *Trans. Am. Math. Soc.* **79**, 103–123.

Miller, J. C. P. 1950 On the choice of standard solutions for a homogeneous linear differential equation of the second order. *Q. Jl mech. appl. Math.* **3**, 225–235.

Miller, J. C. P. 1955 *Tables of Weber parabolic cylinder functions*. London: Her Majesty's Stationery Office.

Miller, J. C. P. 1964 Parabolic cylinder functions. In *Handbook of mathematical functions* (ed. M. Abramowitz & I. A. Stegun), pp. 685–720. National Bureau of Standards Appl. Math. Ser. no. 55. Washington, D.C.: U.S. Government Printing Office.

Moriguchi, H. 1959 An improvement of the WKB method in the presence of turning points and the asymptotic solutions of a class of Hill equations. *J. Phys. Soc. Japan* **14**, 1771–1796.
Olver, F. W. J. 1959 Uniform asymptotic expansions for Weber parabolic cylinder functions of large orders. *J. Res. natn. Bur. Stand.* B **63**, 131–169.
Olver, F. W. J. 1974 *Asymptotics and special functions*. London: Academic Press.
Olver, F. W. J. 1975 Legendre functions with both parameters large. *Phil. Trans. R. Soc. Lond.* A **278**, 175–185.
Whittaker, E. T. 1903 On the functions associated with the parabolic cylinder in harmonic analysis. *Proc. Lond. Math. Soc.* **35**, 417–427.

[175]

LEGENDRE FUNCTIONS WITH BOTH PARAMETERS LARGE

By F. W. J. OLVER
The University of Maryland and *The National Bureau of Standards*

(*Communicated by Sir Harold Jeffreys, F.R.S. – Received* 8 *May* 1974)

CONTENTS

By application of the theory for second-order linear differential equations with two turning points developed in the preceding paper, some new asymptotic approximations are obtained for the associated Legendre functions when both the degree n and order m are large. The approximations are expressed in terms of parabolic cylinder functions, and are uniformly valid with respect to $x \in (-1, 1)$ and $m/(n+\frac{1}{2}) \in [\delta, 1+\Delta]$, where δ and Δ are arbitrary fixed numbers such that $0 < \delta < 1$ and $\Delta > 0$. The values of m and $n+\frac{1}{2}$ are either both real, or both purely imaginary. In all cases explicit bounds are supplied for the error terms associated with the approximations.

1. Introduction and summary

The purpose of this paper is to derive asymptotic approximations for solutions of the associated Legendre equation

$$(1-x^2)\frac{d^2L}{dx^2} - 2x\frac{dL}{dx} + \left\{n(n+1) - \frac{m^2}{1-x^2}\right\}L = 0 \tag{1.1}$$

when the parameters m and n are large, neither necessarily being an integer. For large n and fixed m (or more generally, bounded m) it is possible to construct asymptotic solutions in terms of Bessel functions of order m which are uniformly valid in unbounded x-domains containing one of the singularities ± 1 (see Olver 1974, chapter 12). In a similar way, for large m and fixed n the present writer has shown that it is possible to construct asymptotic solutions in terms of Bessel functions of order $n+\frac{1}{2}$ which are uniform in unbounded x-domains containing either $x = 1$ or $x = -1$. In this case, however, details have not been published.

When both m and n are large the problem is more difficult. The characterizing features of the differential equation in this situation are the regular singularities at $x = \pm 1$ and ∞, and the turning points at $x = \pm a$, where

$$1 - a^2 = m^2/(n+\tfrac{1}{2})^2. \tag{1.2}$$

Thorne (1957*b*) treated the case in which the ratio $m/(n+\frac{1}{2})$ has a given value in the open interval $(0, 1)$. He constructed asymptotic solutions in terms of Airy functions, and also in terms of Bessel functions of order m, these solutions being uniformly valid in a region that includes the interval $0 \leqslant x < 1$. When $m/(n+\frac{1}{2})$ is allowed to approach unity, however, Thorne's approximations no longer apply owing to lack of uniformity with respect to this ratio.

The present paper supplies asymptotic solutions of (1.1) for large values of $n+\frac{1}{2}$, complete with error bounds, which are uniformly valid when $m/(n+\frac{1}{2}) \in [\delta, 1+\Delta]$, where δ and Δ are any fixed numbers such that $0 < \delta < 1$ and $\Delta > 0$. Thus the turning points may be real and distinct $(0 < m < n+\frac{1}{2})$, coincident $(m = n+\frac{1}{2})$ or purely imaginary $(m > n+\frac{1}{2})$. The situation is exactly that covered by theorems I and II of the preceding paper (Olver 1975), and in §§ 2, 3 and 4 of the present paper standard solutions of (1.1) are identified in terms of the uniform asymptotic solutions involving parabolic cylinder functions. The results apply to the interval $-1 < x < 1$.

The remaining sections, §§ 5 and 6, are devoted to the case in which both m and $n+\frac{1}{2}$ are purely imaginary. Again there are either two real or two purely imaginary turning points, which coincide when $|m| = |n+\frac{1}{2}|$. Uniform asymptotic approximations, for the same interval $-1 < x < 1$, are constructed by application of theorems III and IV of the preceding paper.

All results presented are believed to be new.

2. Real parameters and real turning points: preliminary transformations

On removing the term in the first derivative from (1.1) by transformation of dependent variable, we obtain the differential equation

$$\frac{\mathrm{d}^2L}{\mathrm{d}x^2} = \left\{u^2\frac{x^2-a^2}{(1-x^2)^2} - \frac{3+x^2}{4(1-x^2)^2}\right\}L, \tag{2.1}$$

satisfied by $L = (1-x^2)^{\frac{1}{2}}P_n^m(x)$ and $(1-x^2)^{\frac{1}{2}}Q_n^m(x)$, in which

$$u = n+\tfrac{1}{2}, \quad a^2 = 1-\frac{m^2}{u^2} = \frac{(n+m+\frac{1}{2})(n-m+\frac{1}{2})}{(n+\frac{1}{2})^2}; \tag{2.2}$$

compare (1.2). In the present section, and also § 3, we suppose that

$$0 < m \leqslant n+\tfrac{1}{2},$$

so that $u > 0$ and $0 \leqslant a < 1$.

Equation (2.1) has singularities at $x = \pm 1$ and ∞, and turning points at $x = \pm a$. In each of the intervals $(1, \infty)$ and $(-\infty, -1)$, uniform asymptotic expansions of the solutions can be constructed in terms of elementary functions by standard methods.† We shall not pursue the details of those expansions; instead we confine attention to the interval $-1 < x < 1$.

Following Olver (1975, § 2.2), we see that the appropriate Liouville transformation is given by

$$L = \left(\frac{\mathrm{d}x}{\mathrm{d}\zeta}\right)^{\frac{1}{2}} w, \quad \left(\frac{\mathrm{d}\zeta}{\mathrm{d}x}\right)^2 = \frac{x^2-a^2}{(1-x^2)^2(\zeta^2-\alpha^2)}, \tag{2.3}$$

† Given, for example, in Olver (1974, chapters 6 and 10, especially § 5.3 of the former chapter).

where α is defined by

$$\int_{-\alpha}^{\alpha}(\alpha^2-\tau^2)^{\frac{1}{2}}\,d\tau = \int_{-a}^{a}\frac{(a^2-t^2)^{\frac{1}{2}}}{1-t^2}\,dt. \tag{2.4}$$

Evaluation of the last two integrals leads to the equation

$$\tfrac{1}{2}\alpha^2 = 1-(1-a^2)^{\frac{1}{2}}, \tag{2.5}$$

or, in terms of the original parameters,

$$\alpha^2 = 2(n-m+\tfrac{1}{2})/(n+\tfrac{1}{2}). \tag{2.6}$$

Thus α lies in the interval $[0, \surd 2)$, with $\alpha = 0$ corresponding to $a = 0$.

Since the points $\zeta = \pm\alpha$ correspond respectively to $x = \pm a$, integration of the second of (2.3) yields

$$\int_{\alpha}^{\zeta}(\tau^2-\alpha^2)^{\frac{1}{2}}\,d\tau = \int_{a}^{x}\frac{(t^2-a^2)^{\frac{1}{2}}}{1-t^2}\,dt \quad (a \leqslant x < 1), \tag{2.7}$$

$$\int_{\zeta}^{\alpha}(\alpha^2-\tau^2)^{\frac{1}{2}}\,d\tau = \int_{x}^{a}\frac{(a^2-t^2)^{\frac{1}{2}}}{1-t^2}\,dt \quad (-a \leqslant x \leqslant a), \tag{2.8}$$

and

$$\int_{\zeta}^{-\alpha}(\tau^2-\alpha^2)^{\frac{1}{2}}\,d\tau = \int_{x}^{-a}\frac{(t^2-a^2)^{\frac{1}{2}}}{1-t^2}\,dt \quad (-1 < x \leqslant -a). \tag{2.9}$$

From these equations and (2.4) it is evident that ζ is an even function of x, with $x = -1, 0$ and 1 corresponding to $\zeta = -\infty$, 0 and ∞, respectively. Moreover, (2.8) is replaceable by

$$\int_{0}^{\zeta}(\alpha^2-\tau^2)^{\frac{1}{2}}\,d\tau = \int_{0}^{x}\frac{(a^2-t^2)^{\frac{1}{2}}}{1-t^2}\,dt \quad (-a \leqslant x \leqslant a). \tag{2.10}$$

Evaluation of the integrals in (2.7) and (2.10) yields the formulae

$$\tfrac{1}{2}\zeta(\zeta^2-\alpha^2)^{\frac{1}{2}} - \tfrac{1}{2}\alpha^2\operatorname{arcosh}\left(\frac{\zeta}{\alpha}\right) = (1-a^2)^{\frac{1}{2}}\operatorname{artanh}\left\{\frac{1}{x}\left(\frac{x^2-a^2}{1-a^2}\right)^{\frac{1}{2}}\right\} - \operatorname{arcosh}\left(\frac{x}{a}\right) \quad (a \leqslant x < 1), \tag{2.11}$$

and

$$\tfrac{1}{2}\alpha^2\arcsin\left(\frac{\zeta}{\alpha}\right) + \tfrac{1}{2}\zeta(\alpha^2-\zeta^2)^{\frac{1}{2}} = \arcsin\left(\frac{x}{a}\right) - (1-a^2)^{\frac{1}{2}}\arctan\left\{x\left(\frac{1-a^2}{a^2-x^2}\right)^{\frac{1}{2}}\right\} \quad (-a \leqslant x \leqslant a), \tag{2.12}$$

in the case $a > 0$, or

$$\zeta^2 = -\ln(1-x^2) \quad (-1 < x < 1), \tag{2.13}$$

in the case $a = 0$.

The transformed differential equation is given by

$$d^2w/d\zeta^2 = \{u^2(\zeta^2-\alpha^2) + \psi(\alpha,\zeta)\}w, \tag{2.14}$$

where

$$\psi(\alpha,\zeta) = -\dot{x}^2\frac{3+x^2}{4(1-x^2)^2} + \dot{x}^{\frac{1}{2}}\frac{d^2}{d\zeta^2}(\dot{x}^{-\frac{1}{2}}),$$

dots signifying differentiations with respect to ζ; compare Olver (1975, (2.10)). Using the second of (2.3) and carrying out the differentiations, we arrive at the explicit formula

$$\psi(\alpha,\zeta) = \frac{3\zeta^2+2\alpha^2}{4(\zeta^2-\alpha^2)^2} + \frac{\zeta^2-\alpha^2}{4(x^2-a^2)^3}\{(3-4a^2)x^4-(3-6a^2+a^4)x^2-a^2(2-a^2)\}. \tag{2.15}$$

Lemma I of Olver (1975, §3) shows that $\psi(\alpha,\zeta)$ is continuous in the region $\alpha\in[0,\surd 2)$, $\zeta\in(-\infty,\infty)$.

In identifying the solutions of (2.14) we shall need to know the behaviour of $\psi(\alpha, \zeta)$ as ζ tends to its extreme values. When $\zeta \to \infty$ and $x \to 1-$, we derive from (2.11)

$$\zeta^2 - \tfrac{1}{2}\alpha^2 - \alpha^2 \ln\left(\frac{2\zeta}{\alpha}\right) + O\left(\frac{1}{\zeta^2}\right) = (1-a^2)^{\frac{1}{2}} \ln\left\{\frac{2(1-a^2)}{a^2(1-x)}\right\} - 2\ln\left\{\frac{1+(1-a^2)^{\frac{1}{2}}}{a}\right\} + O(1-x).$$

By reversion and use of (2.5) we find that

$$(1-\tfrac{1}{2}\alpha^2)\ln\left(\frac{1}{1-x}\right) = \zeta^2 - \alpha^2 \ln \zeta + \ln \hat{\alpha} + O\left(\frac{1}{\zeta^2}\right) \quad (\zeta \to \infty), \tag{2.16}$$

where

$$\hat{\alpha} = \left(\frac{a}{\alpha}\right)^{4-\alpha^2} \frac{2^{\frac{1}{2}(6-3\alpha^2)}}{(2-\alpha^2)^{2-\alpha^2}} e^{-\frac{1}{2}\alpha^2}. \tag{2.17}$$

Exponentiation of (2.16) yields

$$x = 1 - \left\{\exp\left(-\frac{2\zeta^2}{2-\alpha^2}\right)\right\}\left(\frac{\zeta^{\alpha^2}}{\hat{\alpha}}\right)^{2/(2-\alpha^2)}\left\{1 + O\left(\frac{1}{\zeta^2}\right)\right\}. \tag{2.18}$$

When $x = 1$ the content of the braces in (2.15) vanishes identically. Therefore as $\zeta \to +\infty$

$$\psi(\alpha, \zeta) = \frac{3\zeta^2 + 2\alpha^2}{4(\zeta^2 - \alpha^2)^2} + O\left\{\zeta^{4/(2-\alpha^2)} \exp\left(-\frac{2\zeta^2}{2-\alpha^2}\right)\right\}.$$

Since $\psi(\alpha, \zeta)$ is even in ζ, this result also holds when $\zeta \to -\infty$. Thus we have

$$\psi(\alpha, \zeta) = \frac{3}{4\zeta^2} + O\left(\frac{1}{\zeta^4}\right) \quad (\zeta \to \pm\infty). \tag{2.19}$$

Moreover, from the foregoing analysis it is clear that this result is uniformly valid for $\alpha \in [0, \mathrm{A}]$, where—in this section and the next—A denotes an arbitrary constant in the interval $(0, \sqrt{2})$.

It should be observed that the fact that the order of $\psi(\alpha, \zeta)$ is as small as $O(\zeta^{-2})$ as $\zeta \to \pm\infty$ is not entirely fortuitous. It is a consequence of the choice of the large parameter made at the beginning of this section. Had we made, for example, the quite natural choice $u = \{n(n+1)\}^{\frac{1}{2}}$ in place of the first of (2.2), then in the expression corresponding to (2.15) the content of the braces would not vanish as $x \to 1$, causing $\psi(\alpha, \zeta)$ to behave as a multiple of ζ^2 as $\zeta \to \pm\infty$. In turn, this would have the undesirable consequence of restricting the validity of the resulting asymptotic solutions of the differential equation to bounded values of ζ (compare also Olver 1974, pp. 463–4 and Olver 1975, § 6.3).

3. Real parameters and real turning points: identification of solutions

Let A denote any constant in the interval $(0, \sqrt{2})$. On applying theorem I of Olver (1975, § 6), with $\zeta_2 = \infty$, we obtain the following solutions of equation (2.14):

$$w_1(u, \alpha, \zeta) = U(-\tfrac{1}{2}u\alpha^2, \zeta\sqrt{2u}) + \epsilon_1(u, \alpha, \zeta), \quad w_2(u, \alpha, \zeta) = \bar{U}(-\tfrac{1}{2}u\alpha^2, \zeta\sqrt{2u}) + \epsilon_2(u, \alpha, \zeta), \tag{3.1}$$

valid when $\zeta \in [0, \infty)$ and $\alpha \in [0, \mathrm{A}]$, where U and $\bar{U}$ denote the parabolic cylinder functions defined and discussed in Miller (1955) and Olver (1975). The solutions $w_1(u, \alpha, \zeta)$ and $w_2(u, \alpha, \zeta)$ now need to be identified in terms of standard solutions of (2.1).

As $\zeta \to \infty$, $w_1(u, \alpha, \zeta)$ is recessive and $w_2(u, \alpha, \zeta)$ is dominant. Hence $w_1(u, \alpha, \zeta)$ is a multiple of the solution of (2.1) that is recessive as $x \to 1-$. This solution is $(1-x^2)^{\frac{1}{2}} \mathrm{P}_n^{-m}(x)$, where $\mathrm{P}_n^{-m}(x)$ denotes the Ferrers function of degree n and order $-m$.† Thus from (2.3)

$$w_1(u, \alpha, \zeta) = \{(x^2-a^2)/(\zeta^2-\alpha^2)\}^{\frac{1}{4}} S \mathrm{P}_n^{-m}(x), \tag{3.2}$$

where the coefficient S is independent of x. The value of S can be found by evaluating the ratio of the two sides of this equation as $x \to 1-$. First, we have

$$\mathrm{P}_n^{-m}(x) \sim \frac{1}{\Gamma(m+1)} \left(\frac{1-x}{2}\right)^{\frac{1}{2}m}.$$

Next, as $\zeta \to \infty$

$$U(-\tfrac{1}{2}u\alpha^2, \zeta\sqrt{2u}) \sim (\zeta\sqrt{2u})^{\frac{1}{2}(u\alpha^2-1)} \mathrm{e}^{-\frac{1}{2}u\zeta^2},$$

and from the bound (6.5) of Olver (1975) for $\epsilon_1(u, \alpha, \zeta)$ it is seen that the ratio of this error term to $U(-\frac{1}{2}u\alpha^2, \zeta\sqrt{2u})$ vanishes. Therefore

$$S = \lim_{\zeta \to \infty} \left\{ \Gamma(m+1) \left(\frac{2}{1-x}\right)^{\frac{1}{2}m} \frac{\zeta^{\frac{1}{2}u\alpha^2}(2u)^{\frac{1}{4}(u\alpha^2-1)}}{(1-a^2)^{\frac{1}{4}} \mathrm{e}^{\frac{1}{2}u\zeta^2}} \right\}.$$

On using (2.2), (2.6), (2.16) and (2.17), we arrive at

$$S = \left(\frac{n+\frac{1}{2}}{2}\right)^{\frac{1}{4}} \frac{\Gamma(m+1)}{m^{m+\frac{1}{2}}} \frac{(n+m+\frac{1}{2})^{\frac{1}{4}(2n+2m+1)}}{\mathrm{e}^{\frac{1}{4}(2n-2m+1)}}. \tag{3.3}$$

Combination of (3.2), (3.3) and the first of (3.1) yields the first of the desired approximations:

$$\mathrm{P}_n^{-m}(x) = \left(\frac{2}{n+\frac{1}{2}}\right)^{\frac{1}{4}} \frac{m^{m+\frac{1}{2}}}{\Gamma(m+1)} \frac{\mathrm{e}^{\frac{1}{4}(2n-2m+1)}}{(n+m+\frac{1}{2})^{\frac{1}{4}(2n+2m+1)}} \left(\frac{\zeta^2-\alpha^2}{x^2-a^2}\right)^{\frac{1}{4}} \{U(m-n-\tfrac{1}{2}, \zeta\sqrt{2n+1}) + \epsilon_1\}, \tag{3.4}$$

valid when $0 < m \leqslant n+\frac{1}{2}$ and $0 \leqslant x < 1$. In this result, the variable ζ is given by (2.11) and (2.12), and the error term ϵ_1 bounded by (6.5) of Olver (1975).

To show that $|\epsilon_1|$ is small compared with $|U(m-n-\frac{1}{2}, \zeta\sqrt{2n+1})|$ when n is large, we now obtain an asymptotic estimate for ϵ_1 which is uniform with respect to x and the ratio $m/(n+\frac{1}{2})$ provided that

$$\delta(n+\tfrac{1}{2}) \leqslant m \leqslant n+\tfrac{1}{2}, \tag{3.5}$$

where δ is any fixed number in the interval $(0, 1)$.‡ Then $0 \leqslant \alpha \leqslant \mathrm{A}$, where A is a fixed number in $(0, \sqrt{2})$. Applying the results of § 6.3 of Olver (1975) with $\Omega(x)$ set equal to $|x|^{\frac{1}{3}}$—a substitution that is permissible as a consequence of (2.19)—we derive

$$\begin{aligned} \epsilon_1 &= \mathsf{E}^{-1}(-\tfrac{1}{2}u\alpha^2, \zeta\sqrt{2u})\, \mathsf{M}(-\tfrac{1}{2}u\alpha^2, \zeta\sqrt{2u})\, O(u^{-\frac{2}{3}}) \\ &= \mathsf{E}^{-1}(m-n-\tfrac{1}{2}, \zeta\sqrt{2n+1})\, \mathsf{M}(m-n-\tfrac{1}{2}, \zeta\sqrt{2n+1})\, O(n^{-\frac{2}{3}}), \end{aligned}$$

where E and M are the auxiliary functions introduced in § 5.8 of Olver (1975). Next, on using Stirling's formula we obtain from (3.3)

$$S = (n+\tfrac{1}{2})^{\frac{1}{4}} 2^{\frac{1}{2}(n+m)} \Gamma(\tfrac{1}{2}n+\tfrac{1}{2}m+\tfrac{3}{4}) \{1+O(n^{-1})\}.$$

† For definitions and properties of Ferrers functions, or 'Legendre functions on the cut', the reader is referred to B.M.P. (1953, §3.4), or Olver (1974, pp. 185–9).

‡ The point of the restriction $\delta(n+\frac{1}{2}) \leqslant m$ is to ensure that the turning point at $x = a$ is bounded away from the singularity at $x = 1$. A coalescing turning point and pole cannot be included in the present theory, but could be treated by an extension of the theory of Thorne (1957 *a*, *b*).

Combination of these results with (3.1) and (3.2) yields

$$\mathrm{P}_n^{-m}(x) = \frac{1}{(n+\frac{1}{2})^{\frac{1}{4}}\, 2^{\frac{1}{2}(n+m)}\, \Gamma(\frac{1}{2}n+\frac{1}{2}m+\frac{3}{4})} \left(\frac{\zeta^2-\alpha^2}{x^2-a^2}\right)^{\frac{1}{4}}$$
$$\times \{U(m-n-\tfrac{1}{2}, \zeta\sqrt{2n+1}) + \mathsf{E}^{-1}(m-n-\tfrac{1}{2}, \zeta\sqrt{2n+1})\, \mathsf{M}(m-n-\tfrac{1}{2}, \zeta\sqrt{2n+1})\, O(n^{-\frac{2}{3}})\} \quad (3.6)$$

as $n \to \infty$, uniformly with respect to $x \in [0, 1)$ and values of m satisfying (3.5). Except in the neighbourhoods of zeros of $U(m-n-\frac{1}{2}, \zeta\sqrt{2n+1})$, the relative error of this asymptotic approximation to $\mathrm{P}_n^{-m}(x)$ is uniformly $O(n^{-\frac{2}{3}})$.

Equations (3.4) and (3.6) have been derived by identification of solutions at the singularity $x = 1$. At the other endpoint of the interval of validity it is known that

$$\mathrm{P}_n^{-m}(0) = \frac{\Gamma(\frac{1}{2}n-\frac{1}{2}m+\frac{1}{2}) \cos(\frac{1}{2}n\pi-\frac{1}{2}m\pi)}{\pi^{\frac{1}{2}} 2^m \Gamma(\frac{1}{2}n+\frac{1}{2}m+1)}, \quad (3.7)$$

and it is of obvious interest to compare this result with the form of (3.6) at $x = \zeta = 0$. We have

$$\left(\frac{\zeta^2-\alpha^2}{x^2-a^2}\right)^{\frac{1}{4}} = \left(\frac{\alpha}{a}\right)^{\frac{1}{2}} = \left(\frac{2n+1}{n+m+\frac{1}{2}}\right)^{\frac{1}{4}}.$$

Also, $$U(m-n-\tfrac{1}{2}, 0) = \pi^{-\frac{1}{2}} 2^{\frac{1}{2}(n-m)} \Gamma(\tfrac{1}{2}n-\tfrac{1}{2}m+\tfrac{1}{2}) \cos(\tfrac{1}{2}n\pi-\tfrac{1}{2}m\pi),$$

and $$\mathsf{E}^{-1}(m-n-\tfrac{1}{2}, 0)\, \mathsf{M}(m-n-\tfrac{1}{2}, 0) = \pi^{-\frac{1}{2}} 2^{\frac{1}{2}(n-m)} \Gamma(\tfrac{1}{2}n-\tfrac{1}{2}m+\tfrac{1}{2}).$$

Accordingly, the right-hand side of (3.6) reduces to

$$\frac{\Gamma(\frac{1}{2}n-\frac{1}{2}m+\frac{1}{2})}{\pi^{\frac{1}{2}} 2^m (\frac{1}{2}n+\frac{1}{2}m+\frac{1}{4})^{\frac{1}{4}} \Gamma(\frac{1}{2}n+\frac{1}{2}m+\frac{3}{4})} \{\cos(\tfrac{1}{2}n\pi-\tfrac{1}{2}m\pi) + O(n^{-\frac{2}{3}})\}. \quad (3.8)$$

By means of Stirling's formula it is easily verified that (3.7) agrees with (3.8) within the tolerance of the uniform error term, thus providing a powerful check on the soundness of our asymptotic theory.

To identify the second solution, write

$$w_2(u, \alpha, \zeta) = \{(x^2-a^2)/(\zeta^2-\alpha^2)\}^{\frac{1}{4}} \{S_1 \mathrm{P}_n^{-m}(x) + S_2 \mathrm{Q}_n^{-m}(x)\}, \quad (3.9)$$

where $\mathrm{Q}_n^{-m}(x)$ denotes the second Ferrers function, and S_1 and S_2 are independent of x. A convenient way of determining S_1 and S_2 is to compare the two sides of (3.9), together with their differentiated forms, at $x = \zeta = 0$. On the right-hand side, we have (3.7) and the corresponding expressions

$$\mathrm{P}_n^{-m\prime}(0) = \frac{\Gamma(\frac{1}{2}n-\frac{1}{2}m+1) \sin(\frac{1}{2}n\pi-\frac{1}{2}m\pi)}{\pi^{\frac{1}{2}} 2^{m-1} \Gamma(\frac{1}{2}n+\frac{1}{2}m+\frac{1}{2})},$$

$$\mathrm{Q}_n^{-m}(0) = -\frac{\pi^{\frac{1}{2}} \Gamma(\frac{1}{2}n-\frac{1}{2}m+\frac{1}{2}) \sin(\frac{1}{2}n\pi-\frac{1}{2}m\pi)}{2^{m+1} \Gamma(\frac{1}{2}n+\frac{1}{2}m+1)}, \quad \mathrm{Q}_n^{-m\prime}(0) = \frac{\pi^{\frac{1}{2}} \Gamma(\frac{1}{2}n-\frac{1}{2}m+1) \cos(\frac{1}{2}n\pi-\frac{1}{2}m\pi)}{2^m \Gamma(\frac{1}{2}n+\frac{1}{2}m+\frac{1}{2})}.$$

On the left-hand side,

$$\bar{U}(-\tfrac{1}{2}u\alpha^2, 0) = -\pi^{-\frac{1}{2}} 2^{\frac{1}{2}(n-m)} \Gamma(\tfrac{1}{2}n-\tfrac{1}{2}m+\tfrac{1}{2}) \sin(\tfrac{1}{2}n\pi-\tfrac{1}{2}m\pi),$$
$$\bar{U}'(-\tfrac{1}{2}u\alpha^2, 0) = \pi^{-\frac{1}{2}} 2^{\frac{1}{2}(n-m+1)} \Gamma(\tfrac{1}{2}n-\tfrac{1}{2}m+1) \cos(\tfrac{1}{2}n\pi-\tfrac{1}{2}m\pi),$$

and $\epsilon_2 = \partial\epsilon_2/\partial\zeta = 0$ at $\zeta = 0$. Solution of the two linear algebraic equations for S_1 and S_2 yields the required result:

$$S_1 = 2^{\frac{1}{2}(n+m-2)} (n+\tfrac{1}{2})^{\frac{1}{4}} \sin(n\pi - m\pi)$$
$$\times \{(\tfrac{1}{2}n+\tfrac{1}{2}m+\tfrac{1}{4})^{\frac{1}{4}} \Gamma(\tfrac{1}{2}n+\tfrac{1}{2}m+\tfrac{1}{2}) - (\tfrac{1}{2}n+\tfrac{1}{2}m+\tfrac{1}{4})^{-\frac{1}{4}} \Gamma(\tfrac{1}{2}n+\tfrac{1}{2}m+1)\}, \quad (3.10)$$

$$S_2 = \pi^{-1}2^{\frac{1}{2}(n+m+2)}(n+\tfrac{1}{2})^{\frac{1}{4}}\{(\tfrac{1}{2}n+\tfrac{1}{2}m+\tfrac{1}{4})^{\frac{1}{4}}\Gamma(\tfrac{1}{2}n+\tfrac{1}{2}m+\tfrac{1}{2})\cos^2(\tfrac{1}{2}n\pi-\tfrac{1}{2}m\pi)$$
$$+(\tfrac{1}{2}n+\tfrac{1}{2}m+\tfrac{1}{4})^{-\frac{1}{4}}\Gamma(\tfrac{1}{2}n+\tfrac{1}{2}m+1)\sin^2(\tfrac{1}{2}n\pi-\tfrac{1}{2}m\pi)\}. \quad (3.11)$$

For large n, we find that

$$S_2 = \pi^{-1}2^{\frac{1}{2}(n+m+2)}(n+\tfrac{1}{2})^{\frac{1}{4}}\Gamma(\tfrac{1}{2}n+\tfrac{1}{2}m+\tfrac{3}{4})\{1+O(n^{-1})\}, \quad S_1/S_2 = O(n^{-1}),$$

the O-terms being uniform for values of m satisfying (3.5). Combining these estimates with (3.6) and referring to properties of the auxiliary functions

$$\mathsf{E}(m-n-\tfrac{1}{2}, \zeta\sqrt{2n+1}) \quad \text{and} \quad \mathsf{M}(m-n-\tfrac{1}{2}, \zeta\sqrt{2n+1})$$

given in Olver (1975, §5.8) (including the fact that the former equals or exceeds unity), we see that the term

$$\{(x^2-a^2)/(\zeta^2-\alpha^2)\}^{\frac{1}{4}}S_1\mathrm{P}_n^{-m}(x)$$

on the right-hand side of (3.9) is absorbable in the estimate

$$\mathsf{E}(m-n-\tfrac{1}{2}, \zeta\sqrt{2n+1})\,\mathsf{M}(m-n-\tfrac{1}{2}, \zeta\sqrt{2n+1})\,O(n^{-\frac{2}{3}})$$

for the error term $\epsilon_2(u, \alpha, \zeta)$ associated with the left-hand side. Thus we arrive at the following uniform asymptotic approximation for large n, valid with the same conditions as (3.6):

$$\mathrm{Q}_n^{-m}(x) = \frac{\pi}{(n+\frac{1}{2})^{\frac{1}{4}}2^{\frac{1}{2}(n+m+2)}\Gamma(\frac{1}{2}n+\frac{1}{2}m+\frac{3}{4})}\left(\frac{\zeta^2-\alpha^2}{x^2-a^2}\right)^{\frac{1}{4}}$$
$$\times\{\bar{U}(m-n-\tfrac{1}{2}, \zeta\sqrt{2n+1})+\mathsf{E}(m-n-\tfrac{1}{2}, \zeta\sqrt{2n+1})\,\mathsf{M}(m-n-\tfrac{1}{2}, \zeta\sqrt{2n+1})\,O(n^{-\frac{2}{3}})\}. \quad (3.12)$$

A check on (3.12) is to compare the limiting forms of the two sides of this equation as $x\to 1-$, that is, as $\zeta\to\infty$. Again, it is found that there is agreement within the tolerance of the given error term.

Relations (3.6) and (3.12) apply when $x\in[0, 1)$. Corresponding results for the interval $(-1, 0]$ are derivable from §6.4 of Olver (1975), or by use of the well-known connection formulae for the Ferrers functions. Connection formulae can also be used to derive corresponding approximations for other Ferrers functions, including $\mathrm{P}_n^m(x)$ and $\mathrm{Q}_n^m(x)$. Details are straightforward and need not be recorded.

Some previous asymptotic approximations for associated Legendre functions are included as special cases of the present results. For example, if $m/(n+\frac{1}{2})$ has a *fixed* value in the interval (0, 1), then $m-n-\frac{1}{2}$ tends to $-\infty$ as $n+\frac{1}{2}\to\infty$. In this case the parabolic cylinder functions in (3.6) and (3.12) may be replaced by their asymptotic approximations in terms of Airy functions given, for example, in Olver (1975, §5.3). The results then reduce to Thorne's (1957*b*) asymptotic approximations for Ferrers functions in terms of Airy functions, truncated at their first term.† The unavoidable price of the extensive uniformity of the present results is the complexity of the relations (2.11) and (2.12) between the variables ζ and x.

† Compare also the non-uniform asymptotic approximations given in §5 of Brussaard & Tolhoek (1957) for the similar problem of the eigenfunctions of a symmetric rotator.

4. Real parameters and imaginary turning points

In this section we consider equation (1.1) with the conditions

$$0 < n+\tfrac{1}{2} \leqslant m.$$

Corresponding to (2.2), we set

$$u = n+\tfrac{1}{2}, \quad a^2 = \frac{m^2}{u^2}-1 = \frac{(n+m+\frac{1}{2})(m-n-\frac{1}{2})}{(n+\frac{1}{2})^2}, \tag{4.1}$$

so that $u > 0$ and $a \geqslant 0$. Then we have

$$\frac{d^2L}{dx^2} = \left\{u^2\frac{x^2+a^2}{(1-x^2)^2}-\frac{3+x^2}{4(1-x^2)^2}\right\}L, \tag{4.2}$$

with solutions $L = (1-x^2)^{\frac{1}{2}}\,P_n^m(x)$ and $(1-x^2)^{\frac{1}{2}}\,Q_n^m(x)$.

The turning points of the differential equation (4.2) are located at $x = \pm ia$. The appropriate Liouville transformation is therefore that of § 2.3 of Olver (1975). The results of § 2 above again apply, provided that a^2 and α^2 are replaced throughout by $-a^2$ and $-\alpha^2$, respectively. Thus

$$\tfrac{1}{2}\alpha^2 = (1+a^2)^{\frac{1}{2}}-1 = (m-n-\tfrac{1}{2})/(n+\tfrac{1}{2}), \tag{4.3}$$

and

$$\int_0^\zeta (\alpha^2+\tau^2)^{\frac{1}{2}}\,d\tau = \int_0^x \frac{(a^2+t^2)^{\frac{1}{2}}}{1-t^2}\,dt,$$

that is,

$$\tfrac{1}{2}\zeta(\zeta^2+\alpha^2)^{\frac{1}{2}}+\tfrac{1}{2}\alpha^2\operatorname{arsinh}\left(\frac{\zeta}{\alpha}\right) = (1+a^2)^{\frac{1}{2}}\operatorname{artanh}\left\{x\left(\frac{1+a^2}{x^2+a^2}\right)^{\frac{1}{2}}\right\}-\operatorname{arsinh}\left(\frac{x}{a}\right). \tag{4.4}$$

The last two equations hold when $x\in(-1,1)$, or, equivalently, $\zeta\in(-\infty,\infty)$. When $a = 0$, (4.4) is replaced by (2.13).

With $L = (1-x^2)^{\frac{1}{2}}(\zeta^2+\alpha^2)^{\frac{1}{4}}(x^2+a^2)^{-\frac{1}{4}}w$, the transformed differential equation is given by

$$d^2w/d\zeta^2 = \{u^2(\zeta^2+\alpha^2)+\psi(\alpha,\zeta)\}\,w, \tag{4.5}$$

in which

$$\psi(\alpha,\zeta) = \frac{3\zeta^2-2\alpha^2}{4(\zeta^2+\alpha^2)^2}+\frac{\zeta^2+\alpha^2}{4(x^2+a^2)^3}\{(3+4a^2)\,x^4-(3+6a^2+a^4)\,x^2+a^2(2+a^2)\}; \tag{4.6}$$

compare (2.15). Lemma II of Olver (1975, § 4.1) shows that $\psi(\alpha,\zeta)$ is continuous in the region $\alpha\in[0,\infty)$, $\zeta\in(-\infty,\infty)$. And in the manner of § 2 above it is verifiable that

$$\psi(\alpha,\zeta) = \frac{3}{4\zeta^2}+O\left(\frac{1}{\zeta^4}\right) \tag{4.7}$$

as $\zeta\to\pm\infty$, uniformly for $\alpha\in[0,A]$, where A now denotes *any* fixed positive number.

For large u, approximate solutions of (4.5) are supplied by theorem II of Olver (1975, § 7.1). They are identifiable in terms of the Ferrers functions in the manner of § 3. Corresponding to (3.4), we have

$$P_n^{-m}(x) = \left(\frac{2}{n+\frac{1}{2}}\right)^{\frac{1}{4}}\frac{m^{m+\frac{1}{2}}}{\Gamma(m+1)}\frac{e^{\frac{1}{4}(2n-2m+1)}}{(n+m+\frac{1}{2})^{\frac{1}{4}(2n+2m+1)}}\left(\frac{\zeta^2+\alpha^2}{x^2+a^2}\right)^{\frac{1}{4}}\{U(m-n-\tfrac{1}{2},\zeta\sqrt{2n+1})+\epsilon\}, \tag{4.8}$$

where the error term is subject to the bound (7.3) of Olver (1975), with $\zeta_2 = \infty$. This bound is valid when $0 < n+\frac{1}{2} \leqslant m$ and $-1 < x < 1$.

The asymptotic result corresponding to (3.6) is available from §7.3 of the last reference. Using Stirling's formula we obtain

$$\mathrm{P}_n^{-m}(x) = \frac{1}{(n+\frac{1}{2})^{\frac{1}{4}}\, 2^{\frac{1}{2}(n+m)}\, \Gamma(\frac{1}{2}n+\frac{1}{2}m+\frac{3}{4})} \left(\frac{\zeta^2+\alpha^2}{x^2+a^2}\right)^{\frac{1}{4}} U(m-n-\tfrac{1}{2}, \zeta\sqrt{2n+1})\,\{1+O(n^{-1}\ln n)\} \quad (4.9)$$

as $n \to \infty$, uniformly for $-1 < x < 1$ and

$$n+\tfrac{1}{2} \leqslant m \leqslant (1+\Delta)(n+\tfrac{1}{2}),$$

Δ denoting any positive constant.

A check on (4.9) is to let $\zeta \to -\infty$ and compare the limiting form of the right-hand side with the relation

$$\mathrm{P}_n^{-m}(x) \sim \frac{\Gamma(m)}{\Gamma(m-n)\,\Gamma(n+m+1)} \left(\frac{2}{1+x}\right)^{\frac{1}{2}m} \quad (x \to -1+).$$

Agreement is obtained within the tolerance of the error term in (4.9).

Since the relations (4.8) and (4.9) apply to negative values of x as well as to positive values, they furnish uniform asymptotic approximations for the solutions $\mathrm{P}_n^{-m}(x)$ and $\mathrm{P}_n^{-m}(-x)$ of equation (1.1). With the given conditions these two solutions are independent and numerically satisfactory; therefore similar uniform asymptotic approximations of any other solutions can be constructed by use of connection formulae. In particular, this includes $\mathrm{P}_n^{m}(x)$, $\mathrm{Q}_n^{-m}(x)$ and $\mathrm{Q}_n^{m}(x)$.

Comments similar to those made in the final paragraph of §3 apply also to the results obtained in the present section.

5. Imaginary parameters and real turning points

This case is similar to that of §§2 and 3, the essential changes being that u and m are replaced throughout by $\mathrm{i}u$ and $\mathrm{i}m$, respectively. Thus we are now considering the Ferrers function $\mathrm{P}^{\mathrm{i}m}_{-\frac{1}{2}+\mathrm{i}u}(x)$ with $0 < m \leqslant u$ and $-1 < x < 1$.

Again we write $a^2 = 1-(m^2/u^2)$, and except for (2.6) the transformations (2.3) to (2.13) apply unchanged. Corresponding to (2.14) we have the differential equation

$$\mathrm{d}^2w/\mathrm{d}\zeta^2 = \{-u^2(\zeta^2-\alpha^2)+\psi(\alpha,\zeta)\}\,w, \quad (5.1)$$

with solutions
$$w = \{(x^2-a^2)/(\zeta^2-\alpha^2)\}^{\frac{1}{4}}\, \mathrm{P}^{\mathrm{i}m}_{-\frac{1}{2}+\mathrm{i}u}(\pm x),$$

the function $\psi(\alpha,\zeta)$ being given by (2.15).

For large u, approximate solutions of (5.1), in terms of modified parabolic cylinder functions, are supplied by theorem III of Olver (1975, §9.1), with $\zeta_2 = \infty$. As in §3 above, the first of these solutions is easily related to one of the standard solutions by considering limiting behaviour as $x \to 1-$, that is, as $\zeta \to \infty$. The second approximate solution is identifiable by using the known values of the Ferrers functions and their derivatives at $x = 0$; in this case the analysis is facilitated by the identities given in Miller (1955, §9.3) for Gamma functions of arguments $\frac{1}{4}s+\frac{1}{2}\mathrm{i}a$, $s = 0, 1, 2, 3, 4$. The final results are found to be expressible in the form

$$\mathrm{Re}\{T_1\mathrm{P}^{\mathrm{i}m}_{-\frac{1}{2}+\mathrm{i}u}(x)\} = \frac{m^{\frac{1}{2}}}{(2u)^{\frac{1}{4}}}\left(\frac{\zeta^2-\alpha^2}{x^2-a^2}\right)^{\frac{1}{4}} \{k^{-\frac{1}{2}}(u-m)\,W(u-m,\zeta\sqrt{2u})+\epsilon_1\}, \quad (5.2)$$

and
$$\left(\frac{1}{\mathrm{Re}\,T_2}+\frac{1}{\mathrm{Re}\,T_3}\right)\mathrm{Re}\{T_1\mathrm{P}^{\mathrm{i}m}_{-\frac{1}{2}+\mathrm{i}u}(x)\}+\left(\frac{1}{\mathrm{Re}\,T_2}-\frac{1}{\mathrm{Re}\,T_3}\right)\mathrm{Re}\{T_1\mathrm{P}^{\mathrm{i}m}_{-\frac{1}{2}+\mathrm{i}u}(-x)\}$$
$$= \left(\frac{\zeta^2-\alpha^2}{x^2-a^2}\right)^{\frac{1}{4}}\{k^{\frac{1}{2}}(u-m)\,W(u-m,-\zeta\sqrt{2u})+\epsilon_2\}, \quad (5.3)$$

where W denotes the modified parabolic cylinder function defined and discussed in Miller (1955) and Olver (1975), and T_1, T_2 and T_3 are defined by

$$T_1 = T\exp\{\tfrac{1}{2}\mathrm{i}\,\mathrm{ph}\,\Gamma(\tfrac{1}{2}+\mathrm{i}u-\mathrm{i}m)\},$$

$$T_2 = 2\left\{\frac{2(u+m)\cosh(\pi u-\pi m)}{uk^2(u-m)}\right\}^{\frac{1}{4}} \frac{2^{\frac{1}{2}(u+m)\mathrm{i}}T}{\Gamma(\frac{3}{4}-\frac{1}{2}\mathrm{i}u-\frac{1}{2}\mathrm{i}m)}\exp\{-\tfrac{1}{2}\mathrm{i}\arctan\tanh(\tfrac{1}{2}\pi u-\tfrac{1}{2}\pi m)\},$$

$$T_3 = -4\left\{\frac{\cosh(\pi u-\pi m)}{2u(u+m)\,k^2(u-m)}\right\}^{\frac{1}{4}} \frac{2^{\frac{1}{2}(u+m)\mathrm{i}}T}{\Gamma(\frac{1}{4}-\frac{1}{2}\mathrm{i}u-\frac{1}{2}\mathrm{i}m)}\exp\{\tfrac{1}{2}\mathrm{i}\arctan\tanh(\tfrac{1}{2}\pi u-\tfrac{1}{2}\pi m)\},$$

with
$$T \equiv m^{\mathrm{i}m}\Gamma(1-\mathrm{i}m)\,(u+m)^{-\frac{1}{2}(u+m)\mathrm{i}}\,\mathrm{e}^{\frac{1}{4}(2u-2m+\pi)\mathrm{i}}. \tag{5.4}$$

All functions in the foregoing equations take their principal values, except $\mathrm{ph}\,\Gamma(\frac{1}{2}+\mathrm{i}u-\mathrm{i}m)$, which is zero when $u-m$ is zero and defined by continuity for other values of $u-m$. The error terms ϵ_1 and ϵ_2 are bounded as in (9.4) and (9.5) of Olver (1975), these bounds applying when $0 < m \leqslant u$ and $0 \leqslant x < 1$.

For large u, asymptotic estimates of ϵ_1 in the ζ-intervals $[0,\infty)$ and $(-\infty,0]$ are immediately available from the results of §§ 9.3 and 9.4 of Olver (1975). Thus

$$\epsilon_1 = \mathsf{E}^{-1}(u-m,\zeta\sqrt{2u})\,\mathsf{M}(u-m,\zeta\sqrt{2u})\,O(u^{-\frac{2}{3}}) \quad (0 \leqslant \zeta < \infty), \tag{5.5}$$

and
$$\epsilon_1 = k^{-1}(u-m)\,\mathsf{E}(u-m,\zeta\sqrt{2u})\,\mathsf{M}(u-m,\zeta\sqrt{2u})\,O(u^{-\frac{2}{3}}) \quad (-\infty < \zeta \leqslant 0). \tag{5.6}$$

In each case the O-term is uniform with respect to ζ in the given interval and $m\in[\delta u, u]$, δ again denoting a fixed number in the interval $(0, 1)$. The auxiliary functions E and M are defined as in § 8.6 of the same reference.

It is of interest to notice that the result (5.6) is also derivable from (5.3), as follows. Substituting in (5.4) by means of Stirling's formula, we obtain

$$T = (2\pi m)^{\frac{1}{2}}\,\mathrm{e}^{-\frac{1}{2}m\pi}\,\mathrm{e}^{\frac{1}{2}(u+m)\mathrm{i}}\,(u+m)^{-\frac{1}{2}(u+m)\mathrm{i}}\{1+O(m^{-1})\} \quad (m\to\infty).$$

Then using this result and the definition

$$k(u-m) = \{1+\mathrm{e}^{2\pi(u-m)}\}^{\frac{1}{2}} - \mathrm{e}^{\pi(u-m)},$$

we find that for large u and $m\in[\delta u, u]$ the quantities $\mathrm{Re}\,T_2$ and $\mathrm{Re}\,T_3$ are approximated uniformly by

$$\mathrm{Re}\,T_2 = \frac{2m^{\frac{1}{2}}}{(2u)^{\frac{1}{4}}k(u-m)}\left\{1+O\left(\frac{1}{u}\right)\right\},\quad \mathrm{Re}\,T_3 = -\frac{2m^{\frac{1}{2}}}{(2u)^{\frac{1}{4}}k(u-m)}\left\{1+O\left(\frac{1}{u}\right)\right\}.$$

Accordingly,
$$\frac{1}{\mathrm{Re}\,T_2}-\frac{1}{\mathrm{Re}\,T_3} = (2u)^{\frac{1}{4}}m^{-\frac{1}{2}}k(u-m)\left\{1+O\left(\frac{1}{u}\right)\right\}, \tag{5.7}$$

and
$$\frac{1}{\mathrm{Re}\,T_2}+\frac{1}{\mathrm{Re}\,T_3} = (2u)^{\frac{1}{4}}m^{-\frac{1}{2}}k(u-m)\,O\left(\frac{1}{u}\right). \tag{5.8}$$

From (5.2), (5.8) and the fact that

$$k(u-m)\,\mathsf{E}^{-1}(u-m,\zeta\sqrt{2u}) \leqslant \mathsf{E}(u-m,\zeta\sqrt{2u}),$$

it is perceivable that the whole contribution from the first solution on the left-hand side of (5.3) is absorbable in the O-estimate for the error term on the right-hand side. Then on replacement of x by $-x$ and use of (5.7) we are led to (5.2) with ϵ_1 given by (5.6), as asserted.

Another useful check is to compare the limiting form of (5.2) and (5.6) as $\zeta \to -\infty$ with the known behaviour of the Ferrers functions as $x \to -1$. The writer has confirmed that the agreement is satisfactory, but again details are too lengthy to warrant reproduction.

6. Imaginary parameters and imaginary turning points

This case is related to that of § 4 in the same way that § 5 is related to §§ 2 and 3. After applying the Liouville transformation we obtain the differential equation

$$d^2w/d\zeta^2 = \{-u^2(\zeta^2+\alpha^2)+\psi(\alpha,\zeta)\}\,w, \tag{6.1}$$

with solutions

$$w = (x^2+a^2)^{\frac{1}{4}}\,(\zeta^2+\alpha^2)^{-\frac{1}{4}}\,\mathrm{P}^{\mathrm{i}m}_{-\frac{1}{2}+\mathrm{i}u}(\pm x).$$

Here
$$0 < u \leqslant m, \quad a^2 = \frac{m^2}{u^2}-1, \quad \tfrac{1}{2}\alpha^2 = (1+a^2)^{\frac{1}{2}}-1 = \frac{m}{u}-1,$$

and ζ and $\psi(\alpha,\zeta)$ are given by (4.4) and (4.6), respectively.

Equation (6.1) is in the standard form of theorem IV of Olver (1975, § 10.1). Identifying the solutions by letting $x \to 1-$ in the manner of previous sections, we arrive at

$$\mathrm{Re}\{T_1\,\mathrm{P}^{\mathrm{i}m}_{-\frac{1}{2}+\mathrm{i}u}(x)\} = \frac{m^{\frac{1}{2}}}{(2u)^{\frac{1}{4}}}\left(\frac{\zeta^2+\alpha^2}{x^2+a^2}\right)^{\frac{1}{4}} k^{-\frac{1}{2}}(u-m)\,\{W(u-m,\zeta\sqrt{2u})+\epsilon\}, \tag{6.2}$$

where T_1 is defined in § 5 above. The error term ϵ is bounded as in (10.3) of Olver (1975), this bound being valid when $0 < u \leqslant m$ and $-1 < x < 1$. Equation (6.2) should be compared with (5.2).

For large u,
$$\epsilon = \mathsf{M}(u-m,\zeta\sqrt{2u})\,O(u^{-1}\ln u), \tag{6.3}$$

uniformly for $m \in [u, u+\Delta u]$ and $x \in (-1,1)$, Δ again denoting any fixed positive number.

Corresponding results for other Ferrers functions can be constructed by straightforward use of connection formulae, since $\mathrm{Re}\{T_1\,\mathrm{P}^{\mathrm{i}m}_{-\frac{1}{2}+\mathrm{i}u}(x)\}$ and $\mathrm{Re}\{T_1\,\mathrm{P}^{\mathrm{i}m}_{-\frac{1}{2}+\mathrm{i}u}(-x)\}$ comprise a numerically satisfactory pair of solutions in the present circumstances.

This research was supported by the U.S. Army Research Office, Durham under Contract DA ARO D 31 124 72 G49, and the National Science Foundation under Grant GP 32841X. The writer is indebted to Dr L. Maximon for many suggestions for improving the presentation of the results.

References

B.M.P. 1953 Bateman Manuscript Project. *Higher transcendental functions* (A. Erdélyi, ed.), vol. I. London: McGraw-Hill.

Brussaard, P. J. & Tolhoek, H. A. 1957 Classical limits of Clebsch–Gordan coefficients, Racah coefficients and $D^l_{mn}(\phi,\vartheta,\psi)$-functions. *Physica* **23**, 955–971.

Miller, J. C. P. 1955 *Tables of Weber parabolic cylinder functions*. London: Her Majesty's Stationery Office.

Olver, F. W. J. 1974 *Asymptotics and special functions*. London: Academic Press.

Olver, F. W. J. 1975 Second-order linear differential equations with two turning points. *Phil. Trans. R. Soc. Lond.* A **278**, 137–174.

Thorne, R. C. 1957*a* The asymptotic solution of linear second order differential equations in a domain containing a turning point and a regular singularity. *Phil. Trans. R. Soc. Lond.* A **249**, 585–596.

Thorne, R. C. 1957*b* The asymptotic expansion of Legendre functions of large degree and order. *Phil. Trans. R. Soc. Lond.* A **249**, 597–620.

JOURNAL OF RESEARCH of the National Bureau of Standards—B. Mathematical Sciences
Vol. 80B, No. 4, October–December 1976

Improved Error Bounds for Second-Order Differential Equations With Two Turning Points

F. W. J. Olver*

Institute for Basic Standards, National Bureau of Standards, Washington, D.C. 20234

(July 13, 1976)

New error bounds are given for approximate solutions of differential equations of the form

$$d^2w/d\zeta^2 = \{u^2(\pm\alpha^2-\zeta^2) + \psi(u,\alpha,\zeta)\}w$$

in terms of parabolic cylinder functions. Here ζ is a real variable, α is a bounded real parameter, u is a large positive parameter, and $\psi(u,\alpha,\zeta)$ is a continuous function of α and ζ. Zero values of α are admitted.

Key words: Error bounds; parabolic cylinder functions; turning points; uniform asymptotic approximations; Weber's equation.

1. Introduction and Summary

Recently [2][1], [3], I developed a uniform asymptotic theory of second-order linear differential equations with two coalescing simple turning points, and applied the results to the associated Legendre equation. Subsequently, in the course of writing a paper on connection formulas for multiple turning points [4] it became clear how to effect some improvements in two of the four cases treated in [2]. The purpose of the present note is to describe these improvements. It will be assumed that the reader is familiar with the results presented in [2], and the same notation will be used except where indicated otherwise.

The next section introduces new auxiliary functions for the solutions of the modified Weber equation. The new form of the general approximation theorem is stated and discussed in the third (and concluding) section.

An application of the results to the approximation of Whittaker functions with both parameters large will be published in due course.

2. Auxiliary Functions for the Modified Weber Functions

The modified Weber functions $W(b,x)$ and $W(b,-x)$ are solutions of the differential equation

$$d^2w/dx^2 = (b - \tfrac{1}{4}x^2)w.$$

As in [2], §8.6 we denote by $\sigma(b)$ the smallest positive root of the equation

$$k^{-1/2}(b)W(b,x) = k^{1/2}(b)W(b,-x),$$

* Present address: Institute for Physical Science and Technology, University of Maryland, College Park, Md. 20742.
[1] Figures in brackets indicate the literature references at the end of this paper.

but we now extend this definition to all real values of the parameter b. We already know that

$$\sigma(b) \sim 2b^{1/2}, \qquad b \to +\infty.$$

Next, because $\sigma(b)$ is bounded by the smallest positive zero of $W(b,x)$, the expansion (15.7) of [1] shows that[2]

$$\sigma(b) \to 0, \qquad b \to -\infty.$$

In place of the definition [2], (8.15) of the weight function $E(b,x)$, we now define

$$E(b,x) = \begin{cases} k^{1/2}(b), & -\infty < x \leq -\sigma(b), \\ \{W(b,-x)/W(b,x)\}^{1/2}, & -\sigma(b) \leq x \leq \sigma(b), \\ k^{-1/2}(b), & \sigma(b) \leq x < \infty. \end{cases}$$

Then $E(b,x)$ is continuous for all real values of b and x, and is also a nondecreasing function of x. Furthermore, denoting $1/E(b,x)$ by $E^{-1}(b,x)$, we have

$$E(b,-x) = E^{-1}(b,x).$$

Using the above definition of $E(b,x)$ we now define modulus and phase functions for all real values of b and x by the relations

$$W(b,x) = E^{-1}(b,x)M(b,x)\sin\theta(b,x),$$

$$W(b,-x) = E(b,x)M(b,x)\cos\theta(b,x),$$

$$W'(b,x) = E^{-1}(b,x)N(b,x)\sin\omega(b,x),$$

$$W'(b,-x) = E(b,x)N(b,x)\cos\omega(b,x).$$

Thus when $-\infty < x \leq -\sigma(b)$ we have

$$M(b,x) = \{k(b)W^2(b,x) + k^{-1}(b)W^2(b,-x)\}^{1/2},$$

$$\theta(b,x) = \tan^{-1}\{k(b)W(b,x)/W(b,-x)\},$$

$$N(b,x) = \{k(b)W'^2(b,x) + k^{-1}(b)W'^2(b,-x)\}^{1/2},$$

$$\omega(b,x) = \tan^{-1}\{k(b)W'(b,x)/W'(b,-x)\}.$$

Secondly, when $-\sigma(b) \leq x \leq \sigma(b)$ we have

$$M(b,x) = \{2W(b,x)W(b,-x)\}^{1/2}, \qquad \theta(b,x) = \tfrac{1}{4}\pi,$$

$$N(b,x) = \left\{\frac{W'^2(b,x)W^2(b,-x) + W'^2(b,-x)W^2(b,x)}{W(b,x)W(b,-x)}\right\}^{1/2},$$

$$\omega(b,x) = \tan^{-1}\left\{\frac{W'(b,x)W(b,-x)}{W'(b,-x)W(b,x)}\right\}.$$

Lastly, when $\sigma(b) \leq x < \infty$ we have

$$M(b,x) = \{k^{-1}(b)W^2(b,x) + k(b)W^2(b,-x)\}^{1/2},$$

$$\theta(b,x) = \tan^{-1}\{k^{-1}(b)W(b,x)/W(b,-x)\},$$

$$N(b,x) = \{k^{-1}(b)W'^2(b,x) + k(b)W'^2(b,-x)\}^{1/2},$$

$$\omega(b,x) = \tan^{-1}\{k^{-1}(b)W'(b,x)/W'(b,-x)\}.$$

[2] The actual asymptotic form of $\sigma(b)$ as $b \to -\infty$ can be shown to be $\sigma(b) \sim \frac{1}{2}|b|^{-1/2}e^{\pi b}$, but we shall not make use of this more precise result in the present paper.

Each of the functions $M(b,x)$, $\theta(b,x)$, $N(b,x)$, and $\omega(b,x)$ is continuous in b and x. Reflection properties are given by

$$M(b,-x) = M(b,x), \qquad \theta(b,-x) = \tfrac{1}{2}\pi - \theta(b,x),$$

$$N(b,-x) = N(b,x), \qquad \omega(b,-x) = \tfrac{1}{2}\pi - \omega(b,x).$$

As $x \to \pm\infty$

$$M(b,x) \sim |2/x|^{1/2}, \qquad N(b,x) \sim |x/2|^{1/2}.$$

Compared with [2], §§8.6 and 8.7, the modulus functions $M(b,x)$ and $N(b,x)$ agree with their previous definitions, except when $b < 0$ and $-\sigma(b) < x < \sigma(b)$. On the other hand the phase functions agree only in the case of $\theta(b,x)$ when $b \geq 0$ and $x \geq -\sigma(b)$.

3. Approximation Theorem

The following theorem replaces both Theorems III and IV of [2]. The definitions of E, M, and N are those just given in §2.

THEOREM: *Assume that for* $a \in [0,A]$, $\zeta \in (\zeta_1,\zeta_2)$, *and each value of the positive parameter* u, *the function* $\psi(u,a,\zeta)$ *is continuous and* $\mathcal{V}_{\zeta_1,\zeta_2}(F)$ *is bounded. Then in this region the differential equation*

$$d^2w/d\zeta^2 = \{u^2(\pm a^2-\zeta^2) + \psi(u,a,\zeta)\}w \tag{3.1}$$

has a continuous solution $w(u,a,\zeta)$, *with continuous first and second partial* ζ *derivatives, given by*

$$w(u,a,\zeta) = W(\pm\tfrac{1}{2}ua^2,\zeta\sqrt{2u}) + \epsilon(u,a,\zeta), \tag{3.2}$$

where

$$\frac{|\epsilon(u,a,\zeta)|}{M(\pm\frac{1}{2}ua^2,\zeta\sqrt{2u})}, \quad \frac{|\partial\epsilon(u,a,\zeta)/\partial\zeta|}{(2u)^{1/2}N(\pm\frac{1}{2}ua^2,\zeta\sqrt{2u})}$$

$$\leq E^{-1}(\pm\tfrac{1}{2}ua^2,\zeta\sqrt{2u})\left[\exp\left\{\frac{l(\pm\frac{1}{2}ua^2)}{(2u)^{1/2}}\mathcal{V}_{\zeta,\zeta_2}(F)\right\} - 1\right], \tag{3.3}$$

and

$$l(b) = \sup_{x\in(-\infty,\infty)} \{\Omega(x)M^2(b,x).\ \} \tag{3.4}$$

In this result either the upper sign or the lower sign is to be taken throughout. Upper signs correspond to Theorem III of [2], and lower signs correspond to Theorem IV of [2]. In both cases a numerically satisfactory companion solution to $w(u,a,\zeta)$ may be obtained by applying the theorem to the equation obtained from (3.1) by replacing ζ by $-\zeta$. The proof of the theorem is similar to the proofs of Theorems I, II, III, and IV of [2], and it is unnecessary to record details.

The new result simplifies the general theory by replacing two theorems by a single theorem. The main advantage, however, is that error bounds are now supplied directly for all real values of ζ. In contrast, Theorem III furnishes error bounds only for nonnegative values of ζ; compare [2], §9.4. Theorem IV, however, supplies error bounds for all real values of ζ, and a property of Theorem IV that is not shared by the new theorem is that it is possible to prescribe anywhere in the closure of (ζ_1,ζ_2) a reference point ζ_0 at which $\epsilon(u,a,\zeta)$ and $\partial\epsilon(u,a,\zeta)/\partial\zeta$ both vanish. In consequence, it is conceivable that there may be applications in which Theorem IV is to be preferred. The new theorem should generally be used in place of Theorem III, however.

It should be noted that although the structure of the error bounds (3.3) is independent of the choice of ambiguous sign in (3.1), (3.2), and (3.3), the asymptotic nature of the bounds as $u \to \infty$ differs in the two cases. This is in consequence of the results

$$l(b) = O(b^{1/3}), \qquad b \to +\infty, \tag{3.5}$$

$$l(b) = O(1), \qquad b \to -\infty, \tag{3.6}$$

which correspond to similar results in §§9.2 and 10.2 of [2]. Thus, for example, if $|\psi(u,a,\zeta)/\zeta|$ is independent of u and integrable at $\zeta = -\infty$ and $+\infty$ uniformly with respect to a, then by taking the balancing function $\Omega(x)$ in the definition [2], (6.3) of the error-control function $F(u,a,\zeta)$ to be $1 + |x|$, we may show that

$$\epsilon(u,a,\zeta) = E^{-1}(\tfrac{1}{2}ua^2,\zeta\sqrt{2u})M(\tfrac{1}{2}ua^2,\zeta\sqrt{2u})O(u^{-2/3}\ln u), \tag{3.7}$$

when upper signs are taken, compared with

$$\epsilon(u,a,\zeta) = E^{-1}(-\tfrac{1}{2}ua^2,\zeta\sqrt{2u})M(-\tfrac{1}{2}ua^2,\zeta\sqrt{2u})O(u^{-1}\ln u), \tag{3.8}$$

when lower signs are used. The relations (3.7) and (3.8) apply as $u \to \infty$ and are uniform with respect to $a \in [0,A]$ and $\zeta \in (-\infty,\infty)$; compare [2], §§9.3 and 10.3. Similarly for the derivative $\partial\epsilon(u,a,\zeta)/\partial\zeta$.

This research has been supported in part by the U.S. Army Research Office, Durham, Contract DA ARO D 31 124 73 G204, and the National Science Foundation, Grant GP 32841X2.

4. References

[1] Olver, F. W. J., Uniform asymptotic expansions for Weber parabolic cylinder functions of large orders, J. Res. Nat. Bur. Stand. (U.S.), **63B** (Math. and Math. Phys.) 131–169 (1959).

[2] Olver, F. W. J., Second-order linear differential equations with two turning points, Philos. Trans. Roy. Soc. London Ser. A **278,** 137–174 (1975).

[3] Olver, F. W. J., Legendre functions with both parameters large, Philos. Trans. Roy. Soc. London Ser. A **278,** 175–185 (1975).

[4] Olver, F. W. J., Connection formulas for second-order differential equations with multiple turning points, SIAM J. Math. Anal. Vol. 8, pp. 127–154 (Feb. 1977).

(Paper 80B4–455)